The Science of Grapevines
Anatomy and Physiology

The Science of Grapevines
Anatomy and Physiology

Second Edition

Markus Keller
Irrigated Agriculture Research and Extension Center,
Washington State University, Prosser, WA, USA

AMSTERDAM • BOSTON • HEIDELBERG • LONDON
NEW YORK • OXFORD • PARIS • SAN DIEGO
SAN FRANCISCO • SINGAPORE • SYDNEY • TOKYO
Academic Press is an imprint of Elsevier

Academic Press is an imprint of Elsevier
32 Jamestown Road, London NW1 7BY, UK
525 B Street, Suite 1800, San Diego, CA 92101-4495, USA
225 Wyman Street, Waltham, MA 02451, USA
The Boulevard, Langford Lane, Kidlington, Oxford OX5 1GB, UK

© 2015, 2010 Markus Keller. Published by Elsevier Inc. All rights reserved.

No part of this publication may be reproduced or transmitted in any form or by any means, electronic or mechanical, including photocopying, recording, or any information storage and retrieval system, without permission in writing from the publisher. Details on how to seek permission, further information about the Publisher's permissions policies and our arrangements with organizations such as the Copyright Clearance Center and the Copyright Licensing Agency, can be found at our website: www.elsevier.com/permissions.

This book and the individual contributions contained in it are protected under copyright by the Publisher (other than as may be noted herein).

Notices
Knowledge and best practice in this field are constantly changing. As new research and experience broaden our understanding, changes in research methods, professional practices, or medical treatment may become necessary.

Practitioners and researchers must always rely on their own experience and knowledge in evaluating and using any information, methods, compounds, or experiments described herein. In using such information or methods they should be mindful of their own safety and the safety of others, including parties for whom they have a professional responsibility.

To the fullest extent of the law, neither the Publisher nor the authors, contributors, or editors, assume any liability for any injury and/or damage to persons or property as a matter of products liability, negligence or otherwise, or from any use or operation of any methods, products, instructions, or ideas contained in the material herein.

ISBN: 978-0-12-419987-3

British Library Cataloguing-in-Publication Data
A catalogue record for this book is available from the British Library.

Library of Congress Cataloging-in-Publication Data
A catalog record for this book is available from the Library of Congress.

For Information on all Academic Press publications
visit our website at http://store.elsevier.com/

Typeset by MPS Limited, Chennai, India
www.adi-mps.com

Printed and bound in the United States of America

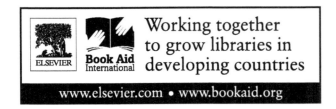

Contents

About the Author ix
Preface to the Second Edition xi

CHAPTER 1 Botany and Anatomy 1
   1.1 Botanical Classification and Geographical Distribution 2
   1.2 Cultivars, Clones, and Rootstocks 10
       1.2.1 Variety Versus Cultivar 10
       1.2.2 Cultivar Classification 17
       1.2.3 Clones 19
       1.2.4 Rootstocks 21
   1.3 Morphology and Anatomy 24
       1.3.1 Root 26
       1.3.2 Trunk and Shoots 33
       1.3.3 Nodes and Buds 41
       1.3.4 Leaves 44
       1.3.5 Tendrils and Clusters 50
       1.3.6 Flowers and Grape Berries 52

CHAPTER 2 Phenology and Growth Cycle 59
   2.1 Seasons and Day Length 59
   2.2 Vegetative Cycle 61
   2.3 Reproductive Cycle 83

CHAPTER 3 Water Relations and Nutrient Uptake 101
   3.1 Osmosis, Water Potential, and Cell Expansion 101
   3.2 Transpiration and Stomatal Action 106
   3.3 Water and Nutrient Uptake and Transport 110

CHAPTER 4 Photosynthesis and Respiration 125
   4.1 Light Absorption and Energy Capture 125
   4.2 Carbon Uptake and Assimilation 130
   4.3 Photorespiration 136
   4.4 Respiration 138
   4.5 From Cells to Plants 141

CHAPTER 5 Partitioning of Assimilates 145
- 5.1 Photosynthate Translocation and Distribution 146
  - 5.1.1 Allocation and Partitioning 156
- 5.2 Canopy–Environment Interactions 163
  - 5.2.1 Light 165
  - 5.2.2 Temperature 174
  - 5.2.3 Wind 177
  - 5.2.4 Humidity 179
  - 5.2.5 The "Ideal" Canopy 180
- 5.3 Nitrogen Assimilation and Interaction with Carbon Metabolism 182
  - 5.3.1 Nitrate Uptake and Reduction 183
  - 5.3.2 Ammonium Assimilation 185
  - 5.3.3 From Cells to Plants 187

CHAPTER 6 Developmental Physiology 193
- 6.1 Yield Formation 194
  - 6.1.1 Yield Potential and its Realization 197
- 6.2 Grape Composition and Fruit Quality 205
  - 6.2.1 Water 209
  - 6.2.2 Sugars 212
  - 6.2.3 Acids 216
  - 6.2.4 Nitrogenous Compounds and Mineral Nutrients 219
  - 6.2.5 Phenolics 223
  - 6.2.6 Lipids and Volatiles 236
- 6.3 Sources of Variation in Fruit Composition 243
  - 6.3.1 Fruit Maturity 244
  - 6.3.2 Light 245
  - 6.3.3 Temperature 249
  - 6.3.4 Water Status 254
  - 6.3.5 Nutrient Status 257
  - 6.3.6 Crop Load 261
  - 6.3.7 Rootstock 264

CHAPTER 7 Environmental Constraints and Stress Physiology 267
- 7.1 Responses to Abiotic Stress 268
- 7.2 Water: Too Much or Too Little 272
- 7.3 Nutrients: Deficiency and Excess 287
  - 7.3.1 Macronutrients 293
  - 7.3.2 Transition Metals and Micronutrients 308
  - 7.3.3 Salinity 317
- 7.4 Temperature: Too Cold or Too Warm 321
  - 7.4.1 Heat Acclimation and Damage 322

		7.4.2 Chilling Stress 326
		7.4.3 Cold Acclimation and Freeze Damage 329

CHAPTER 8 Living with Other Organisms 343
8.1 Biotic Stress and Evolutionary Arms Races 343
8.2 Pathogens: Defense and Damage 352
8.2.1 Bunch Rot 353
8.2.2 Powdery Mildew 357
8.2.3 Downy Mildew 359
8.2.4 Bacteria 361
8.2.5 Viruses 364

Abbreviations and Symbols 369
Glossary 371
References 381
Internet Resources 489
Index 491

About the Author

Markus Keller is the Chateau Ste. Michelle Distinguished Professor of Viticulture at Washington State University's Irrigated Agriculture Research and Extension Center in Prosser. He received his master's degree in agronomy (plant science) in 1989 and a doctorate in natural sciences in 1995 from the Swiss Federal Institute of Technology in Zürich. Having grown up on a diversified farm that included wine grape production among other crops and livestock, he began his research and teaching career in viticulture and grapevine physiology at the Federal Research Station for Fruit-Growing, Viticulture and Horticulture in Wädenswil (now Agroscope Changins-Wädenswil), Switzerland. He then moved to Cornell University in Geneva, New York, and from there to Charles Sturt University in Wagga Wagga, Australia, before coming to eastern Washington. In addition, he has been a regular guest lecturer at the Universidad Nacional de Cuyo in Mendoza, Argentina. He was awarded the Swiss AgroPrize for innovative contributions to Switzerland's agriculture industry. His research focuses on developmental and environmental factors, as well as vineyard management practices, as they influence crop physiology and production of wine and juice grapes.

Preface to the Second Edition

It has been 5 years since *The Science of Grapevines* was first published. Science (Latin *scientia* = knowledge), or the pursuit of knowledge, does not stand still, of course, which poses a challenge for any printed textbook. For the second edition, I have reviewed and integrated novel material in an attempt to present the latest information available from the scientific literature. Additionally, several gaps have been closed by revisiting some of the older, classic literature. The task has not been trivial; the second edition contains more than 750 additional literature references. All chapters have been thoroughly revised and updated. The former large and diverse Chapter 7 has been split: the new Chapter 7, "Environmental Constraints and Stress Physiology," is now focused on responses to abiotic stress, whereas the new Chapter 8, "Living with Other Organisms," addresses responses to biotic stress. Abstracts have been added, summarizing the key points of each chapter in a highly condensed manner. To help readers unfamiliar with specific terminology, a glossary is now included that explains many chemical, anatomical, physiological, and viticultural terms that appear throughout the text. The new list of abbreviations and symbols will provide further assistance in navigating the text.

Grapes were among the first fruit species to be domesticated and remain the world's most economically important fruit crop. According to the Food and Agriculture Organization of the United Nations, grapevines were planted on almost 7 million hectares producing more than 67 million metric tons of fruit in 2012. More than 70% of this crop was used to make wine, 27% was consumed as fresh fruit (table grapes), 2% was consumed as dried fruit (raisins), and less than 1% was processed to grape juice, distilled to brandy, or transformed into vinegar.

The Science of Grapevines explores the state of knowledge of the construction and life of this economically and socially vital plant species. The book is an introduction to the physical structure of the grapevine, its various organs and tissues, their functions, their interactions with one another, and their responses to the environment. It focuses essentially on the physical and biological functions of whole plants rather than the metabolism and molecular biology of individual cells. It is nonetheless necessary to review some fundamental processes at the cell, tissue, and organ levels in order to build up an appreciation of whole-plant function. The book covers those elements of physiology that will enhance our understanding of grapevine function and their implications for practical vineyard management. Although of necessity the text contains a plethora of technical terms and details, I have tried to resist the temptation to dwell in biochemical and molecular-biological jargon. Most physiological processes (water movement through the vine's hydraulic system and evaporation through plant surfaces may be exceptions) are rooted in biochemistry. They are driven or at least facilitated by enzymes which, in turn, are built based on blueprints provided by genes. I have therefore taken it for granted that it is understood that a developmental process or change in chemical composition implies a change in enzyme activity, which in turn implies a change in the activity, or expression, of one or more genes. This does not imply, as used to be thought, that "one gene makes one enzyme," or that "one enzyme makes one chemical," but means merely that all enzymatic processes are rooted in the dynamic expression of certain genes.

Many biochemical and biophysical processes apply to many or even all plants. Perhaps no process is truly unique to grapevines. Chances are if grapes employ a solution to a survival issue, then some or many other species do the same thing, because they share a common ancestor that invented the trick a long time ago. For example, microbes hit upon photosynthesis and respiration long before these discoveries enabled some of them to join forces and evolve into plants. Consequently, although this book is about grapevines, and primarily about the wine grape species *Vitis vinifera*, I have borrowed heavily from research done with other plant species, both wild and cultivated, perennial and annual, woody and herbaceous, including that "queen of weeds"—at least in the fast-paced world of modern molecular biology—*Arabidopsis thaliana*, the otherwise inconspicuous thale cress. I have even taken the liberty of borrowing insights gained using microorganisms such as the yeast *Saccharomyces cerevisiae*, which gives us the wine, beer, and bread that enable us to think about these issues.

This book aims to be global in scale. It covers topics ranging from the physiological aspects of tropical viticulture near the equator all the way to those that pertain to the production of ice wine at the temperate latitudinal margins of grape growing. It moves from vineyards at sea level to vineyards at high altitude. It considers the humid conditions of cool, maritime climates, the moist winters and dry summers of Mediterranean climates, as well as the arid environment, often combined with hot summers and cold winters, typical of continental climates in the rain shadows of massive mountain ranges. Yet a book of this nature is necessarily incomplete, and so is the selection of published information included in the text. No one can read everything that has been and is being published, even in the admittedly relatively narrow field of grapevine anatomy and physiology. The magnitude of the task of reviewing as much of the pertinent literature as possible often forced me to rely on review papers where they were available. I apologize to those friends and colleagues whose work I did not cite or cited incompletely or incorrectly. Science—and scientists—can only ever hope to approximate the truth. This and the simple fact of *"errare humanum est"* will guarantee a number of errors throughout the text. These are entirely my responsibility, and I would be grateful for any feedback that might help improve this book and further our understanding of the world's most important and arguably most malleable fruit crop. After all, the full quote from Seneca the Younger, who was a contemporary of Columella, the Roman author of agriculture and viticulture textbooks, reads *"errare humanum est, sed in perseverare diabolicum"* ("to err is human, but to persevere is devilish").

As with the first edition, many people have made large and small contributions to bring this second edition to fruition. I am indebted to all of these generous individuals. My wife, Sandra Wran, has supported this project throughout and also waited countless long weekends for me to "finish" reading another paper or revising another section. Many of the illustrations were skillfully drawn by Adrienne Mills, and Lynn Mills helped with data collection and some of the most recalcitrant illustrations. Feedback from colleagues and students has helped to eliminate numerous mistakes, some of them more glaring than others. I am also grateful for the encouragement from Nancy Maragioglio and Carrie Bolger at Elsevier, who were always quick to answer my questions *du jour*.

CHAPTER 1

BOTANY AND ANATOMY

CHAPTER OUTLINE

1.1 Botanical Classification and Geographical Distribution 2
1.2 Cultivars, Clones, and Rootstocks 10
    1.2.1 Variety Versus Cultivar 10
    1.2.2 Cultivar Classification 17
    1.2.3 Clones 19
    1.2.4 Rootstocks 21
1.3 Morphology and Anatomy 24
    1.3.1 Root 26
    1.3.2 Trunk and Shoots 33
    1.3.3 Nodes and Buds 41
    1.3.4 Leaves 44
    1.3.5 Tendrils and Clusters 50
    1.3.6 Flowers and Grape Berries 52

ABSTRACT

The diverse species of grapevine belong to the botanical family Vitaceae, which includes mostly shrubs and woody lianas that climb using leaf-opposed tendrils. The vast majority of the thousands of grape cultivars belong to the species *Vitis vinifera*. Some of the other species are used as pest-tolerant rootstocks to which cultivars with desirable fruit properties are grafted. Cultivars are propagated asexually as cuttings so that each individual is a clone of its mother plant. Grapevines comprise vegetative organs (roots, trunk, cordon, shoots, leaves, and tendrils) and reproductive organs (clusters with flowers or berry fruit). All organs are interconnected through the vascular system comprising the xylem for water and nutrient transport, and the phloem for assimilate transport. The roots form the plant–soil interface, while the trunk, cordons, and shoots of a vine form its stem. The shoots carry the leaves, buds, tendrils, and clusters. Leaves are arranged in spiral phyllotaxy in juvenile vines and in alternate phyllotaxy in mature vines. Buds are young, compressed shoots embedded in leaf scales. Tendrils and clusters are modified shoots. After fertilization, the flower pistil develops into the berry fruit. The berry comprises up to four seeds surrounded by the endocarp, the mesocarp or flesh, and the exocarp or skin.

Keywords: Bud; clone; cultivar; grape berry; leaf; root; rootstock; shoot; tendril; *Vitis*

1.1 BOTANICAL CLASSIFICATION AND GEOGRAPHICAL DISTRIBUTION

The basic unit of biological classification is the species. According to the "biological species concept," a species is defined as a community of individuals—that is, a population or group of populations whose members can interbreed freely with one another under natural conditions but not with members of other populations (Mayr, 2001; Soltis and Soltis, 2009). In other words, such communities are reproductively isolated. Although each individual of a sexual population is genetically unique, each species is a closed gene pool, an assemblage of organisms that does not normally exchange genes with other species. Their genes compel the individuals belonging to a species to perpetuate themselves over many generations. Yet all life forms on Earth are interrelated; they all ultimately descended from a common ancestor and "dance" to the same genetic code, whereby different combinations of three consecutive nucleotides of each organism's deoxyribonucleic acid (DNA) specify different amino acids that can be assembled into proteins. Because they are thus interrelated, organisms can be grouped according to the degree of their genetic similarity, external appearance, and behavior. In the classification hierarchy, closely related species are grouped into a genus, related genera into a family, allied families into an order, associated orders into a class, similar classes into a division (plants) or a phylum (animals), related divisions or phyla into a kingdom, and, finally, allied kingdoms into not an empire but a domain. The "evolutionary species concept" recognizes this ancestor—offspring connection among populations that may follow distinct evolutionary paths to occupy separate ecological niches but may continue to interbreed for some time (Soltis and Soltis, 2009). For example, although they have been geographically isolated for over 20 million years, Eurasian and North American *Vitis* species are still able to interbreed readily.

As is the case with many plants, the species of the genus *Vitis* are not very well defined because of the extreme morphological variation among and within populations of wild vines (Currle et al., 1983; Hardie, 2000; Mullins et al., 1992). This implies the following: (i) all *Vitis* species are close relatives that share a relatively recent common ancestor and (ii) evolution is still at work, throwing up new variants all the time (see Section 2.3). Many vine species are actually semispecies—that is, populations that partially interbreed and form hybrids under natural conditions, which is in fact common among plants and may be an important avenue for the evolution of new species (Soltis and Soltis, 2009). Despite some hybridization where their natural habitats overlap, however, the various *Vitis* gene pools usually stay apart so that the populations remain recognizably different. Nonetheless, species that occur in close proximity are more similar than distant species in similar habitats. Grapevines are a good example of the limits of taxonomic systems, demonstrating that there is a continuum of differentiation rather than a set of discrete, sexually incompatible units. As early as 1822, the Rev. William Herbert asserted that "botanical species are only a higher and more permanent class of varieties," and in 1825 the geologist Leopold von Buch postulated that "varieties slowly become changed into permanent species, which are no longer capable of intercrossing" (both cited in Darwin, 2004). A few decades later, Charles Darwin expressed it clearly: "Wherever many closely allied yet distinct species occur, many doubtful forms and varieties of the same species likewise occur" and, furthermore, "there is no fundamental distinction between species and varieties," and, finally, "varieties are species in the process of formation" (Darwin, 2004). Indeed, modern genetic evidence indicates that the various *Vitis* species evolved relatively recently from a common ancestor so that they have not yet had time to develop the complete reproductive isolation that normally characterizes biological species. Thus, *Vitis* species are defined as populations of vines that can be easily distinguished by morphological traits, such as the anatomy of their leaves, flowers, and berries, and that are isolated from one another by geographical, ecological, or phenological barriers; such species are termed *ecospecies* (Hardie, 2000; Levadoux, 1956; Mullins et al., 1992). The following is a brief overview of the botanical classification of grapevines, starting with the domain at the top of the hierarchy and finishing with a selection of species at the base.

Domain Eukaryota

All living beings, making up the Earth's biological diversity or biodiversity, are currently divided into three great domains of life: the Bacteria, the Archaea, and the Eukaryota. The Eukaryota (eukaryotes; Greek *eu* = true, *karyon* = nucleus) include all terrestrial, sexually reproducing "higher" organisms with relatively large cells (10–100 μm) containing a true cell nucleus, in which the DNA-carrying chromosomes are enclosed in a nuclear membrane, and cell organelles such as mitochondria and plastids (Mayr, 2001). They evolved following injections of oxygen into the atmosphere caused by abiotic (i.e., nonbiological) factors such as plate tectonics and glaciation (Lane, 2002). The vast majority of life and the bulk of the world's biomass—the small (1–10 μm), single-celled prokaryotes (Greek *pro* = before) with cell walls composed of peptidoglycans (protein–polysaccharides)—is grouped into other two domains. However, both the photosynthetic organelles (chloroplasts, from cyanobacteria) and the "power plants" (mitochondria, from proteobacteria) of eukaryotic cells have descended from (symbiotic) bacteria that "infected" other single-celled organisms (or were "swallowed" by them) over 1 billion years ago. These organelles still retain some of their own DNA (i.e., genes), although more than 95% of their original genes have since been and are still being lost or transferred (donated) to their host's nucleus (Timmis et al., 2004; Green, 2011). The mitochondria, in turn, have acquired a sizeable fraction (>40%) of genes from the chloroplasts, which may have returned the favor by incorporating some mitochondrial genes (Goremykin et al., 2009). Genetic modification or transformation, resulting in transgenic organisms, is evidently a natural process. Some bacteria cause diseases of grapevines; for example, crown gall is caused by *Agrobacterium vitis* and Pierce's disease by *Xylella fastidiosa* (see Section 8.2).

Kingdom Plantae

The Eukaryota comprise at least five major lineages that are termed supergroups or kingdoms (Green, 2011); the number and associations change as the relationships among organisms become better known. The Plantae have a haplo-diploid life cycle and cell walls composed of cellulose, and are studied in the field of Botany. This supergroup includes the plants and the green and some other algae. There are approximately 500,000 plant species, which are classified into 12 phyla or divisions based largely on reproductive characteristics. The higher or vascular (Latin *vasculum* = small vessel) plants, to which grapevines belong on account of their water conduits, form the subkingdom Tracheobionta (Greek *trachea* = windpipe). The Animalia comprise the multicellular, diploid (having two sets of chromosomes) animals with cells lacking cell walls, except in arthropods (insects, spiders, and the like) which have chitin cell walls, and are the domain of Zoology. The Fungi include the haploid (having one set of chromosomes) mushrooms, molds, and other fungi, with cell walls composed of glucans and chitin; they are studied in Mycology. Because they are more closely related to each other than to any other eukaryotes, the animals and fungi are now grouped together into the Opisthokonta (Green, 2011). Whereas one group of fungi (singular fungus), the yeasts (especially *Saccharomyces cerevisiae*), turns grapes into wine through fermentation, other fungi cause diseases of grapevines (see Section 8.2); for example, gray rot is caused by *Botrytis cinerea* Pers.:Fr. and powdery mildew by *Erysiphe necator* (a.k.a. *Uncinula necator* (Schwein.) Burr.). Animals also can be important pests of grapevines, especially certain insects (e.g., phylloxera, *Daktulosphaira vitifoliae* Fitch), mites, and nematodes. The number and names of the other lineages and the relationships among them remain contested; they include all other "higher-order" organisms, from single-celled microbes or microorganisms (Greek *mikros* = small), including some that resemble unicellular fungi, plants (some algae), and animals (protozoans), to large, multicellular seaweeds (marine algae such as kelps).

Division (phylum) Angiospermae (synonym Magnoliophyta)

The angiosperms (or, in new terminology, the magnoliophytes) are the flowering plants, which include perhaps as many as 400,000 species. They are believed to have evolved from a common ancestor that lived approximately 160 million years ago during the late Jurassic period, and they make up the most evolutionarily successful group of plants. Angiosperms are the plants with the most complex reproductive system: They grow their seeds inside an ovary (Greek *angeion* = pot, vessel) that itself is embedded in a flower. After the flower is fertilized, the other flower parts fall away and the ovary swells to become a fruit, such as a grape berry. Indeed, the production of fruits is what defines the angiosperms and sets them apart from the gymnosperms, with whom they are classed in the superdivision Spermatophyta, or seed plants.

Class Dicotyledoneae (synonym Magnoliopsida)

This class is large and very diverse, and its members are often called dicot plants. The vast majority of plants ($\sim$200,000 species), including most trees, shrubs, vines, and flowers, and most fruits, vegetables, and legumes, belong to this group. Like all members of the Dicotyledoneae, grapevines start their life cycle with two cotyledons (seed leaves) preformed in the seed.

Order Rhamnales (Vitales according to the Angiosperm Phylogeny web)

Grapevines belong to the order Rhamnales, which gets its name from the genus *Rhamnus*, the buckthorns. The order has three families: Rhamnaceae (e.g., *Ziziphus jujuba* Mill., jujube tree), Leeaceae (the oleasters), and Vitaceae. Plants of the family Leeaceae are more recognizable as being related to grapevines than those belonging to the Rhamnaceae, being shrubs or trees with flowers aggregated in inflorescences, black berries, and seeds that resemble grape seeds, also named pips. Some taxonomists have now separated the Vitaceae (Jansen et al., 2006) and Leeaceae from the Rhamnales and placed them in the order Vitales.

Family Vitaceae

The members of this family are collectively termed grapevines. The family contains approximately 1000 species assigned to 17 genera that are typically shrubs or woody lianas that climb by means of their leaf-opposed tendrils—hence the name Vitaceae (Latin *viere* = to attach). Although most species of this family reside in the tropics or subtropics, a single species from the temperate zones has become the world's leading fruit crop grown in almost 90 countries for wine and juice production or as fresh table grapes or dried grapes (raisins). Vitaceae roots are generally fibrous and well branched, and they can grow to several meters in length. The leaves are alternate, except during the juvenile phase of plants grown from seeds, and can be simple or composite. The fruits are usually fleshy berries with one to four seeds. All cultivated grapes belong to either the genus *Muscadinia* ($2n = 40$ chromosomes) or the genus *Vitis* ($2n = 38$ chromosomes). The former classification of *Muscadinia* and *Euvitis* as either subgenera or sections of the genus *Vitis* has fallen out of favor among taxonomists (Mullins et al., 1992). Because of the different numbers of chromosomes, crosses between these two genera rarely produce fertile hybrids. Key morphological characteristics of the two genera include the following:

- simple leaves
- simple or forked tendrils
- generally unisexual flowers—that is, either male (staminate) or female (pistillate)
- fused flower petals that separate at the base, forming a calyptra or "cap"
- soft and pulpy berry fruits.

Genus *Muscadinia*

Members of the genus *Muscadinia* usually have glabrous (hairless) leaves, simple tendrils, non-shredding bark, nodes without diaphragms, and hard wood (Currle et al., 1983; Mullins et al., 1992; Olmo, 1986). Because they do not root from dormant cuttings, they are usually propagated by layering, although they do root easily from green cuttings. The "homeland" of this genus extends from the southeastern United States to Mexico. The genus has only three species, which are all very similar and may not even deserve to be classed as separate species (Currle et al., 1983; Mullins et al., 1992; Olien, 1990).

- *Muscadinia rotundifolia* Small (formerly *Vitis rotundifolia* Michaux): A dioecious plant, although breeding has yielded perfect-flowered and female cultivars, such as Noble, Carlos, or Magnolia, known as "muscadines," that are grown as table, jelly, or wine grapes (Olmo, 1986). The species is native of the southeastern United States. The musky flavor and thick skins of the fruit can be unattractive. The species has coevolved with and therefore resists or tolerates, albeit to varying degrees, the grapevine diseases and pests native to North America, including the fungi powdery mildew and black rot (*Guignardia bidwellii* (Ellis) Viala and Ravaz), the slime mold downy mildew (*Plasmopara viticola* (Berk. and Curtis) Berl. and de Toni), the bacterium causing Pierce's disease, the aphid phylloxera, and the dagger nematode *Xiphinema index* (which transmits the grapevine fanleaf virus), but is sensitive to winter freeze and lime-induced chlorosis (Alleweldt and Possingham, 1988; Olien, 1990; Olmo, 1986; Ruel and Walker, 2006). Although usually incompatible in both flowering and grafting with *Vitis* species, it does produce fertile hybrids with *V. rupestris*, which allows it to be used in modern (rootstock) breeding programs.
- *Muscadinia munsoniana* Small (Simpson): Native to Florida and the Bahamas, with better flavor and skin characteristics than *M. rotundifolia*, but not cultivated.
- *Muscadinia popenoei* Fennell: Native to southern Mexico ("Totoloche grape"), relatively unknown.

Genus *Vitis*

The genus *Vitis* occurs predominantly in the temperate and subtropical climate zones of the Northern Hemisphere (Mullins et al., 1992; Wan et al., 2008a). All members of this genus are perennial vines or shrubs with tendril-bearing shoots. This genus probably comprises 60–70 species (plus up to 30 fossil species and 15 doubtful species) spread mostly throughout Asia (~40 species) and North America (~20 species) (Alleweldt and Possingham, 1988; Wan et al., 2008b,c). The Eurasian species *Vitis vinifera* L. gave rise to the overwhelming majority of grape varieties cultivated today. Plants that belong to this genus have hairy leaves with five main veins, forked tendrils, bark that shreds when mature, nodes with diaphragms, and soft secondary wood. They all can form adventitious roots, a trait that permits propagation by cuttings, yet only *V. vinifera*, *V. riparia*, and *V. rupestris* root easily from dormant cuttings. Although the ancestor of all *Vitis* species may have had perfect (i.e., bisexual or hermaphroditic) flowers (McGovern, 2003), the extant wild species are dioecious (Greek *dis* = double, *oikos* = house), containing imperfect male (i.e., female sterile or staminate) or female (i.e., male sterile or pistillate) flowers on different individual plants, whereas the cultivated varieties of *V. vinifera* have perfect or, in a few cases, physiologically female flowers (Boursiquot et al., 1995; Levadoux, 1956; Negrul, 1936; Pratt, 1971; see also Figure 1.1). Members of this genus are very diverse in both habitat and form. Nevertheless, all species within the genus can readily interbreed to form fertile interspecific crosses called hybrids, which implies that they had a relatively recent common ancestor. Moreover, all *Vitis* species can be grafted onto each other. The genus is often divided into two major groups: the American and the Eurasian groups. The dominant species of the two groups differ greatly in their useful agronomic traits (Table 1.1), which makes them attractive breeding partners (Alleweldt and Possingham, 1988;

FIGURE 1.1

Flower types in the genus *Vitis*: perfect flower (left), female flower (center), and male flower (right).

Illustrations by A. Mills.

Table 1.1 Broad Viticultural Traits of American and Eurasian Grapevine Species		
Trait	**Eurasian Species**	**American Species**
Fruitfulness	Good	Poor or highly variable
Fruit quality	Good	Poor
Usefulness	Highly diverse products	Niche products, rootstocks
Propagation capacity	Good	Variable
Lime tolerance	Good	Highly variable
Phylloxera tolerance	Poor	Good or variable
Disease resistance	Poor	Good or variable

This et al., 2006). Unfortunately, none of the many attempts and thousands of crosses that have been tested to date have truly fulfilled the breeders' hopes of combining the positive attributes while eliminating the negative ones contained in the natural genetic variation of the two groups. Perhaps the genes conferring disease resistance are coupled to those responsible for undesirable fruit composition. Indeed, hybrids have often been banned in European wine-producing countries because of their perceived poor fruit (and resulting wine) quality. Confident predictions of the future availability and spread of newly bred cultivars in the New World (Olmo, 1952) have been wrecked by the reality of taste conservatism of producers and consumers alike. The only unequivocal success story thus far has been the grafting of phylloxera-susceptible European wine grape cultivars to rootstocks that are usually hybrids of tolerant American *Vitis* species (see Section 1.2).

American group

Depending on the taxonomist, this group contains between 8 and 34 species, of which several have become economically important as wine or juice grapes. Because of their varying resistance to the North American grapevine diseases and pests, members of this group are also being used as rootstocks (see Section 1.2) or crossing partners in breeding programs (Alleweldt and Possingham, 1988; This et al., 2006). As an aside, crosses are always listed as maternal parent × paternal parent (i.e., the mother's name comes first). The species of this group generally have thinner shoots with longer internodes and less prominent nodes than the Eurasian species. They also have smaller buds, and the leaves have very shallow sinuses and often a glossy surface. All grape species native to North America are strictly dioecious (i.e., none of them has perfect flowers), and most of them grow near a permanent source of water, such as a river, stream, or spring (Kevan et al., 1985, 1988; Morano and Walker, 1995; Figure 1.2). Following is an incomplete list and brief description of some of the more important species. Much of this information is derived from the Plants Database (plants.usda.gov) of the United States Department of Agriculture Natural Resources Conservation Service:

- *Vitis labrusca* L.: Vigorous climber ("northern fox grape") native to the eastern United States from Georgia to southeastern Canada, with Indiana as its western limit. This species differs from all others in that it usually has continuous tendrils, that is, a tendril at every node. Some of its cultivars (e.g., Concord and Niagara) are commercially grown in the United States for juice, jam, jelly, and wine production. Nonetheless, these cultivars (which, unlike wild *V. labrusca*, have perfect flowers) are of mixed *V. labrusca* and *V. vinifera* descent; they probably arose through natural hybridization, and thus have been classed as *Vitis* × *labruscana* L. Bailey (Mitani et al., 2009; Mullins et al., 1992; Pratt, 1973; Sawler et al., 2013). The distinct foxy flavor (caused by methyl anthranilate) that characterizes this species is popular in the United States but strange to Europeans. The species is cold tolerant, resistant to powdery mildew and crown gall, and tolerant of phylloxera. It is, however, susceptible to downy mildew, black rot, and Pierce's disease and has poor lime tolerance, preferring acid soils. Hybrids of *V. labrusca* were exported to Europe at the beginning of the nineteenth century. Some of these plants carried powdery mildew, downy mildew, black rot, and phylloxera, which drove most populations of wild vines extinct and brought the European wine industry to the verge of destruction.
- *Vitis aestivalis* Michaux: Vigorous climber native to eastern North America, growing in dry upland forests and bluffs. It is very cold hardy (to approximately $-30°C$) and drought tolerant, tolerates wet and humid summers as well ("summer grape"), and is resistant to phylloxera,

FIGURE 1.2

The "bank grape," *Vitis riparia*, growing in a forest in upstate New York (left) and the "canyon grape," *Vitis arizonica* Engelmann, growing up a riverside tree in Utah's Zion National Park (right). Note the large size of the wild vines and the long "trailing trunks" at the bottom right.

Photos by M. Keller.

powdery and downy mildew, and Pierce's disease. The species is very difficult to propagate from cuttings. Its fruits are used to make grape jelly, and the cultivars Norton and Cynthiana are commercially grown as wine grapes in the southern and midwestern United States (Tarara and Hellman, 1991). It seems likely, however, that the two names are synonyms for the same cultivar and that they are hybrids between *V. aestivalis* and *V. labrusca* and/or *V. vinifera* (Reisch et al., 1993; Sawler et al., 2013).

- *Vitis riparia* Michaux: Widespread in North America from Canada to Texas and from the Atlantic Ocean to the Rocky Mountains. This species climbs in trees and shrubs along riverbanks ("bank grape") and prefers deep alluvial soils, but does not do well in calcareous soils (i.e., prefers acid soils), and its shallow roots make it susceptible to drought—a trait it also confers on the rootstocks derived from its crosses with other species. It is the earliest of all the American species to break buds and ripen its fruit, matures its shoots early, is very cold hardy (to approximately $-36°C$), is tolerant of phylloxera, and is resistant to fungal diseases but susceptible to Pierce's disease.

- *Vitis rupestris* Scheele: Native to the southwestern United States from Texas to Tennessee, the species is now almost extinct. It is found in rocky creek beds ("rock grape") with permanent water, and it is vigorous, shrubby, and rarely climbs. It has deep roots for anchorage but is not very drought tolerant on shallow soils, and its lime tolerance is variable. The species tolerates phylloxera and is resistant to powdery mildew, downy mildew, and black rot, but susceptible to anthracnose (*Elsinoë ampelina* (de Bary) Shear).

- *Vitis berlandieri* Planchon: Native to central Texas and eastern Mexico, this species climbs on trees on deeper limestone soils between ridges. It is one of very few American *Vitis* species that have good lime tolerance. Its deep root system makes it relatively drought tolerant, but it is very susceptible to waterlogging. The species breaks buds and flowers much later than other species and is the latest ripening of the American group with very late shoot maturation. It is somewhat tolerant of phylloxera and resistant to fungal diseases and Pierce's disease, but it is very difficult to propagate and to graft (Mullins et al., 1992).
- *Vitis cinerea* Engelmann: Sprawling, vigorous, but relatively low climber native to much of the southeastern United States through Texas. It thrives in moist woodlands and near streams and prefers relatively acid soils. The species is resistant to powdery mildew.
- *Vitis candicans* Engelmann: Very vigorous climber native to the southern United States and northern Mexico. The species is drought tolerant, relatively tolerant of phylloxera, and resistant to powdery and downy mildew and Pierce's disease, but it is difficult to propagate. Other southern species, such as *V. champinii* Planchon and *V. longii* Prince, are probably natural hybrids of *V. candicans*, *V. rupestris*, and other native species (Pongrácz, 1983). They are highly resistant to nematodes.

Eurasian group

There are approximately 40 known species in this group, most of them confined to eastern Asia. Chinese species are particularly diverse, growing in the dry southwest, the northern and southern foothills of the Himalayas, the very cold northeast, and the hot and humid southeast. Although some of them are resistant to fungal diseases and may tolerate high humidity (Li et al., 2008; Wan et al., 2007, 2008a), most of these species are little known, and there may be several additional species that have not yet been described. Most Eurasian species are not resistant to the North American grapevine diseases, and yet one of them has come to dominate the grape and wine industries throughout the world.

- *Vitis vinifera* L.: Native to western Asia and Europe between 30°N and 50°N, but temporarily confined, in more or less isolated populations, to the humid and forested to arid, volcanic mountain ranges of the southern Caucasus between the Black Sea and the Caspian Sea and to the Mediterranean region during the Ice Ages (Grassi et al., 2006; Hardie, 2000; Levadoux, 1956; Maghradze et al., 2012; Zohary and Hopf, 2001). This is the most well-known species of the Eurasian group, as it gave rise to most of the cultivated grapes grown today. The species is highly tolerant of lime, even more so than *V. berlandieri*, and drought. Because the cultivated grapevines have hermaphroditic or, rarely, physiologically female flowers, they are often grouped into the subspecies *V. vinifera* ssp. *sativa* (also termed *V. vinifera* ssp. *vinifera*). But according to many taxonomists, *V. vinifera sativa* is merely the domesticated form of *V. vinifera* (ssp.) *sylvestris*. According to this view, the differences between the two forms are the result of the domestication process (Levadoux, 1956; This et al., 2006), that is, they arose through human rather than natural selection. However, the concept of subspecies (or geographical races) as taxonomically identifiable populations below the species level ought to be abandoned, as it is highly subjective and inefficient and, therefore, biologically meaningless (Wilson and Brown, 1953).
- *Vitis sylvestris* (or *silvestris*) (Gmelin) Hegi: Native to an area spanning central Asia to the Mediterranean region, this group contains the dioecious wild vines (also termed lambrusca vines)

of Asia, Europe, and northern Africa, growing mainly in damp woodlands ("forest grape") on alluvial soils of river- and hillsides (Ghaffari et al., 2013; Maghradze et al., 2012). Taxonomists debate whether this group of vines deserves species status or whether it is a subspecies of *V. vinifera* (ssp. *sylvestris*, but see above) because, apart from their flowers, the two look so similar and interbreed readily (Mullins et al., 1992). Although Levadoux (1956) concluded that they belong to the same species (*V. vinifera*), the wild form is genetically somewhat distinct from *V. vinifera sativa* (De Andrés et al., 2012; Laucou et al., 2011). It has almost disappeared in Europe and is considered an endangered species in some countries, mainly because of the destruction of its natural habitats and its susceptibility to phylloxera and the mildews introduced from North America (Arnold et al., 1998; Levadoux, 1956). The high humidity and periodic flooding in wooded river valleys may have protected the remaining populations from destruction by phylloxera. These wild grapes are thought to be resistant to leafroll and fanleaf viruses (Arnold et al., 1998).

- *Vitis amurensis* Ruprecht: The genetically diverse native of northeastern China and Russian Siberia ("Amur grape") is said to be the cold hardiest of all *Vitis* species (Alleweldt and Possingham, 1988; Wan et al., 2008a). But budbreak occurs about a month earlier than in *V. vinifera*, making it vulnerable to spring frosts. Bloom and fruit maturity are also advanced, which seems to be true for many other Chinese species (Wan et al., 2008c). It is resistant to crown gall, downy mildew, anthracnose, and *Botrytis* but susceptible to phylloxera (Du et al., 2009; Li et al., 2008; Szegedi et al., 1984). Several cultivars with female or, rarely, hermaphroditic flowers and small clusters of small and very dark-colored berries are grown in northeastern China for wine production.
- *Vitis davidii* (Romanet du Caillaud) Foëx: Subtropical Chinese species growing in river valleys. It has large, heart-shaped leaves and is adapted to warm, humid conditions but has poor cold hardiness. The species is resistant to powdery mildew and *Botrytis* but susceptible to downy mildew (Liang et al., 2013). Some cultivars are grown for wine production.
- *Vitis romanetii* Romanet du Caillaud: One of more than 35 native Chinese species. It grows in forests and shrubland, and on hillsides and is resistant to powdery mildew and anthracnose (Li et al., 2008).
- *Vitis coignetiae* Pulliat: Native to Japan and used locally for jam production and some winemaking. The species is dioecious and strongly resembles the American *V. labrusca* (Kimura et al., 1997; Mullins et al., 1992).

1.2 CULTIVARS, CLONES, AND ROOTSTOCKS
1.2.1 VARIETY VERSUS CULTIVAR

The study of the botanical description, identification, and classification of plants belonging to the genus *Vitis* and of their usefulness for viticulture is termed *ampelography* (Greek *ampelos* = vine, *graphein* = to write). The descriptors have traditionally included visual traits such as shoot tips (shape, hairs, and pigmentation), leaves (shape of blade with lobes, sinuses, and serrations), fruit clusters (size and shape), and berries (size, shape, and pigmentation) (Galet, 1985, 1998; Viala and Vermorel, 1901). Not surprisingly, leaf shape in grapevines is rather highly heritable (Chitwood et al., 2014).

The advent of DNA fingerprinting has led to the adoption of this technique for the identification of selections in grape collections and is increasingly being used to uncover the historical origins and genetic relationships of grapevines.

Cultivated grapevines of sufficiently similar vegetative and reproductive appearance are usually called "grape varieties" by growers and "cultivars" by botanists. Botanically speaking, a variety includes individuals of a (wild) population that can interbreed freely, whereas, according to the international code of nomenclature for cultivated plants, "a cultivar is an assemblage of plants that (a) has been selected for a particular character or combination of characters, (b) is distinct, uniform, and stable in these characters, and (c) when propagated by appropriate means, retains those characters" (Brickell et al., 2009). In principle, a cultivar can be produced both sexually (i.e., from a seedling) and asexually (i.e., as a clone), but only the latter method will give grapevine descendants that are genetically identical (i.e., "true to type"), or at least nearly so. Seedlings of the so-called grape varieties are not identical copies of their mother plant. Because grapevines are heterozygous (Greek *heteros* = other, different) across a large number of chromosome positions or loci (singular locus) in their genomes, each seed may give rise to a cultivar with distinct characteristics (Bacilieri et al., 2013; Laucou et al., 2011; Mullins et al., 1992; Pelsy et al., 2010; Thomas and Scott, 1993; see also Section 2.3).

Grapes were among the first fruit crops to be domesticated, along with olives, figs, and dates (Maghradze et al., 2012; Zohary and Hopf, 2001; Zohary and Spiegel-Roy, 1975). Although all four wild fruits had ranges stretching far beyond the eastern Mediterranean region, the deliberate cultivation of grapes for winemaking as well as fresh fruit and raisin production probably started about 7000–8000 years ago in the eastern part of the Fertile Crescent that spans the modern-day countries of the Near East: Egypt, Israel, Lebanon, Jordan, Syria, eastern Turkey, Iraq, and western Iran. This is roughly the same time that farmers invented irrigation and precedes by approximately 2000 years the earliest civilization of Sumer in southern Mesopotamia (McGovern et al., 1996). As an interesting aside, the Sumerian symbol for "life" is a grape leaf, and the "tree of life" of many ancient civilizations is the grapevine. These early societies regarded wine as "nectar of the gods," and wine drinking was considered to be a hallmark of civilization; "savages" or "barbarians" drank no wine. The biblical legend, modeled on the earlier Mesopotamian Epic of Gilgamesh (McGovern, 2003), of Noah's ark stranding near volcanic Mount Ararat on the Turkish/Iranian/Armenian border and his subsequent planting of the first vineyard also reflects the probable origin of viticulture. Indeed, the oldest winery known to date was discovered in southeastern Armenia at an elevation of almost 1000 m and dates to about 6000 years ago (Barnard et al., 2011). Even today, this arid region is home to wild forms of *V. vinifera* (or *V. sylvestris*) with a greater genetic diversity than is found anywhere else (Grassi et al., 2006; Maghradze et al., 2012). Moreover, many cultivated varieties of *V. vinifera* are more closely related to *V. sylvestris* from the Near East than to their wild relatives from Western Europe (Myles et al., 2011). Nevertheless, new archeological discoveries show that wine might have been made from wild grapes in China as far back as 9000 years ago (McGovern et al., 2004). In Europe, grapes were grown at least 4000 years ago (Rivera Núñez and Walker, 1989). Clearly, humans quickly learned how to use fermentation as one of the most important food preservation technologies; they also learned to preserve grapes as raisins by drying them. Domestication of *V. vinifera* was also accompanied by a change from dioecious to hermaphroditic reproduction and an increase in seed, berry, and cluster size (This et al., 2006; Zohary and Spiegel-Roy, 1975); perhaps early viticulturists selected not only more fruitful vines but also plants that were self-pollinated, thus eliminating the need for fruitless pollen donors.

Today, an estimated 10,000 or so grape cultivars are being grown commercially, although DNA fingerprinting suggests that a more accurate figure may be approximately 5000 (This et al., 2006). Many cultivated grapes are closely related to one another; in fact, many of them are siblings or cousins (Cipriani et al., 2010; Di Vecchi Staraz et al., 2007; Robinson et al., 2012). Moreover, many cultivars are known by several or even many synonyms (different names for the same cultivar) or homonyms (identical name for different cultivars). A good example of the latter is Malvasia, a name that has been applied to dozens of related and unrelated cultivars (many with several synonyms) with a range of berry colors and flavor profiles (Lacombe et al., 2007). The vast majority of cultivars belong to the species *V. vinifera*, and comparisons of the DNA contained in chloroplasts suggest that they might have originated from (at least) two geographically distinct populations of *V. sylvestris*: one in the Near and Middle East and the other in a region comprising the Iberian Peninsula, Central Europe, and Northern Africa (Arroyo-García et al., 2006). Most cultivated grapevines have a very large effective population size and evidently did not experience a genetic bottleneck during domestication, which means that domestication and subsequent breeding did little to reduce the genetic diversity in grapes (Fournier-Level et al., 2010; Laucou et al., 2011; Myles et al., 2010, 2011). Such reduction of genetic diversity did, however, occur when phylloxera and mildew pathogens introduced from North America in the second half of the nineteenth century wreaked havoc across European vineyards (Cipriani et al., 2010).

Most current cultivars are not products of deliberate breeding efforts. Rather, they are the results of continuous selection and vegetative propagation over many centuries of grapevines that were spontaneously generated by mutations (Latin *mutatio* = change) and sexual reproduction, and via somatic (Greek *soma* = body) mutations. Seedlings may have been selected that resulted from sexual reproduction following either self-pollination within the flowers of a single mother vine or cross-pollination involving different vines (see Section 2.3). In the latter case, intraspecific hybridization was facilitated by the ancient practice of growing multiple varieties in the same vineyard, a custom that was widespread until the nineteenth century (Cipriani et al., 2010). Somatic mutations are mutations that occur in the dividing cells of the shoot apical meristem (Greek *merizein* = to divide; see Section 1.3). Because somatic mutations typically happen during bud formation (see Section 2.2), clones that result from propagation of affected shoots are also called bud sports. Moreover, since plants, unlike animals, do not separate their somatic cells from their germ line cells (Walbot and Evans, 2003), and because somatic mutations are often nonlethal, such somatic mutations will also be carried forward into the new seeds derived from mutated cell lines (see Section 2.3).

Mutations in the grape genome arise through rare, chance mistakes in DNA replication during the process of cell division. Whenever a cell divides, it first has to double (i.e., copy) its DNA so that each daughter cell gets a complete set of chromosomes. Millions of nucleotides are regularly copied with amazing accuracy, and sophisticated "spell-checking" and repair mechanisms attempt to fix most mistakes that do occur. However, every so often, a nucleotide is inadvertently switched with another one, or a whole group of nucleotides (termed a DNA sequence) is inserted, repeated, omitted, or moved to a different location on the chromosome. Many DNA sequences, called "jumping genes" (a.k.a. transposons, transposable elements, or mobile elements), even move around a chromosome on their own, some of them employing "cut-and-paste" or "copy-and-paste" strategies (Benjak et al., 2008; Carrier et al., 2012; Godinho et al., 2012; Lisch, 2009). Some of these transposons might be remnants of old virus infections following the loss of a virus's ability to move from cell to cell. Most of these repetitive DNA sequences and transposons are silenced, meaning their activity is suppressed by the addition of methyl groups that hinder their transcription (i.e., the copying of the DNA into RNA, which leads to gene expression) (Pecinka et al., 2013). Nonetheless, new genes quite often seem to arise from duplication of an existing gene and subsequent mutation of one of the two copies (Díaz-Riquelme et al., 2009; Firn and Jones, 2009; Guo et al., 2014). Many mutations are also caused by damage to the DNA by so-called reactive oxygen species that are produced during oxidative stress (Halliwell, 2006; Møller et al., 2007; see also Section 7.1).

Thus mutations, if they are not lethal or silenced, can give rise to slightly different forms of the same gene that are termed alleles (Greek *allelos* = each other), just as does the recombination that occurs during sexual reproduction (see Section 2.3). The number of these gene variants at any locus, that is, any gene location on a chromosome, of grapevines varies from as little as 3 to more than 30 (Bacilieri et al., 2013; De Andrés et al., 2012; Laucou et al., 2011). The genome of the *Vitis* species is approximately 500 million DNA base pairs long and comprises an estimated 30,000 genes (Jaillon et al., 2007; Velasco et al., 2007) that form both a vine's "building plan," including its flexible architectural and engineering design, and its "operating manual" by encoding (i.e., prescribing) the amino acid makeup of proteins. For comparison, we seemingly far more complex humans are thought to have fewer than 25,000 genes. Like in humans, though, most of the genetic diversity occurs *within* local populations of (wild) grapevines, rather than *among* them (De Andrés et al., 2012).

Because there are genes that influence growth habit, leaf shape, disease resistance, cluster architecture, berry size and color, flavor, and other quality attributes, some of the mutations also affect these traits (Bessis, 2007; Bönisch et al., 2014; Fernandez et al., 2006, 2010; Pelsy, 2010). For instance, dark-skinned (i.e., anthocyanin-accumulating) fruit is the "default" version in the Vitaceae, and it appears that virtually all *V. vinifera* cultivars with green-yellow fruit (the so-called white cultivars) have a single common ancestor that arose from mutations of two neighboring genes of an original dark-fruited grapevine (Boss and Davies, 2009; Cadle-Davidson and Owens, 2008; This et al., 2007; Walker et al., 2007). These mutations were probably caused by the insertion of a jumping gene that renders grapes unable to "switch on" the anthocyanin assembly line (Hichri et al., 2011; Kobayashi et al., 2004). The chances of both mutations occurring together are extremely small; this might have been a one-time event in the history of grapevines that is thought to have occurred only *after* grapevines were first domesticated (Fournier-Level et al., 2010; Mitani et al., 2009). Although this implies that most white-fruited cultivars are likely to be genetically more closely related to each other than are cultivars with dark fruit, roughly half of all current grape cultivars have fruit with "white" skin. Instability of at least one of these mutations may account for the occasional appearance of dark-skinned variants of white cultivars (Boss and Davies, 2009; Lijavetzky et al., 2006). The same jumping gene that caused the appearance of white grapes, moreover, is also responsible for much of the berry color variation among dark-skinned cultivars (Fournier-Level et al., 2009). For example, Pinot noir has one intact and one mutated allele in each of its two meristematic cell layers (see Section 1.3), whereas Pinot gris lacks the intact version in its inner (unpigmented) cell layer that, moreover, has partly displaced cells from the outer (pigmented) layer, and Pinot blanc carries only the mutated version in both cell layers (Vezzulli et al., 2012; see also Figure 1.3). In a dose-dependent, cultivar-specific manner, the presence of this gene, in concert with a few others, permits more or less of the enzymes that produce anthocyanins to be synthesized, so that cultivars vary greatly in their ability to accumulate anthocyanin pigments.

FIGURE 1.3

Pinot noir (left) growing next to Pinot gris, some of whose berries resemble Pinot blanc due to displacement of the pigmented epidermis cells by underlying nonpigmented cells (right).

Photo by M. Keller.

Mutations also add to the genetic variation within a cultivar, and because mutations accumulate over time, the variation also increases. Jumping genes, of which over 100 are known in *V. vinifera*, are responsible for the majority of mutations that lead to the diversity among grapevine clones (Carrier et al., 2012). In addition, some clonal differences in gene expression may also arise from variation in gene silencing, whereby genetically identical regions may have a different DNA methylation pattern—such differences are, therefore, epigenetic (Pecinka et al., 2013). In other cases, a change in a few of the several hundred amino acids that compose a protein may render that protein ineffective. For example, substitution of only 3 out of the approximately 480 amino acids in a monoterpene glucosyltransferase (an enzyme that attaches a sugar molecule to a monoterpene; see Section 6.2) accounts for the difference between two Gewürztraminer clones, one aromatic and the other nonaromatic (Bönisch et al., 2014). However, although ancient cultivars (e.g., Grenache, Pinot noir) can be quite heterogeneous and are planted as many different clones, the genetic similarity of these clones typically is still on the order of 95–99% (Bessis, 2007; Blaich et al., 2007; Cabezas et al., 2003; Levadoux, 1956; Pelsy, 2010; Wegscheider et al., 2009). One drawback of prolonged vegetative propagation is that the descendant clones tend to accumulate deleterious mutations and systemic pathogens (e.g., viruses and bacteria), both of which will eventually curtail their agronomic performance (McKey et al., 2010). But because the frequency at which mutations occur tends to increase as plants age, and because some of these mutations may increase the plant's adaptive fitness (Thomas, 2013), one might argue that clonal selection programs could gain from focusing on the oldest available vineyards.

Hybridization, on the other hand, occurs by cross-pollination and fertilization (see Section 2.3) of flowers from different plants that are genetically distinct and thus gives rise to a genetically novel individual arising from a seed. Both deliberate interbreeding and natural hybridization have occurred many times in the history of viticulture. Many cultivars were originally selected from domesticated local wild grapes (*V. sylvestris* or *V. vinifera* ssp. *sylvestris*) and further developed by selection of seedlings that arose from intentional or accidental interbreeding with other wild grapes, or by the introduction of exotic varieties (Arroyo-García et al., 2006; Bacilieri et al., 2013;

Grassi et al., 2003; Levadoux, 1956; Maghradze et al., 2012; Myles et al., 2011; Salmasco et al., 2008; Santana et al., 2010; Sefc et al., 2003). Some hybridization also occurs in the opposite direction, whereby pollen from cultivated grapevines finds its way onto the flowers of nearby wild plants (De Andrés et al., 2012; Di Vecchi Staraz et al., 2009). Intriguingly, although the majority of the French wine grape cultivars are genetically closely related to each other and to the wild grapes of western Europe, most Italian and Iberian cultivars are different from these and also more distinct among themselves, and most table grapes are in a different "genetic league" altogether. It appears that the Italian and Iberian peninsulas have long served as "melting pots" and exchange sites for cultivars arriving by sea or land (Bacilieri et al., 2013).

Whereas almost all wine and juice grapes contain seeds, many raisin and table grape cultivars are seedless, a selected trait that seems to originate from a very narrow genetic base, mainly Sultanina (synonyms Sultana, Thompson Seedless, Kishmish), an ancient Middle Eastern grape variety (Adam-Blondon et al., 2001; Boursiquot et al., 1995; Ibáñez et al., 2009). Almost all of the about 20 cultivars with "muscat" aroma, most of them used as raisin or table grapes and many of them with a number of synonyms, are direct descendants of the ancient Muscat à petits grains (synonym Muscat blanc) (Cipriani et al., 2010; Crespan and Milani, 2001). One of these descendants, Muscat of Alexandria, is one parent of the Argentinian Torrontés cultivars—which are unrelated to several different Spanish cultivars named Torrontés (Agüero et al., 2003; Gago et al., 2009). Nevertheless, many "noble" grape lineages are as interwoven as those of their human selectors.

For instance, it is thought that Cabernet franc (which is closely related to Petit Verdot) and Savagnin (synonym Traminer) may have been selected from wild grapes, whereas Cabernet Sauvignon resulted from a natural cross between Cabernet franc and Sauvignon blanc (Bowers and Meredith, 1997; Cipriani et al., 2010; Levadoux, 1956). Cabernet franc also fathered Merlot and Carmenère (Boursiquot et al., 2009; Salmasco et al., 2008). Mutations within Savagnin over time gave rise to what are now considered three distinct cultivars: the original Savagnin blanc with white berries, Savagnin rose with pink berries, and Savagnin rose aromatique (synonym Gewürztraminer) with pink and highly aromatic berries (Pelsy et al., 2010). Syrah (synonym Shiraz) is derived from a cross between Dureza and Mondeuse blanche, whereas Syrah crossed with Peloursin gave rise to Durif (synonym Petite Sirah) (Meredith et al., 1999; Vouillamoz and Grando, 2006). Durif and Peloursin, on the other hand, are closely related to Malbec and Marsanne, which are distinct from the (other) Bordeaux and Burgundy cultivars. Chardonnay, Gamay noir, Aligoté, Auxerrois, Melon, and other French cultivars all originated from the same two parents, Pinot and Gouais blanc (synonym White Heunisch), which means they are full siblings (Bowers et al., 1999; Regner et al., 2000). Pinot blanc and Pinot gris are somatic berry-pigment mutants of Pinot noir that arose independently of each other (Vezzulli et al., 2012; see also Figure 1.3). Similar berry-pigment mutants exist for Garnacha (synonym Grenache), the world's most widely planted grape cultivar (Cabezas et al., 2003). Indeed, variation in berry color seems to be a fairly common outcome of somatic mutation (Bessis, 2007; Furiya et al., 2009; Müller-Stoll, 1950; Regner et al., 2000). The Pinot family in turn is thought to be derived from a spontaneous cross between Traminer and Pinot Meunier (synonym Schwarzriesling) (Regner et al., 2000; Salmasco et al., 2008). Whereas Pinot also appears to be one of the more distant ancestors of Syrah, Traminer features as a parent of Sauvignon blanc, Sylvaner, and many other cultivars (Myles et al., 2011; Sefc et al., 1998). Traminer and Sauvignon blanc, in turn, may be the parents of Chenin blanc (Cipriani et al., 2010). Gouais blanc, crossed with a Traminer variety, probably gave rise to Riesling, and with Chenin blanc it yielded

Colombard and other cultivars. Even Lemberger (synonym Blaufränkisch), which is genetically similar to many southeastern European wine grapes, may have been the result of a Gouais blanc cross with an unknown partner. In fact, based on genetic evidence, the ancient eastern European (Croatian) cultivar Gouais blanc has thus far been proposed to be involved in more than 75 European cultivars. Its prominent position as a parent of many premium wine grape varieties is somewhat ironic because its fruit quality has long been considered inferior. Yet its vigorous growth and high fertility made it a favorite with growers, who grew it alongside selections from wild vines (e.g., Traminer with highly variable yields) for centuries so that it became widespread throughout Europe in the Middle Ages. Moreover, because of their sour berries, Gouais vines were often planted as buffers around vineyards to deter potential grape thieves. Another old cultivar from the Adriatic coast has also attained some popularity in both the Old and the New World, albeit under different names: Primitivo in southern Italy, Crljenak kaštelanski in Croatia, and Zinfandel in California (Fanizza et al., 2005; Maletićet al., 2004).

The case of Gouais blanc demonstrates that obscurity is not unusual for parents of current noble grape cultivars. Further examples of widely grown cultivars, one of whose parents, when finally determined by DNA fingerprinting, turned out to have fallen into oblivion, include Merlot and Tempranillo (Boursiquot et al., 2009; Ibáñez et al., 2012; Salmasco et al., 2008). Moreover, as novel DNA fingerprinting tests show, what breeders get is not always what they set out to develop, despite their best efforts to exclude nondesired pollen from their crosses (Bautista et al., 2008; Ibáñez et al., 2009). The most well-known example is the German wine grape cultivar Müller-Thurgau, which has long been disseminated as a deliberate cross of Riesling flowers with Sylvaner pollen (hence the synonym Riesling × Sylvaner) (Becker, 1976). More than 100 years after its introduction, however, genetic analysis exposed Madeleine Royale, a likely progeny of Pinot noir, as the illegitimate father (Dettweiler et al., 2000). Similarly, Cardinal, a table grape cultivar with worldwide distribution, may be derived from Alphonse Lavallée (a possible progeny of Gros Colman and Muscat Hamburg) and Königin der Weingärten (probably a descendant of Pearl of Csaba) and not, as has been assumed, from Flame Tokay and Alphonse Lavallée (Cipriani et al., 2010; Ibáñez et al., 2009). Natural hybridization can also cross species boundaries, as exemplified by the North American juice grape cultivar Concord, which was selected from a wild *V. labrusca* seedling that was apparently pollinated by an unknown seedling with partial *V. vinifera* parentage (Mitani et al., 2009; Mullins et al., 1992; Sawler et al., 2013). Misidentification of planting material imported to new grape growing regions is another common problem. For instance, the so-called Bonarda in Argentina and Charbono in California are not related to the various Italian cultivars of the same names but are in fact both identical to the old French cultivar Corbeau (Martínez et al., 2008).

Although each cultivar originally began as a single vine that grew from a seedling, most major cultivars grown today have been propagated vegetatively for a long time, some for many centuries and perhaps even millennia (Bessis, 2007; Di Vecchi Staraz et al., 2007; Hardie, 2000; Pelsy et al., 2010). Propagation by cuttings was undoubtedly in use by Roman times 2000 years ago (Columella, 4–ca. AD 70). For instance, Gouais blanc is thought to have been brought from Croatia (then Dalmatia) to France (then Gallia) along with Viognier (the Croatian cultivar Vugava bijela seems to be identical with Viognier and a close relative of the Italian Barbera and the Swiss Arvine) around AD 280 by the Roman Emperor Probus, who encouraged vineyard development to ensure economic stability. In fact, if a map of the distribution of viticulture in Europe, the Near East, and North Africa were to be overlaid with a map of the Roman Empire at its greatest extent, there would be an almost exact geographical overlap. It is believed that the rare extant Swiss variety Rèze is, if not identical with, at least a direct descendant of Raetica, which was considered by Roman writers to be one of the two best wine grape varieties of the Empire (Vouillamoz et al., 2007). A long gap in cultivar description and identification followed the demise of the Roman Empire. The cultivar name Traminer was first mentioned in 1349, Pinot gris (synonym Ruländer) in 1375, Pinot in 1394, Riesling in 1435, Chasselas (synonym Gutedel) in 1523, and Sangiovese in 1590, although it is not certain that the same name was consistently applied to the same cultivar (von Bassermann-Jordan, 1923). Traditional European vineyards were—and in some areas still are—composed of a population of heterogeneous vines, and sometimes several cultivars were planted together in the same block so that many cultivated varieties have no defined origin. The identification and cultivation of "pure" cultivars is quite recent. Nonetheless, the genetic diversity, as well as similarity, among extant *V. vinifera* cultivars agrees well with the idea that viticulture originated in the Near East and spread to Europe via routes around the northern and southern shores of and across the Mediterranean Sea (Bacilieri et al., 2013). More details about the origins, genetic relationships, current distribution, as well as viticultural and enological characteristics, of 1368 extant cultivars can be found in the comprehensive review by Robinson et al. (2012).

1.2.2 CULTIVAR CLASSIFICATION

Following several millennia of cultivation and repeated selection of spontaneous mutants and natural as well as man-made intra- and interspecific crosses in many different regions, there is a vast range of cultivated forms and types of grapevines. Because thousands of grape cultivars are being grown commercially, various attempts have been made to group them into families or tribes. Unlike botanical classifications, these groupings are not always based on phenotypic or genotypic differences, whereby the genotype is the sum of the genetic material of an organism, and the phenotype arises from the interaction of the genotype with the environment during development (Mayr, 2001). The most common methods involve classification on the basis of place or climate of origin, viticultural characteristics, final use, or winemaking characteristics.

Arguably the first cultivar classification was attempted by Plinius (ca. AD 50) in Rome, who created three groups according to berry color and yield:

1. *Anemic cultivars*: Varieties with small, white grapes
2. *Nomentanic cultivars*: Varieties with red grapes, low yielding
3. *Apianic cultivars*: High-yielding varieties, poor quality

During the Middle Ages, European wine grapes were simply divided into two groups according to perceived wine quality (von Bassermann-Jordan, 1923):

1. *Vinum francicum*: "Frentsch" grapes, low-yielding varieties of high quality (e.g., Traminer)
2. *Vinum hunicum*: "Huntsch" grapes, high-yielding varieties of poor quality (e.g., Heunisch)

The division proposed by the Russian botanist Negrul (cited in Levadoux, 1956, and Maghradze et al., 2012) distinguishes three main ecological or ecogeographical groups of varieties, called "proles" (Latin *proles* = offspring), based on their region of origin:

1. *Proles pontica*: Fruitful varieties with small to medium-sized, round berries with seeds ranging from very small to very large that originated from the banks of the Aegean and Black Seas and spread throughout eastern and southern Europe (e.g., Furmint, Clairette, Black Corinth, and Rkatziteli)

2. *Proles occidentalis*: Highly fruitful wine grape varieties of central and western Europe with small clusters of small berries with small seeds (e.g., Riesling, Chardonnay, Sémillon, Sauvignon blanc, Gewürztraminer, the Pinots, and the Cabernets)
3. *Proles orientalis*: Mostly table grape varieties with low bud fruitfulness and large clusters of large, elongate berries with either large seeds or partly to wholly absent seeds (the so-called seedless cultivars) originating from the Near East, Iran, Afghanistan, and central Asia (e.g., Thompson Seedless, the Muscats, and Cinsaut)

Modern genetic research has mostly borne out Negrul's grouping (Aradhya et al., 2003; Bacilieri et al., 2013). Such research also confirmed that many cultivars of the *proles occidentalis* are closely related to the wild *V. sylvestris* (or *V. vinifera* ssp. *sylvestris*), that almost all of the *proles pontica* varieties are closely related to each other, and that the *proles orientalis* are genetically much more diverse than and distinct from the other groups. Yet Negrul's classification is not clear-cut and contains some ampelographic errors. In addition, all of these classifications suffer from the fact that grape cuttings have forever been carried to distant locations and used for (deliberate or natural) interbreeding with locally domesticated and selected variants. Therefore, cultivars have also been grouped according to their viticultural characteristics, although these can be modified by local soil and climatic conditions and cultural practices:

- Time of maturity (Table 1.2): The early maturing Chasselas is used as a basis or standard
- Vigor: Rate of shoot growth
- Productivity: Yielding ability

Table 1.2 Classification of Some Grape Cultivars Based on Their Relative Heat Requirement to Reach Acceptable Fruit Maturity, Subjectively Defined as 18–20°Brix

Group	Red-Wine Cultivars	White-Wine Cultivars
1 (Early)		Madeleine Angevine, Pearl of Csaba
2	Blue Portuguese	Chasselas, Müller-Thurgau
3	Gamay, Dolcetto	Pinot gris, Pinot blanc, Aligoté, Gewürztraminer
4	Pinot noir	Chardonnay, Sauvignon blanc, Sylvaner
5	Cabernet franc, Lemberger	Riesling
6	Merlot, Malbec, Zinfandel, Tempranillo, Cinsaut, Barbera, Sangiovese	Sémillon, Muscadelle, Chenin blanc, Marsanne, Roussanne, Viognier
7	Cabernet Sauvignon, Syrah, Nebbiolo	Colombard, Palomino
8	Petit Verdot, Aramon, Carignan, Grenache, Mourvèdre	Muscat of Alexandria, Ugni blanc
9 (Late)	Tarrango, Terret noir	Clairette, Grenache blanc

Modified from Viala and Vermorel (1909), Gladstones (1992), and Huglin and Schneider (1998).

Classification is also possible in terms of what the grapes will be used for (Galet, 2000), although some cultivars are used for various purposes:

- Table grapes: Large, fleshy or juicy grapes, often seedless, some of them with muscat or foxy aromas. Examples include Cardinal, Cinsaut, Chasselas, and Muscat of Alexandria.
- Raisin grapes: Predominantly seedless grapes. Examples include Thompson Seedless, Flame Seedless, Black Corinth (synonym Zante Currant), and Delight.
- Juice grapes: Some highly aromatic grapes, especially in the United States. Examples include Concord and Niagara.
- Wine grapes: Very sweet, juicy grapes, often low yielding. Examples include Riesling, Chardonnay, Sémillon, Sauvignon blanc, Gewürztraminer, the Pinots and Cabernets, Merlot, Tempranillo (synonyms Aragonez, Cencibel, Tinta Roriz), and Nebbiolo.
- Brandy (distillation) grapes: Generally white grapes producing bland, acidic wines. Examples include Ugni blanc (synonym Trebbiano), Colombard, and Folle blanche.

Of course, divisions can also be made on the basis of winemaking characteristics, but this method is strongly influenced by the market environment and changing consumer demand:

- Grape composition: Basic characteristics (sugar, acid, pH, tannins, flavors, and aroma) important for winemaking.
- Varietal aroma: White grapes can be aromatic (e.g., Riesling, Gewürztraminer, and Muscats) or nonaromatic (e.g., Chardonnay, Sémillon, and Sylvaner).
- Production costs: Value for winemaking reflected in wine and grape price structure and appellation systems.

1.2.3 CLONES

In viticultural parlance, a clone (Greek *klon* = twig) is a group of grapevines of a uniform type that have been vegetatively propagated (usually by cuttings) from an original mother vine that would normally have been selected for a particular desired trait. Such traits include low vigor, high yield, loose clusters, large or small berries, seedlessness, different or more intense fruit pigmentation or other perceived quality attributes, and resistance to a particular disease. Due to vegetative propagation, such clonal traits arise from somatic mutations (see Section 1.3) rather than during sexual reproduction (i.e., in germ cells, see Section 2.3). Nonetheless, somatic mutations may ultimately be propagated both vegetatively and by seeds and result in individual plants of the same cultivar having slightly different genotypes and sometimes phenotypes (Franks et al., 2002; This et al., 2006). This genotypic diversity, which accumulates over time, is termed *clonal variation* (Mullins et al., 1992; Riaz et al., 2002). Even "clonemates" and clones of relatively recent origin are highly unlikely to be strictly genetically identical, because the rules of statistics make somatic mutations of at least some of grapevines' 500 million DNA bases very probable (Lushai and Loxdale, 2002). Most of these mutations, however, occur on the very long stretches of DNA that do not encode any genes. Although the majority of these mutations may thus be "invisible," mutations in coding regions also tend to accumulate over time. Because their cumulative effects may vary among different parts of the plant, grapevines may evolve over their lifetime, and clonal descendants, even when propagated from a single "mother" plant, can be genetically heterogeneous (McKey et al., 2010). If the change is sufficiently distinct (e.g., major change over a short time or many small changes over a long period), clones may come to be called cultivars, as exemplified by the fruit skin color mutants of Pinot: Pinot noir, Pinot gris, and Pinot blanc (Bessis, 2007; Regner et al., 2000). In fact, many clones of the Pinot family are chimeras, which are defined as plants with more than one genetically different cell population that arose from a mutation in only one of the two functionally distinct cell layers or lineages of the shoot apical meristem (Riaz et al., 2002; Thompson and Olmo, 1963; Vezzulli et al., 2012). This includes another old Pinot clone (or Pinot ancestor), Pinot Meunier, whose leaves are densely coated with white hair (Franks et al., 2002; Hocquigny et al., 2004). Chimeras also exist among clones of Chardonnay and Cabernet Sauvignon, which suggests that layer-specific mutations in the apical meristem may be an important source of clonal and varietal variation in grapevines (Moncada et al., 2006; Pelsy, 2010). Therefore, as in the case of species discussed in Section 1.1, there is no clear-cut distinction between clones and cultivars, a fact recognized by Charles Darwin while he developed his revolutionary theory of evolution by (random) mutation and (directional) natural selection (Darwin, 2004).

Although clonal vines were selected and vegetatively propagated for fruitfulness, fruit flavor, and wine quality in Roman times (Columella, 4—ca. AD 70), organized and methodical clonal selection only began in Germany in the late nineteenth century and did not begin in France until the 1960s (Pelsy, 2010). Selection is usually based on the absence of symptoms of virus diseases (e.g., leafroll and fanleaf), healthy growth, and on good performance indicators such as consistent yields and high wine quality, but criteria can also include specific viticultural traits such as fruit set, disease resistance, or drought tolerance. Nevertheless, many so-called clones are actually phenotypes caused by various combinations and degrees of virus infections rather than genuine genetic differences among the plants themselves. Today, clones exist for most major grape cultivars, and their success is based on the ability of some clones to perform differently in diverse environments. Differences among clones, and thus within a cultivar, offer viticulturally relevant diversity and potentially better planting material. For example, some clones of Cabernet franc have less compact clusters with smaller berries and juice with a lower pH than others, or may be less vigorous, while others may be less prone to some fungal diseases (van Leeuwen et al., 2013). Nonetheless, the genetic differences among clones of a cultivar are typically very small. For example, all of the more than three dozen Sangiovese clones analyzed using DNA assays were found to have been derived from the same mother vine (Di Vecchi Staraz et al., 2007; Filippetti et al., 2005).

It is currently fashionable to plant clonal selections in vineyards, although choices are often based on the "fallacy of the perfect clone"—that is, on the false assumption that the "best" clone found on one site will also perform best in a new location. Yet the performance or suitability of a clone on a particular site is strongly modified by the site's environment (soil, climatic conditions, cultural practices, etc.) and also depends on the desired end use of the grapes produced on that site and even regulatory circumstances (e.g., yield restrictions). This makes it nearly impossible to predict clonal performance in new environments. Nevertheless, in some cultivars, certain clonal traits, such as yield, cluster architecture, grape sugar, or acid content, can be quite stable across relatively diverse environments (Huglin and Schneider, 1998). This may be particularly true for old cultivars with very high clonal diversity; examples include the Pinots, Savagnins, and Cabernet franc. Planting several clones of a cultivar in the same vineyard block not only enables growers to conduct their own evaluation at a particular site but also provides some insurance against fluctuations in yield, fruit quality, and disease susceptibility that come with the absence of genetic variation of a single clone.

1.2.4 ROOTSTOCKS

Rootstocks are specialized stock material to which grape cultivars with desirable fruit properties are grafted. The shoot portion of the two grafting partners is termed the *scion*, whereas the rootstock provides the root system to the fused combination of genotypes. The callus cells that form at the graft union initially show an immune response following detection of the nonself nature of the grafting partners (Cookson et al., 2014). For the grafting operation to be successful, the vascular cambiums (Latin *cambium* = change), responsible for cell division (see Section 1.3), of the two grafting partners must make contact with each other so that they can "build" a connection between their separate "plumbing" systems for water and nutrient supply. Grafting of grapevines was common in ancient times; in his *De Re Rustica*, the Roman writer Columella described techniques known to his ancestors. The use of rootstocks derived from North American *Vitis* species saved grape growing in Europe from extinction due to the introduction, on imported planting material, of the aphid-like insect phylloxera in the late nineteenth century (see Section 8.1). On the downside, rootstocks or their progeny that escaped, for example, from abandoned vineyards have also become feral across Europe, and their interbreeding populations are in some areas behaving as invasive species (Arrigo and Arnold, 2007; Laguna, 2003).

Commonly used rootstocks are either individual *Vitis* species or crosses of two or more species (Table 1.3), and due to the dioecious nature of their parents, they are either male or female plants. Examples of male rootstocks include Teleki 5C and SO4 and Riparia gloire de Montpellier, whereas female rootstocks include Kober 5BB, 101-14 Millardet et de Grasset, and Fercal (Meneghetti et al., 2006). The majority of rootstocks in use today are hybrids of three species: *V. riparia, V. rupestris*, and *V. berlandieri* (Galet, 1998). Nevertheless, the genetic basis of the world's rootstocks is extremely narrow because as many as 90% of all *V. vinifera* vines are grafted to fewer than 10 different rootstock varieties. Most rootstocks have remained unchanged following initial breeding efforts before 1900 (Galet, 1998; Huglin and Schneider, 1998). As discussed previously for scion cultivars, this strategy puts vineyards at risk from mutant strains of soil pests, including phylloxera, with potentially devastating effects if a resistance breaks through. An example is the failure in the 1980s, after many years of widespread use, of the *V. vinifera* × *V. rupestris* rootstock Aramon × *rupestris* Ganzin 1 (A × R1) in California. Moreover, with the exception of those derived from *M. rotundifolia*, the so-called resistant rootstocks are not immune to phylloxera or nematodes (microscopic roundworms); they just tolerate them better than do *V. vinifera* cultivars and therefore suffer less from infestation (Grzegorczyk and Walker, 1998; Huglin and Schneider, 1998; Mullins et al., 1992). Tolerant species or cultivars may grow quite well in the presence of these pests, but they do not prohibit pest numbers from building up over time. They can also be symptomless carriers of virus diseases. Infested propagation material can thus be a potential source of inoculum (infectious material) for new vineyards and new viticultural regions.

The technique of grafting combines the tolerance of soil-borne pests of American *Vitis* species with the winemaking quality (i.e., consumer acceptance) of *V. vinifera*. Today, grape rootstock varieties are used not only for their tolerance of or resistance to root parasites, such as phylloxera and nematodes, but also for their ability to influence crop maturity or their tolerance of adverse soil conditions such as drought, waterlogging, lime, and acid or saline soils (May, 1994; Pongrácz, 1983; see also Table 1.3). Rootstocks can also contribute to the management of vine vigor and of grape maturity and composition (Currle et al., 1983; Dry and Coombe, 2004; Galet, 1998).

Table 1.3 Agronomic Characteristics of Important Grapevine Rootstocks

Rootstock	Parent Species	Phylloxera Resistance	Nematode Resistance	Crown Gall Resistance	Phytophthora Resistance	Grafted Scion Vigor	Drought Tolerance	Flooding Tolerance	Lime Tolerance	Salinity Tolerance	Acid Soil Tolerance	Clay Soil Tolerance	Sandy Soil Tolerance	Susceptibility to Mg Deficiency	Susceptibility to K Deficiency	Fruit Maturation	Ease of Bench-Grafting	Ease of Rooting
Riparia Gloire	V. riparia	●	◐		●	●	◐	●	●	●		●	◐	N		A	◐	○
Rupestris St. George	V. rupestris	◐	◐		●	◐	○	●	◐	●	◐	●	●		N	D	◐	○
Rupestris du Lot	V. rupestris	◐	○			◐	○	●	○	●		◐	●			D	◐	●
420A Millardet et de Grasset	V. berlandieri × V. riparia	●	◐		●	◐	○		○			○	◐		Y	D	○	
5BB Kober	V. berlandieri × V. riparia	◐	◐	◐		○	●	◐	○	◐	○	◐	●	Y	Y	D	◐	◐
SO4	V. berlandieri × V. riparia	◐	○			○	●	○	○	●		◐	●	Y	N		●	◐
8B	V. berlandieri × V. riparia	●	◐			◐	◐		◐								●	
5C Teleki	V. berlandieri × V. riparia	●	◐			○	●		○	◐		○	●			A	◐	◐
161-49 Couderc	V. berlandieri × V. riparia	◉	◐			○	●	●	○	●		○	◐				●	●

Rootstock	Parentage													Y/N	Y/N	A/D
99 Richter	*V. berlandieri* × *V. rupestris*	◐	◐			●	◯	◐	●		●	●	◐	Y	Y	D
110 Richter	*V. berlandieri* × *V. rupestris*	◐	◐	●	●	●	◯	◐	●		◐	●	●	Y	Y	D
1103 Paulsen	*V. berlandieri* × *V. rupestris*	◐	◐	◐	◐	●	●	◐	◯	◯	●	◯	◯	N	Y	D
140 Ruggeri	*V. berlandieri* × *V. rupestris*	◐	●			◯	●	●	◯	◐	◐	◐	◐	N	Y	D
44-53 Malègue	*V. riparia* × *V. rupestris*	◐	◯	●	●	◯	◐	●	◐	◯	●	⊙	◐	Y	N	A
3309 Couderc	*V. riparia* × *V. rupestris*	◯	◐	●	●	◯	◐	●	●		●	◐	●	N		A
101-14 Millardet et de Grasset	*V. riparia* × *V. rupestris*	◯	◐			●	◐	●	●	◯	●	●	◯			A
Schwarzmann	*V. riparia* × *V. rupestris*	◯	◯			●	●	◯	●		◐	◐	◯			
Gravesac	*V. berl.* × *V. rip.* × *V. rup.*	◐	◯	●	●	◯	◐	◐			◐	◐	◯			
1616 Couderc	*V. solonis* × *V. riparia*	◯	●		◯	◯	◐	◐	●	●	◯	●	◯			A
Salt Creek (Ramsey)	*V. champinii*	◯	●			◯	◐	●	●	●	◯	●	◯			
Dogridge	*V. champinii*	◯	◐			◯	●	●		●	◯	●	●			
Harmony	*V. cham.* × *V. sol.* × *V. rip.*	◯	◐			●	◐	●	●	●	◐	●	●			
Freedom	*V. cham.* × *V. sol.* × *V. rip.*	◯	◐			●	◐	◐	◐	◐	●	◯	◯			

◐ ⊙ ◯ Excellent/high..............Poorflow ●
A: Advanced; D: Delayed
N: No; Y: Yes
Modified from Currle et al. (1983), Pongrácz (1983), Galet (1998), and Dry and Coombe (2004).

Although the influence on scion vigor is an important consideration in rootstock choice, there are no truly dwarfing, grape rootstocks, unlike in the case of, for instance, apple. On the contrary, many interspecific hybrids grow more vigorously than their parents, perhaps as a result of the effect termed "hybrid vigor" or heterosis and/or of reduced investment in sexual reproduction, which may free up resources for reproductive growth (McKey et al., 2010).

The influence of rootstocks on the composition of the fruit produced by the scion cultivar is discussed in Section 6.3. Although it seems possible for a rootstock and a scion to exchange DNA pieces at the graft union (Stegemann and Bock, 2009), the rootstock effect on fruit composition is indirect and relatively minor. Some of the lesser known eastern Asian species, such as *V. amurensis*, are used increasingly in modern rootstock breeding programs for their cold tolerance (Alleweldt and Possingham, 1988), even though this diminishes the phylloxera resistance of the resulting crosses.

1.3 MORPHOLOGY AND ANATOMY

Many grapevine species are very vigorous, woody climbers named lianas, while others have a more brush-type growth habit (see Section 1.1). All are perennial (Latin *per* = through; *annus* = year), polycarpic (Greek *polus* = many, much; *karpos* = fruit, grain), and deciduous (Latin *decido* = to fall down). These terms are used to describe the vines' typical life form: they live longer than 2 years, they flower and reproduce repeatedly during their life, and they shed their leaves each year. With the aid of their tendrils and flexible trunks, wild vines climb on trees to a height of 30 m or more and spread out their foliage over the tree canopy (see Figure 1.2). As "structural parasites," vines can maximize their leaf area and shoot growth rate while minimizing their investment in a supporting stem structure (Ewers and Fisher, 1991; Wyka et al., 2013). Using the trees for support (i.e., seizing them as a natural trellis system), they can grow to tremendous size; the canopy of a single vine may cover a surface area of dozens of square meters. Grapevines can live to several hundred years of age: Famous examples include the "Great Vine" in London's Hampton Court Palace, which is thought to have been planted before 1770 and continues to produce 200–300 kg of fruit each year, and the "Old Vine" of Maribor (Slovenia), which is believed to be over 400 years old (Vršič et al., 2011). Vegetative propagation, whether naturally by layering or artificially from cuttings, bud grafts, tissue culture, and other means, extends a vine's life span virtually indefinitely. In fact, both rooted cuttings and grafted plants show clear signs of rejuvenation compared to their "mother" plants (Munné-Bosch, 2008). For instance, growth rates, leaf gas-exchange rates, and fruiting of recently propagated vines are like those of young plants and are independent of the age of the vines from which the propagation material was taken.

Like other higher plants, grapevines comprise vegetative organs and generative or reproductive organs or fruiting structures. The vegetative organs include the roots, trunk, shoots, leaves, and tendrils (Figure 1.4), while the reproductive organs include the clusters with flowers or berries. A vine's vegetative part can be divided into a belowground portion (roots) and an aboveground portion (trunk and shoots). The aboveground portion together with the reproductive organs is termed the vine's canopy. The roots, the trunk, and its trained extension, the cordon, together form the permanent structure of the vine, which makes up 50–75% of the biomass (dry matter) of cultivated

1.3 MORPHOLOGY AND ANATOMY

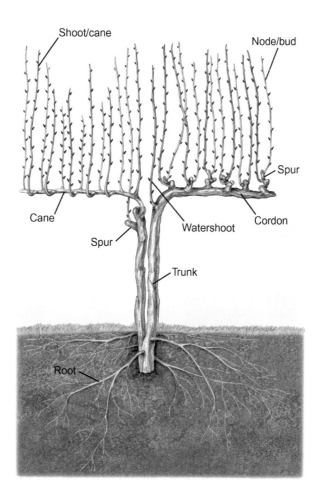

FIGURE 1.4

Own-rooted (ungrafted) grapevine body with vegetative organs, but without leaves and tendrils. The left half is head-trained and cane-pruned, the right half is cordon-trained and spur-pruned.

© Elsevier Inc. Illustration by L. O'Keefe after M. Keller.

grapevines. This proportion is much lower in young plants and increases with age and plant size; it is lower yet in heavily cropped vines and is also lower in humid climates than in dry climates.

The entire plant body is rigid and strong yet flexible and adaptive enough to explore soil resources, intercept sunlight, and develop seeds for propagation. This combination of strength and flexibility is possible because plants employ a "bricks-and-mortar" strategy for growth. The vine's basic building blocks, the cells (bricks), have the ability to divide and expand, and they are joined together by an adhesive matrix, the cell walls (mortar). More precisely, the two cell walls of adjoining cells are "glued" together by the so-called middle lamella, which prevents sliding between cells. Growing cells secrete a thin (<0.1 µm), flexible primary wall, which is reinforced in mature cells that have ceased expanding with a thick (>1 µm), lignified secondary wall for extra mechanical strength (Doblin et al., 2010). The cell walls partly take on the function of bones in animals: They give the plant body form and rigidity, and like bones, cell walls are rich in calcium. The cells differentiate into various types, each "designed" to carry out specific physiological or structural tasks under the direction of different sets of genes; although each cell of the same plant usually contains the same complement of genes (exceptions are discussed in Section 1.2), different genes can be active or inactive—that is, switched "on" or "off." Such similar and specialized groups of cells that form functional units are termed tissues, and different tissues, such as epidermis, parenchyma, phloem, and xylem, in turn make up the plant's organs, which are described in the following sections.

The evolution of such complex multicellular structures from single-celled ancestors required the simultaneous development of innovative solutions to solve the problem of communication between the cells forming a tissue, between different tissues within an organ, between the diverse organs of the plant body, and among plants of a community. As will become apparent throughout this book, plants master the art of intercell communication by cell-to-cell channels named plasmodesmata, whereas a vascular system composed of the xylem and phloem ensures interorgan communication (Lough and Lucas, 2006; Roberts and Oparka, 2003). The roots and shoots "converse" with each other, at least in part, via the production and release of different hormones (Greek *hormaein* = to stimulate); cytokinin and abscisic acid (ABA) are the main root hormones, whereas auxin and gibberellin are the main shoot hormones. Hormones are defined as extracellular signaling molecules that act on target cells distant from their site of production. Perhaps even more fascinating, information exchange between neighboring plants is achieved by emitting and receiving volatile chemicals transmitted through the air and by releasing soluble chemicals into the soil and recruiting symbiotic microorganisms that infect and thereby interconnect the roots of different plants. Of course, successful communication requires not only the production and export of appropriate signaling molecules from transmitting cells (output signals) but also the presence of a sophisticated network of signal perception, decoding, and transformation in the receiving cells (input signals). Such signals integrate genetic programs that perceive and respond to both developmental and environmental cues to direct cell, tissue, organ, and plant growth and physiology (Lough and Lucas, 2006).

1.3.1 ROOT

The root system is the interface between the grapevine and the soil. It provides physical support for the vine in the soil and is responsible for water and nutrient uptake. The roots also serve as storage organs for carbohydrates and other nutrients, which support the initial growth of shoots and roots in spring, and for water. In addition, they are a source of plant hormones (cytokinins and ABA), which modify shoot physiology.

True roots develop from the hypocotyl (Greek *hypo* = below, *kotyledon* = depression) of the embryo (Greek *embryon* = unborn child). In vines grown from seeds, seed germination begins with water absorption, called imbibition, which enables the embryonic root, or radicle (Latin *radicula* = little root), to grow and rupture the seed coat, forming a primary root or tap root, from which multiple secondary roots, branch roots or lateral roots, grow. This type of root morphology is typical of dicotyledons and is called an allorhizic root system (Osmont et al., 2007). In vegetatively propagated vines, roots originate from the cambium layer of woody cuttings (Pratt, 1974). The growth hormone auxin promotes, whereas cytokinin and strigolactone suppress, the initiation of root growth from the cambium (Rasmussen et al., 2013). Most of these so-called adventitious roots form near the nodes, but they also grow on the internodes. These roots are often termed main roots, and they branch off into secondary, tertiary, etc. lateral roots to form a root system with a complicated architecture, which resembles the secondary homorhizic root system typical of monocotyledons (Osmont et al., 2007). In other words, main roots develop directly from a cutting, whereas lateral roots develop from other roots. Unlike the arrangement of shoot lateral organs, the number and placement of lateral roots are not predetermined (and hence inflexible) but depend on the availability of water and nutrients in the soil (Malamy and Benfey, 1997). Because the grapevine is unable to move away from poor soils, it relies on the ability to detect these environmental cues and extend lateral roots to take advantage of favorable soil regions. Moreover, in contrast to many other plant species, lateral root initiation in grapevines is not restricted to the unbranched apical zone. Grapevines are also able to grow new lateral roots on older parts of the roots that have already developed a vascular cambium. Under conditions of very high humidity and high temperature, trunks and other aboveground parts of the vine can also form aerial roots. Whereas this phenomenon is mostly restricted in the genus *Vitis* to greenhouse conditions (and is rare even there), it is quite common in the natural habitat of the *Muscadinia* (Viala and Vermorel, 1909). Nonetheless, *V. sylvestris* vines growing in humid forest habitats readily form aerial roots, too (Levadoux, 1956).

The root system of established grapevines comprises a far-reaching, highly branched structure with a surface area far exceeding that of the leaf canopy it supports. A mature, cultivated grapevine can have more than 100 km of total root length with a surface area greater than 100 m^2, whereas its leaf area is usually less than 10 m^2. The woody roots, whose diameter rarely exceeds 3 or 4 cm, serve to anchor the vine and transport and store soil-derived nutrients, whereas the small absorbing roots ("fine roots," 0.1–1 mm in diameter) are responsible for acquisition of resources such as water and nutrients. The woody roots of mature vines are widely distributed, with horizontal roots exploring the soil for distances of up to approximately 10 m from the trunk (Huglin and Schneider, 1998; Smart et al., 2006; Figure 1.5). Although the majority of roots, especially the fine roots, are normally concentrated in the top 0.5–1 m, roots can grow to a depth of more than 30 m when they encounter no impermeable barriers (Galet, 2000; Lehnart et al., 2008; Morlat and Jacquet, 1993; Pourtchev, 2003; Viala and Vermorel, 1909). Indeed, grapevines are among the most deep-rooted plants, and their root biomass can range from 5 to 40 t ha^{-1}, which may be a reflection of the competition for water and nutrients during the vines' coevolution with their "trellis" trees (Huglin and Schneider, 1998; Smart et al., 2006).

The growing root tip or apex is covered by a slimy root cap whose starch-containing central portion is named the columella. The cap protects the root meristem from abrasion, facilitates penetration of the soil, and contains gravity sensors, which guide the root downward through the soil (see Section 2.2). As the root advances and encounters new regions of moisture and nutrients, the cap is continuously sloughed off and replenished from the inside, whereas the polysaccharide-rich slime, called mucigel or mucilage, secreted by the cap covers the surface of the maturing root. The mucilage harbors microorganisms and probably aids in the establishment of symbiotic mycorrhiza (Greek *mykes* = fungus, *rhíza* = root) in the root tips, which form a network of fungal mycelium that connects plants belowground. This network helps with the uptake of nutrients and water in exchange for carbohydrates supplied to the fungi by the plants (Bais et al., 2006; Marschner, 1995;

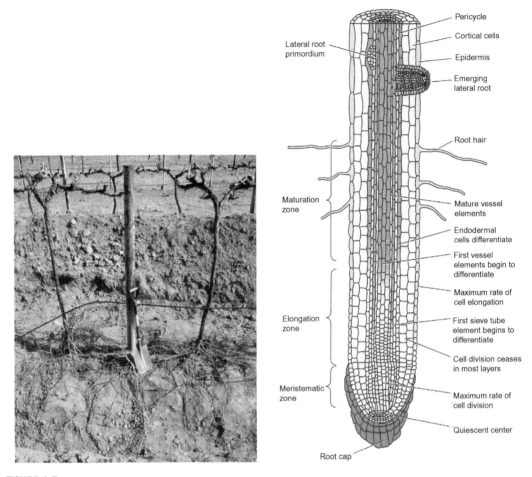

FIGURE 1.5

Distribution of Merlot root system in a drip-irrigated vineyard (left; photo by M. Keller) and diagrammatic longitudinal section of the apical region of the root (right; reproduced from Taiz and Zeiger, 2006).

Smith et al., 2001; Smith and Smith, 2011). In addition, the mucilage and the dead cells lubricate the growing root to reduce the friction between the root surface and the surrounding soil (Jin et al., 2013). The site of active cell division (the apical meristem in the center) and differentiation (the first phloem cells appear to the outside of the meristem) lies just behind the cap and is separated by a transition zone (0.7–1.5 mm behind the tip) from the elongation zone, a short region of cell expansion and further differentiation (the first xylem cells appear; see Figure 1.5). A gradient of competing active oxygen species ($O_2^{\bullet-}$ toward the tip, H_2O_2 toward the base), it seems, determines the location of this transition zone (Chandler, 2011).

The apical meristem contains "stem cells" that supply "founder cells" or "initials" that, following several cycles of cell division, differentiate into the various root tissues according to their position. Cell division is stimulated by the growth-promoting hormone auxin (indole-3-acetic acid and related compounds), which is delivered from the unfolding leaves near the shoot tips via active cell-to-cell transport down the vascular cambium and xylem parenchyma (ground tissue, from Greek *pará* – beside, *enchyma* – infusion) and is then recycled, or refluxed, within the epidermis of the root tip (Aloni, 2013; Friml, 2003; Kramer and Bennett, 2006; Petrášek and Friml, 2009). Cell elongation and differentiation, on the other hand, are promoted by a group of growth hormones termed gibberellins that are produced, upon stimulation by auxin, close to or at their site of action—that is, in tissues with rapidly expanding cells (Ross et al., 2011; Yamaguchi, 2008). The activity of the meristem requires high concentrations of the hormone cytokinin (*trans*-zeatin and related compounds), produced in the root tips, and low concentrations of the "cell-expansion hormone" gibberellin in the endodermis, whereas cell elongation requires the opposite (Wang and Li, 2008; Weiss and Ori, 2007). Although cytokinin is typically regarded as a "cell-division hormone," in root tips, it has the opposite effect. By countering the effect of auxin on cell division, cytokinin controls the rate of cell differentiation in the transition zone, so that the length of the meristem is determined by the balance between the division signal auxin and the differentiation signal cytokinin (Del Bianco et al., 2013). The growth-suppressing hormones ABA and ethylene inhibit differentiation of the stem cells and the initials; they thereby help sustain the meristem during dormancy or during periods of stress that interfere with root growth (Zhang et al., 2010b). Since the phloem (Greek *phloos* = bark) and xylem (Greek *ksulon* = wood) cells differentiate behind the root tip, hormones as well as the water and nutrients feeding the dividing cells must move across the cells after leaving the vascular tissues.

The cells produced in the meristem form a central ring called the endodermis, which divides the root into two regions—the cortex toward the outside and the stele toward the inside. The cortex is responsible for nutrient uptake from the soil and storage of starch and other nutrients, whereas the stele is responsible for nutrient transport up and down the plant (see Section 3.3). The stele differentiates into the pericycle, primary phloem, and primary xylem and is thus also called the vascular cylinder, whereas the cortex develops the exodermis (the hypodermis of the root) and epidermis, which form the dermal ("skin") tissues (Evert, 2006; Galet, 2000). This epidermis, however, is short-lived, and the underlying exodermis forms the boundary between root and soil even a few millimeters behind the root tip (Gambetta et al., 2013; Storey et al., 2003a). The endodermis is the innermost layer of the cortex and consists of a single layer of cells with thickened radial and transverse cell walls, named Casparian strips or Casparian bands, that are impregnated with lignin (Geldner, 2013; see also Section 3.3). In addition, behind the elongation zone, both the exodermis and the endodermis also incorporate the waxy, hydrophobic (water-repellent but not impermeable) biopolyester suberin in their cell walls (Gambetta et al., 2013). The suberization pattern is interrupted by occasional unsuberized cells called passage cells (Geldner, 2013; Vandeleur et al., 2009).

The pericycle beneath the endodermis, rather than the apical meristem, gives rise to the lateral roots by renewed cell division, following dedifferentiation of pericycle founder cells, where the meristem transitions into the elongation zone (Mapfumo et al., 1994a; Nibau et al., 2008; Osmont et al., 2007; Viala and Vermorel, 1909). Lateral root growth is initiated by shoot-derived auxin and aided by locally produced growth hormones termed brassinosteroids (which resemble animal steroid hormones) and small amounts of gaseous ethylene (C_2H_4), which is produced by and released from the differentiating xylem. Apparently, the incoming auxin induces this ethylene production, which in turn blocks further auxin movement, and the resulting local auxin accumulation in some pericycle cells adjoining the primary xylem then induces lateral root growth (Aloni et al., 2006b; Del Bianco et al., 2013). On the other hand, higher concentrations of ethylene and cytokinin moving up from the root tips inhibit lateral root formation (Ivanchenko et al., 2008; Osmont et al., 2007). By blocking the production of auxin transport proteins, cytokinin disrupts auxin accumulation at the appropriate locations, so that lateral roots cannot be initiated (Santner et al., 2009). The fact that cytokinin concentration is highest closest to the site of its production ensures that lateral roots are not formed too close to the tip, which would interfere with its continued growth (Aloni et al., 2006b). Moreover, ABA seems to control the activation of the new lateral root meristem after emergence of the lateral root from its parent root: A high ABA concentration (e.g., due to water stress or high nitrogen availability) inhibits meristem activation and keeps the lateral root in a dormant state (De Smet et al., 2006; Malamy, 2005). Once activated, auxin usually maintains the cell-producing activity of the meristem. Thus, the meristem and elongation zones together form the zone of active root growth in length, and a signaling network consisting of interactions among auxin, ethylene, cytokinin, brassinosteroid, and ABA regulates root architecture in response to environmental (mainly water and nutrient availability) and developmental influences (Osmont et al., 2007).

A few millimeters behind the growth region follows the absorption zone, which is densely covered by root hairs. The colorless root hairs are long protrusions of epidermal cells (epidermis cells that form root hairs are termed trichoblasts) and are mostly responsible for water and nutrient uptake (Gilroy and Jones, 2000; Pratt, 1974). They are very thin (10–15 μm in diameter but can flatten to squeeze through soil pores <2 μm) and can make up more than 60% of the root's surface area, which greatly increases the contact surface between root and soil, or the plant–soil interface, and the exploited soil volume (Clarkson, 1985; Sondergaard et al., 2004; Watt et al., 2006). The initiation and elongation of root hairs is stimulated by auxin and ethylene; the two hormones simultaneously slow down the elongation of the root's epidermis cells (Muday et al., 2012; Strader et al., 2010). As the root grows, new root hairs constantly grow behind the elongation zone, while the older ones are worn away along with the epidermis, the original cortex, and the endodermis. Hence, the absorption zone advances along with the root tip, leaving behind the conducting zone that continues into the trunk and all aboveground organs of the vine.

Procambial cells, which are derived from parenchyma cells and serve as vascular stem cells, generate both the primary phloem and the primary xylem. The xylem cells closest to the root tip are termed protoxylem, its conducting cells (see below) deposit helical thickenings in their secondary cell walls, which enable these cells to stretch as the root elongates (Schuetz et al., 2013). Behind the elongation zone, the xylem cells transition into metaxylem, whose conducting cells are larger than those of the protoxylem and grow thick, pitted secondary cell walls that make them strong yet rigid. At about the same time or slightly later, about 10 cm behind the root tip, the procambium develops into the vascular cambium during the transition from primary to secondary growth (Gambetta et al., 2013; Schuetz et al., 2013). The vascular cambium serves as a lateral meristem and forms a continuous sleeve of one to several cell layers within the root–shoot axis. Once elongation growth has ceased, the cambium forms new, secondary phloem cells toward the outside and secondary xylem cells, which collectively form the wood, toward the inside (Viala and Vermorel, 1909). Thus, if the apical meristem in the root tip is responsible for primary growth, the cambium is responsible for secondary or radial growth of a root and moves outward as the root grows thicker. Its cells can also be regarded as stem cells, and the transcellular flow (i.e., cell-to-cell movement across the cell membranes and cell walls) of auxin from the shoot tip toward the roots, in addition to locally produced strigolactone, cytokinin produced in the root tips, and physical pressure, stimulate cambial cell division and subsequent cell differentiation (Agusti et al., 2011; Aloni, 2001; Aloni et al., 2006b; Dengler, 2001; Matsumoto-Kitano et al., 2008; Ye, 2002). As auxin moves actively and mostly down the cambium "sleeve," its concentration remains highest in these dividing and differentiating cells (Schuetz et al., 2013).

Another meristem, the cork or bark cambium (called phellogen), forms in the pericycle and generates cork (phellem) toward the outside and secondary cortex (phelloderm) toward the inside; these tissues are collectively termed the periderm (Mullins et al., 1992; Pratt, 1974). Because the outermost cell layers, including the exodermis, cortex, and endodermis, rupture and are sloughed off, the mature root consists of periderm on the outside and phloem, vascular cambium, and xylem at the center (Viala and Vermorel, 1909; Gambetta et al., 2013). The deposition of suberin in the cork cell walls and the oxidation of phenolic compounds from dying cells give the mature root a brownish color.

The phloem, vascular cambium, and xylem together are called the vascular tissues, or vascular system, which forms discrete vertical strands of conducting tissues named vascular bundles, separated by rays consisting of at least one layer of compact parenchyma cells (Currle et al., 1983; Stafford, 1988; Viala and Vermorel, 1909). Parenchyma cells are live cells with thin primary walls and are rich in nutrient reserves (e.g., starch and proteins) and tannins. The phloem and xylem are placed in parallel within each vascular bundle (a pattern termed the collateral vascular bundle) and are interconnected by the parenchyma rays, which allows for the transfer of nutrients between the two "circulatory systems." The phloem is composed of sieve elements, companion cells, and phloem parenchyma. In a unique division of labor, the hollow and almost "clinically" dead sieve elements (before they become functional, they resorb their nucleus and vacuole and retain only their cell membrane, a thin layer of cytoplasm, plastids, and a few enlarged mitochondria, tied together so they are not dragged along with the phloem flow; van Bel, 2003) are the conduits for assimilates, whereas the companion cells (which are crammed with mitochondria) carry out the metabolic functions abandoned by the sieve elements and act as transfer cells, loading sugar and other materials from and to the surrounding parenchyma cells into and out of the sieve elements. To this end, pairs of companion cells and sieve elements (which arise from a common mother cell; see Viala and Vermorel, 1909) are interconnected through numerous "lifelines" termed pore/plasmodesma units, and the companion cells and parenchyma cells are connected through membrane-lined channels called plasmodesmata (Roberts and Oparka, 2003; Turnbull and Lopez-Cobollo, 2013). The sieve elements, with a diameter of up to approximately 35 μm, have thicker cell walls than the adjacent parenchyma cells. They are stacked end-to-end and interconnected by large (up to 1 μm diameter) sieve pores derived from plasmodesmata in the end walls (named sieve plates) to form the sieve tubes (Tyree et al., 1974; van Bel, 2003; Viala and Vermorel, 1909). The sieve pores are lined with callose, which enhances their rigidity and prevents collapse under pressure (Barratt et al., 2011; Knoblauch and Oparka, 2012). In addition, extra callose may be deposited in the cell walls surrounding the pores, which restricts passage of molecules through the opening; this callose deposition is reversible in order to regulate plasmodesmal traffic (Burch-Smith and Zambryski, 2012). Circulation of auxin in the sieve tubes, where the hormone is deposited by older leaves, serves to ensure that the amount of callose remains just right (Aloni, 2013). In contrast to the xylem cells, the sieve elements retain their plasma membrane, which is necessary to generate osmotic gradients for assimilate transport in a direction opposite to transpiration (see Section 5.1).

The xylem consists of vessel elements, tracheids, fibers, and xylem parenchyma. The vessel elements and tracheids form strong secondary cell walls that become heavily lignified by the incorporation of lignin (Latin *lignum* = wood) and together are called the tracheary elements. These are arranged longitudinally and embedded in the fibers and parenchyma, the latter forming radial rays (a.k.a. medullary rays) that separate the bundles of tracheary elements (Figure 1.6). The tracheary elements are the actual xylem conduits or water pipes of the vine. Their differentiation is induced by the movement of auxin, which arrives from the shoot tips, in the growing cells that are to become the tracheary elements (Aloni, 2013). Before they can assume their function, however, these cells must die; their organelles and cytoplasm are disassembled, digested, and removed in an orderly manner termed programmed cell death, and the dead cells become the wood (Jones and Dangl, 1996; Plomion et al., 2001; Schuetz et al., 2013; Turner et al., 2007). Their thickened secondary cell walls provide mechanical strength to prevent the pipes from collapsing under the negative pressure generated by transpiration (Ye, 2002; see also Section 3.3). The lignification also renders the cell walls relatively waterproof, although lignin is not completely impermeable to water. Because much of the lignification probably happens postmortem, that is, after the death of the tracheary elements, it is thought that the phenolic lignin building blocks (mostly glucosylated coniferyl alcohol and sinapyl alcohol) are manufactured by the neighboring parenchyma cells and exuded to the cell walls, where the glucose is removed and the resulting monolignols oxidized and assembled into the lignin polymer (Schuetz et al., 2013). The fiber cells, whose differentiation is triggered by gibberellin arriving via the phloem from the leaves, also commit suicide in a tightly controlled manner after the death of the tracheary elements (Aloni, 2013; Courtois-Moreau et al., 2009). The very long (~ 0.5 mm) and thin (~ 20 μm) fibers reinforce and seal their secondary cell walls completely with lignin before they die, and their massive (~ 4 μm) cell walls give the root mechanical support. The only live cells in the mature xylem are those of the parenchyma, and even

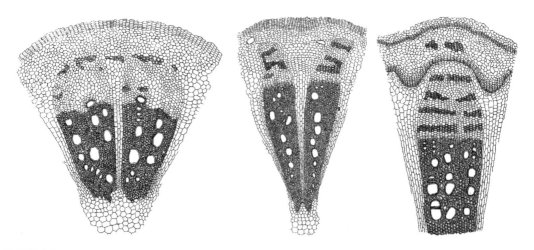

FIGURE 1.6

Cross sections of 1-year-old (left), 2-year-old (center), and 3-year-old *V. vinifera* root (right).

Reproduced from Viala and Vermorel (1909).

these cells die after a few years (Plomion et al., 2001). The parenchyma cells have no well-defined secondary cell walls and serve as storage sites for water and nutrients such as carbohydrates and proteins or amino acids (Schuetz et al., 2013; Zapata et al., 2004), contain raphide crystals to deter herbivores, and are involved in the defense against pathogen penetration.

Multiple vessel elements are stacked end-to-end and interconnected by perforation plates (holes created by targeted removal of the cell walls and membranes) in the end walls to form long hollow columns termed vessels. The perforation plates of the large vessel elements formed early in the growing season are simple (i.e., have a single hole), whereas those of the vessels formed later (latewood) are sometimes scalariform (i.e., have several elongated holes that make the remaining cell wall appear like rungs of a ladder). The much rarer tracheids, which look much like vessels without perforation plates, occur in the latewood and, because they are single cells, are much shorter (1–10 mm) than the vessels; many vessels in grapevines are more than 100 mm long, and some exceed 1 m. Both vessels and tracheids are radially interconnected (although not with each other) through pairs of porous (water-permeable) cell wall depressions named pits. Whereas the connections between individual vessel elements within a vessel are provided by either pits or perforation plates, those between one vessel and another or between neighboring tracheids are provided by pits (Brodersen et al., 2011; De Boer and Volkov, 2003; Sun et al., 2006). The pits are generally scalariform and bordered; bordered pits are pores with overarching secondary cell walls that have a narrower aperture or pit channel opening into a wider pit chamber. The two adjoining vessels or tracheids are separated inside the chamber by a permeable "membrane," which consists of the middle lamella and primary cell walls. Although not a true cell membrane because it does not contain a lipid bilayer, the fine pore size (5–20 nm) of pit membranes tends to slow water flow but also limits the spread of air bubbles (see Section 3.3) and pathogens (Pérez-Donoso et al., 2010). Because of their more open construction, vessels can conduct water more efficiently than can tracheids. Water also moves laterally through pits (more accurately through pit pairs because the pits of the two cells line up to connect the cells), both between parallel vessels or tracheids, and from and to the surrounding parenchyma cells, but not to fibers. Due to their length (~ 0.5 mm) and large diameter (10–150 μm), individual vessel elements can be surrounded by and connected with many dozens of the much smaller parenchyma cells (Sun et al., 2006; Viala and Vermorel, 1909).

1.3.2 TRUNK AND SHOOTS

The aboveground or aerial axis of the grapevine, which comprises the shoots, arms, and trunk, is called the stem by botanists. It provides support for the growing vine and is responsible for water, nutrient, and assimilate transport. The flexible stem (in addition to the roots) also serves as a storage organ for carbohydrates and other nutrients, which support early growth in spring and during stress periods, and for water. Because of their liana nature, cultivated grapevines typically require a trellis system for support, unless they are trained very close to the ground. The trunk is often extended along a horizontal wire to form one or more permanent arms or cordons that support the 1-year-old wood, which in turn gives rise to the fruiting shoots (Figure 1.4). Such vines are said to be cordon trained, whereas vines without cordons are said to be head trained. The shoots, which are called canes after they have matured and the leaves have fallen off, are generally pruned back in winter to either short spurs (the so-called spur pruning, which generally retains one to three buds per spur) or longer canes (the so-called cane pruning, which typically retains eight or more buds per cane). Both spur and cane pruning are usually done manually, which permits maximum control over both bud number and position on the vine. Cordon-trained grapevines may also be pruned by machine, leaving a range of shorter and longer spurs centered on the cordon; this is also called a box cut. Machine pruning is sometimes done as a prepruning operation only, followed by manual touch-up to control bud numbers. Alternatively, very light mechanical pruning, which trims off only the ends of the shoots or canes, may be done in summer and/or winter, a method termed minimal pruning (Clingeleffer, 1984; Possingham, 1994). This practice relies on the vines' self-pruning ability, in which the immature apical portion of the shoots generally dies back and falls off in winter (see Section 5.1). An assortment of combinations of training systems, pruning methods, and other cultural practices are employed in commercial viticulture. They all have as their common goals to constrain the natural vigor of the vines and maintain them at a small, manageable size; to sustain their shape and productivity over the several decades of a vineyard's life; to optimize fruit production and quality depending on the intended end use of the grapes; and to facilitate labor and permit various degrees of mechanization.

Phyllotaxy or phyllotaxis (Greek *phyllon* = leaf, *taxis* = order) describes the regular arrangement of leaves in space or the pattern of lateral organs on the shoot. The points at which these lateral organs connect to the shoot are called nodes. In vines grown from seeds, the epicotyl (embryonic shoot) emerges from the embryo, forming a primary shoot without tendrils, and leaves arranged in a spiral fashion, with leaves offset by 137.5° (the so-called golden angle), displaying 2/5 (sometimes 3/7) phyllotaxy (Galet, 2000; Viala and Vermorel, 1909). This means that two revolutions around the shoot must be made to find a leaf that is located directly above the reference leaf, which is five leaves below. This juvenile phase generally ends when 6–10 leaves have developed and the vine enters adult development (Mullins et al., 1992). All new leaves are now being produced alternately—that is, on two opposite sides of the shoot with a single leaf at each node (distichous or 1/2 phyllotaxy with 180° angles)—and the first tendrils are formed. In vegetatively propagated vines or adult plants, primary shoots usually originate from the buds at nodes of woody canes, and their leaves always display 1/2 phyllotaxy. Shoot growth from buds is sometimes divided into "fixed" growth and "free" growth (Mullins et al., 1992). Fixed growth occurs from leaf primordia and compressed internodes preformed during the previous growing season that overwinter in the dormant bud. It is responsible for the rapid growth of the first 6–12 (occasionally up to 18) leaves in spring; the number increases with increasing temperature during bud development (Buttrose, 1970a; Greer and Weston, 2010a; Morrison, 1991). Free growth, on the other hand, occurs later in the season from the production of new leaf primordia and internodes by the shoot's apical meristem concurrently with internode elongation.

The apical meristem is the site of cell division, where all shoot organs are initiated and where the highly variable pattern of the shoot system is established. This contrasts with the situation in the roots, where lateral roots arise from the pericycle rather than the root apical meristem; the shoot, however, does not have a pericycle. The shoot apical meristem consists of three subpopulations of cells: the central zone at the tip and the rib and peripheral zones below it (Clark, 1997; Evert, 2006; Kerstetter and Hake, 1997). The central zone, which has very small, relatively slowly dividing cells with small vacuoles and is located at the very apex of the apical meristem, serves as a source of cells ("stem cells") for the other two zones, which are involved in cell differentiation. The rib zone at the base of the apical meristem forms the vascular and other tissues in the central part of the shoot. The peripheral (or morphogenetic) zone, which has small, rapidly dividing cells

FIGURE 1.7

Diagrammatic longitudinal section of Concord shoot tip (left; reproduced from Pratt, 1971, reprinted by permission of American Journal of Enology and Viticulture); lateral organs arising from various positions in the dormant bud, illustrating the repeating three-node pattern unit of the shoots of many *Vitis* species (center; reproduced from Carmona et al., 2002); and chimeric Cabernet Sauvignon shoot (right; photo by M. Keller).

with small vacuoles, surrounds the central zone on the flanks of the apical meristem and produces the outer shoot tissues and the lateral meristems. The lateral meristems are produced from a few founder cells that lose their meristem identity and acquire a new organ identity to give rise to the primordia which, in turn, develop into the lateral organs (Sassi and Vernoux, 2013; see Figure 1.7). In the great majority of species of the family Vitaceae (including all species of economic importance), there are two types of lateral meristems: One is responsible for leaf formation and the other for inflorescence and tendril formation (Carmona et al., 2007; Gerrath and Posluszny, 2007).

The various zones overlap with another structural feature of the apical meristem—the horizontal layering of cells divided into the tunica (which usually consists of two parallel cell layers, the outermost of which generates the epidermis) on the outside and the corpus on the inside (Evert, 2006; Thompson and Olmo, 1963). Because the cells of the tunica divide in an anticlinal plane (perpendicular to the surface, so that daughter cells remain in the same layer as their parent cells) and those of the corpus in different directions, the two layers perpetuate separate cell lineages (Clark, 1997; Jenik and Irish, 2000; Kerstetter and Hake, 1997; Sassi and Vernoux, 2013). Mutations occurring in only one of the two layers can therefore give rise to chimeric plants (see Figure 1.7) and to clonal variation within grape cultivars (see Section 1.2).

In contrast to the mature cells of the pith, rays, and cortical parenchyma (Kriedemann and Buttrose, 1971), the meristem cells do not contain chlorophyll, so they cannot perform photosynthesis and must import all carbon and other nutrients. The construction and "installation" of the photosynthetic machinery begins as soon as the cells are released from the meristem, unless those cells remain in darkness. This occurs, for example, in a trunk sucker arising from below the soil surface and leads to the formation of colorless, etiolated shoots that elongate rapidly under the influence of gibberellin in search of light (Hartweck, 2008). The initial cell from which all subsequent cells of a lateral organ are derived is termed a founder cell (Chandler, 2011). This organ initially divides into a set of cells called preprimordium, or "anlage" (German *Anlage* = conception, layout, investment), whose cells proliferate into a primordium. When these primordia cells differentiate, they give rise to a lateral organ.

The lateral organs include leaves, tendrils, clusters (inflorescences), and axillary buds. Axillary buds contain meristem primordia that are called secondary, axillary, or lateral meristems and form several compressed shoot segments before becoming dormant; they can be reactivated later to produce lateral shoots, whose leaves form at right angles to those of the main shoot (Gerrath and Posluszny, 1988a; McSteen and Leyser, 2005, Pratt, 1974; Figure 1.8). Likewise, the inflorescence primordia contain their own flower meristems that later produce the grape flowers and their different floral organs, which are themselves modified leaves—a concept originally proposed by the German writer and philosopher Johann Wolfgang von Goethe (1790). Contrary to their influence in the root tip, cytokinins stimulate cell division in the shoot apical meristem and are necessary for the transition from undifferentiated cells to differentiated primordia (Dewitte et al., 1999; Mok and Mok, 2001; Werner et al., 2003).

Just how the vine "knows" when and where to initiate the various lateral organs is still somewhat of a mystery. Local accumulation of auxin and perhaps cytokinin seems to be the signal for organ formation by activating the cell wall–loosening protein expansin (see Section 3.1) in a founder cell, and the identity of each lateral organ may be determined by hormone gradients (Chandler, 2011; Petrášek and Friml, 2009; Reinhardt et al., 2000; Wang and Li, 2008). Whatever their identity, however, without auxin primordia cannot form (Gallavotti, 2013). Once initiated, the primordia become sinks for auxin, draining auxin from surrounding cells and thereby preventing their growth, which ensures the regular, phyllotactic positioning of lateral organs (Benjamins and Scheres, 2008; Berleth et al., 2007; Kuhlemeier, 2007; Sassi and Vernoux, 2013). Meristems and leaves are in constant communication and clearly influence each other; preexisting leaves are required for the correct positioning of a new leaf, possibly by defining the routes of auxin transport to the meristem via the epidermis (Petrášek and Friml, 2009; Piazza et al., 2005; Reinhardt et al., 2000). The older leaves even influence specific traits of newly developing leaves, such as leaf size and the density and size of stomata, so that the unfolding leaves are adapted to the environment into which they are "born" (Chater et al., 2014; Lough and Lucas, 2006). Moreover, newly formed

FIGURE 1.8

Repeating three-node pattern of a Syrah shoot (left); mistakes do happen in nature—three consecutive tendrils on a *V. vinifera* shoot (center); and dormant bud and lateral shoot in a leaf axil (juncture between petiole and shoot) of a Malbec main shoot (right).

Photos by M. Keller.

axillary buds are not connected to the vascular bundles of the shoot; this connection is established at the time of bud outgrowth. It appears that the buds release auxin, produced mostly from the amino acid tryptophan, just before they break, and this auxin induces differentiation of vascular tissues that connect the buds (and hence the growing lateral organs) to the preexisting vascular network (Aloni, 2001). Thus, all active apical meristems export auxin down the plant toward the roots, whereas inactive meristems (such as those within dormant buds) do not export auxin—or, conversely, meristems that cannot export auxin remain inactive (Leyser, 2010).

The production of leaf-opposed tendrils and clusters appears to be unique to the Vitaceae family and is typically discontinuous; that is, two of every three nodes bear a tendril (Gerrath and Posluszny, 2007; Pratt, 1974). One notable exception is *V. labrusca*, which has a continuous pattern—that is, a tendril at every node. Why the other members of the family leave a "blank" at every third node and how they keep count is still mysterious. The unusual position of clusters and tendrils opposite leaves is a result of the lack of elongation of the internodes separating nodes that bear only rudimentary leaves, termed bracts, from those that bear true leaves (Morrison, 1991; Tucker and Hoefert, 1968). The first (i.e., lowest) two or three nodes usually carry only leaves; the next two nodes are generally the ones with clusters, followed by a node with a leaf only and then a succession of repeating mirror images of three-node units with the clusters replaced by tendrils (see Figures 1.7 and 1.8). The repeating units, or shoot segments, are called phytomers, and each phytomer consists of a node, an internode, a leaf attached to the node, and an axillary bud containing an axillary meristem for branching (McSteen and Leyser, 2005).

Some of the cells produced by the apical meristem do not become primordia. They instead develop into the internodes of the shoot axis. These cells differentiate into the epidermis, cortex, endodermis, phloem, procambium, and xylem (Viala and Vermorel, 1909; Esau, 1948; Galet, 2000). The outermost cell layer, the epidermis, develops a waxy cuticle (Latin *cutis* = skin) on its thick outer cell walls as a protective layer of all aboveground organs. The epidermis also has stomata, various types of hair, or trichomes (Greek *trichos* = hair), and, on young organs, often small and short-lived translucent structures variously named pearls, pearl glands, pearl bodies, or sap balls (Pratt, 1974; Figure 1.9). Pearls also

FIGURE 1.9

Ant feeding on pearls on a Chardonnay shoot.

Photo by M. Keller.

develop on tendrils, inflorescences, petioles, and the abaxial surface of leaves, and they are especially frequent under warm, humid conditions and high nutrient availability promoting vigorous growth (Paiva et al., 2009; Viala and Vermorel, 1909). Although they remain covered by a thin epidermis that sometimes develops a stoma, they are formed of highly vacuolated, thin-walled subepidermal cells accumulating sugars, oils and, perhaps, proteins. Pearls have no glandular activity but instead are also termed food bodies because they ostensibly attract and feed ants, which in turn help defend the vulnerable young plant organs against herbivorous insects (Paiva et al., 2009).

The cortex consists mostly of parenchyma cells that contain chloroplasts, starch, calcium oxalate crystals, and phenolic compounds (including anthocyanins on the sun-exposed side of the shoots of many varieties). Some parenchyma cells about one to four layers below the epidermis differentiate into strands of small collenchyma (Greek *kolla* = glue) cells with thick primary walls, strengthening the growing shoot (Esau, 1948). Shoots that grow on windy sites tend to develop much thicker-walled collenchyma than do shoots at sheltered sites, which is probably why the latter are more easily broken off by passing machinery. The endodermis forms the innermost layer of the cortex, just outside the vascular tissues. It is also called the starch sheath because of its high concentration of amyloplasts. These are specialized plastids responsible for starch production from imported sucrose and its storage in starch grains or granules (Emes and Neuhaus, 1997; Martin and Smith, 1995; Pratt, 1974). In addition, parenchyma cells also accumulate and store protein reserves inside specialized protein storage vacuoles, which are disassembled into transportable amino acids during remobilization in spring. The vascular tissues to the inside comprise yet more parenchyma cells that form the primary rays and alternate radially with the vascular bundles like the spokes of a wheel (Figure 1.10). The vascular bundles, with caps of primary phloem fibers (long, narrow cells with thick, lignified cell walls) on the outside, contain the primary phloem, procambium, and xylem (Bondada, 2012; Esau, 1948). Within the ring of vascular bundles, there is a core of dead, unlignified and rather soft parenchyma cells called the pith.

The differentiation of the vascular bundles is induced by acropetal (toward the tip) polar, active transcellular movement of auxin (at a rate of ~ 1 cm h^{-1}), probably in the epidermis, from developing leaf primordia and young leaves (Benjamins and Scheres, 2008; Berleth et al., 2007; Petrášek and Friml, 2009; Woodward and Bartel, 2005). In addition, gibberellin imported via the phloem and/or produced by the growing xylem cells may regulate xylem cell differentiation and elongation (Israelsson et al., 2005). As in the root tip, the first phloem cells appear before the first xylem cells, providing a conduit for the import of carbohydrates and other material at the earliest stages of leaf development. Each new leaf generated by the apical meristem is connected with preexisting vascular bundles in the shoot by, normally, five divergent bundles called leaf traces. Some leaf traces may arise up to three nodes below their point of departure to a leaf, whereas tendril and axillary bud traces usually begin one node below their point of departure (Chatelet et al., 2006; Gerrath et al., 2001). The shoot vascular bundles that continue their course through the next internode and ensure the continuity of the vine's vascular system are termed sympodial bundles. Their xylem vessels are connected with the vessels of the leaf traces via intervessel pits, whereas the leaf traces ensure continuous vessels between the shoot and the leaf blade. The vessels terminate in the leaf at approximately 50–60% of the leaf length; the xylem conduits beyond this limit are tracheids rather than vessels (Chatelet et al., 2006).

As primary growth ceases, the procambium develops into the vascular cambium, which is divided into intrafascicular cambium and interfascicular cambium. During the ensuing secondary

FIGURE 1.10

Cross section of *V. vinifera* shoot prior to periderm formation (left) and of cane after two layers of phloem have been discarded (center), and longitudinal section through *Vitis* shoot (right: A, pith; B, stele; C, cortex; D, diaphragm).

Reproduced from Viala and Vermorel (1909).

growth, the former produces secondary phloem toward the outside and secondary xylem toward the inside, and the latter forms parenchyma cells that become the secondary rays separating the vascular bundles (see Figure 1.10). The rays' thick-walled but heavily pitted parenchyma cells can accumulate large amounts of storage starch in their amyloplasts (Buttrose, 1969b; Kriedemann and Buttrose, 1971). Within the secondary phloem, tangential bands of conducting tissue composed of sieve tubes, companion cells, and parenchyma cells, collectively termed the soft phloem, alternate with bands of fibers, termed the hard phloem, whose thick secondary walls add strength to the shoot (Bondada, 2012; Esau, 1948). As in the root, the cambium is responsible for radial growth. Strigolactone and cytokinin support auxin in stimulating this cambial activity in the shoot by promoting cell division (Nieminen et al., 2008; Sanchez et al., 2012; Werner et al., 2003). The faster the rate of auxin transport down a shoot, the more active is cell division in the cambium; the "translation" of the auxin signal into secondary growth involves the local production of strigolactone from carotenoids (Agusti et al., 2011). Cytokinins are produced from the purine base adenine in the root tips and unfolding leaves and are transported upward (mostly as zeatin-type cytokinins) in the xylem and downward (mostly as isopentenyladenine-type cytokinins) in the phloem (Aloni et al., 2005; Ha et al., 2012; Matsumoto-Kitano et al., 2008; Nordström et al., 2004). The cambium becomes dormant in late summer and is reactivated in spring by root-derived cytokinin in response to auxin released by the swelling buds (Aloni, 2001). Thus cambium activity results in annual rings of secondary xylem; when we count annual rings, we are counting rings of xylem. The annual rings of grapevines are called diffuse-porous (Mullins et al., 1992; Sun et al., 2006) because the size of the xylem cells produced in spring is similar to that of those produced in summer, before the cambium activity ceases (usually at approximately the time the grapes begin to ripen). Annual rings are composed almost entirely of xylem cells; thus, xylem comprises the bulk of mature shoots, trunks, and roots. Because the vascular cambium moves outward with the growing trunk, it leaves only one or two (occasionally up to four) rings of phloem at the end of each growing season, which are reactivated in spring (Davis and Evert, 1970; Esau, 1948). The xylem remains functional for several years; older, dysfunctional xylem then becomes the heartwood.

Some parenchyma cells form outgrowths into the lumen of tracheary elements, entering via the interconnecting pits; such outgrowths are termed tyloses (Greek *tylos* = knob, knot) and serve mainly to seal injured xylem in the shoots (e.g., due to summer pruning or to freezing), and perhaps to prevent entry of pathogens (Evert, 2006; Sun et al., 2006; Viala and Vermorel, 1909). Mature canes, in contrast to shoots, respond to injury (e.g., due to winter pruning) by secreting a pectin-rich gel (a.k.a. gum) mixed with antimicrobial phenolic compounds to seal the xylem to a depth of at least 1 cm at the site of injury (Sun et al., 2008).

Some secondary phloem cells also form a cork cambium (phellogen), which later cuts off the primary phloem from internal tissues and produces cork (phellem) toward the outside and secondary cortex (phelloderm) toward the inside (Bondada, 2012; Esau, 1948; Galet, 2000; Pratt, 1974). Phellem, phellogen, and phelloderm together form the outer bark, or periderm, whose multilayer combination of dead but elastic and heavily suberized cork cells and waxes, by protecting the shoot (and later the cane, cordon, and trunk) from wounding as well as water and nutrient loss, not only takes over the function of the epidermis/cuticle it replaces but also provides some insulation against temperature fluctuations (Franke and Schreiber, 2007; Lendzian, 2006). The cork layer is interspersed with pores called lenticels that, it seems, assume the role of the vanished stomata, permitting continued gas exchange between the interior tissues and the atmosphere—albeit without the stomatas' ability to control the rate of gas exchange. Additional protection, especially against fungal infection, arises from the incorporation of phenolic compounds termed stilbenes (mainly resveratrol and ε-viniferin), which is one of the reasons woody tissues decay very slowly (Pawlus et al., 2013; see also Section 8.2).

All tissues outside the vascular cambium are collectively referred to as the bark; these include epidermis, cortex, and phloem. The periderm, along with the outermost layers of secondary phloem, dies and turns brown (from oxidation of cellular components) starting at the base and moving up toward the tip as the shoot becomes a cane (Esau, 1948). Thus, shoot browning is caused by the death of the green cortex, rather than by lignification, as is often (mistakenly) believed. The invisible process of lignification instead occurs in all secondary cell walls, but especially in those of the xylem cells, independently of and long before the beginning of browning. Browning is also called shoot "maturation" (French *aoûtement*, from *aoûté* = mature) and is accompanied by deposition of starch in the xylem and phloem parenchyma cells (Eifert et al., 1961; Plank and Wolkinger, 1976). It ends with the death and abscission (Latin *abscissio* = separation) of the apical meristem and the leaves, followed within days by the sealing of the sieve pores and plasmodesmata inside the sieve tubes with callose for overwintering (Davis and Evert, 1970; Esau, 1948; Pouget, 1963; Viala and Vermorel, 1909). Like cellulose ($\beta 1 \rightarrow 4$-glucan) and starch ($\alpha 1 \rightarrow 4$- and $\alpha 1 \rightarrow 6$-glucan), callose ($\beta 1 \rightarrow 3$-glucan) consists of long chains of glucose molecules. It is thought that the circulation of cytokinin in the sieve tubes triggers this callose deposition (Aloni, 2013). In the following spring the callose is degraded under the influence of auxin movement in the phloem, restoring the phloem's transport function for another growing season (Aloni, 2013; Esau, 1948; Pouget, 1963; Pratt, 1974). Phloem reactivation is extremely fast and occurs almost concurrently in the canes and trunk, before any bud swelling can be observed. The reactivated phloem from the previous season is important during the early growth period, before it is replaced with newly formed phloem and discarded around midseason, concurrent with periderm formation (see Figure 1.10). Depending on the species, the cambium resumes its function from approximately 2 weeks (*V. vinifera*) to 2 months (*V. riparia*) after the phloem has been reactivated, and a new phellogen is formed (Davis and Evert, 1970; Esau, 1948). In contrast to phloem reactivation, cambium reactivation is spatially and temporally linked to the swelling and breaking buds. Reactivation begins beneath those buds and progresses downward, so that the cambium of the trunk, and finally that of the roots, resumes its function weeks after the first cell divisions occurred in the canes. The new phellogen isolates most of the secondary phloem produced in the previous growing season, and the dead phloem, together with the old periderm, is eventually sloughed off as strips of dead bark (Davis and Evert, 1970; Esau, 1948). The tissue layers that are eliminated annually (i.e., the dead parts of the bark) are collectively termed the rhytidome (Greek *rhytidùma* = wrinkle) (Viala and Vermorel, 1909).

1.3.3 NODES AND BUDS

The lateral organs of the shoot are attached at nodes, which can be distinguished from the rest of the shoot or cane by their characteristic swelling due to a thicker pith and cortex (Figures 1.10 and 1.11). In all *Vitis* species, a diaphragm consisting of hard, thickened pith cells with sclerified (Greek *skleros* = dry, hard) cell walls divides the pith at the node. It is thought that the diaphragm aids in directing water and nutrients into the leaves. The leaf and bud are inserted just above the node, and each leaf is connected to the shoot via its own vascular bundles inside a petiole. When the petiole falls off, it leaves a leaf scar on the node.

Buds are young, compressed shoots (i.e., the internodes have not yet elongated; see Figure 1.12) enclosed in scales or bracts that are green in summer and turn brown during shoot maturation. Indeed, initiation of bracts is the first sign of bud formation, and their browning is associated with the entry of the buds into dormancy (van der Schoot and Rinne, 2011; van der Schoot et al., 2014). These leaflike structures without petiole or lamina protect the bud from desiccation and freezing. In addition, long, woolly epidermal hairs growing on the inside of the bracts form a down that cushions the bud and gives it a woolly appearance when it opens (breaks). Grapevine buds are always axillary; that is, they arise in the axil (upper angle) between a shoot and a leaf, directly above the petiole insertion point (Galet, 2000; Viala and Vermorel, 1909). All buds are formed by shoot apical meristems. There are three types of buds: prompt buds, dormant buds, and latent buds. Prompt buds are the only true axillary buds of the primary or main shoot and are also called lateral buds (Gerrath and Posluszny, 2007; Pratt, 1974). They develop from the earliest formed (lateral) meristem and can break in the current growing season and give rise to secondary, so-called lateral

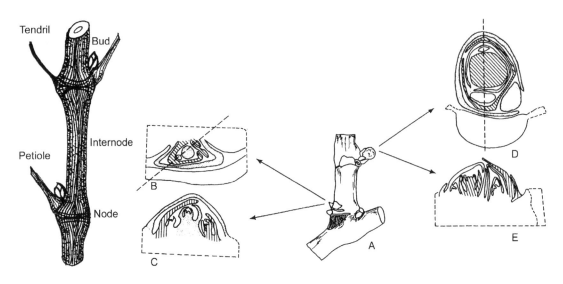

FIGURE 1.11

Location of the main features of a *Vitis* shoot (left; illustration by A. Mills) and one-node Concord spur with one count node and three basal buds (right: A, spur with buds; B, cross section of basal bud; C, longitudinal section of basal bud; D, cross section of compound bud; E, longitudinal section of compound bud; reproduced from Pool et al., 1978, reprinted by permission of American Journal of Enology and Viticulture).

FIGURE 1.12

Longitudinal section of dormant Cabernet Sauvignon bud. A, shoot apical meristem; B, lateral bud primordium; I, inflorescence primordium; L, leaf primordium; U, uncommitted primordium.

Reproduced from Morrison (1991)

shoots. Species such as *V. rupestris*, *V. arizonica*, or *V. monticola* often grow many long laterals, which in turn can give rise to tertiary shoots (Viala and Vermorel, 1909). By contrast, tertiary shoots are rare in *V. vinifera*, and laterals form mainly if the apical dominance of the primary shoot is broken (Alleweldt and Istar, 1969), for example, by removal of the shoot tip (see Section 2.2). Although prompt buds are usually not fruitful, some clusters, referred to as "second crop," can grow on lateral shoots, especially on those growing from the preformed part of the main shoot. Removing the shoot tip, a cultural practice variously named shoot tipping, topping, or hedging in viticultural parlance, appears to promote the development of such second-crop clusters (May, 2000). These clusters are typically small and ripen well after the clusters on the main shoot, if at all. Many lateral shoots fail to form a proper periderm and are abscised in late fall or winter.

Dormant buds are formed later than prompt buds and arise from their own shoot apical meristems in the axil of the first leaf (reduced to a small, scalelike leaf called a prophyll) of lateral shoots, whether or not these shoots emerge from the prompt buds (Gerrath and Posluszny, 1988a, 2007; Pratt, 1974; Viala and Vermorel, 1909). An imaginary longitudinal cut down a shoot between prompt and dormant buds reveals that all prompt buds are located on the same side and all dormant buds on the other; the prompt-bud side is called the dorsal side, and the dormant-bud side is called the ventral side (Huglin and Schneider, 1998). Some of these dormant buds will give rise to next year's crop, but they remain dormant until the following spring, which is why they are also called "winter buds." A dormant bud is in fact an overwintering compound bud (also called an "eye") because it generally contains three separate buds (see Figure 1.11)—a larger primary bud in the center flanked by two smaller secondary buds. The latter are also termed accessory buds and are often separately named a secondary and a tertiary bud because they develop in the axils of the first and second prophyll, respectively, of the shoot primordium that forms the primary bud (Morrison, 1991; Pratt, 1974). Higher-order buds have been reported, but these usually remain rudimentary (Zelleke and Düring, 1994). Just like the shoot apical meristem, the apical meristem of the primary bud generates two types of lateral meristems: One is responsible for leaf production and the other for inflorescence and tendril production (Carmona et al., 2007). These form as many as 12 leaf primordia (8–10 on average) and their associated prompt bud, up to several inflorescence primordia, and several tendril primordia before they become dormant (see Section 2.2). Initially, three or four leaf primordia are produced, each of them flanked by a pair of scales and wrapped in abundant hair (Morrison, 1991; Snyder, 1933; Srinivasan and Mullins, 1981). The bud's apical meristem then starts to produce lateral meristems opposite to leaf primordia. The first two or three of these will normally give rise to inflorescences, and the remainder will become tendrils. Intriguingly, in the majority of cultivars, only the buds produced from fixed growth (i.e., at the first 6–10 nodes), it seems, are able to initiate inflorescences; all buds formed during the shoot's subsequent free-growth phase only initiate tendrils (Morrison, 1991; Sánchez and Dokoozlian, 2005). The inflorescence meristem produces several branch meristems in spiral phyllotaxy, each of them subtended by a bract (Gerrath and Posluszny, 1988b; May, 2004; Srinivasan and Mullins, 1981). The secondary buds generally remain small and are less fruitful. They often grow out when the primary bud has been damaged, for example, by spring frost or insects, but also when vines have been pruned severely so that their shoot system is out of balance with the capacity of the root system (Friend et al., 2011).

Latent buds are buds that remain dormant or hidden for several years and often become part of the permanent structure of the vine. Even decades-old cordons and trunks still carry viable latent buds that can give rise to water shoots or suckers, especially following frost events or on vines that have been pruned too severely (Galet, 2000; Lavee and May, 1997; Sartorius, 1968). Such shoots are usually not fruitful, although exceptions do occur.

Basal buds, located at the base of a shoot or cane (see Figure 1.11), are formed in the axils of last year's bud scales (i.e., in the shoot's prophyll axils), and thus can be prompt buds, dormant buds, or latent buds (Pratt, 1978). Because the basal internodes elongate only insignificantly, these buds appear to be situated in a whorl ("crown") around the shoot (cane) base (Pratt, 1974; Viala and Vermorel, 1909). They are therefore also called "crown buds" and can be quite numerous, though they are typically inconspicuous and do not have leaves associated with them. Although basal buds are usually less fruitful than the other buds, this varies by cultivar: Zinfandel, Sangiovese, Gamay, and Muscat of Alexandria have particularly fruitful basal buds, whereas fruitfulness is very low in the basal buds of Viognier, Gewürztraminer, and Thompson Seedless. Moreover, the fruitfulness of basal buds is much higher in many interspecific hybrids than in *V. vinifera* cultivars (Huglin and Schneider, 1998; Pool et al., 1978). This has implications for the type of winter pruning: cultivars with low basal-bud fruitfulness are better suited for cane pruning, whereas those with high fruitfulness can be pruned to very short spurs.

1.3.4 LEAVES

A leaf differentiates along with a corresponding node from a leaf primordium (Latin *primus* = first, *ordiri* = to begin to weave) produced by the apical meristem in the shoot tip (Figure 1.7). The production of leaf primordia is induced by the "cell-expansion" protein expansin, which in turn is induced by the plant hormone cytokinin, produced either locally near the apical meristem or in the root tips, from where it is imported via the xylem (Mok and Mok, 2001, Pien et al., 2001). The (thermal) time between the formation of two successive leaf primordia is termed the plastochron (Greek *plastos* = formed, *chronos* = time). We distinguish four types of leaves according to their position on the shoot: cotyledons, scales, bracts, and foliage leaves. Cotyledons (embryonic leaves) are preformed in the embryo and are the first two leaves to emerge from the embryo during seed germination. They are short lived and fall off soon after germination. The scales form around buds and protect them from water loss and mechanical injury. Bracts are small, scalelike leaves at branch points on the stem of inflorescences and tendrils. The first two leaves of a shoot grown from a bud ordinarily also develop as bracts and are separated by very short internodes. Subsequent leaves develop into mature leaves termed foliage leaves, which are separated by elongated internodes. Foliage leaves consist of lamina (blade), petiole (stalk), and, at the base of the petiole, a pair of stipules, which are short-lived leaflike structures or sheaths nearly surrounding the shoot's node (Viala and Vermorel, 1909).

The petiole (Latin *petiolus* = little foot) is the leaf stem that connects the lamina to the shoot and contains multiple (typically 12–14) vascular bundles (Viala and Vermorel, 1909). Several vascular bundles are often derived from the same shoot vascular bundle (i.e., leaf trace); this branching is called anastomosis. Depending on species and cultivar, the petiole can be between 2 and 12 cm long and grows toward the light to position the leaf for optimal sunlight interception (Figure 1.13). Where the petiole joins the blade, it divides to form the five main veins of the lamina. An abscission layer forms at each end of the petiole toward the end of the growing season and leads to leaf fall.

FIGURE 1.13

Cross section of a *V. vinifera* leaf with prominent main vein (left; reproduced from Viala and Vermorel, 1909), Zinfandel petiole bending toward the light (center), and pattern of Concord leaf venation (right; photos by M. Keller).

FIGURE 1.14

Chardonnay leaf (left) and Merlot leaf (right).

Illustrations by A. Mills.

The basic function of the lamina (Latin *lamina* = leaf) is to capture sunlight for energy production (in the form of ATP) and carbon dioxide (CO_2) for carbohydrate production to support the vine's metabolism and growth (see Chapter 4). To maximize light absorption, leaves must be as wide as possible, whereas to maximize gas exchange, they must be as flat as possible; both of these properties, however, make leaves inherently vulnerable to overheating and dehydration (Tsukaya, 2006). Different species and cultivars of grapevine differ in their leaf shape, and leaf morphology forms the main basis of ampelography (see Section 1.1). The predominant leaf form of the genus *Vitis* is palmate, in which the five main vascular bundles (main or major veins; see Figure 1.14) serve the leaf's five lobes, which in most *V. vinifera* cultivars are partly separated from each other by more or less deep gaps, termed sinuses (Mullins et al., 1992; Viala and Vermorel, 1909; see

Figure 1.14). The main vascular bundles, which originate from procambial cells, divide into an ever finer, elaborate network whose branches (minor veins) are interconnected or eventually (at the fifth order) end blindly between the leaf mesophyll cells. Interconnected veins form enclosed areas or islets termed areoles, approximately 1 or 2 mm across, and 10 μm of minor veins services up to 10 mesophyll cells (Wardlaw, 1990). The vein network, which can make up several kilometers per square meter of leaf area, acts as an "irrigation/collection system," with the major veins acting as a rapid supply network and the minor veins forming a slow distribution and collection network (Canny, 1993). This system delivers water and nutrients, and it collects assimilates relatively evenly to and from all over the lamina. The veins are enclosed by a set of thickened parenchyma cells termed the bundle sheaths. Grapevine leaves are classified as heterobaric (Greek *baros* = weight), because their prominent bundle sheaths extend all the way to the epidermis, so that they are visible as almost transparent, projecting "ribs," especially on the abaxial side of the leaf (Liakoura et al., 2009; Pratt, 1974; see Figure 1.13). The walls of the bundle sheath cells are waterproofed by deposition of suberin (Sack and Holbrook, 2006). The bundle sheaths function to control the flow of water away from the xylem to the mesophyll cells and serve as "windows," transferring light to the photosynthetic mesophyll cells, in addition to providing mechanical support to protect the leaf blade against collapse during severe dehydration and other stresses.

The leaf margins are serrated (toothed), and both the number of teeth and their sharpness, or sinus depth, vary greatly among *Vitis* species and cultivars (Figure 1.14). It is thought that the sinus depth, but not the teeth number, is inversely related to the rate and/or duration of cell division during leaf formation, so that leaves with greater cell numbers should have shallower sinuses (Tsukaya, 2006), but this association has not been investigated in grapevines. The teeth "funnel" the tracheids of the terminal veins into water pores called hydathodes, which can discharge droplets of xylem sap (Evert, 2006; Pratt, 1974). This process is called guttation (Latin *gutta* = drop) and should not be confused with dew, which is the formation of droplets of atmospheric water that condense on cool surfaces. Guttation is especially pronounced during warm nights with high relative humidity (i.e., when transpiration is minimal and root pressure may force water up the vine; see Sections 2.2 and 3.3). The so-called scavenging cells surrounding the xylem vessels near their endings actively and selectively collect xylem solutes (e.g., nutrient ions and hormones such as cytokinin) and recycle them back to the phloem for reexport (Sakakibara, 2006), whereas undesirable solutes (e.g., excess calcium) may be excreted along with the water. Thus, the hydathodes may function as safety (overflow) valves that can rid the leaves of excess water.

The leaf's shape and the basic architecture of its vascular system are highly correlated and are determined by genetic and hormonal controls during the initial stages of leaf formation (Chitwood et al., 2014). Auxin stimulates cell division at high concentration and cell expansion at low concentration. It may exert its growth-promoting influence via differences in its content among cells (i.e., auxin gradients arising from active cell-to-cell transport away from the auxin source) which, in turn, determines the extent and direction of expansion of these cells by means of sugar and water import (see Section 3.1). As a side note, the herbicide 2,4-D (2,4-dichlorophenoxyacetic acid) is a synthetic, nontransportable auxin that stimulates cell division, but inhibits cell expansion and differentiation; it interferes strongly with patterning, leading to severely distorted leaves (Bondada, 2011). Accumulation of auxin begins in a single epidermal cell at the leaf margin and leads to the initiation and outgrowth of a tooth by promoting local cell division (Kawamura et al., 2010). In addition, a hydathode differentiates at the site of the auxin maximum. Further auxin production in the cells around the hydathode of leaf primordia and unfolding leaves is, at least in part, responsible for the formation of the characteristic leaf shapes (patterning) and differentiation of vascular bundles, whereby cells that experience rapid auxin flow will differentiate into veins (Aloni, 2001; Aloni et al., 2003, 2006b; Benjamins and Scheres, 2008; Berleth et al., 2007; Dengler, 2001). Thus the formation of teeth (serration) in simple leaves is similar to the development of compound leaves and of leaf primordia in general (Kawamura et al., 2010).

Newly formed leaf cells require approximately 2 weeks to expand to their full size (in contrast to root cells, which finish expanding within a few hours), which gives leaves sufficient developmental flexibility to adjust their final size to environmental conditions prevailing during leaf formation. Leaf size and shape can be markedly altered by patterns (gradients) of water availability across the leaf during the period of leaf expansion. The final size of a leaf is usually closely tied to its cell number, although greater cell expansion often partially compensates for lower cell number (Gonzalez et al., 2012; Tsukaya, 2006). Because both cell division and cell expansion are dependent on sugar supply, leaves developing when a vine is photosynthetically "challenged" will produce few and small cells (Van Volkenburgh, 1999). However, while cell division is highly sensitive to sugar availability, cell expansion is more responsive to water availability. As a consequence, the earliest stages of leaf development are mostly under the control of carbon supply, whereas later stages are predominantly controlled by water supply (Pantin et al., 2012). Therefore, leaves can grow to a large size if resources are abundant during their expansion, but they will remain small if they form while environmental conditions are unfavorable.

The lamina grows by cell division and expansion in basal and intercalar (Latin *intercalare* = to insert) meristems and consists mainly of primary tissue; there is no secondary growth to speak of (Evert, 2006). Cell division ceases first in the leaf tips and later toward the base (Gonzalez et al., 2012). Division generally stops when the leaf is only half or less of its final size so that the remainder of leaf growth is caused solely by the expansion of preformed cells, which occurs mainly due to enlargement of their vacuoles (see Section 3.1). In addition to the influence of auxin, both cell division and cell expansion are stimulated by cytokinins produced by the dividing leaf cells as well as by cytokinins imported via the xylem from the root tips. Moreover, cell expansion is also promoted by gibberellins that stimulate loosening of the cell walls by inducing the activity of expansin and other wall-relaxing enzymes (Claeys et al., 2014). Gibberellin production by expanding cells just behind the shoot apical meristem and by the young, rapidly expanding leaves and petioles near elongating internodes is induced by auxin (Hartweck, 2008; Ross et al., 2011).

At the same time as the leaf cells expand and the rate of cell division decreases, they also build up their photosynthetic machinery, namely the chlorophyll-rich chloroplasts (see Section 4.1). By the time the various tissues of the leaf blade begin to differentiate, the leaf is generally only six cell layers thick (Pratt, 1974). The outer surface layers differentiate into the adaxial (upper, top, and sun-exposed) and abaxial (lower, bottom, and shaded) epidermis. The epidermis does not contain chloroplasts but can have various types of hairs called trichomes, particularly along the veins on the abaxial side (Viala and Vermorel, 1909). The number and length of trichomes varies with species and cultivar and gives the underside of some leaves (e.g., *Vitis candicans*; Clairette, Meunier) a matte whitish appearance, whereas others (e.g., *Vitis monticola*; Grenache) are glabrous (Latin *glaber* = hairless, bald). Trichome density is especially high in young leaves, but it decreases after unfolding due to gradual shedding of trichomes, especially on the adaxial side (Karabourniotis et al., 1999). Trichomes probably help reduce water loss (by regulating the leaf surface temperature and increasing the boundary layer resistance; see Section 3.2), deter insects (as mechanical barriers), provide protection from damaging ultraviolet radiation (due to their phenolic compounds absorbing and scattering radiation; see Section 5.2), and may even afford some insulation.

The outside walls of epidermis cells are much thicker than the cell walls of interior tissues (Kutschera, 2008a,b). They contain cutin, a strong, elastic biopolyester composed of polymerized, long-chain hydroxy-fatty acids with 16 (e.g., palmitic acid) or 18 (e.g., stearic acid) carbon atoms, and are covered with a thin, continuous, extracellular membrane termed the cuticular membrane or cuticle (Bargel et al., 2006; Domínguez et al., 2011; Kerstiens, 1996; Schreiber, 2010; Yeats and Rose, 2013). As a specialized modification of the cell wall, the cuticle consists of cutin, polysaccharides (that belong to the cell wall), phenolics (especially hydroxycinnamic acids and flavonoids), and lipids that make up the intracuticular wax embedded in the outer parts of the polymer matrix. It is covered by a layer of overlapping solid, semicrystalline lipid platelets—the epicuticular wax. The intra- and epicuticular waxes are typically referred to as cuticular wax, which seals the plant surface. Although waxes are complex mixtures of different lipids, they are essentially (>90%) made up of very long chains of 20–34 methylene (CH_2) groups that are assembled within expanding epidermis cells and then secreted to the surface along with cutin monomers as the leaf unfolds and expands (Domínguez et al., 2011; Samuels et al., 2008; Schönherr, 2006; Yeats and Rose, 2013). The self-cleaning properties of this water-repellent, or hydrophobic (Greek *hydros* = water, *phobos* = fear), surface film keep the leaves clean by greatly reducing the adhesion of water and particles (Bargel et al., 2006). Together with the thick outer cell walls of the epidermis, the strong and stiff cuticle provides mechanical support that maintains the integrity of leaves and other plant organs, defines the boundaries of these organs (thus preventing fusion of different organs), and determines their growth rate and extent of growth (Javelle et al., 2011). This epidermis/cuticle skin also forms the first mechanical barrier to invading pathogens as well as physical or chemical injury by repelling fungal spores and dust particles thanks to its self-cleaning surface, reduces nutrient leaching by rainfall, and provides some insulation against temperature extremes (Bargel et al., 2006; Domínguez et al., 2011; Kerstiens, 1996; Yeats and Rose, 2013). Its most important function, however, is the protection of the plant from desiccation. The waxy cuticle greatly restricts gas diffusion so that most water vapor and other gases must pass through openings called stomata (Greek *stoma* = mouth; Figure 1.15). These small pores (15–40 μm in length) are formed by degradation of the cell wall joining two specialized cells, the kidney-shaped guard cells, which, in contrast to the surrounding epidermis cells from which they are derived, contain a few chloroplasts (Lawson, 2008). *Vitis* leaves are said to be hypostomatous because the stomata are located almost exclusively in the abaxial epidermis. This arrangement is characteristic of plant species that evolved in warm and dry climates and helps to reduce water loss by transpiration.

Grapevine leaves typically have between 50 and 400 stomata per square millimeter, with *V. vinifera* and most rootstocks at the lower end (<250 mm^{-2}) and *V. labrusca* and *V. cinerea* at the higher end (>300 mm^{-2}) of this range (Düring, 1980; Liu et al., 1978; Scienza and Boselli, 1981). Interestingly, stomatal density is determined by the environmental conditions not of a developing leaf itself but by those that act on the mature leaves (Chater et al., 2014). Mature leaves are able to detect the light and CO_2 environment around them and send a (still unknown) signal to the newly developing leaves that adjusts the number of stomata that will form on those leaves. When the mature leaves experience shade or high CO_2, the stomatal density of the newly developing leaves is reduced (Chater et al., 2014). Stomata act as valves that are responsible for regulating the gas

FIGURE 1.15

Cross section of a leaf (left; © Elsevier Inc. illustration by L. O'Keefe) and stoma on a Chardonnay leaf, surrounded by crystalline platelets of epicuticular wax (right; photo by M. Keller).

exchange (mainly CO_2 for photosynthesis, oxygen for respiration, and water vapor from transpiration) between the leaf and the atmosphere. Their major function is to balance the uptake of CO_2 with the loss of water (see Section 3.2). Stomata open when the guard cells take up water and swell (their volume more than doubles); because they are attached to each other at both ends, they cannot increase in length and therefore bend outward. The aperture (pore width) of open stomata is approximately 8 μm, whereas it is almost zero when the stomata are closed, for example, at night or during severe water stress.

The tissue between the two epidermal layers is called the mesophyll (see Figure 1.15). As the leaf expands, the mesophyll cells stop growing before the epidermis cells do. This pulls the mesophyll cells apart and leads to the formation of a large interconnected network system of intercellular spaces that facilitate the diffusion of gases such as water vapor, O_2, and CO_2 (Van Volkenburgh, 1999). This network is very important for both photosynthesis and respiration, because O_2 and CO_2 diffuse through air about four orders of magnitude more rapidly than they diffuse through water. The mesophyll consists of a single layer of elongated cells termed palisade parenchyma and four to six layers of irregularly shaped, loosely packed cells termed spongy parenchyma (Galet, 2000; Pratt, 1974; Viala and Vermorel, 1909). Both cell types contain large numbers of chloroplasts (Greek *chloros* = light green, yellow), which are the actual centers of photosynthesis and assimilation.

The palisade parenchyma lies below the adaxial epidermis and has small intercellular spaces. It also contains crystals of calcium oxalate, which are arranged inside large cells (typically 50–100 μm long and 20–25 μm wide) in bundles of needlelike crystals called raphides and, in older leaves inside very small cells (~10 μm in diameter) lining the veins, in aggregates of octahedral (star-shaped) crystals called druses (Evert, 2006; Viala and Vermorel, 1909). By scattering the incoming light on their irregular surfaces, the druses may help distribute light evenly to the chloroplasts to maximize the leaf's photosynthetic efficiency (He et al., 2014). The spongy parenchyma follows toward the abaxial side and has large intercellular spaces. Many of the mesophyll cell walls are thus exposed to the air inside the leaf (internal leaf atmosphere) so that the internal leaf surface is approximately 10–40 times the surface area of the leaf. Water evaporates from the mesophyll cell walls into these intercellular spaces, which are continuous with the outside air when the stomata are open so that water vapor is discharged (transpired) into the atmosphere (see Figure 1.15; see also Section 3.2). The CO_2 follows the reverse diffusion route into the leaf (see Section 4.2).

1.3.5 TENDRILS AND CLUSTERS

Tendrils and fruiting clusters of the grapevine are generally considered homologous on the basis of anatomical, morphological, and physiological similarities. Darwin (1875) concluded from observations of grapevines growing in his backyard that "there can be no doubt that the tendril is a modified flower-peduncle." Indeed, studies of gene expression—that is, of the manufacture of RNA and protein from the segment of DNA making up a gene—also suggest that tendrils are modified reproductive organs that have been recruited (i.e., adapted during evolution) as climbing organs (Calonje et al., 2004; Díaz-Riquelme et al., 2009). As sterile reproductive structures, they are prevented from completing floral development by the cell-elongation hormone gibberellin (Alleweldt, 1961; Boss and Thomas, 2002; Boss et al., 2003; Srinivasan and Mullins, 1978, 1980, 1981). Accordingly, the absence of gibberellin production or suppression of its activity at the site of organ initiation gives rise to an inflorescence (flower cluster), whereas the presence of gibberellin results in a tendril. The cell-division hormone cytokinin, in contrast, promotes inflorescence formation over tendril formation (Crane et al., 2012; Srinivasan and Mullins, 1978, 1980, 1981). It seems likely that the response to these hormones is quantitative because there is a continuum of transitional or intermediary structures that are partly tendril-like and partly inflorescence-like. Moreover, the degree of differentiation depends largely on the cultivar and environmental conditions during the time of cluster initiation and differentiation (see Section 2.3). Both tendrils and inflorescences, in turn, are modified shoots (lateral branches), and there is also a variety of intermediary forms between inflorescence and shoot (Figure 1.16). The individual flowers on an inflorescence, again, are modified shoots, with the separate floral organs having evolved from highly modified leaves.

Similar to leaves, tendrils grow in intercalary fashion in addition to growth in their own apical meristem (Tucker and Hoefert, 1968). They also differentiate an epidermis with numerous stomata, a spongy parenchyma, and a large but short-lived hydathode at each of their tips, which soon become suberized (Gerrath and Posluszny, 1988a; Tucker and Hoefert, 1968). Although two tips are most common, vigorously growing shoots often form tendrils with three or more tips (Figure 1.17).

1.3 MORPHOLOGY AND ANATOMY

FIGURE 1.16

Intermediary forms of Syrah inflorescence/tendril/shoot (left; inset: inflorescence that would rather be a tendril; photos by M. Keller) and structure of a grape cluster with berries removed (right; illustration by A. Mills; inset reproduced from Viala and Vermorel, 1909).

FIGURE 1.17

Vitis shoot tip showing tendrils with two and three tips (left) and tips of a tendril coiling around a trellis wire (right).

Photos by M. Keller.

Tendrils enable wild vines to access sunlight at the top of tree canopies with a relatively small investment in shoot biomass per unit height gain. The tendril's tips search for surrounding objects by making sweeping, rotating movements during growth (Darwin, 1875; Galet, 2000; Viala and Vermorel, 1909); this oscillatory growth pattern is termed circumnutation. When one of the tips detects a support (via contact-sensitive epidermis cells), the arms rapidly coil around the support in two directions (see Figure 1.17). Following this so-called thigmotropic (Greek *thigma* = to touch, *trope* = turn) movement, the entire tendril lignifies and stiffens to prevent unwinding (Braam, 2005). Tendrils that fail to find a support die and are abscised at the point of attachment to the shoot.

The fruiting structures of the grapevine are panicles and are known as inflorescences or flower clusters. They are always inserted opposite a leaf and are initially protected by bracts covered with trichomes. After fruit set, the inflorescence is called a cluster or bunch. It is composed of the peduncle (stalk; i.e., the entire branched axis apart from the pedicels of individual flowers) carrying the flowers or, following fruit set, the berries. Each flower or berry is attached to the peduncle via a pedicel (final branch or flower stalk) that widens into the receptacle that carries the flower or berry. The main stem or axis consists of the hypoclade (between the shoot and the first branching) and the rachis (central axis) that is made up of a main axis (also called inner arm) and a lateral wing or shoulder (also called outer arm). Both arms carry multiple secondary branches (and many of these tertiary branches, some of which in turn carry quaternary branches), although the shoulder may be missing or may be a tendril or, occasionally, a shoot. Just as in tendrils, each of the arms and branches is subtended by a bract. The flowers usually occur in groups of three (with two, often slightly smaller, flowers flanking a central, terminal flower) or five; these basic floral units are termed a triad, a dichasium, or a cyme (Gerrath and Posluszny, 1988b; May, 2004; Mullins et al., 1992). A single inflorescence may carry between less than a handful and well over a thousand flowers; this variation contributes to large variations in yield in cultivated grapes (see Section 6.1). The hypoclade, which varies in length from 2 to 10 cm depending on cultivar, connects the cluster with the shoot, similar to the function of the petiole for the leaf, and therefore also contains multiple (typically ~30) vascular bundles that are separated by rays (Viala and Vermorel, 1909). The vascular bundles run the entire length of the cluster framework, then branch in the receptacles to serve the individual berries (Bondada and Keller, 2012; Hall et al., 2011). The parenchyma cells of the rays and to the inside of the vascular bundles accumulate starch and may serve as a transient nutrient storage compartment. The epidermis on the exterior contains numerous stomata and is covered with a cuticle. The stomata, especially those of the pedicels and receptacles, are later converted into suberized lenticels; additional lenticels may also form in stomata-free areas (Bondada and Keller, 2012; Swift et al., 1973; Theiler, 1970). Clusters vary widely in length from 3–5 cm in wild grape species to more than 50 cm in some table grape cultivars, and this is accompanied by varying degrees of branching. The cluster architecture resulting from this branching is probably determined by auxin released from developing flower meristems. Indeed, the development of the inflorescence requires active transport of auxin in the cambium and xylem parenchyma down the inflorescence axis (Morris, 2000).

1.3.6 FLOWERS AND GRAPE BERRIES

As discussed in Section 1.1, wild grapes are dioecious, but most *V. vinifera* cultivars have perfect (hermaphroditic) flowers (Boursiquot et al., 1995). Physiologically female flowers are rare among

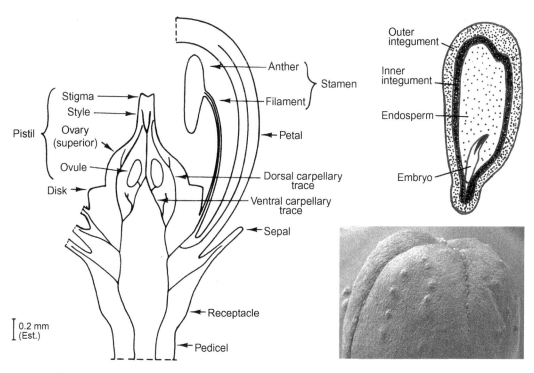

FIGURE 1.18

Diagrammatic longitudinal section of a Muscat Hamburg flower (left; reproduced from Rafei, M.S. 1941. Anatomical studies in *Vitis* and allied genera. I. Development of the fruit. II. Floral anatomy. PhD thesis, Oregon State College), longitudinal section of a *Vitis* seed (top right; illustration by A. Mills), and tip of flower cap with interlocking cells and protruding stomata (bottom right; photo by O. Viret).

V. vinifera cultivars (e.g., Madeleine Angevine and Picolit) and, owing to their propensity to produce large berries, tend to be grown for table grapes, but they are common among *M. rotundifolia* cultivars. Many rootstocks have male flowers, whereas some (e.g., Ramsey, 101-14 Mgt, 5BB, and 41B) have female flowers (Meneghetti et al., 2006). The formation and development of grape flowers is considered in Section 2.3. The flowers (Figure 1.18) are green, inconspicuous, and small; their size varies, depending on species, from 2 mm (*V. berlandieri*) to 6 or 7 mm (*V. labrusca*). They are made up of a receptacle (sometimes called a torus) carrying five sepals (fused to form the calyx), five petals (with their bordering cells interlocked to form a protective cap, termed a calyptra or corolla), five stamens, and the pistil (May, 2004; Meneghetti et al., 2006; Srinivasan and Mullins, 1981; Swanepoel and Archer, 1988). The sepals and petals together form the sterile flower organs, collectively termed the perianth. The stamens are the male reproductive organs with bilobed anthers (microsporangia), each lobe carrying two pollen sacs containing more than 1000 pollen grains (each composed of a large vegetative cell and two sperm cells); their stems are termed anther filaments. The pistil is the female reproductive organ, with a superior ovary surmounted by a short style and a papillate stigma (Considine and Knox, 1979a; Hardie et al., 1996b; Pratt, 1971). The set of stamens of a flower are collectively termed the androecium, and the female parts together are named the gynoecium (Greek *gune* = female, *oikos* = house). The gynoecium of grape flowers generally consists of two seed pockets called carpels (megasporangia) that are fused to form the pistil; the central region in the ovary where the two inner carpel walls meet is termed the septum, which divides the ovary, and later the fruit, into two locules. Female flowers are difficult to distinguish from perfect flowers: they have stamens that look normal but produce sterile pollen, whereas in male flowers the pistil is reduced to a tiny ovary with undeveloped style and stigma (Caporali et al., 2003; see also Figure 1.1).

A network of vascular bundles, or traces, supplies each of the floral organs (see Figure 1.18). The base of the ovary is encircled by a whorl of odor glands termed osmophors but sometimes referred to as nectaries. The ovary comprises two, rarely three, cavities termed locules, each normally containing two seed primordia, called ovules (Nitsch et al., 1960), that are formed by a placental wall to which they remain attached via a funiculus (Latin *funiculus* = thin rope), which serves the same function as the umbilical cord in mammals. Both the calyptra and the pistil (after capfall) have a cuticle-covered epidermis containing chloroplasts and, through their stomata ($10-20$ mm^{-2}; Figure 1.18), are capable of their own gas exchange (Blanke and Leyhe, 1987, 1988, 1989a). The pistil tissues are rich in calcium oxalate crystals, starch, and phenolic compounds, although the starch disappears during berry development (Considine and Knox, 1979a; Hardie et al., 1996a; Sartorius, 1926).

The ovules (female gametophytes or megagametophytes), with their fertilized egg (zygote) and remaining embryo sac, can develop into heart- or pear-shaped seeds whose main role is to protect and nurture the developing embryo (Roberts et al., 2002). Thus, a grape berry with two locules can have a maximum of four seeds, and one with three locules can have six seeds. In practice, the seed number is usually one or two in *V. vinifera* cultivars but two or three in many Chinese *Vitis* species (Wan et al., 2008c). Depending on cultivar and seed number, the seeds make up between 2% and 7% of the fresh weight of mature grapes (Viala and Vermorel, 1909). After fertilization, the pistil develops into the fruit, with the ovary wall becoming the pericarp—that is, the flesh and skin (Figure 1.19); thus, the fruit can be viewed as a swollen or mature ovary (Gillaspy et al., 1993). Grapes are classified as a berry fruit because the thick pericarp encloses the seeds.

The ovule is made up of the outer and inner integuments and the nucellus (Pratt, 1971) and the embryo sac. The latter is composed of seven cells: the egg cell flanked by two synergid cells, the central cell, and the three antipodals. After fertilization, the integuments will form the seed coat, termed testa, and the nucellus will surround the embryo that develops from the egg cell and the endosperm that develops from the central cell (see Figure 1.18). The embryo produces an epicotyl (embryonic shoot tip), two cotyledons (embryonic or seed leaves), a hypocotyl (embryonic shoot base), and a radicle (embryonic root), complete with their vascular system. The epicotyl contains at its tip the shoot apical meristem, and the radicle contains at its tip the root apical meristem. This arrangement defines the apical–basal axis of the embryo and future seedling. The thick outer integument consists of several layers of palisade parenchyma cells and the protoderm, which soon becomes the epidermis and forms a cuticle on the outside that is impregnated with suberin (Javelle et al., 2011; Schreiber, 2010). The outer integument is sometimes further divided into a soft outer integument and a hard middle integument; the outer and inner integuments, as well as the epidermis, accumulate tannins, whereas the middle integument instead becomes lignified as the seed matures (Cadot et al., 2006). The protective testa contains inclusions of calcium oxalate crystals

1.3 MORPHOLOGY AND ANATOMY

FIGURE 1.19

Structure of a ripe grape berry (left; reproduced from Coombe, 1987, reprinted by permission of American Journal of Enology and Viticulture), scanning electron micrograph detail of ruptured Syrah berry surface with epicuticular wax and raphide crystals of calcium oxalate (top right; photo by S. Rogiers), and stylar end of Syrah berry covered with epicuticular wax (bottom right; photo by M. Keller).

and is rich in phenolic compounds such as tannins that protect the seed against premature "consumption" by microorganisms or insects (Dixon et al., 2005). The apparent folding of the integuments leaves a small opening (pore), termed micropyle, through which the pollen tube enters during pollination (Sartorius, 1926; see also Section 2.3); during seed germination, water will enter and the radicle will exit. The endosperm progressively expands at the expense of the nucellus, which may serve as a food supply for the embryo and endosperm during early seed development (Cadot et al., 2006; Nitsch et al., 1960).

The vascular bundles that previously served the ovary give rise to a complex network of vascular traces that supply the seed (ovular bundles) and the pericarp (ventral and dorsal bundles). The dorsal, or peripheral, vascular bundles of most *V. vinifera* cultivars lie approximately 200–300 μm beneath a mature berry's surface and can give some berries a "chicken-wire" appearance (see Figure 1.19), whereas in *V. labrusca* the bundles are approximately twice as far from the surface (Alleweldt et al., 1981). The ventral, or central, vascular bundles and their associated parenchyma cells are termed the "brush" and remain attached to the pedicel when a ripe grape is plucked from the cluster. Individual tracheary elements (mostly vessel elements with some tracheids) in the berry are approximately 250–300 µm long with an average diameter of approximately 3.5–5 µm (Chatelet et al., 2008a). Although existing tracheary elements are stretched at the beginning of ripening, most of them remain intact, and new elements are added throughout berry development. The pericarp consists of three anatomically distinct tissues: the exocarp (or epicarp), mesocarp, and endocarp (Galet, 2000; Viala and Vermorel, 1909).

The exocarp forms the grape's dermal system or "skin," which, depending on thickness (which varies by species and cultivar) and berry size, makes up between 5% and 18% of the fresh weight of mature berries. It is composed of the outer epidermis covered with a waxy cuticle and the underlying outer hypodermis (Bessis, 1972a; Considine and Knox, 1979b; Fougère-Rifot et al., 1995; Hardie et al., 1996b). Epicuticular wax, often referred to as "bloom," covers the surface of the cuticle as overlapping platelets forming a strongly hydrophobic layer that protects the berry from water loss and forms the first barrier against pathogen invasion (Grncarevic and Radler, 1971; Possingham et al., 1967; Rosenquist and Morrison, 1988). The structure of the epicuticular wax changes with age of the berry. Although the wax is normally crystalline, the crystals appear to degrade slowly over time, fusing into amorphous masses and eventually into a more or less continuous layer on top of the normal, uninterrupted wax layer (Casado and Heredia, 2001; Rogiers et al., 2004a). Sun-exposed berries have considerably more amorphous wax than do shaded berries (Rogiers et al., 2004a). This is probably a consequence of solar heating of the berry surface; the transition from crystalline to amorphous wax occurs at approximately 50°C (Khanal et al., 2013). The wax material is rather soft and can be altered or removed by the impact of rain, by abrasion from windblown particles, or by contact with other berries, leaves, or other adjacent objects (Shepherd and Griffiths, 2006). However, because of its vital importance as a desiccation protectant, a damaged or removed wax layer is quickly regenerated, especially in young berries. The cuticle is approximately 1.5–4 µm thick, and the epidermis consists of a single cell layer that is approximately 6–10 µm thick (Alleweldt et al., 1981; Bessis, 1972a; Hardie et al., 1996b). Unlike the hypodermis and mesocarp cells, the cells of the epidermis divide in a strictly anticlinal manner (i.e., normal to the surface), thus maintaining their cell "lineage" (Considine and Knox, 1981). The initially single-layered outer hypodermis (100–250 µm thick) increases to 6–10 layers of thick-walled cells after fruit set and then transiently even to 15–20 layers. These cells contain chloroplasts and tannin-rich vacuoles, interspersed with amyloplasts, which form within a few days after fruit set but disappear during berry development, and the so-called idioblast (Greek *idios* = unique, *blastos* = germ, sprout) cells rich in bundles of needlelike calcium oxalate crystals called raphides (Evert, 2006; Fougère-Rifot et al., 1995; see also Figure 1.19). As an aside, a cell may contain more than one kind of vacuole, so it may be possible for different components to be stored in specialized vacuoles within the same cell and for neighboring cells to differ in the content of their vacuoles (Marty, 1999; Müntz, 2007). The outermost hypodermal cells become stretched and radially compressed, and their cell walls thicken approximately 10-fold as the berry expands (Considine and Knox, 1979b; Hardie et al., 1996b). The innermost cells expand and seem to be progressively converted to mesocarp cells. Thus, the skin becomes thinner (four or five cell layers) in the ripening berry and the vascular bundles, which initially marked the "boundary" between hypodermis and mesocarp, are now situated within the mesocarp (Chatelet et al., 2008a;

Fougère-Rifot et al., 1995). The relative thickness of the skin declines from approximately one-eighth to one-hundredth of the total berry diameter between fruit set and maturity.

The mesocarp, which is commonly called the "flesh" or "pulp" of the grape berry, forms soon after fruit set and consists of 25–30 layers of large, thin-walled, and highly vacuolated storage parenchyma cells that come in a wide variety of sizes (50–400 μm) and shapes (Gray et al., 1999). The diverse vacuoles also vary greatly in size, ranging from 1 to 50 μm in diameter, and can make up as much as 99% of the cell volume in ripe grape berries (Diakou and Carde, 2001; Fontes et al., 2011b; Storey, 1987). They serve as an internal reservoir storing sugars, organic acids, and nutrients (see Section 6.2). Mesocarp cells outside the network of peripheral vascular bundles of the pericarp are termed the outer mesocarp; those toward the inside are the inner mesocarp, which, at maturity, makes up almost two-thirds of the berry volume. At maturity, the mesocarp cells are approximately 75 times larger than the skin cells, and their radial width is approximately 10 times that of the skin cells, whereas the cell walls in the skin are almost 20 times thicker than those in the mesocarp. It is no wonder, then, that the skin sets the upper limit to which a berry can expand.

The innermost tissue of the pericarp, the endocarp, surrounds the seeds and consists of the inner hypodermis with cells rich in star-shaped calcium oxalate crystals (druses) and also the inner epidermis (Fougère-Rifot et al., 1995; Hardie et al., 1996b). The inner epidermis, like the stigma and the integuments, is derived from the same cell layer in the apical meristem that also forms the outer epidermis (Jenik and Irish, 2000). The endocarp is often difficult to distinguish from the mesocarp in grapes (Mullins et al., 1992).

CHAPTER 2

PHENOLOGY AND GROWTH CYCLE

CHAPTER OUTLINE

2.1 Seasons and Day Length 59
2.2 Vegetative Cycle 61
2.3 Reproductive Cycle 83

ABSTRACT

The annual growth cycle of fruiting grapevines is divided into a vegetative and a reproductive cycle. Fruit production extends over 2 years: buds formed in the first year give rise to shoots bearing fruit in the second year. Seasonal growth is driven by day length and temperature, and alternates with winter dormancy. The transition from dormancy to active growth in spring is marked by bleeding of xylem sap from pruning wounds due to root pressure. Root pressure restores xylem functionality and rehydrates the dormant buds. Increasing temperature then leads to budbreak and shoot growth that is marked by apical dominance. Shoots and roots grow as long as the environment permits. The shoots form brown periderm when the days shorten in late summer, enter dormancy, and shed their leaves in autumn. Chilling temperatures release dormancy to resume growth in spring. Flower clusters are initiated in the buds in early summer, and flowers differentiate after budbreak the following spring. Double fertilization during bloom initiates the transition of flowers to berries. Berry growth follows a doublesigmoid pattern of cell division and expansion, seed growth, and final cell expansion concomitant with fruit ripening. Seedless berries have less discernible growth phases. Ripening makes berries attractive for seed dispersers to spread a vine's genes.

Keywords: Anthesis; apical dominance; berry fruit; differentiation; dormancy; phenology; reproductive growth; ripening; root pressure; vegetative growth

2.1 SEASONS AND DAY LENGTH

Phenology (Greek *phainesthai* = to appear, *logos* = knowledge, teaching) is the study of natural phenomena that recur periodically in plants and animals and of the relationship of these phenomena to climate and changes in season. In other words, it is the study of the annual sequence of plant development. Its aim is to describe the causes of variation in timing of developmental events by seeking correlations between weather indices and the dates of particular growth events and the intervals between them. Phenology investigates a plant's reaction to the environment and attempts to predict its behavior in new environments. In viticulture, phenology is mainly concerned with the timing of specific stages of growth and development in the annual cycle. Such knowledge can be used for site and cultivar selection, vineyard design, planning of labor and equipment requirements, and timing of cultural practices as part of vineyard management (Dry and Coombe, 2004).

Grapevines, like other plants, monitor the seasons by means of an endogenous "clock" (Greek *endon* = inside, *genes* = causing) in order to prevent damage by unfavorable environments. Every plant cell contains a clock (perhaps even several clocks), and the clocks of different cells, tissues, and organs act autonomously (McClung, 2001, 2008). The clock is driven by a self-sustaining oscillator consisting of proteins that oscillate with a 24 h rhythm in response to light detected by phytochromes (see Section 5.2) and other light-sensitive proteins that translate the light signal into a clock input that uses this time-encoded information to regulate physiological functions (Fankhauser and Staiger, 2002; Spalding and Folta, 2005). Additional input is provided through feedback loops that arise from various compounds produced by the plant's metabolism (McClung, 2008). Just how plants manage to integrate and transform this environmental and metabolic information into daily and seasonal functions and how they avoid being "fooled" by bright moonlight is still mysterious, but it could involve fluctuations in calcium concentration of the cytosol. Cytosolic Ca^{2+} also oscillates with a 24 h period with a peak before dusk, and it participates in the "translation" of many signals, both internal and external (Hotta et al., 2007; McClung, 2001; Salomé and McClung, 2005; see also Section 7.3). In contrast to the temperature dependence of most biochemical processes, the period of the rhythm is temperature compensated so that it remains constant at 24 h over a wide range of temperatures. The gradual shift in time of sunrise (actually the change between dawn and dusk) as the seasons progress serves to reset the phase of the clock every day and enables plants to "know" when the sun will rise even before dawn (Fankhauser and Staiger, 2002). This synchronization of the circadian (Latin *circa* = about, *dies* = day) clock to the external environment is termed entrainment. It translates day length (i.e., photoperiod) into a vine's internal time as an estimate of both the time of day and the time of year, enabling it to anticipate and prepare for daily and seasonal fluctuations in light and temperature. For example, the genes responsible for "building" enzymes involved in the production of ultraviolet (UV)-protecting phenolics are most active before dawn, photosynthesis-related genes peak during the day (and opening and closing of the stomata can anticipate dawn and dusk), and genes associated with stress are induced in late afternoon (Fankhauser and Staiger, 2002; Hotta et al., 2007).

On a seasonal scale, flowering, onset of fruit ripening, bud dormancy, leaf senescence (Latin *senescere* = to grow old), and cold acclimation are typical responses to day length (the so-called photoperiodism), although each of these developmental processes is also modulated by temperature, and some can be altered by stress factors such as drought (see Section 7.2), nutrient deficiency or excess (see Section 7.3), or infection by pathogens (see Section 8.2). In general, higher temperatures accelerate plant development and advance grapevine phenology (Parker et al., 2011). For example, the time of budbreak of a given *V. vinifera* cultivar grown in a specific location can vary by about 1 month from one year to another (Boursiquot et al., 1995). Even very high temperatures (e.g., average daily mean 35°C) still shorten the time from budbreak to bloom compared with lower temperatures (Buttrose and Hale, 1973). Therefore, one of the most conspicuous consequences of climate change associated with the current and future increase in temperature due to the man-made rise of atmospheric CO_2 is a shift of phenological stages to earlier times during the growing season (Chuine et al., 2004; Wolfe et al., 2005).

Because day length during the growing season is longer at higher latitudes, the effects of day length and latitude on the internal clock are very similar (Salomé and McClung, 2005). In tropical climates, with little or no change in day length (and temperature at lower elevations) during the year, grapevines behave as evergreens with continuous growth and strong apical dominance, and often continuous fruit production, throughout the year (Dry and Coombe, 2004; Lavee and May, 1997; Mullins et al., 1992). With appropriate pruning strategies (and sometimes deliberate defoliation by hand, often in combination with imposed water deficit and sprays of ethephon or other chemicals to induce near-dormancy, followed by applications of hydrogen cyanamide to induce uniform budbreak), two crops can often be harvested each year in the tropics (Bammi and Randhawa, 1968; Favero et al., 2011; Lin et al., 1985). In temperate climates (and at high elevations in the tropics), where winter conditions prevent the survival of leaves, grapevines have a discontinuous cycle with alternating periods of growth and dormancy. Under these conditions, the time of active growth generally occurs from March/April to October/November (Northern Hemisphere) or September/October to April/May (Southern Hemisphere). Several distinct developmental stages or key events have been identified (Figures 2.1 and 2.2) and have been given a variety of names. These stages include dormancy, budbreak (budburst), bloom (anthesis or flowering), fruit set (berry set or setting), veraison (French *vérer* = to change: color change or onset of ripening), harvest (ripeness or maturity), and leaf senescence and subsequent fall (abscission). Although by definition no visible growth occurs during dormancy, metabolism does not rest completely, but high concentrations of the dormancy hormone abscisic acid (ABA) keep it at a minimum necessary for survival of the buds and woody tissues. Although the division of meristematic cells in the buds and cambium is blocked, chromosome duplication and protein synthesis resume during the later stages of dormancy in preparation for the activation of growth when temperature and soil moisture become favorable in spring.

The annual growth cycle of mature, fruiting grapevines is often divided into a vegetative and a reproductive cycle.

2.2 VEGETATIVE CYCLE

In late winter or early spring, grapevines often exude xylem sap from pruning surfaces and other wounds that have not yet been suberized (Figure 2.3). Such sap flow or "bleeding" marks the transition from dormancy to active growth. Initiation of bleeding is related to the restoration of metabolic activity in the roots and is influenced by soil temperature, moisture, and rootstock, but on average it seems to begin when the soil temperature rises above approximately 7°C (Alleweldt, 1965). Indeed, root respiration, as a proxy for metabolic activity in the roots, is closely coupled to soil temperature (Franck et al., 2011). Within days after the first signs of bleeding, the callose that has sealed the sieve plates in the phloem of trunks and canes during the winter begins to be degraded (Pouget, 1963). Bleeding can last for a few days or several weeks (probably depending on whether or not air temperatures are conducive to budbreak); it can also be a stop-and-go process that fluctuates with changes in soil temperature (Andersen and Brodbeck, 1989b; Reuther and Reichardt, 1963). A vine can exude bleeding sap at rates of less than 0.1 L to greater than 1 L per day, with the highest rates occurring on warm and moist soils (Alleweldt, 1965; Currle et al.; 1983; Houdaille and Guillon, 1895). Bleeding is

FIGURE 2.1

Grapevine growth stages according to Baillod and Baggiolini (1993).

2.2 VEGETATIVE CYCLE

FIGURE 2.2

Grapevine growth stages according to Eichhorn and Lorenz.

Reproduced from Coombe, B.G. (1995). Adoption of a system for identifying grapevine growth stages. Aust. J. Grape Wine Res. *1, 104–110, with permission by Wiley-Blackwell.*

FIGURE 2.3

Bleeding grapevine cane (left) and swelling, woolly bud just before budbreak (right).

Photos by M. Keller.

caused by root pressure (Fisher et al., 1997; Priestly and Wormall, 1925; Sperry et al., 1987), whose existence in grapevines was first demonstrated by the English cleric Stephen Hales (1727). Root pressure arises from the remobilization of nutrient reserves from proteins and starch and pumping of amino acids and sugars into the xylem tracheary elements (Reuther and Reichardt, 1963; Roubelakis-Angelakis and Kliewer, 1979). The concentration of sugars apparently increases exponentially as the temperature declines close to and below 0°C; this response occurs within a few hours and is equally rapidly reversible when the temperature rises again (Moreau and Vinet, 1923; Reuther and Reichardt, 1963). Sugars (especially glucose and fructose), however, are only a minor component of the bleeding sap. Organic acids (especially malate, tartrate, and citrate), amino acids (especially glutamine), and mineral nutrient ions (especially potassium and calcium) together contribute a much greater fraction ($>90\%$) of the total solutes present (Andersen and Brodbeck, 1989a; Glad et al., 1992; Marangoni et al., 1986; Priestley and Wormall, 1925). The resulting increase in the osmotic pressure of the xylem sap provides the driving force for osmotic water uptake by the roots from the soil, which generates a positive pressure in the entire xylem of 0.2–0.4 MPa at the trunk base that declines at a rate of 0.01 MPa m^{-1} due to gravity as the sap rises (Scholander et al., 1955; Sperry et al., 1987). Because a pressure of 0.1 MPa can support a water column of approximately 10 m high, root pressure lifts water up the vine (see Section 3.3) and to the swelling buds of even tall, wild grapevines.

Root pressure serves to dissolve and push out air bubbles that have formed during the winter in the xylem and thus restores xylem function (Dixon and Joly, 1895; Scholander et al., 1955; Sperry et al., 1987). This is necessary because when xylem sap freezes, the gas dissolved in the sap is forced out of solution because gases are practically insoluble in ice. This physical process can lead to cavitation (see Section 3.3) upon thawing if the bubbles expand instead of going back into solution. Given that bleeding typically starts before the buds begin to swell, root pressure may also be necessary to rehydrate the dormant buds and reactivate them with cytokinin that is delivered from the roots (Field et al., 2009; Lilov and Andonova, 1976). In fact, the buds' water content quickly rises from less than 50% to approximately 80% during the budswell phase (Fuller and Telli, 1999; Lavee and May, 1997; Pouget, 1963). The delivery of sugars in the absence of phloem flow may enable the buds to resume growth (i.e., cell division in the apical meristem), produce the growth hormone auxin (or release it from storage pools), and break. Accordingly, root pressure−driven water ascent may stimulate bud swelling, and the maximum rate of bleeding occurs a few days before budbreak and generally stops 10−14 days later (Alleweldt, 1965), although root pressure may occasionally remain active for more than 1 month after budbreak (Reuther and Reichardt, 1963). Both bleeding and root respiration decline as soil moisture decreases, and membrane leakage increases sharply if soil moisture becomes too low (Currle et al., 1983; Huang et al., 2005b). This may be the reason why bleeding is minimal or even absent in vines growing in dry soil, which can cause stunted shoot growth (possibly due to cavitation) when shoots begin to grow with stored water from the trunk and root parenchyma in the absence of sufficient soil water to supply the unfolding leaves. Nevertheless, the emerging shoots quickly become dependent on the delivery in the reactivated phloem of sugar and other nutrients remobilized from the permanent vine organs.

Cell division and auxin production in buds, beginning in the basal (closest to the base) leaf primordia of the distal (closest to the tip) buds, starts 1−3 weeks before budbreak (see Figure 2.3), which marks the onset of vegetative growth in spring. At the same time, ABA declines strongly, relieving the growth inhibition that keeps the vine dormant while low temperatures would threaten its survival. The release of auxin and its basipetal (i.e., from the apex toward the base) transport in the cambial region begins suddenly and rushes like a continuous wave down the vine, stimulating the cambium cells to resume division and differentiation into new phloem and xylem cells (Aloni, 2001). The cambium cells become sensitive to auxin when they are exposed to cytokinin, which is supplied from the roots as a result of root pressure. Although it remains to be demonstrated whether the root tips are capable of cytokinin production this early in spring, cytokinins have indeed been found in the bleeding sap of grapevines and, in addition to being influenced by the rootstock (Skene and Antcliff, 1972), higher root temperatures favor cytokinin delivery via the xylem (Field et al., 2009; Zelleke and Kliewer, 1981). As an aside, the same combination of the actions of bud-derived auxin and root-derived cytokinin probably also induces the cambium on either side of new grafts to produce undifferentiated cells termed callus and to form a union, complete with new vascular tissues, between scion and rootstock (Cookson et al., 2013). The differentiation of cells produced in the previous growing season to form the first xylem vessels of the new shoot begins even before the cambium starts producing new cells (i.e., before budbreak), although these vessels mature (i.e., lignify) only after budbreak. Warmer temperatures during vessel differentiation, perhaps by enhancing the supply of remobilized nutrient reserves to the cambium, result in wider vessels that may be more effective at supplying water to the developing shoots later in the season (Fonti et al., 2007); this temperature effect could be an important driving variable for seasonal shoot vigor. Because cambium reactivation occurs well before the new leaves will begin supplying assimilates, the energy and carbon requirements of the dividing cells have to be met by the remobilization of stored starch and supply of sugar from the permanent structure of the vine. This starch breakdown may by induced by gibberellin, whose production is stimulated by the auxin emanating from the buds and shoot tips (Frigerio et al., 2006; Yamaguchi, 2008).

The "chilling requirement" is often defined as the period of low temperature that is necessary to permit 100% of the buds to break (Cooke et al., 2012). Although some grape cultivars (e.g., Thompson Seedless) are believed not to require chilling to resume growth from dormant buds, budbreak generally occurs earlier, more rapidly, and more uniformly on vines that have experienced a period of cool weather (Antcliff and May, 1961; Currle et al., 1983; Kliewer and Soleimani, 1972). Moreover, longer duration of chilling and lower temperatures typically accelerate the rate of budbreak and enhance the uniformity of budbreak once warmer temperatures return (Dokoozlian, 1999; Koussa et al., 1994; Lavee and May, 1997; Pouget, 1963). Thus the buds on a single vine usually break within a few days in areas with cold winters (provided the winter is not so cold as to damage or kill the buds, as discussed in Section 7.4) but may take 10 times longer in areas with mild winters. However, if research on forest trees is relevant for grapevines, it is also possible that the chilling requirement itself may be longer in warmer regions, because higher rather than lower temperatures seem to induce earlier and deeper dormancy (Cooke et al., 2012). As a consequence, budbreak is often erratic in warm temperate and (sub)tropical regions, where the majority of the world's table grapes are grown. This is why growers commonly apply hydrogen cyanamide (H_2CN_2), usually in the form of calcium cyanamide ($CaCN_2$), 2–4 weeks before the intended time of budbreak to end dormancy (the so-called dormancy release) and promote budbreak of many table grapes (Dokoozlian et al., 1995; Lavee and May, 1997). The higher winter temperatures in such warmer areas lead to higher respiration rates (see Chapter 4), and this might be associated with an increase in oxidative stress (e.g., hydrogen peroxide release by the mitochondria) that, under extreme conditions, could result in bud damage, called bud necrosis (Pérez et al., 2007). Hydrogen cyanamide also leads to hydrogen peroxide accumulation by inhibiting the antioxidant enzyme catalase, but the subsequent movement of calcium ions (Ca^{2+}) from the cell walls to the cytoplasm, similar to the movement induced by chilling, then somehow seems to initiate cellular mechanisms culminating in dormancy release of the buds (Pang et al., 2007). One such mechanism may be the activation of glucanase enzymes, which are responsible for the breakdown of winter callose in the phloem and the meristems contained in the buds (Pérez et al., 2009). Like entry into dormancy, dormancy release is a gradual process that occurs over several weeks (Cooke et al., 2012; Pérez et al., 2007).

In cooler, temperate climates, mean daily temperatures above approximately 8–10°C induce budbreak and shoot growth, but the precise base temperature requirement depends on the species (e.g., *V. berlandieri* > *V. rupestris* > *V. vinifera* > *V. riparia*) and cultivar (e.g., Mourvèdre > Syrah > Chardonnay). Cultivars with a lower base temperature break buds earlier than those with a higher temperature threshold (Huglin and Schneider, 1998; Pouget, 1972, 1988). Consequently, the time of budbreak at a particular location and in a given year varies by more than 5 weeks among cultivars of *V. vinifera* and can vary from year to year by about 4 weeks for a given cultivar (Boursiquot et al., 1995; Ferguson et al., 2014; Pouget, 1988). The rate of budbreak increases as the temperature rises above the threshold temperature up to approximately 30°C and then declines again. However, it is the temperature of the buds themselves, rather than the surrounding air temperature, that determines when and how rapidly those buds break (Keller and Tarara, 2010). During the day, the temperature of buds can be several degrees above the ambient temperature, especially when the buds are heated by the sun and when there is little or no wind, yet even at wind speeds greater than 5 m s^{-1} buds may still be approximately 2°C warmer than the air (Grace, 2006). Conversely, at night the bud temperature is often cooler than the air temperature. In contrast to the temperature of the buds, and despite its impact on the rate of bleeding sap flow, root temperature has only a minor influence on the time and rate of budbreak (Field et al., 2009; Kliewer, 1975; May, 2004). Nevertheless, the proportion of buds that do break and subsequent shoot growth both increase considerably as root temperature increases up to a maximum near 30°C (Kliewer, 1975).

FIGURE 2.4

Cane-pruned Cabernet Sauvignon vines displaying strong correlative inhibition in spring. Note that budbreak is not inhibited on spurs adjacent to the canes.

Photos by M. Keller.

Shoot growth starts in the distal (topmost or apical) buds of a cane or vine, proceeding basipetally to the roots. Yet preferential budbreak of distal buds often inhibits the growth of the proximal (basal) buds (Galet, 2000). This phenomenon is often attributed to apical dominance, although it may actually be a case of correlative inhibition. In correlative inhibition, growing shoots may inhibit each other and prevent buds on the same cane from breaking, whereas in apical dominance the shoot tip inhibits the outgrowth of lateral buds (Beveridge, 2006; Ferguson and Beveridge, 2009). Correlative inhibition is most conspicuous in early spring when, in cane-pruned vines, the most distal buds of the cane break early and grow vigorously, whereas those toward the base of the cane grow weakly or not at all (Figure 2.4). This inhibition of basal bud outgrowth is especially prevalent in warm climates, where it can lead to erratic and nonuniform budbreak. It also varies among cultivars; for instance, Carignan and Mourvèdre (synonyms Monastrell and Mataro) show particularly strong correlative inhibition, whereas Syrah and Cinsaut are at the other end of the spectrum. Like the evolution of tendrils, this behavior is an adaptation to the habitat of wild vines, which enables them to access sunlight at the top of tree canopies with a small investment in shoot biomass per unit height gain. This adaptive behavior creates both a viticultural problem (mainly in warm climates) and an opportunity (mainly in cool climates) for cultivated grapevines. The problem arises from the inconsistent budbreak of cane-pruned vines. Viticultural practices aimed at overcoming correlative inhibition and promoting uniform budbreak include spur pruning, bending ("arching") and partial cracking of canes, and application of hydrogen cyanamide (Huglin and Schneider, 1998; Iland et al., 2011; Rosner and Cook, 1983). Spur-pruned vines are typically unaffected by correlative inhibition, and where mixed "populations" of spurs and canes are retained on the same vine, the buds on the spurs break at the same time as the distal buds of the canes (see Figure 2.4; Clingeleffer, 1989; Keller and Tarara, 2010). The opportunity stems from the delay in budbreak on vines that are spur-pruned only after the canes' distal buds have already broken, which may enable vines grown in frost-prone locations to escape spring frost damage (see Section 7.4).

Budbreak is followed by a period of very rapid shoot growth, with a new leaf appearing every few days. Grapevine shoots normally grow upward, although there is considerable variation from erect to droopy growth habits among cultivars. For instance, Cabernet Sauvignon shoots are rather

FIGURE 2.5

Emerging Merlot shoot with pink unfolding leaves and inflorescences days after budbreak (left) and Muscat Ottonel shoot with red young leaves (right).

Photos by M. Keller.

erect, whereas Trebbiano or Concord shoots typically grow downward. Upward growth is guided by a negative gravitropic response, whereby the shoot perceives and responds to gravity (a process termed gravitropism). This response includes sedimentation of starch in specialized amyloplasts, called statoliths (a.k.a. statocytes), in the endodermis, which triggers an increase in auxin and a decrease in extracellular pH on whichever side happens to be lower, causing that side to elongate more than the upper side (Morita and Tasaka, 2004; Petrášek and Friml, 2009). Brassinosteroids, it seems, oppose the action of auxin on gravitropism, perhaps by weakening the cell walls (Vandenbussche et al., 2011); thus, it is possible that the dynamic interplay between auxin and brassinosteroids accounts for the differences in growth habit among cultivars.

Growth seems to adhere more or less to the principle of compound interest: as new leaves become photosynthetically competent, they permit growth of yet more leaves, so that growth is proportional to present biomass (Turgeon, 2010). Consequently, seasonal shoot growth and fruit production may be maximized if conditions favor rapid growth early in the spring (Keller and Tarara, 2010; Keller et al., 2010). In many cultivars, the unfolding leaves, and often the developing inflorescences as well, are reddish before they turn green (Figure 2.5). This transient accumulation of anthocyanin pigments in or just below the epidermis probably protects the developing photosynthetic machinery (e.g., protochlorophyll) from excess light (Chalker-Scott, 1999; Liakopoulos et al., 2006; Rosenheim, 1920). The anthocyanins usually disappear by the time the leaf has produced roughly half of its maximum (i.e., mature) chlorophyll and carotenoid content and the palisade cells have expanded to occupy approximately 50% of the mesophyll (Hughes et al., 2007). In addition, the dense covering of "hair," termed pubescence, of the young leaves also serves the same photoprotective purpose (Liakopoulos et al., 2006). Early growth of both shoots and roots is entirely dependent on nutrient reserves (mainly carbohydrates and proteins or amino acids) stored in the permanent structure of the vine, until new leaves become photosynthetically competent and begin to produce and export carbohydrates. These reserves are thus depleted and reach a minimum around bloom time or even later, making vines vulnerable to stress around the time of bloom (Holzapfel et al., 2010; Lebon et al., 2008;

Weyand and Schultz, 2006b; Williams, 1996; Zapata et al., 2004). This is why grapevines with a low carbohydrate reserve status can experience poor fruit set if photosynthesis becomes limiting during bloom (Keller and Koblet, 1994; Rogiers et al., 2011).

Carbohydrate reserves are used not only as building supplies to construct new plant material in the absence of photosynthesis but also to generate the energy (by respiration) required to fuel this construction (see Section 4.4). Such use, in addition to carbohydrate and other nutrient losses during sap flow, means that grapevines literally lose (dry) weight during the initial spring growth phase, before the photosynthesizing leaves begin to add new biomass from about the five- to six-leaf stage onward (Eifert et al., 1961; Hale and Weaver, 1962; Koblet, 1969; Pradubsuk and Davenport, 2010). Nitrogen reserves, in addition to carbohydrates, are very important for new growth in spring; early season growth and especially leaf chlorophyll correlate closely with the amount of nitrogen reserves stored in the perennial parts of the vine (Cheng et al., 2004b; Keller and Koblet, 1995a; Treeby and Wheatley, 2006). But other macro- and micronutrients are also remobilized from perennial organs and used to support new spring growth (Pradubsuk and Davenport, 2010, 2011). Higher soil temperatures are associated with more rapid starch remobilization in the roots and higher rates of root growth and water and nutrient uptake, which may allow the shoots to grow more rapidly on vines in warm soil than in cool soil (Field et al., 2009; Rogiers et al., 2011, 2014; Skene and Kerridge, 1967; Woodham and Alexander, 1966; Zelleke and Kliewer, 1979).

The initial growth phase is characterized by strong apical dominance, whereby the growing shoot tip suppresses budbreak in prompt (lateral) buds by monopolizing sugar supply for its own growth (Mason et al., 2014). The relatively sugar-deficient prompt buds are unable to initiate cell division for bud outgrowth. In addition to this sugar depletion mechanism, slow but active (i.e., mediated by ATP-consuming pumps) basipetal flow of auxin released by the leaf primordia and young leaves near the shoot apex stimulates internode elongation and contributes to the inhibition of lateral shoot growth (Alleweldt and Istar, 1969; Mason et al., 2014; Müller and Leyser, 2011; Woodward and Bartel, 2005). Intriguingly, however, if auxin flows too rapidly, apical dominance is compromised and lateral buds are able to grow out (Berleth et al., 2007; Dun et al., 2006; Lazar and Goodman, 2006). This might be the case in very vigorously growing shoots that are associated with both rapid main shoot elongation and early and strong lateral shoot growth, which contrasts with the near absence of lateral shoot growth on weak, slow-growing shoots (Huglin and Schneider, 1998; Keller and Tarara, 2010). The suppression of axillary bud outgrowth by auxin is indirect and may only support or reinforce the low-sugar signal, because the hormone does not enter these buds, which themselves start producing and exporting auxin as soon as they begin to grow (Müller and Leyser, 2011; Ongaro and Leyser, 2008). Auxin may exert its remote control and determine which buds will grow out by prohibiting auxin efflux from lateral buds and by inhibiting cytokinin production in the roots and in the shoot tissues adjacent to the buds (Beveridge, 2006; Dun et al., 2006; Ferguson and Beveridge, 2009; McSteen and Leyser, 2005; Prusinkiewicz et al., 2009). In addition, strigolactones or their products, which are produced from carotenoids, also inhibit bud outgrowth, probably by decreasing auxin efflux from buds or by blocking cell division, thus preventing bud activation (Crawford et al., 2010; Dun et al., 2012; Gomez-Roldan et al., 2008; Umehara et al., 2008). Like cytokinins, strigolactones are manufactured in the roots and transported in the xylem in addition to their production in vascular tissues throughout the plant and locally in shoot nodes (Ruyter-Spira et al., 2013). But the influence of strigolactone and of cytokinin on bud outgrowth is antagonistic. Enhanced sugar availability (rapid) and auxin depletion (slow) in the shoot following damage to the shoot tip or its removal (e.g., by summer pruning) activates the prompt buds, releases the repression of cytokinin synthesis in the shoot and delivery from the roots, and may inhibit strigolactone production (Hayward et al., 2009; Mason et al., 2014; Nordström et al., 2004; Tanaka et al., 2006). Sugar and cytokinins that enter the buds stimulate growth of lateral shoots by promoting cell division, especially near the former shoot tip where auxin is depleted first. In a process termed canalization, active export of newly produced auxin from outgrowing axillary buds occurs down polarized files of cells that then develop into new vascular bundles connecting the lateral shoot to the main shoot's vascular system (Müller and Leyser, 2011; Prusinkiewicz et al., 2009; Ruyter-Spira et al., 2013). Many lateral shoots stop growing after they have formed just a few leaves that are ordinarily smaller than the leaves of the main shoot.

Incidentally, removal of shoot tips also temporarily eliminates the strong sugar sink and depletes auxin in the vascular cambium of the shoots, as well as in the trunk and roots, and inhibits their secondary growth and root elongation, at least temporarily (Ferguson and Beveridge, 2009; Fu and Harberd, 2003). Growers in the tropics, where strong apical dominance is notorious, employ continued shoot tipping to encourage lateral growth. In temperate climates, the influence of apical dominance decreases once the main shoot has formed about 18–20 leaves, which usually occurs around the time of bloom (Alleweldt and Istar, 1969). Since auxin concentration declines with increasing distance from the shoot tip, the more basal prompt buds are the first to become activated, and lateral shoots begin to grow out from these buds if the main apex is intact (Prusinkiewicz et al., 2009). Such lateral outgrowth in the presence of an actively growing shoot tip is especially pronounced, and can occur much earlier than the 18-leaf stage, when warm temperatures stimulate rapid growth in spring (Keller and Tarara, 2010). In addition, abundant nitrogen availability also promotes strong and early lateral shoot growth (Keller and Koblet, 1995a). These effects might be related to the stimulation by nitrate of cytokinin production in the roots (Sakakibara, 2006; Sakakibara et al., 2006) and by rapid transpiration on cytokinin delivery via the xylem from the roots to the shoots.

Gibberellins produced upon stimulation by auxin in or near the shoot tip and brassinosteroids produced in the internode epidermis promote cell elongation in the shoot's internodes, perhaps by activating the wall-loosening protein expansin (Cosgrove, 2000; Frigerio et al., 2006; Olszewski et al., 2002; Savaldi-Goldstein et al., 2007; Weiss and Ori, 2007; Yamaguchi, 2008), in addition to promoting production of the membrane water channel protein aquaporin (Maurel et al., 2008; see also Section 3.1). Meanwhile, the thick, cellulose-reinforced outer cell walls and rigid cuticle of the epidermis act like a corset, forcing the internode cells to elongate rather than simply expanding radially in all directions (Kutschera, 2008a,b; Mirabet et al., 2011; Schopfer, 2006). The turgor pressure of the underlying cells is thus transmitted to the epidermis that acts like a tensile skin and becomes stretched in the process (Javelle et al., 2011).

Gibberellins act, it seems, by binding to the so-called DELLA proteins in the cell nucleus and inducing the degradation of these growth-suppressing proteins. Although DELLA proteins bind to and thus inactivate proteins termed transcription factors (whose function, in turn, is to regulate gene expression by binding to specific stretches of DNA in the so-called promoter region of a gene), just how they put the brakes on growth remains to be discovered (Achard and Genschik, 2009; Claeys et al., 2014; Hartweck, 2008; Santner et al., 2009). Mutations that render DELLA proteins unable to bind gibberellins and are thus immune to degradation result in dwarf phenotypes that look like bonsai vines (Boss and Thomas, 2002; Boss et al., 2003); such mutations formed the basis for the short and high-yielding cereal cultivars that gave rise to the agricultural Green Revolution in the second half of the twentieth century (Claeys et al., 2014; Hartweck, 2008).

Both the rate of leaf unfolding and the extension of internodes are influenced by day length and temperature. Long days and high temperatures favor these processes, so that internode length and leaf number increase as day length and/or temperature increase(s) (Alleweldt, 1957). After loss of the shoot tip leads to prompt-bud outgrowth, the most distal lateral shoot often gradually assumes the function of the main shoot, and, at least in *V. vinifera*, the more basal laterals soon stop growing (Alleweldt and Istar, 1969). If water and nutrient availability permits, shoot growth often resumes after fruit set and more or less stops in midsummer. Shortly before or while the grapes begin to ripen, the shoots begin to form a periderm, hence becoming canes and turning from green to yellowish or reddish brown. This shoot "maturation" proceeds acropetally from the base toward the tip and is accompanied by replenishment of storage reserves in preparation for the following growing season (Eifert et al., 1961; Löhnertz et al., 1989). This process is delayed by abundant water and nutrient supply, which also encourages continued or renewed shoot growth beyond veraison. This can cause problems in regions with significant summer rainfall, especially on fertile sites that have high soil water and nutrient storage capacity, where late-season shoot growth can compete with the ripening clusters and storage reserve pools for assimilate supply (Sartorius, 1973).

Unlike the shoots of many other woody perennials, grapevine shoots do not form terminal buds but continue to grow as long as environmental conditions permit. When growth is no longer possible, the shoot tip with the apical meristem dies. The sensitivity of the shoot apex to adverse environmental factors (e.g., drought or low temperature) may be exploited in areas where the soil dries down sufficiently in early summer so that growth can be controlled by appropriate deficit irrigation strategies (see Section 7.2). Decreasing day length, aided by declining temperatures, shortly before the grapes begin to ripen seems to be the trigger that halts cell division in the apical meristem, the prompt buds, and the cambium, thus preventing further lateral shoot emergence and inducing dormancy (Alleweldt, 1960; Alleweldt and Düring, 1972; Davis and Evert, 1970; Fracheboud et al., 2009; Garris et al., 2009; Rohde and Bhalerao, 2007). The synergistic effect of shorter days and cooler temperatures is evident in warm climates, where shoot growth may continue despite a shortening photoperiod so long as temperatures remain permissive. Growth cessation and shoot maturation (i.e., periderm formation) are brought about by a rapid decline in apex-derived auxin in the cambial region and a concomitant increase in the root-derived ABA in the xylem and phloem, where it remains high during the subsequent dormant period (Mwange et al., 2005). In contrast to the buds, the shoot and trunk phloem seems to become dormant only after the first fall frost, and much of the root phloem may not become dormant throughout the winter (Esau, 1948). This can lead to cold injury to the phloem when an unusually cold freeze event follows a warm autumn (Keller and Mills, 2007; see also Section 7.4).

The shoot growth cycle is completed with leaf senescence associated with the recycling of nutrients from the leaves to the canes and permanent parts of the vine (Conradie, 1986; Dintscheff et al., 1964; Löhnertz et al., 1989; Schaller et al., 1990), abscission (shedding) of leaves, and, finally, dehydration and cold acclimation of all woody parts in fall. Before they are shed, the senescing leaves develop their characteristic and often spectacular yellow and/or red autumn color. The coloration occurs because the green chlorophylls are degraded more rapidly than the yellow-orange carotenoids, which are thus "unmasked" in the process. The carotenoids also undergo various

FIGURE 2.6

Brilliant red fall color on Alicante Bouschet leaves, with yellowing Grenache leaves in the background (left) and orange-bronze fall color on Carmenère leaves (right).

Photos by M. Keller.

FIGURE 2.7

Red leaves above the point where a tendril has wrapped itself around a Cabernet franc shoot (left), on undercropped shoots growing from a twisted Cabernet Sauvignon cane (center), and pale yellow leaves on a partly broken Muscat blanc shoot (right).

Photos by M. Keller.

chemical modifications that influence their color. In addition, newly produced red anthocyanins may be accumulated once a leaf has degraded roughly half of its original chlorophyll. Nevertheless, whereas white-skinned grape cultivars are unable to produce anthocyanins, even the leaves of most dark-skinned cultivars normally turn yellow rather than red. The red-fleshed cultivar Alicante Bouschet is a vividly red exception, whereas some other *V. vinifera* cultivars, like Tempranillo, Malbec, Carmenère, or Lemberger, develop orange-red to bronze fall colors (Figure 2.6). Anthocyanin production is triggered by and consequently a sign of sugar accumulation (Gollop et al., 2002; Larronde et al., 1998; Lecourieux et al., 2014; Pirie and Mullins, 1976), which can also be caused by restricted phloem efflux (e.g., due to virus infection, wind-induced or other injury, or girdling; Figure 2.7) or nutrient deficiency (see Section 7.3). After the leaves have been abscised and begin to decompose, the anthocyanins, which seem to be dominated by the glucosides of peonidin and cyanidin, and other phenolic components that have been accumulated in the leaves may be released into the soil. Owing to their phytotoxic nature, these chemicals can be regarded as allelopathic (Greek *allelon* = one another, *pathos* = suffering) compounds that are able to suppress the germination of seeds from other plant species the following spring, at least until they are themselves degraded by soil microbes (Bonanomi et al., 2011; Weir et al., 2004). They thus act as pre-emergence bioherbicides that reduce interspecific competition.

It is thought that decreasing day length triggers leaf senescence. Temperature, by contrast, has no effect on the onset of senescence, although lower temperature may accelerate the rate of senescence once initiated (Fracheboud et al., 2009). In other words, for a given cultivar grown at a particular latitude, leaf senescence should start on approximately the same day each year but progress more slowly in warm than in cool years or on warm than on cool sites. At the other end of the temperature spectrum, heat stress ($\geq 45°C$) may also accelerate the senescence rate (Thomas and Stoddart, 1980). Of course, the critical day length necessary to induce senescence and the interaction with temperature likely vary by species and cultivar. During the carefully orchestrated senescence program, which normally starts with chloroplast degradation in the mesophyll cells at the leaf margins and then progresses inward and toward the base, the leaves transition from photosynthetic carbon assimilation to the breakdown (catabolism) of chlorophyll, proteins, lipids, and nucleic acids and subsequent export of valuable nutrients to other plant parts (Jones and Dangl, 1996; Lim et al., 2007; Masclaux-Daubresse et al., 2010). Before they die, the leaves lose about half of their carbon, much of it due to respiration to generate the energy required for the degradation and export metabolism in the face of declining photosynthesis (Keskitalo et al., 2005; Williams and Smith, 1985). The high respiratory costs of senescence are increasingly met by the degradation of cell membranes and use of their lipids to provide energy through respiration (Buchanan-Wollaston, 1997). Not surprisingly, the nucleus with its complement of genes and the mitochondria with their respiration machinery are the last cell components to be dismantled in the process (Lim et al., 2007). Moreover, chlorophyll is degraded not because its products are reusable but, rather, because its light-absorbing properties make it a strong phytotoxin in the absence of photosynthesis and to facilitate access to more valuable materials such as chloroplast proteins and lipids. The nitrogen and much of the carbon of the chlorophyll molecules, but not their magnesium, remain in the leaf (Hörtensteiner and Feller, 2002), at least until soil microbes decompose the leaf once it has fallen to the ground. Considering that chlorophyll contains only approximately 2% of the leaf's nitrogen, this loss is a rather modest price the plant pays for recycling of proteins and lipids (Hörtensteiner, 2006). Leaves thus manage to remobilize up to 50–80% of their nitrogen and phosphorus, 50% of their sulfur, and 20% of their iron before their cell membranes collapse, marking the end of their life (Keskitalo et al., 2005; Niklas, 2006). Following this resorption process, the phloem is plugged and sealed, and the leaves are shed at predetermined positions of cell separation at the base of the petiole, termed abscission zones, which are put in place (i.e., differentiated) soon after the leaves have fully expanded (Roberts et al., 2002). The formation of abscission zones is induced by the plant hormone ethylene, whereas their position is dictated by auxin, which disappears completely during the subsequent dormancy period. Each year, grapevines invest most of their aboveground carbon and nitrogen in leaf production, which cycles as resorbed material in the perennial plant parts and as dead organic material, the so-called detritus or humus, through the soil.

In viticultural terminology, the rate of shoot growth or shoot elongation over time is referred to as vigor. In addition to genetic effects (species, cultivar, and rootstock), it is strongly influenced by temperature, soil moisture, nutrient availability, vine reserve status, pruning level, and even vine age. Some cultivars (e.g., Thompson Seedless, Cabernet franc, or Grenache) or rootstocks (e.g., 140 Ruggeri, 1103 Paulsen, or St. George) are considered to be vigorous, whereas others are thought to be intermediate (e.g., Cabernet Sauvignon, Syrah, Chardonnay, Teleki 5BB, SO-4, or 3309 Couderc) or weak (e.g., Gamay, 101-14 Mgt, or Riparia Gloire). Based on the discovery of an inverse correlation across genotypes between vigor and concentrations of leaf sugars, organic acids, and glutamine, it has been proposed not only that growth drives metabolism (rather than the reverse), but also that fast growers may be "spenders," investing their resources in growth, whereas slow growers may be "hoarders" saving their resources and accumulating reserves (Hanson and Smeekens, 2009; Meyer et al., 2007). In terms of vineyard management, reducing water supply or increasing the number of buds per vine retained at pruning are often powerful tools to decrease vigor (Freeman et al., 1979).

Growth typically accelerates with increasing temperature up to an optimum of 25–30°C but slows as temperature rises further and ceases at approximately 35–38°C (Buttrose, 1969c; Currle et al., 1983). Temperature has a strong effect on photosynthesis and respiration (see Section 5.2): Warm (25–30°C) daytime temperatures promote photosynthetic CO_2 fixation, whereas somewhat cooler (15–20°C) nighttime temperatures limit respiratory CO_2 loss. Consequently, a sufficient (but not excessive) difference between day and night temperatures, rather than simply a warm daily average temperature, is likely to promote growth, a phenomenon termed thermoperiodic growth (Went, 1944). Nonetheless, the assumption of a linear increase in growth rate with rising mean temperature forms the basis for the calculation of the so-called growing degree days (GDD), thermal time, or heat units. The concept of heat units is based on the recognition that plants measure time as thermal time and pace their development accordingly (Thomas, 2013). Amerine and Winkler (1944) proposed that grape-growing regions could be classified by heat summation or cumulative GDD for a 7-month "standard" growing season (April–October in the northern hemisphere, October–April in the southern hemisphere) as follows:

$$GDD = \sum_{i=1}^{n}(T_i - T_b)$$

where T_i is the mean air temperature of each day for the period April 1–October 31 ($n = 214$ days) and T_b is the base temperature.

Normally, T_i is calculated as the mean of the daily maximum and minimum temperature. Only daily values of $T_i - T_b > 0$ are added to the cumulative GDD, whereas values ≤ 0 are set to zero. Because very little growth and photosynthesis occur in grapevines below 10°C, this value is usually chosen as T_b, but other thresholds down to 0°C are sometimes used (Amerine and Winkler, 1944; Parker et al., 2011). Despite some limitations (Dry and Coombe, 2004) and although variations on this theme have been proposed, including a "heliothermal index" (Huglin and Schneider, 1998; Jones et al., 2010), a "latitude-temperature index" (Jackson and Cherry, 1988), and "biologically effective day degrees" (Gladstones, 1992), the "Winkler regions" (Winkler et al., 1974) are still used as a standard today. For the most part, different indices seem to be highly correlated, which suggests that they describe essentially similar spatial climate patterns (Hall and Jones, 2010; Jones et al., 2010). A lower limit of 850 GDD and an upper limit of 2700 GDD have been proposed for the thermal range that is suitable for wine grape production (Jones et al., 2010), but table grapes and raisins are often produced under considerably warmer conditions. Each 1°C increment in temperature averaged over the period April–October (or October–April) adds 214 GDD to the standard growing season so that vineyards at lower latitude or lower elevation will tend to be classified into higher (warmer) Winkler regions. This applies both across disparate growing regions and within them; whereas latitude is the main driver of differences in GDD between regions (followed by continentality; i.e., distance from an ocean or sea), topography (i.e., changes in elevation) drives most of the GDD differences within a region (Failla et al., 2004; Hall and Jones, 2010; Jones et al., 2010; Keller, 2010). As a rule of thumb, a gain of 100 m in altitude decreases the average temperature by roughly the same extent (0.6–0.7°C), as does an increase in latitude by 100 km (a 1° difference in latitude is ~111 km).

The ongoing rise in temperature associated with global climate change (Intergovernmental Panel on Climate Change, IPCC, 2013) will likely shift many vineyard sites into the next higher Winkler region within the next 50 years. The predicted increase in average spring and autumn temperatures will also lead to longer actual (as opposed to standard) growing seasons, which are determined by the frost-free period (days between the last spring and first fall frosts). Moreover, in addition to its influence on growth rates, a rise in temperature also typically accelerates phenological development, with a consistent trend toward earlier bloom, veraison, and harvest (Alleweldt et al., 1984b; Chuine et al., 2004; Jones and Davis, 2000; Webb et al., 2007; Wolfe et al., 2005). The dependence of plant development on temperature is so persistent that the timing of phenological events, such as bloom or veraison, can be predicted with fairly high accuracy using mathematical models based solely on GDD accumulation (Parker et al., 2011). Nevertheless, although the duration of the phase between budbreak and the time the fruit of a given cultivar attains a specific sugar concentration is shorter in a warmer than in a cooler climate, the cumulative GDD during this period is higher in the warmer climate (McIntyre et al., 1987).

The temperature effect on shoot growth is facilitated, at least in part, by a stimulation of auxin production by high temperature, which in turn promotes cell elongation. Vigor declines with increasing vine size; therefore, shoots of small, young vines usually grow more vigorously than those of large, old vines. This decline arises from the increased mechanical cost of maintaining an older vine's permanent structure, from the hydraulics of water transport, and from the higher number of buds that result in more shoots. Increasing soil moisture and nutrient (especially nitrogen) availability also favor high vigor (due to more abundant resource availability), as does increasing pruning severity, because plants in general try to reestablish a balanced root:shoot ratio (Buttrose and Mullins, 1968; Clingeleffer, 1984; Poorter and Nagel, 2000). Nevertheless, the canopy of lightly pruned vines (e.g., after mechanical or minimal pruning) develops earlier and more rapidly and grows larger than that of heavily pruned vines (e.g., after manual spur or cane pruning) because a higher number of buds leads to more shoots growing simultaneously in spring, using a larger pool of reserves accumulated in the previous growing season (Araujo and Williams, 1988; Intrieri et al., 2001; Possingham, 1994; Sommer et al., 1995; Weyand and Schultz, 2006a,b; Winkler, 1929, 1958). On the other hand, the presence of fruit tends to depress vegetative growth of both shoots, especially lateral shoots, and roots so that the total biomass production per vine remains approximately constant (Eibach and Alleweldt, 1985; Pallas et al., 2008; Petrie et al., 2000b). The direction of growth also affects growth rate (Currle et al., 1983; Kliewer et al., 1989); upright-growing shoots, such as those in a typical vertical shoot positioning trellis system, generally grow more vigorously than shoots forced to grow horizontally, which in turn grow more rapidly than shoots pointing downward, such as those in Geneva double-curtain or Scott–Henry systems. This effect on growth of the shoot angle relative to vertical is termed gravimorphism (Wareing and Nasr, 1961). The effect is associated with variations in xylem vessels, whose number and diameter seem to be tailored to optimize water and nutrient supply to the shoot (Lovisolo and Schubert, 2000; Schubert et al., 1999). Accumulation of auxin near the shoot tip is thought to be responsible for the production of more but narrower vessels in downward-pointing shoots (Lovisolo et al., 2002b). As a consequence, downward-growing shoots have higher hydraulic resistance and smaller leaves with lower stomatal conductance and lower photosynthesis compared with upward-growing shoots (Schubert et al., 1995).

The time of rapid shoot growth coincides with secondary growth (increase in diameter) of the shoots, cordon, trunk, and roots. This secondary growth ends when the decreasing day length after midsummer leads to cessation of cell division in the cambium, well before the temperature becomes too low to limit cell division (Druart et al., 2007). The shortening day length also decreases gibberellin production in the shoot elongation zone, so that growth rates decline (Cooke et al., 2012). The rapid growth period also coincides with the formation of prompt (lateral) and dormant (compound) buds in the leaf axils. However, development of the first compound buds in the basal nodes begins as early as 3 or 4 weeks before budbreak or more than 1 year before these buds will give rise to shoots (Morrison, 1991). Dormancy in the case of these buds means that cell division and expansion are blocked, whereas metabolic processes (e.g., respiration) continue, albeit at a reduced level. Although the buds are under apical dominance from the growing shoot tip, and thus do not normally break, the prompt buds do not become dormant and can break relatively readily to form lateral shoots, which contrasts with the compound buds. Nonetheless, following loss of the shoot tip and removal of lateral shoots or severe defoliation (e.g., due to water stress or physical damage caused by pests), even dormant buds are initially able to resume growth (Alleweldt, 1960; Huglin and Schneider, 1998). This phase is variously referred to as paradormancy, predormancy, conditional dormancy, or summer dormancy, defined as the inhibition of growth by distal organs or, specifically, by auxin released by the shoot tip and young leaves (Galet, 2000; Horvath et al., 2003; Lang et al., 1987; Lavee and May, 1997). However, the same buds that are arrested by incoming (exogenous) auxin will themselves start producing and releasing auxin as soon as they are released from dormancy and begin to grow.

With the slowing down of shoot growth (which starts earlier on less severely pruned vines due to their higher shoot number), the buds progressively lose the ability to break even in the absence of apical dominance, beginning with the most basal buds. In other words, bud dormancy develops gradually over 2–3 weeks, moving from the base toward the tip of the shoot following behind the wave of brown periderm (Alleweldt, 1960; Cooke et al., 2012; Pouget, 1963). While the buds' respiration rate declines gradually after their formation, their water content drops quickly from about 80% to less than 50% as they enter into the phase termed endodormancy, which is also known as true, inherent, organic, deep, or winter dormancy, or rest (Currle et al., 1983; Huglin and Schneider, 1998; Lavee and May, 1997; Mullins et al., 1992). The transition from paradormancy (green and browning buds) to endodormancy (brown buds) is characterized by a sudden, large increase in the temperature and time required for the buds to break (Pouget, 1963). This shift is induced by decreasing day length (i.e., shorter photoperiod) and cooler temperatures, and is associated with a strong increase in ABA in the buds and shoots (Düring and Alleweldt, 1973; Düring and Bachmann, 1975; Koussa et al., 1994; Rohde and Bhalerao, 2007). The para-/endodormancy transition happens around the time of veraison and is thought to occur later, and the endodormancy period to be shorter, in cultivars that break buds early (i.e., at lower temperature) the following spring (Pouget, 1963, 1972). Buds that have become endodormant are now able to acclimate to cool temperatures and acquire cold hardiness (see Section 7.4). The genetic programs for bud dormancy and cold hardiness are superimposed; shorter photoperiods are sufficient to induce dormancy and leaf senescence, whereas low temperatures are required for the dormant organs to acquire cold hardiness (Fennell and Hoover, 1991). Nonetheless, entry into endodormancy appears to be delayed under very warm conditions.

Endodormancy is defined as the inhibition of growth by internal signals. This state of self-arrest prevents budbreak and renewed shoot growth during the favorable growth conditions of late summer and early autumn, which would lead to certain death of the emerging shoots due to fall and winter frosts (Horvath et al., 2003; Lang et al., 1987). It is thought that the inability of buds to grow out arises from the constriction of the plasmodesmata in their apical meristems by callose, whose degradation is induced by gibberellin once the buds have experienced a period of cool temperatures (Burch-Smith and Zambryski, 2012; van der Schoot et al., 2014). In addition, the bud xylem is uncoupled from the shoot xylem, that is, the dormant bud becomes hydraulically isolated from the shoot (Figure 2.8).

Approximately 7 consecutive days with mean daily temperatures between 0°C and 10°C, termed the chilling period, in combination with short days may irreversibly break endodormancy and stimulate the renewed ability of dormant buds to resume growth when favorable (i.e., warm) conditions return (Currle et al., 1983; Galet, 2000; Koussa et al., 1994; Pouget, 1963, 1972). The chilling temperature must be perceived directly by the apical meristem inside each bud (van der Schoot et al., 2014). Because intermittent warm temperatures cancel the effect of earlier cool episodes lasting less than a week, the chilling period ordinarily coincides approximately with leaf fall or soon thereafter in temperate climates but may occur in mid-winter in regions with mild winters. Fascinatingly, while chilling releases the buds from endodormancy, it simultaneously enhances their ability to tolerate freezing temperatures (van der Schoot and Rinne, 2011;

FIGURE 2.8

Uncoupling of xylem flow between Chardonnay buds and shoots, traced by the red xylem-mobile dye basic fuchsin. Paradormant buds are connected to the shoot (left), endodormant buds are hydraulically isolated (center), and swelling buds reconnect to the shoot before budbreak (right).

Photos by L. Mills.

van der Schoot et al., 2014). Thus chilling has a dual purpose: to release buds from dormancy and to increase their cold hardiness (see Section 7.4). Like dormancy induction, dormancy release happens gradually. The release seems to be related to a decline in ABA, and perhaps an increase in gibberellin, in the buds and canes (Düring and Alleweldt, 1973; Düring and Bachmann, 1975; Koussa et al., 1994; see also Section 7.4). The temperature threshold required to break endodormancy is lower for "late" cultivars with a higher base temperature for budbreak and shoot growth in spring and higher for "early" cultivars with lower base temperature. Thus dormancy is less intense and more easily lifted in "early" cultivars than in "late" cultivars; in other words, the extent and duration of dormancy are cultivar specific (Pouget, 1972). Although chilling may not be an absolute requirement to break dormancy in grapevines (at least not in all cultivars), and its effect seems to be quantitative rather than qualitative, no chilling (especially in combination with an extended growing period in late fall) usually leads to delayed, limited, and uneven budbreak (Koussa et al., 1994; Lavee and May, 1997; Mathiason et al., 2009). While 100% of the buds are generally able to grow out after an adequate chilling period, the proportion of buds that do eventually break rarely exceeds 50–70% if the chilling requirement is not met, and the leaves that do emerge from these buds are often malformed and chlorotic (Pouget, 1963). This can be a problem in warm subtropical and tropical climates, where budbreak is often artificially induced by application of H_2CN_2 after pruning. Based on results from studies with such dormancy-breaking agents (Pang et al., 2007; Pérez et al., 2007, 2009), as well as with respiration inhibitors (Pouget, 1963), one might argue that dormancy release may require a temporary decline in bud respiration rates below a cultivar-specific threshold. Perhaps this might also explain the rather startling finding that very hot temperatures ($>40-50°C$), too, seem to be able to break dormancy (Pouget, 1963).

Once endodormancy has been lifted, subsequent low winter temperatures have no additional effect on the ability of buds to break, provided the temperatures are not extreme enough to kill the buds (Pouget, 1963). In fact, the main factor that prevents budbreak after this dormancy release is low temperature (though lack of water can also inhibit budbreak). This last phase is called ecodormancy (or enforced, relative, imposed, or postdormancy or quiescence), which is defined as the inhibition of growth by temporary, adverse environmental conditions (Galet, 2000; Horvath et al., 2003; Lang et al., 1987). In other words, unfavorable growing conditions simultaneously break endodormancy and impose ecodormancy. The latter is related to cold-induced (or drought-induced) accumulation of ABA and coincides with maximal cold hardiness (see Section 7.4). The return of warmer temperatures in spring (or unseasonably warm temperatures during winter) eventually enables the buds to break. Budbreak is preceded by a rapid rise in respiration rate, restoration of the buds' vascular connection to the shoot (Figure 2.8), and gain in the buds' moisture content back up to 80%, along with a concomitant loss of cold hardiness (Ferguson et al., 2014; Lavee and May, 1997; Pouget, 1963). The increase in respiration rate likely indicates renewed meristem activity (i.e., cell division), and the increase in water content is linked to renewed cell expansion in the compressed internodes. The outward sign of this spring awakening is the swelling of the buds that culminates in budbreak. The date of budbreak of a given cultivar is determined by thermal time (cumulative GDD) during the ecodormancy period (Pouget, 1988). Incidentally, the cambium of canes, cordons, and trunks also seems to experience similar periods of rest (endodormancy) and quiescence (ecodormancy) and is reactivated in spring when the cambial cells resume cell division (Begum et al., 2007). While growing apical and cambial meristems may be regarded as being in "online" (active) mode, endodormant meristems are in "offline" (dormant) mode, and ecodormant ones in "standby" (quiescent) mode (van der Schoot and Rinne, 2011).

Dormancy is unknown to roots: low temperatures simply suppress root growth—temporarily and reversibly (van der Schoot and Rinne, 2011). Root growth, especially the formation of lateral roots, is stimulated by auxin that has been delivered acropetally (from its main place of production in the shoot tips and unfolding leaves to the root tip) in the phloem (passive transport, dominant) and parenchyma (active cell-to-cell transport, which is less important except during seed germination) and by gibberellins (promoting cell elongation) produced in the root tips (Friml, 2003; Kramer and Bennett, 2006). Changes in auxin flow usually lead to changes in root growth, and interrupting the auxin flow (e.g., by cutting off the trunk) will inhibit root growth completely. Lateral roots grow from the pericycle (and not the apical meristem, as in the shoot) following the initiation of renewed cell division by basipetal (from the tip to the base) auxin transport in the cortex and epidermis. Following initiation, further division of the "founder cells" leads to the formation of a lateral root primordium, which pushes radially out of the parent root via cell expansion, again stimulated by auxin (and repressed by ABA, e.g., under drought or osmotic stress from excessive nutrient supply or salinity). After emergence, the apical meristem is activated and the new lateral root begins to grow autonomously. Note that whereas dormant buds seem to inhibit adventitious root formation on hardwood cuttings (Smart et al., 2003), the lack of auxin that is normally produced and released by swelling buds may be the reason disbudded cuttings are often unable or struggle to form roots, unless they are dipped in an auxin solution. Indeed, on single-node cuttings, most new roots form directly beneath the bud (Huglin and Schneider, 1998). Cytokinin and strigolactone, by contrast with auxin, normally suppress adventitious rooting; strigolactone might exert its inhibitory role by reducing auxin flow (Rasmussen et al., 2012). Since cytokinin and strigolactone are typical root signals, their absence in cuttings might induce the cambium to replace the missing roots by initiating new ones.

Similar to shoots, roots also display a kind of apical dominance, whereby the growing tip of a root inhibits the outgrowth of lateral roots from its pericycle. In contrast to the shoots, however, the roots use cytokinin rather than auxin to inhibit lateral root initiation too close to the root tip (Aloni et al., 2006b). This enables the root to continue growing down (or across) the soil profile in search of water and nutrients. If the root fails to detect nutrients such as nitrate or phosphate, its tip ceases to produce cytokinin so that lateral roots can proliferate to explore new territory in the upper soil layers. The patchy availability of water and nutrients in the soil therefore forces the roots to move by growing and leads to root systems of varying architecture, length, and density.

Although root tips can extend at rates of several centimeters per day, it seems that root growth generally lags behind shoot growth, peaks around the time of bloom and early fruit development, and tapers off during fruit ripening (Reimers et al., 1994). Early canopy development, such as occurs with lightly or minimally pruned grapevines, is associated with more pronounced early root growth (Comas et al., 2005). The supply of photoassimilates from the leaves is evidently critical to sustain root growth, and loss of leaf area inhibits root growth far more than it does shoot or fruit growth (Buttrose, 1966a). The majority of new roots are produced while the shoots are growing vigorously, but root growth can continue as long as water and nutrient availability are not overly limiting and the soil temperature remains high enough (Comas et al., 2010; Schreiner, 2005; Williams and Biscay, 1991). In warm climates, or in warm growing seasons in cooler climates, which permit continued assimilate production by the leaves, there may be a second flush of new root growth around or just after harvest (Conradie, 1980; Lehnart et al., 2008; McKenry, 1984; Mohr, 1996). Where temperatures are conducive, roots may also grow in the winter after leaf fall; such growth is presumably sustained by internal nutrient reserves (Bauerle et al., 2008b; Schreiner, 2005). Moreover, it appears that vines grafted to drought-susceptible rootstocks (e.g., 101-14 Mgt) produce more new roots in the winter and spring, whereas drought-tolerant rootstocks (e.g., 1103P) grow mostly during the summer by expanding into deeper, moister soil layers (Alsina et al., 2011).

Early canopy development (i.e., larger leaf area) in spring promotes early root growth; thus, root growth of lightly pruned vines peaks earlier than that of heavily pruned vines. The total root biomass increases from growth of permanent roots in length and diameter throughout the growing season from shortly after budbreak until leaf fall in both young and mature vines (Araujo and Williams, 1988; Williams and Biscay, 1991). However, most of the fine roots (<1 mm in diameter), which are the vine's major water- and nutrient-absorbing structures, are short lived, so they have to be replaced continuously, probably because the soil around them quickly becomes exhausted (Comas et al., 2000; Comas et al., 2010; Viala and Vermorel, 1909). Only a few of them live longer and, through secondary growth, may eventually become large, structural roots. These structural roots, however, are extremely long lived and may stay alive in the soil for years even after a vineyard has been removed.

It is estimated that vines allocate between 30% and 60% of total net photosynthesis products (see Section 5.1) to the roots under favorable soil conditions, and this portion is even greater in poor soils or during drought (Anderson et al., 2003). New roots are initially white, but they turn brown after approximately 5 weeks and black after an additional 3–6 weeks. The onset of browning generally indicates cessation of root metabolic activity and thus death of the root as a "functional" organ in terms of water and nutrient absorption and as a source for new laterals (Comas et al., 2000). Black roots shrivel and eventually disappear altogether. Fine roots produced before bloom have an extremely short life span compared with those produced later in the growing season (Anderson et al., 2003). Fine roots also die sooner in dry soil and at shallower soil depth, and so do roots of heavily pruned vines supporting only few shoots. Such an increase in root mortality can be considered a response to herbivory, whereby plants compensate for the loss of shoot biomass (i.e., carbon shortage) and attempt to restore their root:shoot ratio to a balance that is more favorable for growth (Bloom et al., 1985). Therefore, mechanical and, particularly, minimal pruning tend to result in more and longer-lived fine roots than spur or cane pruning. Lightly pruned vines also produce more fine roots (to balance their larger canopy) than heavily pruned vines, even though the former generally support a larger crop. In addition, infection with mycorrhizal fungi greatly prolongs the life span of fine roots, whereas nitrogen addition shortens it.

Grapevines, like other plants, are stationary; unlike animals, they cannot move from place to place to find better food sources. Because uptake of water and nutrients by the roots from the soil quickly depletes readily available resources, the roots must keep elongating throughout the growing season and throughout the vine's life to maintain the supply of these raw materials. This leads to the formation of a branched root system that is elaborated with root hairs and symbiotic mycorrhiza (Gebbing et al., 1977; Possingham and Groot Obbink, 1971; Schreiner, 2005; Stahl, 1900). Root density increases as a vine grows older but appears to settle and remain more or less constant after the vine reaches 5–10 years of age. Although 40–50% of the volume of a typical soil consists of pores, root exploration of the soil ordinarily follows the path of least resistance; roots predominantly grow along natural fracture lines of the soil and reinvade macropores (also called biopores), whose pore diameter is much larger than the size of the soil particles, and which are created by earthworms or old root debris forming microniches of high organic matter (McKenry, 1984; Passioura, 2002; Smart et al., 2006). The decaying dead roots additionally supply nutrients to the growing new roots, provided these mineralized nutrients are not leached by excessive rainfall or irrigation. Roots that penetrate smaller soil pores, whose size is mostly determined by the size distribution of the soil particles, must use turgor pressure to exert a growth pressure that is large enough to deform the surrounding soil (Jin et al., 2013). Of course, the general direction of root growth is downward, guided by a positive gravitropic response.

As in the shoot, the "gravity sensors" are starch-filled statoliths that, in this case, settle at the bottom of the root cap's columella cells (Kutschera and Briggs, 2012; Morita and Tasaka, 2004). When a root is forced away from vertical (by rocks or other obstacles), the resettling of these statoliths triggers a redistribution of calcium and movement of auxin to the new lower side of the root tip and back to the elongation zone via the epidermis, so that the cells on that side elongate more than on the upper side, allowing the root to navigate around the obstacle (Kramer and Bennett, 2006; Muday et al., 2012; Petrášek and Friml, 2009). The gravity-sensing columella cells are also the ones that produce most of the plant's cell division hormone, cytokinin, which is then distributed throughout the vine via the transpiration stream in the xylem, mainly to rapidly transpiring organs (Aloni et al., 2005). Depending on soil conditions, the majority of roots are usually concentrated in the nutrient- and oxygen-rich surface soil, typically in the top 50 cm, with some roots penetrating to several meters of depth (Celette et al., 2008; Huglin and Schneider, 1998; Morlat and Jacquet, 1993; Schreiner, 2005; Smart et al., 2006). Vineyard floor vegetation, such as cover crops or permanent swards, tends to shift the maximum root density of grapevines to 50−100 cm, probably because of competition for water and nutrients (Lehnart et al., 2008; Reimers et al., 1994). Somewhat paradoxically, however, the average rooting depth may be much greater in shallow soils (over bedrock) than in deep soils because the roots in shallow soils can grow down cracks in the rock, following the path of water infiltration. Roots can even grow for several days in air, for example, across large cracks, gaps, or macropores. They pull off this feat by relying on water supplied to the growing tip internally by the xylem, the phloem, and/or by cell-to-cell movement (Boyer et al., 2010).

Root growth and distribution is influenced by species, rootstock, and cultivar. Nonetheless, the vertical and horizontal distribution of the root system of various rootstocks, it seems, is mostly a function of soil texture, composition, and water availability, rather than an inherent trait of the rootstock genotype (Smart et al., 2006). Scion cultivars also alter the root growth of different rootstocks (Erlenwein, 1965a). Such genetic influences and interactions may be partly responsible for the variation in shoot vigor observed with different scion/rootstock combinations because grapevines, like other plants, try to maintain an equilibrium in root:shoot ratio. In fact, partitioning to the roots is generally proportional to that to the canopy, and total leaf mass roughly scales to the power of ¾ with total root mass (Enquist and Niklas, 2002; Price et al., 2010). Therefore, total root length normally correlates closely with leaf area, and root dry weight correlates with aboveground dry weight in grapevines (Petrie et al., 2000b). It has been estimated that a 1-ha vineyard may contain more than 30 t of roots (Huglin and Schneider, 1998). However, whereas root density may be dependent on the genotype, it appears that root distribution is dictated by soil properties (Mullins et al., 1992; Smart et al., 2006). Even genetically identical plants (i.e., clones) can develop vastly different root systems in terms of both size and architecture, depending on microscale variations in soil conditions that the roots encounter during growth. Such differences result from changes in the number, growth rate, distribution, and orientation of lateral roots. Thus, temperature, soil texture, moisture, and nutrient availability also strongly and rapidly affect root growth and distribution.

FIGURE 2.9

Riesling roots grown at 15°C (left) or 25°C (center; reproduced with permission from Erlenwein, 1965b), and Cabernet Sauvignon roots growing down fractures in limestone bedrock (right; photo by M. Keller).

In some cases, the effect of temperature on root growth may be even more pronounced than its influence on shoot growth (Skene and Kerridge, 1967; Figure 2.9); a minimum soil temperature of 6°C seems to be necessary for roots to grow (Richards, 1983), and a 10°C increase in temperature can double or triple the rate of root extension. Although the optimum is close to 30°C, temperatures above that threshold can kill at least the fine roots within days (Huang et al., 2005b). This can result in problems for grapevines grown in pots, where the soil temperature during the day often exceeds the ambient temperature by several degrees. Although under field conditions soil temperatures usually fluctuate much less than aboveground temperatures due to the large heat capacity of wet soils, heavy rainfall or application of cold irrigation water can lead to short-term changes in soil temperature. Moreover, dry and sandy soils are not as well buffered against temperature fluctuations as are wet and loamy soils. Mild water deficit has little effect on lateral root growth, but more severe stress suppresses root growth, although not as severely as it inhibits shoot growth (see Section 7.2).

Roots tend to search out moist soil regions. In other words, they grow toward water or down a moisture gradient in a process called hydrotropism, and this ability is so strong that it can override the roots' gravity response (Eapen et al., 2005). How the roots "sense" the presence of water and how they integrate the resulting signal with the gravity signal is not known, but hydrotropism partly seems to result from the roots' propensity to grow away from regions of high osmotic pressure (Takahashi et al., 2003). Their growth clearly responds to the presence of water tables, the amount and frequency of water application (whether from rainfall or irrigation), and the type of irrigation system (e.g., drip versus furrow). Therefore, grapevines can tolerate considerable drought due to their ability to selectively grow roots in soil patches with available water (Bauerle et al., 2008b). This is why the roots of drip-irrigated vines (especially in sandy soils and dry climates) are often concentrated beneath the drip emitters, whereas in rain-fed vineyards or those irrigated with other methods (e.g., flooding, furrows, and overhead or microjet sprinklers), the root system is typically much more widespread (Stevens and Douglas, 1994), unless root growth in the midrows is prevented by soil compaction from heavy machinery. Roots of mature vines also change their growth pattern accordingly following conversion from flood or sprinkler to drip irrigation (Bowen et al., 2012; Soar and

Loveys, 2007). Similarly, permanent interrow cover crops tend to lead to a concentration of vine roots in the bare, herbicide-treated vine rows (Celette et al., 2008; Morlat and Jacquet, 2003). The roots' ability to find water even below the bedrock underlying the "useful" soil horizons (see Figure 2.9) can also greatly expand the apparent rootzone or rooting depth as estimated from soil pits. Consequently, grapevines planted on apparently shallow soil are not always less vigorous than those planted on deeper soil, and their growth is not necessarily easier to control using deficit irrigation strategies. Restricting the rooting volume in pots, however, does decrease shoot growth, and the reduction is greater the smaller the rooting volume (Wang et al., 2001).

In contrast to its effect on shoot growth, increasing nutrient (especially nitrogen) availability decreases root growth. Soil compaction, leading to a decrease in the soil's pore volume, also reduces root (and consequently shoot) growth (Chapman et al., 2012); pressurizing the soil by 1 bar due to compaction may decrease root elongation by 90%. The resistance compaction imposes on root growth increases as the soil dries down and is associated with increased ethylene and ABA production by the root (Bengough et al., 2011; Jin et al., 2013). Nonetheless, localized soil compaction (e.g., due to heavy machinery or plow pans) reduces shoot growth only when it decreases the total root length (Montagu et al., 2001). When a single root grows through loose soil into compact subsoil, compensatory root growth in the loose soil maintains total root length and thus shoot growth remains unaffected. Such plasticity in root system development in response to heterogeneous soil conditions enables grapevines to acquire soil resources with a conservative investment in root biomass. Constraining roots in pots, however, reduces both root and shoot growth, which is an important yet often overlooked consideration in pot experiments. Roots often become "pot-bound," and as pot size decreases, so does the relative growth rate and overall plant growth (Poorter et al., 2012).

2.3 REPRODUCTIVE CYCLE

Grapevines grown from seeds, like other woody perennials, have a vegetative juvenile phase of 2–4 years before they reach their reproductive phase and become able to produce fruit. It seems that the degree of expression of the gene that controls flowering transition increases gradually during the juvenile phase until it exceeds a threshold that permits the formation of inflorescences (Böhlenius et al., 2006). This mechanism ensures that the vine has ample resources to support fruit production. By contrast, vines grown from cuttings are able to fruit in the first growing season under optimal conditions. Another trait typical of perennial woody species is that they require two consecutive growing seasons for flower and fruit production: Buds formed in the first year give rise to shoots carrying fruit in the second year. Reproductive growth is very similar in the different *Vitis* species and begins with flower formation, which can be divided into three separate processes: inflorescence initiation (or induction), flower initiation, and flower differentiation. The first step involves the formation of lateral meristems as uncommitted primordia or initials by the apical meristem inside the shoot's axial compound buds from spring through early summer (Alleweldt and Ilter, 1969; Carmona et al., 2008; Gerrath, 1992; Gerrath and Posluszny, 1988b; Meneghetti et al., 2006; Morrison, 1991; Pratt, 1971; Srinivasan and Mullins, 1981). Primordia are first formed in the buds at the shoot base, and thereafter they also appear progressively in the buds toward the shoot tip. These primordia (i.e., meristems) can develop (differentiate) into inflorescence primordia (i.e., they form several inflorescence branch primordia or meristems), tendril primordia, or sometimes even shoot primordia (see Figure 1.16). Initiation of uncommitted primordia occurs after four- or five-leaf primordia have been initiated in the buds (i.e., from budbreak to bloom), and subsequent inflorescence differentiation begins around anthesis of the current season's inflorescences and may continue until the buds enter dormancy (Alleweldt and Ilter, 1969; Morrison, 1991; Pratt, 1979; Snyder, 1933; Swanepoel and Archer, 1988). In other words, inflorescences are formed during the year that precedes the flowering and fruiting year, and the fate of the following year's primordia is determined as early as bloom time of the current year. The number of inflorescences is lowest at the shoot base and increases with higher bud position before gradually declining again beyond bud positions 10—12 (Sartorius, 1968; Huglin and Schneider, 1998). Yet in some cultivars only the basal six to eight buds of a shoot (i.e., those preformed in the previous growing season) are able to form inflorescences; the uncommitted primordia initiated at younger node positions (i.e., those produced by the actively growing shoot) always become tendrils, never inflorescences (Morrison, 1991; Sánchez and Dokoozlian, 2005). Thompson Seedless is an exception: Apart from the basal buds, fruitful buds are initiated over the entire length of its shoots (Sánchez and Dokoozlian, 2005). The origin of the shoot is unimportant for the fruitfulness of its buds; shoots arising from 1-year-old spurs or canes produce equally fruitful buds as those growing from older cordons or trunks (Müller-Thurgau, 1883b).

The maximum number of future inflorescences per bud, the so-called bud fruitfulness, appears to be determined by approximately 3 months after budbreak—that is, before veraison (Alleweldt and Ilter, 1969). Depending on species and cultivar, several inflorescence primordia may be formed in the primary buds of most *Vitis* species. Most *V. vinifera* cultivars typically initiate two such primordia, but for some, such as Sémillon, Muscadelle, Gamais, or Carignan, three is more common, whereas the fruitful Gouais blanc forms four, *V. labruscana* cultivars form three or four, and *V. riparia* and *V. rupestris* buds form seven or eight inflorescence primordia (Viala and Vermorel, 1909). Bud fruitfulness is linked to shoot vigor. Weak shoots produce buds of lower fruitfulness than do more vigorous shoots (Huglin and Schneider, 1998). The degree of bud fruitfulness may be associated with auxin production; more auxin seems to lead to more inflorescences per shoot (Costantini et al., 2007). Moreover, environmental conditions strongly modulate the number of primordia. Warm, sunny conditions with adequate water and nutrient availability, along with a sufficiently large, actively photosynthesizing leaf area, are crucial to maximize the number of primordia (see Section 6.1). In addition, the growth direction of the shoot also appears to influence bud fruitfulness, at least in some cultivars: Upright-growing shoots produce buds with more inflorescence primordia than horizontally or downward-growing shoots (Alleweldt and Ilter, 1969).

The commonly accepted view (Carmona et al., 2002, 2008; Considine and Knox, 1979a; Gerrath, 1992; Gerrath and Posluszny, 1988b; May, 2004; Meneghetti et al., 2006; Morrison, 1991; Mullins et al., 1992; Scholefield and Ward, 1975; Snyder, 1933; Srinivasan and Mullins, 1981) of inflorescence and flower differentiation is that the formation of inflorescence meristems and that of flower meristems are seasonally separated (Figure 2.10). According to this view, inflorescence primordia grow rapidly during the summer, producing several branch meristems before the buds enter dormancy in late summer. Further inflorescence development is arrested with the onset of dormancy and does not resume until the buds begin to swell in late winter or early spring. When the buds are reactivated, the inflorescence branch meristems produce additional meristems that give rise to clusters of three or four flower meristems or floral primordia during the period of budswell

FIGURE 2.10

Cluster initiation (1) and differentiation (2—6: numbers indicate increasing orders and positions of branching) and flower development (8: initiation of sepal lobes; 9: initiation of petal lobes; 10: initiation of stamen primordia; 11: initiation of carpel primordia; 12: appearance of locules; 13: ovule initiation; 14: pollen formation; 15: anthesis) of Concord grapevines (left; reproduced with permission of M. Goffinet); inflorescences of the rootstock 3309C emerging after budbreak (right; photo by M. Keller).

that precedes budbreak by several days to weeks. This process is termed flower initiation, and within each of these clusters the floral primordia are initiated "from the top down"; that is, the terminal flower is initiated first and the basal flower last. Flower initiation happens simultaneously in all parts of the inflorescence primordium during the budswell phase and culminates in the formation of a calyx (fused sepals) for each flower. Branching and flower initiation ostensibly cease once a shoot starts growing out of the bud. The subsequent flower differentiation, or the development of the individual organs of each flower, also called floral organogenesis, occurs over approximately 5 weeks during and after budbreak. The corolla (joined petals) and stamens appear successively under the stimulating influence of gibberellin (Cheng et al., 2004a), while the inflorescences become visible on the shoot and then separate. The pistil with its reproductive structures does not develop until the individual flowers become visibly separated on the inflorescence. Separate genes or groups of genes specify the sequence and timing of flower differentiation and the identity of the individual floral organs, although the mechanism by which they exert their influence is only beginning to be elucidated (Díaz-Riquelme et al., 2009; Fernandez et al., 2010). Intriguingly, the genes that specify flower organ identity are activated only *after* the appearance of the organ primordia (Jenik and Irish, 2000).

The dissenting minority opinion of this developmental process differs mainly in the timing of flower initiation (Agaoglu, 1971; Alleweldt, 1966; Alleweldt and Ilter, 1969; Sartorius, 1968). This view holds that a portion of the calyxes are formed before bud dormancy, some during winter dormancy (albeit very slowly), and the remainder during the subsequent budswell and budbreak period. The corolla and stamen also appear before budbreak, followed by the pistil, whose formation is complete within 10—15 days of the emergence of the inflorescence after budbreak. Intriguingly, evidence supporting both of these views comes from both warm and cool climates, so it is not clear what might cause the disparity in observations. Inflorescences that arise on lateral shoots, especially on those laterals on the free-growth portion of a shoot that could not possibly have been initiated in the previous growing season, obviously run through their entire developmental program from inflorescence initiation to fruiting within the same growing season.

The extent of branching before and after the dormancy period and the degree to which the individual flowers develop in spring determine the number of flowers that form on each inflorescence. The flower number is highly variable, even on the same shoot: The basal inflorescence of a shoot typically produces the most flowers, and numbers decline with increasing height of insertion on the shoot (Huglin and Balthazard, 1975; Huglin and Schneider, 1998; May, 2004). Branching of the primordia is induced and controlled by two groups of plant hormones, namely auxins and cytokinins. The production of cytokinins, which occurs predominantly in the root tips, and their transport to the shoots in the xylem are regulated by developmental (genetic) and environmental signals (e.g., water and nitrogen availability). Auxin is produced and released by the floral meristems (i.e., the tip of each floral organ) and controls both formation and differentiation (i.e., morphogenesis) of the flowers; it also induces development of their vascular tissues. The anthers produce and release particularly high amounts of auxin, as well as gibberellin, and thereby synchronize (i.e., delay) the development of the other flower parts so that they become functional just before anthesis (Aloni et al., 2006a; Olszewski et al., 2002; Yamaguchi, 2008). Gibberellins help maintain the inflorescence meristems and favor inflorescence elongation (Alleweldt, 1959). This is why table grape growers sometimes apply gibberellin sprays before bloom to lengthen the rachis and make clusters less compact (Mullins et al., 1992). Application during bloom may lead to lower fruit set, which tends to further promote loose clusters. However, gibberellins act antagonistically to cytokinins and favor tendril formation by inhibiting the branching process so that bud fruitfulness may decrease (Weaver, 1960; Weyand and Schultz, 2006c). This inhibitory role appears to be somewhat unique to grapes because gibberellins are involved in establishing floral primordia in many other plants.

Approximately 2 weeks after the ovules have been formed and 5–10 weeks after budbreak, anthesis marks the beginning of bloom, exposing the male and female floral organs (Figures 2.11 and 2.12). By this time the shoots of *V. vinifera* cultivars have grown to variable lengths depending on vigor but usually have 15–20 unfolded leaves, whereas *V. labruscana* cultivars have fewer than 15 leaves (Pratt and Coombe, 1978). Anthesis theoretically refers to the release of pollen but is commonly regarded as the shedding of the calyptra, or capfall. Similar to budbreak, anthesis generally begins in the shoots growing from distal buds and progresses basipetally toward the trunk. On the other hand, anthesis within an individual inflorescence often starts close to its base, and all flowers normally open within 5–7 days, although this can be delayed during cold and rainy conditions. Only few flowers open at less than or equal to 15°C, and the optimum temperature seems to be in the range of 20–25°C; the rate of capfall slows greatly at 35°C (Galet, 2000; May, 2004). Each day, opening begins around dawn and ends at approximately noon (Staudt, 1999). Different inflorescences on the same vine, or on different vines in the same vineyard, follow the same rhythm but may start it on different days so that the entire bloom period in a particular vineyard may last 2 or 3 weeks. Capfall occurs when the petals detach from the receptacle and are lifted off by the anthers. It is thought that it is the rapid elongation of the anther filaments that pushes the calyptra away from its base once the basal cells have dehisced, which is a rare mechanism among angiosperms. Occasionally, the petals instead open at the top and are spread open and pushed downward by

FIGURE 2.11

Grape flowers immediately before anthesis (left, with cross section center left), during anthesis (center right), and immediately after anthesis (right).

Illustrations by A. Mills.

FIGURE 2.12

Chardonnay inflorescence at the beginning of bloom (left) and in full bloom (right, with insert showing a close-up of a group of flowers).

Photos by M. Keller.

the anthers. Such flowers are termed star flowers and can appear in a range of cultivars (Longbottom et al., 2008; Pratt, 1971).

When the anthers burst open, they release their pollen into the air. The production of pollen is stimulated by the plant hormone cytokinin, and pollen release may be facilitated by deliberate weakening of the cell walls. The latter is achieved by removal of calcium from the cell walls and its sequestration in calcium oxalate crystals (He et al., 2014). The transfer of pollen grains (male gametophytes or microgametophytes) from the anthers to the stigma is called pollination and is followed by abscission of the stamens. During anthesis, the stigma releases a sticky sap that retains and rehydrates the pollen for pollination (Considine and Knox, 1979a; Huglin and Schneider, 1998; Meneghetti et al., 2006). This sap is supplied via the newly formed stylar xylem that has previously been induced by auxin released by the stigma's papillae. Because the tiny, dry pollen grains are highly exposed during this process, the pollen wall contains high concentrations of phenolic compounds called flavonols that act as "sunscreen" (Downey et al., 2003b; see also Section 5.2), protecting the genetic information (DNA) contained in the chromosomes from destruction by UV radiation (Caldwell et al., 1998). It is conceivable, moreover, that the calcium oxalate crystals that are transferred to the stigma along with the pollen grains help scatter excess light. Nevertheless, excessive UVB radiation, such as that resulting from sudden exposure to sunlight of previously shaded inflorescences, especially in conjunction with heat ($>33°C$), can sometimes decrease pollen germination or pollen tube growth (Torabinejad et al., 1998). In other words, high-light stress reduces pollen viability and leads to male sterility.

In contrast to their wild relatives and the few cultivars with female flowers, cultivated grapevines are typically self-pollinated, whereby pollen originates from the flower's own anthers (Mullins et al., 1992; Sartorius, 1926). Wind appears to be responsible for only occasional cross-pollination, where pollen originates from a flower on a different plant than the pollinated one (Scherz, 1939). In fact, the anthers often burst shortly (1–24 h) before the calyptra is shed, releasing the pollen on the flower's own stigma, a process termed cleistogamy (Heazlewood and Wilson, 2004; May, 2004; Staudt, 1999). Despite this self-pollination strategy, the pollen releases sesquiterpene volatiles that were previously accumulated inside the pollen grains (Martin et al., 2009). In addition, the osmophors on the ovary may serve to attract insects by emitting monoterpenes (in the same way that citrus and lavender flowers do), although they ostensibly do not offer the insects any nectar (Bönisch et al., 2014; Brantjes, 1978). An intriguing possibility is that the release of volatiles might be a defense strategy to deter herbivorous insects or fungi that would feed on the highly vulnerable flowers or to attract enemies of these herbivores (Gang, 2005). Insect pollination may be much more common and is perhaps required for commercially acceptable fruit set, in *Muscadinia* than in *Vitis* species (Olien, 1990; Sampson et al., 2001). However, even in the latter, berries originating from cross-pollinated flowers apparently grow larger than those from self-pollinated flowers, which suggests that cross-pollination improves seed formation. In any case, the pollen grains absorb water from the moist stigma to rehydrate (a process that requires approximately 30 min), germinate on the stigma whose surface cell layer subsequently suberizes (Considine and Knox, 1979a), and form a pollen tube. The optimum temperature for pollen germination is between $25°C$ and $30°C$, whereas temperatures below $10°C$ and above $35°C$ inhibit germination. The pollen tube, which carries two small fused sperm cells (male gametes; Greek *gamete* = spouse) surrounded by a large vegetative cell, penetrates the stigma and grows down through the intercellular spaces of the style like a hydraulic drill, delivering its package of chromosomes into the egg inside the ovule (Zonia and Munnik, 2007). The male and female gametes (sex cells) meet when a pollen tube enters the embryo sac and releases the sperm. Their fusion in the ovule is termed fertilization and results in the zygote, which develops into the embryo and leads to seed set (the ovule becomes a seed) and fruit set (the ovary becomes a fruit).

In a typical grape flower, the pollen tubes cover a distance of approximately 2 mm. Both pollen germination and pollen tube growth are induced by auxin, with which the pollen has previously been "loaded" in the anthers and stimulated by the growth hormone gibberellin that is produced in the anthers, perhaps by the pollen itself (Giacomelli et al., 2013; Gillaspy et al., 1993). Growth of the pollen tube represents a more than 1000-fold elongation of the pollen cell and is very rapid; rates of 0.3 mm h^{-1} are not uncommon (Staudt, 1982). Not surprisingly, this process consumes a great deal of energy, and the growing pollen tubes have high respiration rates. Respiration is fueled by the breakdown of starch reserves that are accumulated in the pollen grains from imported sucrose prior to anthesis.

The speed at which the pollen tubes grow toward the ovules is critical for seed and fruit set because the ovules are receptive for only a limited period after anthesis, after which they can no longer be fertilized. Although only one pollen tube from the myriads of pollen grains landing on the stigma needs to reach one of up to four receptive ovules to make fertilization likely, this outcome nevertheless seems to depend on the pollen density present on the stigma. High pollen numbers may promote fertilization, because the tubes from different pollen grains grow at different rates, and higher numbers mean that there may be more fast-growing pollen tubes than when the pollen density is low. In addition, higher pollen numbers are associated with higher rates of pollen germination. Consequently, fertilization, seed number, and fruit set decline if too few pollen grains land on a stigma (Zhang et al., 2010a). The rate of pollen tube growth also depends on temperature; at 25–30°C, fertilization can occur within 12 h, at 20°C after 24 h, and at 15°C after 48 h, whereas even lower temperatures do not permit fertilization because the pollen tubes stop growing before they reach the ovules or they simply reach the ovules too late (Staudt, 1982). Eggs that are not fertilized within 3 or 4 days after anthesis will degenerate (Callis, 1995; Kassemeyer and Staudt, 1981). Therefore, slow pollen tube growth (e.g., due to low pollen "fuel status" resulting from environmental stress interfering with starch accumulation prior to pollination) leads to poor fruit set (Koblet, 1966).

How the pollen tubes find the ovules is not well known, but it seems that they are guided down the pistil by a combination of physical and chemical mechanisms. Among other cues, a glycoprotein (protein attached to a sugar molecule) establishes an extracellular sugar gradient, which the pollen tubes follow by breaking away the sugar from the protein (Cheung et al., 2010). The tubes' tips are then attracted by the synergid cells (which die after they have served their purpose) flanking the egg cell and follow gradients of γ-amino butyrate (GABA) and other attractants (Berger et al., 2008; Chevalier et al., 2011; Palanivelu et al., 2003). GABA is chemically an amino acid, but it is not referred to as such because it is not incorporated into proteins. In plant cells, it is synthesized from glutamate, and in the human brain GABA released from nerve cells is the predominant inhibitory neurotransmitter. In any case, once the pollen tube reaches the ovule, it stops growing, and its volume and turgor pressure increase rapidly. This leads to an explosive rupture of the tip, which ejects the sperm cells into the ovule (Staudt, 1982). Only a single pollen tube usually penetrates the ovule; those that lose the race seem to be repelled and may continue growing in search of another virgin ovule (Cheung et al., 2010; Chevalier et al., 2011). The four ovules in a flower do not all reach the same stage of development simultaneously: Some embryo sacs may fail to develop or may not mature in time to be fertilized, others are not "found" by a pollen tube, and yet others are fertilized but are slow to develop (Kassemeyer and Staudt, 1982; Nitsch et al., 1960; Pratt, 1971). This is why grape berries rarely contain four seeds.

Fertilization in angiosperms is more complex than in other plants or in animals. During the process of double fertilization, one haploid (i.e., having one set of chromosomes) male gamete fuses with the haploid female gamete (the egg cell or simply egg) to form the diploid (i.e., having two sets of chromosomes—one from the mother via her egg and one from the father via his pollen) zygote. The other male gamete (delivered by the same pollen tube) almost simultaneously fuses with the other female gamete, the diploid central cell of the embryo sac, to give rise to the endosperm (Berger et al., 2008; May, 2004; Raghavan, 2003). Similar to the placenta in mammals, the endosperm protects the embryo developing from the zygote, controls nutrient transfer from the berry, and serves as a nutrient store for the developing embryo and the emerging plant organs during seed germination. It seems that double fertilization is an all-or-nothing process: Either both the egg cell and the central cell are fertilized or neither is fertilized (Kassemeyer and Staudt, 1981). The resulting diploid embryo and (initially) triploid endosperm together form the filial (daughter) generation, whereas all other berry tissues (including the seed coat and the nucellus) are part of the maternal (mother) generation or mother plant. Although the contribution of the central cell's maternally derived cytoplasm to the developing endosperm is quite significant (Berger et al., 2006), the endosperm is a terminal tissue; it decays while feeding the embryo and makes no genetic input to the new plant that develops from the embryo.

The embryo should more accurately be called the zygotic embryo because somatic plant cells (e.g., from floral organs, shoot tips, or leaf pieces) can be forced through tissue culture (i.e., *in vitro*) to form the so-called somatic embryos, which can develop into whole plants that are clones of their mother plant (Franks et al., 2002; Martinelli and Gribaudo, 2001; Mullins and Srinivasan, 1976). Somatic embryogenesis or a related technique called shoot organogenesis, in which the embryo stage is bypassed altogether, are exploited in vegetative propagation. Whereas somatic embryogenesis is mostly limited to breeding programs, shoot organogenesis is also used to help eliminate viruses and bacteria by the so-called micropropagation or meristem culture, whereby shoot tips less than 0.5 mm in length are excised under aseptic conditions and grown into full plants (Barlass et al., 1982; Bass and Vuittenez, 1977). Although this practice may occasionally lead to the expression of juvenile leaf characters in the resulting plant, this is normally confined to the lower portion of the new shoots, and cuttings taken from the distal portion will give rise to true-to-type plants (Grenan, 1984). However, such propagation methods are believed to encourage the activation of normally silent jumping genes (see Section 1.2), which can induce occasional somatic mutations and thus introduce additional genetic variation, leading to novel clones, in some of the tissue-cultured vines (Benjak et al., 2008; Carrier et al., 2012).

In contrast to somatic embryogenesis, although they are also derived from somatic cells (and hence may carry somatic mutations), due to genetic recombination in addition to occasional mutations, each egg cell is genetically different from all other egg cells on the same vine, and each sperm cell is genetically different from all other sperm cells on the same vine. The recombination is basically a random shuffling of paternal and maternal genes, just like the shuffling of a deck of cards, and occurs approximately 1–3 weeks before anthesis (depending on cultivar; Lebon et al., 2008) during the production of gametes in a special cell division termed meiosis (Greek *meioun* = to make small). Just before the cell divides, the chromosomes line up alongside each other and randomly exchange pieces of DNA so that the new chromosomes that end up in the gametes constitute a novel and distinctive mix of genes from the mother and the father. This shuffling is the true essence of sex and thus applies only to genes contained in the cell nucleus—that is, on the chromosomes. The other cell organelles with their own genes (i.e., the mitochondria and plastids such as chloroplasts) reproduce asexually (by simple division), and the male organelles are normally excluded from or left behind in the pollen (Strefeler et al., 1992). Only very few of them occasionally "slip" through (Birky, 1995; Timmis et al., 2004), and some of their genes may escape destruction and be "absorbed" by the nucleus (Sheppard et al., 2008). Indeed, this exclusion of the father's organelle genes is the fundamental reason why there are two, and only two, sexes in all sexually reproducing plants (and animals). Building and running an organism as complex as a grapevine (or an animal) requires a great deal of cooperation among the genes. The genes of the plastids and mitochondria have never "learned" how to cooperate, and these organelles would often attack and kill each other when two cells fuse (which, in fact, they do in single-celled algae). Assigning them to two separate sexes and prohibiting them from mixing prevents this problem (i.e., limits the damage) while still permitting the (nuclear) chromosomes to take advantage of the costly sexual recombination.

As a result of the chromosome mixing, every single seed, even within the same berry, will give rise to a genetically unique plant, slightly different from all others, even though it inherited all its plastids and mitochondria from the mother plant. This production of new genotypes by mixing of genes (more accurately, alleles; i.e., slightly different versions of the same gene, one of which may or may not be dominant over the other) in each generation, in addition to occasional mutations (which generate the different alleles in the first place), provides the endless genetic variety that under natural conditions allows evolution (specifically, natural selection) to act by "weeding out" those individuals (phenotypes) less well adapted to the prevailing environmental conditions (Mayr, 2001). As Charles Darwin (2004) stated, "Natural selection acts solely through the preservation of variations in some way advantageous." Because "environment" also comprises other species, including pathogens such as fungi, bacteria, and viruses, the major benefit of recombination is the decreased risk for a plant's offspring to be wiped out by pathogen attack (see Chapter 1). This is the meaning of sex: By producing genetic variation (i.e., by "not putting all its eggs in the same basket"), a species greatly increases the chance that at least some individuals will resist and thus survive pathogen attack (or drought, etc.). Of course, such genetic variation is possible only in wild grapevines and not in cultivated populations of vegetatively propagated (i.e., clonal) plants. Nevertheless, even the various cultivars originally arose this way, whether by deliberate crossing or by natural pollination. Recombination accounts for the vast differences among cultivars in terms of growth, yield formation, fruit composition, pest and disease susceptibility, and reaction to environmental influences. Somatic mutations (bud sports), however, can arise in vegetatively propagated grapevines and may give rise to clonal variation that can be subject to selection by humans, that is, by artificial as opposed to natural selection (see Section 1.2).

Within days after fertilization, the endosperm nucleus begins to divide, whereas the zygote (i.e., the future embryo) does not start dividing until approximately 2 or 3 weeks after fertilization (Nitsch et al., 1960; Pratt, 1971). By this time, the cells of the nucellus have reached their maximum size, and the endosperm, whose nuclei have by now become polyploid, begins to accumulate starch, proteins, and lipids (Cadot et al., 2006; Kassemeyer and Staudt, 1983). The first few divisions of both tissues seem to be mainly under the control of maternal genes, which may give the mother plant some control over which embryos to develop (Raghavan, 2003). The fertilized ovule develops into the seed, whose embryo, endosperm, and seed coat start to produce and release auxin to pericarp tissues, where it stimulates gibberellin synthesis (Giacomelli et al., 2013; O'Neill, 1997; Pandolfini et al., 2007; Pattison et al., 2014; Serrani et al., 2008). In addition, gibberellin may also be produced in and released by the seeds upon stimulation by auxin

(Dorcey et al., 2009; Giacomelli et al., 2013). Auxin reactivates cell division, and gibberellin induces cell expansion (Serrani et al., 2007). Thus, fruit set is dependent on both auxin and gibberellin: These two hormones induce the pistil to develop into the fruit and to differentiate an exocarp (skin) and mesocarp (flesh or pulp). Although auxin induces gibberellin production, auxin synthesis might be stimulated by pollen-derived gibberellins; it is necessary for embryo formation and fruit tissue production and continues until the embryo is mature. Therefore, the interaction, also called cross talk, between auxin and gibberellin plays a key role in the ovary's commitment to initiate fruit development and growth (Gillaspy et al., 1993; Ozga and Reinecke, 2003; Ruan et al., 2012). The promoting effect of gibberellin on cell volume in seedless berries is exploited in table grape and raisin production; growers routinely apply gibberellin sprays at bloom and fruit set to increase berry size (Mullins et al., 1992). In contrast to early seed formation, the seeds later take charge of their own maturation by switching on the embryo's genes; in other words, there is a change from maternal to filial control of seed development (Gutierrez et al., 2007). These genes initiate and coordinate the import of nutrients and accumulation of storage reserves such as starch and proteins in the endosperm and the subsequent desiccation of the seeds.

Although pollination, rather than fertilization, triggers the initial stages of ovary development, perhaps because growing pollen tubes produce gibberellin (Zhang et al., 2010a), subsequent fruit formation depends on a supply of seed-derived auxin to the developing fruit tissues (Kassemeyer and Staudt, 1983; O'Neill, 1997; Pattison et al., 2014). Additionally, the movement of auxin from the berry toward the shoot via the pedicel may prevent premature fruit abscission (Pattison et al., 2014). Therefore, the development of grape berries is usually dependent on pollination, fertilization, and development of at least one seed (Pratt, 1971). Ovaries that do not contain at least one fertilized ovule or whose seed development is aborted early on are abscised from the fruit cluster (Nitsch et al., 1960). Arrested ovule development prior to anthesis prohibits fertilization and also results in flower abortion if all four ovules fail to develop (Kassemeyer and Staudt, 1982). Later degeneration of the embryo and endosperm can result in apparently normal-looking but hollow (and thus nonviable) seeds termed "floaters" because they float in water, unlike normal seeds, which are thus named "sinkers" (Ebadi et al., 1996).

Nonetheless, some cultivars can produce apparently normal seedless berries. This is called parthenocarpic (Greek *parthenos* = virgin) fruit development, if the berries develop without fertilization (although typically not without pollination) and lack seeds completely, or stenospermocarpy, if the berries contain at least one fertilized but subsequently aborted seed so that in many cases all that remains is a small, soft, white rudimentary seed or seed trace (Barritt, 1970; Ledbetter and Ramming, 1989; Olmo, 1936; Pearson, 1932; Roytchev, 1998, 2000; Staudt and Kassemeyer, 1984). Stenospermocarpy is a quantitative trait: The seed traces vary in size and hardness over a wide range (Cabezas et al., 2006; Ledbetter and Shonnard, 1991). Cultivars with parthenocarpic fruit (e.g., Black Corinth) are often used to produce seedless raisin grapes, whereas the most so-called seedless table grapes (e.g., Thompson Seedless) are stenospermocarpic. Intriguingly, the embryos of stenospermocarpic grapes remain alive and can be rescued and grown into full plants using *in vitro* techniques, which permits crossing of seedless table and raisin grapes to breed new cultivars (Cain et al., 1983; Emershad and Ramming, 1984; Emershad et al., 1989). Because the lack of growth hormone production in stenospermocarpic fruit limits cell expansion and hence berry size (Coombe, 1960; Nitsch et al., 1960), gibberellin sprays are often applied after fruit set to increase the size of table grapes with the added benefit of decreased cluster compactness due to elongation of the rachis and reduced fruit set (Weaver, 1960; Weaver and Pool, 1971).

The proportion of flowers that develop into berries following anthesis is typically in the range of 20–50% and is inversely related to the number of flowers per inflorescence. Fruit set is highly variable among cultivars and is strongly modulated by environmental conditions and rootstocks (Alleweldt and Hofäcker, 1975; Huglin and Balthazard, 1975; Keller et al., 2001a; Schneider and Staudt, 1978). Cultivars that are susceptible to environmental conditions leading to poor fruit set include Merlot, Grenache, and Gewürztraminer, whereas the Pinots, Chardonnay, Riesling, and Sylvaner seem to be much less susceptible. The process of meiosis appears to be particularly sensitive to stress conditions that curtail the supply of carbohydrates to the inflorescences. Optimum conditions for flowering and fruit set are thought to be similar to those governing inflorescence initiation (see Section 6.1), namely high light intensity, warm temperature, and adequate soil moisture and nutrient availability. This is the likely reason why years in which the shoots carry more clusters than usual often follow years with above-average fruit set. Adverse environmental conditions (e.g., cloudy, wet, cool, or hot bloom period and water or nutrient stress), insufficient or inefficient leaf area (e.g., due to hail or insect or disease attack), shade created by dense canopies, or excessively vigorous shoot growth (competing with inflorescences for assimilates) often result in poor fruit set and loose clusters (see Section 6.1). Because, as a perennial species, the grapevine does not depend on producing seeds each year, such "self-thinning" enables it to adjust its reproductive output to available resources without jeopardizing survival of the plant. By culling a portion of its reproductive structures, the vine may enable the remaining seeds to attain adequate size and maturity.

Poor fruit set due to excessive abortion of flowers and ovaries is also termed shatter, shedding, or "coulure" (French *couler* = to leak, to fall off). Loss of such ovaries can occur for up to 4 weeks after anthesis (Kassemeyer and Staudt, 1983; Staudt and Kassrawi, 1973). Fertilization of at least one of up to four ovules is required for the ovary to develop into a berry; hence, the abscised ovaries typically do not contain any fertilized ovules, although the vast majority of them have been pollinated (Kassemeyer and Staudt, 1982; Staudt and Kassemeyer, 1984; Pratt, 1973). Normal fruit set despite inadequate seed development in a portion of the fertilized berries, on the other hand, is called "millerandage" and results in clusters having the appearance of "hens and chicks" (Fougère-Rifot et al., 1995; Galet, 2000; May, 2004). The term "hens" refers to large, normal berries with at least one viable seed, and "chicks" describes small berries with tiny, degenerated seeds (i.e., seed traces) that often lack an endosperm (Ebadi et al., 1996; Staudt and Kassemeyer, 1984). Although the expansion of "chicks" is arrested within approximately 3 weeks after anthesis due to seed abortion, these berries apparently ripen normally. By contrast, the so-called "shot berries" are essentially arrested at the ovary stage; remain small, green, and hard; and are high in tannins (Fougère-Rifot et al., 1995). The term is unfortunate: Because these ovaries fail to differentiate a discernible mesocarp and exocarp, shot berries are technically not berries at all (Dry et al., 2010). They usually contain at least one sterile ovule (e.g., due to incomplete or failed meiosis) in addition to degenerated fertile but unfertilized ovules. Perhaps shot berries are prevented from abscising by the initial, although very limited, growth of the nucellus of the sterile ovule (Kassemeyer and Staudt, 1982). Coulure results in very loose clusters, and millerandage leads to highly variable seed numbers and berry sizes, including seedless berries, on the same cluster.

Variations in the degree of fruit set together with the extent of inflorescence branching and variations in the rate and/or extent of rachis cell expansion during cluster development before and after bloom lead to wide variation in cluster size and architecture, both between species and cultivars and within them (e.g., due to environmental influences). Such differences in morphological features have implications not only for yield formation (see Section 6.1) but also for the clusters' susceptibility to fungal attack and consequently for disease severity at harvest (Molitor et al., 2012). For instance, cultivars with tight clusters (e.g., Chardonnay, Riesling, and Pinot) have small rachis cells and a short, compact branching pattern. They are much more susceptible to infection by *Botrytis cinerea* than those with loose clusters (e.g., Cabernet Sauvignon, Syrah, and Thompson Seedless). This is probably because tight clusters have their berries packed into a small volume and hence retain more rain (or irrigation) water and dry down more slowly than loose clusters. In addition, berries touching and rubbing on each other may lose epicuticular wax and hinder uniform fungicide coverage. Yet even cultivars with naturally tight clusters can be manipulated using viticultural practices that tend to decrease cluster compactness, although such practices usually come at the cost of decreased yield. Potential strategies include light or no pruning, resulting in large numbers of clusters and, therefore, relatively poor fruit set (see Section 6.1); removing a portion of the leaves before or soon after bloom, which also diminishes fruit set (Candolfi-Vasconcelos and Koblet, 1990; Palliotti et al., 2011; Poni et al., 2006, 2009; Tardaguila et al., 2010); or cutting through clusters (i.e., removing the distal cluster portion) at or soon after bloom, which often leads to elongation of the remaining rachis (Winkler, 1958). By contrast, cluster thinning before rather than after bloom tends to enhance fruit set and berry size, and thereby increase cluster compactness and *B. cinerea* infection (Smithyman et al., 1998; Winkler, 1958).

In seeded grapes, fertilization is followed by a second period of cell division (the first period is before bloom and determines the cell number of the ovary), which is probably stimulated by a brief increase in cytokinin in addition to auxin produced in the developing seeds. The berry cytokinin concentration, it seems, declines after anthesis throughout berry development (Chacko et al., 1976). During the first 3 weeks after anthesis, the polar nucleus of the seeds' endosperm divides several times before the endosperm becomes cellular so that endosperm nuclei become polyploid (Kassemeyer and Staudt, 1983). By contrast, the zygote cells giving rise to the embryo only begin dividing approximately 2–4 weeks after anthesis. The rate and duration of cell division in the berry pericarp are controlled by the seeds (i.e., the embryos) so that berries containing more seeds will become larger than berries with fewer seeds (Coombe, 1960; Gillaspy et al., 1993). Whereas the cell number successively doubles approximately 17 times before anthesis, only one or two more doublings happen after anthesis (Coombe, 1976). The mesocarp cells cease dividing 3 or 4 weeks after anthesis (the innermost cells stop dividing within 2 weeks), whereas skin cells may continue to divide for up to 5–6 weeks after anthesis (Considine and Knox, 1981; Harris et al., 1968; Nakagawa and Nanjo, 1965; Pratt, 1971; Staudt et al., 1986). Cell division is completed before the berry enters a lag phase of slow or no fresh weight accumulation (Ojeda et al., 1999, 2001). Although it is frequently stated that berry growth is divided into a phase of cell division and one of cell expansion, this is not accurate. Growth is always by cell expansion because plants (like any other organism) cannot add fully grown cells like bricks to a wall. Although dividing cells usually remain small and have small vacuoles, cell division is accompanied and followed by cell expansion (Gillaspy et al., 1993; Pratt, 1971). As the cells expand, the cell wall becomes thinner and the vacuole becomes predominant. Therefore, the vast majority of the volume gain of growing berries is due to cell expansion (i.e., an increase in vacuole volume; see Section 3.1), especially in the mesocarp, which is stimulated by seed-produced auxin. The lag phase is followed by another period of cell expansion so that the volume of mesocarp cells may increase more than 300-fold from anthesis to maturity (Coombe, 1976) or 15-fold between fruit set and maturity. At the same time, the fruit surface area may increase approximately 400-fold from anthesis to veraison and almost doubles again from veraison to maturity (Considine and Knox, 1981). The weight of mature berries varies from less than 0.5 g to more than 10 g among cultivars of *V. vinifera* (Boursiquot et al., 1995; Dai et al., 2011). The berry weight of different clones of the same cultivar can diverge by a factor of two or more (Fernandez et al., 2006). But even within the same fruit cluster on a cultivated vine, the weight of individual berries may vary more than twofold (Dai et al., 2011).

Cell division seems to be mostly under genetic control (i.e., fruit from different cultivars should have different cell numbers), whereas cell expansion is predominantly driven by environmental factors (i.e., berries of the same cultivar grown under different conditions should have different cell sizes). This means that fruit size can vary due to variations in both cell number and cell size. It also means that cell division is comparatively immune to environmental influences so that viticultural attempts to manipulate berry size will have to aim mostly at cell expansion, whereas breeding efforts should address cell division. The maximum extent of cell expansion, and hence berry size, is limited by the elastic properties of the skin, which in turn may be related to the thickness and/or stiffness of the exocarp cell walls or the cuticle (Boudaoud, 2010; Matthews et al., 1987; see also Section 6.2). In ripe grape berries, cell wall thickness ranges from greater than 1 μm in the exocarp to less than 0.1 μm in the mesocarp and, if grapes behave like tomatoes, the resistance of the epidermis (especially the cuticle) against deformation increases strongly during fruit ripening (Bargel et al., 2006; see Section 6.2). As first described by Winkler and Williams (1935), the growth or increase in fresh weight of seeded berries traces a double-sigmoid pattern; that is, the growth curve is bent in two directions, like the letter S, and can be divided into three more or less distinct phases or stages (Figure 2.13).

FIGURE 2.13

Cabernet Sauvignon berry growth over 3 years (left; note 50% anthesis occurred on day 160 in year 1 and on day 150 in years 2 and 3; M. Keller, unpublished data), and beginning of ripening (veraison) in Concord clusters with berries displaying various stages of color change (right; photo by M. Keller).

Stage I: The first rapid increase in size of the seeds and pericarp. The first part of seed development is termed morphogenesis and consists of embryo formation (embryogenesis) and endosperm growth, but there is little growth of the embryo (Ebadi et al., 1996; Nitsch et al., 1960). The seeds are green, and the berry is green and hard and accumulates organic acids (mainly tartrate and malate) but little sugar. The growth hormones auxin, cytokinin, and gibberellin produced by the embryos and released into the pericarp reach high concentrations early on and then decrease during this period (Böttcher et al., 2013; Iwahori et al., 1968; Pattison and Catalá, 2012; Pérez et al., 2000; Sakakibara, 2006; Scienza et al., 1978). Auxin and gibberellin are also exported in the phloem to the pedicel and peduncle, where gibberellin induces cell elongation and auxin induces the production and differentiation of vascular bundles to ensure that the transport capacity of the vascular system does not limit growth of the developing berry. In addition, import from the "mother" vine of the germination-inhibiting hormone ABA prevents seed abortion and promotes normal embryo development (Nambara and Marion-Poll, 2005). Consequently, the ABA concentration in the berry is high during early development but declines as the berry expands (Davies and Böttcher, 2009; Owen et al., 2009). This phase lasts 6–9 weeks and ends when the epidermis cells stop dividing (Staudt et al., 1986). By this time, the berry has attained at least half its maximal size and fresh weight.

Stage II: The lag phase during which the seed enters its maturation stage (just after the embryo's genes take over control from the mother vine). The embryo grows rapidly, and by 10–15 days before veraison the seeds reach their final size, maximum fresh weight, and maximum tannin content (Adams, 2006; Niimi and Torikata, 1979; Ojeda et al., 1999; Pratt, 1971; Ristic and Iland, 2005, Viala and Vermorel, 1909). Moreover, the cell walls of the middle integument thicken and become lignified, and the outermost cell walls of the outer integument thicken to form a cuticle; the hard seed coat now restricts further seed expansion (Cadot et al., 2006). Although the pericarp grows only insignificantly, the concentration of the growth hormone auxin peaks briefly during this phase and then declines sharply (Alleweldt and Hifny, 1972; Alleweldt et al., 1975; Nitsch et al., 1960). The auxin peak is especially pronounced in the seeds, where it may regulate embryo maturation (Gillaspy et al., 1993; Pattison and Catalá, 2012). Because auxin also acts as a ripening inhibitor (Davies and Böttcher, 2009; Davies et al., 1997), it is possible that auxin serves to prevent fruit ripening before the seeds are fully developed (Pattison et al., 2014). The influx of ABA and its production in the berry itself increase toward the end of this phase, suppressing further embryo growth by blocking gibberellin production in the seeds (Davies and Böttcher, 2009; Gutierrez et al., 2007; Nambara and Marion-Poll, 2005; Pérez et al., 2000). ABA also seems to induce changes in the expression of many genes that ultimately bring about fruit ripening (Gambetta et al., 2010; Koyama et al., 2009). At the same time, there may be a fleeting increase in ethylene (though far less pronounced than in the so-called climacteric fruits; Chervin et al., 2004) that is accompanied by heightened sensitivity to ethylene (Chervin et al., 2008) and a striking and persistent rise in brassinosteroids (Pilati et al., 2007; Symons et al., 2006), a class of growth-promoting hormones probably produced in the epidermis. Together ABA, ethylene, and brassinosteroids may trigger the changes in cell wall chemistry that lead to berry softening and expansion. Berry turgor pressure also declines approximately 10-fold during this period (Thomas et al., 2006). The length of the lag phase (1–6 weeks) largely depends on the cultivar and is important in determining the time of fruit maturity; late-ripening cultivars seem to have a long lag phase (Alleweldt and Hifny, 1972; Currle et al., 1983).

Stage III: The ripening period, which lasts 5–10 weeks. Its onset, which typically occurs over a period of 7–10 days within a grape cluster, is termed veraison and signals a fundamental shift from partly photosynthetic to wholly heterotrophic (Greek *trophé* = nutrition) metabolism. Veraison is marked by berry softening and an increase in sugar content, followed by a rapid change in skin color from green to red, purple, or blue in dark-skinned cultivars (see Figure 2.13), and to more or less yellow in some white cultivars. Softening (i.e., a decrease in fruit firmness or increase in deformability) mainly arises from the disassembly of the mesocarp cell walls (Huang and Huang, 2001). Although the cell walls remain functionally intact, they become more open, more hydrated, less flexible, and weaker so that they can no longer hold the cells together very well (Brummell, 2006). Water import via the phloem increases rapidly (water being the solvent for the incoming sugar), and water import through the xylem declines while the berry changes color. The xylem now serves mostly to recycle surplus phloem water back out of the berry (Keller et al., 2006; Patrick, 1997). Berry turgor pressure is low ($P < 0.5$ bar) and varies by cultivar but remains positive and relatively constant throughout ripening (Bernstein and Lustig, 1985; Lang and Düring, 1990; Thomas et al., 2006). This last phase is characterized by a further increase in berry size, which is somewhat unusual because the fruit of most other species reaches their final size before ripening begins (Gillaspy et al., 1993). The volume increase is initially very rapid but slows progressively toward fruit maturity; berry size may plateau or decrease due to evaporative water loss during later stages of ripening. The extensibility of the skin temporarily increases at veraison to accommodate berry expansion, but the skin later restricts further expansion (Matthews et al., 1987). Epidermal cells are stretched until the berry reaches its maximum size (Staudt et al., 1986). Red, purple, or blue pigments called anthocyanins accumulate in the exocarp (and in "teinturier" cultivars, such as Alicante Bouschet, also in the mesocarp), and sugars (glucose and fructose) accumulate throughout the pericarp while organic acids (malate) and chlorophyll are degraded (see Section 6.2). Chloroplasts are converted into chromoplasts (Greek *chróma* = color) that specialize in the accumulation of yellow or orange carotenoid pigments (Fanciullino et al., 2014). Anthocyanin production and other changes, such as invertase and sugar transporter activities, may be triggered by the interplay of sugars and ABA (Davies and Böttcher, 2009; Gambetta et al., 2010; Jia et al., 2011; Koyama et al., 2009; Pirie and Mullins, 1976). ABA acts in a positive feedback loop with sugars, whereby sugars stimulate ABA production and ABA promotes sugar accumulation (Jia et al., 2011). Unlike in climacteric fruits, the "fruit-ripening" hormone ethylene does not play a prominent role in grape ripening, and berry respiration declines during this period (Coombe and Hale, 1973; Geisler and Radler, 1963; Harris et al., 1971). Embryo, but not seed, growth may continue until berry expansion ceases (Nitsch et al., 1960; Pratt, 1971; Staudt et al., 1986). The seeds turn from green to yellow and finally brown due to oxidation of tannins in the parenchyma cells of the outer integument, and the seeds become hard and desiccated (Adams, 2006; Cadot et al., 2006; Huglin and Schneider, 1998; Ristic and Iland, 2005). Storage reserves, such as starch, proteins, and lipids, and mineral nutrients are accumulated in the endosperm until the seeds become dormant (Cadot et al., 2006; Rogiers et al., 2006). Thus, the seed fresh weight declines while the dry weight reaches a plateau or continues to increase until the berry attains its maximum size and weight (Pastor del Rio and Kennedy, 2006; Staudt et al., 1986); seeds that are hard and brown and have attained their maximal dry weight are called "mature." The concentration of ABA increases rapidly after veraison, peaks in the seed and berry flesh during seed maturation, then decreases due to degradation during seed desiccation, and is relatively low in mature seeds and berries (Davies and Böttcher, 2009; Inaba et al., 1976; Lund et al., 2008; Owen et al., 2009; Scienza et al., 1978). ABA, which now accumulates mainly as a result of glucose-induced production by the embryo, helps coordinate seed maturation and is necessary to induce seed dormancy, in which metabolism ceases until the seed is rehydrated for germination (Gutierrez et al., 2007; Nambara and Marion-Poll, 2005; Wobus and Weber, 1999). Intriguingly, the content of cytokinins also increases during this stage (Böttcher et al., 2013). It remains to be seen whether this serves to delay premature senescence of the ripening berry tissues.

The length of each stage of fruit growth and the final berry size depend on the cultivar but are strongly modified by rootstocks and environmental conditions. Optimum conditions for rapid berry development are similar to those governing the other phases of reproductive development: high light intensity, warm temperature, and adequate soil moisture and nutrient availability (see Chapter 6). The rate of development is also modified by the number of seeds in a berry. Increasing seed number appears to delay the transition from one growth phase to another (Staudt et al., 1986). As an extension of this trend, seedless grape berries often do not show clearly discernible phases of growth, and their volume increases much more gradually (Iwahori et al., 1968; Nitsch et al., 1960; Winkler and Williams, 1935). They also tend to remain smaller compared with the berries from seeded cultivars (Boursiquot et al., 1995; Huglin and Schneider, 1998).

Harvest of grapes by birds or mammals, including humans, normally completes the reproductive cycle. If grapes remain on the vine for an extended period, the continued cell wall disassembly, leading to diminished cell wall strength and loss of intercellular adhesion, eventually results in tissue failure, cell separation, and senescence (Considine and Knox, 1979a), which sets the seeds "free"—often aided by infection by opportunistic fungi such as *B. cinerea* and others (Brummell, 2006).

Seeds are ready to germinate at or soon after veraison (Pratt, 1971; Winkler and Williams, 1935); in fact, it is thought that fruit ripening is not initiated until seed maturation has been completed (Cawthon and Morris, 1982; Gillaspy et al., 1993). Whereas seeds may germinate immediately if postveraison berries drop to the ground in a moist environment, they lose this ability within a day upon removal from a berry, when their water content declines from approximately 50% to 20% (Currle et al., 1983). Even though the endosperm may lose cellular integrity and die during seed desiccation, grapevine seeds can tolerate water loss to less than 6% without losing their ability to germinate (Orrù et al., 2012; Zeeman et al., 2010). Indeed, drying may enhance the subsequent ability of seeds to germinate under favorable conditions (Wang et al., 2011). However, it is thought that the change in seed shape from almost spherical with an inconspicuous beak to a pear shape with a pronounced beak that accompanied grape domestication may have reduced the ability of seeds to germinate (Terral et al., 2010).

Once dehydrated, the seeds require a stratification period of 4–12 weeks at chilling temperatures (0–10°C) and at least 40–50% relative humidity to overcome embryo dormancy. The term stratification (Latin *stratum* = blanket) itself is derived from the old practice of spreading seeds between layers of moist, cold soil to break dormancy. Following dormancy release, the base temperature for seed germination, at least in *V. sylvestris*, is around 10°C, and the germination rate increases up to an optimum of at least 25°C (Orrù et al., 2012). Light (especially red light) induces water imbibition and subsequent germination by stimulating the embryo cells to produce gibberellin and expand rapidly, which overcomes seed dormancy by accelerating ABA degradation, facilitates breakage of the seed coat, and induces remobilization of nutrients from the endosperm (Nambara and Marion-Poll, 2005; Olszewski et al., 2002). Seeds do not germinate when there is insufficient water for imbibition, which ensures that the young seedlings do not succumb to drought after germination. The capacity for seed germination may also be influenced by the nutritional status of the mother plant. For instance, high nitrate content during seed formation increases the germination capacity, apparently because nitrate lessens the maintenance of seed dormancy by accelerating ABA breakdown (Alboresi et al., 2005).

The evolution of a fleshy, sweet, and dark-skinned berry during the late Cretaceous period and the subsequent Paleocene epoch about 100 to 56 million years ago was probably an adaptation to a conversion, driven by climate change and the associated demise of herbivorous dinosaurs, from open to closed, wooded habitats and the simultaneous evolution of birds and mammals (Hardie, 2000; Rodríguez et al., 2013; Seymour et al., 2013). Birds are the predominant seed dispersers of European wine grapes, whereas bears and other mammals are major dispersers of American juice grapes, whose distinctive aroma birds find unattractive. The only role grape berries play in the life of a vine is to spread its genes (DNA). The ability to advertise seed maturity by visual and olfactory cues and reward seed vectors with energy-rich sugar greatly increases the chances for gene dispersal. As Charles Darwin (2004) remarked, a fruit's "beauty serves merely as a guide to birds and beasts in order that the fruit may be devoured and the manured seeds disseminated." Berries accumulate pigments and aroma volatiles to advertise to potential seed vectors their content of nutritionally valuable components, such as sugars, amino and fatty acids, vitamins, and antioxidants (Goff and Klee, 2006; Rodríguez et al., 2013). These compounds, in turn, constitute the "ticket price" to cover the costs of transportation by means of the wings and legs of seed dispersers. Low concentrations of (volatile) ethanol produced by yeasts, which have been living on and in fruits since the Cretaceous period, may also form part of the aroma bouquet of ripe berries (Dudley, 2004). Although fruit color is certainly an important attractant during the day, volatiles may be more important at night, when colors are difficult to see. As a rule of thumb, birds rely more on visual cues to detect fruits, while mammals tend to rely more on their sense of smell (Rodríguez et al., 2013). At the same time, tannins and other secondary compounds may deter microbes that also would like to consume the fruit but do not disperse the seeds (Levey, 2004). Likewise, the high acidity and unpleasant aroma (caused by methoxypyrazines and other volatile compounds) of unripe berries prevents the berries from being eaten before the seeds are mature. The seeds themselves are made unpalatable by the accumulation of bitter-tasting tannins, which prevents animals from destroying the seeds by chewing them, and the seeds survive the digestive systems of their dispersers.

The clever strategy of packaging unpalatable seeds in an edible berry enables grapevines to explore new habitats beyond the small area around an existing immobile plant that simply drops its fruit (seeds) to the ground. This strategy ensures that the parent plant does not compete for limited resources with its own offspring. The characteristic accumulation of sugars and other compounds also means that fruit ripening is different from the senescence process in leaves that involves disassembly of proteins (called proteolysis) and recycling of nutrients to other plant parts (Gillaspy et al., 1993).

CHAPTER 3

WATER RELATIONS AND NUTRIENT UPTAKE

CHAPTER OUTLINE

3.1 Osmosis, Water Potential, and Cell Expansion 101
3.2 Transpiration and Stomatal Action 106
3.3 Water and Nutrient Uptake and Transport 110

ABSTRACT

Grapevines consist mostly of water that serves as a solvent for organic and inorganic solutes. Plant cells grow by osmotically attracting water and loosening their rigid cell walls. During osmosis, water moves into a solute-containing compartment, such as a cell, across a semipermeable membrane, increasing the turgor pressure of the cell. The solute and pressure forces together make up the water potential. Growth is extremely sensitive to decreases in water potential due to water deficit. Water evaporates from plant surfaces during transpiration, mostly through stomatal pores, cooling the plant tissues and causing a tension that drives root water uptake and movement up the vine in the xylem to the leaves. The stomata function as valves that regulate water loss to avoid excessive dehydration and cavitation of the xylem, whose low hydraulic resistance provides for effective water transport. Along with water, the roots also take up mineral nutrients from the soil solution in a controlled fashion for accumulation and distribution via the transpiration stream throughout the vine. Nutrient uptake, mostly by ion exchange, and movement to the xylem can occur actively, through ATP-powered transport proteins, or passively, down an electrochemical potential gradient. Shallow roots preferentially take up immobile nutrients, and deep roots preferentially take up mobile nutrients.

Keywords: Apoplast; nutrient uptake; osmosis; stomatal control; symplast; transpiration; turgor; water potential; xylem

3.1 OSMOSIS, WATER POTENTIAL, AND CELL EXPANSION

Between 70% and 95% of the fresh mass of a grapevine's organs consists of water (H_2O). Even the woody tissues of trunk and roots have a water content of approximately 60%. Most of this water serves as a solvent for ions and organic molecules in the interior of the vine's cells. Water can diffuse freely, albeit slowly, across the phospholipid bilayer of cell membranes, which are approximately 5–10 nm thick, and diffusion accelerates with rising temperature. More important, however, H_2O molecules are small enough (0.3 nm diameter) to be able to just barely (i.e., in single file), but extremely rapidly (up to 1 billion H_2O molecules per second), penetrate membrane pores formed by water channel proteins called aquaporins (Katsuhara et al., 2008; Maurel et al., 2008; Tyerman et al., 1999, 2009). In contrast to the diffusion pathway, the rate (but not the direction) of water transport across aquaporins can be fine-tuned by active regulation because these proteins are able to open or close their pores in a process called gating. Additional regulation is provided by varying the number of aquaporin proteins in the membranes. Aquaporins are relatively temperature insensitive and can speed up the passage of water across membranes more than 50-fold; in other words, they strongly raise the membranes' (osmotic) water permeability (also called hydraulic conductivity). At the same time, most of them are impermeable to mineral nutrient ions and even to protons. In fact, cells use protons, in addition to calcium ions (Ca^{2+}) and removal of phosphate groups (the so-called dephosphorylation) from the protein, to close aquaporins (Johansson et al., 2000; Luu and Maurel, 2005; Maurel et al., 2008; Tournaire-Roux et al., 2003). In addition to size exclusion (i.e., filtration), dissolved ions are also repelled by charged residues at the entrance to the pores; thus, cations cannot pass through a "gate" guarded by a positive charge, and anions cannot pass through a gate guarded by a negative charge. In other words, membranes are selectively permeable or semipermeable (Latin *semi* = half). Specialized gates consisting of transport proteins and ion channels are required for molecules other than H_2O to pass through membranes (see Section 3.3).

The concentration of solutes, that is dissolved ions (from dissociated salts) and small organic molecules, inside a cell is almost always higher than that of the exterior (i.e., cell walls and intercellular space). Because these solutes are repelled by the pore openings in the cell membrane, their momentum drags water molecules away from the membrane along with the solutes, which causes water to move passively through aquaporins and across the membrane into the cell (Kramer and Myers, 2013). This water movement is termed osmosis (Greek *osmos* = impulse or thrust). In other words, the presence of solutes in a cell exerts a "pull" on the water molecules surrounding the cell, and this tension is termed osmotic potential. The greater the concentration of solutes, the more water will move into the cell to restore (hydraulic) equilibrium. This leads to a buildup of pressure (force per unit area) due to the incompressibility of water and the rigidity of plant cell walls resisting an increase in volume. Osmotic pressure (π, expressed in MPa) of an aqueous solution is thus caused by dissolved solutes and increases linearly as the solute concentration (c, expressed in mol L^{-1}; i.e., the number of dissolved molecules per unit volume) increases, and also as the temperature (T) increases. This relationship is captured by the following equation, for which its discoverer, the Dutch chemist Jacobus van't Hoff (1887), received the first Nobel Prize in chemistry:

$$\pi = RTc$$

where R is the universal gas constant (=8.31 J K^{-1} mol^{-1} = 0.00831 MPa L K^{-1} mol^{-1}) and T is the thermodynamic (or absolute) temperature in Kelvin (K = °C + 273).

Sugars such as sucrose, organic acids such as malate, and inorganic ions such as potassium (K^+) and chloride (Cl^-) are the major osmotic solutes (osmolytes) of plant cells. Thus, solutes have an osmotic function in addition to their metabolic functions. For instance, cells use K^+ to neutralize the negative charge of inorganic and organic anions (i.e., organic acids) and anionic groups of macromolecules because the total charge of all anions and cations inside a cell must

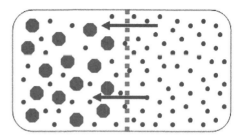

FIGURE 3.1

During diffusion, solutes (red) and water (blue) move in opposite directions, and the pressure remains constant (left). During osmosis, water moves toward the solutes across a semipermeable membrane (brown), increasing the pressure in the solute compartment (right).

balance (Clarkson and Hanson, 1980). When solutes attract water into a cell, the cell swells and causes the cell membrane to exert force on the cell wall. This, in turn, leads to a balancing wall pressure, which raises the energy of the water inside the cell until it equals that of water outside. At this point, the cell's internal hydrostatic pressure (more accurately, the pressure difference between inside and outside of the cell), termed turgor pressure (P) but often abbreviated as turgor (Latin *turgere* = to be swollen), is equal to the difference in osmotic pressure between the cell and its surroundings (i.e., $P = \pi_{inside} - \pi_{outside}$), and net water influx stops (Figure 3.1). Thus, osmotic pressure is defined as the hydrostatic pressure difference required to stop the net flow of water across a membrane separating solutions of differing concentration. In other words, osmosis can only work as a driving force for water flow (see Section 3.3) when two compartments are separated by a semipermeable membrane. Turgor pressure, in turn, is the osmotically maintained pressure inside living plant cells. The turgor pressure of growing cells typically reaches approximately 0.3–1 MPa—that is, 3–10 times atmospheric pressure ($P_a \approx 0.1$ MPa) (Cosgrove, 1997). Opening stomatal guard cells can generate over 4 MPa; nevertheless, the deformation of these cells and their cell walls is completely reversible (Franks and Farquhar, 2007; McQueen-Mason, 2005; Roelfsema and Hedrich, 2005).

The solute and pressure forces in plants and soils are conveniently described as free energy per unit volume (i.e., equivalent to force per unit area, or pressure) that is generated by the random movement and collision of molecules. The free energy of water or, more accurately, of an aqueous solution, per unit molar volume of liquid water (~ 18 mL mol^{-1}), is termed water potential (Ψ, expressed in J m^{-3} or as its pressure equivalent Pa = 10^{-5} bar = N m^{-2}) and is the sum of the component potentials arising from the effects of turgor pressure (i.e., turgor potential, $\Psi_P = P$) and solutes (i.e., solute potential or osmotic potential, $\Psi_\pi = -\pi$) in addition to interactions with matrices of solids (e.g., cell walls) and macromolecules (i.e., matrix potential, Ψ_M) and effects of gravity (i.e., gravitational potential, Ψ_G), as described by the following equation (Boyer, 1969):

$$\Psi = P - \pi + \Psi_M + \Psi_G = \Psi_P + \Psi_\pi + \Psi_M + \Psi_G$$

Although Ψ_M is very important in soils (due to the adhesion of H$_2$O molecules to soil particles) and cell walls, it is very close to zero inside the cells of well-watered plant tissues and is therefore negligible unless the tissue is severely dehydrated (e.g., loss of $\geq 50\%$ of tissue water; see Hsiao, 1973).

The gravitational potential increases with height at a rate of 0.01 MPa m^{-1}. Although this may be of little significance in the relatively short-statured cultivated grapevines that rarely exceed a height of approximately 2 m, the downward pull of Ψ_G is a factor in their taller wild relatives, as well as in the trees serving as their natural trellis system (see Figure 1.2). Yet even in their natural habitats, grapevines never approach the upper limit of about 130 m that gravity imposes on tree height (Koch et al., 2004). Pure water has the highest possible water potential, which by convention equals zero ($\Psi = 0$) at atmospheric pressure and 25°C. Therefore, the water potential of unpressurized aqueous solutions is always negative ($\Psi < 0$). In other words, Ψ is a measure of water concentration, and π is a measure of how much the dissolved molecules, termed solutes, decrease Ψ. The Ψ of a cell describes the potential tension (negative pressure or suction) that the cell solution exerts on surrounding pure water. In the absence of other forces, the movement of water during osmosis is always from a region of lower solute concentration (i.e., greater water potential) to one of higher solute concentration (i.e., lesser water potential). In other words, osmosis is driven by a water potential gradient.

Plant cells use osmotic solutes, especially sucrose and potassium, to lower their Ψ in order to attract water into the cells. Therefore, solutes have a major role in building P, which is necessary for cell expansion and hence plant and organ growth. In fact, cell growth is mainly a function of turgor and the mechanical properties of the cell walls. Cell expansion is made possible by cell wall loosening and accumulation of solutes inside the cell vacuole, which draws water into the vacuole to create turgor pressure. Thus, water movement into cells (i.e., across membranes) is determined by $\Psi_P + \Psi_\pi$ (i.e., it is driven by both pressure and osmotic components). In contrast, water movement outside of cells (i.e., through the apoplast; see Section 3.3) is driven solely by pressure differences (because water does not cross membranes in the apoplast). The P of growing cells is typically approximately 0.3–1 MPa, and this pressure irreversibly extends (i.e., stretches) and "pushes" the plastic cell walls outward (Cosgrove, 1997). Nonetheless, although P causes tension, it does not induce extension of the cell walls, which is controlled by the cross-linking of xylem-delivered calcium with pectic acids produced by the expanding cells (Boyer, 2009; Schopfer, 2006). Although P must be high enough to expand the cell, it must also be low enough to sustain water influx (i.e., maintain a gradient of Ψ). Continued water influx is necessary in expanding cells because the incompressibility of water means that P would otherwise drop to zero as soon as the cell walls become stretched. Indeed, a decrease in P by as little as 0.02 MPa can stop cell expansion but not reverse it (Taiz, 1984).

While P provides the driving force for cell expansion, the cell walls control the rate and direction of expansion (Winship et al., 2010). Cell expansion requires loosening of the cell walls of turgid cells before they can be extended (Boyer, 2009; Braidwood et al., 2014; Hamant and Traas, 2010; Schopfer, 2006). Wall loosening, and thus wall stress relaxation, is thought to be achieved by the "relaxing" activity of proteins known as expansins that are activated at low pH (Cosgrove, 1997, 2000, 2005; Li et al., 2003). According to the "acid growth theory," the cell wall is acidified by protons (H^+) pumped from the cell interior to the cell wall in exchange for potassium ions (K^+), which maintain the electrical charge balance inside the cell (Rayle and Cleland, 1992; Stiles and Van Volkenburgh, 2004; Van Volkenburgh, 1999). As a consequence, the cell wall pH can temporarily decrease from the normal 5.5–6 to as low as 4.5 (compared with a cytosol pH of ~7.5; Kurkdjian and Guern, 1989), which equates to a more than 10-fold increase in H^+ in the cell wall. The ATP-powered H^+ pumps in turn are activated by the growth hormone auxin, which induces phosphorylation of the H^+-ATPase protein (Takahashi et al., 2012) and by light (via phytochromes and other photoreceptors; see Section 5.2). Expansins may temporarily disrupt the chemical associations (hydrogen bonds) between the parallel and inelastic cellulose microfibrils and the cross-linking xyloglucans (hemicelluloses) in the cell walls, which makes the walls flexible enough to allow expansion and insertion of new cell wall material (Braidwood et al., 2014). In addition, auxin also seems to stimulate the production of reactive oxygen species (see Section 7.1), which are then converted by the multipurpose oxidizing enzyme peroxidase into hydroxyl radicals (OH), which in turn may attack the bonds between cell wall polysaccharides (Møller et al., 2007). Slipping of the parallel microfibrils and production and insertion of new wall material is necessary for the expanding cell to enclose its increasing volume; cell walls usually maintain their thickness during expansion (Boyer, 2009; Cosgrove, 2005). Because cells must increase in volume before they can divide, cell expansion, in turn, is also a prerequisite for cell division. As much as 90% of the entire volume gain during plant growth is due to the expansion of cell vacuoles and concomitant stretching and addition of cell wall material. Hence, most of the increase in volume during cell expansion is due to water uptake by the vacuole, whereas the amount of cytoplasm increases only insignificantly.

The term water potential is convenient because it can be applied not only to (cell) solutions but also to soil solution and air—that is, to any medium with variable water content. Water potential indicates the availability of water in any aqueous system. The water potential of air can be estimated from the air's relative humidity (RH; expressed in percentage). The RH is the amount of water vapor in air at a particular temperature in relation to the total amount the air could hold at that temperature (i.e., the saturation vapor density or saturation vapor pressure). The water-holding capacity of air (and thus its saturation vapor pressure) increases exponentially with increasing air temperature. In the 10–35°C range, an increase in air temperature of 12°C doubles the water vapor concentration of saturated air; that is, air at 22°C can hold twice as much water vapor as air at 10°C, and air at 34°C can hold four times as much. In other words, if saturated air (100% RH) is heated by 12°C, its RH drops to 50% (i.e., the air develops a saturation deficit). The relationship between Ψ_{air} (in MPa) and RH (in %) can be approximated by the following equation (Nobel, 2009; Tyree and Zimmermann, 2002):

$$\Psi_{air} \approx 0.46\ T\ \ln(RH/100)$$

where T is the thermodynamic (or absolute) temperature in K and ln is the natural logarithm.

As this equation shows, Ψ of saturated air is always zero ($\ln 1 = 0$), regardless of temperature, whereas Ψ of drying air drops dramatically (Table 3.1). Even relatively humid air exerts an

Table 3.1 Typical Values of Relative Humidity (RH) and Ψ for Leaves and Atmosphere at 25°C

	RH (%)	Ψ (MPa)
Air spaces in leaf	99	−1.4
Boundary layer	95	−7.0
Atmosphere	100	0
	90	−14.5
	50	−95.1
	10	−315.9

enormous tension on the water inside a plant or in the soil. This tension is responsible for transpiration (see Section 3.2). In other words, transpiration is also driven by a water potential gradient. Incidentally, the relationship between temperature and Ψ_{air} is also responsible for the so-called rain shadow effect of the Cascades mountain range in eastern Washington and Oregon and the Andes range in Mendoza (Argentina). As moist air from the Pacific Ocean rises up the western slopes of the mountains, it cools and drops excessive moisture as rainfall or snow so that the saturation deficit (or evaporative demand) of the now drier air increases further as the air warms down the eastern slopes of the mountains.

3.2 TRANSPIRATION AND STOMATAL ACTION

The evaporation of water from plants is called transpiration (Latin *trans* = on the other side, *spirare* = to breathe). Evaporation constitutes a physical phase change of water from the liquid to the gaseous form. The term evapotranspiration that is often found in the literature is more inclusive, as it represents the total amount of water evaporated from a land surface, which includes the water transpired by plants (in a vineyard these are grapevines, cover crops, and weeds) and the water evaporated from the soil and surface water bodies. In a hot climate like central California, such evapotranspiration from well-watered raisin or table grape vineyards can exceed 6 mm d^{-1} (1 mm = 1 L m^{-2}) in the middle of summer and more than 800 mm over the course of a growing season (Williams et al., 2003). Transpiration from a leaf depends on two major factors: the difference in absolute water vapor concentration, or water potential, between the leaf air spaces (Ψ_{leaf}) and the external air (Ψ_{air}) and the resistance (r) of this pathway to diffusion. As water evaporates from the mesophyll cell walls and is discharged as water vapor into the atmosphere (see Figure 1.15), it meets a series of resistances that work in the opposite direction of its movement. A resistance slows the H_2O molecules down and thus limits their rate of diffusion. The major ones are the resistance at the stomatal pore and the resistance due to the thin film of still and relatively moist air at the surface of the leaf (the so-called boundary layer), but the stomatal resistance (r_s) is usually much greater than the boundary-layer resistance (r_b). The boundary layer (also known as the unstirred layer) acts like an additional (cuticular) membrane in series, but wind increases turbulence in the boundary layer, making it thinner and decreasing r_b, which increases transpiration dramatically even as r_s increases (Freeman et al., 1982; Schultz and Stoll, 2010). On the other hand, r_b can be very high in dense canopies so that transpiration of such canopies may be determined mainly by r_b and is not easily influenced by atmospheric conditions. Transpiration rate per unit leaf area (E, expressed in mmol H_2O m^{-2} s^{-1}) increases as the leaf-to-air vapor concentration difference (i.e., the driving force) increases, and it decreases as the total resistance increases. This relationship can be described by the following equation, which is analogous to Ohm's law ($I = Vr^{-1}$, where I is the current and V is the electrical potential difference or voltage), used in electricity and thermodynamics:

$$E = \Delta\Psi(r_s + r_b)^{-1}$$

where $\Delta\Psi = \Psi_{leaf} - \Psi_{air}$.

Water vapor concentration is proportional to water vapor pressure, and the difference between the amount of moisture (water vapor pressure) in saturated air and the actual amount of moisture in the air at a given temperature is called the vapor pressure deficit (VPD). As discussed in Section 3.1, an increase in air temperature strongly increases the air's water-holding capacity and thus results in a greater VPD. The steeper concentration gradient (higher $\Delta\Psi$) stimulates water evaporation from the leaf mesophyll. Therefore, temperature is an important determinant of the transpiration rate, and transpiration increases linearly with increasing temperature (see Figure 5.8).

In a typical grapevine leaf, only 5–10% of the total transpiration occurs across the cuticle, although the cuticular portion can reach up to 30% in old leaves that have low overall transpiration rates compared with younger leaves (Boyer et al., 1997). The bulk of the water vapor escapes through the stomatal pores, even though they cover less than 5% of the leaf surface. Stomata are thus very important for regulating diffusional water loss from a leaf. They function like a pressure regulator, limiting pressure changes (i.e., Ψ_{leaf}) by controlling flow rate (transpiration), and respond very sensitively to environmental variables (Sperry et al., 2002). The most conspicuous of these responses is the light-induced opening of stomata at sunrise and their closing at sunset in response to photosynthetically driven changes in the substomatal CO_2 concentration (Raschke, 1975; Roelfsema and Hedrich, 2005). This minimizes water loss at night, when there is insufficient light energy to fuel photosynthesis (see Section 4.1) and hence no need for CO_2 diffusion into the leaf (see Section 4.2). At night, when the stomatal conductance ($g_s = r_s^{-1}$) less than 50 mmol H_2O m^{-2} s^{-1}, the rate of water evaporation from the leaves is roughly 10% of that during the day (Rogiers et al., 2009; Rogiers and Clarke, 2013). During the course of the day, stomatal opening normally follows the daily change in light intensity, which peaks around midday (Düring and Loveys, 1982; Rogiers et al., 2009). Similarly, the stomatal aperture is wider in bright light than in dim light such as prevails under cloudy conditions or in the interior of dense canopies (Shimazaki et al., 2007). Conversely, stomata close when a leaf runs out of water. Therefore, low Ψ_{leaf} (indicating water stress; see Section 7.2) causes stomata to close partially to protect the xylem conduits from cavitation (see Section 3.3). This effect can override the influence of high light intensity on stomatal opening so that even relatively mild soil water deficit in a vineyard often leads to a midday depression in stomatal conductance (Düring and Loveys, 1982). This temporary decline in g_s is usually not observed in well-watered vines, except for vines growing in hot climates (Downton et al., 1987; Williams et al., 1994). The aperture of the stomatal pores also decreases as air humidity declines (i.e., as the VPD increases), although this could be a response to transpiration rather than to humidity *per se*. As would be expected from the preceding "transpiration equation," this stomatal closure can compensate for variations in humidity, holding transpiration and Ψ_{leaf} relatively constant. Nevertheless, a decrease in relative humidity from 85% to 45% (at constant temperature) can decrease the transpiration rate by more than 80% via partial stomatal closure. A rise in temperature initially increases stomatal aperture due to its stimulating effect on photosynthesis, but very high temperature leads to stomatal closure, possibly due to its stimulating effect on respiration (see Section 5.2). An increase in leaf nitrogen also leads to a wider stomatal aperture because nitrogen also stimulates photosynthesis (see Section 5.3), but this effect levels off at high nitrogen content.

Stomata are hydraulically driven gas valves because, as discussed in Section 1.2, they open when the guard cells' vacuoles take up water and swell, causing their turgor pressure to rise (Franks and Farquhar, 2007). This clever mechanism is possible because the guard cells lack plasmodesmata connecting them to neighboring epidermis and mesophyll cells; that is, they are symplastically isolated (Roberts and Oparka, 2003; Turnbull and Lopez-Cobollo, 2013; Wille and Lucas, 1984). In addition, the guard cells of closed stomata contain many small vacuoles that merge into a few large vacuoles when the stomata open (Martinoia et al., 2012). This allows the stomata to open very rapidly, because the fragmentation in the closed state minimizes the vacuoles' volume while preserving their membrane surface. The osmotic uptake of water is driven by a $\Delta\Psi$ (see Section 3.1) caused by movement of potassium ions (K^+) from surrounding cells into the guard cells across K^+ channels in exchange for hydronium ions (H_3O^+) or protons (H^+), which are driven out by a light-activated pump called H^+-ATPase embedded in the plasma membrane (Roelfsema and Hedrich, 2005; Shimazaki et al., 2007). The reverse exchange (K^+ out/H^+ in) causes water to move out of the guard cells so that the decrease in turgor causes the stomata to close. The H^+ may come from the organic acid malate, which is produced from starch stored in the guard cell chloroplasts and, courtesy of its double-negative charge, also serves as the counterion to K^+. The starch, in turn, is produced by guard cell photosynthesis or from sucrose imported across the cell walls from the underlying mesophyll cells (Lawson, 2008). In addition, sucrose derived from starch breakdown also supplements K^+ as an osmoticum, especially in the afternoon (Roelfsema and Hedrich, 2005; Shimazaki et al., 2007; Talbott and Zeiger, 1996). Thus, guard cells actively, and rapidly, regulate stomatal opening and closing via a dynamic balance of starch/malate and H^+/K^+. The hydraulic effect of Ψ_{leaf} on stomatal opening or closing can be amplified by abscisic acid (ABA), which is produced from carotenoids in dehydrated cells, vascular tissues, and guard cells (Nambara and Marion-Poll, 2005; Wilkinson and Davies, 2002). In addition to the ABA produced in drying leaves as their turgor declines, the amount of ABA arriving from the roots can increase considerably in response to reduced soil moisture (Bray, 1997; Davies and Zhang, 1991; Davies et al., 2002). Thus, ABA acts as a messenger from the roots, indicating water stress when the soil dries out (see Section 7.2). As Ψ_{soil} declines, the roots produce increasing amounts of ABA, which is transported in the xylem sap to the leaves' guard cells (Li et al., 2011; Loveys, 1984; Speirs et al., 2013). ABA triggers stomatal closure by inducing an increase in reactive oxygen species (especially H_2O_2) that, in turn, lead to a brief rise in calcium (Ca^{2+}) in the cytosol, which blocks influx into and enhances efflux from the guard cells of K^+ (Allen et al., 2001; McAinsh and Pittman, 2009). In other words, ABA modifies the permeability of the guard cell membrane by way of a Ca^{2+} signal. The gaseous signaling molecule nitric oxide (NO) also rapidly accumulates in guard cells and, more slowly, in leaf mesophyll cells when the soil dries down (Patakas et al., 2010). However, it remains to be demonstrated whether NO is necessary to induce stomatal closure in an ABA–H_2O_2–NO–Ca^{2+} signaling cascade or whether it serves to detoxify H_2O_2.

The reaction of stomata to ABA is rapid (within ~ 10 min) and independent of Ψ_{leaf} (although ABA is also produced in the leaves in response to decreasing Ψ_{leaf}) and VPD, and it does not involve starch degradation (Roelfsema and Hedrich, 2005). However, stomatal closure is associated with an increase in the pH of the xylem sap (from the "normal" pH 5–6 to 6.5–7), which enhances ABA delivery to the guard cells by reducing uptake of ABA by intervening tissues (Bacon et al., 1998; Hartung and Slovik, 1991; Wilkinson and Davies, 2002). In other words, at higher pH, less ABA becomes "trapped" inside the usually more alkaline parenchyma and phloem cells lining the xylem vessels. It appears that a decrease in the flow rate of xylem sap leads to an increase in the sap's pH, perhaps because there is more time for the adjacent cells to remove protons from the sap (Jia and Davies, 2007). Furthermore, ABA seems to induce closure of aquaporins in the membranes of the bundle-sheath cells lining the xylem in the leaves, thus reducing water flow to the mesophyll cells, a hydraulic effect that enhances the "standard" ABA effect on stomatal closure (Pantin et al., 2013; Shatil-Cohen et al., 2011).

Even when only a portion of the roots experience dry soil, the ABA produced by that portion is sufficient to trigger strong stomatal closure despite the remainder of the roots taking up enough water to maintain high Ψ_{leaf} (Comstock, 2002; Li et al., 2011; Lovisolo et al., 2002a; Stoll et al., 2000). When rainfall or irrigation replenishes soil moisture, the roots stop producing ABA, whose content in the xylem therefore decreases rapidly. Nonetheless, the transpiration-driven rise in water flow up the xylem after such water supply may "flush" residual ABA to the leaves, where it may keep the stomata partially closed for 1 or 2 days, perhaps to enable any embolisms (gas bubbles) that may have formed in the xylem to be repaired (Lovisolo et al., 2008; see Section 3.3). Because the ABA that is present in the leaves is quickly degraded (mainly to phaseic acid), the stomata can then reopen (Nambara and Marion-Poll, 2005; Speirs et al., 2013).

The turnover of water in a rapidly transpiring leaf occurs within 10–20 min; that is, the leaf loses the equivalent of its entire water content every 10–20 min (Boyer, 1985; Canny, 1993). Although transpiration is sometimes described as a necessary evil (due to a plant's dependence on gas exchange), it has some useful side effects. For instance, the negative pressure (tension) in the xylem created by transpiration aids in extracting water from the soil (see Section 3.3) that can be used for plant growth. This, in turn, attracts nutrients, which can then be taken up by the roots. Although the increased flow of water during transpiration leads to a corresponding increase in the rate at which dissolved nutrients move upward in the xylem conduits, transpiration is not necessary for this nutrient movement (Tanner and Beevers, 2001). In fact, plants can absorb and transport nutrients just as well at night when transpiration is minimal as they do during the day. More important, evaporation of water is a powerful cooling process, which is why this principle is also used in air-conditioning. This heat loss prevents leaves exposed to sunlight from overheating. The temperature inside actively transpiring leaves is ordinarily no more than 2°C or 3°C higher than the ambient temperature. Nevertheless, leaf temperature can rapidly fluctuate by more than 10°C in direct sunlight (Sharkey et al., 2008). Water evaporation cannot completely prevent "heat spikes." Grapevines have several cooling tactics to lose (dissipate) excessive heat gained from absorbed solar radiation. The three main pathways are as follows (Mullins et al., 1992; Taiz and Zeiger, 2006):

Radiation: Direct transfer (reradiation) of heat as long-wavelength (infrared) radiation to surrounding objects or the sky. The percentage of incoming solar radiation reflected by a surface is termed albedo. Leaves absorb ultraviolet, visible, and infrared radiation and radiate infrared energy, which is a cooling process. This happens mostly during the night, when the temperature of the surrounding air is cooler than the leaf temperature. Nevertheless, even in bright sunlight, leaves can radiate 50–80% of the energy they absorb.

Convection: Heating or cooling of ambient air via sensible heat—that is, heat that can be felt and measured with a thermometer. The energy of molecules on the leaf surface is exchanged with that of air molecules with which they are in direct contact. The warmed air becomes lighter and rises, which leads to cooling. Convection is driven by the temperature difference between the air and the leaf, and the boundary layer is the main resistance to this process. Wind decreases the thickness of the boundary layer, which increases convective heat transfer and thus convective cooling.

Transpiration: Release of latent heat during the vaporization of water into the air, also referred to as transpirational cooling. Water evaporation consumes energy called the latent heat of vaporization (2.45 kJ g^{-1} at 20°C) because water vapor contains more energy than liquid water.

In other words, evaporation removes heat that cannot be felt or measured with a thermometer. Transpiration is driven by the water potential difference between the inside of a leaf and the outside. This water potential gradient depends on temperature (which in turn depends on solar radiation) and relative humidity.

3.3 WATER AND NUTRIENT UPTAKE AND TRANSPORT

Although the main force driving water movement through plants is water pressure, water uptake by the roots and transport to the leaves is largely brought on by passive forces because the hydraulic system of plants has no moving parts (Tyree and Zimmermann, 2002). The direction of movement is always from high to low pressure or, if the movement is across membranes, from a region with low solute concentration to one with high solute concentration. As discussed in Section 2.2, water absorption and movement in early spring is driven by positive root pressure induced by remobilization of stored nutrients and starch and release of osmotically active ions and organic molecules into the xylem sap. Under these conditions, water moves from the soil into the root by an osmotically generated gradient (difference per unit distance) of water potential and is then pushed up the vine through the xylem. Although an osmotic gradient ($\Delta\pi$)—that is, a gradient in solute concentration—has virtually no effect on the flow rate of water inside the xylem because vessels have no membranes, it drives water inflow into the xylem from the soil across the root cell membranes (Wegner and Zimmermann, 2009). This $\Delta\pi$ usually collapses after leaf expansion (although it can remain important at night or during rainfall or overhead irrigation), and water loss to the atmosphere by transpiration from the unfolding leaves begins to hydraulically produce a negative pressure (tension) in the xylem (P_x), which maintains the upward flow of water in the plant (Dixon and Joly, 1895). At night or under light conditions low enough to lead to stomatal closure, such as during overcast periods, and with plentiful soil water supply, P_x can remain positive, albeit below atmospheric pressure ($P_a \approx 0.1$ MPa at sea level), but drops below zero (i.e., $P_x < 0$ MPa) as soon as the stomata open even slightly to let water vapor escape and declines further as light intensity and transpiration rise (Wegner and Zimmermann, 2009).

Approximately 95–98% of all water absorbed by the roots is lost to the atmosphere in the process of transpiration; very little water is required to build the plant body through cell expansion. Thus, transpiration is the main driving force for water uptake and movement in the xylem up the vine to the leaves. This is why water flow through the trunk of grapevines oscillates with a minimum at night and a maximum at approximately midday and why vines with a large canopy sustain higher flow rates than do small vines (Pearsall et al., 2014; Tarara and Ferguson, 2006). The transpirational pull is possible because water forms a continuous system from evaporating surfaces in the leaves, shoots, tendrils, and flower and fruit clusters to absorbing surfaces in the roots and out into the soil. This system is termed the soil–plant–air continuum. The chemical and physical properties of water (i.e., high surface tension, cohesion or the strong tendency of H_2O molecules to "stick" together by forming hydrogen bonds between the negatively charged oxygen and the two positively charged hydrogen atoms, and strong adhesion of H_2O molecules to the surfaces of xylem conduits) keep it in this continuum so that water exiting the leaves by transpiration is continually replaced by water being pulled up the xylem from the roots (Dixon and Joly, 1895; Steudle, 2001; Tyree and Zimmermann, 2002).

The transpiration—cohesion—tension theory of xylem sap flow states that the evaporation of water from cell walls inside the leaf creates tension, and adhesion of water to the walls of the xylem helps counteract gravity (i.e., the gain in gravitational potential energy with increasing height). Due to cohesion, the tension extends down the xylem, into the roots, and out into the soil. Therefore, plants directly utilize solar energy (which causes water to evaporate from the leaves) to drive water uptake and distribution. Again, the water moves along a gradient of water potential (from high to low). This gradient is intermediate within the root xylem (0.3—2.5 MPa m^{-1} with lower values in well-watered and higher values in water-stressed vines), gentle within a shoot (0.1—0.5 MPa m^{-1}), and quite steep within the petiole (1—5 MPa m^{-1}) of a transpiring leaf (Lovisolo et al., 2008).

For xylem sap to sustain the tension required in a grapevine to pull water from the soil, the hydraulic system has to be airtight. The water's surface tension acts like a seal, keeping water inside the xylem and air out. However, the xylem is a vulnerable pipe, and water is saturated with air at atmospheric pressure (Steudle, 2001; Tyree and Zimmermann, 2002). If the water column inside the xylem is put under too much tension (e.g., excessive transpiration due to drying wind or heat combined with dry soil), if the tension is suddenly relieved (e.g., by shoot, leaf, or cluster removal), or if it receives some other physical shock (e.g., a bump by machinery), it can snap back like a piece of elastic in a rigid tube. This phenomenon is termed cavitation, in which the water column breaks and the conduit lumen suddenly fills with H_2O vapor or air bubbles (Figure 3.2). The air can be sucked to adjacent vessels through the holes of the interconnecting perforation plates and through pits (Brodersen et al., 2013). Such gas blockages are called embolisms and render the vessels nonfunctional; in severe cases, this can lead to leaf wilting and even canopy collapse (Schultz and Matthews, 1988a; Tyree and Ewers, 1991; Tyree and Zimmermann, 2002). For the gases to dissolve and conduits to refill with liquid water, P_x must rise to within approximately 0.1 MPa of atmospheric pressure or above (i.e., $P_x \geq 0$ MPa) (De Boer and Volkov, 2003; Sperry et al., 2002, 2003). More accurately, gas emboli dissolve at $P_x > -2tr^{-1}$ (where t is the surface tension of water and r is the radius of curvature of the air—water interface). If cavitation occurs in the trunk or shoots, repair normally requires positive root pressure or starch remobilization in the surrounding xylem parenchyma cells and active release of sugar and potassium ions to lower the osmotic potential (Ψ_π) in the droplets of xylem sap that remain in the embolized vessels (Secchi and Zwieniecki, 2012). These metabolic processes require an input of energy to proceed. In addition, the parenchyma cells may also release water into the vessels by opening the aquaporins in their cell membranes (Lovisolo and Schubert, 2006; Secchi and Zwieniecki, 2010). The small parenchyma cells in turn are supplied with water by the phloem.

Vitis species are thought to be able to generate root pressure in excess of 0.1 MPa, which is sufficient to push water to a height of over 10 m above ground level (Tibbetts and Ewers, 2000). Transport of solutes and osmotic release of water from xylem parenchyma cells surrounding embolized vessels may occur into any small water droplets that remain along the vessel walls. These droplets thus grow and eventually coalesce, forcing the air back into solution or pushing it out into the pores of the vessel walls (Brodersen et al., 2010). Whereas moderately stressed vines may be able to restore water flow quickly after water supply and while the leaves are transpiring, complete embolism repair in more severely stressed plants may require both a rise in Ψ_{leaf} and a cessation of xylem flow so that it typically occurs only at night, provided soil moisture is high enough, or during rainfall (Holbrook et al., 2001; Secchi and Zwieniecki, 2010). The trunk and shoots, at least after they have produced some lignified secondary xylem, are much less prone to cavitation than the

FIGURE 3.2

Cavitation of xylem vessel by water vapor bubble (left; reproduced from Taiz and Zeiger, 2006), bleeding of xylem sap during budbreak (top right), and guttation at night during the growing season (bottom right; photos by M. Keller).

petioles and minor veins in the leaves (Choat et al., 2005; Lovisolo et al., 2008; Nardini et al., 2001; Zufferey et al., 2011). In the shoots, the primary xylem near the pits is the most vulnerable to cavitation, and cavitation spreads radially outward as drought stress intensifies (Brodersen et al., 2013).

Although the leaves' minor veins can suffer from embolisms even during an average sunny day, the high concentration of inorganic ions and organic molecules in the leaves can generate the positive pressure needed to refill the veins relatively easily (De Boer and Volkov, 2003; Trifilò et al., 2003). Moreover, the vessels and tracheids in leaves and petioles are narrower and much shorter than those in the rest of the plant, and the pits in the adjoining end walls of these tracheary elements normally do not allow air to pass into the next vessel—that is, pits function as hydraulic safety valves (Choat et al., 2008). This prevents spreading of embolisms in the xylem network following leaf damage and sacrifice, a protective mechanism dubbed the "lizard's tail" strategy (Tyree and Zimmermann, 2002; Zufferey et al., 2011). Finally, the high number of parallel tracheary elements makes it unlikely that the entire transport pathway is interrupted at any one time. Indeed, the parenchyma cells in the rays act like a physical barrier that tends to contain cavitated areas within sectors that look like spokes of a wheel (Brodersen et al., 2013). The fact that roots, trunks, and shoots are not simply hollow pipes ensures that some residual water flow persists under all but the most severe drought conditions (see Section 7.2), and this flow can rapidly rehydrate the leaves upon restoration of soil moisture (Lovisolo et al., 2008). Consequently, Ψ_{leaf} recovers within hours of a rainfall or irrigation event, regardless of the severity of drought stress that existed prior to rewatering.

The root xylem of grapevines is much more vulnerable to cavitation than their shoot xylem (Lovisolo and Schubert, 2006; Lovisolo et al., 2008). This can be a problem with shallow root systems because roots in the surface soil are the first to experience dry soil, and their failure can trigger leaf wilting and canopy collapse. If deeper roots are present, cavitation in the surface roots will simply shift water uptake down to wetter soil layers so that shoot damage can be avoided (Sperry et al., 2002). Moreover, the expansive mycelium network created by symbiotic mycorrhizal fungi also facilitates water uptake into the roots (Stahl, 1900) and helps avoid cavitation by osmotic water "lifting," which is possible due to the high carbohydrate concentration in the fungi supplied from the plants. Plants, it seems, can also reciprocate the "favor" at night by transferring water from the roots to the symbionts (Querejeta et al., 2003). In a "compensation game," fungal colonization of the roots tends to increase as soil moisture, and hence root growth, decreases and vice versa (Schreiner et al., 2007). The fungi can even transfer water, as well as sugar and nutrients, between neighboring plants (including grapevine−weed transfer) either directly (through the mycelium connecting the roots of two plants) or indirectly (through uptake of water that was released by the roots of another plant) (Egerton-Warburton et al., 2007).

Grapevines that grow under conditions of moderate water deficit develop narrower xylem vessels than vines growing with abundant water supply (Lovisolo and Schubert, 1998; Mapfumo et al., 1994b). Narrower vessels might be a consequence of curtailed supply of assimilates to the cambium resulting from lower photosynthesis due to higher r_s. Although the smaller vessel diameter increases r_h, which may lead to reduced water loss and increased cavitation resistance, this effect cannot be very large because the few large vessels typically present in grapevines dominate the overall r_s (Tibbetts and Ewers, 2000). The vulnerability to cavitation also varies among species and cultivars (Schultz, 2003); less vulnerable cultivars (e.g., Syrah) probably have denser and stronger wood and less permeable pits (i.e., more resistant xylem) than more vulnerable cultivars (e.g., Grenache, which nevertheless thrives in hot, arid, and windy conditions).

The flow of water from the soil toward the roots is maintained by suction at the root surface, which is caused by root pressure (i.e., osmotically) or transpiration (i.e., hydrostatically). As transpiration removes water, the soil adjacent to the roots (rhizosphere) becomes drier than the more distant (bulk) soil. Thus, water flow to the roots, and from the roots to the leaves, is caused by a water potential gradient. In analogy to Ohm's law (and to the transpiration equation discussed previously), the flow rate (F, volume per unit time) of water from the soil to the leaves can be described by the following equation:

$$F = \Delta\Psi r_h^{-1} = \Delta\Psi \, l_h \text{ (because } l_h = r_h^{-1})$$

where $\Delta\Psi = \Psi_{soil} - \Psi_{leaf}$, r_h is the hydraulic resistance, and l_h is the hydraulic conductance.

Inside the xylem conduits (vessels and tracheids), the osmotic component of Ψ is insignificant because there are no membranes in the pathway; hence, water flow follows a pressure difference over distance (i.e., $\Delta\Psi \approx \Delta P_x$). Sometimes a negative sign is added on the right side of the equation to indicate that water flows in the opposite direction to that of $\Delta\Psi$—that is, from high to low Ψ (Tyree and Zimmermann, 2002). Although this equation is surely an oversimplification, it is quite convenient to explain most phenomena involved in water flow through plants. The hydraulic conductance is a measure of transport capacity or water permeability. At the whole-plant level, it determines Ψ_{leaf} at any specific transpiration rate.

The term "conductance" should not be confused with "conductivity" (L_h or K), which refers to the water flow normalized to ΔP over a specific length (x) of the flow path (thus, $l_h = L_h x^{-1}$). In addition, "specific conductivity" (L_s) applies to the cross-sectional area (A) of a unit flow path ($L_s = L_h A^{-1}$) and is roughly proportional to the number of xylem conduits passing through that cross section and the fourth power of their radius (r^4; the rate of water flow per unit cross-sectional area is termed flux, $J = FA^{-1}$) (Tyree and Ewers, 1991; Tyree and Zimmermann, 2002). In other words, a doubling of vessel diameter results in a 16-fold increase in L_s. Therefore, L_h and L_s are independent of the total length of the pathway, whereas l_h decreases (and r_h increases) with increasing path length (e.g., increasing plant height). The term r_h reflects the resistance to water flow due to friction between water and conduit walls and between the H_2O molecules. It is a measure of how much a component of the flow path hinders the free flow of water and thus slows water flow. There are both short-distance radial and long-distance axial or longitudinal components of r_h; the axial resistance to water flow is 2–3 orders of magnitude lower than the radial resistance. In other words, the axial r_h behind the elongation zone of grapevine roots accounts for only 0.1–1% of the total resistance (Gambetta et al., 2013).

The radial r_h is the resistance water encounters when it flows across the root into the xylem and out of the xylem into leaf and fruit cells, much of it imposed by cell membranes whose permeability is regulated by aquaporins (Maurel et al., 2008; Perrone et al., 2012; Tyerman et al., 2009; Vandeleur et al., 2009). Most of the axial r_h is imposed by the vine's "plumbing layout," termed hydraulic architecture, which is mainly determined by the number, shape, size, and arrangement of xylem conduits and their interconnections (e.g., pits), as well as the total length of the flow pathway and the number and shape of bends in the pathway (i.e., plant size and shape) (Tyree and Zimmermann, 2002). Due to their small size and fine pore meshwork, however, the pits connecting individual xylem conduits usually account for more than 50% of the xylem system's total r_h (Choat et al., 2008; Nardini et al., 2011). The importance of the pits for overall r_h also means that longer vessels decrease r_h because the flowing water has to cross fewer high-resistance vessel end walls (Lens et al., 2011). The increase in r_h with increasing plant height partly accounts for the tendency of photosynthesis to decrease as plants grow taller, because their leaves must close their stomata earlier (i.e., at less severe water deficit) to maintain Ψ_{leaf} above a species- and cultivar-specific minimum (Hubbard et al., 2001; Ryan and Yoder, 1997). In other words, photosynthesis in tall vines, such as those climbing in trees, may be at least partly hydraulically limited. Note, however, that growth is more sensitive to water stress than is photosynthesis, so it may be more correct to say that growth, rather than photosynthesis, is hydraulically challenged (Ryan et al., 2006; see also Section 5.2). Moreover, the larger vines' more extensive root system and their usually greater trunk diameter tend to counter this hydraulic limitation trend.

As in electricity, resistances encountered in series (e.g., in the roots, trunk, cordon, shoots, and leaves) are additive (i.e., $r_h = r_{h1} + r_{h2} + \cdots + r_{hn}$), whereas for those acting in parallel (e.g., several shoots on the same cordon), it is the individual conductances that are additive (i.e., $r_h^{-1} = r_{h1}^{-1} + r_{h2}^{-1} + \cdots + r_{hn}^{-1}$). Nevertheless, whole-plant r_h is not constant for a given vine but varies for several reasons. In the long term, r_h usually increases with increasing plant size and age (as does r_s), but it decreases with increasing shoot number on otherwise similar vines. In the short term, a rise in temperature decreases r_h (by ~2–2.5%/°C due to the decrease in water viscosity), which enables more rapid water delivery to the leaves for transpirational cooling (see Section 3.2) during heat episodes. Moreover, transport of nutrient ions, especially of cations such as K^+, may decrease r_h by increasing pit permeability. Cations may induce shrinking of the gel-like polymer matrix ("hydrogel") in the so-called pit membranes, which are specialized modifications of the primary cell wall, rather than proper cell membranes, with pores of 5–20 nm in diameter (Nardini et al., 2011; Pérez-Donoso et al., 2010; van Ieperen et al., 2000; Zwieniecki et al., 2001). The cations may moreover diminish the extent of the diffuse electrical double layer that lines the pit membrane pores because of the attraction of cations to the negative charges of the wall polymers and slows water flow akin to the boundary layer on leaf surfaces (van Doorn et al., 2011). Nitrate (NO_3^-), but not other anions such as phosphate ($H_2PO_4^-$) or sulfate (SO_4^{2-}), also decreases r_h, perhaps by altering the status of aquaporins either directly or via a rise in pH brought about by NO_3^- assimilation (Gloser et al., 2007; Gorska et al., 2008; see also Section 5.3). In addition, higher NO_3^- content is also associated with higher K^+ content in the xylem sap (Keller et al., 2001b). This means that increasing the concentration of some nutrients in the xylem sap increases the sap flow rate (and hence water uptake by the roots) and, conversely, that nutrient deficiency strongly reduces sap flow even under well-watered conditions. It also means that vines may be able to actively, rapidly, and reversibly modify r_h by altering the concentration of ions in the xylem sap, for instance, by remobilizing/resorbing nutrients in the surrounding parenchyma cells. It is conceivable that the xylem is able to enhance the delivery of nutrient elements to the leaves by taking advantage of these nutrients' influence on r_h, thereby favoring more rapid water flow.

Grapevines sustain very rapid water flow up their trunks that matches the canopy transpiration rate plus the water "bound" by the growing tissues. The flow rate in field-grown vines has been estimated to fluctuate daily from low values (<0.1 L h^{-1}) at night to high values (1–10 L h^{-1}) during days with clear sky and high VPD, but it decreases to 10–20% of clear-sky values during rainy or overcast days with low VPD (King et al., 2014; Pearsall et al., 2014; Tarara and Ferguson, 2006; Williams et al., 2003). Aquaporin activity in the root cell membranes, and hence radial r_h, approximately matches the diurnal oscillations in flow rate to enable appropriate water uptake by and flow across the roots (Luu and Maurel, 2005; Maurel et al., 2008; Tyerman et al., 2009; Vandeleur et al., 2009). The rate of sap flow also varies with canopy size and water availability: For instance, maximum values at midday may be approximately 1 or 2 L h^{-1} in deficit-irrigated Cabernet Sauvignon wine grapes (declining to <0.5 L h^{-1} as the soil dries down), but can reach 3–10 L h^{-1} in large, well-watered Thompson Seedless table grapes or Concord juice grapes

(Dragoni et al., 2006; King et al., 2014; Pearsall et al., 2014; Tarara and Ferguson, 2006; Williams et al., 2003). A pressure gradient greater than 10 kPa m^{-1} is required to maintain such flow rates in the trunk. Rapid transpiration favors hydrostatic (and hence apoplastic) water flow through the root and thus reduces r_h (Steudle, 2000, 2001; Steudle and Peterson, 1998). Conversely, when there is little or no transpiration, such as at night or with dry soil, r_h increases substantially because water flow is driven by osmotic gradients (i.e., across membranes). Of course, cavitation in the xylem also greatly increases r_h—to infinity in the affected vessel. Cavitation may be a primary reason for the increase in r_h in leaves under drought stress (Johnson et al., 2012). If cavitation is to be avoided, the flow of water from the soil to the leaves must balance the water lost through the stomata. By combining the transpiration equation with the flow equation, this can be written as follows:

$$(\Psi_{leaf} - \Psi_{air})(r_s + r_b)^{-1} = (\Psi_{soil} - \Psi_{leaf})r_h^{-1}$$

As long as transpiration rate and soil-to-leaf flow rate remain constant, Ψ_{leaf} will remain constant. For a given microclimate (Ψ_{air}), which determines the VPD, and a given soil water status (Ψ_{soil}), r_b and r_s determine the transpiration rate, whereas whole-vine r_h determines Ψ_{leaf} at that transpiration rate (Tyree and Zimmermann, 2002). In other words, the difference between Ψ_{soil} and Ψ_{leaf} is determined by the transpiration rate and r_h, which in turn is correlated with r_s. Thus, r_h defines how wide the stomatal pores can be open without desiccating the leaves (Brodribb and Holbrook, 2003; Jones, 1998; Zufferey et al., 2011). A larger r_h will result in a greater decrease in Ψ_{leaf} when transpiration increases. Water will flow from the soil to the vine as long as Ψ_{xylem} is lower than Ψ_{soil}, and it will flow from the xylem to the leaf as long as Ψ_{leaf} is lower than Ψ_{xylem}. Even in well-watered vines, Ψ_{leaf} can fluctuate widely during the day (by as much as 1.5 MPa, depending on cultivar; see Section 7.2), following both opening and closing of the stomata at dawn and dusk and depending on the evaporative demand (i.e., VPD) of the air that drives transpiration (Williams and Baeza, 2007). These changes in Ψ_{leaf} (i.e., changes in Ψ of the cell walls) must be balanced by changes in turgor pressure for the leaf cells to maintain their volume and solute concentration, which is achieved due to vacuoles and rigid but elastic cell walls. Thus P_{leaf} declines from a nighttime maximum to a minimum around midday and recovers in the afternoon (Rüger et al., 2010). Because transpiration-driven water flow is minimal at night, Ψ_{leaf} reaches a maximum as well (Schultz and Matthews, 1988a). In other words, soil water status determines the baseline Ψ_{leaf} in the near absence of transpiration (Tardieu and Simonneau, 1998).

As a soil dries, the extraction of water by the roots and its transport to the shoots become increasingly more difficult for grapevines. To maintain the water potential gradient that drives water flow, the Ψ in the shoot must decrease. Therefore, the predawn Ψ_{leaf} can be used as a robust indicator of the Ψ_{soil} to which the roots are exposed. Transpiration during the day decreases Ψ_{leaf} and P_{leaf} below the predawn value so that Ψ_{leaf} and P_{leaf} at a particular time of day are the result of both soil water status and transpiration driven by evaporative demand (Rüger et al., 2010; Smart, 1974; Tardieu and Simonneau, 1998; Zufferey et al., 2011). Thus while the absolute values of Ψ_{leaf} and P_{leaf} decrease in drying soil, their daily amplitudes tend to increase, because the decrease is greater during the day than at night. At the same time, the influence of the VPD on Ψ_{leaf} diminishes in drying soil, because the stomata close, so that Ψ_{soil} becomes increasingly dominant in altering Ψ_{leaf} (Williams and Baeza, 2007).

The water potential equation introduced in Section 3.1 also applies to the soil solution, although, in contrast to plant cells, the Ψ_π of the soil water is generally negligible (approximately −0.01 MPa), except in saline soils (see Section 7.3). Consequently, Ψ_{soil} is determined mainly by Ψ_M, which is close to zero in wet soils but can decrease to -3 MPa in dry soils due to the surface tension resulting from expanding air spaces. In drying soil, a point will eventually be reached at which the resistance to water flow is so great that the vine can no longer maintain a sufficient $\Delta\Psi$ to sustain transpiration and becomes drought stressed (see Section 7.2). Drought stress can also be induced by low soil temperature, which increases the resistance to water uptake and flow inside the roots (i.e., it increases the root's radial r_h), ostensibly due to closure of aquaporins (Aroca et al., 2012; Maurel et al., 2008). Consequently, cool soils are associated with higher leaf r_s, leading to lower photosynthesis, compared with warm soils (Rogiers and Clarke, 2013). Nevertheless, this can be a problem when cold irrigation water is applied to rapidly transpiring vines, especially at high atmospheric VPD (see Section 3.2). The resulting imbalance between water uptake by the roots and water loss from the leaves can induce cavitation in the xylem, which may lead to leaf wilting and injury (Scheenen et al., 2007). When vines are grown in pots, especially in dark pots heated by the sun, whose soil temperature can be several degrees above air temperature, applying cold irrigation water can inhibit growth (Passioura, 2006). But even field-grown grapevines can experience significant fluctuations in root temperatures, especially in sandy soils and dry climates, where the root zone of irrigated vines is quite shallow (Stevens and Douglas, 1994).

Water influx is proportional to the surface area of the root system (Steudle, 2001). Grapevines often have dense root systems in the topsoil and can extract water effectively from the surface soil layers. Because extracting water from lower soil layers is less effective, the surface soil dries more quickly than the subsoil. Under nonirrigated conditions, roots continue to grow into deeper, wetter soil layers, whereas the roots of irrigated plants proliferate mostly in the topsoil; in both cases, the Ψ of the advancing root tips remains high as long as the roots can find water (Hsiao and Xu, 2000). Root water uptake shifts to deeper soil layers as the soil dries and back to the upper layers after rainfall or irrigation.

Because plant-available nutrient ions are dissolved in the soil solution, nutrient uptake also depends on water flow through the soil–root–shoot pathway. Although nutrients are often concentrated in the biologically active surface soil, water and nutrient availability varies greatly in both space and time. To complicate matters further, different nutrients are often not readily available in the same location. For example, NO_3^- moves through the soil approximately 10 times faster than K^+ and 500 times faster than $H_2PO_4^-$, so percolating water drains (i.e., leaches) nitrate down to the subsoil much more rapidly than other nutrients. As a consequence, shallow roots often take up soil-immobile nutrients (e.g., potassium and phosphorus) from the topsoil, whereas deeper roots tap water and soil-mobile nutrients (e.g., nitrate) that leach deeper into the soil profile. These deep roots can also transport water to the surface roots in a process called hydraulic lift or hydraulic redistribution, which occurs mostly at night when transpiration is minimal (Bauerle et al., 2008a; Bleby et al., 2010). Deep roots need not even be grapevine roots; some cover crops might also be able to lift water to the topsoil, where it may be available to the vines, although this does not apply to vineyard floor covers dominated by grass species (Celette et al., 2008; Patrick King and Berry, 2005). Hydraulic redistribution keeps the surface roots alive, at least so long as they do not lose contact with the drying soil, so that they are ready to take advantage of water arriving from rainfall or irrigation (Bleby et al., 2010). By locally increasing soil moisture due to water loss from the roots when $\Psi_{soil} < \Psi_{root}$, hydraulic redistribution even enables the fine roots to sustain nutrient uptake in drying soil. In addition, roots can passively move water laterally (horizontally) and

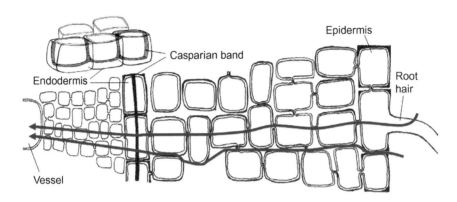

FIGURE 3.3

Water and nutrient flow from the soil through a root into the xylem via the symplastic (top blue line) and apoplastic (bottom blue line) pathways.

Illustration by A. Mills and M. Keller.

downward (e.g., when rainfall or irrigation make the surface soil wetter than the subsoil) according to local gradients of Ψ_{soil} (Bauerle et al., 2008a; Smart et al., 2005; Stoll et al., 2000). Water redistribution by roots tends to even out local differences in soil moisture (e.g., due to variations in soil texture or organic matter), delays soil drying, and may assist in maintaining symbiotic mycorrhiza. The redistribution process even continues in dormant plants through the winter (Bleby et al., 2010). However, water that is lost from vine roots by diffusion can also be reabsorbed by competitors (e.g., some weeds or cover crops).

As shown in Figure 3.3, water that enters the root initially moves radially through the epidermis and cortex tissues along both symplastic (or symplasmic) and apoplastic (or apoplasmic) routes (Steudle, 2000, 2001; Steudle and Peterson, 1998). The symplast is everything inside the plant that is bound inside a membrane (i.e., the interconnected cytoplasm of the cells); thus, the symplastic pathway is an intracellular pathway. Within the symplast, water and solutes move from cell to cell via small connecting pores called plasmodesmata (Lough and Lucas, 2006; Roberts and Oparka, 2003). Short-distance cell-to-cell movement of solutes within the symplast thus occurs by diffusion but is also influenced by osmotic gradients across membranes, and long-distance movement between symplastic domains occurs by mass flow through the interconnecting phloem (see Section 5.1). The apoplast, on the other hand, comprises everything that is outside the membrane and includes cell walls, intercellular spaces, and xylem conduits. Cell walls consist of a porous framework of the hydrophilic (i.e., wettable; from Greek *hydros* = water, *philos* = loving) polymers cellulose and hemicellulose (matrix glycans) embedded in a polymer matrix of pectin, proteins, and in some cells lignin, which is imbued with water like a sponge so that approximately 65–75% of the cell wall is water (Carpita and Gibeaut, 1993; Cosgrove, 1997; Doblin et al., 2010; Steudle, 2001). In other words, cell walls resemble dense aqueous gels. Nonetheless, because cell walls comprise only approximately 5% of a tissue's total volume, they have very limited water storage capacity. In addition, the cell walls' dense network of microfibrils also acts like a sieve (pore sizes of 4–14 nm) because it permits passage only to small molecules such as nutrient ions, amino acids, and sucrose.

Water on an apoplastic route flows around the cell membranes; hence, the apoplastic pathway is an extracellular pathway. Because there are no membranes along the apoplast, hydraulic, nonselective water flow dominates along this pathway (Steudle, 2000, 2001). Rather than moving exclusively in one compartment or the other, a third possibility for water flow is straight across cell walls, membranes, and through cells; in other words, water moves alternately in the apoplast and the symplast. This route is termed the transcellular pathway and is probably the preferred pathway for water on its way to the xylem. However, the endodermis cells that form behind the growing root tip and separate the cortex from the stele have thickened, hydrophobic radial and transverse cell walls called Casparian strips or Casparian bands, which act like gaskets due to their impregnation with lignin (Clarkson, 1993; Geldner, 2013; Perrone et al., 2012; Schreiber, 2010; Steudle, 2000). In addition, access of water and nutrients to channel and transport proteins (see below) in the endodermis cell membranes is also blocked by the deposition of suberin in the secondary cell walls, so that nutrients can only move into and out of these cells through plasmodesmata (Geldner, 2013). This waterproofing blocks diffusion and hydraulically separates the apoplast in the cortex from that in the stele, forcing water and dissolved nutrient ions to pass directly through the cell membranes into the cell interior—that is, into the symplast. Thus, on their way to the xylem, water and solutes must pass at least two cell membranes (i.e., in and out of at least one endodermis cell), regardless of whether the initial route is symplastic or apoplastic (Tester and Leigh, 2001).

Unlike cell walls, cell membranes are semipermeable (see Section 3.1); that is, they are, in principle, impermeable to solutes, including even small ions such as protons. In reality, membranes act like selective sieves, permitting entry to some ions and not to others. The degree of selectivity or semipermeability is termed the reflection coefficient (σ), which can vary from 0 to 1. Membranes with $\sigma = 0$ behave nonselectively and are equally permeable to water and solutes, whereas at $\sigma = 1$ membranes retain solutes completely and have high water but no solute permeability (Steudle and Peterson, 1998). A high σ in the endodermis ensures, for instance, that nutrients taken up by the roots do not leak back out into the soil solution, which also permits the buildup of root pressure under conditions of slow transpiration. Pores created by special channel and transport proteins embedded in the membranes regulate the passage of water and nutrient ions, and each cell type or tissue seems to be equipped with a unique set of these proteins. Many channel proteins, especially in the parenchyma tissue around vascular bundles, are specific for water (aquaporins; see Section 3.1) and control water flow by opening or closing (Maurel et al., 2008; Steudle, 2000). Thus, they control the rate but not the direction of the flow, which simply and passively follows the $\Delta\Psi$ (Tyerman et al., 1999). For instance, aquaporins close at night but open during the day; this opening decreases the radial r_h, which enables the roots to meet the water demand for transpiration.

In contrast with the bidirectional aquaporins, many ion channels are said to be either inward rectifying or outward rectifying. Inward-rectifying channels allow ions to move into cells, whereas outward-rectifying channels facilitate ion movement out of cells (Hedrich, 2012; Tyerman et al., 1999). Such one-way movement allows channels to function like valves, which prevents unnecessary nutrient loss. Because of their ability to control the passage of ions, membranes also act as electrical isolators, which leads to electrical potential gradients across membranes (negative charges typically dominate inside the cell) in addition to concentration gradients. The sum of the gradients of concentration and the electric charge of an ion is called the electrochemical potential gradient.

Some channels, termed ion channels, are specific for cations in general, and some ion channels are specific for particular cations or anions, whereas solute carriers are responsible for the transport across membranes of neutral solutes. Some neutral solutes (e.g., urea and boric and silicic acid) and gases (e.g., CO_2 and NH_3) may also move through aquaporins (Maurel et al., 2008). A channel is called specific if it is selective, or discriminates, for a particular molecule or group of molecules such that other molecules cannot pass through the channel, regardless of whether their movement is passive or active. Ions moving passively simply diffuse across channels down the electrochemical potential gradient. As with water flow, the direction of movement is from high to low potential, typically from high to low concentration. Ions moving actively must be transported (pumped) across membranes by transport proteins called transporters or carriers against their electrochemical potential, ordinarily from low to high concentration (Grossman and Takahashi, 2001). Active movement typically draws on a proton (H^+) gradient across the membrane and requires an input of energy in the form of ATP, which is hydrolyzed to ADP and phosphate (Gilroy and Jones, 2000; Lalonde et al., 2004). However, rather than burning ATP directly for nutrient transport, the ATP is used to power pumps, termed H^+-ATPases, which export H^+ from the cell interior (cytosol) to the apoplast at a rate of approximately 1 H^+ per "consumed" ATP (Britto and Kronzucker, 2006; Sondergaard et al., 2004). The resulting decrease in apoplast pH (to pH 4 or 5) generates an electrochemical gradient of -100 to -150 mV that, in turn, fuels nutrient uptake via the membrane's channels and carriers (Hedrich, 2012). Thus, H^+-ATPases act as energizers that convert the chemical energy released by ATP hydrolysis into chemiosmotic energy; the resulting ADP is then recharged by respiration. Whereas in the so-called antiport, one proton is pumped out of the cell for each charge of an incoming cation, two protons are needed per charge of an incoming anion that then enters together with the backflowing protons in the so-called proton cotransport or symport (Amtmann and Blatt, 2009).

Active transport enables grapevines to concentrate nutrient ions inside the roots well above their concentration in the surrounding soil solution (Keller et al., 1995, 2001b), although some concentration is also possible simply due to the electrical potential gradients (Amtmann and Blatt, 2009). Most ions occur in the soil water at much lower concentration than would be required for plant cellular functions, and without a concentration mechanism nutrient uptake would not be possible. At the same time, concentrating nutrient ions inside the root cells also generates an osmotic driving force for water uptake by the roots.

Whereas many anions (e.g., NO_3^-, SO_4^{2-}, or Cl^-) are always taken up actively against their electrochemical potential gradient, some cations (e.g., K^+, NH_4^+, or Na^+) are taken up passively across relatively generic (i.e., nonselective) ion channels when their availability in the soil solution is high and actively and very selectively across carriers when availability is low (Tester and Leigh, 2001; Véry and Sentenac, 2003). For most macronutrients, the carriers become saturated at approximately 1 mM in the soil solution, and the channels are switched on when the nutrient concentration rises above this threshold (Britto and Kronzucker, 2006). Such high nutrient concentrations may be reached after nitrogen fertilizer applications and are relatively common for potassium; moreover, the passive uptake of NH_4^+ and Na^+ can sometimes result in ion toxicity (see Section 7.3). Some transporters (e.g., those for NO_3^- and K^+) can function in both active and passive mode, depending on the external concentration, and anion channels can also transport organic acids (Britto and Kronzucker, 2006; Dechorgnat et al., 2011). Although controlled ion uptake is also important to maintain a neutral electrical charge balance in the plant, uptake rates of cations and anions are rarely precisely equal; roots restore neutrality by releasing H^+, HCO_3^-, or OH^- into the soil immediately surrounding them, termed the rhizosphere (Clarkson and Hanson, 1980). In addition, the high uptake capacity of the passive channels is usually accompanied by nutrient efflux from the roots. Whereas anions can leak back passively in exchange for protons, cations must be pumped back out actively (Britto and Kronzucker, 2006). Therefore, because either uptake or efflux is active, even passive nutrient uptake could be called active in terms of energy expenditure for the related active efflux. For most nutrient ions save phosphate, this efflux increases with rising external concentration and becomes almost equivalent to ion uptake at high soil nutrient concentration, resulting in considerable—and apparently futile—nutrient cycling across the membranes (Britto and Kronzucker, 2006).

Once inside the stele, most water and dissolved nutrients ultimately return to the apoplast when the water is released into the xylem conduits across the pit membranes separating the vessels and tracheids from the surrounding parenchyma cells (see Section 1.3), which themselves have acquired the solutes via their plasmodesmata from other parenchyma cells or from the cell wall apoplast (Sattelmacher, 2001; Sondergaard et al., 2004). Because the cell walls of the xylem conduits are waterproof around the pits, release into the xylem is another bottleneck of high radial r_h in the vine's hydraulic system after the passage through the endodermis. In addition, the presence of aquaporins and selective channel or transport proteins in the parenchyma cell membranes provides a chance to actively control access to the xylem (De Boer and Volkov, 2003; Dechorgnat et al., 2011; Tester and Leigh, 2001). Such controlled release and reversible exchange between parenchyma cells and xylem conduits also enables the vine to modify the xylem sap composition and hence to fine-tune nutrient supply with the changing demand of the shoots. Inside the xylem conduits, the tension (negative P_x) induced by transpiration and the very low axial r_h of the tracheary elements facilitate rapid transport of water (up to 2 m h^{-1} during a sunny day) and its dissolved nutrients to the shoots, where they are delivered initially to the apoplastic compartment of transpiring organs.

Even with the concentration mechanism discussed previously, xylem sap is very dilute (i.e., it has a low π) compared with phloem sap (see Table 5.1) and the contents of shoot and leaf cells. Nevertheless, due to the transpiration stream, it transports large amounts of nutrient ions by mass flow, although the nutrient concentration can fluctuate severalfold between day and night due to the diurnal change in transpiration rates. Lower concentrations are associated with higher flow rates, which in turn are associated with greater water uptake by the roots; despite the lower concentration, however, the total nutrient flow to the shoots is greater because transpiration-driven water influx is strongly coupled with nutrient uptake (Wegner and Zimmermann, 2009). In addition to nutrient ions, the xylem can also transport sugars, especially during springtime remobilization of storage reserves as discussed previously, as well as other organic molecules such as amino acids as nitrogen carriers (see Section 5.3), organic acids as carriers of metal ions (see Section 7.3), and hormones such as ABA and cytokinins. Solutes can also diffuse in the nonxylem apoplast, but this diffusion is at least one order of magnitude slower than the movement in the xylem (Kramer et al., 2007). This is because cations interact with, and hence are slowed by, the negatively charged carboxyl groups of the cell wall pectins, which function like cation exchangers, and also because the pores in the apoplast hinder the passage of solutes and increase their path length (Sattelmacher, 2001). In addition, xylem-sap water and solutes can readily diffuse to and from adjacent parenchyma tissue en route to the leaves, which helps to buffer fluctuations in xylem sap concentrations (Metzner et al., 2010a,b).

The long-distance transport of water and solutes in the xylem, the so-called transpiration stream, occurs by bulk flow (a.k.a. mass flow), in which large numbers of H_2O molecules move together and drag the dissolved ions along, and is driven by a hydrostatic pressure gradient (ΔP_x) caused by transpiration, which "sucks" water up the vine against gravity and friction. The flow rate increases exponentially (actually to the power of 4) with increasing diameter of the xylem conduits, and so does the velocity of the flow (actually to the power of 2), due to the decreasing r_h with increasing diameter (Tyree and Ewers, 1991). Thus, the large vessels formed in spring can transport water much more rapidly than the narrower ones formed during summer. Moreover, older vines with more annual rings and a larger trunk diameter, and hence more large vessels, are able to transport much more water to support a larger leaf area than can young vines, and their generally deeper and more extensive root systems are able to access soil moisture at deeper layers, which is especially important later in the season as evapotranspiration dries the soil down from the top. The roots, trunks, and cordons also act as important water storage compartments; that is, they behave as hydraulic capacitors. These reservoirs, which are also called hydraulic capacitance, are depleted in the morning, when the stomata open and transpiration rises, and replenished in the afternoon (Schultz and Matthews, 1988a; Steppe and Lemeur, 2004). Depletion and replenishment of the internal water reserves leads to daily cycles of shrinkage and expansion of the trunk diameter. Older vines can store more water due to their higher number of xylem (and phloem) parenchyma cells; this buffering capacity could make them less vulnerable to xylem cavitation so that they can cope with drought stress better than young vines. Moreover, the xylem parenchyma cells may also be able to absorb nutrient ions such as nitrate and potassium during periods of abundant supply and release them back into the xylem in times of starvation. In other words, these cells may serve as short-term storage reserves of both water and nutrients.

The xylem constitutes more than 99% of the total length of the transpiration stream; this proportion increases with increasing plant size (Sperry et al., 2003). Nevertheless, a few dozen micrometers in the roots and in the leaves exert the vast majority of whole-vine r_h, even though the entire transpiration stream can be several meters long, and r_h increases with increasing transport distance (i.e., with increasing plant height) (Sperry et al., 2002). Thus, one of two major resistances to water flow through a grapevine is not in the trunk or shoots but in the roots, where the radial (i.e., nonxylem) rather than axial (i.e., xylem lumen) resistances dominate (Steudle and Peterson, 1998); within the xylem, the pit membranes between adjoining vessels and tracheids can account for half or more of the total r_h (Choat et al., 2008; Hacke et al., 2006). The other main resistance is in the leaves, at the terminal section of the transpiration stream, where the water flows through orders of veins in series and in parallel, leaves the xylem network at the veins across and around the bundle sheath parenchyma encircling the veins, and moves into and around the mesophyll cells before it evaporates into the air spaces, changing from the liquid to the vapor phase, and diffuses out of the stomata (Comstock, 2002; Sack and Holbrook, 2006; Sack et al., 2003; see also Section 1.3). Similar to the endodermis in the roots, suberin deposition in the cell walls of the bundle sheath forces water to cross membranes, so that the rate of water flow can be regulated by aquaporins (Shatil-Cohen et al., 2011; Tyerman et al., 2009). In a transpiring grapevine, r_h of the leaves accounts for approximately one-fourth of the whole-plant resistance to water flow, and within the leaf the post-vein (i.e., nonxylem) resistance predominates. This means that the distance between the veins and the evaporative surfaces is directly correlated with, and controls, leaf r_h, which in turn limits photosynthesis (Brodribb et al., 2007). Uptake of water and solutes into the leaf mesophyll cells from the xylem vessels occurs initially through pits (from the apoplast back into the symplast of neighboring parenchyma cells) and then through plasmodesmata or, across the cell membranes, via aquaporins (for water) and energy-dependent carrier proteins (for solutes) that are stimulated by the growth hormone auxin. Therefore, the mesophyll is somewhat hydraulically isolated from the remainder of the transpiration stream (Shatil-Cohen et al., 2011).

During the day, grapevines lose most of the absorbed water in transpiration, but a small fraction ($\sim 1\%$) is used for cell expansion, cell metabolism, and phloem transport. At night, when the stomata are essentially closed and transpiration is drastically reduced, those small amounts may become the dominant component of water flow. In fact, water is circulated between the phloem and xylem, and this circulation maintains water flow in the xylem in the absence of any significant transpiration, even at night (Köckenberger et al., 1997; Windt et al., 2006). The water taken up to balance this "growth water" and phloem counterflow amounts to a transpiration-independent water flow that is sufficient to transport nutrients absorbed by the roots; hence, nutrient uptake and long-distance transport in the xylem are not dependent on transpiration (Tanner and Beevers, 2001). Of course, the increase in water flow due to transpiration results in a corresponding increase in the speed at which dissolved molecules move up the xylem (Peuke et al., 2001; Wegner and Zimmermann, 2009). Rapid transpiration thus increases nutrient uptake by the roots, especially when nutrient availability in the soil is high (Alleweldt et al., 1984a; Hsiao, 1973; Keller et al., 1995). Over the course of a day, however, root water uptake must exceed transpiration sufficiently to also supply water for growth because plant growth is mainly caused by cell expansion, which results almost entirely from an increase in the cell's water content (Boyer, 1985; Hsiao and Xu, 2000; Schopfer, 2006). This is why growth is not the same thing as a gain in dry mass. Water import into cells is driven by cell wall loosening and accumulation of solutes inside the cell and thus depends on a water potential gradient between the cell and the supplying xylem ($\Delta\Psi = \Psi_{xylem} - \Psi_{cell}$; see Section 3.1). Therefore, expanding tissues normally have lower Ψ_{cell} than mature (nongrowing) tissues. Because growth competes with transpiration for xylem water, cells must also maintain lower Ψ_{cell} during the day than at night if they are to sustain expansion during the day (Boyer and Silk, 2004). Indeed, grape berries, which must remove water from the transpiration stream surging up the shoot and past the fruit clusters, grow mostly at night, especially before veraison and when the water demand of the transpiring canopy is high (Greenspan et al., 1994, 1996; Matthews and Shackel, 2005).

Growth rates of all tissues change rapidly with fluctuating Ψ_{xylem} (i.e., with extracellular water status), and growth is extremely sensitive to water deficit (Boyer, 1985; Boyer and Silk, 2004; Hsiao and Xu, 2000). Shoot growth of grapevines declines with decreasing Ψ_{xylem} and stops completely when Ψ_{xylem} around midday reaches approximately -1.0 to -1.1 MPa. Because water stress decreases Ψ_{xylem}, a major cause for growth inhibition under water deficit may simply be the smaller $\Delta\Psi$, which reduces water uptake by the expanding cells (Nonami and Boyer, 1987; Nonami et al., 1997). Moreover, due to the small, immature xylem conduits near the shoot tip and in young petioles and expanding leaves, the r_h in the "growth zone" is much greater than in the mature portion of the shoot so that a steeper $\Delta\Psi$ is required to sustain growth (Schultz and Matthews, 1993b). If water ceases to move out of the xylem, shoot and leaf growth stop immediately because cells must take up water to expand, whereas root growth is less affected (Boyer and Silk, 2004; Hsiao and Xu, 2000; Wu and Cosgrove, 2000). Lower Ψ_{xylem} is also responsible for the smaller berry size of water-stressed vines. Moreover, it explains why pre-veraison water deficit has a larger effect on berry size than post-veraison water deficit: Water influx into the berry changes at veraison from predominantly via the xylem to predominantly via the phloem (see Sections 1.3 and 6.2).

Photosynthesis of mature leaves is less affected by decreasing Ψ than is expansion of young leaves (Hsiao and Xu, 2000), and solute transport in the phloem can continue at decreasing Ψ. As a consequence, the import of solutes may exceed their use so that solutes often accumulate in sink organs, such as roots or berries, of vines that experience mild water deficit. Therefore, berries grown on mildly water-stressed vines are not only smaller but also have higher sugar content. Yet the decrease in photosynthesis under more severe water stress leads to a reduction in berry sugar (see Section 7.2). As in the berries, some of the water for root growth comes from the phloem in addition to direct influx from the soil (Boyer and Silk, 2004). Moreover, because xylem cells are differentiated behind the growing root tip, water uptake for transpiration also occurs behind the expanding cells. Due to this hydraulic isolation, the root tip usually experiences slightly higher soil moisture than the mature root. All of this makes root growth somewhat less vulnerable to fluctuations in Ψ_{xylem} and ensures that it is favored over shoot growth when the soil dries (Boyer and Silk, 2004; Hsiao and Xu, 2000).

CHAPTER 4

PHOTOSYNTHESIS AND RESPIRATION

CHAPTER OUTLINE

4.1 Light Absorption and Energy Capture 125
4.2 Carbon Uptake and Assimilation 130
4.3 Photorespiration 136
4.4 Respiration 138
4.5 From Cells to Plants 141

ABSTRACT

Leaves capture sunlight using chlorophyll pigments to generate chemical energy that powers transformation of inorganic carbon dioxide and water into sugar during photosynthesis. Two chlorophyll-rich photosystems work in series to intercept and convert light energy into the chemical energy of ATP (adenosine triphosphate) and NADPH (nicotinamide adenine dinucleotide phosphate), capturing electrons from water and releasing oxygen as a waste product. Carbon dioxide diffuses into the leaves through stomatal pores, permitting copious amounts of water vapor to escape. Incorporation of CO_2 into carbohydrates occurs via the ATP- and NADPH-powered Calvin cycle whose three-carbon product is used to make sugars and starch. Starch can be stored for nighttime consumption, and sucrose is exported via the phloem throughout the vine. Because the first Calvin cycle enzyme, rubisco, can work in opposing directions, it converts a portion of the carbohydrate back to CO_2 during photorespiration, thus reducing the efficiency of photosynthesis. Some sucrose is burnt to CO_2 in a series of reactions during respiration to fuel ATP generation. By-products of this reaction sequence give rise to most of a vine's diverse organic compounds. Respiration is the vine's powerplant, supplying energy for metabolism, nutrient uptake, transport, growth, and cell maintenance.

Keywords: ATP; NADPH; photosynthesis; photorespiration; respiration; rubisco; starch; sugar

4.1 LIGHT ABSORPTION AND ENERGY CAPTURE

Photosynthesis is the most important way of obtaining energy for all plants including grapevines. As the name suggests, photosynthesis (Greek *photos* = light, *synthesis* = building a whole) is

FIGURE 4.1

Portion of the electromagnetic spectrum.

Reproduced from Taiz and Zeiger (2006).

a process by which sunlight is converted into chemical energy that is used to produce (synthesize) organic compounds within the plant from inorganic compounds acquired from outside the plant. Sunlight is a form of electromagnetic radiation that has properties of both waves and particles. The wave nature of light can be visualized with a prism separating a ray of visible or white light into a continuous array of different colors according to their wavelengths (Figure 4.1). Light's particle nature enables us to describe light energy as a stream of energy-carrying particles called quanta (singular: quantum) or photons.

The window of the spectrum called visible light, because our eyes are sensitive to it, also happens to be the range of wavelengths ($\sim$ 400–700 nm) that is important for photosynthesis. This region is only a small portion of the entire spectrum of electromagnetic radiation, and other organisms can see light of other wavelengths. Insects and birds, for instance, often distinguish between objects such as flowers or eggs according to their color patterns in the ultraviolet (UV) range, and plants can detect both UV and infrared radiation in addition to visible light (see Section 5.2). All forms of electromagnetic radiation travel at the speed of light ($c \approx 300{,}000$ km s^{-1}), but the energy content (E) of photons is higher the shorter their wavelength (λ), or the higher their frequency (v), within the spectrum. This can be written as follows:

$$E = h\upsilon = hc\lambda^{-1}$$

where h is the Planck's constant (6.63×10^{-34} Js).

The energy contained in the photons is exploited by plants during photosynthesis, which proceeds in two stages. The first is the photochemical capture of solar energy, transported in the form of photons, and its temporary storage in high-energy chemical bonds of adenosine 5'-triphosphate (ATP; the universal biological energy currency) and nicotinamide adenine dinucleotide phosphate (NADPH). The second stage uses this energy to enzymatically convert carbon dioxide (CO_2) and water (H_2O) to sugar (carbohydrate), which in turn is used to produce all other organic compounds throughout the plant. Of course, in the absence of light, there is no energy source for

4.1 LIGHT ABSORPTION AND ENERGY CAPTURE

photosynthesis, and there is no carbohydrate production. The two stages of photosynthesis occur in different regions of specialized organelles termed chloroplasts that contain the photosynthetic machinery. Although the cells of all green plant organs and tissues have chloroplasts, the leaves are their main residence, and they are well designed to intercept the maximum amount of light (see Section 1.3). The conversion of light energy into chemical energy happens in the membranes of small, disk-shaped, stacked sacs called thylakoids (Greek *thylakos* = sac) inside the chloroplast, whereas the conversion of CO_2 (which is discussed in Section 4.2) occurs in the fluid chloroplast matrix referred to as stroma. This chapter provides a brief summary of the most important aspects of the photosynthetic processes. Because of their importance for life on Earth, the majority of these processes are virtually identical in all plants and are covered in great detail in general texts on plant physiology, from which most of the following information is derived (Atwell et al., 1999; Salisbury and Ross, 1992; Taiz and Zeiger, 2006).

Before the light energy can be used, the light particles must first be absorbed and translated into a flow of electrons that are derived from the splitting of water. The green chlorophyll is the main light-absorbing pigment that makes light energy capture possible. It is the world's most abundant pigment. Chlorophyll is green because it absorbs predominantly blue (430 nm) and red (680 nm) light and reflects most of the green light with wavelengths intermediate between blue and red (Figure 4.2). In plants, chlorophyll comes in two slightly different versions called chlorophyll *a* and chlorophyll *b*, both of which, along with various other pigments called accessory pigments (e.g., carotenoids such as carotenes and xanthophylls), are located in the thylakoid membranes. When a chlorophyll molecule absorbs a photon, it is converted from its low-energy state, called ground state, to an excited state by abruptly shifting one of its electrons from a lower energy atomic orbital, or shell, close to the atomic nucleus to a more distant, higher energy orbital. This instantaneous electron "jump" from one energy level to another is termed a quantum leap. The excited chlorophyll is extremely unstable and returns to its ground state within nanoseconds (10^{-9} s) by releasing its available energy as either heat or fluorescent radiation (in which the photon is transferred to xanthophylls and reemitted), or by transferring either the energy or an electron to another pigment molecule.

Approximately 250 chlorophyll molecules are grouped together in the so-called photosystem II (PSII), draped on the framework of a multiprotein complex in the thylakoid membrane. The majority of these chlorophylls, called antenna pigments, are responsible for absorbing photons (i.e., for "harvesting" light) and transferring their energy from one pigment to another until it arrives at a single specialized chlorophyll molecule called the reaction center. The antenna pigments transfer only their excitation energy so that the reaction center is the only chlorophyll molecule that actually releases an electron in a mechanism called photochemistry. It wins out in the race of possible energy-releasing mechanisms, and thus makes photosynthesis possible, because it occurs at a rate approximately 1000 times faster (i.e., in picoseconds, 10^{-12} s, making it one of the fastest known chemical reactions) than the other processes. The chlorophyll molecule replaces the lost electron (e^-) by capturing an electron from water, which is thereby oxidized (i.e., split), releasing oxygen as a waste product and a proton (H^+). This water-splitting reaction ($2H_2O \rightarrow 4H^+ + 4e^- + O_2$) is catalyzed by a protein complex that contains manganese, calcium, and chloride ions at a ratio of 4:1:1 (Nelson and Yocum, 2006). Thus, water is the source of all electrons (and protons) involved in photosynthesis (both e^- and H^+ are later incorporated into a sugar molecule by the enzyme rubisco; see Section 4.2); for each electron that is released, one water molecule is consumed and its

FIGURE 4.2

Structure of the chlorophyll *a* molecule (left) and absorption spectra for different light-absorbing pigments (right).

oxygen released to the air. The oxygen in photosynthetically produced sugar therefore comes from CO_2, not from H_2O.

The electron released from the reaction center chlorophyll is transferred to an acceptor molecule in a process termed photochemical quenching because it "quenches" the excitation energy of the chlorophyll (Baker, 2008). The acceptor in turn passes the electron on to a secondary acceptor and so on down a cascade of steps called the electron transport chain until it arrives at the reaction center of a second photosystem, confusingly named photosystem I (PSI), operating in series with PSII.

PSI is an extremely efficient nano-photoelectric machine that is also composed of a multiprotein complex and contains approximately 175 chlorophyll molecules, which also absorb light and transfer its energy (Nelson and Yocum, 2006). The electron is transferred from the excited PSI reaction center to yet more acceptors until it is accepted by an iron sulfur-containing protein called ferredoxin (Latin *ferrum* = iron), which passes it to $NADP^+$ that is thereby reduced to NADPH.

The protons produced during the oxidation of water are released into the thylakoid's interior, the so-called thylakoid lumen (Baker, 2008). In addition, the electron transport chain between PSII and PSI also "pumps" protons from the stroma into the thylakoids using some of the energy released by the electrons "flowing" down the chain. This results in a relative shortage of protons in the stroma (pH 8), creating an electrochemical potential (i.e., a pH difference equating to a 3000-fold difference in protons) across the thylakoid (pH 4.5) membrane. This so-called proton motive force constitutes a source of energy that powers another protein complex in the thylakoid membrane, called ATP synthase or ATPase, to produce ATP from ADP and inorganic phosphate (P_i): $ADP + P_i \rightarrow ATP$. Working like a turbine in a hydroelectric power plant, each rotation of the enzyme ATP synthase releases three molecules of ATP while transferring 14 protons back out to the stroma (i.e., the "waterfall" or flow of protons drives the "turbine"). This process, which recycles all protons back to the chloroplast stroma, is called photophosphorylation.

In essence, energy transfer and photochemistry convert light energy into chemical energy in the form of NADPH and ATP, which provide the energy for carbon assimilation, as well as for the assimilation of nitrogen (see Section 5.3), phosphorus, and sulfur, and other metabolic processes in the chloroplast (Baker, 2008). The proportion of absorbed photons that are used for photosynthesis is termed quantum yield or quantum efficiency, and it decreases as incident light intensity increases (Baker, 2008). Although inevitable, the release of oxygen so close to the PSII reaction center is problematic because excited chlorophyll can cause this oxygen to be transformed into the very reactive singlet oxygen ($^1O_2$), which can damage membranes and cause mutations or cell death by oxidizing lipids, proteins, and DNA (Apel and Hirt, 2004; Halliwell, 2006; Møller et al., 2007). Not surprisingly, many herbicides (e.g., diuron, diquat, paraquat, simazine) used to kill unwanted plants in vineyards act by blocking the photosynthetic electron transport chain, leading to "self-poisoning" of plants by producing excessive amounts of $^1O_2$ in PSII reaction centers. Spray drift of such herbicides on grapevine leaves thus causes necrotic lesions wherever a spray droplet has landed. As an aside, whereas vegetatively propagated grapevines would have great difficulty evolving resistance to such herbicides, sexually propagated weeds can do so much more readily. This is because even minute mutations that decrease the transport of herbicide molecules or their binding to the thylakoid membrane can markedly reduce herbicide efficiency (Powles and Yu, 2010). Hence, the repeated use of similar herbicides may encourage herbicide resistance by increasing the selection pressure on weed populations.

Chlorophyll, then, is a "Jekyll and Hyde" molecule, enabling the crucial light absorption for photosynthesis but acting as a phototoxin when it becomes overly excited (Hörtensteiner, 2009). Excessive oxygen is toxic due to the formation of free radicals termed active (or reactive) oxygen species (see Section 7.1). Fortunately, under normal conditions, the carotenoids (especially the yellow xanthophyll pigments violaxanthin, antheraxanthin, and zeaxanthin) associated with the reaction center and antenna complex quickly "scavenge" singlet oxygen in addition to dissipating excess light energy absorbed by the antenna pigments and converting it into heat. This dissipation is termed nonphotochemical, energy-dependent quenching or thermal dissipation, and it employs the so-called xanthophyll cycle, which converts violaxanthin via antheraxanthin to zeaxanthin when there is too much light but runs in the reverse direction when light intensity declines (Baker, 2008; Demmig-Adams and Adams, 1996; Fanciullino et al., 2014; Noctor and Foyer, 1998). This is why the amount of antheraxanthin and zeaxanthin in grapevine leaves often rises in the morning and declines in the evening, whereas violaxanthin follows the opposite trend (Düring, 1999). In addition, the carotenoids also assist with light harvesting by absorbing photons at wavelengths not covered by chlorophyll (Bartley and Scolnik, 1995; see Figure 4.2).

When a leaf absorbs more light energy than can be converted to chemical energy and used in CO_2 assimilation, the "energy overload" will damage the photosynthetic apparatus (especially PSII) by "knocking out" manganese ions (Mn^{2+}) from the water-splitting protein complex and inactivating the reaction center. Such damage is an unavoidable ancillary cost of the business of doing photosynthesis and increases in proportion to the light (especially high-energy UV light) intensity (Takahashi and Badger, 2011; Takahashi and Murata, 2008). It reduces the efficiency of photosynthesis and is called photoinhibition, which also is a form of nonphotochemical quenching (Baker, 2008; Gamon and Pearcy, 1990b). The damaged proteins are normally degraded (i.e., they undergo proteolysis) and recycled to build new PSII proteins; thus, photoinhibition results from the balance between damage and repair. Excess light, however, suppresses the repair cycle (Takahashi and Badger, 2011). When the photosynthetic fixation of CO_2 is limited, the demand for ATP and NADPH declines. Oxygen can seize the resulting surplus electrons from PSI, especially in strong light, and the ensuing superoxide is converted to hydrogen peroxide (H_2O_2). Such reactive oxygen species strongly inhibit the repair process by interfering with the assembly of proteins (Takahashi and Murata, 2008), which can cause problems under conditions of environmental stress that curtail CO_2 fixation (see Section 7.1).

4.2 CARBON UPTAKE AND ASSIMILATION

Just as it does for other plants, uptake of CO_2 by leaves presents a special dilemma for grapevines. Because CO_2 is available in the atmosphere around plants and photosynthesis takes place inside the leaf cells, CO_2 must move from the atmosphere to the leaf interior. This happens by way of diffusion through the stomata (see Figure 1.15), then through the intercellular air spaces, cell walls, and finally across membranes into the cells and chloroplasts. As discussed in Section 3.3, the rate of diffusion is dependent on a driving force (a potential gradient or concentration gradient) and the resistance (the inverse of conductance) to diffusion. The diffusion path for CO_2 is almost entirely through the stomata, whereas the path for water vapor involves both the stomata and the cuticle (Boyer et al., 1997). CO_2 diffusion through the stomata follows the same path as H_2O during transpiration (E) but in the reverse direction (see Section 3.2), but the path for CO_2 is longer because it must cross additional cell walls and membranes (i.e., the plasma membrane and the chloroplast membranes) before it can be assimilated in the chloroplasts. Thus, although CO_2 may be able to use certain aquaporins (which are normally reserved for water transport and hence have been renamed cooperins if they are permeable to CO_2) to cross membranes, it meets more points of resistance than does water (Evans et al., 2009; Katsuhara et al., 2008; Maurel et al., 2008; Nobel, 2009). About half of the total resistance in the mesophyll is imposed by the cell walls alone (Terashima et al., 2011).

FIGURE 4.3

Relationship between photosynthetic CO_2 uptake and transpirational H_2O loss in mature leaves of two grapevine cultivars.

M. Keller, unpublished data.

Moreover, the concentration gradient of CO_2 from the outside air to the leaf's interior is very small. The current atmospheric CO_2 concentration (C_a) is approximately 0.039% (= 390 µmol mol^{-1} = 390 µbar = 390 ppm), whereas the concentration inside the leaf (C_i) cannot be less than 0%. Grapevines literally have to extract their main source of food and energy out of thin air. By contrast, air at 20°C and 50% relative humidity contains approximately 1.25% H_2O, whereas the air inside the leaf is at almost 100% relative humidity and contains approximately 2.5% H_2O. This means that the gradient for H_2O diffusion out of the leaf is more than 40 times steeper than that for CO_2 diffusion into the leaf. To make matters worse, H_2O molecules are smaller than CO_2 molecules, so H_2O diffuses approximately 1.6 times more easily through air than does CO_2. As a consequence, the "exchange rate" of mature leaves is only approximately 1 CO_2 molecule fixed for every 200–600 H_2O molecules lost; the current value depends on cultivar and soil conditions (Figure 4.3). This exchange rate is also termed water use efficiency (WUE = AE^{-1}). Leaves develop fewer stomata when they unfold under elevated atmospheric CO_2 concentrations (Woodward, 1987); the unfavorable exchange rate between CO_2 and H_2O might be the cause of this adaptive response.

The net uptake of CO_2 by the leaf equals the net rate of photosynthesis per unit leaf area (A, usually expressed in µmol CO_2 m^{-2} s^{-1}) and can be approximated by the following formula:

$$A \approx g_l \Delta C$$

where $\Delta C = (C_a - C_i)$ and g_l is the leaf conductance.

The term g_l reflects the combined gas conductances of the boundary layer (g_b) and the stomata (g_s) and increases with increasing stomatal aperture (see Section 3.2). As with water flow, conductance is the reciprocal of resistance ($g = r^{-1}$). The stomata are open in the light, and CO_2 diffuses into the leaf and into the mesophyll cells to be assimilated in photosynthesis. In the dark or under drought conditions, the stomata are almost closed ($g_s < 50$ µmol CO_2 m^{-2} s^{-1}), and photosynthesis ceases.

However, as discussed previously, opening of the stomata not only allows CO_2 to diffuse into the leaf but also allows water vapor to escape from the leaf. Thus, the dilemma for the vine is that it cannot simultaneously maximize CO_2 uptake and minimize H_2O loss. Consequently, g_s continually adjusts in a manner that optimizes CO_2 acquisition relative to H_2O loss; in other words, the stomata's "job" is to optimize the marginal water cost of carbon gain (Franks et al., 2013). Once CO_2 has entered the chloroplast, the chemical energy generated by the chloroplast pigments is used to convert (assimilate) CO_2 and water into carbohydrate (e.g., the hexose sugar glucose $C_6H_{12}O_6$). In the overall process of photosynthesis, the leaf absorbs CO_2 molecules and releases O_2 molecules as a waste product of the process (see Section 4.1), as shown in the following equation:

$$6CO_2 + 12H_2O \rightarrow C_6H_{12}O_6 + 6O_2 + 6H_2O$$
$$\Delta G = +2840 \text{ kJ mol}^{-1} (1 \text{ mol glucose} = 180 \text{ g})$$

The incorporation of CO_2 into carbohydrate is achieved by the action of 13 enzyme proteins in the Calvin cycle (named after one of its discoverers, the American biochemist Melvin Calvin, but also called photosynthetic carbon reduction cycle or reductive pentose phosphate pathway) located in the chloroplast stroma. The Calvin cycle proceeds in three successive stages (Figure 4.4). The first stage is termed carboxylation and combines CO_2 and protons derived from water with an initial carbon acceptor (ribulose-1,5-bisphosphate) to generate two intermediate molecules (3-phosphoglycerate), thereby "converting" the inorganic CO_2 into an organic molecule. As its name suggests, this initial organic compound, like all other intermediates of the Calvin cycle, is

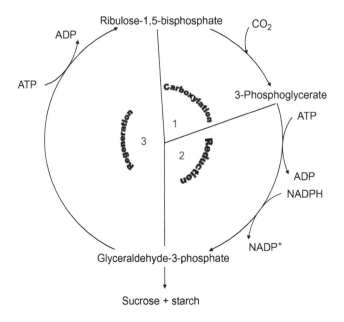

FIGURE 4.4

The three stages of the Calvin cycle that converts CO_2 into carbohydrates.

Illustration by M. Keller.

a sugar phosphate; thus, despite the previous simplified and convenient equation, glucose is not a direct product of the Calvin cycle. Because 3-phosphoglycerate is the first stable product of CO_2 fixation and contains three carbon atoms, grapevines are grouped with most other crop plants as the so-called C_3 plants. The carboxylation step is catalyzed by the enzyme ribulose bisphosphate carboxylase/oxygenase (rubisco). Rubisco occurs at very high concentration (often in substantial excess). It represents approximately one-third of a leaf's total protein complement, appropriating up to half of the leaf's nitrogen, which arguably makes it the most abundant protein nature ever invented (Spreitzer and Salvucci, 2002). Rubisco commandeers an extremely prominent position in grapevine physiology because almost all of the carbon that is assimilated is initially "captured" by this protein. As its full name suggests, the enzyme "works" in both directions so that CO_2 competes with oxygen for the same acceptor. Although under normal conditions the "forward" reaction (carboxylation) is much faster than the "reverse" reaction (oxygenation), the fact that O_2 is present in the atmosphere at a much higher concentration ($\sim 21\%$) than CO_2 (0.039%) means that the competition results in a loss of some of the CO_2 that has entered the cells (Spreitzer and Salvucci, 2002; see also Section 4.4).

The second stage, called reduction, forms a triose phosphate carbohydrate (glyceraldehyde-3-phosphate) using ATP and NADPH from photochemistry. The two reactions involved in this reduction oxidize ATP to ADP (i.e., one phosphate group is taken from ATP) and NADPH to $NADP^+$, which then have to be returned to the thylakoid membranes to be "recharged." During the final stage of the cycle, termed regeneration, the initial CO_2 acceptor molecule is restored in a further series of enzymatic reactions involving 10 of the 13 Calvin cycle enzymes and various intermediate sugar phosphates. The last of these steps also uses ATP. Thus, the cycle consumes two NADPH molecules and three ATP molecules for every CO_2 molecule fixed into carbohydrate. The regeneration stage allows continued uptake of CO_2, but it also means that only one out of every six triose phosphates produced can be either used for starch production within the chloroplast or exported to the cell's cytosol for the synthesis of sucrose (Paul and Foyer, 2001). Three turns of the Calvin cycle are required to form one 3-phosphoglycerate molecule, and six turns are required to produce one glucose molecule. However, the cycle operates very rapidly; free sugars such as sucrose appear within less than 30 s of "feeding" CO_2 to a leaf. Note that animal cells carry out all reactions of the Calvin cycle with the exception of the first and the last step. Because the two enzymes necessary to carry out those steps are missing, animals cannot convert CO_2 into sugar.

The organic product of photosynthesis, photosynthate, is often given as $(CH_2O)_n$, which is the basic component of carbohydrates. The simple sugars (usually glucose or fructose) are termed hexoses or monosaccharides and are produced by putting six of the basic units together (i.e., $n = 6$). They can be formed either directly inside the chloroplast or in the cytosol from triose phosphates exported from the chloroplast. Several enzymes are involved in hexose synthesis, some of which remove the phosphate groups (P_i), which can then be used to recharge ADP into ATP.

Combining glucose and fructose gives the disaccharide sucrose ($n = 12$), which is the most abundant sugar found in nature. Sucrose, a glycoside, is the major end product of photosynthesis and the predominant organic transport compound in the phloem of grapevines. Sucrose is produced in the cytosol rather than the chloroplast. The glucose phosphate molecule is energized (as UDP-glucose) before it is linked with the fructose phosphate molecule. In this case, the required energy is not provided by ATP but instead comes from a related compound, uridine triphosphate (UTP). The P_i released during sucrose synthesis is transported back into the chloroplast for continued

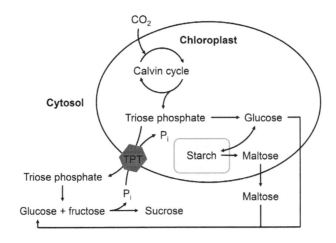

FIGURE 4.5

Simplified diagram of the dynamics of sugar production, export, and starch accumulation inside a leaf cell. The triose phosphate transporter (TPT) exports triose phosphate in exchange for inorganic phosphate (P_i) released during sucrose synthesis.

Illustration by M. Keller.

photophosphorylation (Paul and Foyer, 2001). A special protein embedded in the chloroplast membrane, the phosphate translocator or triose phosphate transporter (Figure 4.5), is responsible for exchanging triose phosphate (out) for P_i (in). Sucrose is exported from the cell to the phloem and distributed throughout the plant (see Section 5.1) for use in the production of other organic components, maintenance processes, and growth of the various sink organs.

In contrast to sucrose synthesis, the production of starch inside the chloroplast requires ATP to form ADP-glucose. The starch "assembly line" links thousands (5000–500,000) of glucose molecules together, resulting in the long, osmotically inert chains of glucose that are called starch. Depending on the nature of the bonding between the glucose molecules, the chains can be linear ($\alpha 1 \rightarrow 4$-glucans), in which case they are termed amylose, or branched ($\alpha 1 \rightarrow 4$- and $\alpha 1 \rightarrow 6$-glucans), in which case they are called amylopectin. Starch is organized into semicrystalline grains or granules of alternating amorphous and crystalline regions that can grow by adding layers (resulting in growth rings) from less than 1 µm to more than 100 µm in diameter (Emes and Neuhaus, 1997; Martin and Smith, 1995; Tetlow et al., 2004; Zeeman et al., 2010).

The Calvin cycle feeds carbon into starch production when the export of sucrose from the cell cannot keep pace with photosynthesis (Paul and Foyer, 2001). This is necessary because osmotic pressure is proportional to the number of dissolved particles (see Section 3.1): n glucose molecules cause n times the osmotic pressure of a polymer composed of n identical glucose units. Accumulation of the osmotically active sucrose or glucose (i.e., many small solute particles) would therefore cause water to flood the cell and it would burst. To avoid this potentially lethal problem, sucrose that accumulates in the leaf cells slows its own formation in a process called feedback inhibition. The P_i released during sucrose production assumes the function of "regulator"; a high P_i concentration indicates a high rate of sucrose export and promotes triose phosphate export from the chloroplast (Woodrow and Berry, 1988). Conversely, a low P_i concentration indicates that sucrose export is slow (e.g., due to low sucrose demand of sink organs) and promotes starch accumulation for temporary storage in the chloroplast. Therefore, phosphate deficiency resulting from low P_i uptake by the roots also leads to starch accumulation and prevents starch mobilization for export, and may lead to feedback inhibition of photosynthesis (Hendrickson et al., 2004b; see also Section 7.3). Starch also accumulates in leaves when the phloem of a shoot is obstructed or injured. This can happen due to physical injury from strong wind or trellis wires, crown gall or virus infection (e.g., leafroll virus), or tendrils coiling around a shoot. Such shoots can be easily recognized in a vineyard because the leaves turn bright red due to anthocyanin production from surplus sugar (see Figure 2.7). Accumulation of light-absorbing anthocyanin pigments may occur in order to protect the leaves' photosynthetic machinery from damage due to excessive light absorption because sugar accumulation leads to feedback inhibition of photosynthesis to bring the supply of sugar back in balance with the demand for sugar (Ayre, 2011).

The rate of sucrose export depends on the rate of photosynthesis: The greater the rate of photosynthesis, the greater the rate of concurrent export (Wardlaw, 1990). As discussed previously, however, grapevines, like other plants, often use starch production as an overflow mechanism when sucrose formation exceeds the leaves' export capacity or the demand for sucrose by the various plant organs (Johnson et al., 1982). The starch can be degraded (remobilized) by removing glucose units simultaneously from the many chain ends in a starch granule and converting them, partly via intermediary maltose, back to sucrose following maltose and glucose export out of the chloroplast (Smith, 2012; Smith et al., 2005; Zeeman et al., 2010). Little or no starch turnover normally occurs in leaves during the day, but degradation is switched on at night when there is no photosynthesis to provide carbon for continued sucrose export and generation of energy and reducing agents in the leaf (Geiger and Servaites, 1991; Geiger et al., 2000; Tetlow et al., 2004). In the absence of environmental limitations, the starch pool is depleted by sunrise and replenished by sunset, so that sucrose export remains relatively constant over the course of a day—night cycle. An intricate, albeit poorly understood, control mechanism that involves the circadian clock ensures that this starch accumulation and turnover is adjusted such that it compensates for changes in day length during the growing season to maintain a steady supply of sucrose over a 24 h period (Geiger et al., 2000; Graf and Smith, 2011; Stitt et al., 2007). Because more starch builds up during long days than during short ones, more is broken down during the ensuing short nights. In other words, the degradation of leaf starch reserves is matched exactly to the length of the night (Graf and Smith, 2011). This means that the total amount of sucrose export from a leaf increases as day length increases so that more sugar is potentially available for growth, yield formation, and fruit ripening at higher latitudes. Daytime starch degradation, on the other hand, apparently occurs only when dense clouds or canopy shade prevent net carbon assimilation (Smith et al., 2005). Due to its ability to channel assimilated carbon into the fairly large short-term storage pools of starch and sucrose, the vine can insulate the rate of sucrose supply to other organs from the changes in the rate of photosynthesis that occur during the course of day and night (Woodrow and Berry, 1988).

Glucose molecules (from split sucrose) can also be linked together as linear $\beta1\rightarrow4$-glucans forming chains (polymers) of 2000–25,000 glucose units called cellulose. Cellulose is arranged in approximately 35 parallel chains that form a network of long and very strong crystalline microfibrils held together by hydrogen bonds (Doblin et al., 2010). These inert, inextensible fibrils, whose tensile strength is greater than that of steel, form the basic framework (i.e., the stiff structural elements) of all plant cell walls, which make up a large portion of a vine's dry weight. In fact, cellulose is the most abundant organic polymer on Earth, making up approximately half of the total organic matter. For instance, dry wood consists of 40—50% cellulose (in addition to 25% hemicellulose and 25—35% lignin; Plomion et al., 2001), and cotton fibers used in the textile industry are made up of approximately 95% cellulose; cellulose is also the stuff that makes the paper of this book. Whereas the cellulose microfibrils are responsible for the stiffness and mechanical strength of cell walls, the wall matrix forms the plastic element that holds the microfibrils together and enables cell walls to move and expand (Hamant and Traas, 2010; see also Section 3.1). The viscous matrix is also composed of polysaccharides (e.g., the hydrophilic pectin and the cellulose-binding matrix glycans, also called hemicellulose, which is mainly composed of xyloglucan) in addition to structural and enzyme proteins (Cosgrove, 2005; Doblin et al., 2010; Reiter, 2002; Wolf et al., 2012). During periods of rapid growth, cellulose production can be a major drain on a cell's resources because both new (dividing) and expanding cells require the insertion of new (primary) cell wall material. Indeed, most of the carbon fixed by photosynthesis may be incorporated into cell walls (Reiter, 2002). In secondary cell walls, which fully grown cells insert between their primary cell wall and the cell membrane, the carbohydrate polymers are complemented by the three-dimensional, water-repellent polymer lignin. Lignin consists of simple phenolic building blocks that originate from the aromatic amino acid phenylalanine and are linked together in the process of lignification that often coincides with or even follows cell death and employs enzymes such as peroxidase along with hydrogen peroxide as an oxidizing agent (Boerjan et al., 2003; Schuetz et al., 2013). By displacing water, lignin fills the spaces between the polysaccharides, adding extra mechanical support and waterproofing, especially to the xylem cells of the wood. Its importance as a stiffener and strengthener of the cell wall structure makes lignin the world's second most plentiful polymer, comprising approximately 30% of all organic carbon (Boerjan et al., 2003). Once lignified, cells become unable to expand further and are glued to their cell neighbors by the pectin-rich middle lamella.

A chloroplast in a young, expanding leaf is basically a biosynthetic organelle manufacturing compounds for cell growth and development. In a mature leaf, on the other hand, a chloroplast fixes CO_2 for the production and export to other parts of the plant. Photosynthesis therefore increases with leaf age and in grapevines reaches a maximum approximately 5 or 6 weeks after leaf unfolding, gradually declining thereafter, as senescence sets in (Bertamini and Nedunchezhian, 2003; Williams and Smith, 1985). Senescence is the end phase of differentiation and eventually ends in death and abscission of the leaf.

4.3 PHOTORESPIRATION

As previously discussed, rubisco, the first enzyme in the Calvin cycle, can catalyze the addition of both CO_2 and O_2 to ribulose-1,5-bisphosphate. The latter is called oxygenation and is the main reaction occurring during a process known as photorespiration (Bauwe et al., 2010; Ogren, 1984; Spreitzer and Salvucci, 2002). Because the earth's current atmosphere contains vastly more O_2 (21%) than CO_2 (0.039%) and due to the release of O_2 in the water-splitting reaction (see Section 4.1), the O_2:CO_2 ratio in the chloroplast fluid is approximately 24:1 at 25°C. Moreover, though CO_2 is much more water soluble than is O_2, the solubility of CO_2 decreases with increasing temperature; thus, the O_2:CO_2 ratio increases so that higher temperatures favor photorespiration.

Photorespiration is a problem for grapevines because the process works in the opposite direction of photosynthesis and therefore competes with it for ATP and NADPH, produces toxic phosphoglycolate instead of phosphoglycerate, and results in a loss of CO_2 that could otherwise be used for sugar production. Part of this CO_2 can be recovered in a series of reactions called the photorespiratory carbon oxidation cycle or glycolate pathway after the initial product of the oxygenation reaction. This process is quite complex and involves interconversion and transport of several amino acids and cooperation among three distinct cell organelles—the chloroplast, peroxisome, and mitochondrion (Bauwe et al., 2010; Foyer et al., 2009; Ogren, 1984). It also costs energy in the form of ATP and results in heat loss. In addition, the release of toxic ammonium (NH_4^+) from amino acids requires yet more ATP for detoxification (i.e., reassimilation). Although this reassimilation of NH_4^+ is costly, losing the NH_4^+ and then replacing it with soil-absorbed nitrate (NO_3^-) would be even more expensive. The flux of NH_4^+ due to photorespiration is at least 10-fold higher than that due to root NO_3^- uptake and assimilation (see Section 5.3).

The glycolate pathway recovers approximately 75% of the lost carbon and returns it as phosphoglycerate to the Calvin cycle. Consequently, photorespiration reduces the efficiency of photosynthesis, and approximately one out of every four CO_2 molecules that enter the chloroplast is released back to the atmosphere (Bauwe et al., 2010; Düring, 1988). The biological function, if any, of photorespiration is not understood. One possible function could be the use of electrons when energy supply exceeds demand, such as in very bright light, or during drought conditions when the photosynthetic pigments capture more photons than the Calvin cycle can use to fix CO_2 (Apel and Hirt, 2004; Foyer et al., 2009; Lawlor and Cornic, 2002; Niyogi, 2000; Ogren, 1984). The "venting" of surplus electrons substantially reduces a potential "energy overload" that would otherwise damage the photosynthetic machinery and lead to photoinhibition (Takahashi and Badger, 2011; see also Section 4.1).

Nonetheless, it is likely that this seemingly wasteful pathway may be an evolutionary "hangover." Photosynthesis, and hence rubisco, evolved very early in the existence of life on Earth (more than 3 billion years ago) when the atmosphere was rich in CO_2 but almost devoid of O_2 (Bauwe et al., 2010; Foyer et al., 2009; Xiong and Bauer, 2002). As a remarkable aside, it was the "invention" of photosynthesis with oxygen as its waste product (in addition to injection of oxygen into the air from abiotic sources) that enabled the evolution of all oxygen-breathing life forms, including humans (Xiong and Bauer, 2002). This innovation was a momentous breakthrough event in the history of life because it enabled the photosynthetically endowed organisms to exploit limitless supplies of water and solar energy. Of course, not all life depends on oxygen for respiration, as viticulturists and winemakers should know. Brewer's or baker's yeast (*Saccharomyces cerevisiae*) is one of many anaerobic microorganisms that cannot or prefer not to use oxygen. Although yeast is a facultative anaerobe and thus is not poisoned by oxygen, unlike obligate anaerobes, it relies on energy-inefficient fermentation rather than respiration of sugar for energy generation (Rolland et al., 2006). This is why Louis Pasteur described fermentation as "life without oxygen." Ethanol production occurs only under anaerobic conditions; in the presence of oxygen, ethanol is oxidized to acetic acid (i.e., vinegar) by other (aerobic) yeasts or *Acetobacter* bacteria.

The rising atmospheric CO_2 concentration ($[CO_2]$) that is a consequence of rapid burning of fossil fuels, forest clearing, and soil cultivation and is responsible for global climate change (Huang et al., 2000; IPCC, 2013) will, to some extent, turn the clock back, as it were, to benefit the carboxylation function of rubisco and favor photosynthesis over photorespiration (Foyer et al., 2009; Paul and Foyer, 2001). Whereas during most of the last million years, that is, at least five times as long as humans of our species, *Homo sapiens*, have roamed Earth, [CO_2] never exceeded 300 ppm, and was often below 200 ppm during the repeated glacial periods (ice ages), it has increased from approximately 270 ppm prior to the beginning of industrialization to greater than 390 ppm today (Franks et al., 2013; Petit et al., 1999). If the rise continues at the average annual rate of approximately 2 ppm that occurred during the first decade of the twenty-first century, [CO_2] will reach close to 600 ppm by the end of the century (Franks et al., 2013; IPCC, 2013).

The increase in ambient [CO_2] results in a steeper CO_2 concentration gradient into the leaves, which favors carboxylation by rubisco and photosynthesis, while decreasing g_s. Indeed, across all C_3 species investigated thus far, an increase in [CO_2] from approximately 370 ppm to approximately 570 ppm was associated with a one-third increase in both light-saturated and daily photosynthesis that was accompanied by a greater than 80% increase in leaf starch content, whereas rubisco content and g_s declined by approximately 20% (Ainsworth and Rogers, 2007; Long et al., 2004). At least some of the decrease in g_s is a consequence of the lower stomatal density of leaves grown at higher [CO_2] (Franks et al., 2013; Woodward, 1987). Grapevines are no exception to this general trend (Düring, 2003; Schultz, 2000; Schultz and Stoll, 2010; Tognetti et al., 2005). All other plant processes affected by elevated [CO_2] are thought to be an outcome of these two basic responses (Long et al., 2004). In the long term, however, photosynthesis can only increase if the vine can utilize the extra sugar produced, for example, by increasing shoot and root growth or yield, all of which increase under higher [CO_2] (Bindi et al., 1996; Johnson et al., 1982).

Some plant species long ago, and often independently, "invented" solutions to the problem posed by high [O_2]: The most prevalent of these are the so-called C_4 plants, which were named for the four-carbon oxaloacetate, rather than the usual three-carbon 3-phosphoglycerate, as the product of CO_2 fixation (Foyer et al., 2009; Lundgren et al., 2014; Ogren, 1984; Sage et al., 2012). Among other adaptations, the C_4 plants use a biochemical "pump" to concentrate CO_2 near the rubisco enzymes, which almost eliminates photorespiration and speeds up the rates of photosynthesis and growth, especially in high light and at high temperature. Because of their competitive advantage in warm climates, these plants include not only a few important commercial crops (e.g., maize, sorghum, and sugarcane) but also many highly successful weed species (e.g., crabgrass, switchgrass, pigweed, and nutgrass). Photosynthesis in C_4 species benefits less from the current rise in atmospheric [CO_2] so that they can be expected to become somewhat less competitive in the near future—although it is far from clear whether this will make weed control any easier.

4.4 RESPIRATION

Respiration is the series of three processes (glycolysis, citric acid cycle, and electron transport chain) responsible for releasing the energy stored in carbohydrate and incorporating it into a form (ATP) that can be readily used to power most energy-requiring metabolic processes in the plant. In addition, it produces a number of intermediate organic components ("carbon skeletons") that can be used for the manufacture of other compounds. In the process, glucose is completely oxidized to CO_2 and its electrons are transferred to oxygen, which is reduced to water. In other words, respiration is the oxidation of carbon reduced by photosynthesis. Therefore, the chemical equation that summarizes this process, which occurs in all living tissues, is basically the reverse of that used to describe photosynthesis:

$$C_6H_{12}O_6 + 6O_2 + 6H_2O \rightarrow 6CO_2 + 12H_2O$$
$$\Delta G = -2840 \text{ kJ mol}^{-1}$$

Sugars serve as the main carriers of energy that the leaves "extract" from the sun. Because sucrose, rather than glucose, is the transport sugar in grapevines, and because respiration also occurs in nonphotosynthetic tissues that rely on sucrose import, sucrose is the sugar used as a starting point (substrate) for respiration. However, sucrose must be broken down into its two-component hexoses, glucose and fructose, both of which can then enter the glycolysis pathway in the cells' cytosol. Stored starch can also be used for respiration after it has been hydrolyzed to glucose (Figure 4.6). When such carbohydrates are not readily available, tissues can rely on alternative substrates to fuel respiration (Zell et al., 2010). These include organic acids (e.g., malate), lipids, and even proteins. The nature of these substrates can be guessed at by measuring the so-called respiratory quotient (RQ = CO_2 released/O_2 consumed). Only sugars will be respired to one molecule of CO_2 for each molecule of O_2 consumed, which results in RQ = 1. For the comparatively more oxidized organic acids RQ ≈ 2, and for the more reduced lipids and proteins, RQ < 0.5.

Although the oxidation of one sucrose molecule with its 12 C atoms yields a net energy equivalent of 60–64 ATP molecules, the maximum respiration efficiency is probably slightly less than 5 molecules of ATP per molecule of CO_2 released (Amthor, 2000). Glycolysis (Greek *glykos* = sugar, *lysis* = splitting) degrades (oxidizes) glucose in a series of enzymatic reactions to the three-carbon compound pyruvate (Figure 4.6). This process initially requires input of energy in the form of ATP but later regenerates twice as many ATP molecules in addition to NADH. Pyruvate still contains more than 75% of the energy contained in glucose and can be used as the starting material for the citric acid cycle.

The next steps of respiration are carried out by the citric acid cycle (also called the Krebs cycle after its discoverer, the German physician Hans Adolf Krebs; the tricarboxylic acid cycle; or the TCA cycle) and an electron transport chain known as oxidative phosphorylation. They occur in specialized organelles termed mitochondria (singular: mitochondrion), which can be regarded as the cell's power plants. The number of mitochondria per cell varies and is directly related to the metabolic activity of the tissue to which the cell belongs (i.e., energy generation is driven by demand), but there are hundreds of mitochondria scattered through most cells. This is possible because mitochondria have their own genome and replicate independently of their "mother cell" by simple division. In other words, they reproduce asexually like bacteria, and they can do so many times while the cell (and its nucleus) does not divide. Nevertheless, most of the mitochondrial enzymes are now encoded by genes in the nucleus, partly in concert with the few remaining mitochondrial genes (Millar et al., 2011). Not surprisingly, the enzymes carrying out glycolysis are attached to the outside of the mitochondria (Fernie et al., 2004). Pyruvate is transported into the mitochondria, where it is oxidized to acetic acid and linked to a sulfur-containing coenzyme called acetyl coenzyme A (acetyl-CoA). Acetyl-CoA is then combined with oxaloacetate already present in the cycle or supplied from the glycolysis intermediate phosphoenolpyruvate by the enzyme phosphoenolpyruvate carboxylase (Figure 4.6). In a series of three enzymatic reactions, the TCA cycle reduces the resulting molecule completely to CO_2 and H_2O, generating a substantial amount of

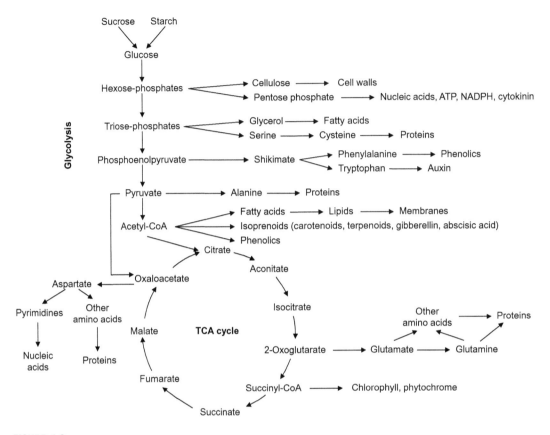

FIGURE 4.6

Glycolysis and TCA cycle are of central importance for the manufacture of many organic compounds throughout the plant.

Modified after Salisbury and Ross (1992).

chemical energy or reducing equivalents (10 NADH per glucose) in the process. The remaining three steps of the cycle restore oxaloacetate, which is then ready to accept the next transformed pyruvate (i.e., acetyl-CoA), allowing the continued operation of the cycle. Yet, many steps of the TCA cycle can also be performed by enzymes present in the cytosol, and the cycle is often incomplete in that it interacts with the cytosol by means of membrane transporters that permit the bypassing of various steps in the mitochondria (Millar et al., 2011).

The electron transport chain transfers electrons from NADH to oxygen in a stepwise manner involving several carrier proteins bound to the inner membrane of the mitochondrion. This mitochondrial electron transfer chain is structurally very similar to the photosynthetic electron transfer chain in the chloroplast and also involves a transfer of protons from the interior of the mitochondria to the cytosol (see Section 4.1). The free energy released during this process (actually the potential gradient resulting from the movement of protons across the mitochondrial membrane) is used to power an enzyme termed ATP synthase that recharges ADP to ATP using P_i (Fernie et al., 2004; Millar et al., 2011). This ATP synthase is life's universal power generator and is responsible for almost all of the ATP produced by living cells. The production of 15 ATP equivalents per pyruvate molecule by oxygen-consuming mitochondria (burning sugar as fuel) is by far the most efficient means of biological energy generation. However, because respiration also produces a small amount of highly reactive and damaging oxygen intermediates (the so-called reactive oxygen species; see Section 7.1), it may slowly but surely lead to oxygen poisoning, which in turn results in aging and, eventually, senescence (Logan, 2006). In addition to the "standard" electron transport chain, the mitochondria employ an alternative electron transport pathway, which uses the enzyme alternative oxidase to reduce oxygen to water but does not produce ATP, in an attempt to minimize the production of reactive oxygen species under stress conditions such as low temperature (Millar et al., 2011).

In addition to the formation of ATP, respiration generates various intermediate carbon compounds and thus is central to the production of a wide variety of components of the plant's metabolism (Fernie et al., 2004; Sweetlove et al., 2010; see also Figure 4.6). These include cellulose, nucleotides (and thus nucleic acids, cytokinins, etc.), amino acids (and thus proteins, flavonoids, anthocyanins, lignin, etc.), fatty acids (and thus lipids, etc.), isoprenoids (and thus terpenoids, carotenoids, cytokinins, abscisic acid, gibberrellins, etc.), and porphyrins (and thus chlorophylls, phytochromes, etc.). For example, all 20 standard amino acids formed in plants for the manufacture of proteins are assembled from the carbon skeletons provided by the organic acid products of glycolysis, the TCA cycle, and the Calvin cycle. Therefore, a large portion of the sugar that is metabolized by glycolysis and the TCA cycle is not oxidized to CO_2 but instead is diverted to biosynthetic purposes. Indeed, it seems that the TCA "cycle" operates as a full cycle only when a cell has a high demand for ATP. Otherwise it may take various noncyclic "shortcuts" depending on the cell's need for metabolic precursors or intermediates (Sweetlove et al., 2010). Moreover, both ATP and ADP themselves are also used as the original precursor molecules for the assembly of all compounds that comprise the group of cell division hormones called cytokinins (Sakakibara, 2006).

4.5 FROM CELLS TO PLANTS

Although photosynthesis is the ultimate source of carbohydrates and supplies some ATP and NAD(P)H in photosynthetic tissues, respiration is the powerhouse of plants, providing the driving force for biosynthesis (production of new materials), cell maintenance, and active transport (including nutrient uptake). Respiration is generally broken down into two components: growth respiration and maintenance respiration (Amthor, 2000; Penning de Vries, 1975; Penning de Vries et al., 1974). In growth respiration, reduced carbon is processed to fuel the accumulation of new biomass ("construction cost"). Sugar export from the leaves in the phloem to other growing organs also requires energy ("transport cost"), as do uptake, assimilation, and long-distance transport of mineral nutrients. These processes are sometimes viewed as comprising a third respiratory category that supports growth (Kruse et al., 2011). A substantial portion of growth respiration involves the production of carbon skeletons for nitrogen assimilation (see Section 5.3). The cost of production for plant tissues varies with the nature of the chemical compounds that are being produced (Table 4.1). For instance, producing proteins or phenolic compounds is much more expensive than producing

Table 4.1 Average Construction Costs of Major Chemical Plant Components

Component	Cost[a]
Carbohydrates	1.17
Organic acids	0.91
Lipids	3.03
Amino acids, proteins	2.48
Phenolics	2.60
Inorganic compounds (minerals)	0

[a]*Expressed as g glucose used to produce 1 g of each compound class.*
From Vivin et al. (2003).

carbohydrates; thus, plant parts with a high protein content, such as leaves, are relatively expensive to build (Vivin et al., 2003). The rate of growth respiration increases as the rate of growth increases. The other component, maintenance respiration, is needed to keep existing, mature cells in a viable, functional state and is coupled to energy generation, internal carbon and nutrient transport, protein and lipid turnover (i.e., production−breakdown−recycling−resynthesis), and adjustment or acclimation to changing environmental conditions. Protein turnover and adjustment/acclimation are linked so that plants do not have to maintain a complete set of enzymes and transport proteins at all times just to be ready for changing conditions (Araújo et al., 2011). It is much more economical for them to produce specialized proteins, or groups of proteins, "on demand" and recycle them when not needed so that their amino acid building blocks can be reassigned to other temporary jobs. Scientists do not have a good understanding of energy utilization by maintenance respiration, but estimates suggest that it may represent more than 50% of the total respiratory CO_2 release.

Growth is not possible without respiration, but respiration consumes some of the carbon that could otherwise be used to "build" the plant body. Therefore, the net amount of carbon accumulated in a vine's biomass during a growing season equals the total amount of carbon fixed in photosynthesis minus that "lost in respiration." Respiration can consume a significant portion (30−80%) of the photosynthetically fixed carbon each day over and above the losses due to photorespiration, and it releases large amounts of CO_2 back to the atmosphere (Amthor, 2000; Atkin and Tjoelker, 2003; Atkin et al., 2005). Grapevine leaves respire at a rate of approximately 5−10% of the rate of photosynthesis. However, all plant tissues, whether photosynthetic or not, respire, and they do so 24 h a day. At least the green shoots and, to a much lesser degree, even the older woody organs (especially the relatively chlorophyll-rich inner bark) are able to refix some of the lost CO_2 by photosynthesis and thereby contribute to their own construction and maintenance cost (Kriedemann and Buttrose, 1971; Pfanz, 2008; Palliotti et al., 2010; Saveyn et al., 2010). Respiration rates vary with the age of the plant and differ from one organ to another depending on the availability of carbon substrates, stage of development, and temperature. For instance, a temporary 10°C rise in temperature usually leads to at least a doubling of the respiration rate (i.e., $Q_{10} \geq 2$). Young grapevines lose approximately one-third of their daily photosynthate to respiration, and this loss increases in older plants as the amount of nonphotosynthetic tissue (e.g., in cordon, trunk, and roots) increases. The total annual respiration of a vineyard soil has been estimated to exceed 5 tonnes of carbon per hectare, with about half of this attributed to grapevine roots (Franck et al., 2011). Overall, both whole-plant respiration and growth are approximately proportional to the three-quarter power of whole-plant mass (Price et al., 2010).

As a general rule, the greater the metabolic activity of a given tissue or organ and/or the higher its growth rate, the higher its respiration rate. Meristems have the highest respiration rates so that young, rapidly growing plant parts, such as developing buds, shoot and root tips, unfolding leaves, or flowers, can respire up to 10 times more rapidly than older parts that rely mainly on maintenance respiration. Between 70% and 90% of imported carbon is incorporated in new material in growing organs, which contain about 45% carbon (i.e., 0.45 g C g^{-1} dry matter; Keller et al., 1995); the remainder is released as CO_2 (Amthor, 2000). This means that during and after budbreak, vines have a negative carbon balance: They will literally lose weight (dry matter) until the new leaves become photosynthetically competent and begin to replace the lost carbon (Buttrose, 1966b; Holzapfel et al., 2010; Weyand and Schultz, 2006a). It appears that respiration constitutes a compromise between energy generation and carbon gain. A high respiration rate means that more ATP is produced (and thus more energy is available for metabolism), but this comes at the expense of carbon lost to accumulation of biomass. Conversely, a low respiration rate improves the carbon balance but reduces the energy available to utilize this additional carbon. The increase in photosynthesis associated with the elevated atmospheric [CO_2] that is driving global climate change (see Section 4.3) may also stimulate respiration by increasing leaf sugar and starch concentrations (Amthor, 2000; Leakey et al., 2009). Although this will diminish the net plant carbon balance, it may nevertheless lead to greater carbon export for growth and yield formation, which is associated with higher growth and maintenance respiration. The rise in temperature that accompanies the higher atmospheric [CO_2] may also stimulate respiration and growth, although probably not as much as do short-term, temporary increases in temperature, due to the effects of acclimation to long-term temperature changes (Amthor, 2000).

Although nutrient ions *per se* do not cost any glucose for their construction (they are inorganic chemicals), grapevines spend much of the energy generated during respiration on nutrient acquisition—that is, on the uptake of nutrient ions from the soil solution and their assimilation into organic compounds. Nitrate and sulfate acquisition incurs particularly high respiratory costs and causes an increase in respiration rates to supply carbon skeletons for the production of amino acids (see Section 5.3). Vigorous species or cultivars are characterized by spending less respiratory energy for nutrient acquisition and more for growth. In addition, their maintenance costs also seem to be slightly lower, although this difference is small compared with the difference in energy allocated to nutrient acquisition. Thus, rather than having higher photosynthetic rates, vigorous plant species often respire a lower proportion of their acquired carbon than do slow-growing species; that is, the respiration:photosynthesis ratio is smaller in vigorous species (Loveys et al., 2002). However, differences in maintenance respiration become more important when vines are grown under unfavorable conditions. For example, low soil nitrogen availability decreases the absolute rates of whole-plant respiration but increases the proportion of carbon fixed during the day that is lost during respiration. This happens because vines with a low nitrogen status have reduced photosynthesis and invest more of their photosynthate in the growth of roots, which have higher respiration rates than shoots.

CHAPTER 5

PARTITIONING OF ASSIMILATES

CHAPTER OUTLINE

5.1 Photosynthate Translocation and Distribution	146
5.1.1 Allocation and Partitioning	156
5.2 Canopy–Environment Interactions	163
5.2.1 Light	165
5.2.2 Temperature	174
5.2.3 Wind	177
5.2.4 Humidity	179
5.2.5 The "Ideal" Canopy	180
5.3 Nitrogen Assimilation and Interaction with Carbon Metabolism	182
5.3.1 Nitrate Uptake and Reduction	183
5.3.2 Ammonium Assimilation	185
5.3.3 From Cells to Plants	187

ABSTRACT

Grapevines manufacture carbohydrates and other assimilates in source organs and transport them via the phloem to sink organs for further metabolism or storage. All organs start out as sinks, but leaves transition to sources as they grow, and woody storage tissues may variously serve as sources or sinks. The phloem transports mainly sucrose and some amino acids and inorganic nutrients such as potassium dissolved in water. Phloem flow follows a pressure gradient that is generated osmotically by loading of solutes in sources and unloading in sinks. Assimilate partitioning to the various sinks of a vine depends on their relative priority, which changes with position, seasonal development, and environmental influences. Microclimatic variables such as light, temperature, wind speed, and humidity alter a vine's canopy and are altered by it. This interaction is sensed by the leaves and used to determine canopy architecture and photosynthesis, and thus vine growth and productivity. In addition to the photosynthates produced in the leaves, assimilates are also produced from mineral nutrients taken up by the roots. Nitrogen is taken up mostly as nitrate and converted into amino acids for metabolism, storage, or export. Photosynthesis and nitrogen assimilation are interdependent and influence each other via positive and negative feedback loops depending on environmental and developmental factors.

Keywords: Allocation; microclimate; nitrogen assimilation; partitioning; phloem; pressure flow

5.1 PHOTOSYNTHATE TRANSLOCATION AND DISTRIBUTION

Grapevines, like all other plants, employ a division of labor between different tissues and organs that is coupled to their structural framework and relies on an efficient transport system connecting the various parts. The organic compounds produced during photosynthesis and nutrient assimilation (photosynthates and assimilates) and certain mineral ions must be transported from their place of manufacture or storage ("source": supply) to places of intermediate or final use or storage ("sinks": demand). A source is any plant organ that exports material. The typical source organ is the mature leaf that produces more photosynthate than it needs for its own growth and metabolism, but all green (i.e., chlorophyll-containing) tissues (including those in the shoot and cluster) can contribute to a vine's total assimilate production. Other sources of grapevines include the woody structures (canes, cordon, trunk, and roots), which function as storage organs. However, all sources begin their life as sinks. A sink is a nonphotosynthetic plant organ or an organ that produces insufficient photosynthate to supply its own needs. Sinks can be growing vegetative organs (expanding leaves and root tips), storage organs (shoots or canes, trunks, and roots), or reproductive organs (flowers and developing fruits and seeds).

All leaves are initially sinks because they need to build up their photosynthetic machinery before they can start producing and exporting their own sugar. The construction phase entails an investment of carbon (and other nutrients) in both building material and energy (respiration). The time period required for a leaf to make the transition from sink to source is important because it determines how quickly the leaf starts paying dividends on the carbon invested to construct it. It seems as though leaves unfolding in spring pass through this period more rapidly than leaves unfolding in fall; thus, late-season leaf growth may be too slow to make a significant contribution to fruit ripening or carbohydrate storage. This is one reason why viticultural practices aim at suppressing shoot growth late in the growing season. Such growth may compete for resources with the ripening fruit as well as with the replenishment of the reserve pool and cold acclimation of the vine. The sink-to-source transition happens gradually, beginning at the leaf tip; a grapevine leaf generally starts exporting assimilates when it has reached approximately one-third its final size but continues to import carbon up to half its final size, which requires bidirectional assimilate transport in the petiole (Koblet, 1969; Taiz and Zeiger, 2006). The oldest, most basal leaf on a shoot typically begins exporting assimilates when the shoot has approximately five or six leaves (Hale and Weaver, 1962; Koblet, 1969; Koblet and Perret, 1972; Yang and Hori, 1979, 1980). Export is initially directed toward the shoot tip, but as soon as the next leaf above makes the transition from sink to source, the older leaf also begins to export a portion of its assimilates toward the shoot base and into the permanent structure of the vine. Only a few days later the direction of export from the older leaf becomes exclusively basipetal. This pattern is repeated as new leaves unfold from the growing shoot tip, although export of leaves below a cluster becomes bidirectional again after fruit set. Following the sink–source transition, a leaf will no longer import substantial amounts of sugar, even under conditions of heavy shading (i.e., the leaf acts at most as a very weak sink). Approximately 40 days after unfolding, the leaf reaches maturity. Thereafter, its chlorophyll content and photosynthetic rate often begin to decline, and the leaf gradually exports less photosynthetically fixed carbon as it ages (Hunter et al., 1994; Intrieri et al., 1992, 2001; Poni et al., 1994a, b). Under nonlimiting conditions, however, many leaves, especially those that unfold early in the growing season, can remain fully photosynthetically active for more than 100 days; indeed, leaf senescence and the associated decline in photosynthesis appear to set in at about the same time of year in all leaves in a canopy (Schultz et al., 1996). This means that leaves that are formed late in the growing season have a progressively shorter window of opportunity to "pay back" their construction costs. Old leaves, though still photosynthetically active, do not seem to export assimilates but, rather, retain them for their own metabolic demands (Koblet, 1969). Instead, the leaves become a major source of mineral nutrients (particularly nitrogen, phosphorus, and potassium) toward the end of their life as a result of the resorption of proteins and other constituents during senescence (Thomas and Stoddart, 1980).

The woody organs of the vine are generally regarded as sinks because they do not photosynthesize except to refix a portion of the CO_2 lost during respiration (Pfanz, 2008; Saveyn et al., 2010). They rely on imported photosynthate and nutrients for growth and metabolism. However, sugar and other nutrients imported and stored as starch and other material (e.g., amino acids and proteins) in these tissues during the growing season are remobilized in spring to support budbreak and the initial shoot and root growth before the new leaves start to export assimilates (Holzapfel et al., 2010; Wardlaw, 1990; Williams, 1996). This starch remobilization (i.e., hydrolyzation to glucose and glucose phosphate that are then converted back to sucrose for export) is stimulated by the plant hormone gibberellin. Although the oldest leaf on a new shoot begins exporting some of its assimilates to the woody parts of the vine as early as 2 or 3 weeks after budbreak, remobilization and export from the wood to the developing shoots continues through bloom (i.e., transport is bidirectional). In addition, stored reserves also serve as a buffer by providing a temporary supply of carbon that can be mobilized during stress periods, such as overcast conditions or loss of leaf area, that interfere with photosynthesis (Bloom et al., 1985; Holzapfel et al., 2010).

The transport must be sufficiently flexible to supply the growing and metabolizing tissues of roots, shoots, leaves, flowers, and fruits. It therefore must be relatively rapid to keep up with the demand for "fuel" and "building blocks" by these growing tissues, which can be far away, and must be able to reverse the direction of flow depending on the relative needs of different organs at different times. This translocation of solutes occurs in the phloem tissue (more precisely in the phloem's sieve elements; see Section 1.3), and because solute concentration is usually higher in the source phloem than in the sink phloem, it follows a gradient of concentration (i.e., chemical potential).

By far the major compound (besides water) translocated in the phloem of grapevines is sucrose (Koblet, 1969; Swanson and El-Shishiny, 1958), which accounts for 50—70% of the sap osmotic pressure (Table 5.1). The mass flow resulting from the steep osmotic gradient created by sucrose also drags along numerous other solutes, of which potassium (K^+), amino acids, glucose, fructose, and malate make up much of the remainder of the osmotic components (Patrick, 1997; Peuke, 2010; Peuke et al., 2001). The concentration of amino acids and organic acids (e.g., malate and ascorbate) varies widely but is normally much lower than the sucrose concentration. Nonetheless, as the leaves change from carbon sources to mineral nutrient sources during senescence, phloem sap composition shifts from predominantly sucrose to mainly amino acids (Thomas and Stoddart, 1980). Some nutrient ions can also be transported in the phloem. For instance, most of a grape berry's K^+ is imported via the phloem, and phosphate can be retrieved from old leaves and recycled to young, expanding leaves under conditions of phosphorus deficiency. The phloem also contains short pieces of RNA, termed microRNA (or miRNA), and several peptides and proteins.

Table 5.1 Composition of Phloem and Xylem Sap

Solute	Xylem Sap	Phloem Sap
Sugars (mostly sucrose, some glucose and fructose)	≤ 5 g L^{-1} ≤ 15 mM	100–300 g L^{-1} 300–900 mM
Amino acids (mostly glutamine, some glutamate, asparagine, and aspartate)	≤ 2 g L^{-1} ≤ 15 mM	5–40 g L^{-1} 30–270 mM
Proteins (mostly glycolytic, structural, and stress-related proteins)	Traces	0.2–2 g L^{-1}
Organic acids (mostly malate)	0.1–0.5 g L^{-1} 0.7–4 mM	1–3 g L^{-1} 7–20 mM
Inorganic ions	0.2–4 g L^{-1}	1–5 g L^{-1}
Potassium (K$^+$)	1–10 mM	40–100 mM
Sodium (Na$^+$)	0.1–2 mM	0.1–5 mM
Magnesium (Mg^{2+})	1–2 mM	1–6 mM
Calcium (Ca^{2+})	1–5 mM	0.1–2 mM
Phosphate (H$_2$PO$_4^-$/HPO$_4^{2-}$)	0.1–3 mM	7–15 mM
Nitrate (NO$_3^-$)	0.01–20 mM	Traces
Total solute concentration	10–100 mM	250–1200 mM
Osmotic pressure (π)	0.02–0.2 MPa	0.6–3 MPa
pH	5.0–6.5	~7.5

Modified from Buchanan et al. (2000), Peuke et al. (2001), and Keller et al. (2001b).

Phloem-specific proteins, manufactured in the companion cells, are involved in, among other things, sugar metabolism and transport (sucrose carriers, ATPases), nutrient transport (K$^+$ and Ca^{2+} channels), and water transport (aquaporins) across membranes. Other proteins are responsible for the almost instantaneous sealing of the phloem's sieve plate pores upon injury (e.g., by pruning or removal of leaves or clusters), whereas long-term plugging is achieved within minutes by deposition of callose (see Section 1.3), whose production is induced by a sudden increase in (cytosolic) calcium in the sieve elements (Furch et al., 2007; Kühn et al., 1999; Oparka and Santa Cruz, 2000; Van Bel, 2003). Because the sieve elements lack the necessary machinery to manufacture proteins, RNA pieces, which even include base sequences that act as a "zip code" that specifies the recipient tissue and ensures proper delivery, might be used as long-distance signals (even across graft unions) that promote or, more frequently, inhibit the production of certain proteins in the target sink organs (Kehr and Buhtz, 2008; Lough and Lucas, 2006; Turgeon and Wolf, 2009; Turnbull and Lopez-Cobollo, 2013). Such inhibition is called gene silencing because it essentially "switches off" specific genes by preventing their RNA from being translated into proteins. Furthermore, almost all plant hormones (auxin, gibberellins, cytokinins, and abscisic acid) are also translocated in the phloem in addition to the xylem. By acting as chemical signals, these hormones exert remote controls on physiological processes in organs and tissues distant from their site of production. Even electrical signals may be transmitted through the phloem, which gives the phloem properties of a neural network (Fromm and Lautner, 2007; Mancuso, 1999). Electrical signals may also be transmitted through the xylem, where they arise locally in response to hydraulic pressure waves or transported compounds due to physical injury. Such signals are generated in response to sudden environmental stimuli (e.g., temperature, light, touch, and wounding) and exert an additional regulation of physiological functions. The long-distance information transmission of electrical signals is much more rapid (>10 m h^{-1}) than that of chemical signals (~ 1 m h^{-1}). Unfortunately, some undesired hitchhikers, such as viroids, viruses, and some bacteria, also use the phloem-generated mass flow for transport and are able to slip through the intercellular plasmodesmata, enabling them to spread easily throughout the vine (Christensen et al., 2004; Lough and Lucas, 2006; Oparka and Santa Cruz, 2000; Roberts and Oparka, 2003; Van Bel, 2003). Finally, the ability of the phloem to transport fungicides, insecticides, and herbicides within the plant is exploited in viticultural practice to combat pathogens and weeds.

The composition of the phloem sap not only depends on species, cultivar, and rootstock but also varies with changing physiological conditions, location within the plant, and time of the growing season. Although the composition also varies with time of day, the amount of transported solutes remains more or less constant over a 24 h period (Peuke et al., 2001). Solutes can also be exchanged between the phloem and xylem (Lalonde et al., 2004; Metzner et al., 2010b). Xylem-to-phloem transfer occurs particularly in leaf traces (i.e., in the vascular bundles branching away from the shoot into a leaf) and in the minor veins of a leaf, but it can also happen in the shoot, especially at nodes. It enhances the delivery of nutrients to rapidly growing, but slowly transpiring, sinks (e.g., meristems, fruits, and seeds). Phloem-to-xylem transfer is important in roots and shoots for redirecting some solutes delivered by the phloem (e.g., amino acids and K$^+$) back to the shoots and for buffering short-term fluctuations in nutrient concentrations in the xylem.

Phloem sap typically moves at flow velocities of 0.2–1.5 m h^{-1} (Koblet, 1969; Windt et al., 2006), which is much faster than would be expected from simple diffusion. Phloem flow can be bidirectional within a single internode but is unidirectional within a given sieve element. In contrast to the xylem, the rate and direction of solute movement in the phloem are under metabolic control and change over time in response to source and sink development. Thus, the direction of phloem flow can be reversed over time and often is against the direction of the transpiration stream in the xylem. The pressure flow theory of phloem sap movement developed by Münch (1930) states that sap moves from places of high (turgor) pressure ($P = \Psi_P$) to places of low pressure within the phloem (Figure 5.1). The volume flow rate (F_v) varies with the resistance of the flow path (r_h) and can be described by the following equation (Patrick, 1997):

$$F_v = \Delta P r_h^{-1}$$

where $\Delta P = P_{source} - P_{sink}$, and r_h is the hydraulic resistance of the phloem path.

It is important to realize that P_{source} refers to the pressure inside the source phloem and P_{sink} refers to the pressure inside the sink phloem; that is, the hydrostatic pressure difference or pressure gradient (ΔP) occurs within the sieve tubes rather than between a source organ and a sink organ. In other words, the sieve tubes are pressurized (to between 0.5 and 1.5 MPa; Knoblauch and Peters, 2010), and the buildup and dissipation of pressures in the phloem are isolated from the turgor pressures in the source and sink cells to permit continued phloem flow and translocation of solutes. It follows that the flow rate increases as P_{source} increases and/or as P_{sink} decreases. This means that sinks compete for assimilates by lowering the pressure in their sieve elements (Wardlaw, 1990). Lower P_{sink} is also thought to translate to lower pressure throughout the entire phloem and thus serves as a signal to increase phloem loading in the source (Thompson, 2006; Thompson and Holbrook, 2003). Note that

FIGURE 5.1
Schematic representation of the movement of phloem sap from regions of high pressure in a source to regions of low pressure in a sink. The purple circles represent solute transporters. All values are in MPa.

Modified after Evert (2006), Nobel (2009), and Taiz and Zeiger (2006).

although water influx progressively dilutes the phloem sap along the transport path (Tyree et al., 1974), inside a sieve tube the $\Delta\pi$ between source and sink does not oppose the driving force (ΔP) for sap flow. This is because the sieve pores are lined with membranes only in the axial direction. Despite the universal and longstanding acceptance of Münch's pressure flow theory, however, clear experimental evidence for ΔP is still lacking, and r_h is sometimes positively—rather than negatively—correlated with flow velocity; the latter may depend more on loading and unloading capacity than on r_h (Knoblauch and Oparka, 2012). Regardless of pressure flow, the bulk flow rate of solutes (F_s) through the phloem can be written as follows (Lalonde et al., 2003; Patrick, 1997):

$$F_s = F_v c$$

where c is the solute concentration.

In other words, the transport rate in the phloem is dependent on solute (assimilate) concentrations and a ΔP. The radius of a sieve tube, and hence the cross-sectional area (A), is important insofar as it determines the hydraulic conductance ($l_h = L_h A = r_h^{-1}$, where L_h is the hydraulic conductivity; Patrick et al., 2001). Unlike the xylem (see Chapter 3), the phloem cannot take advantage of transpiration as the driving force for mass flow but instead relies on a pressure-driven mass flow in the opposite direction (Lalonde et al., 2004; Van Bel et al., 2002). Ingeniously, sucrose is not only the major cargo (i.e., the main transport form of carbon) of the phloem system but also its fuel (i.e., the osmotic driving force for mass flow) (Hellmann et al., 2000, Münch, 1930; Van Bel, 2003). The source cells "load" sucrose into the sieve elements, which results in an osmotic gradient ($\Delta\Psi_\pi$) that generates a water potential gradient ($\Delta\Psi$). This draws water from the xylem into the sieve elements, increasing their turgor pressure (P). In other words, phloem transport depends on the osmotic generation of pressure in the source phloem. The sap moves toward the sink because sucrose is being removed (unloaded) there, which leads to a corresponding loss of water (which passively follows the sugar by osmosis) from the sieve element so that P is lower at the sink end of the sieve tubes. Water movement into and out of the phloem is driven by $\Delta\Psi$, whereas water movement inside the phloem (from source to sink) is driven by ΔP ($=P_{source} - P_{sink}$), which in turn is generated by loading (at the source) and unloading (at the sink) of sugar and other solutes. Whereas most of the water entering the phloem in the minor veins of source leaves is absorbed from the xylem, the water leaving the phloem in the sink tissues is used in part for growth of the sink cells, and the remainder enters the xylem to be recycled back to the source (Bradford, 1994; Münch, 1930; Patrick, 1997; Van Bel, 2003; see also Figure 5.1). In transpiring sinks, such as expanding leaves and young grape berries, some water also evaporates to the atmosphere, although berry transpiration rates decline during the course of berry development (Blanke and Leyhe, 1987; Düring and Oggionni, 1986; Rogiers et al., 2004a).

The phloem system is composed of three sequential sectors: collection phloem, transport phloem (also called path phloem), and release phloem (Lalonde et al., 2003; Patrick et al., 2001; Van Bel, 2003). Each of these sectors executes a specific task. Sucrose produced in the mesophyll cells of source leaves moves (i.e., diffuses) symplastically via plasmodesmata to the bundle sheath cells of the leaf's minor veins. From there, it is loaded into the companion cells and sieve elements of the collection phloem (Lalonde et al., 2003, 2004). The way in which grapevines load sucrose into the phloem is controversial. Some researchers argue that loading occurs via the apoplast—that is, across cell walls—because the transport sugar in grapevines is sucrose (rather than the larger raffinose and/or stachyose) and because the sucrose concentration inside the phloem is much higher than in the surrounding cells (diffusion against a concentration gradient is not possible). If this is true, then sucrose must leak or be pumped out to the cell walls (which are permeable to small organic molecules such as sugars and amino acids) and then be pumped across the cell membranes of the companion cells and sieve elements by proton cotransport (or symport), a process that requires an input of energy in the form of ATP (Ayre, 2011; Carpaneto et al., 2005; Chen et al., 2012; Geiger, 2011; Lalonde et al., 2003; Patrick et al., 2001). Hexose sugars that may have leaked to the apoplast might be transported back into the cells using hexose transporters (Hayes et al., 2007). The average energy demand for phloem loading has been estimated at roughly one-third of the entire respiratory energy demand of mature leaves (Amthor, 2000). Despite this high energy demand, active loading may be advantageous, because it can increase P_{source} while keeping the sugar concentration in the leaves low, thus increasing carbon export to sinks and reducing feedback inhibition of photosynthesis (Fu et al., 2011; Turgeon, 2010).

Other scientists believe that in contrast to most other crop species but similar to many tree species, this sucrose transport occurs entirely within the symplast rather than across the apoplast. According to this passive loading model, sucrose in the leaves moves by diffusion or mass flow (generated by a ΔP between mesophyll and phloem cells) along a continuous, membrane-bound pathway (through plasmodesmata, of which there may be about 10,000 per cell) from the mesophyll to the phloem, which is therefore called "open phloem" (Gamalei, 1989; Rennie and Turgeon, 2009; Turgeon, 2000; Turgeon and Wolf, 2009). The plasmodesmata are thought to operate as pressure-sensitive "valves" that close when two neighboring cells experience a $\Delta P > 0.2$ MPa; this enables them to quickly isolate wounded cells from the rest of the symplast (Oparka and Prior, 1992). Provided the concentration of sucrose inside the minor-vein phloem cells is lower than that in the surrounding mesophyll cells (unlike in many other species), there is no need for active, energy-dependent transport. This implies, however, that the flow direction of the phloem is reversible, depending on the direction of the pressure gradient. Despite the proposed symplastic continuity, hexoses do not enter the phloem in significant amounts (Koblet, 1969). Although there is plenty of glucose and fructose in the mesophyll cells, these sugars are mostly sequestered in the vacuoles (Turgeon, 2010). One costly drawback of the open phloem design is that viral pathogens can also pass through the numerous plasmodesmata and thus become systemic once a vector (usually an insect or a nematode) has deposited them near the phloem (see Section 8.2). It is also possible that both modes of loading operate in tandem in the leaves (Lalonde et al., 2003; Turgeon, 2000; Turgeon and Wolf, 2009).

Amino acids can be produced in the mesophyll cells or delivered from the roots in the xylem along with K^+ and other nutrient ions (see Section 5.3). These compounds can be loaded into the phloem symplastically (leaf-produced amino acids) or apoplastically (xylem-delivered amino acids and K^+) (Lalonde et al., 2003). The apoplastic step requires channels (for K^+) and transporters (for amino acids). The energy-dependent transporters are driven by ATPases that move a proton across the membrane along with, or in exchange for, an amino acid, depending on whether a transporter functions in symport or antiport mode (Lalonde et al., 2004). It appears that loading of K^+ and that of sucrose are coupled such that K^+ may stimulate sucrose loading into the phloem (Gajdanowicz et al., 2011; Geiger, 2011). Other compounds, such as organic acids and hormones, probably enter the phloem mostly by diffusion, although auxin may also be actively loaded, and are swept along the pathway by mass flow. The minor veins are also a rerouting system for water; water entering the leaves from the xylem is carried away again in the phloem if it is not used for transpiration or cell expansion (see Section 3.2).

Once the solutes have entered the phloem, they are exported away from the source in the transport phloem of the main veins, petioles, shoots, trunk, and roots. This long-distance transport occurs by mass flow rather than by diffusion because the pressure inside the phloem is lower in the distant sink organs than in the leaves, with water being provided by the xylem. Because phloem flow is driven by the ΔP between the sieve elements in the source and in the sink, the export rate of sugar depends on both the sucrose production in the source leaf (which in turn depends on net carbon assimilation; see Section 4.2) and the capacity of sink tissues to remove (and use) sucrose from the phloem. In contrast to the xylem (see Section 3.3), and although it is often symplastically isolated from the surrounding cells, the transport phloem is rather "leaky" (Ayre, 2011; De Schepper et al., 2013). A considerable portion of the solutes are passively or actively released to the phloem parenchyma along the way to supply axial sinks (e.g., for maintenance respiration, radial growth by the cambium, and reserve storage in woody organs), from where they can also be actively and rapidly retrieved back into the phloem, for example, via starch hydrolysis and ATP-fueled proton/sucrose symport (Hafke et al., 2005; Lalonde et al., 2003; Martens et al., 2006; Patrick, 1997; Van Bel, 2003).

Sucrose retrieval may be supported by efflux from the phloem of K^+ ions that are sequestered by the surrounding tissues that invest energy to acquire K^+ from the apoplast. Contrary to the normal K^+/sucrose symport in source leaves (phloem loading) and sink organs (phloem unloading), in this K^+/sucrose antiport system the energy from the K^+ gradient is used to reload sucrose back into the phloem (Gajdanowicz et al., 2011). "Pathway storage" can buffer diurnal and environmentally induced fluctuations in assimilate supply from the leaves (Wardlaw, 1990) so that the sugar concentration in the phloem entering distant sinks varies little compared with that in the minor veins of the source leaves. Symplastic net release along the transport phloem is favored by high source:sink ratios (i.e., under sink limitation, such as with light crop loads), whereas under low source:sink ratios (i.e., under source limitation) the plasmodesmata close and solutes can only leave the phloem apoplastically and hence slowly (Patrick, 1997; Patrick et al., 2001). Sucrose that has diffused to the apoplast along the transport phloem may also be transported into xylem parenchyma cells for the production of fortified cell walls and to keep sucrose out of the transpiration stream (Schmitt et al., 2008).

The transport phloem is in local water potential equilibrium with the apoplast: Ψ_{phloem} and Ψ_{xylem} are tightly coupled because water moves across the separating membranes (Thompson and Holbrook, 2003; see also Section 3.1). Water enters the transport phloem osmotically from the surrounding apoplast, including the xylem, along the pathway and progressively dilutes the sieve element contents. The sucrose concentration in the phloem consequently declines with increasing transport distance and has been estimated to be less than 50 mM in the pedicel of ripening grape berries (Zhang et al., 2006). This dilution gradually raises Ψ_{phloem} and increases the transport rate (Patrick et al., 2001). If the increase in flow rate cannot compensate for the decrease in solute concentration, however, then more distant sinks will receive less sugar and will be at a competitive disadvantage compared with more proximal sinks.

Upon arrival in the sink tissue (e.g., shoot tips, root tips, fruit, and seeds), the solutes are unloaded from the phloem (release phloem) in reverse order of the loading sequence and metabolized or stored in the sink cells (Ayre, 2011; Lalonde et al., 2004; Patrick, 1997; Patrick et al., 2001). In contrast with the transport phloem, this release is final and irreversible—that is, retrieval back into the phloem stops in the release phloem. The sucrose concentration in the phloem is almost always much higher than in the surrounding sink parenchyma cells (but lower than in the source), so unloading again passively follows a concentration gradient. In most sinks (e.g., root and shoot tips and expanding leaves), the unloaded sucrose by and large follows a symplastic pathway through plasmodesmata to the site of usage, although there may also be some additional apoplastic unloading (Ayre, 2011; Lalonde et al., 2003; Patrick, 1997; Ruan et al., 2012; Van Bel, 2003). The symplastic pathway is one of passive diffusion and bulk flow and is characterized by its large transport capacity and low hydraulic resistance. Nonetheless, the majority of the resistance that water and solutes encounter between the source and sink cells is imposed by the narrow plasmodesmata that connect the phloem symplast with the surrounding parenchyma symplast. The ability to open or close the plasmodesmata provides sink cells with a mechanism to control solute import (facilitated diffusion) by changing the hydraulic resistance of the pathway. In young, expanding leaves, assimilate import seems to be restricted to the larger veins (the minor veins are probably responsible for assimilate collection for export) and is maximal at approximately 15–20% of the leaf's final size. The transition from sink to source starts in the leaf margins so that during the transition phase (from 30% to 50% of full leaf size, corresponding roughly to leaf positions four to six from the shoot tip; Koblet, 1969) there can be simultaneous import into the leaf base and export from the leaf tips, although import and export probably occur in different vascular bundles (Wardlaw, 1990). The unloading pathway is turned off after a leaf has completed its sink–source transition so that even heavily shaded mature leaves in the interior of dense canopies do not import assimilates, unless there is no competition from sink organs (Lalonde et al., 2003; Quinlan and Weaver, 1969, 1970; Turgeon, 1984).

Grape berries are special in the sense that although unloading initially follows mostly the standard symplastic route, the apoplastic (extracellular) path of unloading (in which sucrose is released to the apoplast) comes to predominate at or just before veraison (Zhang et al., 2006). The pericarp cells seem to become symplastically isolated from the phloem inside the fruit by partial blockage (probably by callose) of the plasmodesmata that connect the sieve element/companion cell complex with the surrounding parenchyma cells inside the vascular bundles and with the berry's flesh cells. Even before veraison, however, there are no plasmodesmatal connections between tissues of different genomes, such as a seed's filial tissues (embryo and endosperm) and its maternal tissues (which include the seed coat). Therefore, although unloading into the seed coat may be symplastic, unloading into the embryo and endosperm is always apoplastic (Bradford, 1994; Patrick, 1997; Walker et al., 1999). Apoplastic unloading also occurs at the interface between leaf or berry cells and biotrophic pathogens, such as the powdery mildew fungus *Erysiphe necator* (see Section 8.2). In addition to the facilitated diffusion mentioned previously, apoplastic unloading and postphloem transport require energy-dependent active transport to "pump" (again using ATPases) the solutes across the membranes of the phloem and pericarp cells (Carpaneto et al., 2005; Giribaldi et al., 2007; Martinoia et al., 2012; Sarry et al., 2004; Zhang et al., 2008). These ATPases power sucrose as well as hexose transporter proteins (Ageorges et al., 2000; Davies et al., 1999; Fillion et al., 1999; Hayes et al., 2007; Lecourieux et al., 2014; Vignault et al., 2005) and, along with the invertases discussed later, are activated by sugars and abscisic acid, whose concentration inside the berry increases at veraison (Agasse et al., 2009; Pan et al., 2005). Nutrient ions, such as K^+, may be unloaded as well, and may then be taken up by berry cells by passive diffusion through channel proteins (Cuéllar et al., 2013).

As illustrated in Figure 5.2, the importing berry cells can retrieve sucrose from the apoplast unaltered, again with the (active) aid of sucrose transporters in the cell membranes, or sucrose can be split into hexoses (glucose and fructose) by the enzyme invertase located in the cell walls (apoplast). The hexoses are then imported into the cell and then into the vacuole by membrane-based hexose transporters or directly into the vacuole (i.e., bypassing the cytosol) via intracellular vesicles—a mechanism termed endocytosis. Sucrose can also be split inside the cell by a cytoplasmic invertase or enter the vacuole unaltered. Inside the vacuole, it can again remain unaltered or be split by a vacuolar invertase (Famiani et al., 2000; Robinson and Davies, 2000). It appears that most of the sucrose released from the phloem to the apoplast is broken down to hexoses there before being pumped into the cytoplasm, where some of them may be reassembled to sucrose for import into the vacuole, only to be split into glucose and fructose again (Sarry et al., 2004; Terrier et al., 2005). During ripening, the "surplus" hexoses—that is, sugars that are not used in further metabolism—are accumulated for storage inside the vacuoles. If grapes are similar to tomatoes, then cultivars that accumulate different amounts of hexoses and therefore reach different degrees Brix at maturity should mainly differ in their rate of sugar transport across the mesocarp cell membranes (Patrick et al., 2001). Moreover, due to the dependence of phloem flow on ΔP, berries that are able to

FIGURE 5.2

Possible pathways for sugar from the phloem to the vacuole of a grape berry pericarp cell (Glc, glucose; Fru, fructose; HT, hexose transporter; Inv, invertase; ST, sucrose transporter; Suc, sucrose). Specialized forms of invertase work in the cell wall, cytoplasm, and vacuole, respectively.

Illustration by M. Keller.

rapidly use or accumulate sugars are more competitive than their slower siblings and will tend to develop and ripen more rapidly.

Despite the continued import of sucrose into most sink tissues, its concentration in these tissues remains low because it is "consumed" by respiration (for growth and metabolism), used as building blocks in the construction of new cell walls and other cell components, or converted to storage material such as starch (in shoots, trunks, and roots). However, this is not the case in postveraison grape berries, which accumulate large amounts of hexose sugars to concentrations that exceed that in the phloem of the pedicel by severalfold. The switch from symplastic to apoplastic phloem unloading before veraison isolates the sink symplast (berry pericarp cells) from the vascular symplast (phloem) and enables the berries to accumulate osmotic solutes to high concentrations without inhibiting phloem flow. The sucrose gradient generated by the activity of cell wall invertase may drive the efflux from the phloem of sucrose along with protons against the steep phloem—apoplast proton gradient (Ayre, 2011; Carpaneto et al., 2005; Keller and Shrestha, 2014). In addition, the deposition of solutes, such as hexoses and K^+, in the apoplast lowers the apoplast's osmotic potential (Ψ_π), leading to water efflux from the phloem (Keller and Shrestha, 2014; Wada et al., 2009). This decreases phloem pressure (i.e., turgor potential, Ψ_P), which in turn increases the berry's sink strength and thus the flow rate of phloem sap to the berry. Because sugar and other solutes are actively pumped into the berry vacuoles, they remain there for good or until they are consumed by other organisms. Under conditions that limit sink activity, however, less sucrose is unloaded from the release phloem, which consequently slows the flow rate in the transport phloem. This leads to sucrose "congestion" in the collection phloem, and sucrose becomes "stuck" in the leaf apoplast (with apoplastic loading) or the mesophyll cells (with symplastic loading), which ultimately leads to feedback inhibition of photosynthesis (Ayre, 2011; see also Section 4.2). Such conditions may include water deficit, nutrient deficiency, or low temperatures (see Chapter 7).

Nutrient ions and other small solutes probably move from the phloem to most sink tissues passively by diffusion through the plasmodesmata or by bulk flow. The pressure flow mechanism requires water to flow out of the phloem along with the unloaded solutes. Therefore, in addition to delivering nutrients, the bulk flow of phloem sap provides a mechanism to meet the water demands of expanding sink cells (Lalonde et al., 2003; Patrick, 1997; Patrick et al., 2001) because cells, and therefore tissues and organs, grow as water moves across the cell membrane down a water potential gradient (see Section 3.1). This is especially true of expanding grape berries (see Section 6.2), but even in growing roots a considerable portion of water is derived from phloem import. This is necessary because transpiration rates in sink tissues are typically too slow to drive sufficient water import via the xylem to sustain growth of these organs.

5.1.1 ALLOCATION AND PARTITIONING

The amount of photosynthate available for export from a source leaf depends on the leaf's carbon balance, which is determined by its photosynthetic rate and metabolic activity. The process by which the leaf regulates the distribution of fixed carbon to the various metabolic pathways within the leaf is termed allocation. Of course, allocation is also important in the sink organs that import carbon. Allocation can be grouped into three main categories (Taiz and Zeiger, 2006):

1. *Utilization*: Fixed carbon is consumed in respiration for energy (ATP) "generation" or used as building blocks for the production of other components (mainly amino acids) needed by the cell for metabolism and growth.
2. *Storage*: Fixed carbon is converted to starch during the day and stored within the chloroplast for remobilization at night and when environmental constraints limit photosynthesis.
3. *Transport*: Fixed carbon is converted to sucrose for temporary storage in the vacuole (mainly as a buffer against short-term fluctuations in sucrose formation) or export to sink organs. Most of the sucrose produced by a source leaf is allocated to transport.

As discussed in Section 4.2, carbon allocation is regulated by the availability of triose phosphate sugars and inorganic phosphate. The coordination of the allocation of fixed carbon to form either sucrose or starch is of particular importance because only sucrose is available for immediate export. This coordination is at least partly driven by the demand for sucrose of all the sinks on a vine. High sink demand removes sucrose from the source, which favors sucrose production over starch production. Consequently, once the source's own needs have been met (allocation to utilization), the proportion of currently produced assimilate allocated to storage in the leaf is determined by the sink demand for sucrose (allocation to transport). Thus leaf starch accumulates if assimilate production exceeds total sink demand (Johnson et al., 1982). On the other hand, an abundance of fixed carbon in the source can also stimulate sink growth, apparently by inducing the production of additional sucrose transporters. Moreover, both the proportion and the total amount of carbon stored as starch in a leaf also depend on day length—that is, season and latitude (Stitt et al., 2007). Although a greater portion of the fixed carbon is allocated to starch during short days, the total amount is smaller than that produced during long days. This is followed by slower rates of starch degradation during the longer nights so that the overall sucrose export over a 24 h period is less in short days. This adjustment of starch turnover to the amount of carbon fixed during the day leads to slower growth rates during short days, which balances the utilization of carbon with its supply (Stitt et al., 2007).

The distribution of exported assimilates to the various sink organs is called partitioning, and it occurs via source—sink turgor gradients (ΔP) within the phloem (Taiz and Zeiger, 2006). Leaves that have recently become sources initially export their assimilates to the growing shoot tip and unfolding leaves. As new leaves are added on the shoot, the older leaves find themselves at increasing distance from the shoot tip and begin exporting toward the shoot base (Figure 5.3). Assimilate transport is a dynamic process that can be adapted to the developmental status of the vine and to changing requirements imposed by the environment. In other words, although sources usually supply materials to nearby sinks, they do not do so equally to all sinks on a plant (Wardlaw, 1990). Certain sources preferentially supply specific sinks, but the supply pathways are flexible. Moreover, as the various organs of a vine develop, they compete with each other for space, light, and nutrients. Therefore, there is a hierarchy of relative priorities among the sinks, and this hierarchy is dynamic and sensitive to environmental variables. Flowers are generally poor competitors (especially under source-limiting conditions), whereas after fruit set the berries and seeds dominate the shoots, which in turn often outcompete the roots (Buttrose, 1966a; Hale and Weaver, 1962; Wardlaw, 1990). How much of the available assimilates a particular sink organ receives (imports) is a matter of supply and demand. The import rate depends on the organ's sink strength (i.e., assimilate demand) relative to the strength of all other sinks on the same vine and also on the total amount of assimilate available from the various sources (i.e., assimilate supply). Sink strength is defined as the product of

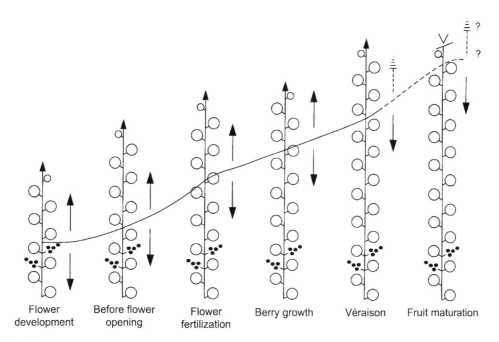

FIGURE 5.3

Source—sink relations along a grapevine shoot change over the course of a growing season. Arrows indicate direction of movement.

Reproduced from Koblet (1969).

sink size (total weight of a sink) and sink activity (rate of assimilate import per unit sink weight) and is often estimated as an organ's growth rate. However, small sink strength is not always associated with low sink priority, which refers to the preferential supply of available assimilates between competing sinks (Minchin and Lacointe, 2005). For example, seeds are too small to be strong sinks but are generally the top-priority sinks (i.e., they are ranked higher than the surrounding berry flesh), whereas reserve storage pools (e.g., in the wood of canes, trunks, and roots) often have the lowest sink priority (but see later discussion). The relative priority of a particular sink depends on its ability to lower the concentration of sucrose in the phloem—that is, on the rate of phloem unloading—and thereby maintain a favorable pressure gradient to the source (Patrick, 1997; Wardlaw, 1990). The sinks usually convert most of the imported sucrose to the hexose sugars glucose and fructose, using invertase and/or sucrose synthase, before they can use the sugar for their own metabolic processes, including the reassembly of starch (Hawker et al., 1991). Modifying either the size or the activity of a sink leads to changes in assimilate transport patterns in a vine. The following patterns are important in partitioning (modified from Taiz and Zeiger, 2006):

Proximity: Although leaves on one shoot can export assimilates to fruit clusters on other shoots, even if they are located several meters away (Koblet and Perret, 1972; Meynhardt and Malan, 1963), the closer a source is to a specific sink, the more likely it is to supply materials to that sink (Wardlaw, 1990). The mature leaves closest to the shoot tip generally export their assimilates to the growing tip, whereas the more basal leaves preferentially supply the grape clusters and (beginning around bloom) the permanent parts of the vine. The intermediate leaves export in both directions (Hale and Weaver, 1962; Koblet, 1969).

Connection: A source leaf favors a sink with which it is directly connected via vascular bundles. Any leaf on a shoot is usually connected with the leaves and clusters above and below it (i.e., on the same side of the shoot). Therefore, a flower or fruit cluster is supplied mostly by the leaves located on the same side of the shoot as itself (Koblet, 1969; Motomura, 1990, 1993; Yang and Hori, 1980). This "unilateralism" even applies to the supply of clusters with assimilates derived from lateral shoots (Koblet, 1975; Koblet and Perret, 1971).

Interference: The "normal" pathways of translocation can be altered by wounding or pruning, which interrupts the direct connection between a source and a sink. Following such interruption, vascular interconnections (anastomoses) can provide alternative connections (e.g., cross-transfers). This has implications for canopy management. Hedging (shoot tipping or topping) stimulates cross-transfer of assimilates to clusters on both sides of the shoot and induces even young leaves to switch from upward (acropetal) export to downward (basipetal) export that can extend to neighboring nontopped shoots (Koblet and Perret, 1972; Quinlan and Weaver, 1970). This is probably why shoot tipping during bloom often improves fruit set (Coombe, 1959, 1962; Vasconcelos and Castagnoli, 2000), although the subsequent outgrowth of lateral buds may later reverse assimilate flow again. Defoliation of a shoot triggers compensatory assimilate import into that shoot's clusters from neighboring shoots (Quinlan and Weaver, 1970). Leaf removal or shade in the cluster zone force more distal leaves to export their assimilates to the clusters (but not to the shaded leaves), whereas loss (or shading) of younger leaves induces older leaves to reverse the export direction toward the shoot tip at the expense of the perennial parts of the vine. Fruit removal on a shoot induces that shoot to export assimilates to neighboring shoots with clusters (Quinlan and Weaver, 1970). Girdling (phloem removal) of a shoot near its base leads to the leaves exporting assimilates to clusters on both the same and opposite sides of that shoot (Motomura, 1993).

Communication: Grapevines carefully balance their investment in growth and reproduction (reproductive productivity). Vegetative growth also must be balanced between shoot growth (photosynthetic productivity) and root growth (water and nutrient uptake). As Goethe stated, "in order to spend on one side, nature is forced to economize on the other side" (cited in Darwin, 2004). This requires interaction between centers of supply and demand, which happens by way of pressure gradients (low sink pressure stimulates import via the phloem), nutrients, and hormones. Cytokinins produced by root tips or seeds and transported to the leaves in the xylem may signal sink strength and integrate assimilate supply with sink demand (Ha et al., 2012; Lalonde et al., 2003; Paul and Foyer, 2001). The strongest and most competitive sinks may be those with the highest amount of cytokinins (Thomas, 2013). But the same may be true for auxin produced in shoot tips and seeds and exported in the phloem and parenchyma cells. A developing grape cluster can attract assimilates from nearby shoots, especially when assimilate production exceeds those shoots' own demand (Currle et al., 1983; Intrigliolo et al., 2009; Koblet and Perret, 1972; Quinlan and Weaver, 1970). Mechanical elimination by shoot hedging of the auxin signal required for root growth may further enhance delivery to the clusters because lack of auxin flow may temporarily inhibit root growth, so long as outgrowth of lateral buds does not generate new shoot apices.

Competition: Competition determines the priority of each sink relative to all other sinks on the plant in terms of assimilate distribution. The greater the ability of a sink to store or metabolize imported assimilates (i.e., the higher its sink capacity), the better its ability to compete for exported assimilates. A rapidly growing sink is competitive because the consumption of assimilates lowers sink pressure. A competitive sink is said to have high sink strength. Moreover, the greater the number of sinks competing for assimilates, the lower the availability of assimilates for each individual sink. This has implications for winter pruning and canopy management. Leaving more buds per vine increases the number of shoots and clusters but decreases the vigor of each shoot and the proportion of flowers that set fruit. Moreover, increasing the number of berries per plant or per unit leaf area can limit berry expansion (Keller et al., 2008) or slow the rate of fruit ripening. Conversely, applying gibberellin to table grape clusters increases not only berry size but also the amount of imported assimilates (Weaver et al., 1969).

Development: The relative importance of different sinks changes during plant development. Shoot tips have high priority after budbreak, when the vine needs to establish its new canopy (Hale and Weaver, 1962). By contrast, clusters are low-priority sinks until bloom. This is why shoot topping or trunk girdling during bloom may improve fruit set, whereas loss of leaf area reduces set (Coombe, 1959, 1962; Poni et al., 2009). After fruit set, however, the clusters become powerful sinks and dominate the sink hierarchy during grape development, especially for adjacent and nearby leaves (Hale and Weaver, 1962; Williams, 1996). After veraison, the clusters are initially very strong sinks, but the vine then gradually shifts its priorities to the woody perennial organs to replenish storage reserves and acquire cold hardiness (Candolfi-Vasconcelos et al., 1994).

In summary, for an organ or a tissue to be a strong sink, it pays to be large, close to a source, and have good connections. However, many other factors also modulate sink strength and assimilate partitioning. Environmental variables play an important role in modifying the pattern of partitioning. For instance, soil water or nutrient deficits decrease overall vine growth but increase the proportion of assimilates partitioned to the roots (see Sections 7.2 and 7.3).

Source and sink organs are parts of a single inseparable system, and an effect on one part is bound to have a consequential and concurrent influence on the other. For instance, removing a sink improves the availability of assimilates to adjacent sinks (Quinlan and Weaver, 1970). In viticulture, we are mainly interested in maximizing or optimizing the vine's investment in reproduction (i.e., fruit production). In mature vines, the proportion of total biomass (dry matter) partitioned to the fruit can range from 10% (e.g., in nonirrigated, relatively water-stressed, cluster-thinned Pinot noir) to 70% (e.g., in heavily irrigated, minimally pruned Concord). The dilemma for grapevines as perennial plants is to maximize annual seed dispersal (short-term reproductive output) while at the same time ensuring long-term survival of the plant (for long-term reproductive output). The vine must coordinate the supply of photosynthate to seed production and, hence, fruit development such that it does not occur at the expense of other essential processes and structures. Early shoot growth during and after budbreak is completely dependent on remobilized storage reserves from the "permanent" parts of the vine, such as canes, spurs, cordons, trunks, and roots, which are therefore depleted beginning before budbreak (Eifert et al., 1961; Loescher et al., 1990; Williams, 1996; Zapata et al., 2004). Even unfolding leaves, before becoming sources, must compete with other sinks for assimilates, yet they are stronger sinks than the cambium of the cane, trunk, and roots (Wardlaw, 1990). As more leaves are added on the shoot, they increasingly support the shoot's further growth, which soon becomes independent of the parent plant. The growing shoots also begin to contribute to radial growth via the cambial activity in the permanent parts of the vine. Beginning at approximately the time of bloom, assimilate export from the shoots normally begins to replenish the parent plant's storage reserves and also sustains secondary growth of the cordons, trunk, and roots.

The ability to refill the storage pools also depends on the amount of fruit the vine has to support because the berries generally dominate the hierarchy of sink priorities between fruit set and seed maturity (Hale and Weaver, 1962; Williams, 1996). Although the replenishment of reserves is often described as an overflow mechanism that directs "surplus" assimilates to storage only after all other assimilate-requiring processes, such as growth and fruit production, have been satisfied (Lemaire and Millard, 1999), this may depend on the vine's developmental stage. The woody parts of the vine may become high-priority sinks late in the season (Loescher et al., 1990), when the plant "loses interest" in the fruit and focuses on the replenishment of storage reserves and on cold acclimation (see Section 7.4). In other words, in addition to the overflow mechanism (the so-called passive storage), accumulation of reserves may be an active storage process when carbon is diverted to storage at the expense of other processes (Wiley and Helliker, 2012). This appears to happen after about midripening (Candolfi-Vasconcelos et al., 1994; Wample and Bary, 1992) and can clash with a grower's or winemaker's desire to improve fruit quality by delaying harvest (i.e., prolonged "hang time"), especially when adverse environmental conditions limit the availability of resources.

The net assimilation rate of a grapevine canopy determines the export rate from the canopy, and the export rate increases with increasing source—sink ΔP in the phloem. Therefore, the best way to increase assimilate translocation is to increase the rate of photosynthesis in source leaves because an increase in the amount of photosynthate available for export increases the turgor in the source phloem, which results in a greater ΔP between source and sink. Leaves that unfold under intense competition from other sinks often achieve higher photosynthetic rates than leaves that do not experience competition during development. Even in fully grown leaves, however, photosynthesis can still be influenced by natural or manipulated changes in sink demand (Downton et al., 1987; Flore and Laksø, 1989; Paul and Foyer, 2001). The presence of fruit on a vine tends to increase the leaves' photosynthetic rate, especially in vines with small leaf area that leads to source limitation (Edson et al., 1993, 1995a; Eibach and Alleweldt, 1984; Hofäcker, 1978). The fruit also induces a shift in assimilate supply to reproductive growth at the expense of vegetative growth, especially root growth, keeping whole-vine biomass relatively constant (Edson et al., 1995a,b; Petrie et al., 2000b; Williams, 1996). This shift is particularly pronounced at low soil moisture (Eibach and Alleweldt, 1985). Destruction or removal of leaves (i.e., source removal, for example, to improve fruit exposure to sunlight) also leads to a compensatory increase in photosynthesis of the remaining leaves and a delay of their senescence (Buttrose, 1966a; Candolfi-Vasconcelos and Koblet, 1991; Currle et al., 1983; Hofäcker, 1978; Hunter and Visser, 1988; Iacono et al., 1995). However, this compensation is incomplete, and severe defoliation may retard fruit development and ripening (Petrie et al., 2000a). A late spring frost, for example, by killing leaves and flowers that have been "built" using remobilized storage reserves, can severely deplete the reserve pool and restrict renewed canopy development even in the absence of fruit (Gu et al., 2008). If defoliation occurs late in the season, it can result in poor carbohydrate reserve status and in nitrogen deficiency the following growing season (Loescher et al., 1990; Sartorius, 1973). Nitrogen and other nutrients that could be remobilized from the leaves and transported to the perennial parts of the vine for storage are lost when leaves are removed before normal abscission. This problem also arises when leaves are killed by an early fall frost; nutrients cannot be recycled from dead leaves. Alternatively, inadequate carbohydrate reserves in the roots, by curtailing new root growth and nutrient uptake and assimilation, can also lead to nitrogen deficiency (Loescher et al., 1990). However, even before such nitrogen deficiency may limit plant growth, it is possible that an inadequate reserve status may interfere with the plant's ability to generate adequate root pressure prior to budbreak (Améglio et al., 2001; see also Section 2.2). This would prevent the vine from purging air bubbles from the xylem, potentially leading to poor budbreak or early canopy collapse.

In contrast, excessive fruit removal (i.e., sink removal, for example, cluster thinning), especially if done early, may result in an assimilate surplus due to a shortage of demand and can decrease photosynthesis because of feedback inhibition from accumulating sugar (Currle et al., 1983; Downton et al., 1987; Iacono et al., 1995; Naor et al., 1997; Paul and Foyer, 2001; Roitsch, 1999). In addition, some of the surplus sugar may be used for vegetative growth, especially of lateral shoots, and for root growth (Morinaga et al., 2003; Pallas et al., 2008; Petrie et al., 2000b). Whereas fruit clusters cannot compete with shoot growth before bloom because of their low sink strength, shoots may continue to grow vigorously after fruit set on vines with low amounts of fruit (Keller et al., 2008). This may result in dense, shaded canopies, which can be detrimental to disease incidence and fruit quality (see Section 6.2). Vines without fruit also may begin to "shut down" earlier in autumn, which is visible as the leaf chlorophyll content declines in response to decreasing photosynthesis, and the leaves are shed before those on similar plants with fruit (Keller et al., 2014; see also Figure 5.4). Such premature leaf senescence may be triggered by the accumulation of sugar in the leaves, which signals a surplus of uncommitted assimilates (Lim et al., 2007; Rolland et al., 2006; Thomas, 2013; Wingler et al., 2009). It is possible that such sugar accumulation also serves as an internal signal for nutrient recycling during leaf senescence.

FIGURE 5.4

Senescing Cabernet Sauvignon leaves in early November (northern hemisphere) from plants that had all fruit removed at veraison in mid-August (left) or with the fruit still on the plant (right).

Photo by M. Keller.

Whereas the presence of fruit influences the physiology of the rest of the vine, fruit growth often compensates for changes in sink number, and this is called the yield component compensation principle (see Section 6.1). Thus, early loss of fruit, for example due to poor fruit set or early thinning, is partly compensated by increased growth of the remaining fruit, which tends to lead to an increase in berry size (Dokoozlian and Hirschfelt, 1995; Keller et al., 2008; Kliewer et al., 1983; Winkler, 1958). If the loss in sink number occurs before bloom (e.g., due to cold injury), then the proportion of the remaining flowers that set fruit often increases in addition to berry growth, which can result in compact clusters with large berries. Such growth compensation is greatly reduced or absent if the sink number decreases only later in the growing season, for example during the lag phase or at veraison, and the remaining fruit may instead accumulate sugar more rapidly (Dokoozlian and Hirschfelt, 1995; Nuzzo and Matthews, 2006). Conversely, vines usually self-adjust to a large number of sinks (e.g., due to high bud numbers following mechanical or minimal pruning) by lowering the percentage of fruit set and decreasing berry growth so that there may be many small, loose clusters with small berries (Clingeleffer and Krake, 1992; Intrieri et al., 2001; Miller and Howell, 1998; Possingham, 1994; Winkler, 1958).

A puzzling phenomenon in grapes is that different berries on the same cluster seem to develop independently at different rates (Coombe, 1992; Coombe and Bishop, 1980) and accumulate very different amounts of sugar, even though they may be supplied by the same phloem strand through the peduncle. This shows that individual berries, rather than the cluster as a whole, are able to control delivery of phloem solutes. Berries containing more seeds than others seem to be better able to attract assimilates, possibly because each seed produces and exports auxin (see Section 2.3). Because auxin stimulates cambial activity and xylem and phloem development, more auxin export increases the production of vascular tissues in the pedicel so that more water and nutrients can be imported by the fruit (Else et al., 2004). Indeed, fruit size and pedicel cross-sectional area are strongly correlated, and the (dry) weight of a cluster can be estimated from the diameter of its peduncle (Castelan-Estrada et al., 2002). Therefore, berries that begin development with a head start (differences in the rate of cell division before bloom can result in differences in ovary size among berries) usually remain more competitive throughout their growth and ripening (Coombe, 1976).

Grapevines often form large numbers of meristems (shoot and root tips and berries) capable of growth under favorable conditions. Thus, they respond strongly to variations in the availability of resources: They are said to be very "plastic" (Lawlor, 2002). Environmental conditions, therefore, also play an important role in the regulation of partitioning. Low irradiance stimulates carbon supply to the shoots, whereas low nutrient or water availability favors supply to the roots (Keller and Koblet, 1995a). In the case of water, this does not mean that root growth actually increases in response to water deficit. On the contrary, root growth decreases, but less so than shoot growth. These relationships are explored in subsequent chapters.

5.2 CANOPY—ENVIRONMENT INTERACTIONS

Grapevine productivity (growth, yield, and fruit composition) ultimately depends on the photosynthetic capacity of the vine's canopy, integrated over the growing season. A canopy in the viticultural sense is defined as the aboveground parts of a vine (i.e., shoots, leaves, fruit, trunk, and cordon). Compared with leaves, shoot and berry photosynthesis is minor and never exceeds respiration. Therefore, canopy photosynthesis is the sum of the photosynthetic activity of all the leaves minus the respiratory activity of the leaves and all nonphotosynthetic tissues. Although canopy photosynthesis is to a large extent determined by the availability of resources, internal factors also play a role. For example, the length and layout of the water-flow pathway from the roots to the leaves may limit photosynthesis via the effect of hydraulic resistance and gravity on stomatal conductance (see Section 3.3). Because the xylem pressure declines with increasing height, the leaves near the top experience relatively more water stress than those closer to the ground. This influence of a plant's "plumbing" design and length has been put forward as the main reason photosynthesis and vigor tend to decline as plant height and age increase (Koch et al., 2004; Ryan and Yoder, 1997). Alternatively, the decrease in gas exchange in tall plants may be a consequence, rather than the cause, of slower growth because growth is more sensitive to water deficit than is photosynthesis (Ryan et al., 2006). In this case, hydraulics and gravity limit the water potential gradient from the xylem to the growing cells and, hence, the turgor necessary for cell expansion, which decreases sink strength (see Section 3.1; Bond et al., 2007). In support of this sink limitation hypothesis, taller plants tend to accumulate greater amounts of reserve carbohydrates than do shorter plants (Sala and Hoch, 2009). Alternatively again, a lower water potential gradient from the xylem to the phloem might curb phloem export from the leaves (Barnard and Ryan, 2003). The outcome in the latter two scenarios is the same: Sugar accumulates in the leaves and curtails photosynthesis by feedback inhibition. Whatever its direct and indirect consequences, hydraulic limitation is probably of little importance in cultivated grapevines whose size—in contrast with their tall, tree-climbing wild relatives—is strictly limited by trellis design and annual pruning practices.

Resource availability is determined by meteorological or climatic (Greek *klima* = surface of the Earth, region) conditions in addition to edaphic (Greek *edaphos* = ground, soil) factors. The impact of variables related to the soil, such as water and nutrient availability, is discussed in Sections 7.2 and 7.3; this section provides an overview of the importance of aboveground climatic variables. Whereas the term "weather" refers to the daily, seasonal, and annual fluctuations of temperature, precipitation, humidity, and wind, "climate" constitutes the summarized and averaged weather situation over a long time. A "long time" is defined by the World Meteorological Organization as 30 years or more. In viticulture, we often distinguish three levels of climate:

1. *Macroclimate*: The climate of a region, which is ordinarily described by data collected at one or several weather stations. It is sometimes viewed as the mean of all the microclimates in a region. It is mainly determined by the geographic location (i.e., latitude, altitude, and distance from large bodies of water) but is independent of local topography, soil type, and vegetation. The size of the region may extend for hundreds of kilometers.
2. *Mesoclimate*: The climate of a site or large vineyard. It is a local variant of the macroclimate modified by topography, and hence also called local climate or topoclimate (Greek *topos* = place). It may differ from the macroclimate because of altitude or elevation from a valley floor. The extent of a particular mesoclimate may be from hundreds of meters to several kilometers. This is the climate that is relevant for vineyard site selection.
3. *Microclimate*: The climate within and immediately surrounding the canopy or within a vineyard. It may differ from the mesoclimate because of aspect, slope, and even soil type. Due largely to the presence of leaves, differences in microclimate may occur over as little as a few centimeters or over hundreds of meters. This is the climate that can be manipulated by vineyard cultural practices.

Compared with the situation close to the equator, regions at higher latitude experience longer periods of summer daylight between the spring and autumn equinoxes (the dates of equal lengths of day and night). Greater day length enables leaves to be photosynthetically active during a greater portion of each day, but this benefit is partially offset by the lower intensity of solar radiation. This is because from a grapevine's viewpoint, the sun "passes" lower on the horizon—that is, at a greater zenith angle, whereby the zenith is directly overhead, at 0°, and the horizon is at 90°. Thus, at the summer solstice, or the "longest" day, the total daily global irradiance is approximately equal between 30° and 50° latitude, where most of the world's grapes are grown. However, at the two equinoxes the daily irradiance received at 50° latitude is only about 75% of that at 30°. Altitude (elevation above sea level), aspect (slope direction), and grade (slope steepness) of a vineyard site also determine the radiation received by the vines (Failla et al., 2004). The effect of altitude at the same latitude is particularly noticeable as a strong and predictable decrease in temperature with increasing elevation ($\sim 0.65°C$ for each 100 m), provided the atmosphere is well mixed, which occurs courtesy of solar heating and wind. Nonetheless, at night and during the winter (especially in sheltered valleys), such mixing is often insufficient (especially during clear, calm weather) to prevent the formation of the so-called temperature inversions. During such inversions, cold air settles in valley floors and local depressions so that the temperature is coldest at ground level and increases (2.5–3°C per 100 m) up to the inversion top (typically 200–300 m above ground level), above which it begins to decrease normally (Daly et al., 2008).

A 10% south-facing slope at 45°N receives as much radiation, and hence energy, during the April–October growing season as a horizontal plain at 40°N. Grape-growing regions near the 45th parallel, the midpoint between the equator and the poles, include Bordeaux or the Rhône Valley in France, Italy's Piemonte, Croatia, Crimea, northern China, Oregon's Willamette Valley, and, in the Southern Hemisphere, New Zealand's Otago region. Examples of regions near the 40th parallel include central Spain and Portugal, southern Italy, northern Greece and Turkey, China's Beijing region, northern California, and near 40°S Hawke's Bay in New Zealand. During the summer, a

50% equator-facing slope receives approximately 25% more solar energy than a north-facing slope, but the difference is much greater in the winter because of the lower elevation of the sun (Holst et al., 2005). Not only does the disparity in radiation translate into warmer, drier soils and earlier budbreak on equator-facing slopes but also, because the energy difference due to vineyard aspect increases during the ripening period, pole-facing slopes can be at a serious disadvantage in marginal climates. Differences are also much larger under clear skies; clouds filter out the direct radiation from the sun so that during overcast days there is only diffuse radiation. Thus, in regions with frequent cloud cover (few sunshine hours) the incident radiation is similar regardless of aspect. The leaves (and the roots) perceive differences in solar radiation mainly as differences in temperature, with more radiation leading to higher daytime air and soil temperature, whereas light intensities above the canopy are similar regardless of aspect, at least during the growing season (Holst et al., 2004; Mayer et al., 2002).

In addition to the overall climate, annual climate variation and short-term weather fluctuations are also very important in viticulture. Whereas the long-term climatic averages are pertinent to site selection and choice of cultivars and rootstocks when establishing a new vineyard, climate variation among growing seasons and weather variation within seasons (see Sections 7.2, 7.4, and 7.5) determine the weather-associated risks involved in grape production and often influence management decisions in established vineyards. Climatic conditions that are relevant for individual vines are part of the (canopy) microclimate, which is strongly influenced by the presence of leaves. These conditions include light, temperature, wind, and humidity (Smart, 1985). Leaves alter all of these from the exterior of the canopy to the interior and from the top to the bottom.

5.2.1 LIGHT

Light has a more profound effect on plant development than does any other climatic factor or signal. Grapevines, like all plants, use light both as a source of energy (in fact, as their only energy source) and as a source of information. They can accurately perceive fluctuations in quantity (intensity), quality (spectral composition), directionality, and periodicity (day length) of the incoming light (Fankhauser and Staiger, 2002). The amount of incident light on a grapevine canopy varies with latitude, season, time of day, and cloud cover. Leaves effectively absorb sunlight in the visible and ultraviolet (UV) region of the electromagnetic spectrum (Blanke, 1990a). More precisely, the epidermis absorbs most of the potentially damaging UV light but is transparent to visible light, which penetrates the chloroplast-rich mesophyll cells (see Section 1.3). The visible wavelength range from 400 to 700 nm is utilized in photosynthesis (see Section 4.1), and this light is called photosynthetically active radiation (PAR). In other words, plants are mainly sensitive to the same spectral "window" of light that we see as the colors of the rainbow. Approximately 85−90% of incident light in the PAR range falling on a grape leaf is absorbed by the leaf; the rest is either reflected at its surface (6%) or transmitted through the leaf (4−9%) (Smart, 1985).

Because absorbing an optimal amount of light is so important for leaves, they have evolved anatomical and physiological strategies that allow them to adapt to a range of light environments. Leaves grown in the shade (e.g., in the canopy interior) are often larger but thinner than leaves grown in full sunlight, mainly because of their shortened palisade cells and limited investment in structural biomass (Keller and Koblet, 1995a; Palliotti et al., 2000). As a consequence of their larger size, shade leaves have fewer stomata per unit leaf area than do sun leaves. Shade leaves also have more total chlorophyll per reaction center and per unit nitrogen, but they have less rubisco protein and carotenoids (especially xanthophylls) and up to 50% lower respiration rates than sun leaves (Cartechini and Palliotti, 1995; Evans, 1989; Schultz, 1991; Seemann et al., 1987). These adaptations enhance light absorption and energy transfer in the shade but make these leaves considerably less efficient at higher light intensity (Ortoidze and Düring, 2001; Palliotti et al., 2000; Schultz et al., 1996). Moreover, such leaves are highly light sensitive, and when they are suddenly exposed to the sun, they can suffer from severe oxidative stress and photoinhibition and may even die in extreme cases (Iacono and Sommer, 1996; Triantaphylidès et al., 2008; see also Section 7.1).

In complete darkness (i.e., at night), a leaf's net CO_2 assimilation is negative because there is only respiration but no photosynthesis (see Section 4.4), and the leaf's metabolism and continued sugar export depends on the breakdown of starch accumulated during the day. As the irradiance or light intensity or photon flux (number of photons per unit surface area per unit time, expressed in $\mu mol\ m^{-2}\ s^{-1}$) increases, photosynthetic CO_2 uptake increases. The light intensity at which the leaf's net CO_2 assimilation is zero (i.e., photosynthetic CO_2 uptake balances respiratory CO_2 release) is called the light compensation point. This point depends on species, cultivar, and developmental conditions, but in typical grapevine leaves it is reached at an irradiance of approximately 10–70 $\mu mol\ m^{-2}\ s^{-1}$ (Cartechini and Palliotti, 1995; Düring, 1988; Keller and Koblet, 1994; Rühl et al., 1981). A further increase in photon flux leads to a concomitant, almost linear increase in photosynthesis until it starts to level off and reaches light saturation. Grapevine leaves generally reach light saturation between 700 and 1200 $\mu mol\ m^{-2}\ s^{-1}$ (Figure 5.5), which is well below the photon flux of full sunlight (which can exceed 2000 $\mu mol\ m^{-2}\ s^{-1}$) but above that under cloud covers (100–1000 $\mu mol\ m^{-2}\ s^{-1}$, depending on cloud type and density). The irradiance at which individual leaves reach light saturation varies widely, depending again on species, cultivar, developmental stage, and nutrient status, but it generally reflects the maximum irradiance to which a leaf was exposed during its growth. Moreover, at any given stomatal aperture width (i.e., stomatal

FIGURE 5.5

Relationship between light intensity and photosynthesis (left) and between stomatal conductance and light-saturated photosynthesis (right) of mature grapevine leaves. Note that the influence of cultivar is minor compared with that of light and nitrogen status.

M. Keller, unpublished data.

conductance), light-saturated photosynthesis "runs" faster in vines with high nutrient (especially nitrogen) status than in those with low nutrient status (see Figure 5.5).

Photosynthesis below light saturation is referred to as light limited (i.e., insufficient light for maximum photochemistry), and that above light saturation is referred to as CO_2 limited (i.e., enzymatic reactions cannot keep pace with photochemistry). Leaves respond readily to changes in light conditions, such that transient decreases in light due to, say, passing clouds, are associated with equally transient decreases in photosynthesis (Carbonneau and de Loth, 1985). The fact that less than 10% of the PAR falling on a grape leaf will reach the leaves beneath has implications for canopy photosynthesis. Only the leaves on the exterior of a canopy are exposed to saturating irradiance, so only these exterior leaves will achieve maximum rates of photosynthesis (Schultz et al., 1996; Smart, 1974). The interior leaves receive much less light (often less than 10 μmol m^{-2} s^{-1} inside a dense canopy) that is either transmitted through other leaves or passes through gaps in the canopy (Dokoozlian and Kliewer, 1995a,b; Mullins et al., 1992). Most of these gaps are not static due to wind moving shoots and leaves, which together with the absorption and reflection of light by leaves and soil surface creates an irregular patchwork of highly variable irradiance within the canopy. Light incident on individual leaves, or only on sections of a leaf, can fluctuate within milliseconds to minutes, often resulting in fleeting sunflecks (Kriedemann et al., 1973; Rascher and Nedbal, 2006). The contribution of these sunflecks to the daily photon flux at any given point within a typical canopy varies from 20% to 90%. Leaves that experience frequent but brief sunflecks benefit most in terms of photosynthesis, but because shaded leaves can heat up very rapidly upon sudden exposure to sunlight, they are prone to heat damage and wilting from water stress (Pearcy, 1990).

Depending on trellis and training system, and on vine vigor, light in the fruiting zone can range from less than 1% of ambient (e.g., nonshoot-positioned single-curtain trellis) to approximately 10% (e.g., vertically shoot-positioned systems) and more than 30% (e.g., double-curtain systems) (Cartechini and Palliotti, 1995; Dokoozlian and Kliewer, 1995b; Gladstone and Dokoozlian, 2003; Williams et al., 1994). Moreover, the shaded side of a canopy can receive as little as 3—6% (40—100 μmol m^{-2} s^{-1}) of the light intercepted by the sunlit side (Smart, 1985). As a consequence, whole-canopy photosynthesis is almost never light saturated (Flore and Lakso, 1989; Intrieri et al., 1997; Petrie et al., 2009; Poni et al., 2003, 2008). On the other hand, shade leaves inside the canopy have 30—50% lower respiration rates than sun leaves at the canopy surface, irrespective of temperature and leaf age. This adaptation (together with their lower light compensation point) enables many shade leaves to maintain a positive daily net carbon balance, but this balance is necessarily much lower than that of sun leaves. Particle films (e.g., clay minerals such as kaolin) applied to a canopy in areas with high solar radiation to protect the vines from heat and water stress decrease light absorption by the leaves (Shellie and King, 2013). Nonetheless, whole-canopy photosynthesis can remain unaffected or even increase, because the particles also reflect light. This reflection may improve the light distribution within the canopy, which tends to compensate for the decrease in photosynthesis of the exterior leaves. Similarly, the contribution of interior leaves to whole-canopy photosynthesis is greater in loose, open canopies than in dense canopies, where the outermost leaves absorb almost all of the available light (Smart, 1974).

Grapevines also adjust the distribution of proteins, chlorophyll, and photosynthetic capacity to canopy density so that well-exposed leaves have high nitrogen and photosynthetic capacity per unit leaf area (Bowen and Kliewer, 1990; Kriedemann, 1968b; Palliotti et al., 2000; Prieto et al., 2012; Williams, 1987). When photosynthesis in some leaves drops below a certain threshold for some time, the vine sheds these leaves by initiating the processes of senescence and abscission in order to prevent a wasteful situation in which the leaves' demand for water and nutrients outweighs their supply of fixed carbon (Flore and Lakso, 1989; Poorter et al., 2006; Taylor and Whitelaw, 2001). In other words, the plant disposes of those leaves that can no longer "pay" for their share of the respiratory costs of the shoots and roots that support them (Reich et al., 2009). Because a decline in photosynthesis is accompanied by a decline in stomatal conductance and consequently in transpiration rate, the delivery to these leaves of root-derived cytokinins via the transpiration stream also decreases, which may serve as a signal to initiate the early senescence program (Boonman et al., 2007, 2009; Buchanan-Wollaston, 1997; Pons et al., 2001). The ratio of red:far-red (FR) light incident on shaded leaves is also lower than at the canopy surface, and this may provide an additional senescence signal (Boonman et al., 2009; Rousseaux et al., 2000). Leaf senescence therefore depends more on a leaf's position than on its age, and abscission due to canopy shading or during stress may be viewed as the elimination of surplus leaves that do not contribute to canopy photosynthesis (Hikosaka, 2005). Senescence is accompanied by remobilization of carbon, proteins, and mineral nutrients from these leaves, although approximately half of a leaf's resources cannot be recycled and are lost (Bertamini and Nedunchezhian, 2001; Hikosaka, 2005). This adaptive strategy enables the plants to survive episodes of limited supply of photosynthates by remobilizing buffer reserves from older, sequentially senescing leaves and permanent parts of the plant and translocating them to organs with high sink priority to temporarily supply carbon for maintenance and/or growth processes (Geiger and Servaites, 1991; Hunter et al., 1994). Leaf senescence does not cause any change in the direction of phloem transport in the leaf: There is no switch from export to import. There is only a progressive change in the nature of the exported materials from predominantly sucrose to mostly amino acids and other nutrients (Masclaux-Daubresse et al., 2010; Thomas, 2013). In other words, shaded leaves in the interior of dense canopies are not "parasitic" on a vine (Koblet, 1975; Quinlan and Weaver, 1970; Wardlaw, 1990). Such leaves still maintain a positive daily carbon balance and are simply discarded following recycling of their accessible resources: A senescing leaf is still a source, albeit now mostly one of nitrogen instead of carbon (Reich et al., 2009; Thomas, 2013). Where the recycled compounds end up depends on the relative strength of the various sink organs. When the shoots are actively growing, remobilized nutrients are redistributed to young, better exposed leaves, but when shoot growth has stopped (e.g., due to water deficit), the fruit clusters or the permanent structures may be the main recipients of recycled nutrients.

Grapevines are quite shade tolerant. They adapt to low light by altering leaf and shoot growth, producing longer and thinner internodes, although this occurs at the cost of greater shoot hydraulic resistance (Cartechini and Palliotti, 1995; Schultz and Matthews, 1993b). When entire shoots or vines experience low light, such as during overcast periods, grapevines produce new leaves, especially lateral leaves, rather than maintain the source capacity of old leaves; however, this response is at the expense of fruit production and root growth (Keller and Koblet, 1995a). Such vines tend to have elevated concentrations of tissue nutrients and low amounts of reserve carbohydrates (Keller et al., 1995).

Wild vines are very well adapted for maximum light capture. Their elaborate branching structure, with a large number of short shoots armed with tendrils, enables them to spread their foliage over tree canopies, which results in an enormous increase in sunlight absorption. This option is not available to cultivated grapevines that are often confined to small trellis systems with foliage concentrated within a more or less defined canopy volume; this is especially true for shoot-positioned vines. In fact, fewer than 20% of the leaves often account for more than 80% of a vine's total carbon assimilation. Therefore, the canopy surface area of a vineyard, rather than its total leaf area, is important because the more solar energy that is intercepted by foliage, the greater the biomass production and the yield potential. Canopy surface area is referred to in terms of exposed surface area or effective surface area and, for fully developed canopies, varies from 30% to 85% of the total leaf area depending on trellis design and row spacing. Vines with a small proportion of exposed surface area have many shaded (interior) leaves and may produce less assimilate for export to be available for fruit ripening, root growth, nutrient uptake and assimilation, and replenishment of storage reserves for cold hardiness and spring growth.

The fact that photosynthetic CO_2 assimilation shows light saturation while the absorption of photons continues to increase with rising irradiance is a potential source of trouble for leaves. When a leaf absorbs more light than it can utilize, some of the excess energy must be driven away (dissipated) as heat, or else it will cause photoinhibition (see Section 4.1). Photoinhibition in grapevines occurs frequently around midday, when incident light and temperature are at a maximum, and leads to a temporary reduction in CO_2 assimilation (Chaves et al., 1987; Correia et al., 1990; Düring, 1999; Iacono and Sommer, 1996). Over the course of a growing season, this temporary depression of photosynthesis, known as "dynamic" photoinhibition, can result in a roughly 10% loss of potential biomass production. Leaves acclimate to high light conditions by moving the chloroplasts, which are normally aligned at the abaxial surface of mesophyll cells (and thus perpendicular to the solar rays to maximize light absorption), to anticlinal positions in the cells (and thus parallel to the solar rays). This reversible chloroplast relocation is regulated by a protein photoreceptor termed phototropin that is activated by blue light (450 nm) and is an attempt to avoid photoinhibition (Li et al., 2009; Spalding and Folta, 2005; Takahashi and Badger, 2011). The effect of chloroplast relocation may be reinforced by relocation of druses (calcium oxalate crystals). Under low light the druses "sit" near the bottom of the leaf's palisade cells, but under excess light they move to the top of these cells where they reflect some of the incoming light (He et al., 2014). Both low and high temperatures exacerbate the effect of strong light and photoinhibition (Gamon and Pearcy, 1990a,b). This can be problematic in cool-climate regions, which often experience large diurnal temperature fluctuations so that vines can be exposed to below-optimum temperatures and high light intensities in the morning (especially on the east side of the canopy), particularly in spring and fall. Off-season cold spells (combined with a clear sky) can result in irreversible damage due to degradation of photosynthetic pigments; this condition is called chronic photoinhibition (see Section 7.4).

Leaf layers alter not only the quantity of light but also its quality. The spectral characteristics of chlorophyll make leaves strong absorbers of photons in the blue (400—500 nm) and red wavebands (600—700 nm) of the solar spectrum (Figure 5.6; see also Section 4.1). Absorption of green (500—600 nm) and particularly FR (700—800 nm) light is weaker (Blanke, 1990a), and many photons of these wavelengths are reflected or transmitted (and scattered) in the form of diffuse radiation. Thus, light that is reflected from leaves is enriched in the FR region, which lies in the infrared portion of the electromagnetic spectrum, whereas light that is transmitted through leaves is depleted in the red (R) region of the spectrum. The outcome in both situations is identical: The R:FR ratio declines. These changes in the spectral composition of solar radiation are detected by various photoreceptors or photosensors that absorb light and translate the light signals via signaling networks into physiological responses. Perhaps the most important photosensor is a pigment system

FIGURE 5.6

Light reflection, absorption, and transmission of a typical leaf (left; reproduced from Taiz and Zeiger, 2006), and interior of a dense grapevine canopy (right; photo by M. Keller).

called phytochrome that works like a light-regulated switch (Rockwell et al., 2006; Smith, 2000). Like other plants, grapevines have at least five different phytochrome proteins that are abbreviated phyA–phyE, and each exists as a mixture of two reversible forms—an inactive form (P_r), which absorbs red light (660–670 nm) and has a half-life of approximately 100 h, and an active form (P_{fr}), which absorbs FR light (725–735 nm) and has a half-life of only approximately 1 h. When P_r absorbs red light, it alters its configuration to become P_{fr}, which induces other proteins to carry P_{fr} from the cytosol to the nucleus. Inside the nucleus, P_{fr} interacts with yet other proteins termed transcription factors to modify gene expression (Bae and Choi, 2008; Fankhauser and Staiger, 2002; Franklin, 2008; Smith, 2000). In other words, red light photoactivates phytochrome which then indirectly switches light-regulated genes on. Conversely, in the dark or when P_{fr} absorbs FR light, it changes back to P_r, switching the light-regulated genes off, as shown in the following diagram (Chory, 1997):

$$P_r \underset{\text{FR light}}{\overset{\text{Red light}}{\rightleftarrows}} P_{fr} \longrightarrow \text{Shade-avoidance response}$$

Although light quantity during the day varies as much as 10-fold due to variations in cloud cover, weather conditions, and time of year, these conditions have only a minor effect on R:FR, which is remarkably constant at approximately 1.2 (Holmes and Smith, 1977a,b). However, during dawn and dusk R:FR drops temporarily, especially at higher latitudes with longer twilight periods, and the magnitude and duration of this decline inform plants about seasonal changes (Franklin, 2008). The largest variation in R:FR occurs due to sunlight interacting with leaves. As vines grow and their canopy size increases, the R:FR of the light reaching individual leaves (even on the exterior) decreases. When the plant spacing is wide and/or canopy density low, vines do not shade each other, and the R:FR decreases mainly due to the increase in reflected FR. In grapevines with a loose canopy and vertical shoots, this increase in FR modifies the light environment of the internodes without greatly affecting the spectral balance of the leaves, which is dominated by the contribution of direct sunlight. Nonetheless, the decrease in R:FR is large enough to be sensed by phytochrome molecules located in the shoot tissue and to reduce the proportion of P_{fr}. Above a grapevine canopy or on its sunlit side, R:FR is similar to the ambient value (1.1–1.2), whereas on the shady side, values of 0.3–0.5 have been recorded. Similarly, inside a dense canopy R:FR can be reduced to values as low as 0.1 (Dokoozlian and Kliewer, 1995a,b; Smart et al., 1982). In this case, the decrease is due to the selective absorption of R by the leaves. This effect may be more important in vineyards of table and raisin grapes (e.g., Thompson Seedless) and of juice grapes (e.g., Concord), which are commonly grown with considerably larger and denser canopies than are wine grapes.

Together with blue- and UVA-light receptors named cryptochromes, the phytochrome system enables plants to monitor the quantity and quality of light, as well as its duration and direction. This information allows them to adjust their development in ways that optimize the capture of energy for photosynthesis and synchronize vegetative and reproductive development (Whitelam and Devlin, 1998). Generally, red light induces phytochrome-controlled responses, whereas FR light inhibits these responses. The detection of a lower R:FR by phytochrome (translated into less P_{fr}) provides a clear signal of light transmitted through, or reflected from, nearby plants and thus indicates the proximity of potential competitors. Plants can detect the presence of neighbors very early, well before they begin shading each other. To avoid being shaded, they respond to low R:FR signals by several morphological changes, collectively termed the shade avoidance syndrome (Morelli and Ruberti, 2002; Smith and Whitelam, 1997; Whitelam and Devlin, 1998). These changes mainly include stronger apical dominance (less lateral shoot outgrowth), accelerated shoot elongation rates (vigor), decreased leaf size and thickness, altered shape and more horizontal orientation of leaf blades, and more vertical shoots. Shoots that do grow in the shade tend to be thinner and lighter, have longer internodes and thinner leaves, and accumulate less carbohydrate reserves than shoots that grow in the sun (Buttrose, 1969c; Cartechini and Palliotti, 1995; Kliewer et al., 1972; May, 1960; Morgan et al., 1985).

Such alterations of vegetative growth often occur at the expense of reproductive growth. Therefore, prolonged low R:FR signals, which indicate that the competitors cannot be outgrown, also lead to developmental responses, such as early flowering or even inhibition of flower (inflorescence) initiation, reduced seed and fruit set, truncated fruit development, and often decreased seed viability (Morelli and Ruberti, 2002). These changes are probably brought about by the stimulating effect of phytochrome on auxin production and its distribution in the shoots and roots, and by increased ethylene production, which may augment brassinosteroid and gibberellin action and tissue sensitivity to gibberellin, leading to cell wall loosening and enhanced cell elongation in the shade (Franklin, 2008; Gallavotti, 2013; Morelli and Ruberti, 2002; Pierik et al., 2004). The regulation of auxin transport also enables phytochrome to coordinate shoot and root growth, whose reciprocal adjustment allows plants to better utilize the available light. Thus, low R:FR enhances shoot elongation but reduces root growth, whereas high R:FR has the opposite effect. In addition to its involvement in the control of photomorphogenesis and photoperiodism, the phytochrome system also induces changes in the composition of chemicals such as mineral nutrients, chlorophylls, anthocyanins, and other metabolites. Phytochrome seems to play a role in regulating the activities of certain enzymes, such as nitrate reductase (important for nitrogen assimilation; Smart et al., 1988; see also Section 5.3), phosphoenolpyruvate carboxylase (producing oxaloacetate for the TCA cycle; see Section 4.4), and phenylalanine ammonia lyase and chalcone synthase (two key enzymes for the synthesis of phenolic compounds, including anthocyanins and tannins; see Section 6.2). For example, P_{fr} may lead to activation of the genes that code for the enzymes that make anthocyanins; thus, a decrease in R:FR leads to a decrease in anthocyanin production, which impairs fruit color.

The regulation of growth and development by phytochrome in response to changing R:FR has implications not only for planting density and trellis design in vineyards but also for research studies conducted with artificial light sources. The cool white fluorescent tubes and high-pressure sodium lamps that are typically used in growth chambers and greenhouses have light spectra that differ markedly from the spectra of direct sunlight and even of natural daylight under overcast conditions and are almost devoid of FR (Hogewoning et al., 2010). This and the compounding fact that these lamps generally emit irradiances that are far below photosynthetic light saturation (often as little as $100\ \mu mol\ m^{-2}\ s^{-1}$) make it doubtful that such studies mimic plant growth under field conditions.

In addition to the effects of visible light, the UV range of the electromagnetic spectrum, which is commonly divided into UVA (320–400 nm), UVB (280–320 nm), and UVC (<280 nm), is also important. Much of the UVB and all of the UVC are absorbed by the ozone (O_3) layer in the stratosphere and never reach the surface of the Earth. Just as in the case of visible light, the UVB radiation a vineyard receives depends mainly on the position of the sun (latitude, altitude, season, and time of day) and on cloud cover. In addition, UVB has been increasing at Earth's surface due to the depletion of stratospheric O_3 (Madronich et al., 1998; McKenzie et al., 1999), although this process is thought to be close to its maximum and is expected to recover by approximately 2050—provided all countries continue to implement the so-called Montreal Protocol that limits the emission of O_3-depleting chemicals, such as chlorofluorocarbons. The main problem with UV light is that as the wavelength of electromagnetic radiation declines, its energy content increases. Although it comprises only approximately 0.5% of the total solar irradiance, the high energy of UVB makes it damaging to membranes, proteins, and DNA; thus, plants have evolved mechanisms to screen out UV radiation at or near the surface of their organs (Jansen et al., 1998; Rozema et al., 1997).

Leaf hairs, which are particularly dense on the young, expanding leaves of some cultivars, scatter and attenuate a large fraction of the UV radiation (Karabourniotis et al., 1999). The epicuticular wax, although itself not a strong UV absorber, also reflects and scatters some of the incident UV light (Figure 5.7) (Kerstiens, 1996; Rozema et al., 1997; Shepherd and Griffiths, 2006). In addition, the vacuoles of the epidermis cells of leaves and fruits accumulate phenolic compounds known as flavonols and cinnamic acids, which act as an optical filter or "sunscreen" that absorbs the damaging UVB radiation and, being strong antioxidants, detoxify the reactive oxygen species generated as a consequence of UV-induced damage (Bachmann, 1978; Egger et al., 1976; Hernández et al., 2009; Kolb and Pfündel, 2005; Kolb et al., 2001; Rozema et al., 1997; Yamaskai et al., 1997).

FIGURE 5.7

Cabernet Sauvignon leaves grown in full sunlight (right) and with the UV portion of the spectrum filtered out (left) display differences in epicuticular wax and hence in leaf surface glossiness.

Photo by M. Keller.

Additional phenolics are incorporated into the leaf cell walls (Weber et al., 1995). Together these phenolics effectively screen UV radiation so that virtually none is transmitted through the leaves. Particularly high concentrations of flavonols, in addition to anthocyanins that also absorb UV light, are produced in the unfolding leaves at the shoot tips, giving the shoot tips of many grape cultivars a pink or red appearance (see Figure 2.5). This protects the UV-susceptible and highly sun-exposed young leaves from photooxidative damage while the photosynthetic apparatus is being assembled in their cells. When the leaves senesce at the end of the growing season, they again accumulate flavonols and, in some cultivars, anthocyanins to protect their cells from sun damage, this time during the disassembly of the photosynthetic apparatus and retrieval of nutrients from the leaves. Flavonols are also effective inhibitors of auxin transport (Agati and Tattini, 2010; Peer and Murphy, 2007; Winkel-Shirley, 2002). Because basipetal auxin flow from the shoot tip is associated with apical dominance and suppression of lateral shoot growth, UV light may also influence shoot, and hence canopy, architecture. Although the extent of this influence is currently unclear, high UV exposure is indeed associated with a stimulation of lateral shoot growth in grapevines (M. Keller, unpublished data). This effect seems to be especially pronounced if soil moisture does not limit shoot growth, which may explain why overirrigation in climates with high solar radiation is very effective in promoting dense canopies.

Grape flowers and berries at the exterior of the canopy may be exposed to high intensities of UV radiation, most of which is absorbed by the berry skin (Blanke, 1990a). Accumulation of phenolic compounds, especially flavonols, in the skin is greater in these berries than in berries in the interior of the canopy, which has implications for fruit quality (see Section 6.2). It also has implications for yield formation (see Section 6.1). Similar to the situation in shoots, the need for sun-exposed flowers to produce flavonols for UV protection may interfere with auxin transport. Because auxin flow is necessary for fruit set (see Section 2.3), high UV exposure might lead to poor fruit set, especially in combination with nitrogen deficiency, which further promotes flavonol formation (M. Keller, unpublished data).

5.2.2 TEMPERATURE

Leaf temperature changes rapidly in response to fluctuations in radiation and air turbulence. Exterior leaves and grape berries are heated by the sun, whereas shaded leaves and berries in the interior of the canopy are generally close to air (ambient) temperature (Vogel, 2009). The heat load on a leaf exposed to full sunlight is so great that the leaf would heat up by $1-2°C\ s^{-1}$ to temperatures that would denature proteins and kill the tissues within less than 1 min if no heat was lost to the environment. This is indeed a problem when there is no wind at all, whereas even slow air movement can prevent lethal leaf heating (Vogel, 2009). Although leaves on the sunlit side of a canopy are often approximately $2-3°C$ warmer than leaves on the shaded side, the leaves are generally less than $5°C$ warmer than the surrounding air because of the emission of longwave radiation, sensible heat loss, and evaporative cooling (latent heat loss) due to transpiration (see Section 3.2). Sensible heat loss is the process whereby air circulation around the leaf removes heat from the leaf surface, as long as the leaf is warmer than the air. Evaporative heat loss occurs because water evaporation requires energy. Water stress decreases transpiration and therefore increases leaf temperature. Accordingly, the leaf temperature of grapevines that are experiencing water deficit tends to be higher than that of fully irrigated vines, regardless of whether the leaves are sunlit or shaded (Grant et al., 2007). Despite these cooling processes, however, direct sunlight can occasionally heat leaves by $10°C$ above air temperature (Sharkey et al., 2008). Conversely, at night leaves may be up to $10°C$ cooler than the surrounding air.

Growing grapevines in a manner that maximizes the canopy surface area for maximum light interception creates a dilemma for the vines, especially in a warm climate. Although they appear to be able to survive brief temperature "spikes" up to $60°C$, grape leaves are killed at temperatures above approximately $45°C$ due to disintegration of the cell membranes, which leads to membrane leakage and loss of cell contents, and due to thermal denaturation of proteins (see Section 7.4). In a movement termed paraheliotropism, grapevine leaves change their angle to the sun over the course of a day and a growing season, aligning the leaves in parallel with the solar rays during hot periods, to avoid overheating and keep light intensity on the leaf at or slightly below light saturation to maintain photosynthesis while avoiding photoinhibition (Gamon and Pearcy, 1989; Takahashi and Badger, 2011). In addition, transpiration rates increase within a range of less than $10°C$ to more than $40°C$, although stomata usually start closing at approximately $35°C$ (Figure 5.8). The increase occurs because warmer air can hold more moisture, thereby increasing the vapor pressure deficit, and protects the leaf from overheating. The stomatal closure at high temperature seems to be a response to heat-induced reduction in photosynthesis rather than the cause of it. Therefore, high temperatures can result in excessive water loss of the canopy while simultaneously reducing CO_2 assimilation. It is possible that the higher transpiration rate of Chardonnay compared to Cabernet Sauvignon at high temperature (see Figure 5.8) is related to the differences in leaf shape between the two cultivars (see Figures 1.14 and 5.7). Local leaf temperature increases approximately with the square root of the distance from the edge, and lobing improves heat transfer (Vogel, 2009), so Chardonnay's more-or-less entire leaves may need to evaporate more water to keep them from overheating.

As would be expected from a process that involves biochemical (enzymatic) reactions, CO_2 assimilation is strongly influenced by temperature (Geiger and Servaites, 1991). Moreover, other photosynthetic processes are also sensitive to temperature (e.g., electron transport has a pronounced

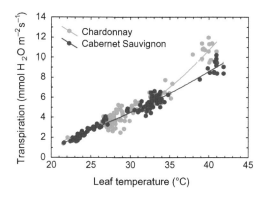

FIGURE 5.8

Relationship between leaf temperature and transpiration of mature leaves of two grapevine cultivars.

M. Keller, unpublished data.

optimum at 30°C), especially at high irradiance. Below approximately 15°C, photosynthesis is strongly curtailed by an inhibition of sucrose synthesis, which leads to accumulation of phosphorylated intermediates and prevents the release of phosphate (P_i) for regeneration of ribulose-1,5-bisphosphate (Hendrickson et al., 2004b; see also Section 4.2). This so-called "end-product limitation" or "feedback inhibition" occurs because low temperature restricts cell division in sink organs more than photosynthesis; the time it takes for a cell to divide increases exponentially with declining temperature (Körner, 2003). This decrease in sink activity results in surplus sugar accumulating in the leaves and in the perennial organs of the plant. As in the case of very high temperatures, the stomata will partially close in response to the reduced photosynthesis, rather than photosynthesis declining because of lower stomatal conductance.

Similar to light, therefore, rising temperature initially stimulates photosynthesis. However, instead of reaching a saturation point, the temperature response shows a relatively broad optimum, with very high temperatures resulting in a reduction of carbon fixation. In grapevine leaves, there is very little photosynthesis below 10°C, the optimum ordinarily falls between 25°C and 30°C, and photosynthesis typically declines sharply above approximately 35°C (Currle et al., 1983; Downton et al., 1987; Gamon and Pearcy, 1990b; Hendrickson et al., 2003; Kriedemann 1968b; Williams et al., 1994). Yet sometimes the photosynthetic rate at 45°C may approach half the rate at 30°C (Greer and Weedon, 2012; Mullins et al., 1992). The optimum temperature range depends on species, cultivar, light intensity, and developmental stage, but it also reflects the maximum temperature experienced during leaf growth. For example, low irradiance leads to a flatter and broader temperature response curve so that shaded leaves have a lower and less pronounced temperature optimum than exposed leaves (Berry and Björkman, 1980; Gamon and Pearcy, 1990b). Furthermore, in north—south-oriented rows, east-facing leaves typically reach lower daily maximum temperatures than west-facing leaves, which can limit photosynthesis and growth on cool sites (Hendrickson et al., 2004a). Nevertheless, the west-facing leaves often contribute less photosynthate than their east-facing counterparts because the high vapor pressure deficit (not the higher temperature) in the afternoon may lead to partial stomatal closure. On the other hand, high temperatures during leaf development can shift the optimum to as much as 35°C due to photosynthetic temperature acclimation that makes Calvin cycle enzymes and membranes more heat tolerant. Conversely, low temperatures will shift the optimum downward (leaves produce more photosynthetic proteins). In general, the optimal temperature for photosynthesis tends to increase by approximately 1°C for each 2–3°C increase in growth temperature—up to the maximum (Berry and Björkman, 1980; Hikosaka et al., 2006; Schultz, 2000). This means that the temperature optimum decreases with increasing vineyard elevation due to the decrease in average temperature (by ~0.65°C per 100-m elevation gain) and at a given site increases from spring to summer and declines in autumn. Leaves developing during the hot summer months have a considerably higher temperature optimum for photosynthesis than leaves developing in spring or autumn. However, even mature leaves can acclimate to changes in temperature within 1 or 2 weeks. The seasonal shift in temperature optimum occurs in both irrigated vines and vines growing under natural water stress conditions, but water stress (see Section 7.2) can become the dominant factor determining photosynthetic performance, largely overriding the temperature response.

Clouds alter not only the light intensity but also the temperature. Although clouds are usually associated with cooler days than would be the case under clear skies, they also prevent nighttime inversions and thereby dampen the amplitude of the diurnal temperature swings. In other words, clouds keep days cooler and nights warmer. This has implications for the daily carbon balance of grapevines. Because canopy net photosynthesis is light limited and thus lower during cloudy days, whereas whole-plant respiration is higher during cloudy nights, less carbon is available for growth, fruit production, and ripening. Frequent cloud covers, which are indicated by relatively few sunshine hours in climate records, in many cool and maritime growing regions are one reason for the marked interannual fluctuations in yield and fruit quality in such areas.

Both the increasing temperature and rising CO_2 content of the air ($[CO_2]$) associated with global climate change will also influence photosynthesis. Models as well as empirical observations generally show a relatively steep increase in net CO_2 assimilation rates above approximately 20°C when $[CO_2]$ rises from 300 to 600 ppm (current ambient values are ~390 ppm) and a shift of the optimum temperature for photosynthesis from 25–35°C to 35–40°C (Sage and Kubien, 2007). Below 20°C, there appears to be little effect of rising $[CO_2]$, and above approximately 40°C photosynthesis drops rapidly, although at 45°C it is still twice the rate (and similar to the rate at 20°C) at 600 ppm CO_2 compared with 300 ppm. Although heat effects will certainly be important in a warmer world with higher $[CO_2]$, the influence of climate change will be more important at the lower limits of temperature because current and predicted temperature increases will be higher at night, at higher latitudes, and in the winter (IPCC, 2013).

The temperature response of CO_2 assimilation partly reflects a conflict of interests in addition to the dilemma created by the need for evaporative cooling at high temperatures. Enzyme activities are stimulated by increasing temperature, and rubisco is no exception. Unfortunately, rubisco's oxygenation rate (releasing CO_2) increases faster with increasing temperature than its carboxylation rate (fixing CO_2) (Berry and Björkman, 1980; Foyer et al., 2009; Woodrow and Berry, 1988). This is due to temperature-induced increases in both the specificity of rubisco for oxygen and the solubility of oxygen. In addition to its effect on photorespiration, temperature also influences mitochondrial respiration—that is, glycolysis and the TCA cycle and the associated electron transport chain. The proportionality factor or temperature coefficient (Q_{10}) of respiration is approximately 2 or 3; in other words, the respiration rate roughly doubles or triples for every 10°C rise in temperature, although respiration declines rapidly above approximately 30°C (Kruse et al., 2008; Mullins et al., 1992; Schultz, 1991; Williams et al., 1994). The Q_{10} value is not constant and must be viewed as an oversimplification of the temperature response of respiration (Kruse et al., 2011). Nonetheless, because of its exponential increase with rising temperature, the respiration rate of grapevine leaves at 10°C is close to zero, whereas at 25°C it can be as high as 2 μmol CO_2 m^{-2} s^{-1}. The temperature effect on respiration is especially important because photosynthesis "rests" at night, whereas respiration "works" 24 h a day in all plant parts, not just the leaves. This temperature effect also reduces net CO_2 assimilation because at higher temperatures vines use a greater proportion of their daily fixed carbon for respiration. Therefore, even at modestly high temperatures (25–30°C) there may be less carbon available for vine growth and fruit ripening than at cooler temperatures (15–20°C). Incidentally, water-stressed plants often have a lower Q_{10} than plants with abundant water supply (Atkin et al., 2005). In other words, water deficit decreases the temperature sensitivity of respiration, although this is often complicated by an increase in soil temperature as the soil dries. Because the Q_{10} declines when the temperature increases briefly (Atkin and Tjoelker, 2003), a hot day may stimulate respiration more in vines growing in cool climates than in vines growing in warm climates. However, respiration acclimates somewhat to above- or below-average temperatures within days to compensate for the change in temperature. Thus, heat waves lasting several days decrease the respiration rate, whereas cold waves increase the respiration rate (Atkin and Tjoelker, 2003). Above approximately 35°C, the capacity of the electron transport system becomes limiting to photosynthesis, and at very hot temperatures (above ~45°C) membranes become increasingly fluid and leaky so that respiration declines again (Sage and Kubien, 2007). The fact that the Q_{10} is higher when there is abundant (saturating) sugar for respiration suggests that a low crop load might stimulate respiratory CO_2 release more in a warm climate than in a cool climate.

The temperature that a leaf experiences during its growth also affects leaf growth. Meristem temperature, through its effect on cell division, is an important determinant of the rate of leaf appearance and the rate of leaf expansion, as long as the temperature causes neither chilling stress nor heat shock (see Section 7.4). Leaf area development depends on three separate processes: leaf initiation, leaf expansion, and outgrowth of lateral shoots. Increasing temperature accelerates all three of these processes so that higher temperature leads to more rapid canopy development, longer shoots, and denser canopies. Indeed, a fundamental response of plants to high temperature appears to be a shift in carbon partitioning (see Section 5.1) to favor shoot growth at the expense of fruit growth and ripening and, probably, storage reserve accumulation (Richardson et al., 2004).

5.2.3 WIND

Strong wind (>6 m s^{-1}) can induce physical damage in addition to a reduction of shoot length, leaf size, and stomatal density. However, even if it is not strong enough to induce visible damage (Figure 5.9), wind mechanically disturbs shoot growth by repeated bending (often at rates of dozens to hundreds of times per minute), leading to shorter but thicker shoots. This so-called thigmomorphogenetic response intensifies with increasing height above the ground (Braam, 2005; Niklas and Cobb, 2006). Wind during the day seems to inhibit shoot growth more than at night (Hotta et al., 2007). The reduction in shoot growth is also more severe on the windward (into the wind) side of the canopy than on the leeward (away from the wind) side and intensifies with increasing number of wind perturbations rather than with increasing wind speed in each single

FIGURE 5.9

Physical wind damage on Merlot canopy from shoot being knocked against a foliage wire (left) and leaf being twisted at the petiole junction (right).

Photos by M. Keller.

event (Tarara et al., 2005; Williams et al., 1994). Moreover, the shoots are often displaced away from the wind so that the canopy becomes lopsided, which has consequences for fruit exposure to sunlight. Clusters on the windward side of the canopy may receive more intense light for longer duration than do clusters on the leeward side of the canopy (Tarara et al., 2005). Thigmomorphogenetic responses that resemble responses to wind may also be triggered by other physical influences, such as repeated bending or touching by passing vineyard workers, animals, or machinery. Shoot elongation ceases within minutes of a mechanical stimulus and is accompanied by cessation of phloem flow due to rapid callose deposition in the sieve tubes. This response is followed by a temporary increase in radial growth, callose removal, and recovery of elongation growth over the next few days (Coutand, 2010).

As discussed in Section 3.3, wind decreases a leaf's boundary layer resistance, which increases transpiration and enables evaporative cooling. This is an advantage under warm conditions because leaves tend to heat up rapidly when the wind speed drops below approximately 0.5 m s^{-1} (Vogel, 2009). Under otherwise similar conditions, the temperature of sun-exposed leaves tends to vary inversely with wind speed, whereas shaded leaves track the air temperature (Vogel, 2009). To avoid excessive water loss and dehydration in stronger wind, however, grapevines respond to wind speeds greater than 2.5 m s^{-1} by partly closing their stomata (Campbell-Clause, 1998; Freeman et al., 1982; Williams et al., 1994). Although this strategy may conserve water, it also reduces photosynthetic

CO_2 assimilation and increases leaf temperature. This may at least partially explain why vines growing in areas with frequent strong winds often produce fewer and smaller clusters and lower fruit-soluble solids (Williams et al., 1994). Conversely, reduced wind speed and air mixing in sheltered vineyards could also decrease photosynthesis because leaves may deplete the CO_2 in the air surrounding the foliage. Moreover, wind speeds less than 0.5 m s^{-1} result in humid canopies, which favors disease development. For instance, the powdery mildew fungus *E. necator* requires only 40% relative humidity for germination (see Section 8.2); this threshold is easily exceeded within the leaf boundary layer, where the fungus resides (Keller et al., 2003a).

Wind moving down the rows in a vineyard creates less turbulence and movement of leaves than wind moving across rows (Weiss and Allen, 1976). This may decrease water loss, especially under dry conditions, and reduce the negative effect of stomatal closure on photosynthesis. In addition, wind speed at the center of a canopy is often less than 20% of the speed at the exterior. Although this may not be important in terms of stomatal effects on CO_2 assimilation (which is light limited), it has implications for drying of leaves and fruits after rain: interior surfaces dry more slowly than exposed surfaces. Nevertheless, even mild winds can move leaves inside a canopy sufficiently to briefly increase their light exposure. The resulting sunflecks on these leaves can account for a temporary rise in photosynthesis, improving the leaves' photosynthetic efficiency and the overall carbon balance of the canopy (Intrieri et al., 1995; Kriedemann et al., 1973).

5.2.4 HUMIDITY

Intuitively, one would expect transpiration by leaves and berries to increase the humidity inside the canopy, with subsequent implications for the development of fungal diseases (see Section 8.2). This is a subject that has been very little studied. Increases in humidity of less than 10% have been recorded, and the significance of these increases is not well understood. However, a decrease in relative humidity from 95% to 50% increases the vapor pressure deficit of the atmosphere surrounding a leaf more than 10-fold (see Section 3.1). Moreover, humidity strongly depends on air temperature because increasing temperature also sharply increases the vapor pressure deficit (see Section 3.2). Therefore, as the vapor pressure deficit increases (e.g., on west-facing leaves in the afternoon), stomatal conductance tends to decline in an attempt to control excessive water loss by transpiration, which will limit photosynthesis. This means that an increase in relative humidity effectively reduces the transpiration rate and increases CO_2 assimilation. Due to the modulating effect of leaf layers, this benefits exterior leaves more than interior leaves.

Humidity also affects leaf growth. A high vapor pressure deficit reduces the growth rate of leaves by decreasing the rate of cell division and cell expansion, even when there is no soil water deficit. Therefore, leaves growing in low humidity remain smaller than leaves grown in high humidity. Vines growing in dry climates tend to have more open canopies than vines growing in more humid climates, even when they are equally well watered.

In addition to air humidity, even small changes in vine water status can have an effect on canopy microclimate. Water-stressed vines (see Section 7.2) often have higher canopy temperatures than fully irrigated vines because there is less transpirational cooling of water-stressed leaves and because such vines have a sparser, more open canopy.

5.2.5 THE "IDEAL" CANOPY

The upper limit on vineyard productivity is set by the total seasonal amount of PAR intercepted by the vines planted in the vineyard. Canopy structure, and especially the spatial distribution of leaves, has important consequences for canopy light interception and hence productivity (Prieto et al., 2012). As discussed previously, the acquisition of energy and carbon by canopies depends on total leaf area, leaf surface distribution, canopy structure, and photosynthetic capacity of individual leaves. A canopy that has an ideal microclimate in terms of maximum light interception for vine productivity has the following features (modified from Smart, 1985 and Smart et al., 1990):

- Rows should be oriented from north to south to maximize light interception by both sides of the canopy for some part of the day. However, row direction is of relatively little importance for wine quality compared to the features described next, and the north–south requirement may well be overridden by other considerations (e.g., topography and economic row length). In a high-irradiance environment, it may be beneficial to deviate from the north–south orientation because rows shifted somewhat to the northeast–southwest may protect grape berries from overheating on the vulnerable west side of the canopy.
- The ratio between canopy height and row width (distance between rows) should be 1:1. For any one canopy height, as row width decreases, the percentage light interception by the canopy increases. However, decreasing row width increases the likelihood of one row shading the lower canopy of the neighboring row. Similarly, for any one row width, as the height of the canopy increases, the percentage of light interception by the canopy increases, but again the potential for cross-row shading also increases. The 1:1 ratio is a compromise between the canopy's intercepting as much light as possible and avoiding one row shading another.
- Foliage should be trained vertically to avoid shading on one side of the canopy and promote light interception by leaves and grapes (fruit exposure). However, this requirement may be less important in regions with high temperature and high irradiance during the ripening period than in regions with frequent cloudy and cool conditions.
- The canopy surface area should be approximately 21,000 m^2 ha^{-1}, and 80–100% of the leaves should be on the outside of the canopy. The larger the surface area, the more light is intercepted, and hence the potential for photosynthesis and yield production is increased. However, if the surface area is too large, then the canopy height:row width ratio of 1:1 is not adhered to.
- The canopy should be approximately 30–40 cm wide with 1–1.5 leaf layers. This leaf layer number is the number of leaves from one side of a canopy to another. Higher values are associated with shading and reduced fruit quality, whereas lower values are associated with incomplete light interception.
- There should be approximately 15 shoots per meter of canopy. A higher figure means that shoots may be crowded and shading may occur. A lower figure means that light interception is suboptimal. The value varies according to cultivar and vine vigor and can be considerably higher without detriment in warm regions with plenty of sunshine compared with cool regions where light is often limiting. Maintaining ideal shoot spacing is the underlying theme of canopy management.
- A total of 20–40% of the canopy should be composed of gaps. Too many gaps reduce the yield potential, whereas too few gaps indicate that shading in the canopy is likely.

- Shoots should stop growing when they are approximately 15 nodes long (~1 m) to provide sufficient leaf area to ripen the fruit (10–15 cm^2 g^{-1}). If shoots grow beyond this length and are not supported or trimmed, they will fall across each other and create shade. Excessive trimming, however, wastes the vine's resources and is an indication of too much vigor.
- Lateral shoot growth should be limited to less than 10 lateral nodes per main shoot. For example, 10 lateral nodes could be made up of 5 lateral shoots of 2 nodes each or 2 lateral shoots of 5 nodes each. Excessive lateral growth indicates high shoot vigor, and growing shoot tips compete with the fruit for carbohydrates.
- The fruit zone should be near the top or the outside of the canopy so that 50–100% of the fruit is exposed to the sun. This promotes anthocyanin and tannin production and also improves disease control. High temperature ($>33°C$) and the UV radiation of bright sunlight can inhibit the spread of powdery mildew colonies, and exposed fruit also dries more quickly after rain, which reduces bunch rot infections.
- Fruit exposure on the east side of north–south-oriented rows should be near 100%, whereas on the west side it should be closer to 50% because of the increased heat load in the afternoon. Excessive fruit exposure can result in heat damage, sunburn, and impaired anthocyanin accumulation, while increasing phenolics other than anthocyanins beyond desirable levels, especially in white grapes (see Section 6.2).
- The renewal zone, which is the part of the shoot that will become the fruit-bearing unit in the following year, should also be near the top or the outside of the canopy to promote inflorescence initiation (i.e., bud fruitfulness).
- The pruning weight (total weight of canes removed during winter pruning) should be in the range 0.3–0.6 kg m^{-1} of canopy length with canes weighing 20–40 g each. Values higher than 1 kg m^{-1} indicate overly vigorous vines.
- The ratio of yield to pruning weight per vine should be in the range of 5–10. This is thought to be a measure of the balance between reproductive and vegetative growth. Values higher than 10 may be associated with overcropping and delayed ripening, whereas values lower than 5 are associated with low yield and high shoot vigor.

Vineyard sites with low to moderate vigor potential tend to produce canopies that are close to this "ideotype." Such sites are often characterized by relatively shallow rootzones and well-drained soils with somewhat limited water and nutrient storage capacity. Provided vines are not spaced too closely at planting and are not grafted to high-vigor rootstocks, they can be easily trained to a vertical shoot positioning system (especially in cool climates) or sprawl trained with little or no shoot positioning (especially in warm climates). Deep and fertile soils with high water storage capacity are less desirable because they may result in overly vigorous vines and dense canopies (Cortell et al., 2005). Although achieving balanced vines with ideal canopy characteristics is more difficult on such sites, it is not impossible. Vines can be grafted to low-vigor rootstocks (see Section 1.2) and spaced at wider planting distances that accommodate lighter pruning levels (see Section 6.1). In contrast, close spacing often results in shoot crowding and competition among shoots for access to light rather than in decreased growth due to competition between roots for water and nutrients, which tends to exacerbate the canopy problem (Falcetti and Scienza, 1989). In fact, high planting density could result in a so-called "tragedy of the commons," whereby plants that compete with each other for the same soil resources may increase both root and shoot growth—thereby keeping the root:shoot ratio constant—at the expense of fruit growth (and hence crop yield and quality) rather than limiting their root production to match resource availability (O'Brien et al., 2005). Vines growing on sites with high vigor potential can also be trained to trellis systems that are vertically divided (e.g., Scott Henry, Smart–Dyson, Ballerina) or horizontally divided (e.g., Geneva Double Curtain, Lyre). Such canopy division increases the length of the canopy per plant and is advisable when the pruning weight exceeds 1 kg m^{-1} canopy length. This will allow greater numbers of shoots to be spaced more ideally, which in turn will improve light interception and fruit exposure and accommodate higher yields without sacrificing fruit quality (Reynolds and Vanden Heuvel, 2009; see also Section 6.2). The impact of shoot number per unit canopy length may also depend on when the shoot density is established. Removing young shoots (a practice called shoot thinning) often leads to compensatory growth of the remaining shoots such that the total leaf area per vine is similar in shoot-thinned and nonthinned plants (Bernizzoni et al., 2011). For example, shoot thinning approximately 3 weeks before bloom led to compensatory growth of lateral shoots, which negated the beneficial influence of lower shoot density on canopy microclimate (Reynolds et al., 1994a), whereas shoot thinning at veraison had no such effects (Smart, 1988). However, delaying shoot thinning or not thinning dense canopies at all can nevertheless be associated with delayed fruit ripening (Bernizzoni et al., 2011; Reynolds et al., 2005).

5.3 NITROGEN ASSIMILATION AND INTERACTION WITH CARBON METABOLISM

Nitrogen (N) assimilation is the process by which inorganic N acquired by the vine is incorporated into the organic carbon compounds necessary for growth and development. Nitrogen is generally the fourth most abundant element in grapevines after hydrogen (H), carbon (C), and oxygen (O), although calcium (Ca) can exceed N in vines grown on calcerous, high-pH soils. Typical concentrations in dry matter (biomass) are roughly 60 mmol H g^{-1}, 40 mmol C g^{-1}, 30 mmol O g^{-1}, and 1 mmol N g^{-1}, and the production of 1 kg biomass requires between 20 and 50 g N (Xu et al., 2012). Therefore, N is the mineral nutrient for which plants have the highest demand and the nutrient that most often limits growth. This makes N the most important mineral nutrient of grapevines. Moreover, when grapes are harvested, some of the fixed N is permanently removed from the vineyard soil. This loss amounts to 2 or 3 kg t^{-1} of fruit removed but can be reduced to less than 1 kg t^{-1} if stalks (grape peduncles after destemming) and pomace (solid remains of grapes after pressing of juice or must) are recycled back to the vineyard. Any such loss has to be replaced by addition of organic or mineral fertilizer or by biological N fixation using leguminous cover crops.

Although N makes up almost 80% of the atmosphere, grapevines, in contrast to legumes whose roots "employ" symbiotic bacteria called rhizobia for the task of fixing atmospheric nitrogen gas (N_2), cannot directly utilize this N_2. Instead, they generally rely on uptake by their roots of nitrogen ions—mostly in the form of nitrate (NO_3^-)—dissolved in the soil water. These nitrate ions are then reduced to ammonium (NH_4^+) and assimilated into amino acids in the roots and leaves for translocation as both nitrate and amino acids throughout the vine to be used in growth, metabolism, or storage. Amino acids are N-containing organic acids, the units or "building blocks" from which protein molecules are manufactured by cellular ribosomes. Ribosomes, composed of proteins and RNA, utilize the DNA's genetic code to assemble amino acids into proteins, thereby translating the RNA template that is transcribed from the plant's genes. Each 1 g protein typically contains 0.16 g N, and plants invest approximately 55% of their total N content in proteins (Niklas, 2006).

5.3.1 NITRATE UPTAKE AND REDUCTION

Close to 99% of all the nitrogen present in the soil is bound in organic matter, which cannot be taken up directly by the roots. Due to rapid nitrification of NH_4^+ derived from organic sources to NO_3^- by microbes in most aerobic and reasonably moist soils, NO_3^- is the primary source of N for grapevines (Bair et al., 2008; Keller et al., 1995, 2001b). However, *Vitis* species are also capable of taking up NH_4^+ (Jiménez et al., 2007) and, almost certainly, amino acids as well, especially in acidic soils that are rich in organic matter (Fischer et al., 1998; Grossman and Takahashi, 2001). In fact, where they are available, roots prefer amino acids over NO_3^- for uptake (Miller et al., 2008). Nitrate is dissolved in the soil water and taken up across the root's epidermis and cortex (see Sections 1.3 and 3.2). The concentration of NO_3^- in the soil water can vary from 10 μM to 100 mM, but it is generally orders of magnitude lower than the concentration inside the xylem stream (Crawford, 1995; Crawford and Glass, 1998; Keller et al., 1995, 2001b; Robinson, 1994). Although the removal of NO_3^- by nitrogen assimilation and xylem transport should favor passive diffusion into the root, the concentration gradient in the "wrong" direction is far too great for this to occur under most conditions. Therefore, roots absorb NO_3^- actively by means of proton/nitrate (H^+/NO_3^-) cotransport using an ATP pump, a protein called H^+-ATPase embedded in the cell membrane (Barbier-Brygoo et al., 2011; Crawford, 1995; Crawford and Glass, 1998). In this process, energy from ATP consumption is used to "pump" protons out of the root cells into the soil water, generating a proton gradient across the membrane. The protons diffuse back into the cells, carrying negatively charged nitrate molecules with them. The micronutrient boron (B) is essential to keep the ATP pump going so that insufficient B availability results in a strong decrease in the roots' ability to absorb NO_3^- (and K^+) and, consequently, may lead to secondary N deficiency and accumulation of sugars and starch in the leaves (Camacho-Cristóbal and González-Fontes, 1999). The so-called low-affinity nitrate transporters operate when the external NO_3^- supply is high (in the mM range), whereas under low NO_3^- availability (in the μM range) the high-affinity transporters are switched on (Barbier-Brygoo et al., 2011). Some nitrate transporters can even switch back and forth between low-affinity mode and high-affinity mode, using a phosphorylation switch that removes (low-affinity) or adds (high-affinity) a phosphate group from or to the transport protein (Tsay et al., 2011; Wang et al., 2012).

Once inside the root cells, nitrate can be moved to the vacuoles for temporary storage or loaded into the xylem and transported to the shoots and leaves as shown in Figure 5.10. Much of the NO_3^-, however, is assimilated into organic N compounds—that is, amino acids (Loulakakis et al., 2009; Roubelakis-Angelakis and Kliewer, 1992; Tester and Leigh, 2001; Xu et al., 2012). Nitrogen assimilation in grapevines occurs in both the roots and the leaves (Keller et al., 1995; Llorens et al., 2002; Perez and Kliewer, 1982; Stoev et al., 1966; Zerihun and Treeby, 2002). The following overview of nitrogen assimilation mainly summarizes aspects of the detailed reviews provided by Loulakakis and Roubelakis-Angelakis (2001) and Loulakakis et al. (2009). The first step of this

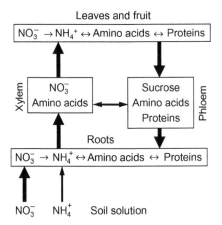

FIGURE 5.10

Simplified diagram of nitrogen uptake, assimilation, and circulation in grapevines.

Modified from Keller (2005).

process is the reduction of NO_3^- to nitrite (NO_2^-) by the enzyme nitrate reductase (NR) using two electrons provided by NADH or NADPH:

$$NO_3^- + 2\,H^+ + 2\,e^- \rightarrow NO_2^- + H_2O$$

There are two slightly different forms of NR. One is located in the cytosol of root epidermis and cortex cells and of leaf mesophyll cells. The other, which appears to be restricted to roots, is bound to the outer surface of the plasma membrane (and thus located in the apoplast). The cytosolic NR is active only during the day, whereas the apoplastic NR "operates" day and night but peaks during the night. The membrane form of NR also normally prefers electrons from succinate over those from NADH. NR is the main molybdenum (Mo)-containing protein in plants (Schwarz and Mendel, 2006), and one symptom of Mo deficiency is the accumulation of nitrate due to diminished NR activity (see Section 7.3). NR is stimulated by the presence of nitrate but requires light (in the leaves, where activation is mediated by phytochrome; see Section 5.2) or carbohydrates (mainly sucrose, in the roots) for full induction. The plant hormone cytokinin promotes but the amino acid glutamine suppresses NR activity.

Grapevines can distribute large amounts of nitrate or store it for later use in the cell vacuoles (which usually occupy more than 90% of the volume of mature cells) without deleterious effects (Roubelakis-Angelakis and Kliewer, 1992). This vacuolar storage pool buffers the nitrate concentration in the cytosol and may also participate in maintaining cell turgor (Dechorgnat et al., 2011). In contrast, nitrite is a toxic ion that can induce mutations in plant tissues. It is therefore immediately transported from the cytosol into plastids (in roots) or chloroplasts (in leaves), where it is reduced to ammonium by the enzyme nitrite reductase (NiR). This reaction uses six electrons provided by reduced ferredoxin (Fd_{red}), which is thus oxidized (Fd_{ox}), in the following process:

$$NO_2^- + 8\,H^+ + 6\,e^- \rightarrow NH_4^+ + 2\,H_2O$$

5.3 NITROGEN ASSIMILATION AND INTERACTION

The protein ferredoxin is produced in the photosynthetic electron transport chain in the chloroplasts (see Section 4.1) or by NADPH generated by the oxidative pentose phosphate pathway in the nonphotosynthetic plastids (Crawford, 1995). Because nitrite is toxic, vines maintain an excess of NiR whenever NR is present. In the roots, nitrate is sufficient to activate NiR, whereas in the leaves light is required in addition to nitrate.

5.3.2 AMMONIUM ASSIMILATION

Like nitrite, ammonium (NH_4^+) is toxic to plants and is either rapidly incorporated into amino acids (i.e., assimilated) or stored (in case of excess supply) in the cell vacuoles. Ammonium assimilation (Figure 5.11) is normally catalyzed by the two enzymes glutamine synthetase (GS) and glutamate synthase, also known as glutamine-2-oxoglutarate aminotransferase (GOGAT), in a cycle consisting of two sequential reactions:

$$\text{Glutamate} + NH_4^+ \rightarrow \text{Glutamine}$$

$$\text{Glutamine} + \text{2-oxoglutarate} \rightarrow \text{2 Glutamates}$$

FIGURE 5.11

Structures and reactions of compounds involved in ammonium metabolism.

Reproduced from Taiz and Zeiger (2006).

The first step of the GS/GOGAT cycle requires energy provided by ATP and involves a divalent cation (Mg^{2+}, Mn^{2+}, Ca^{2+}, or Co^{2+}) as a so-called cofactor of GS (Roubelakis-Angelakis and Kliewer, 1983). There are two main forms (isoenzymes or isozymes) of GS: GS1 is located in the cytosol of all plant organs and in the phloem companion cells, and GS2 is located in the plastids of photosynthetic tissues and roots (Grossman and Takahashi, 2001). The cytosolic GS1 is central to ammonium assimilation in the roots, and its activity increases with increasing sugar content. In contrast, GS2 predominates in ammonium assimilation in the leaf mesophyll.

The glutamine produced by GS in the first reaction stimulates the activity of the iron- and sulfur-containing protein glutamate synthase. This enzyme produces two molecules of glutamate in a reaction that consumes two more electrons. Plants contain two types of glutamate synthase: One is called NADH-GOGAT because it accepts electrons from NADH, and the other is termed Fd-GOGAT because it accepts electrons from ferredoxin (Temple et al., 1998). The NADH type is located only in plastids of nonphotosynthetic tissues such as roots or vascular bundles of developing leaves, whereas the ferredoxin type is located in both chloroplasts and nonphotosynthetic plastids but dominates in leaves and is very important in the recapture of ammonium released during photorespiration (see Section 4.3) in addition to its role in assimilating ammonium derived from nitrate reduction (the so-called primary ammonium assimilation). Up to 90% of the ammonium that flows through the GS/GOGAT cycle is generated during photorespiration (Stitt, 1999; Stitt et al., 2002). Both glutamine synthetase and glutamate synthase are stimulated by light and sucrose and inhibited by amino acids (Grossman and Takahashi, 2001). The carbon "backbone," 2-oxoglutarate, also known as α-ketoglutarate, is provided by the TCA cycle (see Section 4.4) and from stored organic acids, such as malate and citrate (Gauthier et al., 2010).

One of the two glutamate molecules produced by GOGAT is used to regenerate the cycle, and the other is used to supply amino acids for general metabolism. In the roots, glutamate can also be transported back to the cytoplasm, where it is converted back to glutamine by a slightly different form of GS for export in the xylem to the shoots. Glutamine thus provides N groups, either directly or via glutamate, for the production via amino transferases of virtually all organic nitrogenous compounds in the vine. Some such groups might also be derived from asparagine, which can apparently be produced by the addition of ammonium to aspartate by the ATP-powered asparagine synthetase (Masclaux-Daubresse et al., 2010).

In an alternative pathway for ammonium assimilation (see Figure 5.11), the enzyme glutamate dehydrogenase (GDH) catalyzes a reversible reaction that can either form (i.e., aminate) or degrade (i.e., deaminate) glutamate:

$$\text{2-Oxoglutarate} + NH_4^+ \leftrightarrow \text{Glutamate} + H_2O$$

GDH also comes in two main forms, one localized in the mitochondria of all organs (but especially in phloem companion cells) and the other in leaf chloroplasts. The mitochondrion version uses electrons supplied by NADH, whereas the chloroplast version uses electrons donated by NADPH. As a stress-related enzyme GDH, activated by Ca^{2+}, may participate in ammonium assimilation in tissues with excessive ammonium concentration and in senescing (dying) leaves, where it is thought to recycle and thereby detoxify the ammonium that is released during protein remobilization (Loulakakis et al., 2002; Masclaux et al., 2000). Thus, GDH may complement the GS/GOGAT cycle, whose activity declines during senescence in parallel with the decrease in rubisco, photosynthesis, and chlorophyll. But the main role of GDH seems to be very different. The enzyme can "work" in the reverse direction, oxidizing glutamate when fixed carbon is depleted, for example, as a result of restricted photosynthesis. Under these conditions, GDH can participate in the remobilization and breakdown of proteins (termed proteolysis) and in the subsequent degradation of amino acids to supply carbon skeletons back to the TCA cycle (see Section 4.4) for continued energy (ATP) regeneration (Aubert et al., 2001; Miyashita and Good, 2008; Robinson et al., 1991). This may be the main reason that NH_4^+ accumulation is sometimes found in organs that suffer from carbon starvation (Keller and Koblet, 1994).

5.3.3 FROM CELLS TO PLANTS

Nitrogen is a structural component of nucleic acids (i.e., of the organic bases in DNA, RNA, and ATP), chlorophyll, hormones, and amino acids. Many amino acids are ultimately assembled into proteins such as enzymes (which catalyze or increase the rates of biochemical reactions) and are essential to plant metabolism and energy generation. Other proteins become channels and transporters that carry molecules across membranes, and others store nitrogen for future use. Nitrogen uptake and assimilation "cost" large amounts of energy to convert stable, low-energy inorganic compounds present at low concentration in the soil water into high-energy organic compounds with high concentrations inside the plant. Nitrate assimilation uses 2.5 times the NADPH required for CO_2 assimilation. The reduction of one molecule of nitrate to nitrite and then to ammonium requires the transfer of 10 electrons and often accounts for 10—25% of the total energy expenditures in both roots and shoots. This means that a vine may use as much as one-fourth of its energy to assimilate N, a constituent that comprises less than 2% of total plant dry weight.

The distribution of available N in the soil is extremely heterogeneous. Nitrate concentrations in the soil solution can vary by several orders of magnitude, both temporally and spatially, even over short distances (Crawford, 1995; Keller et al., 1995, 2001b; Robinson, 1994). Because N is so important to them, plants have evolved mechanisms that remove nitrate from the soil solution as quickly as possible. The root system can react to a patch of nitrate by rapidly activating nitrate uptake and, more slowly, initiating lateral root proliferation within the nitrate-rich zone (Forde, 2002; Gastal and Lemaire, 2002; Nibau et al., 2008; Scheible et al., 1997). These responses are especially pronounced when vine N status is low (see Section 7.3). To support such localized nitrate uptake, those roots in contact with the rich N source also augment water influx, whereas in a compensatory response the roots that happen to grow in N-poor soil patches simultaneously take up less water (Gloser et al., 2007; Gorska et al., 2008). It seems obvious that this mechanism would improve nitrate delivery to those roots that are best positioned for immediate uptake by accelerating mass flow to the root surface. Accordingly, the rates of nitrate uptake and respiration decline rapidly as newly formed roots age (Volder et al., 2005).

When photosynthetic electron transport generates more energy than is needed by the Calvin cycle (see Section 4.2), some of this energy becomes available for N assimilation. Because the production of amino acids also requires a supply of carbon through the TCA cycle, the incorporation of inorganic N into amino acids is a dynamic process that is regulated by both internal (e.g., availability of carbon and N metabolites) and external factors (e.g., light and availability of inorganic N). When a grapevine takes up nitrate through the roots, this nitrate functions as a signal that induces the plant to switch on the pathway of N assimilation and divert carbon away from starch production toward the manufacture of amino acids and organic acids, such as malate and citrate, that substitute for nitrate to maintain charge neutrality following its reduction (Foyer et al., 2003; Huppe and Turpin, 1994; Jiménez et al., 2007; Stitt, 1999; Stitt et al., 2002). Consequently, export from the leaves of malate in the phloem increases as nitrate supply in the xylem increases. At the same time, K^+ flow in both the xylem and phloem also increases, because K^+ acts as a counterion that helps to neutralize the negative electrical charge of both nitrate and malate (Peuke, 2010).

Nitrogen assimilation proceeds rapidly in vines with high carbohydrate status and slows as the photosynthetic sugar supply declines (Perez and Kliewer, 1982). A decrease in carbon status can occur due to poor weather conditions, drought stress, loss of leaf area due to hail or insects, or infection by pathogens and can result in nitrate accumulation in the vine's tissues (Keller, 2005). Although energy-dependent nitrate uptake also declines in vines with low carbon status, uptake is less limited than N assimilation (Keller et al., 1995; Oaks and Hirel, 1985; Rufty et al., 1989). In contrast, when nitrate becomes limiting, N metabolism slows, which leads to accumulation of starch and, ultimately, suppression of photosynthesis through feedback inhibition. The increase in carbohydrate content in response to N deficiency is often greater in roots than in leaves.

Both roots and shoots of grapevines have the capacity to assimilate nitrate to amino acids. The relative extent to which nitrate reduction takes place in the roots or the leaves depends on a number of factors, including species, cultivar and rootstock, weather conditions, and the amount of nitrate the roots take up (Alleweldt and Merkt, 1992; Keller et al., 1995, 2001b; Llorens et al., 2002; Zerihun and Treeby, 2002). Species native to temperate regions usually rely more heavily on root N assimilation than do species of tropical or subtropical origin. Despite the variation among grape cultivars in their capacity to assimilate nitrate, neither total N nor nitrate concentrations in the leaf blades vary much among cultivars, although nitrate does fluctuate in the petioles (Christensen, 1984). Petioles, it seems, serve as temporary nitrate storage sites, from where the leaves retrieve it on demand (Wang et al., 2012). When environmental factors permit rapid photosynthesis that is associated with high vine carbon status, and when nitrate availability in the soil water is relatively low, nitrate absorbed by the roots is rapidly assimilated in the roots and transported in the xylem as metabolically active glutamine. As the supply of N increases, so does root uptake, and the roots' capacity for assimilation becomes increasingly saturated so that increasingly more nitrate is transported to the shoots in addition to glutamine (Alleweldt and Merkt, 1992; Keller et al., 1995, 2001b; Pate, 1980; Stoev et al., 1966).

This makes sense for the vine because reduction and assimilation of one nitrate molecule in the roots costs the equivalent of 12 ATP molecules, but it is substantially cheaper in the leaves. The reason for this is that N assimilation in the roots is dependent entirely on sucrose transported in the phloem for both carbon skeletons and energy (Huppe and Turpin, 1994; Oaks and Hirel, 1985). The capacity of the roots to assimilate N is directly related to their carbohydrate status, and the carbon skeletons for glutamine production are derived from recently imported sucrose or from stored starch (Emes and Neuhaus, 1997). Therefore, high N availability, such as after heavy fertilizer application or tilling of cover crops (Bair et al., 2008; Guerra and Steenwerth, 2012; Müller, 1985), can lead to lower starch reserves in the vine's permanent organs, especially under conditions that limit photosynthesis (Cheng et al., 2004b; Keller et al., 1995). Because nitrate reduction and subsequent assimilation are more sensitive to sugar supply than are nitrate uptake or long-distance transport, low carbohydrate status leads to export of nitrate to the shoots. Under such conditions, nitrate may accumulate in the leaf blades and petioles (Perez and Kliewer, 1982).

In contrast to roots, leaves can use excess photosynthetic energy for N assimilation (Huppe and Turpin, 1994). Indeed, increasing N supply and leaf N status stimulates light-saturated photosynthesis in leaves (Keller et al., 2001b; see also Figure 5.5), but it also increases the respiration rate due to the energy demands of N assimilation. The qualifier "light-saturated" is important because light intensities below saturation limit photosynthesis directly (see Sections 4.1 and 5.2), and N has no effect. Moreover, the total amount of carbon fixed per day saturates at high N levels so that there is an optimal leaf N content that maximizes carbon gain. The optimal N concentration is higher in leaves grown at high irradiance than in those grown under cloudy conditions or in the shade (Hikosaka et al., 2006). Excessive N supply reduces the amount of sucrose available for export, worsening the situation for the roots. Leaves also gradually acquire an increased capacity for N assimilation as they unfold and mature along with an increase in photosynthesis. Accordingly, the amounts and activities of GS and GOGAT initially increase with leaf age but decrease in senescing leaves. This is especially true for GS2, whereas GDH follows the opposite trend. At least some of the NH_4^+ that forms by the action of GDH during senescence or that escapes the action of GS during photorespiration may be volatilized to the atmosphere (Farquhar et al., 1979; Xu et al., 2012).

As long as sufficient photosynthate and energy are available, vines can assimilate most of the absorbed nitrate. Because it depends on photosynthetic energy, the production of glutamine and glutamate in the leaves occurs mainly during the day, and all the necessary enzymes are induced by the active form of phytochrome (see Section 5.2). In the roots, on the other hand, N assimilation proceeds both day and night. The "primary" amino acids glutamine and glutamate can be converted to many other amino acids by the action of enzymes called transaminases. These glutamate-utilizing enzymes recycle 2-oxoglutarate so that it can be used again in ammonium assimilation, and the reactions can proceed at night. Under favorable conditions that are associated with high rates of N assimilation, the vine converts surplus glutamine ($\sim 19\%$ N) to arginine. Arginine has the highest N:C ratio of all amino acids ($\sim 32\%$ N) and is the major and most efficient N storage compound in grapevines, both as soluble amino acid and incorporated in proteins (Kliewer, 1967a; Kliewer and Cook, 1971; Schaller et al., 1989; Xia and Cheng, 2004). The amount of stored arginine in the perennial parts of the vine increases as external N supply increases (Perez and Kliewer, 1982). These reserves are available as a buffer during periods of low N supply and to support new growth in spring. In fact, reserve N may be at least as important as reserve carbohydrates in supporting spring growth of grapevines (Cheng et al., 2004b; Keller and Koblet, 1995a; Schaefer, 1981). Up to half of the developing canopy's N demand is supplied from such reserves. Because of the rapid shoot growth in spring, the nutrient demand of the canopy is greatest between budbreak and bloom, even though most of the N uptake from the soil occurs after bloom (Conradie, 1986; Löhnertz et al., 1989; Peacock et al., 1989; Pradubsuk and Davenport, 2010), provided there is sufficient soil moisture (Keller, 2005). Limited availability of N reserves due to inadequate refilling in the previous growing season can restrict early shoot growth and canopy development (Celette et al., 2009; Keller and Koblet, 1995a) and may lead to poor fruit set (Keller et al., 2001a). The reserve pool is depleted during and after budbreak, reaching a minimum by bloom time or, sometimes, as late as veraison, and it is replenished later in the growing season (Eifert et al., 1961; Kliewer, 1967a; Löhnertz et al., 1989; Pradubsuk and Davenport, 2010; Schaller et al., 1989; Schreiner et al., 2006; Zapata et al., 2004). Insufficient N availability at this time—for example, because of late-season growth of a competing cover crop—may not allow the vine to fully replenish its N reserves (Celette et al., 2009). Accumulation of storage reserves occurs mostly when other plant requirements, such as growth and fruit production, have been satisfied—that is, when supply of resources exceeds demand (Lemaire and Millard, 1999; see also Section 5.1). In warm climates, where leaves remain photosynthetically active for several weeks or even months after harvest, it appears that much of the N taken up by the roots after harvest is directly incorporated into the reserve pool (Conradie, 1986; Peacock et al., 1989). While such postharvest uptake may not occur in cool climates (Pradubsuk and Davenport, 2010), late-season foliar application of N fertilizer may nevertheless increase the reserve N pool, albeit at the expense of carbohydrate reserves (Xia and Cheng, 2004). Clearly, then, late-season fertilizer application or cover crop tillage can be used to increase the amount of reserve N for the following spring. Increasing the size of the reserve pool—for instance, by N fertilizer application—will increase the N content of the emerging leaves and early season growth, which enhances photosynthesis and may lead to greater seasonal demand for N due to the larger canopy (Cheng et al., 2004b; Keller and Koblet, 1995a; Treeby and Wheatley, 2006).

Much of the N that continuously arrives, mainly in the form of glutamine and nitrate, via the xylem in the leaves is redistributed as glutamine via the phloem to the shoot tips, fruit clusters, and woody tissues for further use in growth and metabolism or storage. Since the phloem can transport only small amounts of nitrate (Peuke, 2010), most nitrate has to be assimilated to glutamine once the xylem has delivered it to the leaves or stored as nitrate in the leaf vacuoles to buffer temporary shortages in N supply. Nonetheless, some nitrate may be remobilized and exported to sink organs from older leaves via the phloem, especially under conditions of N deficiency (Fan et al., 2009; Wang et al., 2012). In addition, nitrogenous compounds, as well as other mineral ions, are remobilized from proteins and nucleic acids in senescing leaves at the end of the growing season and exported to the perennial parts of the vine for storage (Conradie, 1986; Dintscheff et al., 1964; Kliewer, 1967a; Loulakakis et al., 2002). This process is also called resorption and is associated with a decline in photosynthesis and a transition from nutrient assimilation to nutrient remobilization (Hoch et al., 2003; Lim et al., 2007). As much as 90% of leaf N may be recycled during senescence (Löhnertz et al., 1989), and the GS1 form of glutamine synthetase produces glutamine for export from the leaves. Some of the remobilized protein is not broken down completely and may be exported by the phloem in the form of small peptides two or three amino acid units in length. Remobilization also occurs from shaded leaves, and from flower clusters prior to fruit set, in the interior of dense canopies (see Section 5.2). The recovered N is recycled to unshaded leaves and to the shoot tips for production of new leaves that are better exposed to sunlight (Hikosaka, 2005). Therefore, the leaf N content declines with both decreasing light and increasing leaf age, which is why the (typically older) leaves in the interior of a canopy contain less N than the leaves at the exterior (Prieto et al., 2012).

The ability of GDH to oxidize glutamate during periods of carbon starvation may have implications for yield formation in grapevines. When adverse environmental conditions, such as overcast skies or unseasonably cool weather, restrict photosynthesis (see Section 7.4), vines can experience severe carbon shortage. Meanwhile, N assimilation is reduced when photosynthesis declines, which leads to an inhibition of amino acid production. Consequently, plants may become starved for both carbon and N (Stitt et al., 2002). This may be especially critical during bloom, when storage reserves in the permanent organs are at their lowest. In addition, local carbon starvation can also occur inside dense canopies or due to stunted shoot growth limiting leaf area. In such situations, GDH participates in the resorption of proteins from relatively unimportant sinks that may subsequently be discarded. The amino acids derived from this proteolysis can serve as alternative electron donors for respiration in the absence of sufficient carbohydrates (Araújo et al., 2011). It is possible that this mechanism may account for the observed accumulation of ammonium in grape flowers during the development of inflorescence necrosis (a.k.a. early bunchstem necrosis). This syndrome is an extreme form of poor fruit set that culminates in abscission of portions of or whole inflorescences and is probably induced by carbon starvation (Jackson, 1991; Jackson and Coombe, 1988; Keller and Koblet, 1994; Keller et al., 2010; Rogiers et al., 2011; see also Section 6.1). The accompanying buildup of toxic ammonium likely is a consequence, not a cause, of the senescence process associated with organ abscission. Such abscission, preceded by nutrient recycling, can be viewed as a "sacrifice" by the vine of comparatively less important tissues or organs that will permit more important sinks to survive episodes of stress-induced carbon starvation. For instance, if the incident light is insufficient to sustain growth of all vine organs, leaf growth, especially on lateral shoots, may be enhanced to capture more light at the expense of inflorescence survival (Keller and Koblet, 1994, 1995a; Keller et al., 2001a).

The rate of photosynthesis in grapevine leaves increases as the supply of N in the xylem increases (Keller, 2005). The leaves' photosynthetic capacity—that is, the light-saturated rate of photosynthesis—also seems to increase linearly with increasing leaf N content, mainly because rubisco, the enzyme responsible for CO_2 fixation, comprises approximately one-third of a leaf's protein and up to 25% of its N (Evans, 1989; Seemann et al., 1987; see also Section 4.2). Thus, the optimal leaf N content required to maximize photosynthesis increases with increasing light intensity. Plants therefore tend to distribute leaf N so that it closely follows the light distribution within a canopy in order to maximize whole-canopy photosynthesis (Gastal and Lemaire, 2002). This non-uniform distribution of N has to be taken into account when leaves are sampled for nutrient analysis to determine vineyard fertilizer requirements. On the other hand, when N becomes limiting, rubisco activity declines more than electron transport capacity (see Section 4.1). To avoid excessive light absorption that would damage the photosynthetic apparatus via oxidative stress, the leaves decrease their chlorophyll content (i.e., turn pale green or even yellow) and activate their energy-dissipation and antioxidant systems (Chen and Cheng, 2003a; Keller, 2005). In addition, they decrease the angle between the leaf blade and the petiole, which further reduces the number of light photons captured by a leaf (see Section 7.3).

Warm conditions stimulate organ growth (e.g., leaf expansion) more than N assimilation and use (e.g., protein synthesis), so the N:C ratio in the vine's organs decreases compared to that seen in cool conditions. Low temperature reduces protein production and enzyme activity so that amino acids and nitrate accumulate in the leaves because nitrate uptake can continue under low temperature (Lawlor, 2002). Cell division appears to be particularly sensitive to N status, and inadequate N supply generally suppresses growth more than photosynthesis so that carbohydrates accumulate in the leaves and the N:C ratio is low (Gastal and Lemaire, 2002). However, under the elevated nitrate concentration in the soil water that occurs after fertilizer applications (Müller, 1985), root absorption can exceed the capacity of a vine for assimilation because nitrate reduction and assimilation are among the most energy-intensive reactions. This is especially true under conditions that favor rapid transpiration, such as warm and sunny days, and therefore enhance nutrient uptake. Under high N availability and high light intensity, glutamine and even nitrate can be transported back from the leaves to the roots in the phloem as signals to limit nitrate uptake (Dechorgnat et al., 2011; Gessler et al., 1998; Lemaire and Millard, 1999; Wang et al., 2012). This feedback regulation normally coordinates nitrate uptake with the vine's demand for N. However, when grapevines have access to large amounts of soil nitrate during periods of low solar radiation, nitrate assimilation cannot keep pace with uptake because of a sucrose and energy shortage, and nitrate accumulates in the leaves and other organs once available starch reserves have been exhausted (Keller et al., 1995, 2001b; Perez and Kliewer, 1982; Scheible et al., 1997; Smart et al., 1988; Zerihun and Treeby, 2002). High leaf nitrate reduces the amount of sucrose available for export, depleting root starch reserves and inhibiting root growth, which strongly decreases the root:shoot ratio (Keller and Koblet, 1995a; Keller et al., 1995; Scheible et al., 1997). This has special implications during the establishment phase of a vineyard. Heavily fertilizing and irrigating young vines in order to harvest a crop in the year after planting may curb root development, which might be detrimental to vine performance in the long term. The increase in leaf nitrate under cool or overcast conditions also has implications for the interpretation of the results from analysis of leaf samples collected in the vineyard to estimate fertilizer requirements: If samples are collected during cool or cloudy periods, the vine N status may appear higher than it is in reality.

During drought or salt stress, and possibly also with N deficiency, a larger proportion of glutamate appears to be converted to the amino acid proline. Much of this proline synthesis results from glutamine produced by the GS1 form of glutamine synthetase in the phloem's companion cells. Proline is highly soluble and can accumulate in cells to high levels without disrupting their metabolism. This allows plants to lower their tissue water potential while maintaining turgor pressure during periods of drought (see Section 7.2) or salinity (see Section 7.3).

CHAPTER 6

DEVELOPMENTAL PHYSIOLOGY

CHAPTER OUTLINE

- 6.1 Yield Formation 194
  - 6.1.1 Yield Potential and its Realization 197
- 6.2 Grape Composition and Fruit Quality 205
  - 6.2.1 Water 209
  - 6.2.2 Sugars 212
  - 6.2.3 Acids 216
  - 6.2.4 Nitrogenous Compounds and Mineral Nutrients 219
  - 6.2.5 Phenolics 223
  - 6.2.6 Lipids and Volatiles 236
- 6.3 Sources of Variation in Fruit Composition 243
  - 6.3.1 Fruit Maturity 244
  - 6.3.2 Light 245
  - 6.3.3 Temperature 249
  - 6.3.4 Water Status 254
  - 6.3.5 Nutrient Status 257
  - 6.3.6 Crop Load 261
  - 6.3.7 Rootstock 264

ABSTRACT

Plant size and the environment limit the capacity of grapevines to grow shoots and fruit, and vines self-regulate to balance their annual vegetative and reproductive growth. Grapevine yield is made up of several components, namely the numbers of buds per vine, shoots per bud, clusters per shoot, and berries per cluster, in addition to berry weight. Yield formation occurs over two growing seasons, and developmental and environmental factors interact to determine the final crop through their variable impact on the different yield components. The berries' physical and chemical composition at harvest determines their quality for fresh consumption or processing. Organic acids, tannin precursors, some minerals such as calcium, and some aroma precursors and green aroma components accumulate during early berry growth, while amino acids, flavonols, and other minerals such as potassium accumulate throughout berry development. Fruit ripening begins at veraison and is associated with berry softening, skin stiffening, accumulation of sugars, anthocyanins, tannins, and flavor and aroma components, increase in pH, and decrease in chlorophyll, organic acids, and green aroma components. Harvest fruit composition varies considerably both within and between vineyards and between years due to differences in fruit maturity and environmental impacts dominated by light, temperature, and water and nutrient status, with some variation also being due to crop load and rootstock.

Keywords: Anthocyanin; aroma; capacity; flavor; organic acid; ripening; sugar; tannin; vigor; yield

6.1 YIELD FORMATION

The amount of fruit production in a given year and over the lifetime of grapevines determines both their reproductive success as a species and their agronomic trait of yield potential. Viticultural yields are determined by the amount of carbohydrate (sugar) partitioned to the fruit rather than to other organs. Yield formation is often referred to as cropping, with the crop being the amount of fruit borne on a vine or produced by a vineyard. The "crop level" is the amount of fruit (sometimes the number of clusters) per shoot or per unit of canopy length, the "crop size" is the yield per vine or per unit of land area, and the "crop load" is the crop size relative to vine size (estimated as pruning weight or leaf area) and is a measure of the sink:source ratio. The terms crop level and crop size are often used synonymously, and "overcropping" refers to the production of more crop than a vine can bring to acceptable maturity by normal harvest time (see Section 6.2).

Grapevine yield is made up of a number of different components. Yield components are those factors in grapevine reproduction that, multiplied together, total the yield obtained from a single vine or an entire vineyard. For a single vine, this can be written as follows (Coombe and Dry, 2001):

$$\text{Yield} = \frac{\text{buds}}{\text{vine}} \times \frac{\text{shoots}}{\text{bud}} \times \frac{\text{clusters}}{\text{shoot}} \times \frac{\text{berries}}{\text{cluster}} \times \text{berry weight}$$

The vineyard yield is the sum of the yield of all individual vines and depends on the planted land surface, the number of bearing vines per unit land area (i.e., planting density), and the trellis and training system (e.g., single versus divided canopy), which together with the pruning method (e.g., spur, cane, or minimal) define the size of each vine. The upper limit on the number of fruitful shoots per vine is largely determined by the number of buds left after winter pruning (i.e., pruning severity), but the maximum number of fruit clusters available for harvest is limited by the number of inflorescences initiated in those buds during the previous growing season. The weight of each cluster depends on the number of berries on the cluster and their final weight. The berry number, in turn, is determined by the number of flowers that set fruit.

The discussion of the reproductive cycle in Section 2.3 is mainly concerned with anatomical and temporal aspects involved in the initiation and differentiation of grape clusters and their subsequent development to fruit maturity. However, yield formation also depends on a number of both internal and external factors and the interactions among them. They include the genetic makeup (genotype) of the plant (species, cultivar, clone, and rootstock), the vineyard site (soil, water and nutrient availability, and climate), seasonal weather patterns (light, temperature, rainfall, and humidity), the trellis and training system, cultural practices (pruning, canopy management, irrigation, nutrition, and pest and disease management), as well as legal aspects and market-driven actions (wine style, winery demand, and yield regulation).

Although interactions between genotype and environment lead to tremendous spatial and temporal variation in grape yield, both within and between vineyards, not all components of yield respond equally to environmental conditions. The planting density may decrease over the life of a vineyard because a grower may choose not to replace some of the vines that die for one reason or another. This change constitutes the difference between number of vines per unit of land area and number of bearing vines per unit of area. It is often said that the potential yield of a vineyard increases as the planting density increases, but this is not necessarily true (Winkler, 1959). The total number of shoots growing on a vine is determined primarily by planting density, trellis and training system, and pruning level (i.e., the number of buds retained at pruning). For a given planting density and pruning level, the berry weight is relatively highly conserved, and the number of clusters per shoot ordinarily varies far less than the number of berries per cluster (Currle et al., 1983). This is true for both the variation between cultivars and the variation between growing seasons for the same cultivar.

A few wine grape cultivars have been grouped according to their "reproductive performance" (Dry et al., 2010). In this classification, Cabernet Sauvignon and Merlot were characterized by high flower numbers per inflorescence (>300) and low fruit set (~30%); Syrah, Tempranillo, Pinot noir, Chardonnay, and Sauvignon blanc were characterized by low flower numbers (~200), high fruit set (~45%), and low berry numbers per cluster (~100); whereas Sangiovese, Nebbiolo, and Zinfandel were characterized by low cluster numbers, high flower numbers (>300), high fruit set (~45%), and high berry number (>200). Despite this variation, average berry weight (~1 g) and yield (3–4 kg m^{-1} of cordon) were similar among these cultivars. Vine growth and maintenance (i.e., survival) on the one hand, and reproduction on the other hand, both require resources that the environment provides in limited supply. Therefore, there is an optimal balance between a vine's vegetative and reproductive growth that is the outcome of a trade-off between survival and reproduction. The interaction between these two goals is related to the interdependence of vine capacity and vigor (Huglin and Schneider, 1998; Winkler, 1958; Winkler et al., 1974).

Capacity is defined by the total annual growth of the vine, including its production of fruit, leaves, shoots, and roots in one growing season. Capacity, which is termed annual net primary production in the general plant ecology literature, is thus a measure of the net resource gain from the environment, analogous to profit in economic theory (Bloom et al., 1985). Although accurate estimates require measurements of total plant dry mass at the beginning and at the end of the growing season (but before leaf fall), capacity can be approximated from a vine's total weight of fruit and canes at the end of the growing season—that is, for manually pruned vines, yield plus pruning weight. In practice, pruning weight is often incorrectly used as the sole indicator of vine capacity. Vine capacity increases with increasing size of the root system, number of shoots, and leaf area. In fact, this total annual growth is roughly proportional to the vine's ability to intercept sunlight, which depends on its total and sun-exposed leaf area. Capacity is also thought to be approximately proportional to the three-fourths power of total plant dry mass ($C \sim m^{3/4}$), which simply states that although large plants are able to grow more than small plants, the extra gain diminishes as plant size increases further (Niklas, 2006; Niklas and Cobb, 2006).

Vigor, on the other hand, is defined as the rate of shoot growth and can be measured by the change in shoot length over time. Alternatively, vigor is sometimes defined as the vine's annual vegetative biomass. Ignoring the secondary growth of a vine's perennial structure, the latter definition integrates the growth of all shoots over an entire growing season and can be estimated from a vine's pruning weight. Pruning weight is, however, not an appropriate measure of vine vigor for mechanically or minimally pruned grapevines, because insufficient amounts of cane material are pruned off to permit accurate estimates.

Vigor and capacity are interrelated and are ultimately determined by vine size, pruning level, the amount of stored reserves present at the beginning of the growing season, seasonal weather patterns, and availability of water and nutrients in the rootzone. Thus, a young vine may grow vigorously but have a low capacity, whereas a mature vine with many buds can give rise to shoots of low vigor but can have a high capacity. The main factors involved in the regulation and interdependence of capacity, vigor, and crop load (i.e., vine balance) are as follows (Coombe and Dry, 2001; Downton and Grant, 1992; Huglin and Schneider, 1998; Winkler et al., 1974):

- Grapevines have a finite capacity because of the limitations imposed by the amount of external resources available. Thus, vines can only support a certain number of shoots and ripen a certain amount of fruit in any one growing season.
- Vine capacity is proportional to total potential growth. Thus, older and/or larger vines (with greater capacity) can support more shoots and more fruit than younger and/or smaller vines.
- Pruning tends to depress vine capacity because it reduces the number of shoots and leaves per vine, delays canopy development (delay in the development of maximum leaf area), and reduces the total amount of fruit produced. Vines typically respond to more severe pruning with increased vigor and production of more total 1-year-old wood.
- The production of fruit tends to depress vine capacity by reducing vigor and decreasing the amount of stored carbohydrates, which can lead to lower capacity during the following year.
- The vigor of a shoot is inversely proportional to the number of shoots per vine and to crop load. Thus, an increase in shoot number and an increase in crop load both lead to a decrease in vigor.
- The direction of shoot growth influences vigor; that is, shoots growing upward are most vigorous, and shoots growing downward are least vigorous.
- Fruitfulness increases with shoot vigor to a maximum and then levels off, and it may even decline as vigor increases further due to deterioration in canopy microclimate.
- Vines can self-regulate by adjusting budbreak to bud number. Severe pruning stimulates budbreak of buds not deliberately retained at pruning (noncount buds) and thus results in a large number of double shoots, watershoots, and suckers. Greater than 100% budbreak (i.e., more than one shoot per retained bud) is an indication that the vine has the capacity to support more growth.

Siquidem luxuriosa vitis nisi fructu compescitur, male deflorescit, et in materiam frondemque effunditur; infirma rursus, cum onerata est, affligitur.
(Indeed, a vigorous vine, unless restrained by cropping, aborts its flowers and teems with vegetative growth; a weak vine, on the other hand, declines when burdened with fruit.)
—**Columella: De Re Rustica**

These relationships have important implications for the choice of trellis design and training system to be used, as well as pruning and other cultural practices. For example, the balance-pruning concept is based on the assumption that larger vines can support more shoots with a larger crop. This concept was obviously known to the Romans, as the quotation from Columella (4–ca. AD 70) shows, and it eventually led to the recommendation to retain approximately 30 buds for each kilogram of canes removed during winter pruning in order to make optimal use of the capacity of most wine grape cultivars. In addition, irrigation and nutrition can also modulate the relationship among vine balance, vigor, and capacity because vegetative growth often reacts more strongly to water or nutrient supply than does reproductive growth, at least after fruit set (see Sections 7.2 and 7.3).

Grapevines have an inherent propensity to self-regulate the balance among the growth of roots, shoots, and fruit. For example, effects of high planting density are partly offset by production of fewer buds per vine and (sometimes) fewer shoots per bud. Similarly, leaving twice as many buds per vine during winter pruning will normally not double the yield because the number of berries and their average size will decrease. Generally, as the number of clusters per vine increases, the number of berries per cluster and the average weight of these berries decrease. Consequently, the yield component compensation principle states that changing the level of one yield component will induce compensatory changes of other yield components, resulting in a subproportional change in yield.

6.1.1 YIELD POTENTIAL AND ITS REALIZATION

The yield potential of a crop plant can be roughly defined as the yield of a cultivar when grown in environments to which it is adapted, with nonlimiting external resources, and with pests, diseases, weeds, and other stresses effectively controlled. In viticulture, the yield potential is typically regarded as the maximum amount of fruit a vine or vineyard could potentially produce once the numbers of clusters and flowers are set at the beginning of each growing season. Taking all external factors into account, the largest yield potential is often achieved on vines of medium vigor because excessive vigor tends to impair bud fruitfulness, whereas very low vigor is generally a sign of limitation by some stress. A grapevine bud is said to be fruitful if it contains at least one inflorescence primordium that develops into a cluster. Such clusters "in the making" can be seen (using appropriate magnification) in longitudinal cuts of winter buds, and this enables growers to conduct an initial assessment of a vineyard's yield potential prior to winter pruning. Based on the observed average number of clusters per shoot, the pruning strategy can be adapted to compensate for annual changes in bud fruitfulness and bud injury. Bud fruitfulness can also be estimated retroactively by counting the number of clusters per shoot during the growing season or at harvest, but this ignores clusters that have been abscised due to environmental stress. Fruitfulness is an inherited characteristic that is modulated by environmental factors at the time of inflorescence initiation, which occurs before bloom, and subsequent differentiation, beginning around bloom (see Section 2.3). Yield per node is different from bud fruitfulness because it is a function of the number of shoots per node, the number of clusters per shoot, the number of berries per cluster, and the size of these berries.

Once the maximum number of potentially fruitful shoots has been determined by the number of buds retained at winter pruning (pruning level), bud fruitfulness sets the vine's yield potential for the following growing season. It must be kept in mind, however, that leaving more buds does not lead to proportionally more shoots, because fewer of the extra buds will break (Jackson et al., 1984). The extent to which this initial yield potential is realized by the vine depends on how many flowers develop on each inflorescence and how many of these flowers set fruit and develop into berries (Alleweldt, 1958).

Conditions required for the formation of the maximum number of inflorescence primordia include high irradiance, warm temperature, and adequate water and nutrient supply (Currle et al., 1983; Meneghetti et al., 2006; Mullins et al., 1992; Sartorius, 1968; Sommer et al., 2000; Vasconcelos et al., 2009). Long days also ostensibly favor the induction of inflorescence primordia in American *Vitis* species and their cultivars (Mullins et al., 1992) and in some *V. vinifera* cultivars (e.g., Riesling and Muscat of Alexandria). However, the influence of photoperiod (day length) on bud fruitfulness is far less important than the impact of irradiance (Srinivasan and Mullins, 1981). The effect of light intensity on fruitfulness is independent of the effect of temperature, although the two are usually linked: cloudy or shady conditions are often associated with cool temperatures. Overcast conditions during the bloom—fruit set period decrease bud fruitfulness (Keller and Koblet, 1995a; May and Antcliff, 1963). A marked reduction in fruitfulness occurs when daily total photosynthetic photon flux is reduced to one-third or less of that reached under clear skies (which can have $>2000\ \mu mol\ m^{-2}\ s^{-1}$). The primary reason for this pronounced light effect is probably the influence of light on leaf photosynthesis; low light limits the amount of assimilates produced by the leaves and thus the amount supplied to the developing buds.

Furthermore, buds that develop inside dense canopies are less fruitful than the buds near the exterior of the canopy (Buttrose, 1974a; Dry, 2000; May et al., 1976; Sánchez and Dokoozlian, 2005; Shaulis et al., 1966; Sommer et al., 2000). This effect may be part of the shade avoidance syndrome that is mediated by phytochromes (Morelli and Ruberti, 2002; see also Section 5.2). In extreme cases, such shading may even cause bud necrosis, which looks similar to—and thus is sometimes misdiagnosed as—winter freeze damage (Perez and Kliewer, 1990; Vasudevan et al., 1998). However, bud fruitfulness appears to be independent of the light exposure of the buds themselves; instead, fruitfulness depends strongly on the supply of assimilates to the buds (Keller and Koblet, 1995a; May and Antcliff, 1963; Sánchez and Dokoozlian, 2005). These assimilates may come from stored reserves or from leaf photosynthesis, but they are mostly supplied by the leaves on the same side of the shoot (Hale and Weaver, 1962). Consequently, low light on the leaves above and below a bud reduces bud fruitfulness. This has implications for the design of trellis and training systems in viticulture (Reynolds and Vanden Heuvel, 2009). A renewal zone (the zone where buds are located that produce next year's shoots and crop) with well-exposed leaves, ideally situated close to the top of the canopy, and a canopy that is not too dense will maximize the potential crop.

In addition to high light intensity, relatively high temperatures are also required for maximum inflorescence initiation and development (Buttrose, 1969a, 1970b, 1974a; Sommer et al., 2000; Srinivasan and Mullins, 1981). Warm temperatures promote the induction of inflorescences, whereas cool temperatures ($<20°C$) promote tendril formation (Buttrose, 1969a, 1970b). Fruitfulness increases as the temperature during the cluster initiation phase in the buds increases from $10°C$ to almost $35°C$, although some cultivars (e.g., Thompson Seedless and Muscat of Alexandria) appear to be unfruitful below $20°C$ (Vasconcelos et al., 2009). Excess heat ($>35°C$), however, may reduce the primordia number to zero, rendering the buds unfruitful.

Severe loss of leaf area (e.g., due to pest or disease infection or hail) can reduce inflorescence initiation and differentiation (i.e., branching), resulting in fewer and smaller inflorescences being formed in each winter bud (Bennett et al., 2005; Candolfi-Vasconcelos and Koblet, 1990; Lebon et al., 2008; May et al., 1969). Because of the importance of functional leaf area for inflorescence development, the number and size of the future inflorescences decrease as the severity of leaf destruction increases and the earlier the loss occurs, although a negative effect still persists if loss of leaf area occurs as late as 2 months after bloom.

Water deficit normally reduces bud fruitfulness (Buttrose, 1974b; Srinivasan and Mullins, 1981), but this is not always the case. A mild water deficit reducing canopy density can actually increase fruitfulness due to the improved canopy microclimate and thus better light exposure (Keller, 2005; see also Section 7.2). Mineral nutrition also influences bud fruitfulness, but little is known about what these effects are. It is likely that they also involve changes of canopy microclimate. Soil nitrogen (N) availability during bloom is important because the response of cluster initiation to vine N status shows a clear optimum, with both N deficiency and N excess resulting in a reduced number of inflorescences per shoot (Alleweldt and Ilter, 1969; Currle et al., 1983; Keller and Koblet, 1995a; Srinivasan and Mullins, 1981). Phosphorus and potassium deficiency also decrease bud fruitfulness (Vasconcelos et al., 2009).

Because both the number of inflorescences and the extent of branching of these inflorescences are often lower in the buds produced at the shoot base than at higher bud positions, cane-pruned vines tend to have a greater yield potential than spur-pruned vines with similar bud numbers (Vasconcelos et al., 2009). This may be partly compensated by greater numbers of flowers per inflorescence on shoots arising from spurs compared with those from the same bud positions on canes (Huglin and Schneider, 1998). Although the number of flowers per inflorescence of a particular cultivar can vary from less than 10 to more than 1000 in the same vineyard, the environmental causes of this variation are mostly unknown. It seems logical, however, that much of this variation may be introduced during the period of inflorescence differentiation in the buds. In addition, high temperatures during the budswell and budbreak periods are associated with fewer flowers per inflorescence (Keller et al., 2010; Petrie and Clingeleffer, 2005; Pouget, 1981), perhaps because with the faster development at higher temperatures there is more competition from the emerging shoots for remobilized storage reserves (Bowen and Kliewer, 1990). High temperatures and warm, moist soil conditions, on the other hand, accelerate the rate of reserve mobilization in spring and favor both shoot growth and flower development (Keller et al., 2010; Rogiers et al., 2011). Flower differentiation and early development during and after budbreak are entirely dependent on the availability of stored nutrient reserves and the rate of their remobilization because the unfolding leaves act as sinks during this period. The need to supply carbon and other nutrients to both shoots and flowers may be one reason that fruit-bearing vines tend to grow much less vigorously than do vines without fruit (Greer and Weston, 2010a). The more shoots that are growing on a vine, the more their demand for stored nutrients may compete with the formation of flowers. The amount of reserves stored during the previous growing season is therefore critical for maximum flower formation. Overcropping, excessively cool or hot seasons, untimely loss of active leaf area, and poor water and nutrient management all contribute to limiting (and sometimes depleting) reserves in the vine's permanent organs and thus interfere with flower formation. In other words, the number of flowers formed per inflorescence and per vine increases as vine reserve status increases (Bennett et al., 2005; Holzapfel et al., 2010). Adequate reserve status may be particularly important to sustain productivity of high-yielding vineyards, where functional leaves seem to be required after fruit harvest to replenish the reserve pool (Holzapfel et al., 2006, 2010). This means that heavily cropped and vigorously growing vines are more dependent than lightly cropped vines and vines whose growth has ceased by veraison on favorable weather conditions and sufficient leaf area late in the growing season to refill their nutrient reserves before dormancy (Holzapfel et al., 2006; Zufferey et al., 2012).

As discussed in Section 2.3, bloom occurs within 5—10 weeks after budbreak. The actual timing depends on the cultivar and weather conditions. Anthesis typically does not occur until the mean daily temperature exceeds approximately 18°C. Under favorable conditions (high light and high temperature), an inflorescence can complete flowering within a few days, and bloom lasts approximately 7—10 days for individual vines (Currle et al., 1983). Unfavorable environmental conditions (e.g., low temperature or rain) can extend the bloom period to more than 1 month, and the abscised calyptra may remain on the flower. This does not prevent pollination, but it can interfere with fertilization by reducing pollen viability or germination rate, leading to excessive abscission of flowers (shatter) and poor fruit set, thereby decreasing the number of berries per cluster (Koblet, 1966).

The proportion of flowers that set fruit is often between 20% and 50% (Currle et al., 1983; Mullins et al., 1992). Limitation in assimilate supply is the main cause of flower abortion and poor fruit set (Ruan et al., 2012). Thus the final fruit yield is primarily dependent on the photosynthetically active leaf area early in the growing season. Loss of leaf area during this period generally leads to poor fruit set due to low pollen viability resulting from carbon starvation of the flowers, which is a consequence of their low sink strength before fruit set (Candolfi-Vasconcelos and Koblet, 1990; Caspari et al., 1998; Coombe, 1959; Winkler, 1929; see also Section 5.1). A leaf area of less than about 5 cm^2 per flower can severely limit fruit set (Keller et al., 2010). Although the total requirement for assimilates by flowers and small berries is quite low, the developing flowers cannot compete with the growing shoot tips if resources become limiting. As a consequence, the flowers are very susceptible to environmental stress. Ovaries use glucose derived from the breakdown of phloem-imported sucrose as a signal: High amounts of glucose promote cell division and thus fruit set and growth, whereas low amounts of glucose activate the process of programmed cell death, culminating in flower or fruit abortion (Ruan et al., 2012).

Ideal conditions during bloom, leading to maximum fruit set, are practically identical to those required for maximum inflorescence initiation. The optimum temperature range for bloom and fruit set in grapevines appears to be 20—30°C, and both high (>35°C) and low (<15°C) temperatures reduce fruit set (Currle et al., 1983; Kliewer, 1977a). At more than 40°C, fruit set can be completely inhibited, at least in vulnerable cultivars (e.g., Sémillon), which may lead to abortion of inflorescences (Greer and Weston, 2010b). Overcast, cool, or hot weather and water or nutrient deficits limit the amount of photosynthate available to the flower clusters and can lead to embryo abortion and hence poor fruit set (Keller, 2005; Lebon et al., 2008; Müller-Thurgau, 1883a; Roubelakis and Kliewer, 1976). Water deficit and other stresses that lead to low sugar supply to and starch accumulation by the developing pollen before anthesis also interfere with meiosis and result in male sterility, which also curtails fruit set (De Storme and Geelen, 2014; Dorion et al., 1996; Goetz et al., 2001; Saini, 1997; Zinn et al., 2010). Even moderate water deficit at this time can significantly reduce fruit set and final yield (see Section 7.2). Fruit set can also suffer in the interior of dense canopies, which might be associated with the shade avoidance syndrome that is mediated by phytochromes (Morelli and Ruberti, 2002; see also Section 5.2). The vulnerability to stress of grape flowers, and especially of the anthers, could be an adaptation to domestication and vegetative reproduction (McKey et al., 2010). Because sexual reproduction may be a metabolically expensive process, clonally propagated plants are thought to reduce over time their investment in male function. It is possible that such a bias against male function was aided unconsciously by early viticulturists who selected hermaphroditic variants from their dioecious progenitors during *V. vinifera* domestication.

Unfavorable conditions and carbon shortage may also result in a high number of the so-called "shot berries" (ovaries that fail to develop properly) or, in extreme cases, a syndrome termed inflorescence necrosis or early bunch stem necrosis, whereby whole inflorescences or portions of them die and are abscised (Figure 6.1) (Caspari et al., 1998; Jackson, 1991; Jackson and Coombe, 1988; Keller and Koblet, 1994; Keller et al., 2001a, 2010; Winkler, 1929). The term "necrosis" (Greek *nekrós* = dead) may be unfortunate because it implies an unprogrammed, chaotic cell death without nutrient recycling (Müntz, 2007), whereas the death and abscission of portions of an inflorescence is likely to be organized and genetically programmed and to involve mobilization and recycling of nutrients to other sinks. Deliberate shedding of reproductive organs allows the vine to adjust its reproductive output according to the prevailing external conditions. Such self-thinning avoids a situation whereby resources are (over)invested in reproduction at the expense of the long-term survival of the vine. Removing the shoot tips, especially those of vigorously growing shoots, or cane or trunk girdling during bloom can sometimes ameliorate the impact of such adverse conditions by temporarily eliminating strong competitors for assimilates, whereas removal of leaves before bloom accentuates fruit loss (Caspari et al., 1998; Coombe, 1959, 1962; Koblet, 1969; Müller-Thurgau, 1883a; Vasconcelos and Castagnoli, 2000). Thus, the number of berries per cluster is most susceptible to environmental stress at or just before anthesis.

In addition to its effect on fruit set, bloom-time temperature might also somehow influence bunch stem necrosis, a physiological disorder that is associated with necrotic lesions on the rachis (see Section 7.3). The lesions develop only after the berries begin to ripen, apparently often beginning around stomata or dead bracts at branching points or beneath the epidermis, and may interrupt assimilate supply to the berries distal to a lesion, which interferes with ripening and may lead to abortion of the afflicted cluster portion (Bondada and Keller, 2012; Currle et al., 1983; Hifny and

FIGURE 6.1

Inflorescence necrosis on Müller–Thurgau following a cool and overcast bloom period (left) and on nitrogen-deficient Cabernet Sauvignon following water deficit during bloom, with normal fruit set on nitrogen-sufficient vines (right).

Photos by M. Keller.

Alleweldt, 1972; Theiler, 1970). Compared with nonaffected clusters, even on the same vine, berries on symptomatic clusters are typically lighter, often due to shriveling, and have lower amounts of sugars, potassium, and anthocyanin but higher amounts of organic acids (especially tartrate) and calcium, and lower pH (Morrison and Iodi, 1990; Ureta et al., 1981). However, although a negative correlation has been found in some circumstances between the average daily maximum temperature during the bloom period and the incidence of bunch stem necrosis (Theiler and Müller, 1986, 1987), a satisfactory explanation for this apparent phenomenon has yet to be found.

Although plants (and animals) regulate organ size and shape at the organ level rather than at the level of individual cells (Anastasiou and Lenhard, 2007), the final size of grape berries that set after bloom is determined by the number of cell divisions before and after bloom, the extent of expansion of these cells, and the degree of weight loss (shrinkage or shrivel) prior to harvest. According to Harris et al. (1968), the ovary contains approximately 200,000 cells compared with 600,000 in the mature berry. This implies that there are many more cycles of cell division before bloom than after bloom. Consequently, there should be an ample opportunity for early season environmental factors to influence berry size (Coombe, 1976; Keller et al., 2010). Because cell division is dependent on carbon supply, stress factors that diminish assimilate supply will also limit the cell number and hence berry size (Caspari et al., 1998; Keller et al., 2010; Van Volkenburgh, 1999). The initial postbloom rates of cell division and cell expansion are controlled by the number of seeds that have been fertilized, perhaps because more seeds make more auxin and stimulate more gibberellin production, making the berry a stronger sink (Cawthon and Morris, 1982; Scienza et al., 1978). Thus, the final berry volume increases with increasing number of seeds per berry (Gillaspy et al., 1993; Huglin and Schneider, 1998; Olmo, 1946; Winkler and Williams, 1935). Although this assertion has not been tested, it is possible that a loss of ovule viability at temperatures above approximately 35°C might be associated with fewer seeds per berry and thereby constrain berry size (Williams et al., 1994). The effect of seed number is most pronounced at the lower end so that the increase in volume when seed number increases from three to four is only small. The absence of seeds in many table grape cultivars limits berry growth, which conflicts with the general consumer demand for large table grapes. Growers therefore often stimulate berry expansion by applying gibberellin sprays at bloom and fruit set and/or by removing a strip of bark around the canes or trunk at fruit set (Currle et al., 1983; Williams and Matthews, 1990). Whereas gibberellin may stimulate the berries' sink strength (Weaver et al., 1969), bark removal, called girdling, eliminates phloem flow to the perennial parts of the vine so that more sugar is available to the developing fruit clusters (Roper and Williams, 1989). In addition, girdling increases leaf water status due to a decrease in stomatal conductance that may be associated with feedback inhibition of photosynthesis (Williams et al., 2000). However, girdling typically constitutes only a temporary interruption of phloem flow because callus, ostensibly formed by nearby phloem parenchyma cells, heals the wound (Sidlowski et al., 1971). Application of gibberellin to seeded cultivars often decreases fruit set and can also lead to lower bud fruitfulness and even bud necrosis (Huglin and Schneider, 1998). Although this carryover effect is typically regarded as undesirable, it may be exploited as a means to limit the crop size of large, minimally pruned grapevines (Weyand and Schultz, 2006c).

Cell enlargement is generally more sensitive to temperature than is cell division. Nevertheless, maximum cell division occurs at approximately 20–25°C, and temperatures below 15°C or above 35°C lead to a marked reduction in the rate of cell division, resulting in smaller berries

(Kliewer, 1977a). Even a few days with temperatures greater than 40°C after fruit set can limit final berry size (Matsui et al., 1986). The decrease in cell division due to both cold and heat stress may be caused by the thermosensitivity of cytokinin production and transport. It is believed that cytokinins are inactivated at low temperatures by conversion of the active form to a storage form (Mok and Mok, 2001), whereas during a heat period, the seeds ostensibly produce less cytokinin and the xylem transports less cytokinin from the roots (Banowetz et al., 1999). It is possible that the need for leaf cooling, which causes rapid transpiration rates in leaves during heat stress (see Section 5.2), reduces the availability of cytokinin for fruit development. Heat also decreases cell expansion; however, whereas high temperature before the berry enters the lag phase of growth leads to an irreversible reduction in berry size, the temperature experienced after veraison often has no noticeable effect on berry size (Hale and Buttrose, 1974; Kliewer, 1973). In sensitive cultivars, however, heat stress (≥ 40°C) during ripening may also restrict berry expansion (Greer and Weston, 2010b).

Because heat has a negative effect on both fruit set and berry growth, poor set and small berries are typical of years that experience a "heat wave" during the bloom−fruit set period (Kliewer, 1977a). The temperature sensitivity of berry cell division and expansion also has implications for vineyard management: removing leaves in the cluster zone early during berry development may result in smaller berries due to solar heating of the sun-exposed fruit. Warm temperatures in the range of 25−32°C between bloom and the lag phase of berry growth are furthermore thought to promote the capacity of seeds to germinate (Currle et al., 1983). In addition to their effect on berry size, both cool and hot temperatures also delay fruit development, especially by prolonging the lag phase of berry growth (Hale and Buttrose, 1974). Under cool conditions, the duration of phase I of berry development also has been found to be inversely correlated with the average temperature during this period (Alleweldt et al., 1984b).

Although the berries and their seeds are predominantly sink organs, it should be remembered that they are green—at least before veraison—and their skin absorbs much of the incident light (Blanke, 1990a). The cells of both the seeds of preveraison berries and the berry pericarp contain chloroplasts with functional (and shade-adapted) photosynthetic machinery that participates in berry and seed metabolism (Blanke and Leyhe, 1989b; Famiani et al., 2000; Goes da Silva et al., 2005; Kriedemann, 1968a; Martínez-Esteso et al., 2011). Berry and seed photosynthesis may refix some of the CO_2 that would otherwise be lost in respiration and may produce some of the NADPH and ATP required for energy-intensive metabolism in addition to the accumulation of malate as a by-product of CO_2 refixation (Koch and Alleweldt, 1978; Palliotti and Cartechini, 2001; Palliotti et al., 2010). Moreover, the epidermis of the developing berry has a few stomata (~ 1 or 2 mm^{-2}, compared with $100-400 \text{ mm}^{-2}$ in leaves) for gas exchange. As the berry expands, the stomatal density on the berry surface decreases because no new stomata are produced after anthesis, and the stomata are progressively occluded by epicuticular wax, suberin, and other components so that they become nonfunctional by veraison (Bessis, 1972b; Currle et al., 1983; Rogiers et al., 2004a; Swift et al., 1973). Only early in its development does the berry fix an appreciable amount of carbon, which it can use for growth and metabolism; photosynthetic enzymes and chlorophyll are dismantled as the berry begins to ripen (Famiani et al., 2000; Geisler and Radler, 1963; Martínez-Esteso et al., 2011). Consequently, although immature berries may recycle approximately 20−30% of the daily respiratory carbon loss (up to 65% in the light) via photosynthesis, the photosynthetic rate declines to near 10% of the respiration rate during ripening (Koch and Alleweldt, 1978; Ollat et al., 2002). The respiration rate, too, decreases throughout berry development and ripening (Geisler and

Radler, 1963; Palliotti et al., 2010). Therefore, grape berries rely heavily on continued import of sugar for proper development and metabolism.

Any limitation in assimilate supply (e.g., due to loss of active leaf area, poor weather conditions, and water or nutrient stress) during pollination and early berry growth can lead to abortion of embryos. Such early abortion of some fruits may enable the survival of others; that is, by limiting its investment in the number of reproductive organs, the vine has a greater chance to support the remaining fruits to seed maturity. Thus, early stress also leads to fewer seeds per berry and strongly reduces the size of mesocarp cells but does not impair fruit ripening (Ollat et al., 2002). The amount of photosynthates that are partitioned to the fruit, especially early during fruit development, is critical for final fruit size (Candolfi-Vasconcelos and Koblet, 1990; May et al., 1969; Ollat and Gaudillere, 1998). Insufficient photosynthate supply due to inadequate leaf area or low rates of photosynthesis limits both cell division and cell expansion, with very little compensation later in the season so that final berry size is irreversibly reduced. Therefore, berry size at harvest also tends to decline as the number of berries per vine increases (Keller et al., 2008; Kliewer et al., 1983). Conversely, limiting assimilate supply late in the growing season (i.e., after veraison) has little to no effect on berry size but restricts ripening and may curtail the realization of the following year's yield potential by limiting the replenishment of storage reserves (Holzapfel et al., 2010; Zufferey et al., 2012).

Water deficit or nutrient deficiency typically reduce yield, particularly if a deficit occurs early in the growing season, when the inflorescences are not competitive in comparison with the shoot tips (Williams and Matthews, 1990). Inadequate water or nutrient supply before bloom can lead to abortion of entire inflorescences (limiting cluster number), stress during bloom results in poor fruit set (limiting berry number), and stress during the early stages of fruit development restricts berry growth (limiting berry size). Although the rates of both cell division and cell expansion may be reduced, cell division is much less sensitive to changes in water status than is cell expansion (Ojeda et al., 2001; see also Section 7.2). If the stress occurs later, then it will limit only cell expansion, which still may result in somewhat smaller berries. However, in contrast to the weak sink strength of the flowers prior to fruit set, the developing seeds and berries are stronger sinks than the shoots and, especially, the roots. Therefore, although stress does reduce seed and berry size, the decrease is less than would be expected from the reduction in photosynthate availability. This is probably due to remobilization of stored reserves in both the leaves and the permanent parts of the vine.

Grape berries grow mostly at night, and they often lose some weight during the day, especially during periods of water deficit (Greenspan et al., 1994, 1996; Keller et al., 2006). Whereas such daytime berry shrinkage is common before veraison, it seems to occur only under severe water deficit after veraison, at least until the mesocarp cell membranes of some cultivars begin to fail during the late stages of ripening (Tilbrook and Tyerman, 2008). The same extent of water deficit occurring before the lag phase of berry growth normally curtails berry size more than if it occurs during or after that phase. This means that the maximum berry size is mostly determined before veraison and that increasing the supply of irrigation water later in the growing season cannot compensate for the limitation imposed on berry size by an early deficit (Figure 6.2). Although late-season irrigation may lessen berry shrinkage due to drought-induced dehydration, irrigation and even rainfall shortly before harvest alter berry size and composition far less than is often feared (Coombe and Monk, 1979; Huglin and Schneider, 1998; see also Section 6.2).

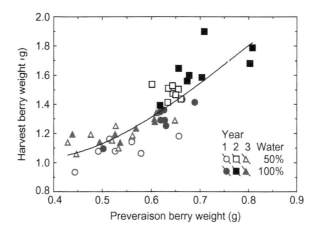

FIGURE 6.2

Weight of Sauvignon blanc berries during the preveraison lag phase and at harvest in 3 years. The closed symbols represent vines that were irrigated throughout berry development to replace 100% of the evaporative water loss from the vineyard, and the open symbols represent vines irrigated at 50% of that amount between fruit set and veraison and 100% during ripening (Keller, unpublished data).

At every step during the yield formation process, there are management tools available to maximize or optimize final yield according to the desired specifications. For example, the timing and rate of water or nutrient application may be varied according to the intended end use of the grapes produced in a particular vineyard block. Seasonal weather patterns, as well as economic and legal considerations, often determine whether and when these cultural practices are applied. Growers also always have to keep in mind the yield component compensation principle when deciding on strategies aimed at optimizing yield and fruit quality. In addition, because berry growth is most sensitive to stress factors during the early stages of development and because final berry size is largely predetermined by veraison, yield predictions are generally much more accurate after this stage than before it.

Cultivars, clones, and rootstocks differ in their sensitivity to changes in environmental conditions. Such genetic differences also influence yield formation and consequently yield variation among vineyard sites and between years. This has implications for the selection and layout of planting material during vineyard establishment. Because genetically heterogenous plants typically experience lower yield variation than homogenous plants, it may be prudent to block vineyards into smaller units of different clones, cultivars, and rootstocks, rather than maximizing block size for each clone—rootstock combination.

6.2 GRAPE COMPOSITION AND FRUIT QUALITY

The grape berries' physical and chemical composition at harvest is responsible for the fruit quality characteristics and, consequently, the quality attributes of the wine or grape juice produced from the fruit. Much of what we call "quality" in grapes, juice, and wine has to do with our perception of taste and flavor, which is the outcome of the processing in the brain of stimuli received by taste receptors in the tongue and olfactory receptors in the nose. Compounds ranging from sugars to acids and phenolics to volatile chemicals all contribute interactively to the overall taste and flavor perception (i.e., sensory quality). As berries ripen, they undergo a multitude of both physical and chemical changes, although many changes and processes important to fruit quality also occur long before ripening begins (Brummell, 2006; Conde et al., 2007; Coombe, 1992; Deytieux et al., 2007; Hrazdina et al., 1984; Kanellis and Roubelakis-Angelakis, 1993; Sarry et al., 2004). These developmental changes are in turn coordinated by the interplay and cooperation of hundreds or thousands of genes (Boss et al., 1996; Castellarin and Di Gaspero, 2007; Davies and Robinson, 2000; Deluc et al., 2007; Goes da Silva et al., 2005; Pilati et al., 2007; Robinson and Davies, 2000; Terrier et al., 2005). The expression of many of these genes increases at the beginning of ripening, whereas that of many other genes decreases, and it is thought that abscisic acid (ABA), probably in concert with sugars, plays a central role in bringing about these changes (Gambetta et al., 2010; Jia et al., 2011; Koyama et al., 2009; Lijavetzky et al., 2012; see also Section 2.3). A general overview of the changes during berry development is shown in Figure 6.3.

The beginning of grape ripening (veraison) is easily recognized by the change of skin color and the sudden softening of the berry (see Section 2.3). The change in color occurs due to the degradation of the green chlorophylls (Hardie et al., 1996a; Uhlig and Clingeleffer, 1998), which unmasks the previously invisible yellow to orange-red carotenoids, and due to the simultaneous accumulation of red, purple, or blue anthocyanin pigments in dark-skinned cultivars. White grapes, of course, do not make anthocyanins (see Section 1.2); they owe their skin color to a combination of chlorophylls and carotenoids (both diminishing during ripening), and the pale yellow, but barely visible, flavonols. A minor portion of the chlorophylls may be converted in the skin and, to a lesser extent, in the pulp, into colorless tetrapyrroles, which are identical to those produced in senescing leaves and are strong antioxidants (Müller et al., 2007).

Actually, softening, which can be measured as deformability, coincides with the beginning of sugar accumulation but precedes by several days the change in skin pigmentation and resumption of berry growth (Coombe, 1992; Coombe and Bishop, 1980; Creasy et al., 1993; Terrier et al., 2005; Wada et al., 2009). Softening occurs due to the gradual disassembly of the mesocarp cell walls (Brummell, 2006; Huang and Huang, 2001) and the decline in mesocarp cell turgor during the preveraison lag phase of berry growth (Thomas et al., 2008). Subsequent alterations include expansion (and later shrinkage) of the berry volume; structural changes in the skin and pulp; a switch in metabolic pathways; rapid accumulation within cell vacuoles of solutes leading to dry mass gain; and decrease in acidity and increase in pH. The loosening of the skin cell walls (aided by the release of calcium and modifications in polysaccharides such as cellulose, hemicellulose, and pectin) and the incorporation of proteins (e.g., hydroxyproline-rich extensins for added tensile strength) enable the skin cells to become tangentially stretched while the pulp cells resume expansion after the lag phase (Chervin et al., 2008; Huang et al., 2005a; Nunan et al., 1998; Ortega-Regules et al., 2008; Vicens et al., 2009). Cell wall loosening proteins such as expansin, whose activity peaks at the end of the lag phase, and various enzymes modify the cell wall polysaccharide components by "unlocking" the hydrogen bonds between the stiff cellulose microfibrils and the hemicellulose matrix just before and during ripening to enable the berry to soften (Deluc et al., 2007; Ishimaru and Kobayashi, 2002; Nunan et al., 2001; Pilati et al., 2007; Rose and Bennett, 1999; Rose et al., 1997, 1998; Schlosser et al., 2008). In addition to this enzyme action, the hydroxyl radical (OH), produced from other

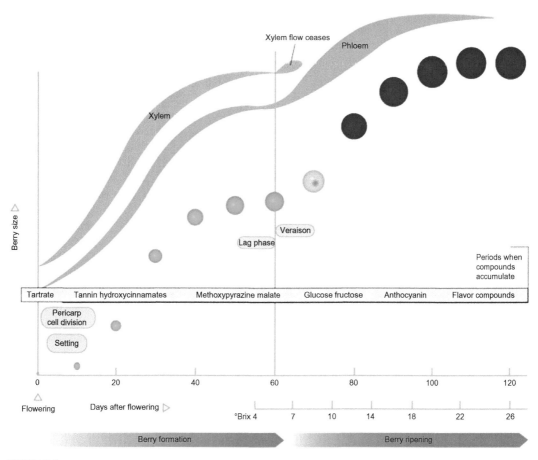

FIGURE 6.3

Diagrammatic representation of the relative size and color of grape berries at 10-day intervals after bloom and of the major changes occurring during berry development. Note that "Xylem flow ceases" refers to inflow only; outflow continues. Also note that "Tannin" refers to tannin precursors (tannins are formed during ripening).

Reproduced from Coombe (2001).

reactive oxygen species with the help of apoplastic peroxidase enzymes, is thought to sever the bonds between cell wall polysaccharides nonenzymatically, which may enhance the cell wall loosening and degradation process required for cell expansion and fruit softening (Møller et al., 2007; Rose et al., 2004). Peroxidase appears to be activated in the skin and, to a lesser extent, in the mesocarp at veraison (Calderón et al., 1994; Ros Barceló et al., 2003).

Cell wall loosening in the pulp seems to be responsible for berry softening, and soon afterward cell wall loosening in both the pulp and the skin enables berry expansion (Huang and Huang, 2001; Huang et al., 2005a), which thus occurs despite the more than 10-fold decline in mesocarp turgor pressure just before veraison (Thomas et al., 2006). This turgor loss occurs due to the accumulation of sugars and other solutes not only in the mesocarp cell vacuoles but also in the surrounding apoplast (mainly cell walls), which starts slightly before veraison due to the switch from symplastic to apoplastic phloem unloading (Keller and Shrestha, 2014; Wada et al., 2008, 2009; Zhang et al., 2006). Solute accumulation apparently begins in the stylar end of the berry and is accompanied by a decrease in xylem influx that also progresses from the stylar to the pedicel end (Castellarin et al., 2011; Keller et al., 2006). Transport of solutes into the vacuoles occurs via an assortment of active transporters and passive channels embedded in the vacuolar membrane (tonoplast), many of which are specific for a particular compound or for closely related compounds (Fontes et al., 2011b; Martinoia et al., 2012). For instance, some transport proteins are responsible for hexose movement, others transport malate, others carry cations such as potassium (K^+), sodium (Na^+), or calcium (Ca^{2+}), or anions such as phosphate (HPO_4^{2-}), and still others move anthocyanins. Some of these transporters (e.g., hexose carriers) move their cargo into the vacuole along with protons (H^+), whereas many others move H^+ out of the vacuole in exchange for their cargo (Hedrich, 2012). To maintain the acidic environment required inside the vacuoles and permit H^+ export to power the transporters, proton pumps continuously relocate H^+ from the cytosol into the vacuoles. These proton pumps are named PPase (inorganic pyrophosphatase) and ATPase (Fontes et al., 2011b; Terrier et al., 1998).

Although the berry volume increases during ripening, it cannot do so indefinitely because the skin sets a limit to mesocarp expansion (Matthews et al., 1987). Following a temporary increase at veraison, the extensibility of the skin later decreases (i.e., its stiffness, the so-called elastic modulus or modulus of elasticity, increases) in parallel with the stiffening of the epidermis and, especially, the cuticle. The cuticle stiffness in mature fruit is similar to that of the common plastic polyethylene yet lower than that of leaf cuticles (Wiedemann and Neinhuis, 1998). Cuticle stiffening occurs partly due to a decrease in the unsaturated and elastic C_{18} cutin monomers and a concomitant increase in the saturated and rigid C_{16} monomers (Bargel et al., 2006; Marga et al., 2001). In addition, the intracuticular wax embedded in the cutin matrix acts to constrain the cuticle's ability to expand (Khanal et al., 2013). Although the activity of pectin methylesterase loosens cell walls, this effect may be reversed due to the binding of Ca^{2+} ions by the newly exposed carboxyl groups; the resulting calcium-pectate gel then stiffens the (epidermal) cell walls (Goldberg et al., 1996; Mirabet et al., 2011). This incorporation of Ca into the cell walls can deplete the soluble apoplastic Ca during fruit expansion—at a time when Ca influx via the xylem declines (de Freitas et al., 2012; Rogiers et al., 2006). Furthermore, additional polysaccharides and phenolics such as hydroxycinnamic acids and flavonoids, including tannins, may be deposited in the skin cell walls (Amrani Joutei et al., 1994; Lecas and Brillouet, 1994; Schlosser et al., 2008). The phenolics are bound to cell wall polysaccharides and proteins by peroxidase, which further stiffens the cell walls, especially in the epidermis. This limits cell expansion and thus counteracts the initial wall-loosening properties of peroxidase (Bargel et al., 2006; Potters et al., 2009; Thompson et al., 1998). Note that because many environmental stresses increase the amount of phenolics in the apoplast, one way by which such stresses often limit berry size might be due to the influence of these phenolics on cell wall stiffness. By the same logic, the bonding of phenolics to cell wall polymers would presumably diminish their extractability during fermentation.

At the same time, protein incorporation and polysaccharide modifications also occur in the mesocarp cell walls (Ortega-Regules et al., 2008). Peroxidase-mediated inclusion of soluble proteins, especially extensin glycoproteins, reinforces the cell walls (Nunan et al., 1998; Ros Barceló et al., 2003). Despite these massive changes, the thickness of the epidermis, hypodermis, and mesocarp cell walls remains essentially constant during ripening (Hardie et al., 1996b; Nunan et al., 1998). Moreover, the ultrastructure, cell walls, and membranes of the berry cells remain intact (Diakou and Carde, 2001; Fontes et al., 2011a; Fougère-Rifot et al., 1995; Hardie et al., 1996b); cell compartmentation is retained, mitochondria remain functional, and plasmodesmata maintain their cell-to-cell connections (except the ones in the phloem membranes, which are plugged to facilitate apoplastic phloem unloading; see Section 5.1). Only the locule area around the seeds appears to be subject to some degradation, especially toward the end of the ripening phase (Krasnow et al., 2008). In the remainder of the pericarp, senescence and loss of cell compartmentation only seem to set in once berries have attained their maximum weight, at least in some cultivars, and are sometimes associated with subsequent berry shrinkage (Hardie et al., 1996a; Tilbrook and Tyerman, 2008). Large berries, which have a large surface:volume ratio, evaporate more water and thus shrink more rapidly than do small berries (Rogiers et al., 2004b). Nonetheless, shrinkage due to dehydration occurs only slowly, which is why the production of raisin grapes is essentially restricted to hot climates.

6.2.1 WATER

The final size of the berries of a particular grape variety is mainly influenced by cell enlargement because cell division determining the number of cells is comparatively little affected by environmental conditions. The rapid growth of the berry during ripening is mostly due to water import and retention by the mesocarp vacuoles. Because by mid-ripening these vacuoles may occupy 99% of the mesocarp cell volume (Diakou and Carde, 2001; Storey, 1987); grape juice essentially consists of vacuolar sap of mesocarp cells. Thus, water is by far the dominant chemical component of the ripening berry. Nonetheless, due to the accumulation of sugars and other solutes during ripening, the water content is highest before veraison (~90%), declining to approximately 75−80% at maturity and sometimes to less than 70% in overripe grapes (>30°Brix). Consequently, 1 kg of grapes yields approximately 0.6−0.8 L of juice or wine, depending on species, cultivar, berry size, maturity, and extent of pressing (Viala and Vermorel, 1909). However, the water content can decrease to less than 15% when grapes are dried to produce raisins (~68°Brix); thus, 1 kg of grapes yields less than 250 g of raisins (Bongers et al., 1991; Parpinello et al., 2012).

Xylem sap is the main source of water for the berry before veraison; it is thought to account for approximately 75% of the total water influx (Ollat et al., 2002). Xylem flow into the berry declines at veraison, and phloem sap concomitantly becomes the primary or only source of berry water (Greenspan et al., 1994; Keller et al., 2006) because excess water is delivered to the berry in the phloem along with assimilates such as sucrose and amino acids. The rate of carbon import is thought to increase three- to fourfold at veraison, and virtually all of this carbon is imported via the phloem (Ollat et al., 2002).

Berry transpiration represents the bulk of water loss from the berry, and the evaporation rate from the berry surface increases with increasing berry size. However, transpiration rates of grape berries are 10−50 times lower than those of leaves and decline during berry development (Becker and Knoche, 2011; Blanke and Leyhe, 1987; Düring and Oggionni, 1986; Rogiers et al., 2004a). Indeed, it appears that grape berries are designed to minimize evaporative water loss. First, they are typically round or nearly so, a geometry that translates into the least amount of surface area relative to volume;

this in turn translates into a large water storage capacity compared with the evaporative surface. Second, berries have far fewer stomata than do leaves, and these stomata become nonfunctional by veraison (Blanke et al., 1999; Swift et al., 1973). Finally, their cuticle is impregnated with approximately 10 times more wax than the leaf cuticle (Possingham et al., 1967; Radler, 1965). The berry cuticle, with its intra- and epicuticular waxes, is a formidable barrier to water diffusion (Becker and Knoche, 2011; Leide et al., 2007). Yet the decrease in transpiration during fruit development is not caused by an increase in the amount of cuticular waxes (Grncarevic and Radler, 1971; Rogiers et al., 2004a). Instead, changes in the composition of these waxes, such as the incorporation of long-chain ($>C_{28}$) n-alkanes, seem to account for the developmental increase in cuticular resistance to water vapor (Leide et al., 2007).

In addition to evaporating from the berry surface, water can also flow back through the xylem from the berry to the shoot (Keller et al., 2006; Lang and Thorpe, 1989). Due to the low transpiration rate of the berry, much of the water received by a preveraison berry via the phloem may be recycled back to the vine, especially in the afternoon, when Ψ_{leaf} may be lower than Ψ_{berry}. The decreasing Ψ_{berry} after veraison due to sugar import changes the water potential gradient ($\Delta\Psi$) in favor of the berry so that water that has already entered the berry cell vacuoles becomes increasingly protected from being "pulled back" by the leaves. Instead, surplus phloem water, which moves to the berry cell walls osmotically during apoplastic phloem unloading (see Section 5.1), may be recycled out of the berry via the xylem without entering the berry cells (Keller et al., 2006). It appears that cultivars differ in the extent of such xylem backflow, perhaps due to variation in the hydraulic resistance of the pedicel or berry xylem (Tilbrook and Tyerman, 2009; Tyerman et al., 2004).

The (usually) reversed xylem flow in the postveraison berry provides some independence from the fluctuations in Ψ experienced by the leaves due to transpiration. Although the berry is not hydraulically isolated from the vine (i.e., the berry xylem remains functional after veraison; Bondada et al., 2005; Chatelet et al., 2008b), the sensitivity of berry water status to soil and plant water status declines greatly after veraison, and the daily cycle of berry shrinkage during the day and expansion at night becomes much less pronounced (Greenspan et al., 1994, 1996; Matthews and Shackel, 2005). Accordingly, whereas preveraison berries shrink and expand readily with fluctuating vine water status, postveraison berries are much less subject to such fluctuations. If the results obtained with algae are applicable to grape berries, then the increase in solute concentration, leading to "osmotic dehydration," should lead to closure of aquaporins, which would decrease membrane water permeability (Ye et al., 2004). This not only could contribute to the high resistance to dehydration of ripening berries but also may support the changes in skin stiffness in limiting berry growth. Although the berries will still shrink during prolonged episodes of water stress (Keller et al., 2006), shrinkage becomes limited because phloem water supply is much less prone to changes in vine water status than is xylem water supply. On the other hand, berry expansion becomes limited because the increasing cutinization of the thick epidermal cell walls and the increasing stiffness of the cuticle decrease the skin's elasticity as the fruit matures (Bargel and Neinhuis, 2005; Bargel et al., 2006; Matthews et al., 1987). Cuticle stiffening is at least partly due to incorporation of phenolics (flavonoids) and polysaccharides into the cutin matrix, both of which have the effect of reinforcing the rigidity imparted by the cuticular wax (Domínguez et al., 2009, 2011).

A rigid (i.e., unelastic) skin may be necessary for grapes to limit cell expansion while sugars accumulate to high concentrations (i.e., high π). Simultaneous accumulation of solutes in the cell walls and recycling of surplus phloem water in the xylem keep the mesocarp turgor pressure (P) low. But the increase in surface area of the expanding berry causes mechanical stress due to tension in the skin tissues (whose stiff cell walls are easily fractured) that also bear the P that is transmitted to them from the interior mesocarp cells (Considine and Brown, 1981, Considine and Knox, 1981). Because the skin cell walls are more than 10-fold thicker than the mesocarp cell walls, the mechanical structure of the berry is often approximated as an inflated elastic shell, in which the tensional stress (s) is needed to balance the outward P (Boudaoud, 2010; Lustig and Bernstein, 1985). The balance of forces in a berry can be stated as follows:

$$s = Pdh^{-1}$$

where d = berry diameter, h = skin thickness.

In other words, given constant P during ripening (Thomas et al., 2006), the tension in the skin cells increases as the berry diameter increases and as the skin thickness decreases. As the berry expands, the skin would have to become thicker to support the same P and keep its cells from being pulled apart (Boudaoud, 2010). Thus, the stiffening of the cuticle and epidermis and the decrease in skin extensibility and thickness seem to come at the cost of increased vulnerability to cracking (splitting), which can occur during periods of high humidity or rainfall when water may be absorbed through the skin (Considine and Kriedemann, 1972; Lang and Thorpe, 1989; Meynhardt, 1964). The permeability of the berry cuticle for water uptake appears to be severalfold higher than that for transpirational water loss, and the resistance to osmotic water uptake through the skin declines as fruit sugar concentration increases (Becker and Knoche, 2011; Beyer and Knoche, 2002). In addition, perhaps even predominantly, water uptake can also occur through the berry pedicel and receptacle, especially in the corky scar area left from the abscised flower cap (Becker and Knoche, 2011; Becker et al., 2012). Cracking can ruin the quality of table and juice grapes and increase the incidence of bunch rot (see Section 8.2). It is also a common problem in tomato, cherry, and blueberry production (Ehret et al., 1993; Marshall et al., 2007; Sekse, 1995). Warm and moist conditions may favor cracking because the water permeability of the cuticle increases with rising temperature, while the cuticle's strength decreases (Becker and Knoche, 2011; Beyer and Knoche, 2002; Lopez-Casado et al., 2010; Schönherr et al., 1979). Moreover, there seems to be a relatively abrupt transition from high to low cuticle stiffness as the temperature rises above 25–30°C (Domínguez et al., 2011). Cuticular permeability also increases with increasing relative humidity of the atmosphere, probably because absorbed water molecules lead to swelling of the cuticle (Schönherr, 2006; Schreiber, 2005). In addition, moisture increases the cuticle's vulnerability to cracking, perhaps because water acts as a plasticizer that softens the cuticle and increases its flexibility (i.e., decreases stiffness) but decreases its strength (Domínguez et al., 2011; Lopez-Casado et al., 2010; Wiedemann and Neinhuis, 1998).

The uptake of surface water by moist grape berries, or even berries in a very humid atmosphere, is one reason why removal of leaves around the fruit clusters is a key canopy management practice, especially in regions that typically experience rainfall during the ripening phase. The increased air flow and, once sunshine returns, higher berry surface temperature are conducive to rapid drying of the clusters after rainfall, which limits the amount of water that can be absorbed through the skin and receptacle, as well as the extent of dilution of berry solutes. However, cracking may also ensue under conditions of root pressure (e.g., when irrigation suddenly relieves a drought episode) that may prohibit recycling of phloem-derived water out of the berry via the xylem (Keller et al., 2006).

Either way, the pressure generated inside the berry by the incompressible water is transmitted from the mesocarp cells to the epidermal cells. Berries are generally thought to crack when the mesocarp P exceeds the cell wall resistance of the skin cells, a relationship that is cultivar specific (Bernstein and Lustig, 1985; Considine, 1981; Lang and Düring, 1990; Lustig and Bernstein, 1985). At advanced stages of maturity, the vulnerability to cracking may decline in cultivars that are prone to shrinking (e.g., Syrah) and especially once the mesocarp cell membranes begin to break down, and hence lose osmotic competence (Clarke et al., 2010).

The slow transpiration rate of grape berries strongly limits transpirational cooling, which results in marked day−night fluctuations of berry temperature and permits solar radiation to raise the skin temperature of sun-exposed berries by 10−15°C, and occasionally even more, above ambient temperature (Cola et al., 2009; Millar, 1972; Smart and Sinclair, 1976; Spayd et al., 2002). The postveraison decrease in berry transpiration also slows the drying rate of raisin grapes, especially if the grapes are harvested late in the season, when temperatures and vapor pressure deficits decline. To facilitate drying and mechanical harvesting, growers often cut off the fruiting canes with clusters still attached to the shoots in late summer, a practice known as "drying on the vine" or "harvest pruning." Yet such drying takes much longer than drying manually harvested clusters spread on the ground because the temperature in the canopy is lower than that on the ground, so late-ripening and slow-drying cultivars such as Thompson Seedless are not well suited to this practice (Ramming, 2009). Because severing canes eliminates photosynthesizing leaf area at a time when inflorescence primordia may still be differentiating, this practice can sometimes come at the cost of some reduction in yield potential for the subsequent year (May, 2004; Scholefield et al., 1977a,b, 1978).

6.2.2 SUGARS

Sugars, especially glucose and fructose, are among the principal grape ingredients determining fruit and wine quality because they are responsible for the sweet taste of the fruit (and thus of grape juice and wine); they decrease (i.e., mask) the perception of sourness, bitterness, and astringency (which is the main reason many people add sugar to tea and coffee); and they enhance the "mouthfeel" (a.k.a. "texture"), "body," or "balance" of wines (Hufnagel and Hofmann, 2008b). Most important for wine, (anaerobic) fermentation converts hexose ($C_6H_{12}O_6$) sugars into alcohol (ethanol: CH_3CH_2OH) and CO_2 (two molecules each for each molecule of hexose sugar):

$$C_6H_{12}O_6 \rightarrow 2\ CH_3CH_2OH + 2CO_2$$

This process is accomplished by various species of yeast (but mostly by *Saccharomyces cerevisiae*), which live on the surface of grape berries (Barata et al., 2012b; Belin, 1972). The sugar content (more accurately, the concentration of total soluble solids) is measured in °Brix (g/100 g of liquid, or % w/w soluble solids, as estimated by refractometry at a standard temperature of 20°C). As a general rule, yeast converts sugar into ethanol at a ratio of approximately 0.6% (v/v) ethanol per 1°Brix (Jones and Ough, 1985; Ough and Amerine, 1963). Ethanol is important for wine quality partly because most aroma volatiles are more soluble in ethanol than in water, so that their aroma impact declines with increasing ethanol content, and partly because it can suppress the perception of a wine's overall "fruity" character (Escudero et al., 2007). As the grape sugar concentration increases during fruit ripening, there is not only more sugar available to be converted into ethanol, but the influence of sugar (and other berry constituents) on yeast metabolism also alters the content in wine of yeast-derived metabolites, such as higher alcohols, esters, and sulfur-containing volatiles (Hufnagel and Hofmann, 2008b; Ribéreau-Gayon et al., 2006; Noble and Bursick, 1984; Robinson et al., 2014).

Unlike under the anaerobic conditions that predominate during fermentation, under the aerobic (i.e., oxygen-rich) conditions of a vineyard, yeasts meet their energy requirements by (aerobic) respiration rather than fermentation so that the sugars present in unharvested grapes are turned into CO_2 and water. The little ethanol that is being produced is further respired (i.e., oxidized) to acetic acid (CH_3COOH, and eventually to CO_2) and water by aerobic bacteria of the genus *Acetobacter*:

$$CH_3CH_2OH + O_2 \rightarrow CH_3COOH + H_2O$$

Employing the same process in the presence of oxygen, these bacteria can also turn wine into vinegar (French *vin* = wine, *aigre* = sour), especially at high temperature and high pH.

Sugar is imported into the developing berries as sucrose via the phloem. Although grape flowers and young berries may contain some starch due to their photosynthetic activity, starch does not accumulate to any great extent (Fougère-Rifot et al., 1995; Hardie et al., 1996a; Kriedemann, 1968a; Palliotti et al., 2010; Swift et al., 1973). Starch does, however, accumulate in the pedicel and rachis (Amerine and Bailey, 1959; Amerine and Root, 1960). In addition, the seed endosperm accumulates starch that serves as a source of energy and carbon for seed germination. Sugars typically make up more than 90% of the soluble solids in mature berries, with much of the remainder being organic acids. In berries of most *Vitis* species and their cultivars, 95—99% of these sugars are present in the form of the hexoses glucose and fructose, with the remainder predominantly made up of sucrose (Dai et al., 2011; Hawker et al., 1976; Hrazdina et al., 1984; Kliewer, 1967b; Liu et al., 2006; Lott and Barrett, 1967). The two hexoses have been estimated to be responsible for more than 80% of the osmotic pressure (π) in the mesocarp and almost 60% in the exocarp (Storey, 1987). In contrast, sucrose may account for up to 30% of the sugar content in mature *Muscadinia* berries (Carroll et al., 1971; Carroll and Marcy, 1982). With regard to the taste of sweetness, fructose > sucrose > glucose; the fact that fructose is almost twice as sweet as sucrose may explain why inexpensive high-fructose corn syrup is so widely used by the food industry. At low temperatures (~10°C), pre-veraison *V. vinifera* berries contain approximately equal amounts of glucose and fructose, but the glucose:fructose ratio rises as the temperature increases, and at 35°C it can reach 3—10 (Currle et al., 1983; Findlay et al., 1987; Harris et al., 1968; Kliewer, 1964, 1965a). Irrespective of temperature, however, the glucose:fructose ratio declines to approximately 1 around veraison and remains more or less constant or decreases slightly thereafter, whereas the proportion of sucrose increases at low temperatures (Kliewer, 1964, 1965a, 1967b,c; Viala and Vermorel, 1909).

A few days before a berry begins to change color at veraison, hexose sugars begin to accumulate in the cell vacuoles of the mesocarp and, with a slight delay, in the exocarp (Castellarin et al., 2011; Coombe, 1987). Sugar accumulation is dependent on import of sucrose from photosynthesizing leaves or woody storage organs and tends to proceed more rapidly in berries that transpire more rapidly (Rebucci et al., 1997). Up until veraison, most of the imported sucrose is used for energy generation (respiration; see Section 4.4) and production of other organic compounds but does not enter the vacuole. By the time they begin changing color, the berries generally have reached a soluble solids concentration of 9°Brix or 10°Brix. During ripening, there is a rapid increase in hexose sugars due to import of sucrose and its subsequent breakdown by invertases (see Section 5.1); it is thought that the activation of sugar storage triggers malate efflux from the vacuoles (Sarry et al., 2004). Sugar storage in the vacuoles may deplete sugars in the cytosol, so the respiratory carbon flux through glycolysis (see Section 4.4) also declines at veraison, at least in the mesocarp (Giribaldi et al., 2007; Harris et al., 1971). In contrast, the rate of glycolysis in the skin may actually increase during ripening (Negri et al., 2008). In the mesocarp, malate compensates for the lack of sugar as the substrate for berry respiration. Thus, hexose storage and malate breakdown are activated simultaneously at veraison.

Sugar accumulation can continue for some time after the stiffening cuticle puts an end to berry expansion. Depending on species and cultivar, the natural sugar content in grape berries reaches a maximum at about 25°Brix, when the berries contain about 0.7 M each of glucose and fructose. This results in an ethanol concentration of approximately 15% (v/v) in wine made from such grapes, which also happens to be close to the upper limit of survival for the most ethanol-tolerant yeasts. Berries at 25°Brix have a total solute concentration of approximately 2 M, which is equivalent to an osmotic pressure of 5 MPa. Although import of sugar ceases around this stage, the sugar concentration can increase further due to berry dehydration and shrinkage (Figure 6.4). Thus, the final sugar concentration at harvest varies widely; table grapes and grapes destined for juice production may be harvested at 15–17°Brix, whereas overripe, late-harvest wine grapes may reach twice that sugar concentration. Dried raisin grapes or wine grapes that undergo a postharvest drying phase for the production of straw or passito wines and similar products are much more concentrated still. The solute concentration tends to increase by roughly 2°Brix for every 10% loss in berry weight (Muganu et al., 2011).

Sugars are carbohydrates, a class of chemicals that also includes the mostly insoluble polysaccharides contained in the cell walls. The contribution of these polysaccharides to a berry's total

FIGURE 6.4

Syrah berries reach a maximum amount of sugar when their sugar concentration approaches 25°Brix, but the concentration may continue to increase as berries shrink due to water loss; notice the large variation in sugar content due to variation in berry size (left; Keller, unpublished data) and mature Cabernet Sauvignon berries shrinking from dehydration (right; photo by M. Keller).

sugars is small, for cell walls make up only 1.5% of the berry weight at veraison, decreasing to 1% during ripening (Barnavon et al., 2000). The bulk of the cell wall polysaccharides in grape berries is composed of cellulose and pectins with smaller amounts of hemicelluloses, with most of the latter made up of xyloglucans (Hanlin et al., 2010; Lecas and Brillouet, 1994; Nunan et al., 1997; Vidal et al., 2001). Hundreds to thousands of glucose units are assembled in the long cellulose chains, whereas pectins and hemicelluloses also contain other sugar "building blocks," such as arabinose, xylose, galactose, mannose, and rhamnose in addition to glucose, as well as their oxidized forms, or sugar acids, such as glucuronic acid and galacturonic acid (Brummell, 2006; Doco et al., 2003; Vidal et al., 2001). The latter forms the "backbone" of pectin that is interspersed with rhamnose to which other sugars are attached as side chains. In contrast, the highly branched hemicellulose does not have a backbone proper and consists of a mixture of different sugar units.

As the berries begin to soften, β-galactosidases, aided by other enzymes, pry a portion of the pectin away from its bondage in the cell wall and cut its sugar chains into shorter pieces, making them water soluble; this depolymerization process intensifies due to the activity of polygalacturonases and pectate lyases in the late stages of fruit ripening (Barnavon et al., 2000; Nunan et al., 2001; Robertson et al., 1980; Rose et al., 1998; Vicens et al., 2009). Cell expansion during the early ripening phase is accompanied by production of additional pectin so that the total amount of pectin increases while its concentration declines (Silacci and Morrison, 1990). At the same time, the xyloglucans are also shortened (depolymerized), and their content decreases under the influence of enzymes that degrade and modify the cell walls (Brummell, 2006; Deytieux et al., 2007; Doco et al., 2003; Ishimaru and Kobayashi, 2002; Rose and Bennett, 1999; Schlosser et al., 2008). Due to the very large volume and thin walls of the mesocarp cells, most of the cell wall material is in the berry skin, where xyloglucans tightly bind the cellulose microfibrils and pectins, forming an extensive and tough network of polysaccharides (Doco et al., 2003). In addition to their cross-linking function, xyloglucans may also keep the cellulose fibrils properly spaced and porous (Hamant and Traas, 2010; McCann et al., 1990; Wolf et al., 2012). Thus, hemicellulose depolymerization and pectin disassembly enable berries to soften at veraison.

Although the mesocarp accounts for approximately 75% of the fresh weight of a mature berry, 75% of the berry's total pectin is located in the skin (Vidal et al., 2001). In contrast to the simple sugars, cell wall polysaccharides do not taste sweet, but they are nonetheless important for fruit and wine quality. The cell wall matrix, especially the shorter pectin fragments that result from cell wall disassembly and the integrated extensin proteins, may bind other berry ingredients, such as tannins and anthocyanins, during fruit ripening and postharvest processing (Cadot et al., 2006; Hanlin et al., 2010; Kennedy et al., 2001). The pectin binding capacity arises from the multitude of negative charges of pectins and their degradation products, which confer considerable cation binding and exchange capacity on these cell wall materials (Nobel, 2009; Sattelmacher, 2001). The binding capacity probably exerts its major influence after berries are damaged and their cell compartmentation is broken to release the juice from the vacuoles during the juicing and winemaking processes. In addition, the binding of tannins to pectin fragments might be responsible for the decrease in the perception of astringency in ripening fruit because such binding may increase tannin solubility and thus make tannins less available to bind to salivary proteins when berries are chewed (Macheix et al., 1990; Soares et al., 2012). Thus, alterations in cell wall polysaccharides may account for some of the changes in extractability of quality-relevant compounds observed during ripening that are discussed later. Moreover, yeast enzymes can degrade and extract some of the noncellulose cell wall polysaccharides, especially pectins, during fermentation.

Because most white grapes are fermented *after* pressing—that is, after the skins and seeds have been removed—they result in wine with much lower polysaccharide content than red wines, which are typically fermented "on the skins" (O'Neill et al., 2004).

6.2.3 ACIDS

Acidity plays an important part in the perception of fruit and wine quality because it affects not only tartness (sour taste) but also sweetness by masking the taste of sugars and thus gives grapes, juice, and wine their fresh, crisp taste. In addition, acids (especially malate) also taste astringent and increase the overall perception of astringency (Hufnagel and Hofmann, 2008b), which is why acidity in wine is sometimes difficult to distinguish from astringency due to tannins. Wines with high acidity (and low pH) are often perceived as "sour" (unless the acidity is masked by residual sugar), whereas wines with low acidity (and high pH) are described as "flat," because the tongue functions essentially as a pH meter, detecting the concentration of hydronium ions (H_3O^+) as a sour taste. Similarly, the acidity of grape juice is estimated by titration (hence the term titratable acidity) against an alkaline solution (usually sodium hydroxide, NaOH). This volumetric analysis, more accurately, measures the concentration of the acids' rather weakly bound hydrogen ions or protons (H^+) that are released by the addition of the strong base NaOH. The high concentration of organic acids in grape berries and the active transfer into the fruit vacuoles of protons by H^+ pumps (which, in turn, drives the membrane-bound pumps for many other components including acids) are responsible for the low pH of grape juice. In other words, the pH falls as the H^+ concentration rises as defined by the relationship $pH = -\log_{10}[H^+]$ (where $[H^+]$ is the H^+ concentration in mol L^{-1}). Metal cations, especially potassium (K^+) and sodium (Na^+), tend to counter the influence of organic acids by substituting for H^+ and thereby raising the juice pH (Boulton, 1980a,b). Although the pH is generally inversely related to acid concentration (Figure 6.5), there is no simple relationship between titratable acidity and pH nor (especially)

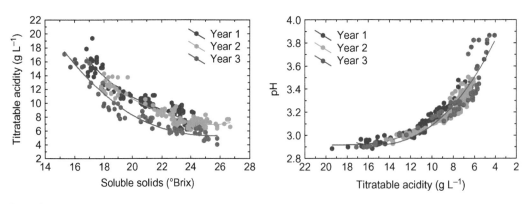

FIGURE 6.5

The acidity declines as the concentration of sugars (expressed as soluble solids) increases (left), and the pH rises as the acidity decreases (right) during ripening of Cabernet Sauvignon berries over 3 years; year 2 had the coolest and year 3 the warmest ripening period (Keller, unpublished data).

between "total acidity" and pH (Iland et al., 2011; Smith and Raven, 1979). In fact, the organic acids tend to buffer and hence stabilize the pH (Shiratake and Martinoia, 2007).

Although plants generally regulate their cell's cytoplasmic pH within quite narrow limits, the pH inside the vacuoles can vary considerably. Because grape juice consists mostly of the sap from mesocarp vacuoles, the pH of the juice is somewhat influenced by changes in soil conditions, vine nutrition, and solute transport (Conradie and Saayman, 1989). Nevertheless, juice from mature grapes harvested from vines grown on a soil with pH 8.5 usually has a pH of approximately 3.5, much like that from vines grown on a soil with pH 5.5, although the H^+ concentration in the two soils differs by a factor of 1000. The pH increases during fruit ripening due to both a decrease in organic acids and an increase in metal cations. The latter effect becomes dominant toward the end of the ripening phase because the rate of malate respiration decreases (most malate has already been respired, and the temperature typically declines in autumn), whereas K^+ may continue to move into the berries until phloem influx ceases. Consequently, juice expressed from harvested fruit typically has a pH between 3.0 and 3.5, but it can sometimes exceed 4.0 in overripe fruit. Values of pH in excess of approximately 3.6 are undesirable because they lead to decreased color intensity and microbial stability (i.e., increased potential for spoilage) and increased susceptibility to oxidation in wine and other grape products.

Juice from mature grapes typically contains between 5 and 10 g L^{-1} of organic acids. A titratable acidity concentration of $6.5-8.5 \text{ g L}^{-1}$ is considered optimal for the production of well-balanced wines (Conde et al., 2007). Although acidity is commonly expressed as tartaric acid equivalents, the organic acids are present as a mixture of several free acids (e.g., malic acid and tartaric acid) and their salts (e.g., malate, tartrate), which are usually ionized (Etienne et al., 2013; Kliewer et al., 1967; Smith and Raven, 1979). For example, malic acid ($C_4H_6O_5$) also occurs in the interconvertible forms $C_4H_5O_5^-$ (the malate anion) and $C_4H_4O_5^{2-}$ (the malate dianion), and tartaric acid ($C_4H_6O_6$) may also exist as $C_4H_5O_6^-$ (the tartrate anion, a.k.a. bitartrate) or $C_4H_4O_6^{2-}$ (the tartrate dianion). The relative proportions of the various forms are dependent on the pH: ionization increases with rising pH. Though this book uses the terms malate and tartrate for all forms of these acids, almost all of the malate and much of the tartrate are present as the free acids in the low-pH grape vacuoles (Iland and Coombe, 1988).

In most fruit species, malate and citrate are the dominant organic (carboxylic) acids. Grapes of both *Vitis* and *Muscadinia* species are unusual because tartrate is much more abundant than citrate. In fact, tartrate and malate are the major soluble organic components that accumulate in the preveraison grape berry and together comprise 70—90% of the berry's total acids (Dai et al., 2011; Jackson, 2008; Keller and Shrestha, 2014; Kliewer, 1966, 1967b; Ruffner, 1982a). Other organic acids mainly include oxalate, citrate, succinate, and fumarate, as well as phenolic acids, amino acids, and fatty acids. Citrate is an important flavor precursor in grapes because it can be converted during malolactic fermentation of wines to diacetyl, a compound imparting a "buttery" or "butterscotch" flavor. The tart malate, on the other hand, can be produced or degraded by various yeasts (which also independently produce succinate) and is converted to the much milder-tasting lactate (and CO_2) by lactic acid bacteria (e.g., *Oenococcus oeni*) during (anaerobic) malolactic fermentation. In contrast, tartrate is rather metabolically and microbially stable, which is why it is often added in winemaking to adjust the juice and wine pH. Nonetheless, the fungus *Botrytis cinerea* (see Section 8.2) and some lactic acid bacteria are able to degrade tartrate, especially at high pH (>3.5). Also, in the presence of iron or copper molecules, tartrate can be oxidized to glyoxylate (Monagas et al., 2006).

Most organic acids are accumulated early in berry development and stored in the vacuoles of the mesocarp and skin cells from synthesis inside the berry (Hale, 1962; Ruffner, 1982a,b; Sweetman et al., 2009), although malate can be transported in both the phloem and, in small amounts, the xylem (Hardy, 1969; Peuke, 2010). The acids serve important roles in regulating osmotic pressure, at least before sugars assume this role at veraison, and as charge-balancing anions for K^+, Na^+, Ca^{2+}, and Mg^{2+} and complexing agents for essential but toxic heavy metals, such as Fe^{2+}, Zn^{2+}, and Cu^{2+}, in the vacuoles (Meyer et al., 2010).

Tartrate and oxalate are both products of the breakdown of ascorbate (vitamin C), which occurs at least partly in the cell walls (DeBolt et al., 2004, 2006; Green and Fry, 2005; Saito et al., 1997). Ascorbate can be produced by oxidation of sugars, such as glucose, mannose, and galactose, or of galacturonic acid derived from pectin breakdown in the berries or imported from the leaves via the phloem; some may even be produced within the phloem (DeBolt et al., 2007; Melino et al., 2009). After veraison, the tartrate concentration decreases somewhat due to dilution from water import, whereas the amount of tartrate per berry generally remains fairly constant (Hale, 1977; Possner and Kliewer, 1985). Some tartrate turnover or respiration does, however, seem to occur (Drawert and Steffan, 1966a; Takimoto et al., 1976). Contrary to widespread opinion, tartrate does not form crystals with K^+ or Ca^{2+} in intact grape berries; thus, it does not become insoluble and precipitate during ripening. Despite the continued influx of K in the phloem, crystals are formed between oxalate (not tartrate) and Ca rather than K (DeBolt et al., 2004; Fougère-Rifot et al., 1995). Of course, once the cell compartmentation breaks down when grapes are crushed, tartrate may indeed crystallize with both K and Ca, but this process is very slow. In fact, the discovery of calcium tartrate in an ancient jar was used as evidence that Neolithic people made wine at least 7000 years ago (McGovern et al., 1996). The insoluble products of the biomineralization that takes place in the berry are deposited as bundles of needle-like Ca-oxalate crystals (raphides) in the vacuoles of hypodermis cells called idioblasts that specialize in Ca-oxalate production (Franceschi and Nakata, 2005; Loewus, 1999). The idioblasts probably serve as a storage compartment for surplus Ca and can release Ca on demand, which is important after veraison, when Ca influx into the berry through the xylem has mostly ceased. The granular Ca-oxalate crystals (druses) accumulated in the endocarp, on the other hand, may serve to store and later release oxalate. It is possible that, similar to their function in the leaves (He et al., 2014), the druses might also help scatter and thereby distribute sparse light to the interior of the berry. Whereas the production of tartrate and malate ceases by veraison, oxalate continues to accumulate throughout ripening, at least in some cultivars (Keller and Shrestha, 2014). Its (concurrent) breakdown releases H_2O_2 (and CO_2) after veraison, which in turn (and aided by the enzymes polyphenol oxidase and peroxidase) oxidizes the phenolics in the seed coat so that the seeds gradually turn brown (Braidot et al., 2008; Hanlin et al., 2010). Raphides are thought to play an antifeeding role by discouraging insects and mammals from eating unripe grapes (or leaves). Raphides are also a cause of contact dermatitis among people harvesting agave for tequila production, Ca-oxalate is a major component of kidney and gall stones, and both oxalate and tartrate are believed to be the dominant pain-inducing toxins of stinging nettles.

Malate is produced from phosphoenolpyruvate as a by-product of glycolysis and as a by-product of the photosynthetic refixation of CO_2 released during berry respiration or, in a reversible reaction, from oxaloacetate in the TCA cycle and is initially stored in the berry's vacuoles (Etienne et al., 2013; Martínez-Esteso et al., 2011; Moskowitz and Hrazdina, 1981; Ruffner, 1982b; Sweetman et al., 2009; Terrier et al., 1998). Preveraison malate accumulation is greatest at temperatures between 20°C and 25°C and declines sharply above 38°C (Kliewer, 1964; Lakso and Kliewer, 1975, 1978). Like tartrate, most malate is formed during the preveraison period (Kliewer, 1965b; Ruffner and Hawker, 1977). There is very little acid synthesis after veraison, although this "production stop" seems to be somewhat delayed in the skin (Gutiérrez-Granda and Morrison, 1992; Iland and Coombe, 1988). Fruit acidity therefore peaks immediately before veraison, a fact that is exploited in the production of verjuice or verjus (French *vert* = green, unripe; *jus* = juice). Verjuice has been used as an acidifying cooking ingredient and substitute for vinegar since ancient times and is made by pressing unripe (and unfermented) grapes harvested around veraison for this purpose or collected during cluster thinning for crop load adjustment. The amount of malate per berry declines after veraison due to inhibition of glycolysis (see Section 4.4) and re-metabolism of stored malate for respiration through the TCA cycle and gluconeogenesis (Drawert and Steffan, 1966b; Martínez-Esteso et al., 2011; Ruffner, 1982b; Ruffner and Hawker, 1977; Ruffner et al., 1975; Steffan and Rapp, 1979). It appears that the activation of hexose storage in the vacuoles triggers efflux from the vacuoles and breakdown of malate in the outer mesocarp in the area of the peripheral vascular bundles (Gutiérrez-Granda and Morrison, 1992; Sarry et al., 2004). Thus, the berry switches its carbon supply for respiration from glucose to malate at veraison so that the juice acidity declines as the sugar content rises (Figure 6.5). Provided that low temperatures do not slow down malate catabolism, most of the decrease in acidity occurs early in ripening, slowing around 16–18°Brix, and often becoming insignificant above 20–21°Brix—even as the pH continues to rise (Keller et al., 2012; Keller and Shrestha, 2014). In contrast with the situation in the mesocarp, and consistent with the persistence of glycolysis in the skin (Negri et al., 2008), there seems to be little, if any, decrease in malate in the skin during ripening (Gutiérrez-Granda and Morrison, 1992; Iland and Coombe, 1988; Possner and Kliewer, 1985; Storey, 1987).

Gluconeogenesis is the formation of glucose through a reversal of the glycolytic pathway. Such a reversal can occur if the cells' energy level exceeds demand so that the carbon of pyruvate can be recycled back to glucose. Gluconeogenesis has a temperature optimum at approximately 20°C and decreases to half the maximum rate at both 10°C and 30°C (Ruffner et al., 1975). Respiratory malate degradation increases with increasing temperature up to 50°C, whereas gluconeogenesis is more important at lower temperatures, when respiration, energy demands, and sugar import are low. Nonetheless, the contribution of gluconeogenesis to total sugar accumulation is minor and seems unlikely to exceed approximately 5% (Ruffner and Hawker, 1977).

6.2.4 NITROGENOUS COMPOUNDS AND MINERAL NUTRIENTS

Nitrogen-containing compounds are important sources of food for yeast during fermentation, and the nitrogen content of grape must greatly impacts fermentation rate. Musts with less than approximately 150 mg yeast-assimilable N L^{-1} are prone to sluggish or stuck fermentations and associated problems with hydrogen sulfide (H_2S) production (Bell and Henschke, 2005). H_2S is liberated from proteins that are broken down by the starving yeast cells and has a "rotten egg" odor. Consequently, taste intensity and aroma quality of wines tend to decrease as the nitrogen content of the grapes decreases (Rapp and Versini, 1996). Grape musts are therefore often supplemented with diammonium phosphate ((NH_4)$_2$$HPO_4$) to enhance yeast nutrition.

Although the nitrogen content of grape berries is extremely variable, they may contain approximately half the total canopy nitrogen at maturity (Pradubsuk and Davenport, 2010; Wermelinger and Koblet, 1990; Williams and Biscay, 1991). The majority (50−90%) of the mesocarp nitrogen is present in the form of free amino acids (Roubelakis-Angelakis and Kliewer, 1992). The remainder is made up mostly of proteins, ammonium, and nitrate. Whereas amino acids predominate in the mesocarp, proteins are much more abundant in the skin and, especially, the seeds (Yokotsuka and Fukui, 2002). The various berry tissues contain hundreds of different proteins, most of which are enzymes that are involved in the construction and transport of the broad range of berry metabolites and in respiration that provides the energy required for these activities (Goes da Silva et al., 2005; Martínez-Esteso et al., 2011; Sarry et al., 2004). The seeds accumulate large amounts of storage proteins (up to 90% of the total seed protein) that serve as a source of energy, nitrogen, and sulfur during and after seed germination. Moreover, proteins are also embedded in the pectin matrix of mesocarp and skin cell walls (Lecas and Brillouet, 1994; Saulnier and Brillouet, 1989; Vidal et al., 2001). Grape proteins can be extracted during fermentation and make up as much as 90% of the total protein content of wine; the remainder of the proteins is derived from yeast (Fukui and Yokotsuka, 2003).

Import of nitrogenous compounds into the berry before veraison can occur in both the xylem, predominantly in the form of glutamine and nitrate, and the phloem, mostly as glutamine. After veraison import becomes essentially restricted to the phloem. Nevertheless, more than half of a berry's total nitrogen is normally imported after veraison, and nitrogen accumulation may continue throughout ripening (Conradie, 1986; Kliewer, 1970; Löhnertz et al., 1989; Pradubsuk and Davenport, 2010; Rodriguez-Lovelle and Gaudillère, 2002; Schaller et al., 1990). Although grape berries can assimilate nitrate (Schaller et al., 1985), glutamine is the major nitrogen transport compound (see Section 5.3); enzymes called aminotransferases convert glutamine in the berry into other amino acids. Of these, arginine and proline together account for 60−80% of the free amino acids in mature grape berries (Bell and Henschke, 2005; Goes da Silva et al., 2005; Hernández-Orte et al., 1999; Roubelakis-Angelakis and Kliewer, 1992; Stines et al., 2000; van Heeswijck et al., 2001). However, the concentrations of arginine and proline may vary by an order of magnitude among grape cultivars grown under similar conditions (Huang and Ough, 1991; Kliewer, 1970). Arginine can be accumulated in the skin, pulp, and seeds throughout fruit development, although in some cultivars (e.g., Chardonnay and Cabernet Sauvignon) accumulation appears to cease at veraison, whereas in yet others (e.g., Müller-Thurgau) it may not begin until veraison (Kliewer, 1970; Schaller and Löhnertz, 1991). In contrast, most of the proline is accumulated after veraison, and accumulation seemingly continues throughout ripening (Coombe and Monk, 1979; Kliewer, 1968; Lafon-Lafourcade and Guimberteau, 1962; Miguel et al., 1985; Stines et al., 2000). Proline often serves to protect cells from excessive osmotic stress; thus, proline accumulation during ripening could be related to the osmotic pressure caused by the accumulating hexose sugars (van Heeswijck et al., 2001). Proline accumulation might also contribute to the rise in juice pH because its production from glutamate may involve hydroxide (OH^-) release (Smith and Raven, 1979). Furthermore, the release of amino acids associated with protein breakdown during ripening may contribute to the increase in amino acids after veraison (Martínez-Esteso et al., 2011). Arginine and other amino acids and protein residues, but not proline, can react with sugars in the so-called Maillard reaction to produce brown polymers termed melanoidins, which leads to slow nonoxidative, nonenzymatic browning during grape processing—for instance, during drying for raisin production and subsequent storage (Frank et al., 2004). Melanoidins can have pleasant (malty, bread crust−like, caramel, and coffee) or unpleasant aromas (burnt, onion, solvent, rancid, sweaty, and cabbage) and may taste bitter. Note that beer, coffee, chocolate, toasted bread, and caramel owe much of their color and aroma to melanoidins.

At very high concentrations, proline and some other amino acids (threonine, glycine, serine, alanine, and methionine) impart a sweet taste, whereas arginine and other amino acids (lysine, histidine, phenylalanine, valine, etc.) appear to taste bitter, and still others (glutamine, glutamate, asparagines, and aspartate) have an umami (Japanese *umami* = savory) taste (Hufnagel and Hofmann, 2008b). However, with the exception of proline and, perhaps occasionally, glutamate, amino acids are present in wine at concentrations that are far too low to have any sensory impact (Hufnagel and Hofmann, 2008b). Moreover, the amino acid profile of a wine differs considerably from that of its grape constituents and may reflect metabolic activities of yeasts and other microorganisms. For instance, arginine is the major yeast-assimilable amino acid, whereas proline cannot be readily used by yeast (Bell and Henschke, 2005). Some of the "fermentation aroma" of wines—that is, the bouquet caused by flavor and aroma compounds generated during fermentation—stems from the conversion by yeasts of amino acids to the so-called higher alcohols, which are defined as alcohol molecules with more than the two carbon atoms of ethanol (Robinson et al., 2014). For example, leucine yields isoamyl alcohol with a "harsh," "burnt," or "whiskey" odor, and valine yields ethyl isobutyrate with an aroma of "strawberry." Several minor amino acids in grapes have implications for fruit quality and human health.

The key position of phenylalanine as a chemical precursor for phenolic compounds is discussed in Section 6.2.5. Phenylalanine is an aromatic amino acid, in the chemical rather than the sensorial sense, produced by the shikimate pathway. Its aromatic sibling tryptophan not only is one of the key precursors for the plant hormone auxin but also can be degraded to the aroma volatile methyl anthranilate (Wang and De Luca, 2005) or be further metabolized to the neurotransmitter serotonin and the human hormone and antioxidant melatonin (Iriti et al., 2006). Fermentation yeast can metabolize phenylalanine and tyrosine, another shikimate offshoot, into phenylpropanoid alcohols and acetates with aromas of "honey" or "rose" (Robinson et al., 2014). In addition, lactic acid bacteria may convert tyrosine during malolactic fermentation into the neurotransmitter tyramine, which may lead to increased heart rate and higher blood pressure in humans, although the amounts of tyramine found in wine are rarely high enough to cause problems. Histidine is important because wine spoilage bacteria can convert it to histamine, a biogenic amine that at high concentrations is suspected to trigger headaches (including migraine) and asthma attacks. Unfortunately, ethanol inhibits histamine degradation in the body, which tends to exacerbate the reaction. Nonetheless, the scientific evidence for a link between biogenic amines and adverse reactions in humans appears to be inconclusive (Jansen et al., 2003).

In addition to proline, certain proteins also accumulate after veraison and continue to accumulate throughout ripening, probably in response to the osmotic stress arising from increasing sugar concentration, and they occupy an increasing fraction of the total berry proteins (Giribaldi et al., 2007; Monteiro et al., 2007; Murphey et al., 1989; Negri et al., 2008; Salzman et al., 1998). These are the chitinases and thaumatin-like proteins, which are members of the so-called pathogenesis-related (PR) protein families (see Section 8.1) and can make up half or more of the total soluble protein in mature grape berries, with roughly equal amounts in the skin and the pulp. These PR proteins seem to be responsible for haze formation or clouding in wine made from white grapes, so they may contribute to protein instability in wine (Pocock et al., 2000; Tattersall et al., 2001). Due to the rise in protein content, the potential for protein instability increases with advancing grape maturity. This is exacerbated by the concomitant increase in mesocarp pH, which makes PR proteins progressively easier to extract. In addition, protein extraction into the juice from damaged machine-harvested grapes is much greater than that from hand-harvested grapes, especially if the grapes are left standing for prolonged periods or transported over great distances before processing (Pocock et al., 1998). Red wines, however, are not normally subject to this protein haze problem because the tannins extracted from the grape skins during fermentation precipitate these and other proteins.

Macronutrients other than nitrogen also accumulate in the berry. Potassium (K^+) is by far the most important of these and is the major cation in grapes; its concentration in expressed juice often exceeds 1 g L^{-1} and can be substantially higher in the skin (Coombe, 1987; Donèche and Chardonnet, 1992; Hrazdina et al., 1984; Mpelasoka et al., 2003; Possner and Kliewer, 1985; Rogiers et al., 2006; Storey, 1987). Because it can substitute for protons in grape juice, K^+ impacts the juice pH: A 10% increase in K^+ concentration is associated with a rise in pH by approximately 0.1 units (Boulton, 1980a). Although K^+ can be transported in both xylem and phloem, its concentration in the xylem is orders of magnitude lower than that in the phloem, and import into the berry probably occurs predominantly in the phloem (see Section 5.1). Import occurs throughout berry development, but it sometimes increases markedly at veraison. Along with K^+, the amount of sodium (Na^+) also increases in both the pulp and the skin during ripening, but the Na^+ content per berry is at least an order of magnitude lower than the K^+ content (Etchebarne et al., 2010; Iland et al., 2011; Iland and Coombe, 1988). Most of the K^+ is stored in the vacuoles of the inner hypodermis and the mesocarp. It is possible that K^+ contributes to berry cell expansion by lowering the cells' osmotic potential, at least before sugars assume this role at veraison (Davies et al., 2006; Mpelasoka et al., 2003; Rogiers et al., 2006).

Free K^+ ions, liberated when the cell membranes are ruptured during grape processing, can form crystals with tartrate, and the resulting precipitation of the poorly soluble K-tartrate ($KC_4H_5O_6$, termed potassium bitartrate, potassium hydrogen tartrate, or cream of tartar) will decrease juice and wine acidity and may increase the pH (although the fact that K^+ substitutes for H^+ on the tartrate molecule, so H^+ remains in solution, could result in a decrease in juice pH upon K-tartrate precipitation). Extraction of K^+ from the skins during fermentation amplifies this effect in red wines (Mpelasoka et al., 2003). Note that cream of tartar (along with other components such as anthocyanins and other phenolics) is also sometimes extracted industrially from pomace remaining after pressing of juice or must.

Free calcium ions (Ca^{2+}) also combine with tartrate to form insoluble calcium tartrate in grape juice and wine, but this crystallization is very slow and may occur over a period of months or years. In contrast to K^+, Ca^{2+} import is largely restricted to the xylem. Consequently, 75–100% of the berry's final Ca^{2+} content is accumulated before veraison (Creasy et al., 1993; Esteban et al., 1999; Rogiers et al., 2006). Magnesium (Mg^{2+}), on the other hand, can be transported in both the xylem and the phloem (see Section 5.1), and approximately half the Mg is imported into the berry after veraison (Bertoldi et al., 2011; Esteban et al., 1999; Pradubsuk and Davenport, 2010; Rogiers et al., 2006). The accumulation of K^+, Mg^{2+}, and especially Ca^{2+} is higher in berries that have higher transpiration rates, even after veraison (Düring and Oggionni, 1986). Windy conditions and other factors that favor berry transpiration could therefore be associated with higher fruit mineral nutrient content.

Both Ca and Mg are important components of cell walls, and Ca also accumulates to high concentrations in the cell vacuoles in the form of Ca-oxalate crystals (see Section 7.3). The highest concentrations of these ions are found in the seeds, which is also true for phosphorus (P) and sulfur (S). However, due to its much larger volume, the mesocarp can sometimes be an equally important storage site for Mg, P, and S in terms of total amount per berry (Bertoldi et al., 2011; Rogiers et al., 2006). Like K, P accumulation in grape berries occurs throughout berry development and ripening (Bertoldi et al., 2011; Pradubsuk and Davenport, 2010).

Micronutrients, or trace metals, such as iron (Fe), copper (Cu), manganese (Mn), and zinc (Zn), are mostly concentrated in the seeds, whereas boron (B) also accumulates in the skin (Bertoldi et al., 2011; Rogiers et al., 2006), which probably reflects the importance of B as a cell wall component (see Section 7.3). However, similar to macronutrients, the mesocarp and, in some instances, even the skin may also contribute substantial amounts on a per berry basis. The amount of most of these ions generally increases throughout berry development and ripening, but most of the Mn and Zn is accumulated before veraison (Pradubsuk and Davenport, 2011; Rogiers et al., 2006). Import in the phloem of Cu and B may be so strong that not only their amount per berry but also their concentration continue to increase during ripening, whereas the concentration of other micronutrients tends to decline. In contrast to the pericarp, however, the seeds essentially stop accumulating phloem-mobile nutrient ions other than K at veraison (Rogiers et al., 2006). Trace metals may be important in altering the color hue of grape juice and wine. For instance, a small increase in Fe concentration can lead to an increase in the blue tint and a corresponding decrease in red. Moreover, dark pigments can be formed through interactions of proline and other amino acids with phenolics in the presence of Fe, whereas Cu, as a component of polyphenol oxidase, is involved in the enzymatic oxidation of phenolics. In addition, fungicide-derived Cu may lead to an increase in H_2S formation by yeast during fermentation (Eschenbruch and Kleynhans, 1974).

Minute quantities of rare earth elements have also been detected in grape berries (Bertoldi et al., 2009, 2011; Yang et al., 2010). Most of these belong to the lanthanoid series and are thought to have biological properties that resemble those of Ca, with whose effects they may interfere. In grapes, the rare earth elements are dominated by cerium (Ce), neodymium (Nd), and lanthanum (La). Most of them are present in the skin, with slightly smaller amounts in the mesocarp and only traces in the seeds. Rare earth elements accumulate predominantly before veraison, but accumulation may continue at a lower rate during ripening (Bertoldi et al., 2011).

6.2.5 PHENOLICS

Compounds other than water, sugars, acids, and minerals account for only a small proportion of the berry weight but contribute significantly to fruit and wine quality. Among these, the class of phenolic compounds is one of the most important contributors. All phenolic compounds are made up of an "aromatic ring" consisting of six carbon atoms with one or more hydroxyl (OH) groups or derivatives of this basic structure (Table 6.1). They are very important for the color and astringency of red wine, contribute to grape and wine flavor and aroma, and are the main substrates for juice and wine oxidation (Macheix et al., 1991; Singleton, 1992). Their susceptibility to oxidation, which they owe to their hydroxyl groups and unsaturated double bonds, is what makes phenolics such good antioxidants (Danilewicz, 2003; Rice-Evans et al., 1997).

Phenolics are produced inside the berry by several different routes (Figure 6.6) and thus form a diverse group from a metabolic standpoint, but they are usually classified into nonflavonoids that accumulate mainly in the berry pulp and flavonoids that accumulate mainly in the skin, seeds, and stem.

Table 6.1 The Main Classes of Phenolic Compounds in Grapes

Class Name	Basic Chemical Structure	Examples
Flavan-3-ols		Catechin (left) Epicatechin (right)
Tannins		Hypothetical proanthocyanidin tetramer with (from top to bottom) epigallocatechin, epicatechin, catechin, and epicatechin gallate (note: the subunits shown are a combination of those found in both seeds and skins)
Anthocyanins		Cyanidin-3-GLU[a] (R_1 = OH, R_2 = H) Delphinidin-3-GLU (R_1 = OH, R_2 = OH) Peonidin-3-GLU (R_1 = OCH$_3$, R_2 = H) Petunidin-3-GLU (R_1 = OCH$_3$, R_2 = OH) Malvidin-3-GLU (R_1 = OCH$_3$, R_2 = OCH$_3$)
Flavonols		Kämpferol-3-GLU (R_1 = H, R_2 = H) Quercetin-3-GLU (R_1 = OH, R_2 = H) Myricetin-3-GLU (R_1 = OH, R_2 = OH) Isorhamnetin-3-GLU (R_1 = OCH$_3$, R_2 = H) Laricitrin-3-GLU (R_1 = OCH$_3$, R_2 = OH) Syringetin-3-GLU (R_1 = OCH$_3$, R_2 = OCH$_3$)
Hydroxycinnamic acids		Cinnamic acid (R_1 = H, R_2 = H) Coumaric acid (R_1 = H, R_2 = OH) Caffeic acid (R_1 = OH, R_2 = OH) Ferulic acid (R_1 = OCH$_3$, R_2 = OH)
Hydroxybenzoic acids		Protocatechuic acid (R_1 = H, R_2 = OH) Gallic acid (R_1 = OH, R_2 = OH) Syringic acid (R_1 = OCH$_3$, R_2 = OCH$_3$)
Stilbenes		Resveratrol (R_1 = OH, R_2 = OH) Pterostilbene (R_1 = OCH$_3$, R_2 = OCH$_3$) Piceid (R_1 = OH, R_2 = GLU) Viniferins (resveratrol polymers)

Chemical structures courtesy of J. Harbertson.
[a]*GLU, glucose.*

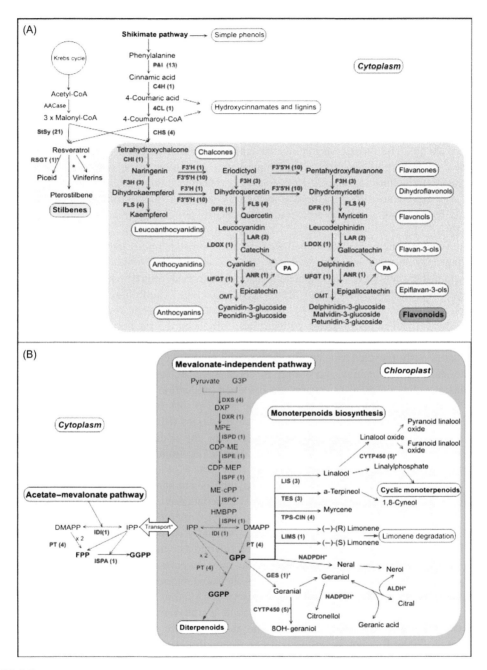

FIGURE 6.6

Biosynthetic pathways for the production of phenolics (A) and terpenoids (B) in grapevines.

Reproduced from Velasco et al. (2007).

Like other biochemical processes throughout the plant, the location, timing, and extent of the production of the multifarious phenolic compounds is tightly regulated by transcription factors (genetic "switches" and "throttles") that control the activity of the genes involved in the various biosynthetic pathways (Guo et al., 2014; Hichri et al., 2011). In addition to their production in grape berries, some flavonoids might also be imported via the phloem from the leaves (Buer et al., 2007; Zhao et al., 2010). The two main "assembly lines" are the shikimate pathway (named after one of its intermediates) and the malonate pathway. The shikimate pathway is responsible for the synthesis of most plant phenolics, whereas the malonate pathway is less important (although it is essential in fungi and bacteria). The former operates within chloroplasts and converts simple carbohydrate precursors derived from glycolysis (phosphoenolpyruvate) and the pentose phosphate pathway (erythrose-4-phosphate) into aromatic amino acids (phenylalanine) while releasing phosphate (Weaver and Herrmann, 1997). The well-known systemic broad-spectrum herbicide glyphosate kills plants by blocking a step in this reaction sequence (Holländer and Amrhein, 1980; Powles and Yu, 2010).

The subsequent phenylpropanoid pathway converts phenylalanine into cinnamic acid by the deaminating enzyme phenylalanine ammonia lyase (PAL), the pathway's key enzyme. Sitting at a branch point between protein and phenolic metabolism, PAL exerts a kind of metabolic traffic control by routing phenylalanine away from the manufacture of amino acids and proteins toward the phenolics assembly line. Whereas 30–40% of a plant's fixed carbon flows down the phenylpropanoid pathway (mainly because this pathway also produces the phenolic building blocks of the cell wall structural polymer lignin; Humphreys and Chapple, 2002), the ammonium released by PAL is continuously recycled via the nitrogen-assimilating GS/GOGAT cycle (see Section 5.3) to regenerate more phenylalanine, according to tissue-specific demands (Weaver and Herrmann, 1997). This recycling mechanism ensures the sustained production of phenolics from an amino acid even in the face of nitrogen deficiency. It is also possible that additional phenylalanine might be derived from the breakdown of proteins under such conditions (Margna et al., 1989).

Cinnamic acid is further modified and activated by the addition of a coenzyme A thioester (*S*-CoA, itself assembled from ATP and pantothenate, a.k.a. vitamin B_5) to yield 4-coumaryl-CoA. The slightly volatile cinnamic acid has a "floral" and "honey" odor and forms part of the flavor of cinnamon. The first step specific to flavonoid biosynthesis and hence to the flavonoid pathway, catalyzed by the enzyme chalcone synthase (CHS), is the condensation of 4-coumaryl-CoA with three molecules of malonyl-CoA, which is derived from acetyl-CoA produced during respiration (Yu and Jez, 2008). The product of this reaction, chalcone, is rapidly isomerized to the flavanone naringenin, which is the precursor of a large number of flavonoids with a basic chemical structure of C6–C3–C6 (Table 6.1). The flavonoids are the most important and most extensive group of phenolics. Alternatively, stilbene synthase—using the same substrates with the same stoichiometry as CHS, from which it may have evolved—catalyzes the synthesis of resveratrol (Schröder, 1997; Tropf et al., 1994), which is used for the formation of the various stilbene derivatives, such as piceid (resveratrol-glucoside), pterostilbene, and the polymeric viniferins (Jeandet et al., 2002; Monagas et al., 2006; Romero-Pérez et al., 1999; Schmidlin et al., 2008). These nonflavonoids are produced in the skin in response to injury (e.g., wounding or pathogen infection) and are important components of the vine's defense against attacking pathogens (see Section 8.2). They have also been implicated in numerous beneficial effects on human health. For instance, besides its antimicrobial properties, resveratrol appears to be capable (albeit in some cases at very high doses) of reducing heart disease, preventing many types of cancer (by acting as an antioxidant and antimutagen among other activities) and memory loss, delaying (type II) diabetes, and even delaying aging and diseases related to aging (Baur and Sinclair, 2006; German and Walzem, 2000; Guilford and Pezzuto, 2011; Pezzuto, 2008).

In addition to the stilbenes, nonflavonoid phenolics also include the phenolic acids (hydroxybenzoic and hydroxycinnamic acids). High concentrations of tartrate esters of hydroxycinnamic acids (e.g., caffeic, coumaric, and ferulic acids) are accumulated in the vacuoles of the berry mesocarp and exocarp during and after bloom, but their concentration often decreases during berry development (Monagas et al., 2006; Singleton et al., 1986). They may, in part, be used to manufacture lignin for the berry's vascular bundles and the seed coat (Braidot et al., 2008; Currle et al., 1983; Humphreys and Chapple, 2002). Some ferulic acid also becomes linked to the glycerol/fatty acid polyester suberin (Franke and Schreiber, 2007) that plugs the berry's stomata before ripening begins (and is the same compound that seals wine bottles in the form of cork derived from the bark of the cork oak, *Quercus suber* L., after which the chemical is named). Nonetheless, hydroxycinnamic acids are the major phenolic constituents of free-run grape juice and white wines and, at high concentrations, seem to contribute some astringency to grape juice and wine (Hufnagel and Hofmann, 2008a,b; Ong and Nagel, 1978; Sáenz-Navajas et al., 2010). Free hydroxycinnamic acids may also contribute to the heart disease-decreasing properties of grapes and grape products, but their esterification with ethanol may render hydroxycinnamates somewhat bitter (Hufnagel and Hofmann, 2008a,b). However, their main enological importance may lie in the fact that *Brettanomyces* and some other yeasts can convert them during fermentation to volatile phenols such as ethyl or vinyl guaiacol and eugenol, which are odor active and at low concentrations smell "smoky," "woody," "leathery," or "peppery" (Chatonnet et al., 1992; Rapp and Versini, 1996). The common esterification with tartrate, however, seems to protect hydroxycinnamates from this unfavorable conversion by *Brettanomyces* (Schopp et al., 2013).

Guaiacol ("smoky"), eugenol ("clove"), and related compounds are also extracted into wine from toasted oak barrels as breakdown products of lignin (Chatonnet and Dubourdieu, 1998; Maga, 1989), and eugenol and other phenylpropanoids are important aroma components of basil leaves and of wines made from grapes of *Vitis cinerea* and, to a lesser extent, *Vitis riparia* (Sun et al., 2011). At higher concentrations, the odor of volatile phenols becomes unpleasantly "pharmaceutical," "medicinal," or "burnt" at the expense of varietal fruit aroma. This poses a problem not only following contamination with *Brettanomyces* during the winemaking process but also when grapes have been exposed during ripening to smoke from wildfires. Smoke-derived phenols can be absorbed by grape berries and bound to various sugar molecules, perhaps in the cell walls, which leads to accumulation of glycosylated phenols in the skin and mesocarp. These phenols may subsequently be released (turning them into volatiles) during alcoholic and malolactic fermentation and wine storage (Hayasaka et al., 2010; Kennison et al., 2008). Just like in toasted barrels, the volatile phenols in the wildfire smoke originate from the thermal (oxidative) degradation of lignin (Edye and Richards, 1991; Fine et al., 2001). Grapes appear to be most sensitive to the adverse influence of smoke exposure during the initial stages of fruit ripening (Kennison et al., 2009, 2011). In addition, rapid oxidation of hydroxycinnamic acids during grape drying or processing results in browning of both raisins and grape juice, including the grapes used for the sweet, fortified Spanish dessert wine Pedro Ximenez (Sapis et al., 1983; Serratosa et al., 2008; Singleton et al., 1985). Such (enzymatic) oxidation becomes possible when physical disruption of cell compartmentation (due to mechanical damage before or after harvest or bunch rot infection) brings the phenolics into contact with oxygen and two groups of enzymes termed polyphenoloxidases and peroxidases that convert phenolics to quinones, which can then polymerize with other phenols to form brown pigments (Hernández et al., 2009; Macheix et al., 1991; Pourcel et al., 2007; Singleton, 1992). This reaction is particularly dominant if the grapes contain plenty of hydroxycinnamic acids and little glutathione (Greek *thion* = sulfur) that would otherwise protect the phenolics, or at least the colorless quinones, from oxidation by binding to them (Kritzinger et al., 2013; Monagas et al., 2006).

Most other phenolics are glycosylated—that is, they are attached to a sugar molecule—to increase their solubility, prevent their free diffusion across membranes, and decrease their reactivity toward other cellular components (glycosylation also decreases their antioxidant activity), which permits accumulation and storage in the cell vacuoles (Bowles et al., 2006; Rice-Evans et al., 1997). Glucose, following its activation by uridine diphosphate to the so-called UDP-glucose, is usually the sugar of choice for attachment to an oxygen atom (hence the term *O*-glucoside) of the phenolics by a family of enzymes called glycosyltransferases. These phenolics include the flavonoids such as tannins, anthocyanins (glucosides of the anthocyanidins malvidin, peonidin, delphinidin, petunidin, and cyanidin), and flavonols (glycosides of quercetin, kämpferol, myricetin, isorhamnetin, syringetin, and laricitrin) (Table 6.1). Note that although grapes produce at most traces of pelargonidin-glucoside (He et al., 2010a,b), which is the principal anthocyanin of strawberries, the major flavonols are the same in the two species. Similar to other phenolics, flavonoids have also been implicated in the protective activity of grapes and their products against heart disease among other benefits. Indeed, they have important pharmacological properties, reducing oxidative stress by acting as antioxidants that are more powerful than ascorbate and α-tocopherol (a.k.a. vitamins C and E; Rice-Evans et al., 1997), which is one of the supposed main reasons why eating grapes and drinking wine and grape juice may be good for our health (Dixon et al., 2005; German and Walzem, 2000; Guilford and Pezzuto, 2011). Nonetheless, human physiologists have noted that our bodies normally balance reactive oxygen species (see Section 7.1) and antioxidants so carefully that consuming antioxidants may not decrease oxidative damage (Halliwell, 2006, 2007).

Among enologically important phenolic compounds, tannins are the most abundant class; the lignin formed in the seed coat from cinnamic acids is enologically unimportant because it is not extracted during fermentation. Tannins are divided into two classes: hydrolysable and condensed (nonhydrolysable) tannins. The major building blocks or subunits of the hydrolysable tannins are gallic and ellagic acids, with small amounts of hydroxycinnamic acids, bound to glucose. The much longer condensed tannins are composed of flavan-3-ols (a.k.a. flavanols) such as catechin and epicatechin, both of which taste bitter when they occur as monomers (Dixon et al., 2005). Grapes contain only condensed tannins; the hydrolysable tannins present in some wines are extracted from oak wood, for example, during barrel aging (Puech et al., 1999). Condensed tannins are also called proanthocyanidins because they break down to anthocyanidins when heated in acids. Indeed, grape berries manufacture these tannin precursors via a branch in the anthocyanin-forming pathway by enzymes called leucoanthocyanidin reductase and anthocyanidin reductase. The former synthesizes catechin from the cyanidin precursor leucocyanidin, and the latter produces epicatechin directly from cyanidin and epigallocatechin from delphinidin (Bogs et al., 2005; Dixon et al., 2005; Terrier et al., 2009; Xie et al., 2003). In other words, the same anthocyanidins can be converted either to tannins (by reduction and subsequent polymerization) or to anthocyanins (by glucose addition). Upon extraction from the grapes, tannins can polymerize with anthocyanins and among themselves and thus are important for color stability in grape juice and wine (Jackson, 2008; Zimman and Waterhouse, 2004). Mostly, however, they are responsible for the bitter taste and the astringent tactile sensation—or "drying," rough feeling on the tongue, which is also called "texture"—of grapes and wines because they bind to and precipitate proteins that are much larger than the tannins (Gawel, 1998; Robichaud and Noble, 1990). Precipitation of salivary proteins increases the friction of the tongue gliding over the inner surfaces of the mouth, a property that gave rise to the term mouthfeel. In fact, the name tannin derives from these molecules' ability to tan leather—that is, to convert animal hide into leather by interacting with the hide's proteins. The perceived quality of red wines tends to increase as their tannin content increases (Mercurio et al., 2010; Sáenz-Navajas et al., 2010), but tannins are typically undesirable in white wines.

The colorless tannins are assembled from simple phenolic building blocks in the chloroplasts, wrapped in outgrowths of the chloroplast membrane, and "shuttled" to the vacuoles and cell walls of the seed coat and, to a lesser extent, the skin cells of the berries of both red and white cultivars (Amrani Joutei et al., 1994; Braidot et al., 2008; Brillouet et al., 2013; Gagné et al., 2006; Gény et al., 2003; Joslyn and Dittmar, 1967; Souquet et al., 1996; Thorngate and Singleton, 1994). Low concentrations of tannins may also occur in the mesocarp (Terrier et al., 2009). In addition, tannins and tannin precursors are also found in the rachis tissues (Souquet et al., 2000; Sun et al., 1999), but this source normally contributes only a minor portion of the tannins extracted during winemaking because grapes are typically destemmed prior to fermentation. Sealing off of the forming tannins within a membrane at their production site in the chloroplasts ensures that they do not precipitate—and thereby denature—enzymes on their way to and inside the vacuoles and also prevents them from reacting with other vacuolar components in intact berries (Brillouet et al., 2013).

Tannins are the true polyphenols of grapes; by definition they consist of at least two flavonoid subunits and can be composed of long, often branched, chains of almost identical subunits, varying in length over two orders of magnitude. Tannins with two to four subunits are usually classified as oligomers, and tannins with more than four units are classified as polymers. Though catechin is the major flavan-3-ol in grape skins, epicatechin and, to a lesser extent, epigallocatechin provide most of the extension subunits for the polymeric skin tannins. Epicatechin dominates in the seeds, followed by smaller portions of epicatechin gallate (Bogs et al., 2005; Fournand et al., 2006; Hanlin and Downey, 2009; Hanlin et al., 2011; Prieur et al., 1994; Souquet et al., 1996). In the seeds, but generally not in the skin, catechin and epicatechin are often esterified with gallic acid; this so-called galloylation increases their astringency or "coarseness" (Brossaud et al., 2001; Dixon et al., 2005; Vidal et al., 2003). Epicatechin gallate is usually only a minor compound in the skin, whereas, in a twist of confusing terminology, epigallocatechin is normally absent in the seed (Downey et al., 2003a; Gagné et al., 2006; Hanlin et al., 2010; Mattivi et al., 2009). The same tannin subunits are present in the seeds of diverse *Vitis* species, albeit in somewhat variable proportions (Liang et al., 2012a).

The shortest of these flavonoid polymers taste bitter but are usually confined to the seeds. Skin tannins on average are longer (4 to more than 100 subunits) than seed tannins (2−20 subunits) and are thought to provide a better mouthfeel (Hanlin et al., 2011; Kennedy et al., 2001; Monagas et al., 2006; Souquet et al., 1996). Overall, bitterness decreases and astringency increases with increasing degree of polymerization—that is, with increasing molecule size or chain length (Cheynier et al., 2006). The perceived astringency also increases as the tannin concentration rises and varies noticeably depending on the relative composition of the various subunits. Higher proportions of epicatechin extension subunits and of gallocatechin in terminal positions are believed to enhance astringency, whereas higher proportions of epigallocatechin in both extension and terminal positions may decrease astringency (Quijada-Morín et al., 2012). Nevertheless, although the total amount of tannins present in a wine is certainly somewhat related to the wine's astringency, small (and mostly unknown) chemical changes, especially during wine aging, may account for the perception of tannins as "fine," "soft," "green," or "harsh." For instance, the incorporation of galloylated procyanidins extracted from grape seeds seems to increase astringency (Brossaud et al., 2001). The perception of astringency, but not bitterness, also increases with increasing acidity, whereas ethanol has the opposite effect.

Although the amount of tannins contained in the seeds is typically several times higher than that in the berry skin, grape cultivars differ in the amount of both seed and skin tannins they produce. For example, the seeds of Cabernet Sauvignon, Merlot, and Syrah seem to contain much lower tannin concentrations than those of Cabernet franc, Pinot noir, Grenache, or Tempranillo. On the other hand (and perhaps surprisingly), not only Tempranillo and Nebbiolo but also many table grape cultivars (e.g., Red Globe, Flame Seedless, and Ruby Seedless) may have far higher skin tannin concentrations than Cabernet Sauvignon, Merlot, Syrah, Pinot, and even Tannat, which are all relatively similar. Barbera and Malbec skins, along with those of more table grapes (e.g., Muscat Hamburg), are at the low end of the spectrum. Although most of them lack the ability to produce anthocyanins, white grapes accumulate the same amount of tannins as their dark-skinned siblings (De Freitas and Glories, 1999; Rodríguez Montealegre et al., 2006; Viala and Vermorel, 1909).

Because the genes responsible for the biosynthesis of tannin subunits in seeds are switched on soon after fertilization (Dixon et al., 2005), seed tannins accumulate from fruit set through veraison or soon thereafter, and then their concentration, or at least their extractability, declines (Bogs et al., 2005; Downey et al., 2006; Kennedy et al., 2000). They also become increasingly polymerized (and thus less bitter) during ripening and seed desiccation. Polymerization is accompanied by oxidation, which is responsible for the browning of the seed coat after veraison (Adams, 2006; Braidot et al., 2008; Kennedy et al., 2000; Ristic and Iland, 2005). Oxidized tannins bind strongly to cell walls, providing a measure of "stress-proofing" to the seed coat and limiting the extractability of seed tannins (Cadot et al., 2006; Gény et al., 2003; Pourcel et al., 2007). The suberized cuticle of the seed coat further limits tannin extraction during winemaking. Consequently, the amount of seed tannins per berry and the amount extracted into wine depend mainly on the number of seeds (rather than the amount per seed) and fruit maturity but are relatively unaffected by environmental factors.

The tannin subunits in the skin are mostly produced and assembled in the developing flowers and young berries up to, or somewhat before, veraison (Adams, 2006; Bogs et al., 2005; Downey et al., 2003a; Hanlin and Downey, 2009; Harbertson et al., 2002; Kennedy et al., 2001). In other words, the production of flavanols from which tannins can be assembled occurs during the warmest part of the growing season and before anthocyanin accumulation begins, a developmental "switch" that grapes share with strawberries (Fait et al., 2008). Indeed, anthocyanins cannot be produced while the production of tannin precursors is in full swing, because the competing flavanol-forming enzymes suppress the attachment of glucose to anthocyanidins (Fischer et al., 2014). It is this competition among biosynthetic enzymes that makes flavonoid production a biphasic process: flavanols for tannins before veraison, and anthocyanins for (red) pigmentation beginning at veraison. As a result, the concentration, though not the absolute amount, of skin tannins declines during the postveraison berry expansion phase. At the same time tannin polymerization, which is thought to be an antioxidant mechanism, increases at veraison and may or may not continue through ripening (Bachmann and Blaich, 1979; Bindon and Kennedy, 2011; Hernández et al., 2009; Kennedy et al., 2001). In addition, a portion of the anthocyanins, or of their nonglycosylated anthocyanidin precursors, produced from veraison onwards might be incorporated into tannins. Such changes may account for most of any apparent tannin accumulation in the skin after veraison and for the decrease in free anthocyanins later in the ripening phase (Fournand et al., 2006).

With structural differences, such as different chain length, charge, and stereochemistry, come differences in functional properties among tannins and hence in their reactivity toward other molecules (Kraus et al., 2003). This may be one reason why a considerable fraction of the skin tannins can be bound to the insoluble matrix, consisting of cell wall pectins and glycans, of grape berries (Amrani Joutei et al., 1994; Kennedy et al., 2001; Lecas and Brillouet, 1994; Pinelo et al., 2006). The binding capacity of cell walls, which reduces tannin extractability into wine, seems initially to increase dramatically after veraison and then to decrease as the fruit approaches maturity and cell walls are dismantled and pectins degraded (Robertson et al., 1980). However, while some studies suggested a general decrease in tannin and anthocyanin extractability during grape ripening (Downey et al., 2006), others found no change in extractability (Fournand et al., 2006), and still others found an increase in extractability (Bindon et al., 2013). Nonetheless, longer-chain—and hence more astringent—tannins, it seems, are not as readily extracted as their shorter-chain counterparts (Fournand et al., 2006). The larger tannins may be increasingly bound to de-esterified cell wall pectins as berry ripening progresses (Bindon et al., 2012). It would be surprising if variations in functional properties among the tannins of different cultivars or arising during fruit ripening did not account for at least some of the differences observed in tannin extractability, subsequent reactions during winemaking, and, ultimately, astringency. Depending on maturity and binding capacity, between 25% and 75% of the berry tannin may be extracted during winemaking, 50–80% of which is derived from the skins (Cerpa-Calderón and Kennedy, 2008). Such a wide range of extractability means that the tannin content of a wine often cannot be predicted from the tannin content of the grapes. Indeed, the tannin concentration in red wines of the same cultivar varies more than two orders of magnitude, although the variation in skin tannins is within less than one order of magnitude (Harbertson et al., 2002, 2008; Ristic et al., 2010; Seddon and Downey, 2008). Because an increase in temperature weakens the association between tannins and cell walls (Hanlin et al., 2010), it is possible that high fermentation temperatures lead to more complete tannin extraction than do low fermentation temperatures.

Due to this incomplete extraction, which also applies to other phenolic compounds, organic acids, and other ingredients as well, grape pomace of both red and white cultivars contains considerable amounts of nutritionally valuable organic and inorganic compounds (Arvanitoyannis et al., 2006; González-Centeno et al., 2013; Kammerer et al., 2004; Lu and Foo, 1999; Mazza and Miniati, 1993; Monagas et al., 2006). These chemicals can be further extracted by distillation (i.e., production of brandy, such as marc or grappa) and pressing (e.g., grape seed oil). Also, they can be used for the production of dietary supplements or phytochemicals, for composting, or may be recycled back to the vineyard.

Anthocyanins (Greek *anthos* = flower, *kyanos* = blue) are the second class of phenolics of major sensory and enological importance after the tannins. They are responsible for the red, purple, or blue coloration of dark-skinned grapes and of grape juice and wine produced when such skins are extracted. Anthocyanin accumulation in grape berries starts at veraison. It is probably triggered by the increasing sugar concentration in the berry skin and mediated by calcium signals; the threshold for switching on the genes involved in anthocyanin production seems to be approximately 9°Brix or 10°Brix (Figure 6.7) (Castellarin et al., 2007a,b; Keller and Hrazdina, 1998; Larronde et al., 1998; Lecourieux et al., 2014; Pirie and Mullins, 1977; Vitrac et al., 2000). In other words, accumulation of anthocyanins begins *after* the onset of sugar accumulation and berry softening and after the production of the flavanol tannin subunits in the skin has ceased (Bogs et al., 2005; Castellarin et al., 2011; Fournand et al., 2006; Pirie and Mullins, 1980; Sadras and Moran, 2012). Note that although usually used to describe the beginning of ripening, the term veraison refers to a population of berries in a cluster or vineyard (e.g., meaning that 50% of the berries have changed color) rather than to an individual berry. At veraison, each cluster can have berries ranging from green to blue, often spanning a threefold range in sugar concentration (Figure 6.7), which implies that the different berries of a cluster are at different developmental stages and that the profound changes in the expression of ripening-related genes, whereby many genes are being "switched on" while others are being "switched off," occur rapidly but asynchronously in these berries.

Anthocyanins are generally confined to the vacuoles of the outer hypodermis, except in teinturier (French *teinturier* = dyer) cultivars whose mesocarp is red, and in overripe grapes with senescing skin cells that leak vacuolar components into the mesocarp (Fontes et al., 2011b; Moskowitz and Hrazdina, 1981; Viala and Vermorel, 1909). But even in teinturier grapes (e.g., Alicante Bouschet or syn. Garnacha Tintorera, Gamay Fréaux, Dunkelfelder, Rubired), the anthocyanin content in the skin is several fold higher than that in the mesocarp (Falginella et al., 2012). Storage inside the acidic cell vacuoles protects anthocyanins from oxidation, courtesy of the vacuoles' low pH, and ensures that they function as red pigments; at the neutral pH of the cytosol they would be mostly colorless or blue and unstable. More accurately, however, anthocyanins accumulate mostly within special, membrane-bound structures or vesicles named anthocyanic vacuolar inclusions

FIGURE 6.7

Change in the pigmentation of Merlot (left and center) and Pinot noir (right) grape skins due to anthocyanin accumulation with increasing sugar concentration at veraison (Biondi and Keller, unpublished data; photos by M. Keller).

(formerly called anthocyanoplasts) that trap and stabilize the pigments and may also contain long-chain tannins (Conn et al., 2003, 2010; Falginella et al., 2012; Markham et al., 2000). These vesicles originate from the endoplasmic reticulum, on whose surface the anthocyanins are produced, and may serve as transport vessels that carry the pigments to their final destination in the vacuoles with which the vesicles fuse (Gomez et al., 2011).

The amount and relative composition of anthocyanins varies significantly among *Vitis* species and cultivars. As a rule, anthocyanidins of *V. vinifera* cultivars are bound to one glucose molecule (the so-called 3-glucosides), and only traces of anthocyanidins bound to two glucose molecules (3,5-diglucosides) are sometimes present (Liang et al., 2011). In contrast, most other *Vitis* species, whether from North America or Asia, as well as the *Muscadinia* species, contain considerable amounts of 3,5-diglucosides (Dai et al., 2011; Hrazdina, 1975; Liang et al., 2012b, 2013; Mazza and Miniati, 1993; Ribéreau-Gayon and Ribéreau-Gayon, 1958). Hybrids (e.g., Concord) contain both the 3-glucosides and 3,5-diglucosides as well. The inability of most *V. vinifera* cultivars to attach a second glucose to anthocyanidins arises from two mutations that disrupt the gene that prescribes the necessary 5-*O*-glucosyltransferase (Jánváry et al., 2009).

The glucose stabilizes the metabolically unstable anthocyanidins, enhances their solubility, and turns them into anthocyanins; grapes cannot accumulate anthocyanidins (Bowles et al., 2006; He et al., 2010b). The glucose, in turn, can bind to acetic, coumaric, or caffeic acid to form acylated anthocyanins, but these may be less stable. Thus, the stability in solutions (e.g., grape juice or wine) is highest for the 3,5-diglucosides and decreases for the 3-glucosides, then the acylated 3,5-diglucosides, and, finally, the acylated 3-glucosides. In addition, anthocyanin stability is also influenced by the type and number of functional groups: more methoxy groups (-OCH$_3$) enhance stability, whereas more hydroxyl groups (-OH) reduce stability. Functional groups, moreover, are important for the color of anthocyanins: Addition of a hydroxyl group at the R position (Table 6.1) or addition of methoxy groups result in a red color, whereas hydroxyl groups at both the R and R, positions turn anthocyanins blue (He et al., 2010b; Jackson, 2008; Lillo et al., 2008; Tanaka et al., 2008). The genes encoding the enzymes that are responsible for adding these hydroxyl groups, the flavonoid hydroxylases, therefore determine the color of anthocyanins. Like the tannin-forming flavanols, two or three anthocyanin molecules can also bond together to form oligomeric pigments in grape berries (Vidal et al., 2004).

The two independently produced "starter" compounds of the anthocyanin-specific metabolic pathway are cyanidin and delphinidin (Figure 6.6). Once glucosylated, cyanidin-3-glucoside can be methoxylated to become peonidin-3-glucoside, whereas methoxylation turns delphinidin-3-glucoside into either petunidin-3-glucoside or malvidin-3-glucoside (Boss and Davies, 2009). Thus, the dark-skinned *V. vinifera* cultivars typically have between 5 and 20 anthocyanins, while the number may approach 40 in other *Vitis* species that accumulate diglucosides in addition to the monoglucosides (He et al., 2010a,b; Liang et al., 2012b; Mazza and Miniati, 1993). In dark-skinned grapes, the pigments based on malvidin are normally more abundant than those based on other anthocyanidins, and this dominance tends to increase during fruit ripening (Keller and Hrazdina, 1998; Liang et al., 2011; Pomar et al., 2005; Wenzel et al., 1987). Nonetheless, the relative proportion of the different anthocyanins varies by cultivar (Cantos et al., 2002; Carreño et al., 1997; Dai et al., 2011; Mattivi et al., 2006; Mazza and Miniati, 1993; Pomar et al., 2005). For instance, malvidin derivatives dominate in Malbec, Pinot noir, Syrah, Tempranillo, and many red table grapes; peonidin derivatives hold the majority in Nebbiolo; cyanidin and/or peonidin derivatives prevail in

Pinotage and other table grapes (and in the mesocarp of Alicante Bouschet); and delphinidin and cyanidin derivatives are dominant in Concord. The distribution is more "balanced" in Merlot, Mourvèdre, and Sangiovese. Moreover, Pinot noir and some varieties of *Muscadinia rotundifolia* ostensibly lack the acylated forms of anthocyanins that most other cultivars contain; the pigment profile of Pinot gris is very similar to that of Pinot noir, albeit on a much lower level (Castellarin and Di Gaspero, 2007; He et al., 2010b). Even clones of a cultivar may vary somewhat in their relative anthocyanin composition (Muñoz et al., 2014; Revilla et al., 2009). Anthocyanins are furthermore responsible for the reddish or bluish berry skin of many Muscat cultivars and Gewürztraminer, although it is possible that the bronzelike color of Pinot gris and Gewürztraminer arises from an interaction between anthocyanins and carotenoids (Forkmann, 1991; Liang et al., 2011). Among the Muscats, cyanidin-glucoside often dominates in the more lightly colored cultivars (as it does in Gewürztraminer), whereas malvidin- and peonidin-based pigments tend to be more abundant in the blue/black-skinned cultivars (Cravero et al., 1994). The berries of some dark-skinned cultivars (e.g., Syrah, Durif, Petit Verdot, and Tempranillo) accumulate pigments to much higher concentrations than those of other cultivars (e.g., Malbec, Pinot noir, Nebbiolo, and many table grapes), with still others (e.g., Cabernet Sauvignon, Merlot, and Barbera) being intermediate (Ferrandino et al., 2012). However, there is no close relationship between the amounts of anthocyanins and tannins in the skin.

Anthocyanins are probably involved in photoprotection, and the skin of red cultivars absorbs much more visible and, especially, ultraviolet (UV) light than the skin of white cultivars (Blanke, 1990a). A dark (red, purple, blue, or black) berry skin is the "default" version for grapes of all *Vitis* species (Cadle-Davidson and Owens, 2008; This et al., 2007; Viala and Vermorel, 1909). Most so-called white grapes, whose skin is green or yellow, have evolved from a dark-skinned relative by two mutations that prevent the activation of the gene that prescribes glucosyl transferase, the enzyme that attaches a glucose molecule to the anthocyanidins (Boss and Davies, 2009; Kobayashi et al., 2004; Walker et al., 2007). The same skin color mutation is also responsible for the conversion of Pinot noir to Pinot blanc. The inability to attach glucose probably impedes the entire branch of the phenylpropanoid and anthocyanin-producing pathways through feedback inhibition, which prevents the accumulation of hydrophobic and possibly toxic anthocyanidins (Yin et al., 2012).

Flavonols, which are present at much lower concentrations than tannins and anthocyanins, are produced in the epidermis throughout flower and berry development and act as "sunscreen," protecting the berry from harmful UVB radiation (Downey et al., 2003b; Keller and Hrazdina, 1998; Keller et al., 2003b; Kolb et al., 2003; Ribéreau-Gayon, 1964). This is why sun-exposed grapes contain far higher amounts of flavonols than shaded grapes, and the sun-exposed side of a berry can have a much higher flavonol concentration than its shaded side (Downey et al., 2004; Lenk et al., 2007; Price et al., 1995). In addition to the skin, the seed coat also contains flavonols (Liang et al., 2012a). Flavonol glycosides may contribute some astringency to wines (Hufnagel and Hofmann, 2008a,b; Sáenz-Navajas et al., 2010). Grape cultivars with dark-skinned fruit contain flavonols based on quercetin and myricetin, with small amounts of kämpferol, laricitrin, isorhamnetin, and syringetin. White grapes, however, apparently produce only quercetin, kämpferol, and traces of isorhamnetin (Ferrandino et al., 2012; Mattivi et al., 2006), which indicates that an entire branch of the flavonol pathway is switched off in white grapes. Intriguingly, despite the dominance of malvidin-based compounds in the anthocyanin profile of many dark-skinned cultivars, the structurally corresponding flavonol, syringetin, usually occurs only in trace amounts. Conversely, quercetin derivatives typically dominate the flavonol profile (although in some cultivars myricetin is more prominent than quercetin), whereas the "matching" cyanidin often contributes little to the anthocyanin pool (Adams, 2006; Cheynier and Rigaud, 1986; Downey and Rochfort, 2008; Liang et al., 2011, 2012b; Mattivi et al., 2006). However, cyanidin is the preferred substrate for anthocyanidin reductase and, thus, for tannin formation (Dixon et al., 2005).

Crushing the grapes during or after harvest brings anthocyanins in contact with other juice components, such as tannins and organic acids, with which they react to form polymeric pigments (Monagas et al., 2006; Wollmann and Hofmann, 2013; Zimman and Waterhouse, 2004). Between 50% and 90% of the total anthocyanins present in the skins are extracted during fermentation. The large variation in anthocyanin extraction is partly caused by differences in fruit maturity (extractability seems to decline as grapes mature) and amount and composition of cell wall material (e.g., cellulose content and degree of pectin methylation), which are themselves influenced by maturity and probably sun (especially UV) exposure (McLeod et al., 2008; Ortega-Regules et al., 2006b). It is also partly due to differences in skin contact time, fermentation temperature, and additions such as macerating enzymes. Despite the incomplete extraction during winemaking, the variation in wine color is preset by the composition of the grapes at harvest, and winemaking practices can do little to change this variation.

All phenolic compounds extracted into wine are subject to a variety of enzymatic (mainly oxidation processes that dominate during the early stages) and chemical (increasingly dominant during later stages) reactions that modify wine color and decrease astringency during aging (Cheynier et al., 2006; Fulcrand et al., 2006; Jackson, 2008; Monagas et al., 2006; Ribéreau-Gayon, 1973; Somers, 1971; Waterhouse and Laurie, 2006). The type and quantities of the different pigments influence intensity, hue, and stability of wine color. Moreover, anthocyanins come in five main molecular forms; they coexist in a dynamic equilibrium, but most of them are actually colorless at juice or wine pH (Jackson, 2008). Above pH 3, less than 30% exists as the red-colored flavylium cation, and its proportion declines further with increasing pH to almost nothing at pH 4. Even more important, sulfur dioxide (SO_2), used in winemaking to control microbial growth, effectively bleaches anthocyanins. As a consequence, the colorless forms account for 75—95% of the anthocyanins in wine.

Once extracted from the grapes into juice or wine, flavonols can be partially hydrolyzed in wine, whereby their glucose "tail" breaks away (Monagas et al., 2006). The resulting flavonol aglycones, together with cinnamic acids and catechin, can serve as the so-called cofactors, which stack together with anthocyanins in wine. This short-lived association of anthocyanin pigments with colorless phenolics is termed copigmentation and shifts the color toward purple or blue and intensifies it (Scheffeldt and Hrazdina, 1978; Tanaka et al., 2008). Copigmentation is responsible for the temporary purpleness and color stability (i.e., protection of the unstable anthocyanins from SO_2 bleaching and high pH) of young red wines and is said to account for 30—50% of their color (Boulton, 2001). Indeed, the color of young wine seems to be limited more by cofactors than by anthocyanins, although anthocyanins can also join together with other anthocyanins in the so-called self-association, which also stabilizes color. The reaction of anthocyanins with acetaldehyde or pyruvate extracted from the grapes or produced during fermentation to form orange-red pyranoanthocyanins confers additional stability. Yet red wines owe most of the gradual change in their color from red-purple to tawny during aging to conversion of anthocyanins to polymeric pigments that form by condensation between anthocyanins and tannins and other components (Bindon et al., 2013; Monagas et al., 2006; Ribéreau-Gayon, 1973; Somers, 1968). Contrary to popular opinion, however, polymerization does not always increase color stability. Thus, aged red wine color is determined by the grapes' anthocyanin and tannin composition, and both of these phenolic classes can limit wine color. In other words, if grapes are rich in anthocyanins but poor in tannins, wine color will be no better than if the reverse is the case. Tannins also polymerize with other tannins, and there is no upper limit to this polymerization (Jackson, 2008; Wollmann and Hofmann, 2013). Even long polymeric tannins, as long as 70 subunits, remain in solution; tannins precipitate only when they bind to proteins in wine. Moreover, polymerization is a dynamic process; polymeric tannins are susceptible to cleavage so that their degree of polymerization can increase or decrease during wine aging.

The grapes' anthocyanin, flavonol, and tannin profile determines the color potential of the wine made from these grapes, and color is a main driver of perceived red wine quality. Color hue and intensity are obviously important for quality in their own right, and they also influence the (subjective) perception of quality in other ways. Intriguingly, people associate certain flavors with specific colors, and the perception of taste and/or flavor intensity and complexity tends to increase as the color intensity increases. In other words, if two otherwise identical grape juices or wines differ in (red) color, the more deeply colored one is perceived as being more complex and as having more "body" and more dark-fruit aromas (Delwiche, 2003; Morrot et al., 2001). In the brain, it is clearly all about perception. In contrast, phenolic compounds are far less important for white grape and wine quality, which is determined mainly by the grapes' flavor and aroma potential. In white wines, phenolics may even contribute to undesirable astringency and bitterness—traits that are particularly unattractive in sparkling wines. Because the phenolics are concentrated in the skin and seeds, skin contact is usually minimized in white winemaking to limit extraction of hydroxycinnamic acids, flavanols, and tannins (Singleton and Trousdale, 1983).

6.2.6 LIPIDS AND VOLATILES

Lipids—consisting primarily of epicuticular and intracuticular waxes, fatty acids, membrane lipids, and seed oils—are present in the grape berry throughout development. Contrary to widespread belief, plant membranes also contain cholesterol in addition to many other sterols and their derivatives, called steroids or phytosterols (Fujioka and Yokota, 2003), although the contribution of cholesterol to the total lipids is about two orders of magnitude less than in animals. For example, cholesterol and other sterols are common, albeit minor, constituents of fruit cuticular waxes (Leide et al., 2007). Grape seeds contain up to 17% (fresh weight basis) polyunsaturated fatty acids (mainly linoleic acid) in addition to the monounsaturated oleic acid characteristic of olives, and the saturated palmitic and stearic acids (Currle et al., 1983; Miele et al., 1993; Rubio et al., 2009). Similarly, linoleic acid accounts for up to half of the skin's and mesocarp's fatty acids, followed by palmitic, linolenic, and oleic acids (Roufet et al., 1987). In most grape cultivars, the concentration of these lipids in the skin is up to three times higher than in the mesocarp and changes little during fruit ripening. When seeds and skins are damaged during the postharvest destemming and crushing process, the ensuing catabolism of fatty acids involves the formation of C6 compounds with a "fresh green," "grassy," or "herbaceous" aroma. For instance, linoleic acid may be oxidized to hexanal, and linolenic acid to hexanol. These C6 compounds are also called "green leaf" volatiles because they are produced by leaves as a generic wound response and because of their characteristic smell of freshly cut grass (Dudareva et al., 2013; Rodríguez et al., 2013). They are typical aroma components of tomatoes and strawberries and contribute to the aroma of unripe grapes. Such C6 compounds are also present in the rachis in addition to the berries, but their extraction during fermentation is very limited (Hashizume and Samuta, 1997; Schwab et al., 2008). Alcohol dehydrogenase enzymes, some of which are particularly active after veraison, might convert some of the aldehydes (such as hexanal) to their more "fruity" alcohol derivatives (especially hexanol) during ripening (Kalua and Boss, 2009; Tesniere et al., 2006). During fermentation, yeast may further convert these C6 compounds to hexyl acetate with an odor of "fruit," "apples," or "herb" (Dunlevy et al., 2009). Yeast also converts lipids to esters and acetates, which impart an overall "sweet-fruity" aroma (Robinson et al., 2014).

However, the most important group of lipids in terms of fruit and wine quality are the carotenoids, especially β-carotene (provitamin A) and lutein, in addition to minor compounds, such as the β-carotene-derived xanthophylls neoxanthin, violaxanthin, and zeaxanthin. Carotenoids belong to the class of isoprenoids, which are also called terpenoids or terpenes. Most carotenoids are C40 compounds; that is, they contain a "backbone" of 40 carbon atoms and are thus tetraterpenes assembled from isopentenyl pyrophosphate (a.k.a. isopentenyl diphosphate), which is derived from glyceraldehyde-3-phosphate and pyruvate through the methyl-erythritol 4-phosphate (MEP) pathway (Cazzonelli and Pogson, 2010; Della Penna and Pogson, 2006; Hirschberg, 2001; Tanaka et al., 2008). They are hydrophobic and, similar to their role in leaves (see Section 4.1), are accumulated in (skin) chloroplasts of young grape berries as part of the photosynthetic machinery to protect berry tissues from oxidative stress, especially stress caused by high light (Düring and Davtyan, 2002). Carotenoids have pharmacological properties as antioxidants and provitamins (Bartley and Scolnik, 1995; Della Penna and Pogson, 2006) and contribute to the yellow to reddish skin color (especially in the so-called white cultivars) when the chlorophylls are degraded during and after veraison (Uhlig and Clingeleffer, 1998). During this transformation of chloroplasts into chromoplasts, carotenoids become confined to oil bodies termed plastoglobuli, which enhances their light stability (Cazzonelli and Pogson, 2010). Nonetheless, the concentration of most carotenoids declines during grape ripening, which is in sharp contrast to the situation in tomatoes that owe their skin color at maturity mainly to high concentrations of red lycopene and orange β-carotene (Fanciullino et al., 2014; Fraser et al., 1994; Razungles et al., 1988). This decrease is at least partly due to degradation of carotenoid pigments into glycosylated norisoprenoids (C13 ketones) that are precursors of potent grape and wine aroma and flavor components (Baumes et al., 2002; Razungles et al., 1996). Carotenoid cleavage dioxygenases are the enzymes that (oxidatively) "slice" the various carotenoids, thus converting them into norisoprenoids (a.k.a. apocarotenoids) whose further conversion products, following hydrolysis of the glucose unit in juice or wine, contribute to floral and fruity aroma attributes (Dunlevy et al., 2009; Mathieu et al., 2005; Robinson et al., 2014; Winterhalter and Schreier, 1994). Like other volatile organic compounds, norisoprenoids have poor water solubility (i.e., they are lipophilic, which allows them to easily cross membranes and evaporate), relatively low molecular weight, and high vapor pressure in the ambient temperature range. When such volatiles escape into the air, they can be detected by the olfactory receptors of the nose so that we can smell them as an odor; they sometimes impart taste as well (Schwab et al., 2008). Our perception of a specific aroma volatile's odor, however, may change depending on its concentration, which is why the same volatile is often given varying sensory descriptors.

It is thought that the composition of the carotenoids determines the resulting profile of aroma volatiles they produce (Klee, 2010). Because grape cultivars differ in their fruit carotenoid composition, they also differ in their bouquet of flavor and aroma compounds produced from these carotenoids or their breakdown products. In other words, part of the "varietal aroma" may be derived directly from variations in skin color, some of which arise from the amounts and relative proportions of different carotenoids (Lewinsohn et al., 2005; Schwab et al., 2008). Examples of norisoprenoids (apart from the hormone ABA) include β-damascenone ("rose," "dried fruit," "exotic fruit," or "tropical flowers" aroma) and β-ionone ("violet" or "raspberry" odor contributing to "floral" or "fruity" aromas) which are found in most, if not all, grape cultivars (Robinson et al., 2014; Sefton et al., 2011). Chardonnay, in particular, is a good example of a cultivar whose varietal aroma is dominated by norisoprenoids, although in wine these are often masked by components derived from the oak wood that is used during fermentation and/or wine ageing (Gambetta et al., 2014; Sefton et al., 1990, 1993; Simpson and Miller, 1984). In addition, both β-damascenone and β-ionone strongly enhance the overall fruity character and reduce or mask vegetative notes of wines (Escudero et al., 2007; Pineau et al., 2007). Though these two volatile components have little aroma on their own, at low concentrations they may enhance the "berry fruit" aroma and at high concentrations lead to "plum" and "dried fruit" aromas. β-Damascenone is also an important aroma component of brandies, beer, rose petal oil, tobacco, and tea leaves (Sefton et al., 2011), whereas β-ionone is one of the most important volatiles contributing to tomato fruit flavor and also to the fragrance of many flowers. The production of these norisoprenoids appears to be directly dependent on the amount of their respective carotenoid precursor formed during fruit development. Although the carotenoids are largely accumulated before veraison, the production of the resulting apocarotenoids does not peak until late in ripening. Moreover, the bulk of these norisoprenoids are glycosylated (Sefton et al., 1993; Mathieu et al., 2005), and only their release from the sugar molecule during fermentation and wine storage leads to volatilization and "smellability." Owing to competing chemical processes of norisoprenoid production and destruction, however, the concentration of these fruity aroma compounds may increase or decrease during wine ageing (Sefton et al., 2011).

Other volatile isoprenoid compounds such as the highly aromatic monoterpenoids (C10; e.g., geraniol, linalool, nerol, and citronellol and its derivative rose oxide), which are reminiscent of "rose," "lilac," "pine," or "citrus," accumulate in postveraison grapes of Muscat cultivars, Gewürztraminer, and, to a lesser extent, Riesling, Viognier, and Chenin blanc (Bayonove et al., 1975b; Gunata et al., 1985; Guth, 1997a,b; Ribéreau-Gayon et al., 1975; Robinson et al., 2014; Wilson et al., 1984). Although monoterpenes are important in Riesling, the typical Riesling aroma may be dominated by the norisoprenoids vitispirane ("floral-fruity"), trimethyl-dihydronaphthalene (TDN, which smells of kerosene), and β-damascenone (Simpson and Miller, 1983; Skinkis et al., 2008; Strauss et al. 1987). Many, if not all, other cultivars (e.g., Sauvignon blanc, Cabernet Sauvignon, Syrah, and the Pinots) also produce monoterpenes, but at much lower concentrations than these aromatic cultivars. The "muscat" aroma is sought after in table grapes, and the same (and other) monoterpenoids also contribute to the characteristic fragrances (and varietal differences) of many flowers (e.g., rose, lilac, and jasmine), mint leaves, and conifer needles, whereas citrus fruits and basil leaves owe their aroma predominantly to mono- and sesquiterpenoids (C15). Although grape leaves are capable of producing monoterpenes, the berries appear to manufacture their own complement of these aroma compounds (Gholami et al., 1995; Luan and Wüst, 2002). The production of geraniol and nerol is mainly restricted to the berry skin, whereas linalool and rose oxide synthesis occurs in both the skin and the mesocarp, especially in the Muscats

(Gunata et al., 1985; Luan and Wüst, 2002; Luan et al., 2005; Park et al., 1991; Wilson et al., 1986). The sesquiterpene rotundone, which is the key aroma compound of both black and white pepper, also imparts a "peppery" and "spicy" aroma to grapes (and the resulting wines) of Syrah, Mourvèdre, Durif, and, to a lesser extent, Cabernet Sauvignon (Wood et al., 2008). Like the mono terpenes, rotundone accumulates mostly in the berry skin, beginning at veraison and continuing through even advanced stages of fruit maturity (Caputi et al., 2011).

All volatile grape terpenoids are mono-, sesqui-, or norisoprenoid terpenes (the diterpenoid hormone gibberellin is not volatile) produced from the simple isoprene building block isopentenyl pyrophosphate (McGarvey and Croteau, 1995; Tanaka et al., 2008). Two compartmentally separated biochemical assembly lines operate in parallel with some interaction to produce a wide variety of terpenoids (Dudareva et al., 2013; Dunlevy et al., 2009; Schwab et al., 2008): the mevalonate pathway in the cytosol and the deoxy-xylulose 5-phosphate/methyl-erythritol 4-phosphate (DOXP/MEP) pathway in the plastids (Figure 6.6). Although the mevalonate pathway uses citrate derived from malate breakdown in the TCA cycle as a starting substrate, this pathway is probably limited to the production of sesquiterpenes and a few other compounds (Dudareva et al., 2013; Sweetlove et al., 2010). Grape berries make monoterpenoids mainly through the mevalonate-independent DOXP/MEP pathway (Luan and Wüst, 2002), which also produces carotenoids, the chlorophyll "backbone," the hormones gibberellin and ABA, and other compounds. The berries of different cultivars manufacture specific bouquets of monoterpenes from the original geraniol, and much of this interconversion occurs only late during the ripening phase (Luan et al., 2005, 2006). Most isoprenoids (e.g., 80–90% of the terpenes) found in grapes are glycosylated and may be regarded as potential aroma or flavor precursors that can be released enzymatically or chemically during winemaking (Bönisch et al., 2014; Park et al., 1991). The sugar of choice for the glycosylation is glucose, which in turn is often bound to other, minor sugar molecules to form disaccharide glycosides (Dunlevy et al., 2009). Many such glycosylated precursors continue to release aroma volatiles in aging wines until they are exhausted.

Among the more than 20 red-colored Muscat cultivars, the amount of terpenes (especially linalool) tends to vary inversely with the anthocyanin content (Cravero et al., 1994). In other words, the more deeply colored cultivars tend to be less aromatic. Indeed, it appears that the production of terpenoids (i.e., the terpenoid pathway) and the accumulation of phenolics (i.e., the shikimate/phenylpropanoid pathway) are in competition for access to carbon substrates (Dudareva et al., 2013; Xie et al., 2008), although the carbon itself is abundant due to the accumulation of hexoses during ripening. Interestingly, carvone, a limonene-derived monoterpene isolated from the oil of caraway or dill seeds, is used in biochemical experiments as an inhibitor of PAL. It is possible, therefore, that cultivars that generate and accumulate high amounts of terpenoids may tend to produce lower amounts of phenolics and vice versa and that viticultural attempts to enhance the production of one class of compounds might come at the cost of lower production of the other. For example, tomato that was genetically engineered by introducing a gene from basil to accumulate very high concentrations of terpenoids (geraniol and its derivatives) produced lower amounts of carotenoids and norisoprenoids derived from them (Davidovich-Rikanati et al., 2007). There may be trade-offs with aroma potential that concern yield formation as well: Among cultivars with muscat aroma, the more aromatic ones seem to be much more vulnerable to conditions that cause poor fruit set than are the less aromatic cultivars (Huglin and Schneider, 1998).

Both grapes and wines contain hundreds of different volatiles in addition to the isoprenoids. Although these belong to many different classes of chemicals, including alcohols, esters and aldehydes, ketones, acids, lactones, and furanones, the majority of them (the sugar-derived furanones are one exception) are produced by oxidation of fatty acids by enzymes such as lipoxygenase, isomerases, alcohol dehydrogenase (which converts aldehydes into alcohols), and alcohol acyl transferases (which convert alcohols into esters) (Jackson, 2008; Kalua and Boss, 2009; Robinson et al., 2014; Schwab et al., 2008). For example, fatty acids from cell membrane lipids are degraded during advanced stages of fruit maturity and may subsequently be converted to volatile esters and aldehydes. Smaller portions of the same chemical classes are also contributed by the catabolism of amino acids (often by yeast during fermentation)—for instance, the leucine-derived methylbutanal and methylbutanol, which are also important flavor compounds of apples and strawberries. Cultivars differ greatly in the type and amount of volatile constituents they produce, and although most volatiles have no odor and many esters simply smell fruity, these differences largely account for the characteristic varietal aroma and flavor (i.e., the bouquet) that can be directly attributed to volatiles present in grapes (Conde et al., 2007; Jackson, 2008). It is thought that the more diverse the mix of volatiles, the less important will be the role of the individual components and the more complex will be the overall aroma and flavor.

Like the phenolics and isoprenoids, these components are normally accumulated in glycosylated form, which prevents these otherwise volatile compounds (as aglycones) from prematurely escaping to the atmosphere. However, although terpene glycosides may add a bitter taste (Noble et al., 1988), it is the volatile components that are central to the distinctive fruity aroma (and flavor), and many of these are slowly released during the winemaking and aging processes in part due to the activities of glycosidase and other metabolic enzymes of yeast and lactic acid bacteria (Gambetta et al., 2014; Jackson, 2008). For example, yeast can transform geraniol into rose oxide (although some transformation also occurs in grape berries). In other words, the majority of potentially important wine flavor and aroma components cannot be perceived in the fruit because human saliva lacks the glycosidase enzymes necessary to break away the sugar molecules (Kennison et al., 2008). Although grape berries do contain glycosidase, their high glucose content greatly slows the enzyme's activity (Aryan et al., 1987; Robinson et al., 2014). This is why the nonvolatile glycosylated compounds (glycosides) are odorless and are generally called aroma (flavor) precursors or potential aroma (flavor) compounds.

When harvest decisions are based on tasting fruit in the vineyard, they often (consciously or unconsciously) become decisions on the basis of the perception of sweetness and sourness in addition to the disappearance of "veggie" aroma (discussed later). Moreover, there is substantial fruit-to-fruit variation within a cultivar due to, among other variables, differences in fruit location (e.g., sun exposure), growth temperature and light, nutrition, harvest date (i.e., "ripeness"), and postharvest handling and storage. For instance, the continued metabolism of harvested grapes, especially machine-harvested grapes that are damaged in the process, can lead to substantial aroma and flavor loss if the fruit is picked, transported, or left standing in the heat for prolonged periods. On the other hand, the continuation of berry metabolism, much of it altered in response to dehydration-induced osmotic stress, together with the concentration effect from water loss in harvested grapes can also be exploited during a deliberate and prolonged postharvest drying phase to produce distinct wine styles such as the dry (e.g., Amarone) or sweet (e.g., Recioto, Vin Santo, and Vins de Paille) straw wines (Costantini et al., 2006; Zamboni et al., 2008).

In contrast to the glycosylated components discussed previously, the distinctive "foxy" *Vitis labrusca* aroma components methyl anthranilate, amino acetophenone, and the furanone furaneol are produced in the free volatile form during ripening and are emitted into the air, contributing to the characteristic smell of ripening Concord vineyards (Shure and Acree, 1994). Methyl anthranilate is also an important aroma component of orange flowers and Chinese jasmine green tea and, along with furaneol, of old strawberry varieties (before their aroma was sacrificed in the process of breeding transport-resistant varieties). Birds dislike the aroma and thus generally avoid Concord grapes; therefore, synthetic forms of methyl anthranilate are used in bird repellents in a variety of crops. Methyl anthranilate seems to be assembled in the outer mesocarp from a breakdown product of tryptophan and methanol (Wang and De Luca, 2005). Methanol may be either directly released from the cell walls or produced from methane (CH_4) released from cell wall pectins when the berries soften during and after veraison (Keppler et al., 2008; Lee et al., 1979). Tryptophan is thought to enable this reaction by conferring on pectin the ability to emit CH_4 upon UV irradiation via the production of reactive oxygen species, thus acting as a photosensitizer (Messenger et al., 2009).

A close relative of methyl anthranilate, 2-amino acetophenone (acetophenone itself is a simple isoprenoid), is present in Concord and other cultivars with *V. labrusca* parentage in both free and glycosidically bound form and is reminiscent of "acacia flower," "naphthalene," or "floor polish." Like methyl anthranilate, it repels birds (Mason et al., 1991). Although amino acetophenone is a regular aroma component of beer and corn chips (and of the pheromone of honey-bee queens), its formation, along with that of the ascorbate-derived furanone sotolon, during wine storage is thought to contribute to the unpleasant "atypical aging" off-flavor of *V. vinifera* wines (Kritzinger et al., 2013; Robinson et al., 2014).

Because of its volatile nature, the concentration of methyl anthranilate begins to decline in mature grapes, as does that of other nonglycosidically bound aroma compounds, such as the "veggie," "herbaceous," "grassy," "bell pepper," or "asparagus"-like methoxypyrazines typical of Sauvignon blanc, Sémillon, and Carmenère berries, but also of Cabernet Sauvignon, Cabernet franc, Merlot, and several other red and white cultivars—as well as of bell pepper and asparagus (Bayonove et al., 1975a; Hashizume and Samuta, 1999; Koch et al., 2010; Lacey et al., 1991; Ryona et al., 2010). In contrast to many other *V. vinifera* cultivars, grapes from *V. riparia* and *V. cinerea* contain particularly high amounts of methoxypyrazines (Koch et al., 2010; Sun et al., 2011). It is possible that such differences among species and cultivars arise from the presence of several transposons (jumping genes) that lead to differences in the expression of genes encoding enzymes involved in methoxypyrazine production (Guillaumie et al., 2013). Unlike methyl anthranilate, however, the cyclic, nitrogen-containing methoxypyrazines, which are probably produced from the amino acids leucine and valine, are not released into the air but are accumulated as free volatiles that are released only when the berry cells are disrupted by frugivores, which they help discourage from eating unripe berries. Therefore, methoxypyrazines accumulate early during berry development to a maximum before veraison and are then degraded during ripening to less than 10% of their maximum concentration (Hashizume and Samuta, 1999; Koch et al., 2010; Lacey et al., 1991; Ryona et al., 2008, 2010). Nevertheless, the methoxypyrazine concentration at or just before veraison is thought to be a very good predictor of its concentration at harvest (Ryona et al., 2008). The postveraison decrease, it seems, is similar to that of malate: Most of the decline occurs during the initial ripening phase, with only a slow further decrease above approximately 20°Brix

(Lacey et al., 1991; Roujou de Boubée et al., 2000). In other words, grapes of the same cultivar that contain high amounts of malate at harvest also tend to be high in methoxypyrazines. Although the leaves produce methoxypyrazines as well, the clusters manufacture their own complement, almost exclusively in the skin and seeds of the berries and in rachis tissues (Koch et al., 2010; Roujou de Boubée et al., 2002). Methoxypyrazines are extracted extremely rapidly during juice processing, before fermentation even starts, and are stable in wine (Hashizume and Samuta, 1997; Roujou de Boubée et al., 2002). Consequently, the methoxypyrazine content of wine is closely correlated with that of the grapes used to make the wine (Ryona et al., 2009).

Certain sulfur-containing compounds, namely volatile thiols (i.e., compounds with a sulfhydryl, or -SH, group), termed mercaptans, also contribute to the aroma complex in wine but probably not in grapes. Although they are partially bound by saliva proteins (Starkenmann et al., 2008), they usually have a very unpleasant smell of rotten eggs; in fact, skunks use sprays of volatile thiols as a defense method (Wood et al., 2002). At very low concentrations, however, such thiols are reminiscent of "black currant" (mercaptobutanol and mercaptopentanol)—characteristic of ripe Sauvignon blanc and Cabernet Sauvignon berries—or "passion fruit" and "grapefruit" (mercaptohexanol), especially in Sauvignon blanc and Gewürztraminer but also in Merlot, Muscat, Riesling, Sémillon, and other cultivars (Dunlevy et al., 2009; Guth, 1997a; Tominaga et al., 1998, 2000). These compounds may be produced from breakdown products of fatty acids and are bound to a nonvolatile thiol termed glutathione and its breakdown product, cysteine, in the grape skin and, to a lesser extent, the mesocarp (Kritzinger et al., 2013; Peña-Gallego et al., 2012; Peyrot des Gachons et al., 2002a,b; Robinson et al., 2014). The amount of glutathione itself increases in concert with sugars during grape ripening (Adams and Liyanage, 1993; Okuda and Yokotsuka, 1999), but the concentration of the bound aroma precursors in the berries seems to be subject to diurnal fluctuations, with peak concentrations occurring near sunrise (Kobayashi et al., 2012). Although the binding to glutathione and cysteine renders these thiols odorless in grapes, the odor-active thiols are released by enzymes of bacteria in the human mouth (Starkenmann et al., 2008) and by yeast enzymes during fermentation (Dubourdieu et al., 2006). The concentration of volatile thiols in wine is directly related to the concentration of their precursors in the grapes, even though only a minor portion ($<5\%$) of these precursors is actually converted to aroma compounds.

Yeast also seems to be able to produce another S-containing compound, termed dimethyl sulfide (($CH_3)_2S$, often abbreviated as DMS), perhaps from the S-containing amino acids cysteine and methionine or from the cysteine-containing peptides glutathione and cystine (de Mora et al., 1986). The DMS concentration further increases during wine aging and may enhance the fruity character, especially of black currant and "strawberry/raspberry," and add "truffle" (of which it is a major aroma component) and "olive" notes (Escudero et al., 2007; Segurel et al., 2004). At higher concentrations, however, DMS has an unpleasant "cabbage" odor, which is considered an off-flavor.

Other aroma compounds do not enter wine from the grapes at all. Instead, they have their origin in storage or packaging materials. One example is the infamous TCA (2,4,6-trichloroanisole), which imparts a musty or moldy odor at extremely low concentrations (parts per trillion) and is responsible for the majority of cork taints in wine. Most of the TCA is produced by microorganisms in corks that have been bleached with chlorinated compounds or in oak wood that has been treated with chlorine-containing insecticides or fungicides.

6.3 SOURCES OF VARIATION IN FRUIT COMPOSITION

Uniform fruit composition is often described as a critical factor for premium winemaking and is equally desirable for the production of grape juice, table grapes, and raisins. Nevertheless, when put to the test, no differences in fruit uniformity were found among vineyards producing fruit that sold for high, medium, or low prices (Calderon-Orellana et al., 2014). It could be argued, moreover, that less uniform fruit may sometimes result in more complex wine, and this might be one of the rationales behind blending wines from different vineyard sources. In many European wine regions, such blending was, and in some cases still is, traditionally done in the field, where a range of clones or even cultivars were interplanted, sometimes systematically and sometimes randomly. Of course the genotype is a major source of differences in fruit composition (Dai et al., 2011; Kliewer, 1967b; Kliewer et al., 1967). Whereas traits related to color, astringency, acidity, and flavor are key quality attributes of wine grapes, table grape quality depends more on sugar:acid ratio, firmness as a sign of freshness, texture for crunchiness or crispness, juiciness, and visual appearance including color (Mullins et al., 1992; Piva et al., 2006; Rolle et al., 2012; Winkler, 1958), although Muscat flavor is often sought after as well. Seedlessness is important in some table grape markets and is usually desirable for raisin grapes. High sugar content, moreover, is much more important for raisin grapes than it is for table grapes, not only because consumers prefer sweeter raisins, but also because a higher sugar concentration (20–22°Brix is considered ideal) at harvest cuts down on drying time (Christensen et al., 1995; Parpinello et al., 2012; Ramming, 2009).

Fruit composition changes over time during berry development and ripening as part of the grapevine's developmental program and is therefore under genetic control. In addition, like phenology and yield formation, the extent of these changes may also be modulated by environmental factors and by the interaction of these factors with the genotype of the vine. It is often difficult to separate the influence of one of these factors from that of another. For example, solar radiation affects both incident light and tissue temperature. An increase in water supply leads to an increase in shoot growth, which can decrease the proportion of exposed leaves, which in turn alters both light quantity and light quality and may also lead to changes in canopy microclimate (see Section 5.2). Furthermore, the sugar concentration is higher and the malate concentration lower in berries that have fewer seeds, apparently because veraison occurs earlier in such berries (Currle et al., 1983; Rapp and Klenert, 1974). Thus, even within a grape cluster, there is often a natural variation of sugar concentration in the range of 5–7°Brix that results from asynchronous development of individual berries (Coombe, 1992; Pagay and Cheng, 2010). This range may be twice as large across berries harvested on the same day within a vineyard (Kasimatis et al., 1975; Singleton et al., 1966). Due to such asynchronous development, neither time nor thermal time after bloom adequately reflects the developmental stage and maturity of individual berries in a sample from the entire population of berries on a vine, let alone in a vineyard. However, the variation in fruit composition generally decreases as the grapes ripen and more and more of the berries approach their sugar maximum and acid minimum (Rankine et al., 1962).

Any factor that influences vine growth and metabolism either directly or indirectly impacts fruit composition, and this leads to large variation among growing seasons in terms of fruit quality. Wine grape production is especially sensitive to climate variability. This sensitivity is the source of considerable variation in vintage quality, which in turn forms one of the bases for the existence of an entire "industry" of wine judges, consumer magazines, and related services. The majority of the fluctuations in fruit composition are caused by climate variability. In fact, weather differences among years, in addition to vineyard location rather than, say, within-vineyard differences in soil composition, are by far the strongest determinants of fruit composition and wine quality (Downey et al., 2006; Noble, 1979; Reynolds et al., 2007, 2013). Such climate variation often trumps differences in soil moisture (except at the far low and high ends of the moisture spectrum) in both dry-farmed and irrigated vineyards (Bowen et al., 2011; Keller et al., 2008; Pereira et al., 2006b; van Leeuwen et al., 2004). For instance, annual variation accounts for an almost twofold range in sugar concentration, a two- or threefold variation in acid and anthocyanin content, and at least a 10-fold range in aroma components (Cacho et al., 1992; Castellarin et al., 2007b; Downey et al., 2004; Viala and Vermorel, 1909; Wood et al., 2008). The differences in anthocyanin profiles—that is, the relative proportion of individual pigments for a given cultivar—varies more among years than from veraison to harvest in the same year (Pomar et al., 2005). The concentration of carotenoids, it seems, varies less among diverse grape cultivars than it does among growing seasons for the same cultivar (Oliveira et al., 2004). Even the amount of skin and seed tannins varies as much between years as it does between cultivars (Harbertson and Keller, 2012; Liang et al., 2012a). These differences also hold true at the level of molecular biology: Differences in gene expression in grape berries between growing seasons can be greater than those between pre- and postveraison berries within a season (Pilati et al., 2007). Cultural practices can at best be used to fine-tune what nature imposes in any given season. The same applies to clones of a cultivar and to the rootstock used as a grafting partner: Growing season and vineyard site still vastly outweigh the influence of these on fruit composition (Harbertson and Keller, 2012; Keller et al., 2012; Schumann, 1974). Even grapes harvested at identical concentrations of soluble solids in different years or different vineyard sites can have very different amounts of anthocyanins and aroma-active compounds. Naturally, this vintage variation results in different sensory properties in wines made from these grapes (Forde et al., 2011). For a specific clone of a cultivar, variation in grape composition occurs within one cluster, within one vine, between vines, between vineyards, and between vintages (Rankine et al., 1962). This variation arises due to the factors discussed in the following sections (see also Jackson, 2008), as well as to other factors (e.g., some effects of pathogen infection are discussed in Section 8.2).

6.3.1 FRUIT MATURITY

Contrary to widespread misunderstanding, grape berries are physiologically mature when their seeds attain the ability to germinate (i.e., at or immediately after veraison; see Section 2.3), not at some arbitrarily defined point thereafter. Subsequent alterations in fruit composition may nevertheless be beneficial for the seeds and hence for the species because they help attract seed dispersers. As argued in Section 2.3, the accumulation of red pigments and aroma volatiles serves as an "advertisement," signaling to potential seed vectors the availability of food material of high nutritional and health value. In fact, the majority of pigments and volatiles are manufactured from nutritionally valuable components such as sugars, amino acids, fatty acids, and carotenoids, and thus serve as positive nutritional signals (Goff and Klee, 2006). Many of the associated changes during grape ripening were discussed in Section 6.2 for individual compounds or specific groups of compounds. The changes in the concentrations of sugars, organic acids, amino acids, pH, and phenolic compounds (especially anthocyanins in red grapes) are rather well known. Sugars and anthocyanins accumulate and malate declines rapidly during the early ripening phase, and the sugar concentration may continue to increase due to berry shrinkage well beyond the natural maximum of less than or equal to 25°Brix. The amount of anthocyanins, however, tends to decline at these late stages of ripening (Holt et al., 2010). In addition to these major compounds, flavor and aroma components also undergo changes as grapes ripen. For instance, methoxypyrazines, whose production can vary by an order of magnitude among vineyard sites, decline rapidly during and after veraison, but their degradation slows considerably later in the ripening phase (Lacey et al., 1991; Marais et al., 1999; Roujou de Boubée et al., 2000; Ryona et al., 2008). Whereas esters may dominate the volatile profile of preveraison berries, grassy-herbaceous aldehydes (e.g., hexenal and hexanal) peak during the early ripening phase but may be converted to more fruity alcohols (e.g., hexanol) during later stages of fruit maturation (Kalua and Boss, 2009). Norisoprenoids and most other terpenes, or their glycosylated aroma precursors, appear to continue to increase even at very advanced stages of maturity (>30°Brix) (Fang and Qian, 2006; Hardie et al., 1996a; Park et al., 1991). Such changes during extended ripening or "hang time" of the grapes form the basis for delaying harvest to minimize vegetal aromas and maximize terpenoid-based and other fruity aroma compounds (Kalua and Boss, 2009; Ryona et al., 2008; Wilson et al., 1984). Whereas the concentration of some volatile phenols (e.g., guaiacol) increases strongly during maturation, perhaps due to conversion of hydroxycinnamic acids, others (e.g., eugenol) seem to remain constant or even decrease (Fang and Qian, 2006). The precursors of some aroma-active esters also appear to decrease over time (resulting in wines with less fruity aroma), but the majority show no consistent trend. It is not clear whether changes late in the growing season are indeed due to the continued production or modification of these quality-relevant compounds in the grape berries or are simply brought about by the slow dehydration of the berries (i.e., concentration effect, weather permitting). Moreover, physical and chemical changes in fruit composition also lead to changes in the number and composition of microorganisms that call the fruit home. As the sugar concentration and the pH increase, both the number and the diversity of this microflora increase. Yet the consequences of these population dynamics for grape juice and wine composition are not well understood and little appreciated.

Clearly, some of the alterations in fruit composition that occur with increasing grape maturity are desirable (e.g., more sugar, less acid and bitter tannin, and more "good" and fewer "bad" flavors), whereas others are not (e.g., even more sugar, high pH, and less "good" and more "bad" flavors). Moreover, the desired level of fruit maturity depends on the intended use of the grapes. Taking soluble solids as a simple measure of maturity, table and juice grapes may be harvested when they have accumulated just 16–18°Brix, which is immediately after they have fully turned color. Grapes destined for sparkling wine production are considered optimally ripe at approximately 18–20°Brix (Iland et al., 2011). And while the typical harvest "window" for many white wine grapes ranges from 20–24°Brix, red wine grapes and white wine grapes destined for sweet "dessert" wines are often left on the vines well beyond the physiological sugar maximum of 24°Brix or 25°Brix—weather permitting. The factors described in the following sections exert much of their influence on fruit composition by accelerating or retarding grape ripening and hence maturity.

6.3.2 LIGHT

As the source of energy for photosynthesis, sunlight is the most important climatic factor affecting grape composition. The effect of light is often expressed in terms of the effects of shade, especially as it refers to the within-canopy microclimate (Jackson and Lombard, 1993; Smart, 1985). Apart from weather conditions, this microclimate, and hence the extent of sun exposure of the fruit clusters, is a function of planting distance, trellis design, canopy density, shoot length, and shoot architecture (e.g., plentiful, overhanging lateral shoots) and can be modified by canopy management practices such as shoot thinning and positioning, hedging, or leaf removal (Reynolds and Vanden Heuvel, 2009). It is important to remember, however, that shade decreases both light intensity and temperature during the day, and it is very difficult to disentangle the separate effects of each of these variables. By reducing the vapor pressure deficit, the cooling effect of shading also reduces fruit transpiration and consequently water loss from the fruit.

Shade on the leaves depresses the rate of photosynthesis, thus limiting the rate of sugar export to grape berries, which may curtail berry growth and sugar accumulation (Cartechini and Palliotti, 1995; Rojas-Lara and Morrison, 1989; Smart et al., 1988). When shaded leaves die, their nutrients are recycled to important sinks. If the fruit happens to be important while leaves are senescing (e.g., around veraison), they may receive a dose of nitrogen, potassium, and other nutrients, which may eventually lead to an increase in juice pH (Morrison and Noble, 1990; Smart et al., 1985a,b). In situations that favor high vigor and dense canopies, hedging or removal of leaves and/or lateral shoots in the fruit zone soon after fruit set may open up the canopy and reduce K^+ accumulation and the associated pH increase in the berries (Coniberti et al., 2012). Although increases in K^+ and pH do not occur in shaded fruit as long as the leaves are exposed to sunlight (Crippen and Morrison, 1986), severe cluster shading generally tends to delay fruit ripening, including sugar accumulation. Berry growth can also be slower in shaded clusters than in sun-exposed clusters, especially if shade occurs early during berry development (Dokoozlian and Kliewer, 1996). Exposing shaded clusters to sunlight at veraison or later does not lead to compensatory berry growth but may increase the incidence of sunburn on the berries. Where fruit exposure is associated with high temperatures, however, sun-exposed berries may remain smaller than shaded berries (Bergqvist et al., 2001; Crippen and Morrison, 1986; Reynolds et al., 1986).

Tartrate seems to be relatively insensitive to low light intensity; if anything, its production decreases in low light, perhaps because the synthesis of its precursor, ascorbate, is a light-sensitive process (DeBolt et al., 2007; Kliewer and Schultz, 1964). In contrast, shaded fruit often has higher amounts of malate, which is partly due to malate production being favored over tartrate production in low light and partly due to malate catabolism being slowed by the lower temperatures that accompany shade (Kliewer and Schultz, 1964; Pereira et al., 2006a; Viala and Vermorel, 1909). Regarding amino acids, although shaded berries have less total nitrogen, they may contain more arginine and less proline than sun-exposed berries (Pereira et al., 2006a). Because light (partly via phytochromes) stimulates CHS and many other enzymes of the pathways involved in the production of phenolics, shading reduces the production of phenolic components. However, different classes of phenolics react differently to specific portions of the light spectrum. Anthocyanins and hydroxycinnamic acids are stimulated primarily by visible light, whereas flavonols accumulate mainly in response to UVB (Berli et al., 2011; Gregan et al., 2012). Anthocyanin accumulation is not only responsive to light intensity but also to day length: Longer days, it seems, are generally associated with more intense anthocyanin production than are shorter days (Jaakola and Hohtola, 2010). This photoperiod effect—or the cumulative effect of more light each day—may be one reason (in addition to the cooler temperatures) that anthocyanins tend to accumulate more rapidly and to higher concentrations at higher than lower latitudes, at least before the autumnal equinox. But although light is necessary for anthocyanin production, direct sun exposure of the fruit is not nearly as important for anthocyanin accumulation as for flavonol production; indeed, flavonols may be useful markers of light exposure (Downey et al., 2004; Haselgrove et al., 2000; Keller and Hrazdina, 1998; Macheix et al., 1990; Pereira et al., 2006a; Price et al., 1995; Spayd et al., 2002). However, high amounts of flavonols may lead to noticeable astringency or even bitterness in grapes and wines (Gawel, 1998).

Because the synthesis of all acids and phenolics is dependent on imported sucrose, episodes of cloudy or overcast conditions slow anthocyanin accumulation and may result in fruit with lower anthocyanin content (Cartechini and Palliotti, 1995; Keller and Hrazdina, 1998; Smart et al., 1988). Clouds may also induce a shift in the relative proportion of individual anthocyanins. In grapes grown under overcast conditions, malvidin-based anthocyanins may come to dominate the anthocyanin profile because their production seems to be less responsive to low solar radiation than that of other anthocyanins (Keller and Hrazdina, 1998). Moreover, acylated anthocyanins may be less sensitive to canopy shade than are their nonacylated counterparts; thus the relative proportion of acylated anthocyanins is higher in heavily shaded grapes than in partially or fully sun-exposed grapes (Ristic et al., 2010). Shading of clusters by leaves only interferes with pigmentation if it is severe (e.g., inside dense canopies) and apparently suppresses the genes that code for anthocyanin biosynthesis (Jeong et al., 2004; Koyama and Goto-Yamamoto, 2008; Ristic et al., 2010). It appears that anthocyanin formation in the skin reaches a plateau at a photon flux of approximately 100 $\mu mol\ m^{-2}\ s^{-1}$ in the fruit zone; above this light intensity (under clear-sky conditions there can be >2000 $\mu mol\ m^{-2}\ s^{-1}$), temperature becomes the dominant factor in berry coloration (Bergqvist et al., 2001; Downey et al., 2006; Spayd et al., 2002; Tarara et al., 2008). Too much radiation (of UVB) can even inhibit anthocyanin production or induce degradation, perhaps due to the formation of hydrogen peroxide (H_2O_2) as a result of oxidative stress. The H_2O_2 in combination with the enzyme peroxidase might degrade anthocyanins (and other phenolics) in the vacuoles. Nonetheless, all else being equal (including latitude), grapes produced at higher altitudes tend to have higher concentrations of total phenolics and anthocyanins than their lower elevation counterparts, although it is not clear if this is due to the increase in UVB with increasing altitude (Berli et al., 2008).

Cultivars differ in their fruit's susceptibility to shade; compared with sunlit fruit, the anthocyanin content is often considerably lower in shaded Pinot noir, Cabernet Sauvignon, or Malbec grapes than in Merlot, whereas coloration in varieties such as Syrah, Petit Verdot, or Nebbiolo appears to be almost unaffected by shade. Taken together, it appears that sun exposure of grapes may enhance the color of (young) red wines mainly by increasing the amount of cofactors (flavonols) and less so that of anthocyanins. Nevertheless, sun exposure may also improve the extractability of anthocyanins during fermentation (Cortell and Kennedy, 2006).

Visible light may enhance tannin accumulation in the skin, but usually not in the seeds, of sun-exposed berries up to veraison and increases the average length of tannin polymers (Cortell and Kennedy, 2006; Downey et al., 2006; Koyama and Goto-Yamamoto, 2008). This light effect seems to be more prominent in cultivars that produce low amounts of tannins than in cultivars with high tannin content (e.g., Nebbiolo). As in the case of anthocyanins, only severe shade may reduce tannin production in the skin (Ristic et al., 2010). Contrary to its influence on anthocyanins, however, sun exposure appears to decrease the extractability of tannins during ripening, which tends to cancel out at least part of the prior gain in tannin content (Downey et al., 2006). The nature of this change in tannin extractability is unknown, but considering that rising temperature and UV radiation enhance methane release from cell wall pectins, possibly via the generation of reactive oxygen species (Keppler et al., 2008; McLeod et al., 2008), it is conceivable that the resulting pectate may bind some of the tannin. Nevertheless, grapes grown on less vigorous vines, which normally have higher fruit exposure, and the wine produced from these grapes tend to have greater amounts of skin tannins and anthocyanins and, therefore, ultimately higher concentrations of stable polymeric wine pigments, than those from more vigorous vines (Cortell et al., 2005, 2007a,b). However, anthocyanin accumulation can be impaired in vines with very low vigor due to insufficient leaf area and excessive fruit exposure.

Because carotenoids act as photoprotectants whose synthesis is regulated by phytochrome (Cazzonelli and Pogson, 2010), their preveraison production is induced by light exposure, whereas episodes of low light (e.g., due to overcast or rainy conditions) are associated with a temporary decrease of xanthophyll carotenoids (Düring and Davtyan, 2002). Given that xanthophyll production apparently does not respond to temperature, one might therefore conclude that grape berries are most vulnerable to sunburn (see Section 6.3.3) if bright, hot days suddenly follow a period of low light conditions or if shaded berries are suddenly exposed to the sun. Sudden exposure occurs, for example, following leaf removal in the cluster zone too late in the growing season in an attempt to improve the microclimate in the fruit zone. Another reason for this vulnerability is that the shaded berries have not produced sufficient amounts of UV-protective flavonols in their skin (Kolb et al., 2003). Moreover, if grapes behave similarly to apples, then only cultivars resistant to sunburn should respond to high light exposure with elevated carotenoid production, whereas carotenoids should decrease in exposed fruit of susceptible cultivars (Merzlyak et al., 2002).

The DOXP/MEP pathway that produces isoprenoid compounds in grape berries does not operate in the dark (Mongélard et al., 2011). Therefore, it is conceivable that long days favor the production of carotenoids and terpenoids. The productivity of the pathway also increases with increasing light intensity (Mongélard et al., 2011). Sun exposure not only enhances the preveraison production of carotenoids but also accelerates their postveraison degradation, and this is associated with increased conversion to aroma-active norisoprenoids (Baumes et al., 2002; Razungles et al., 1998; Schultz, 2000). Although this may result in high norisoprenoid concentrations in exposed fruit, removing leaves to increase light exposure can sometimes—for unknown reasons—diminish the amount of norisoprenoids (Lee et al., 2007). In some cases, grape norisoprenoids were found to be insensitive to canopy shade and light exposure (Ristic et al., 2010). Light might also promote methyl anthranilate formation in sunlit Concord berries because higher UVB radiation favors the release of its precursor, methane, from cell wall pectins, possibly as a result of oxidative stress (McLeod et al., 2008; Messenger et al., 2009). Furthermore, UVB light seems to enhance the production and emission of terpenes (Loreto and Schnitzler, 2010). Although the net effect on aroma precursors in harvested grapes is poorly understood, the concentration of glycosylated monoterpenoids is reduced in shaded fruit (Reynolds and Wardle, 1989b). In contrast, shading below about 50% of full sunlight substantially increases methoxypyrazine accumulation in young berries, and differences induced early during berry development often persist through most or all of the ripening period (Koch et al., 2012; Marais et al., 1999; Ryona et al., 2008). Consequently, growers often remove leaves in the fruit zone to increase sun exposure of the developing clusters and thereby enhance berry monoterpenoids and reduce methoxypyrazines (Gregan et al., 2012; Reynolds and Wardle, 1989a; Reynolds et al., 1996; Scheiner et al., 2010). It is not clear whether this is due to a direct effect of light on aroma synthesis or degradation or an indirect effect of higher berry temperature, or a combination of light and temperature effects. Although postveraison fruit exposure cannot compensate for inadequate exposure before veraison (Koch et al., 2012), even relatively brief periods of low light, such as overcast weather, during ripening may be detrimental to the flavor quality of the fruit. The sensory perception in wine of "fruitiness" is positively correlated and that of "vegetative/asparagus/green pepper" is negatively correlated with the average light intensity during the ripening period (Marais et al., 2001; Ristic et al., 2010). In fact, it is quite common for "fruity" and "vegetal" characters to be inversely correlated in wines (Escudero et al., 2007; Heymann and Noble, 1987; Noble and Ebeler, 2002).

In addition to the direct effects discussed previously, light exposure of grape clusters could also have indirect effects on wine quality. It appears that yeasts are among the fungi that are most susceptible to UVB radiation (Newsham et al., 1997). Although this premise has yet to be tested, it is possible that light exposure alters the yeast microflora on the surface of grapes. Indeed, it is known that the distribution of yeast species and strains, as well as that of other fungi and bacteria, on grape berries can be influenced by environmental effects (Bokulich et al., 2014; Longo et al., 1991; Stefanini et al., 2012; Vezinhet et al., 1992). This could have implications for wine flavor because even in musts inoculated with single-strain yeast cultures, the native yeasts may be important during the initial stages of fermentation before the inoculated strain takes over.

6.3.3 TEMPERATURE

Sunlight heats plant tissues. The dependence of cellular integrity and enzymatic reactions on conditions that are neither too cold nor too hot makes temperature a very important factor influencing grape composition. In general, higher temperatures accelerate plant development and are associated with a more rapid succession of phenological stages, including earlier veraison (see Section 2.2). In cool regions, low temperatures often limit photosynthesis and sugar production in the leaves, although growth and sink activity of the fruit decrease more than photosynthesis under low temperature (Klenert et al., 1978; Körner, 2003; Wardlaw, 1990). Conversely, in hot regions, temperatures often exceed the photosynthetic optimum during a large part of the day; thus, the temperature for a considerable proportion of the growing season can be higher than the optimum range for photosynthesis. In addition, high nighttime temperatures in hot regions increase the proportion of assimilated carbon that is lost through respiration, which reduces the total amount of sugar available to the clusters. Global climate change is expected to be associated with higher temperature increases at night than during the day (IPCC, 2013). Indeed, on a global scale, the average growing season temperature for individual wine regions has been found to correlate positively with average wine vintage ratings, with a clear temperature optimum that varies by region and wine style (Jones et al., 2005).

The long-distance transport of assimilates in the phloem is relatively insensitive to temperature but is inhibited by prolonged periods above 40°C, probably due to temporary blockage of sieve plate pores by callose. Heat summation (usually expressed as growing degree days; see Section 2.2) has been found to be a very good predictor of fruit maturity, especially if a lower temperature threshold of 18°C is applied (rather than the 10°C threshold generally used in grapevine growth models). Temperature, in addition to seasonal rainfall, is thought to be the main factor driving fluctuations in grape quality between years; sugars may accumulate most rapidly in the temperature range 20–30°C, provided soil moisture and other factors are not limiting

(Alleweldt et al., 1984b; Hofäcker et al., 1976; Kliewer, 1973). Nevertheless, the link between temperature and fruit quality is not straightforward, in part because the temperature of individual berries is also important. In fact, photosynthesis is less restricted by low temperatures than are most other plant processes, so lack of sugar supply is probably not the primary reason for the slow ripening during unseasonably cool or hot periods or the slower sugar accumulation in cool compared with warm regions; sink activity and its direct control by temperature seem to be much more important. Although both low and high temperatures do indeed limit photosynthesis (see Section 5.2), this limit might also be due to feedback inhibition from sugars accumulating in the leaves, because berry expansion and sugar accumulation in the berries may stall at temperatures less than 10°C and greater than 40°C (Greer and Weston, 2010b; Matsui et al., 1986; Sepúlveda and Kliewer, 1986; Hall and Keller, unpublished data). Perhaps the ripening berries are lower in the sink hierarchy of vines that experience cold or heat stress. Such stress seems to have a much less marked effect on shoot growth, which implies that assimilates are preferentially supplied to the shoot tips under such conditions (Greer and Weston, 2010b; Sepúlveda et al., 1986).

Berries in different positions in a canopy experience different temperatures. The surface of green berries exposed to full sunlight can be 12°C warmer than the surrounding air, and that of dark-colored berries can be as much as 17°C warmer, whereas shaded berries are usually close to air (ambient) temperature (Bergqvist et al., 2001; Smart and Sinclair, 1976; Spayd et al., 2002; Tarara et al., 2008). Berries facing the sun on a sun-exposed cluster are also heated much more than the nonexposed berries on the same cluster (Kliewer and Lider, 1968). In addition to shade, increasing wind speed also reduces the difference between skin and ambient temperature (i.e., wind has a cooling effect, at least during the day) due to the influence of wind on the boundary layer resistance and thus convective heat loss from the berries (Cola et al., 2009). Moreover, berries on loose clusters are heated less than those on tight or compact clusters because berries that touch one another conduct more heat and lose less to convection (Smart and Sinclair, 1976). A viticultural strategy employed to diminish cluster compactness is the practice of cutting through inflorescences at bloom, which in some cultivars (e.g., Riesling) results in compensatory stretching of the remaining portion of the cluster but also diminishes yield. Another strategy, used mainly for some table grape cultivars (e.g., Thompson Seedless and Flame Seedless), is the application of gibberellin sprays at bloom, which leads to decreased fruit set ("berry thinning") and elongation of the rachis and also increased berry size (Currle et al., 1983; Weaver and Pool, 1971; Williams, 1996). The quality of table grapes may be negatively impacted by excessive temperature due to discoloration and fruit shriveling (Mullins et al., 1992).

Vineyard row direction has a major influence on heating of exposed fruit (Figure 6.8). In east–west-oriented rows, the berries on the south side of the canopy (north side in the Southern Hemisphere) can be at higher than optimal temperature during much of the day (Bergqvist et al., 2001). In north–south-oriented rows, berries on the west side can be considerably warmer than those on the east side because ambient temperatures are generally highest after solar noon (Reynolds et al., 1986; Spayd et al., 2002). Strong solar heating of berries may be advantageous in cool climates but may hinder fruit ripening in warm climates, so the practice of removing leaves in the fruit zone may have to be applied judiciously depending on vineyard location. Whereas fruit sugar content is relatively insensitive to temperature, the berry temperature affects metabolic processes that convert sugars to the various acids and secondary substrates that are important for fruit quality and that degrade (malic) acid.

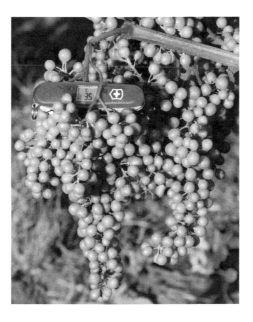

FIGURE 6.8

Sun-exposed grape berries may heat up considerably above the ambient temperature.

Photo by M. Keller.

Both low (<10°C) and high (>40°C) temperatures can limit berry size and delay fruit ripening. In warm, sunny climates, berries on the west side of the canopy in north−south-oriented rows can remain considerably smaller than berries on the east side, likely because the high afternoon temperature of west-exposed berries limits their growth, especially if seasonal water deficit leads to high sun exposure (Shellie, 2011). Very high fruit temperature limits sugar accumulation in sun-exposed berries but may have little effect on sugar concentration in shaded berries, which also seem to have less variation in sugar than their exposed counterparts (Kliewer and Lider, 1968). Conversely, although acidity in mature grape berries is also lower if the grapes were ripened at higher temperature because of the marked impact of temperature on malate respiration (Kliewer, 1973), this effect is much more pronounced in shaded berries than in exposed berries. Nevertheless, due to the pronounced temperature difference between shaded and sun-exposed berries, the latter tend to have much lower amounts of malate, though not of tartrate, and higher pH at maturity (Kliewer and Lider, 1968; Reynolds et al., 1986). The temperature optimum for preveraison malate accumulation in immature berries is 20−25°C, and there is a sharp drop in accumulation above approximately 40°C (Kliewer, 1964; Lakso and Kliewer, 1975, 1978). As a consequence, grapes grown in warm climates or warm seasons may have higher acidity at the beginning of ripening compared with grapes grown in cooler climates or cool seasons, unless excessive sunlight exposure heats the berries above the optimum in the warm climate (Klenert et al., 1978). However, acidity declines rapidly after veraison in warm-climate regions or warm growing seasons, whereas the decrease is much more gradual in cool-climate regions or cool growing seasons. Malate degradation also seems to start earlier in warm conditions (Buttrose et al., 1971; Ruffner, 1982b). Therefore, the concentration of organic acids at harvest is usually higher and the pH lower during cool growing seasons than during warm seasons, mainly because increasing temperature stimulates malate respiration (Currle et al., 1983; Hofäcker et al., 1976; Jackson and Lombard, 1993; Klenert et al., 1978).

The French saying, "*C'est l'août qui fait le goût*" ("It's August that makes the taste"), along with a neat linguistic detail (French *août* = August, *aoûté* = mature), points to the importance of temperature conditions during veraison and early ripening for high harvest fruit quality. Temperature is one of the main climatic variables affected by the human-induced rise in the amount of atmospheric CO_2. The forward shift of phenological development resulting from current and future climate change (see Section 2.3) may be of particular importance because earlier veraison implies that the critical ripening period shifts toward the hotter part of the growing season. That this has already occurred during the second half of the twentieth century has been documented in France, where the period between budbreak and harvest has become shorter and ripening is occurring under increasingly warm conditions (Duchêne and Schneider, 2005; Jones and Davis, 2000). Although this trend has been correlated with greater fruit sugar and lower acid concentrations, and generally better wine quality, even warmer does not necessarily mean even better. For one thing, if veraison occurs earlier, there is less time for the preveraison production of flavanols that can be assembled into tannins before and during ripening. Moreover, with early and rapid sugar accumulation that is not necessarily paralleled by accumulation of anthocyanin pigments and flavor components, growers—or their winemaking customers—may be tempted to wait for flavors to "catch up" so that the grapes may be harvested with very high sugar concentrations that result in wines with high alcohol contents. This tendency may be exacerbated by an adherence to the adage of "less is more" with respect to a perceived yield—quality dichotomy.

In *V. vinifera* cultivars, anthocyanin production increases up to an optimum berry temperature near 30°C but is thought to be inhibited above 35°C (Kliewer, 1977b; Kliewer and Torres, 1972; Spayd et al., 2002), whereas the optimum seems to be somewhat lower in *V. labrusca* and its hybrids with *V. vinifera* (Yamane et al., 2006). Excessively high temperatures, similar to excessive light, may also induce oxidative stress, which could even lead to anthocyanin degradation in addition to the inhibition of anthocyanin production (Mori et al., 2007). Consequently, if the air temperature is 25°C and berries are at 40°C, red pigmentation will be drastically impaired. This sensitivity to daytime temperature (night temperatures seem to be much less important) at least partially explains the poor skin color often observed in warm climates or in warm growing seasons, especially when the grapes are directly exposed to the sun (Haselgrove et al., 2000; Ortega-Regules et al., 2006a; Shellie, 2011). It also explains why, at the same sugar concentration, red coloration is lower under hot compared with cool conditions (Sadras and Moran, 2012). Cultivars with *V. labrusca* parentage appear to be especially sensitive to heat; pigmentation stalls at high temperatures, which is why Concord production for grape juice is mostly confined to cool-climate areas.

High temperature also leads to shifts in the relative amounts of individual anthocyanins. Because the formation of malvidin-based anthocyanins seems to be far less responsive to temperature than that of other anthocyanins, the former become dominant in grapes grown under hot conditions (Mori et al., 2005; Ortega-Regules et al., 2006a; Tarara et al., 2008). Maximum skin coloration often occurs in exposed fruit at cool ambient temperature, and minimum coloration in shaded fruit at high or very low temperatures, but (daytime) temperature, rather than light, appears to be the main driver of anthocyanin accumulation (Spayd et al., 2002; Tarara et al., 2008). Incidentally, cool autumn temperatures also lead to more intense coloration of the leaves, especially in combination with high light that results in energy overload in the leaves. This effect is particularly intense in grapevines infected with leafroll viruses (Gutha et al., 2010). Unfortunately, the virus interferes with grape ripening and leads to poor fruit color (see Section 8.2). The formation of tannins in grape berries appears to increase with increasing temperature, whereas flavonol production may be relatively unaffected by temperature but instead is more responsive to light (especially UV light). Temperatures below approximately 20°C are thought to stimulate flavonol accumulation in plant tissues, an effect that is apparently enhanced by low nitrogen status (Olsen et al., 2009). Consequently, cool growing seasons, especially those with lower temperatures before veraison, are associated with lower amounts of tannins and higher amounts of flavonols in grape skins than are warm growing seasons (Ferrandino et al., 2012; Pastor del Rio and Kennedy, 2006).

The concentration of amino acids, especially proline and γ-aminobutyric acid, rises with increasing temperature, but arginine is less responsive to temperature (Buttrose et al., 1971; Kliewer, 1973). Thus, warm growing seasons may be associated not only with lower acidity (especially malate) but also with higher amino acid contents (Pereira et al., 2006b). Similarly, the K^+ content also rises with temperature, which in combination with reduced malate has the follow-on effect of higher juice pH. Sun-exposed berries are also subject to sunburn or sunscald (dead, brown patches on the skin) and subsequent shriveling (see Section 7.4). Such damage may be due to overheating (above 42°C) and excess UV and/or visible light leading to oxidative stress (collectively termed photooxidative damage) to the epidermis and hypodermis. The impact of temperature on aroma and flavor compounds is not well understood and probably varies depending on the chemical nature of the various volatile compounds and their precursors. For instance, methoxypyrazine concentrations are highest when the grapes develop under very cool conditions and, thus, in more northerly latitudes and/or higher altitudes, especially in cool growing seasons. Even if grapes from a warm and a cool region (or site) were harvested at the same concentration of soluble solids (i.e., sugar), the "cool" grapes would still typically contain severalfold more methoxypyrazines than their "warm" counterparts (Lacey et al., 1991). The temperature prior to veraison may be more important for methoxypyrazine production (less at higher temperature) than is the effect of temperature on the degradation (more rapid at higher temperature) of these compounds after veraison.

Although the activity of the DOXP/MEP pathway increases strongly with increasing temperature, even above 35°C (Mongélard et al., 2011), the accumulation of terpenoids in grape berries shows a relatively broad temperature optimum from approximately 10°C to 20°C. Nonetheless, fruit monoterpene concentrations may be negatively correlated with the average daily maximum temperature over the ripening period (Marais et al., 2001). It is possible that this is because terpenes are increasingly volatilized as the temperature increases: terpene emission has a $Q_{10} = 2-4$ in the range 20–40°C (Loreto and Schnitzler, 2010). Thus, although the production of terpenes increases as the temperature rises, so does their loss through volatilization. Above about 45°C, however, terpene biosynthesis becomes rapidly inhibited. Norisoprenoids such as β-damascenone or β-ionone appear to be relatively insensitive to temperature, although they may be masked by elevated amounts of methoxypyrazines under cool conditions. The "peppery" rotundone, on the other hand, may accumulate faster and to higher concentrations when grapes ripen under cooler conditions (Caputi et al., 2011).

In hot regions, veraison occurs early and the temperature is often too high during ripening for optimal fruit quality. Strategies of double-pruning are sometimes employed, whereby green shoots that grew after the normal winter pruning are hedged severely and leaves and laterals (and clusters) are removed soon after fruit set to break paradormancy (see Section 2.2) and induce renewed budbreak that will result in a crop that ripens during the cooler fall season (Gu et al., 2012; Iland et al., 2011). Conversely, in climates close to the latitudinal or altitudinal limits of grape production, early fall frosts may occur before the grapes are ready for harvest. Some growers, however, deliberately delay harvest until the grapes are frozen for the production of ice wine. It appears that the freezing process has little effect on subsequent juice or wine composition, except, of course, when grapes are pressed while frozen or are left on the vine for prolonged periods during which they may be invaded by pathogens. The sugar concentration is rather immune to freezing and thawing, while titratable acidity decreases somewhat, and K^+ and pH increase (Spayd et al., 1987). The concomitant rise in K^+ and pH might be mostly a result of K^+ extraction from the skin during thawing.

6.3.4 WATER STATUS

Water supply from both the soil and the atmosphere is very important for fruit composition due to its influence on vine growth, yield formation, and fruit ripening. Most of the differences in grape composition and perceived wine quality caused by soil-related differences in the so-called terroir may be attributable to differences in soil moisture rather than, for example, parent rock or soil type and composition (Huglin and Schneider, 1998; van Leeuwen et al., 2004). Nonetheless, deep, fine-textured soils not only hold more water than do shallow, coarse-textured soils, but they also tend to be more fertile, because clay has a higher capacity for nutrient storage and exchange than does sand. At a given vineyard site, soil moisture depends on the annual amount and temporal distribution of rainfall and on temperature-mediated evaporation. Site-specific and local variation in soil moisture due to differences in effective rootzone, water-holding capacity, and drainage can modulate the response of individual grapevines to climatic factors and has a pronounced impact on vine-to-vine variation within a vineyard. In fact, the variation in vigor and yield among vines in the same vineyard is often related to the variation in plant-available water, both in hot and in relatively cool climates (Cortell et al., 2005; Hall et al., 2002; Lamb et al., 2004). However, the spatial variation in fruit quality is not necessarily the same as the spatial variation in yield. Water availability influences shoot vigor and thus canopy microclimate. Increasing soil moisture stimulates vigor, which can lead to a denser canopy and shaded fruit. A larger total leaf area also has a greater water demand due to transpiration, which in turn increases the vine's susceptibility to drought. By contrast, severe water deficit may cause senescence and subsequent abscission of older leaves, which will increase sun exposure of the fruit (Romero et al., 2010).

Abundant water supply favors shoot and berry growth but delays veraison and slows the rate of fruit ripening (Alleweldt et al., 1984b; Hepner et al., 1985). Although this is generally true for both water supplied by rainfall and that coming from irrigation, it is important to remember that rainfall, unlike irrigation, is also normally associated with both lower temperature and lower light intensity due to cloud cover. Moreover, rainwater adversely affects the canopy microclimate by leading to surface wetness on leaves and fruit, and by increasing the humidity around them. Surface wetness and high humidity are associated with increased disease pressure (see Section 8.2), and uptake of water across the berry skin can lead to fruit cracking (see Section 6.2). Moisture also hinders drying of raisin grapes. Thus, at comparable amounts of (excess) water supply, rainfall should be expected to be more detrimental to fruit composition than is irrigation, unless the irrigation water is supplied via overhead sprinklers.

Large, dense canopies that result from abundant water and nutrient availability are associated with reduced fruit sugar, high acidity, and poor color (Dry and Loveys, 1998; Jackson and Lombard, 1993; Salón et al., 2005). Water deficit, in contrast, typically reduces yield (see Section 7.2) and can increase or decrease berry sugar content, acidity, pH, and color, depending on the extent and timing of the deficit. Some degree of water deficit is regarded as beneficial for fruit composition and wine quality. Grapes, especially wine grapes, have been grown successfully on rather marginal sites with infertile, shallow soils of low water-holding capacity for many centuries, and controlled application of stress is the premise of modern deficit irrigation strategies (Dry and Loveys, 1998; Dry et al., 2001; Keller, 2005; Kriedemann and Goodwin, 2003; see also Section 7.2).

Mild to moderate water deficit generally seems to be most effective if applied during the first phase of rapid berry expansion by limiting shoot and berry growth and canopy density (Dry et al., 2001; Huglin and Schneider, 1998; Iland et al., 2011; Salón et al., 2005). Considerable improvements in fruit composition may be attributable to indirect effects of smaller, more open canopies and, sometimes, reduced crop load, but water deficit may also have more direct effects on berry size and composition. When water deficit is associated with lower canopy density, the improved light penetration into the fruit zone may also raise berry temperature (Santos et al., 2005). Whereas any plant water deficit almost always limits berry size (see Section 7.2), its influence on sugar accumulation is less pronounced (Hepner et al., 1985; Hofäcker et al., 1976; Matthews and Anderson, 1988; Santesteban and Royo, 2006; Stevens et al., 1995; Williams and Matthews, 1990). This apparent difference might arise from a decrease in the ability of berries growing under limited sugar supply to accumulate water so that the sugar concentration, albeit not the amount per berry, remains constant. The same principle may apply to the accumulation of mineral nutrients such as K^+ and Ca^{2+} in the berries. Although the amount of these nutrients may be reduced under water deficit, the concentration in the juice may remain unaffected due to the concomitant decrease in berry size (Esteban et al., 1999; Etchebarne et al., 2009; Freeman, 1983; Keller et al., 2008). Indeed, the juice K^+ concentration may even increase with decreasing plant water status, leading to an increase in pH (Dundon and Smart, 1984; Keller et al., 2012). In addition, the decrease in berry size may be mostly due to a smaller mesocarp, whereas the weight of the skin and seeds seems to be less affected by water deficit (Roby and Matthews, 2004). The resulting increase in the skin:juice and seed:juice ratios increases the relative contribution of skin- and seed-derived compounds during winemaking (Casassa et al., 2013).

The decrease in photosynthesis and sugar export from the leaves under more severe stress can curtail berry sugar accumulation, especially if the deficit occurs during ripening (Currle et al., 1983; Dry et al., 2001; Hardie and Considine, 1976; Matthews and Anderson, 1988; Peyrot des Gachons et al., 2005; Quick et al., 1992; Rogiers et al., 2004b; Santesteban and Royo, 2006). Thus, although mild or moderate water deficit may increase sugar accumulation, limit berry size, and improve fruit composition by restricting shoot growth or by reducing canopy density (Kennedy et al., 2002; Ojeda et al., 2002; Shellie, 2014; van Leeuwen et al., 2004), severe stress, contrary to winemaking lore and European regulations, can delay berry development and ripening because of a reduction in photosynthesis or even leaf abscission (Romero et al., 2010, 2013; Williams and Matthews, 1990; Williams et al., 1994). Although the size of ripening berries is much less responsive to soil water status than is the size of preveraison berries, postveraison berries also shrink, even if slowly, due to dehydration under severe water deficit (Keller et al., 2006; Ojeda et al., 2001; Salón et al., 2005). Such weight loss leads to a concentration effect on berry solutes, so that there is an apparent gain of about 2°Brix for every 10% loss in berry weight (Muganu et al., 2011).

Small berry size is less important for most white wine grapes than for red wine grapes. Skin and seed components are not usually extracted during white winemaking, and the yellow-green color of the so-called white wines is rarely viewed as a limiting trait that requires improvement in the vineyard. This is why in the grape and wine quality literature the term "color" is typically, though not necessarily correctly, used as an acronym for red color. Constituents of the skin, moreover, are not nearly as critical for the quality of nonwine grapes, whose surface-to-volume ratio is therefore of little concern. On the contrary, large rather than small berries are typically desired for juice, table, and raisin grapes. Consequently, insufficient water supply, especially before veraison, to vines producing raisin grapes can lead to undesirable loss of berry size and sugar content (Christensen, 1975).

Soil moisture has little effect on tartrate content per berry, although water deficit early during berry development has been found to limit tartrate accumulation (Eibach and Alleweldt, 1985; Esteban et al., 1999). Preveraison malate accumulation also tends to decline with a decrease in soil moisture so that the concentration of titratable acidity is often lower at harvest (Esteban et al., 1999; Keller et al., 2008; Salón et al., 2005; Stevens et al., 1995; van Leeuwen et al., 2009). Because ABA was found to inhibit enzymes involved in malate respiration and gluconeogenesis (Palejwala et al., 1985), it is possible that ABA mediates the response to water deficit in ripening berries. However, although the amount of ABA in grape berries correlates positively with the concentration of ABA in the xylem sap at all stages of berry development (Antolín et al., 2003), the ABA effect on malate seems to be in the "wrong" direction: More ABA would be expected to preserve malate. At least part of the decrease in malate with increasing water deficit may be associated with higher berry temperatures due to a decrease in canopy density (Romero et al., 2010). Nevertheless, the reduction in malate content is more pronounced when water deficit occurs before veraison than after veraison (Matthews and Anderson, 1988; Williams and Matthews, 1990). The concentration of arginine is lower in berries of water-stressed vines, whereas the proline concentration may increase or decrease, perhaps depending on the extent or timing of the stress (Coombe and Monk, 1979; Freeman and Kliewer, 1983; Matthews and Anderson, 1988). Despite these changes in amino acid composition, the amount of PR proteins in grape berries seems to be rather insensitive to water deficit (Pocock et al., 2000). It is thought that the elevated protein concentrations that are sometimes found in juice made from water-stressed grapes is due to the smaller size of those berries rather than an increase in protein accumulation.

The improved red pigmentation often observed with mild to moderate water deficit, especially preveraison water deficit, is to some extent simply due to smaller berry size (see Section 7.2), which increases the skin:pulp ratio (Freeman, 1983; Kennedy et al., 2002; Romero et al., 2010; Salón et al., 2005; Shellie, 2014). In addition, there is also a more direct influence of water deficit enhancing the production of anthocyanins (Dry et al., 2001). This effect may arise as a combination of enhanced fruit exposure due to lower canopy density and of the stimulating effect of root-derived ABA on the genes stipulating flavonoid (especially anthocyanin) biosynthesis and, hence, on the activity of the corresponding enzymes (Castellarin et al., 2007a,b; Jeong et al., 2004; Romero et al., 2010). Exploiting this effect of ABA, table grape growers sometimes apply the hormone as a spray at or just before veraison to boost red coloration of the fruit (Kataoka et al., 1982; Peppi et al., 2008). Even wine grapes, grown in either hot or cool climates that limit anthocyanin production in the skin, may benefit from ABA sprays applied at veraison, which vines may interpret as a perceived water deficit (Balint and Reynolds, 2013, Gu et al., 2011). Application of the synthetic compound ethephon also improves red color formation (Szyjewicz et al., 1984), perhaps because its breakdown product, ethylene, stimulates ABA production. Water deficit may also shift the relative proportion of individual anthocyanins, favoring the production of pigments with higher degrees of hydroxylation and methoxylation, notably malvidin- and petunidin-based anthocyanins (Castellarin et al., 2007a,b). On the other hand, irrigation using partial rootzone drying (see Section 7.2) may leave malvidin-based anthocyanins unchanged while enhancing accumulation of the glucosides of the other four anthocyanins (Bindon et al., 2008).

Severe water stress, especially before veraison, can lead to uneven ripening with some berries remaining green or coloring poorly. The accumulation of tannins, flavonols, and cinnamic acids, on the other hand, appears to be rather insensitive to changes in plant water status, both before and after veraison (Downey et al., 2006; Gény et al., 2003), unless water deficit enhances fruit exposure to sunlight. Carotenoid concentration is often lower in grapes grown at low soil moisture; much of this effect can probably be attributed to higher sun exposure due to the reduced canopy density, which accelerates carotenoid degradation. It appears that this degradation is associated with enhanced conversion of carotenoids to aroma compounds such as norisoprenoids: Water deficit often increases the concentration of β-damascenone in wine (Ou et al., 2010). Mild water deficit is thought to enhance the grapes' aroma potential, for instance, by increasing the amount of volatile thiol precursors, irrespective of the influence of water status on berry size (Peyrot des Gachons et al., 2005). More severe stress, however, seems to curtail this aroma potential. This finding may be particularly relevant for white wine grapes, for which a high aroma potential is much more desirable than a high amount of potentially bitter phenolics. Lower plant water status and lower vigor tend to be associated with lower methoxypyrazine concentrations in the fruit (Scheiner et al., 2012). Indeed, water deficit may accelerate or advance by a few days the postveraison breakdown of methoxypyrazines (Sala et al., 2005). Clearly, wine grape quality benefits from carefully controlled irrigation in dry climates or dry growing seasons.

6.3.5 NUTRIENT STATUS

Of all mineral nutrients, nitrogen is the most potent in its ability to influence vine growth, morphology, fruit production, and tissue composition (Bell and Henschke, 2005; Jackson and Lombard, 1993). Increasing soil nitrogen availability enhances photosynthesis, which means that more sugar is available for growth and fruit ripening. Yet it is commonly accepted that, just as with water, there can be too much of a good thing. High nitrogen supply favors vegetative growth, including growth of lateral shoots (Keller and Koblet, 1995a; Keller et al., 1998, 1999, 2001b), which can result in dense canopies associated with reduced fruit sugar, high acidity, and poor color. Growing shoot tips may also compete with the clusters for supply of assimilates, and fruit ripening is delayed (Keller et al., 2010; Lebon et al., 2004; Pallas et al., 2008; Sartorius, 1973; Spayd et al., 1994). Moreover, high vine nitrogen status may sometimes increase berry size, especially if nitrogen is applied late enough that it does not enhance fruit set (Choné et al., 2006; Keller et al., 2001a).

In addition to this indirect influence on fruit composition, the effects of nitrogen can also be more direct because nitrate uptake leads to a reprogramming of the expression of many genes involved in metabolism (Scheible et al., 2004). Thus, high nitrogen status stimulates the production but not the respiration of organic acids including malate, which substitutes for nitrate to prevent tissue alkalinization due to nitrate reduction (Scheible et al., 2004; Stitt et al., 2002). Although this is sometimes associated with higher malate concentrations in grape berries, juice from berries grown on a high nitrogen "diet" nevertheless tends to have a higher pH, possibly because high rates of nitrate (NO_3^-) uptake by the roots are often associated with high K^+ uptake (Keller et al., 1999, 2001b; Ruhl et al., 1992; Spayd et al., 1994). Since K^+ helps to neutralize the negative electrical charge of both NO_3^- and malate, K^+ uptake, transport in the xylem, and reexport in the phloem along with malate are all correlated with NO_3^- uptake (Peuke, 2010). Moreover, carbon skeletons are necessary to assimilate nitrate into amino acids. Increased formation of organic acids and amino acids in turn diverts carbon away from sugar production and inhibits the formation of phenolic components, such as anthocyanins, tannins, flavonols, and hydroxycinnamic acids. Even more directly, whereas nitrogen deficiency induces enzymes of the shikimate pathway (Weaver and Herrmann, 1997), nitrate supply suppresses the expression of genes involved in phenolics production (Fritz et al., 2006; Olsen et al., 2009; Scheible et al., 2004). In other words, abundant NO_3^- keeps these secondary-metabolism genes "switched off" or, if they were "on" before NO_3^- supply, it switches them off, preventing them from designing and building the necessary enzymes. In contrast, carotenoid accumulation and terpenoid release are increased by high nitrogen status, but it is not known whether this enhances or suppresses isoprenoid aroma production and accumulation. It is, however, clear that high vine nitrogen status strongly stimulates the production of cysteine- and glutathione-bound thiols in grape berries (Choné et al., 2006). Glutathione concentrations themselves are also much higher in high-nitrogen grapes, and this may stabilize both aroma volatiles and phenolics in (white) wine by limiting their oxidation (Choné et al., 2006; Monagas et al., 2006). Moderate nitrogen availability is thought to maximize the flavor and aroma potential of white grapes, whereas both nitrogen deficiency and nitrogen excess ostensibly diminish the aroma potential (Peyrot des Gachons et al., 2005).

The shade problem created by excessive nitrogen supply cannot be overcome by the popular canopy management technique of leaf removal in the cluster zone in an attempt to improve fruit exposure to sunlight because the high nitrogen and low flavonol and carotenoid (e.g., xanthophylls) contents in the berries make them more susceptible to sunburn (Düring and Davtyan, 2002; Kolb et al., 2003). Another common "Band-Aid" action is to tip or hedge (i.e., cut off) the excess shoot tips, which can temporarily improve assimilate supply to the fruit clusters and eliminate shade from overhanging shoots. However, repeated hedging on fertile sites or in areas that permit continued shoot growth during the growing season may only make matters worse because it wastes the vine's resources and eliminates young, photosynthetically active leaves while leaving old, less efficient leaves behind (Coombe, 1959; Keller et al., 1999). Moreover, the generation of multiple new shoot apical meristems by (repeated) stimulation of lateral shoot growth due to the elimination of apical dominance may draw resources away from the clusters and the storage pool in the permanent organs of the vine.

High soil nitrogen also inhibits root growth, which could make vines more vulnerable to water stress in subsequent growing seasons—a consideration that is especially important during establishment of young vines. Nitrogen deficiency, on the other hand, generally reduces fruit set and bud fruitfulness, leading to loss of yield potential. Unless the yield is drastically reduced, however, fruit sugar concentration may be inadequate because there is insufficient nitrogen available for efficient photosynthesis (Keller et al., 1998; Schreiner et al., 2013). Whereas soil nitrogen status apparently has little effect on berry tartrate, malate concentration tends to be lower at low nitrogen availability, but this response may be modified by other environmental variables (Choné et al., 2006; Keller et al., 1998, 1999; Spayd et al., 1994).

A typical reaction of plants to low nitrogen status is the accumulation of phenolic compounds. It appears that the stimulation of phenolics production is a response to low nitrate content rather than low amino acid status. Therefore, pigmentation of red grapes during veraison is maximized at low to moderate nitrogen availability, and it is minimized when high vine nitrogen status, such as due to earlier fertilizer applications, coincides with overcast conditions (Hilbert et al., 2003; Keller and Hrazdina, 1998; Kliewer and Torres, 1972). This nitrogen–light interaction affects not only total grape color but also the distribution of individual anthocyanins. It seems that conditions favoring color accumulation also lead to the most balanced distribution of pigments (Keller and Hrazdina, 1998; King et al., 2014). Because the formation of malvidin-based anthocyanins is less susceptible to stress conditions than that of other anthocyanins, the former can become dominant in grapes grown with high nitrogen, particularly in combination with poor light or excessive heat. Under nonlimiting light conditions, and in contrast to the situation with water deficit, low nitrogen status favors the production of pigments with lesser degrees of hydroxylation and methoxylation (i.e., cyanidin- and peonidin-based anthocyanins). Vine nitrogen status therefore has a direct influence on the production of individual pigments in the grape skin in addition to the indirect effect brought about by its influence on vigor and fruit set. Moreover, low nitrogen status also favors the accumulation of flavonols in preveraison berries (Keller and Hrazdina, 1998). Similarly, the amount of fruit tannins and cinnamic acid derivatives may increase in response to low nitrogen supply (Schreiner et al., 2014).

The optimum supply of nitrogen in a vineyard also depends on the intended use of the grapes. Because of the potential of phenolic compounds to increase the perception of astringency and bitterness, accumulation of phenolics is much less desirable in table, raisin, and juice grapes, and in grapes destined for white wine production than in those destined for red wine production. Higher nitrogen availability limits phenolics accumulation, which may be advantageous for white wine quality (Choné et al., 2006).

Nitrogen supply influences the transport of nitrate and amino acids to the berries and amino acid production and accumulation inside the berries (Bell and Henschke, 2005). It has been estimated that approximately two-thirds of the berries' total nitrogen is derived from the leaves after veraison, and that the crop's demand for nitrogen is mainly a function of the number of berries per vine (Treeby and Wheatley, 2006). Because nitrate transport in the phloem is very limited, virtually all of the leaf-derived nitrogen would have to be in the form of amino acids (see Section 5.3). Considering that fruit nitrogen at harvest correlates positively with leaf nitrogen at veraison (Holzapfel and Treeby, 2007), increasing the nitrogen status of the leaves at veraison may be a possible strategy to enhance the nitrogen content of the berries at harvest.

Arginine accumulation in grapevines is restricted at low soil nitrogen availability but responds strongly to nitrogen supply (King et al., 2014; Kliewer and Cook, 1971; Schreiner et al., 2014). Arginine increasingly accumulates in the berries as nitrogen supply rises, whereas the production of proline and other amino acids may be somewhat less responsive to nitrogen status (Bell and Henschke, 2005; Bell et al., 1979; Kliewer and Torres, 1972; Spayd et al., 1994). Thus, yeast-assimilable nitrogen increases strongly as soil nitrogen status increases; nitrogen supply during ripening may be more effective at raising fruit nitrogen status than is nitrogen supply earlier in the season (Choné et al., 2006; Rodriguez-Lovelle and Gaudillère, 2002; Schreiner et al., 2013). This has implications for winemaking: Because yeast cells cannot metabolize proline, nitrogen-deficient grapes often lead to sluggish or stuck fermentations (Bell and Henschke, 2005; Spayd et al., 1995). Such fermentations can be associated with malate production by yeast and often result in the formation of H_2S from sulfur (S), SO_2, and the S-containing amino acids cysteine and methionine. H_2S is perceived as a "reduced" smell at low concentration but worsens to a smell of cabbage or even rotten eggs at high concentration. Its production increases in proportion to the amount of S present in the grapes as well (Schütz and Kunkee, 1977). In addition, during fermentation of must of either low or very high nitrogen concentration, yeast may also form acetaldehyde and acetic acid (i.e., volatile acidity). Moreover, low plant nitrogen status, especially in combination with water stress, is suspected to be one reason for premature aging of white wines, although increasing nitrogen fertilizer application has sometimes exacerbated the problem (Linsenmeier et al., 2007). This phenomenon is termed atypical aging and is associated with the development of an off-odor reminiscent of "acacia flower," "naphthalene," or "floor polish" at the expense of varietal aroma.

A threshold of approximately 150 mg N L^{-1} of yeast-assimilable nitrogen (which includes a number of other amino acids and ammonia in addition to arginine) seems to be required for grapes to readily complete fermentation (i.e., to <2 g L^{-1} of residual sugar) (Bell and Henschke, 2005). Higher concentrations of amino acids in the berries often result in higher concentrations of fruity and floral esters (e.g., ethyl acetate, ethyl butyrate, and other similar compounds) and thiols and in lower concentrations of higher alcohols (the so-called fusel alcohols, such as propanol, butanol, hexanol, and related compounds) in wine, which typically enhances the wine's sensory properties. On the other hand, high concentrations of nitrogen in grape juice often result in the yeast producing more SO_2 during fermentation. Because SO_2 inhibits bacterial growth, malolactic fermentation by lactic acid bacteria (*O. oeni*) in such wines can be difficult to achieve.

When grapes from vines with high nitrogen status are dried during raisin production, they tend to turn dark brown due to the involvement of arginine in Maillard reactions, which produce pigmented melanoidins from sugars. Grape proteins (e.g., PR proteins) also increase at high plant nitrogen status, which can increase the requirement for bentonite fining in the resulting wines (Spayd et al., 1994).

Indirect effects of high plant nitrogen status include heightened susceptibility to fungal diseases, such as *Botrytis* bunch rot and powdery mildew (both partly related to vigor and canopy density), which may be detrimental to grape and wine quality (Bell and Henschke, 2005; Christensen et al., 1994; Valdés-Gómez et al., 2008; see also Section 8.2). The optimum availability of soil nitrate to supply the vine's demands and satisfy fruit quality requirements depends on cultivar, intended use of the grapes, and climatic conditions. In a warm, dry, sunny climate or growing season, the optimum nitrogen supply may be higher than that under cool, humid, and cloudy conditions. Nitrogen availability also depends on soil moisture because the roots can take up only nitrogen ions dissolved in the soil water (see Section 5.3). Whether the nitrogen that vine roots take up is derived from fertilizer applications or from organic amendments such as manure or compost makes little difference to the vine. High rates of organic amendments have the same deleterious consequences for fruit composition as do high rates of nitrogen fertilizer: low sugars, tannins, and anthocyanins, and high K^+, pH, and vegetal aroma components (Morlat and Symoneaux, 2008).

Abundant K^+ availability in the soil may increase the pH of grape berries due to enhanced K uptake and transport to the berries. This tends to increase the juice pH, and consequently the wine pH (Jackson and Lombard, 1993; Mpelasoka et al., 2003; Walker and Blackmore, 2012). Thus incorporation into the soil of excessive rates of K fertilizer (up to 900 kg K ha^{-1} for 5 years) may sometimes but not always lead to a higher juice pH (Dundon and Smart, 1984; Huglin and Schneider, 1998; Morris et al., 1980, 1983, 1987). One reason for this inconsistency might be that higher K status is often associated with higher berry malate concentrations (Hale, 1977; Hepner and Bravdo, 1985; Ruhl et al., 1992). In any case, the presence of high amounts of K in grape juice or wine may lead to precipitation of K-tartrate during storage (Morris et al., 1980, 1983). In contrast to nitrogen, K supply may have little impact on the amount of amino acids in grape berries (Schreiner et al., 2014).

High rates of K uptake by the roots may decrease Mg uptake (Morris et al., 1983). In contrast, high soil Mg or Ca may decrease K uptake by the roots and sometimes lead to lower juice and wine pH. The effect of high Ca and Mg availability on juice pH appears to be enhanced by a concomitant increase in grape tartrate and malate content, although the reasons for this increase are unclear. Therefore, contrary to intuition, high soil pH, which is associated with high soil Ca status, may be coupled with high rather than low juice and wine titratable acidity (Noble, 1979), which may or may not be associated with a lower pH.

Little is known of the influence of soil nutrients other than N on the accumulation of phenolic components and flavor volatiles in grape berries. The stimulation of anthocyanin production in leaves of plants deficient in K, P, or Mg (see Section 7.3) suggests that low nutrient status in general might also enhance the production of phenolics in the fruit. Nevertheless, anthocyanins, tannins, and cinnamic acid derivatives in grape berries seem to be rather insensitive to K or P supply (Schreiner et al., 2014). High rather than low amounts of Mg (or Mn) apparently protect anthocyanins from degradation in cell vacuoles, perhaps by stabilizing the pigment molecules via formation of blue anthocyanin–Mg complexes (Sinilal et al., 2011). High vine sulfur status (e.g., as a result of foliar application for powdery mildew control) may increase the concentration of volatile thiols in wine (Lacroux et al., 2008).

6.3.6 CROP LOAD

The classical view of the relationship between grape yield and quality is that of a linear decrease in quality with increasing yield per vine (Currle et al., 1983). However, this is an oversimplification, and there are many instances in which the quantity and the quality of the crop are not related (Bowen et al., 2005; Bravdo et al., 1985a,b; Hofäcker et al., 1976; Keller et al., 2005, 2008) or vary in the same instead of opposite directions (Chapman et al., 2004b). In a study conducted over 4 years with 80 individual Riesling vines, the between-vine variation in yield was approximately fivefold higher than that in either juice soluble solids or titratable acidity (Geisler and Staab, 1958). A meta-analysis of Riesling clonal trials conducted at 16 locations over 37 years found that yield as well as soluble solids increased over time, whereas titratable acidity decreased (Laidig et al., 2009). Much of the variation in yield was attributed to site effects, whereas the change in fruit composition was clearly linked to the rise in average temperature during the same period. Although yields and mean temperatures continued to vary considerably from year to year, the variation in fruit composition declined over time, indicating that composition was not greatly affected by yield and that improved vineyard management contributed to the change in fruit composition. Similarly, seasonal weather, vineyard site, or other factors were found to be much more important than yield in determining fruit composition and wine quality of Chardonnay, Cabernet Sauvignon, or Merlot (Bowen et al., 2005, 2011; Ough and Nagaoka, 1984).

It is not so much the crop size or yield *per se* that is important but, rather, the crop load, which is a reflection of a vine's sink:source ratio (Bravdo et al., 1984, 1985a; Jackson and Lombard, 1993; Santesteban and Royo, 2006; Winkler, 1958; see also Section 6.1). For instance, an increase in planting distance is typically associated with a higher yield per vine, but vine size, and thus the leaf area per vine, also increases, so fruit composition and wine quality may be completely unaffected (Winkler, 1969). Conversely, when a higher planting density, which is associated with lower yields per vine, leads to competition among shoots for sunlight, the fruit may have higher titratable acidity while at the same time the juice pH is also higher (Falcetti and Scienza, 1989), possibly due to recirculation of potassium from shaded, aging leaves to the clusters. Vines planted at higher density also tend to produce fruit that is higher in methoxypyrazine content (Sala et al., 2005). Furthermore, water availability may be far more important than crop load in determining fruit and wine quality. Excess water supply can compromise quality even in vines whose yield is severely reduced by cluster thinning after bloom, whereas quality may be high in high-yielding vines that experience seasonal water deficit (Bravdo et al., 1985b).

As a general rule, a leaf area of $1-1.5$ m^2 is required to fully ripen 1 kg of fruit, and this normally results in a yield:pruning weight ratio of approximately $5-10$ (Kliewer and Dokoozlian, 2005; Kliewer and Weaver, 1971). If the crop load is lower than this (i.e., undercropping or sink limitation), then the vine will invest comparatively more resources in vegetative growth, which can, in extreme cases, delay ripening and reduce fruit quality via the follow-on effects of a dense canopy (Bravdo et al., 1985b). Moreover, berry size may increase to compensate for the low number of berries relative to leaf area (Keller et al., 2008; Kliewer et al., 1983; Santesteban and Royo, 2006). This increase may be a result of elevated rates of sugar import due to a low sink:source ratio. There may also be a strong increase in the import of amino acids from the leaves via the phloem as a result of sink limitation (Do et al., 2010). This leads to a marked increase in the concentration of arginine and, especially, proline in grapes of lightly cropped vines (Kliewer and Ough, 1970; Kliewer and Weaver, 1971). Conversely, if there is insufficient leaf area to ripen the fruit (i.e., high crop load, overcropping, or source limitation), then the rate of ripening will decline. This is why vines of medium vigor often produce both higher yields and better-quality fruit than vines at either end of the vigor spectrum. Such vines are said to be balanced—that is, their crop size matches their vegetative growth and leaf area development. Furthermore, the leaf area required to ripen 1 kg of fruit also depends on the trellis and training system; it appears to be higher in a typical vertically shoot-positioned canopy (>0.8 m^2 kg^{-1}) than in vines that are sprawl trained or trained to divided canopies (<0.8 m^2 kg^{-1}; Kliewer and Dokoozlian, 2005).

Overcropping ordinarily delays fruit maturation and therefore decreases grape sugar, pH, amino acids, and color, and often increases titratable acidity if harvest cannot be delayed (Bravdo et al., 1984; Huglin and Schneider, 1998; Intrieri et al., 2001; Kliewer and Ough, 1970; Miller and Howell, 1998; Weaver and Pool, 1968; Weaver et al., 1957; Williams, 1996; Winkler, 1954, 1958). Such differences may arise because berries on heavily cropped vines begin ripening later (i.e., veraison is delayed) or because they ripen more slowly (i.e., the rate of ripening is low) compared with lightly cropped vines (Nuzzo and Matthews, 2006). Delayed ripening can sometimes be beneficial, especially for white wine grapes in warm climates or warm growing seasons, which may benefit from enhanced acid retention, lower pH, and improved production of aroma precursors in the cooler autumn weather. However, a late harvest also increases the risk of rainfall, the amount and frequency of which typically increase in autumn. It should be noted, moreover, that if equal numbers of berry or cluster samples are collected from each of a set number of vines in a nonuniform vineyard to assess fruit maturity, then the sample will overestimate the true maturity status of the vineyard (Rankine et al., 1962). This is because the fruit of lightly cropped vines tends to be more mature at a specific time point than the fruit of heavily cropped vines. To avoid this bias, the sampling strategy should take into account the differences in crop loads among vines (i.e., sampling proportionally more berries from high-yielding than from low-yielding vines).

The effect of crop load on berry composition depends on how a difference in crop load is achieved. An outbreak of pests or diseases, or a hail storm, can reduce photosynthetically active leaf area after the yield potential has been established, which results in reduced sugar accumulation. If an increase in yield is accompanied by deterioration in the canopy microclimate, then the changes in fruit and wine composition are negative (Reynolds et al., 1994b). In other words, the so-called overcropping effects are often actually shade effects caused by poor pruning practices or other vineyard management errors. For example, if pruning is too light (i.e., too many buds left), then there may be too many shoots, which leads to dense canopies. If pruning is too severe (i.e., too few buds left), however, then the few remaining shoots may grow too vigorously and produce too many laterals, which leads to shade in the fruiting zone. In contrast, where an increase in yield is accompanied by an improvement in the canopy microclimate or by a decrease in berry size, fruit composition may be improved as well. For example, the concentration of undesirable methoxypyrazines in grapes grown on high-yielding minimally pruned vines can be dramatically reduced compared with lower yielding spur-pruned vines. Indeed, when yields are increased by decreasing the pruning level (i.e., leaving more buds), the resulting wines may have more intense (red) color and are often fruitier and less vegetative, perhaps due to the reduced shoot vigor and smaller berry size (Chapman et al., 2004a,b; Freeman, 1983; see also Section 6.1). If, on the other hand, yields are stimulated by abundant water supply, the result is usually the opposite. Similar to heavy pruning, excessive early cluster thinning may also increase both berry size and methoxypyrazine concentration (Dunlevy et al., 2013). Carotenoids are another class of compounds whose production may be higher at higher rather than lower crop loads (Fanciullino et al., 2014). It is unknown, however, whether this results in differences in norisoprenoid accumulation during fruit ripening.

Balancing shoot growth and fruit production is an important viticultural goal. For an individual vine, the crop load/fruit quality relationship generally follows an optimum curve with increasing quality as crop load is increased from a very low level, followed by a plateau, and finally a reduction in quality when crop load is further increased. Under changing external conditions (including cultural practices), this curve can be shifted upward or downward. Rather than setting a specific, inflexible target yield, economically minded vineyard managers aim to achieve the highest possible yield without compromising quality. Moreover, where weather conditions permit, delayed ripening can sometimes be compensated by delayed harvest (Keller et al., 2005; Nuzzo and Matthews, 2006).

On overcropped vines, cluster thinning is typically employed to reduce the crop load and enhance ripening. This can be especially beneficial in cultivars that are prone to overcropping due to their large clusters, such as Syrah, Mourvèdre, Grenache, or Zinfandel. In addition to such wine grape cultivars, heavily cropped table grapes may also benefit from cluster thinning, especially if done soon after fruit set (Dokoozlian and Hirschfelt, 1995). One form of cluster thinning is cutting away the distal portion (one-third or half) of flower clusters during or shortly after bloom, which not only decreases yield but also often leads to less compact clusters due to the compensatory stretching of the rachis (Winkler, 1958). It is possible that the advanced maturity observed following such early thinning may be attributed partly to the tendency for berries in the proximal portion of a cluster to ripen more rapidly than those in the distal portion (Weaver and Ibrahim, 1968). In table grapes, where large berries are desirable, cluster thinning (or even flower cluster thinning) is sometimes supplemented by berry thinning. A different, more indirect approach, developed for wine grapes, exploits the relationship between leaf area and fruit set as discussed in Section 6.1, while avoiding the compensatory berry growth that often results from early cluster thinning. Leaf removal around the clusters before or during bloom may lead to lower fruit set, less compact clusters, smaller berries, and, as a consequence, more rapid accumulation of sugars and anthocyanins and higher titratable acidity in the berries (Gatti et al., 2012; Palliotti et al., 2011; Poni et al., 2006, 2009; Tardaguila et al., 2010).

The size of individual berries is often more important than the crop level or even the crop load in determining fruit composition. Berry size and development are influenced by the number of seeds per berry. While higher seed numbers usually result in larger berries, they also tend to slow down berry development and delay the beginning of ripening (Staudt et al., 1986). Large increases in berry size after veraison may be coupled with concomitant increases in berry sugar and K due to phloem import (Rogiers et al., 2006). The concentration of K in the berries decreases as the crop load increases (Hepner and Bravdo, 1985). However, because tartrate synthesis inside the berry ceases at veraison, an increase in berry size can lead to a substantial decrease in tartrate due to a "dilution effect." Of course, more K and less tartrate in the berries will result in a corresponding increase in juice and wine pH. In addition, larger berries have a relatively smaller skin:pulp ratio, which has implications for red wine composition and quality due to the importance of the extraction of skin-derived compounds (mainly anthocyanins, tannins, and flavonols) during fermentation. In contrast to grapes used for (red) winemaking, large size and crispness are important quality traits of table grapes.

6.3.7 ROOTSTOCK

Grafting does not directly affect the quality-relevant traits of the grapes produced by the scion cultivar, because the chemicals responsible for these traits are produced inside the berry and are therefore determined by the genotype of the scion (Gholami et al., 1995; Koch et al., 2010). This is true even though a rootstock and a scion may be able to exchange DNA pieces at the graft union (Stegemann and Bock, 2009). Because long-distance transfer of such genetic material beyond the immediate contact zone has never been observed, the only way in which such hybrid cells could become part of a new shoot apical meristem, and thus heritable, is through the emergence of shoots from the graft union itself (Bock, 2010; Stegemann and Bock, 2009). However, because the rootstock to which a scion cultivar is grafted may alter water and nutrient uptake and distribution, plant growth, and yield formation, it seems logical to expect the rootstock to influence fruit composition too. Thus an indirect effect on fruit composition, especially on acidity, is possible due to the potential influence of the rootstock on scion vigor, canopy configuration, and yield components (Keller et al., 2001a; Ruhl et al., 1988; Schumann, 1974). Rootstock effects on yield and its components, however, are often minor and variable from year to year (Di Filippo and Vila, 2011; Nuzzo and Matthews, 2006; Stevens et al., 2008, 2010; Williams, 2010). Moreover, despite considerable variation in yield, the effects of climate variation, vineyard location, and soil properties on fruit composition still trump the variation introduced by rootstocks (Harbertson and Keller, 2012; Keller et al., 2012; Kidman et al., 2014; Schumann, 1974). For example, in a soil with very high nematode population, Chardonnay vines grafted to 15 different rootstocks were found to produce up to 7 times more fruit than own-rooted vines, with considerable variation from year to year, and yet there was no difference in fruit soluble solids at harvest (McCarthy and Cirami, 1990).

Rootstocks may vary in their potential to take up mineral nutrients from the soil solution (Keller et al., 2001b). Because of the association between juice K^+ concentration and pH, it is possible that rootstocks with a lower capacity for K^+ uptake could be used to control wine pH. The main difference, however, may be between own-rooted vines and vines grafted to rootstocks in general. The pH and K^+ are often higher in grapes and wines from own-rooted than from grafted vines (Harbertson and Keller, 2012; Keller et al., 2012; Ruhl et al., 1988; Walker et al., 1998, 2000). In contrast, the ranking of rootstocks among different studies, sometimes even for the same scion cultivar in the same environment, has been anything but consistent (Walker and Blackmore, 2012). Nevertheless, the rootstock may alter amino acids, especially arginine, in the berries of its grafting partner. For example, 140 Ruggeri and 101−14 Mgt may sometimes lead to considerably lower amino acid concentrations in the fruit of their grafting partner than some other rootstocks and, particularly, own-rooted vines (Treeby et al., 1998).

One of the more important uses of rootstocks to alter fruit composition is in areas affected by salinity (see Section 7.3). Because rootstocks such as Ramsey, 1103 Paulsen, 110 Richter, 140 Ruggeri, or 101−14 Mgt can exclude much of the salt dissolved in the soil solution from root uptake and xylem transport, the scions grafted to them accumulate less Na^+ and Cl^- in the fruit (Walker et al., 2000, 2010).

ENVIRONMENTAL CONSTRAINTS AND STRESS PHYSIOLOGY

CHAPTER 7

CHAPTER OUTLINE

- 7.1 Responses to Abiotic Stress ... 268
- 7.2 Water: Too Much or Too Little ... 272
- 7.3 Nutrients: Deficiency and Excess ... 287
  - 7.3.1 Macronutrients ... 293
    - Nitrogen ... 293
    - Phosphorus ... 298
    - Potassium ... 300
    - Sulfur ... 303
    - Calcium ... 304
    - Magnesium ... 306
  - 7.3.2 Transition Metals and Micronutrients ... 308
    - Iron ... 308
    - Zinc ... 311
    - Copper ... 312
    - Manganese ... 313
    - Molybdenum ... 314
    - Boron ... 314
    - Nickel ... 316
    - Silicon ... 316
  - 7.3.3 Salinity ... 317
- 7.4 Temperature: Too Cold or Too Warm ... 321
  - 7.4.1 Heat Acclimation and Damage ... 322
  - 7.4.2 Chilling Stress ... 326
  - 7.4.3 Cold Acclimation and Freeze Damage ... 329

ABSTRACT

Grapevines are often exposed to suboptimal growing conditions that cause environmental or abiotic stress. Such resource limitation leads to acclimation processes throughout the plant that shift investment to those organs or processes that favor acquisition of the limiting resource. Abiotic stress is often associated with accumulation of damaging reactive oxygen species, which can culminate in organ or plant senescence. Excessive soil water leads to insufficient oxygen being available for proper root function, whereas water deficit limits growth and photosynthesis and, if severe, may result in xylem cavitation and canopy collapse. Deficiency or excess of mineral nutrients, or excessive salt levels, also limit growth and photosynthesis, but other plant responses and symptoms depend on the mineral element in question. Grape growers exploit water and nutrient deficit principles to control shoot growth and enhance fruit quality. High temperatures accelerate grapevine development, but heat stress impairs photosynthesis and damages fruit tissues. Low but nonfreezing temperatures may cause chilling stress and limit growth and fruit production. They may also trigger cold acclimation, enabling vines to survive freezing temperatures. Cold injury to grapevine tissues results mostly from intracellular ice formation and depends on the interaction between tissue temperature and temperature-driven acclimation and deacclimation processes.

Keywords: Cold hardiness; drought stress; heat stress; nutrient deficiency; salinity; temperature; toxicity

7.1 RESPONSES TO ABIOTIC STRESS

Grapevines, like other plants, require just three categories of resources to grow and produce fruit: carbon, water, and mineral nutrients (Bloom et al., 1985). Nonetheless, they are often exposed to suboptimal growing conditions that cause environmental stress. Such stress is said to be abiotic because it arises from nonbiological constraints, such as an overcast or too-bright sky, heat or cold, water surplus or deficit, and nutrient deficiency or toxicity. These conditions limit either the availability of one or several resources to the plant or the plant's ability to put these resources to use. Although there is a general perception that "stressing" grapevines in the vineyard will improve fruit quality, stress adversely affects plant growth, development, and productivity. However, the environment *per se* does not constitute abiotic stress; external conditions are neutral. It is the ability of plant mechanisms to function in a particular set of environmental conditions that determines whether these conditions are "stressful." Depending on the severity and duration of a stress event and on when it starts and ends, stress can trigger acclimation processes in the entire vine. Shoot and root apical and lateral meristems give vines the flexibility to integrate developmental decisions with continuous changes in the environment. Their high degree of developmental and morphogenic plasticity enables grapevines to maximize their photosynthetic efficiency, long-term survival, and reproductive potential. Such plasticity is characteristic of plants that evolved in resource-rich environments (Bloom et al., 1985).

The optimum resource allocation hypothesis holds that plants respond to insufficient resource availability by investing biomass in those organs and processes that enhance the acquisition of the resource that most strongly limits growth, often at the expense of investment in plant parts that have a large requirement of that resource (Bloom et al., 1985; Poorter and Nagel, 2000). In general, plants exposed to carbon limitation (e.g., due to cloudy weather) often increase partitioning to the shoots (especially the leaves), whereas plants exposed to nutrient deficiency typically increase root growth or at least limit shoot growth more than root growth (Chapin, 1991; Keller and Koblet, 1995a). Consequently, if a nutrient deficiency is relieved by fertilizer application, the vine may respond by shifting its carbon investment to favor shoot growth (to expand the leaf area in order to enhance the now more limiting carbon acquisition) at the expense of the roots. The same response would be expected when irrigation water is applied to relieve drought stress (provided nutrients are not limiting). Moreover, the number and size of sinks competing for carbon during stress periods, and their developmental stage or relative priority, are important because they determine which sink is preferentially supplied with the remaining resources and which ones are abandoned in order to guarantee the survival of the plant (Geiger and Servaites, 1991). Thus, many plant reactions to stress involve morphogenic responses that are characterized less by an overall cessation of growth and more by a redirection of growth that entails inhibition of cell expansion, local stimulation of cell division, and changes in cell differentiation (Potters et al., 2007, 2009). Plant hormones are important mediators of these responses, and the various organs of a grapevine respond differently to different hormones.

The ratio of auxin to cytokinin, rather than their absolute concentrations, may determine which organs are to be favored during a particular stress, and tissue auxin and cytokinin concentrations are usually inversely correlated. For example, a low auxin:cytokinin ratio stimulates shoot growth over root growth, whereas a high auxin:cytokinin ratio favors (lateral) root growth and results in shorter shoot internodes and smaller leaves—but probably stronger apical dominance. In addition, auxin also appears to suppress cytokinin biosynthesis and to promote gibberellin and ethylene biosynthesis, but ethylene interferes with auxin transport toward the roots and instead promotes transport toward the shoots (Negi et al., 2008; Woodward and Bartel, 2005). Whereas auxin produced by the expanding leaves near the shoot tip and transported basipetally from cell to cell and in the phloem inhibits lateral shoot growth, a local increase in shoot-derived auxin in the roots prompts pericycle cells to resume cell division and produce lateral and adventitious roots while at the same time inhibiting root elongation. Conversely, cytokinins produced in the root tips and transported acropetally in the xylem inhibit lateral root initiation but activate dormant lateral buds and stimulate lateral shoot growth, thus counteracting apical dominance (Ha et al., 2012). Cytokinins apparently can also be produced in very young, expanding leaves, probably by the dividing cells (Nordström et al., 2004), and in general increase sink strength by promoting cell division and differentiation (except in roots, where they inhibit cell division) and delay senescence. The ability of auxin to promote root formation and that of cytokinin to induce shoot formation are used to regenerate plants from undifferentiated callus in tissue culture. In addition, the stress hormone abscisic acid (ABA) normally inhibits growth by inhibiting production of the cell elongation hormone gibberellin and by stiffening the cell walls, although at low concentration ABA may instead activate growth (del Pozo et al., 2005; Hartweck, 2008). An organ's sensitivity to ABA may be modulated by cytokinins: low cytokinin contents increase ABA sensitivity and high cytokinin contents reduce it (Ha et al., 2012). As leaves expand and mature, their sensitivity to ABA increases markedly (Chater et al., 2014); perhaps this is because their cytokinin content decreases.

Various abiotic stresses (e.g., drought, nutrient surplus, salinity, and chilling) cause cell dehydration and hence osmotic stress. One universal mechanism by which plants cope with this challenge is the accumulation of so-called compatible solutes inside their cells (Bohnert et al., 1995; Bray, 1997; Zhu, 2002). Compatible solutes are relatively small (i.e., low molecular weight), hydrophilic (i.e., highly water-soluble), stable organic compounds that cannot be easily metabolized and do not disrupt cell functions. These compounds include sugars (mainly sucrose and fructose), sugar alcohols (e.g., mannitol or glycerol), and amino acids (e.g., proline), and they probably serve to reduce the cells' osmotic potential (Ψ_π). This osmotic adjustment supports continued water uptake or prevents excessive water loss and helps plant tissues to maintain a higher water potential (Ψ).

Compatible solutes also act as osmoprotectants, protecting enzymes and membranes from osmotic stress (and thereby stabilizing them), and also seem to play a role in protecting the tissues against oxidative stress.

When grapevines are exposed to drought, cold or heat, intense light, nutrient deficiency, salinity, and other environmental stresses, they generally suffer from oxygen toxicity, termed oxidative stress (Apel and Hirt, 2004). Oxidative stress normally is a secondary stress that develops as a consequence of the effects of primary stresses, namely osmotic and ionic stresses. Such oxidative stress can result from an inability to use photosynthetic energy because light capture proceeds while carbon fixation declines so that stresses that curtail photosynthesis typically increase the leaves' susceptibility to bright light. The excess energy then interacts with oxygen to form so-called reactive (or active) oxygen species. Whereas molecular (atmospheric) oxygen (O_2) is rather inactive (if this were not so, we would all burst into flames), most of the aptly named reactive oxygen species are highly reactive. They are formed as intermediates and by-products of the reduction of O_2 to water (H_2O), in which electrons (e^-) are added in a stepwise manner. This can be written in simplified form as follows [modified from Lane (2002) and Apel and Hirt (2004)]:

$$O_2 + e^- \rightarrow O_2^{\cdot -} + e^- + 2H^+ \rightarrow H_2O_2 + e^- \rightarrow {}^{\cdot}OH + e^- + H^+ \rightarrow H_2O$$

The reactive agents of decay include free radicals (e.g., superoxide, $O_2^{\cdot -}$; hydroxyl radical, ${}^{\cdot}OH$; and singlet oxygen, ${}^1O_2$) and somewhat less unstable nonradicals (e.g., hydrogen peroxide, H_2O_2). The production in the photosynthetic chloroplasts of O_2 and its associated dangers are discussed in Section 4.1. In the mitochondria, the reduction of O_2 to H_2O is carried out by the enzyme cytochrome oxidase as part of ATP generation (see Section 4.4) without releasing any reactive oxygen species, but other respiratory reactions are not so "electron-tight." Thus, reactive oxygen species are normal by-products of photosynthesis and, to a small extent, respiration (Halliwell, 2006; Møller et al., 2007; Noctor and Foyer, 1998). Plants also use H_2O_2 and $O_2^{\cdot -}$ as key components of the cascade of signals involved in the adaptation to changing environments (Foyer and Noctor, 2005; Yang and Poovaiah, 2002). However, the formation of reactive oxygen species increases under stress, which can lead to oxygen toxicity (Apel and Hirt, 2004; Mittler, 2002). In contrast to the comparatively "lazy" O_2, these highly reactive molecules crave electrons because their oxygen atom is just one electron short of filling its outer shell. Consequently, they damage and, at excessive concentrations, even kill cells by rapidly oxidizing (i.e., taking away electrons from) the cells' components such as membrane lipids, proteins, cell wall polysaccharides, and DNA (Lane, 2002; Møller et al., 2007). Oxidative damage to DNA is a major cause of mutations, whereas damaged membranes become leaky and damaged proteins (e.g., enzymes and transporters) are inactivated. In leaves, it appears that regardless of the initial source and type of reactive oxygen species, they all ultimately result in overproduction of ${}^1O_2$ by excess light energy, which oxidizes membrane lipids and triggers the processes culminating in cell death (Triantaphylidès et al., 2008). Therefore, failure to quickly dispose of (scavenge) the reactive culprits leads to inhibition of photosynthesis and development of chlorotic and necrotic leaves. Note that the contact herbicide paraquat acts by inducing $O_2^{\cdot -}$ and H_2O_2 production by "stealing" electrons from photosystem I (Halliwell, 2006).

To cope with oxygen toxicity, plants have evolved a suite of antioxidant defense systems designed to prevent oxidative damage by detoxifying active oxygen species (Apel and Hirt, 2004; Halliwell, 2006; Lane, 2002; Mittler, 2002; Noctor and Foyer, 1998). These defense systems employ antioxidants, which are defined as compounds that "consume" or quench active oxygen species without themselves being converted to damaging radicals in the process. Several of these systems use the amino acid cysteine, which is a component of proteins and peptides and is easily and reversibly oxidized (Møller et al., 2007). One system employs the enzymes superoxide dismutase and catalase to convert superoxide into water. Another, the so-called ascorbate–glutathione cycle, turns H_2O_2 into H_2O by allowing it to oxidize the antioxidant ascorbic acid (commonly known as vitamin C) and then recovering (reducing) its oxidized form, partly with the help of the peptide glutathione. Indeed, plants detoxify many electrophilic (i.e., oxidizing) chemicals by attaching glutathione to them. Because some plant species have "learned" to use this mechanism to detoxify triazine herbicides, they have become herbicide-resistant weeds (Powles and Yu, 2010).

Chemically, oxidation typically means loss of an electron from a compound, whereas reduction means gain of an electron so that the compound's charge is reduced; reactions involving reduction and oxidation are termed redox reactions. The so-called redox state of a cell or tissue is often estimated by the proportion of reduced ascorbate and/or glutathione relative to their total amount (Potters et al., 2009). Furthermore, the production of carotenoids and flavonoids (e.g., anthocyanins) in addition to compatible solutes and stress proteins also contributes to the detoxification process by scavenging active oxygen species or preventing them from damaging the cells' structures. Flavonoids may be oxidized and hence degraded by peroxidase, thereby consuming H_2O_2 (Pérez et al., 2002; Ros Barceló et al., 2003). Oxidative stress, therefore, results from an imbalance between production and degradation of reactive oxygen species.

Although many stresses reduce the leaves' photosynthetic rate or the total photosynthetic leaf area, the causes of these decreases and the impact of a stress factor vary with the type of stress. For instance, whereas low nutrient availability decreases photosynthetic rates in the leaves, this decrease is not as great as and starts later than the reduction in growth (Chapin, 1991). This is because nutrients are needed to produce building blocks (e.g., proteins) for the addition of new biomass by the vine's sinks, whereas the photosynthetic apparatus in the source leaves is already established. Therefore, although vines grown under conditions of nutrient deficiency show diminished growth, they experience a relative excess of photosynthate rather than a shortage, and they accumulate sugar and starch in their leaves. Low water availability also decreases photosynthesis, but this does not result in starch accumulation in the leaves because the decrease in growth due to water deficit approximately matches the decrease in photosynthesis (Poorter and Nagel, 2000).

Eventually, a stress may induce a depletion of assimilates available for export. Carbon depletion in turn stimulates processes involved in photosynthesis and reserve mobilization ("famine" response), whereas abundant availability of carbon favors carbon utilization, export, and storage ("feast" response) (Hellmann et al., 2000). When they are starved for sugar, plant cells initially consume their available starch reserves and then sacrifice their cell membrane phospholipids (mainly linoleic and linolenic acids) to recycle fatty acids and other metabolites (Aubert et al., 1996). Under prolonged and severe stress that occurs as a consequence of a deficit or surplus of water or nutrients, or extremes of temperature, senescence of cells and, eventually, whole organs sets in. Senescence is defined as the ordered (organized or programmed) degradation of cell constituents that ends in death and is generally followed by shedding (abscission) of the dead organs by breakdown of the cell walls in the preformed abscission zones. The moment of death of a cell occurs when the vacuolar membrane ruptures (van Doorn and Woltering, 2004). When a grapevine sacrifices an organ or part of its structure, this constitutes an adaptive strategy to survive a stress period.

Several plant hormones interact to bring about and coordinate the (premature) abscission of leaves, flowers, or fruit. ABA stimulates the senescence process, and ethylene operates as an accelerator of abscission, whereas auxin acts as a brake (Roberts et al., 2002). Therefore, ABA and ethylene can trigger senescence and abscission only in organs whose auxin concentration is low. For instance, although a variety of soil-related stresses result in ABA production in the roots and transport to the shoots, the developing leaves and flowers (but not fully developed leaves and flowers), as well as the grape berries during the period of seed development (but not berries with mature seeds), are usually protected from shedding by their high auxin levels. However, the original trigger (i.e., initiating signal) and partial executioner of the senescence program again appears to be oxidative stress, namely in the form of $^1O_2$ (Triantaphylidès et al., 2008).

Different tissues within the same organ do not die at the same time. In senescing leaves, the mesophyll cells are the first to be disassembled, followed by the epidermis and, finally, the vascular bundles. The earliest and most dramatic change in the cellular structures during senescence is the breakdown of the chloroplasts, which contain the cell's photosynthetic machinery and the majority of the leaf protein and are the site of major biosynthetic processes. This means that photosynthesis declines due to rubisco and chlorophyll degradation. Rubisco thus serves a dual role: it is both the main photosynthetic enzyme and the main storage protein for nitrogen remobilization (Thomas, 2013). Meanwhile, the chloroplasts in the guard cells remain green and functional, probably to keep the stomata closed to prevent desiccation before nutrient recycling is complete (Tallman, 2004). The carbon, nitrogen, and other nutrients stored in the leaf's (or inflorescence's) proteins, nucleic acids, polysaccharides, and lipids are rapidly remobilized during senescence and can be used to sustain the growth and metabolism of important sink organs such as young leaves, clusters, or roots. Of course, the partitioning of resorbed nutrients from leaves that undergo senescence depends on the relative strength or importance of the various sinks of the vine. Moreover, the presence of strong sinks (e.g., high crop load) may increase this nutrient recycling, whereas weak sinks will decrease it. In other words, strong sinks that cannot satisfy their appetite for nutrients (especially nitrogen) by import from the roots will induce remobilization from the reserve pools in older tissues (Thomas, 2013). But overall lack of sink demand can also induce leaf senescence, which can be triggered by a surplus of uncommitted assimilates (Thomas, 2013; see also Figure 5.4).

7.2 WATER: TOO MUCH OR TOO LITTLE

Water availability influences canopy development, vine microclimate, yield, and fruit composition, and lack of water constitutes the overwhelming limitation to plant growth and yield formation (Kramer and Boyer, 1995). Under natural conditions, water is supplied by snow and rainfall, and the portion that is not drained by gravity is temporarily stored in the soil for extraction by plant roots or surface evaporation. Following each recharge, the stored water reserves decrease progressively as water evaporates from the soil surface or is transpired by plants (see Section 3.2); this evapotranspiration (ET) increases as the temperature rises. The total water use of a fully irrigated vineyard over the course of a growing season in a cool climate or short season varies from 300 to 600 mm (3–6 ML ha^{-1}), depending on cultivar, planting density, canopy size and configuration, and seasonal temperature patterns (Evans et al., 1993; Williams, 2014). Vineyard water use in a warm climate with longer growing seasons ranges approximately from 400 to 800 mm, or 4–8 ML ha^{-1} (Williams and Baeza, 2007).

The amount of rainfall varies greatly from region to region and from year to year, which may drastically impair vine performance and the economics of viticulture in some regions and some years. Water availability depends not only on how much rainfall a vineyard receives but also on when the rain falls and how rapidly it evaporates. In addition, soil water-holding capacity and hence the amount of plant-available water varies with soil depth, texture (i.e., the relative proportions of clay, silt, and sand particles), and organic matter content. For example, a loamy soil can hold up to 25% water (0.25 g H_2O per gram soil), a clay soil even 40%, whereas a sandy soil may contain only 12% water at field capacity. The field capacity is also called drained upper limit, whereby the soil pores with a diameter of less than 30 μm retain water against gravity after all the water in the larger pores has drained away and the soil $\Psi_M \approx -0.01$ MPa. The loam's higher water-holding capacity compared with sand is counteracted to some extent by the fact that a larger fraction of the water is held too tightly by the loam to be accessible to the roots. Although roots can extract water more easily from water-saturated sand than from loam, a coarse soil dries much more rapidly so that water extraction becomes much more difficult in the sand as soil water potential (Ψ_{soil}) decreases (Chapman et al., 2012). This is because the hydraulic resistance to water movement in the soil (r_{soil}) is approximately 10 times lower in a water-saturated sand (due to its large pore size) than in a loam but increases by the same mechanism that causes cavitation in the xylem (see Section 3.3): The pores of a drying soil fill with air, which results in failure of capillary forces. Just as larger xylem vessels are more vulnerable to cavitation, so too are larger soil pores (i.e., coarser soil) more susceptible to capillary failure. The coarser the soil type, the more limiting the soil is to plant water use, whereas in a fine-textured soil the vulnerability of the xylem to cavitation is more limiting (Sperry et al., 1998). Grapevines growing in sandy soils therefore often invest more resources in root growth than vines growing in loamy soils to increase the root:shoot ratio in order to compensate for the lower soil water extraction potential. Similarly, a coarse soil favors greater rooting depth than a fine-textured soil because less water is available for extraction at low Ψ_{soil} in the surface layers of a coarse soil. But even for the same soil type there can be large variation in pore size and hence physical properties depending on soil management practices (Chapman et al., 2012). Soil compaction, for example, decreases pore size and increases r_{soil}, which curtails the soil's ability to transport water and nutrients and constrains root growth. Variation in soil moisture due to differences in water-holding capacity and effective rootzone strongly influences vine performance both between and within vineyards (Hall et al., 2002; Lamb et al., 2004).

Waterlogging occurs when the soil moisture is well above field capacity. Grapevines can be grown hydroponically, which shows that they do not suffer from an excess of water during waterlogging. However, at high soil moisture, such as during flooding due to heavy rainfall or excessive irrigation, especially on poorly drained soils and soils with high ground water levels, water drives gases out of the soil pores so that there is insufficient oxygen (O_2) for proper root function and growth (Mancuso and Boselli, 2002). This condition is termed hypoxia (i.e., O_2 deficiency) or, in extreme cases, anoxia. This is because gases (including O_2) diffuse four orders of magnitude less rapidly in water than in air. Due to the dependence of respiration on O_2 supply, waterlogged roots can become so O_2 deficient that they cannot respire and switch to anaerobic fermentation of pyruvate to recycle NADH to NAD for continued ATP production via glycolysis, which is far less efficient than respiration and uses up storage carbohydrates while producing lactate and, later, volatile ethanol (Bailey-Serres and Voesenek, 2008; Zabalza et al., 2009). The depletion of storage reserves is aggravated by the susceptibility of phloem transport to low ATP availability, which impairs phloem unloading (van Dongen et al., 2003). Inhibition of respiration is accompanied by an increase in reactive oxygen species and a decline in pH of the root cells' cytosol, and the excess protons in turn block the aquaporins in the cell membranes so that the roots become less water permeable (Tournaire-Roux et al., 2003). Consequently, but somewhat paradoxically, flooding-induced anoxia rapidly increases root hydraulic resistance and decreases water uptake, which is associated with a decline in stomatal conductance and transpiration rate (Aroca et al., 2012; Flore and Lakso, 1989; Keller and Koblet, 1994; Striegler et al., 1993). Thus, similar to water deficit, an excess of water can lead to partial closure of the stomata, which over time diminishes photosynthesis and may damage the photosynthetic apparatus (Else et al., 2009; Stevens and Prior, 1994). In addition, energy-intensive processes such as cell division come to a halt. Cell division also ceases when uncontrolled acidification of the cytosol, due to leakage of acids from the vacuoles and slow escape of CO_2 produced during fermentation, leads to death of the root meristem cells (Roberts et al., 1984).

Vitis species vary in their susceptibility to waterlogging (Mancuso and Boselli, 2002; Mancuso and Marras, 2006); for instance, *Vitis riparia* can tolerate root flooding for several days by rapidly suspending root metabolism, especially nutrient uptake and assimilation, in order to conserve energy (i.e., to reduce ATP demand). On the other hand, susceptible species such as *V. rupestris* are unable to adapt their root metabolism and literally run out of energy. In addition, the inability of susceptible species to suppress nutrient uptake can lead to ion (especially K^+) poisoning, membrane leakage, and death. Excessive soil water, similar to the effect of soil compaction, also seems to increase ethylene production by the roots and prevents the diffusion of this gas away from the roots (Feldman, 1984; Sharp and LeNoble, 2002). Small doses of ethylene produced by the roots from the amino acid methionine normally promote root growth, perhaps by increasing tissue responsiveness to auxin and gibberellin (Kende and Zeevaart, 1997; Potters et al., 2009). Together with auxin, ethylene also induces the formation of lateral roots and root hairs, as long as the ethylene gas can easily diffuse away from the roots. If diffusion is inhibited, however, ethylene interferes with cell division in the root meristem and delays the inactivation by gibberellin of the growth-suppressing DELLA proteins so that root growth slows (Dugardeyn and Van Der Straeten, 2008; Feldman, 1984; Mancuso and Boselli, 2002). If only a portion of the root system experiences waterlogging, then vines may compensate by increasing root growth in the nonaffected portion, which is typically in the surface soil (Kawai et al., 1996; Stevens et al., 1999; Striegler et al., 1993). Anoxia also increases the xylem sap pH, which is a response similar to that in vines experiencing water deficit, and leads to reduced shoot growth. The leaves of waterlogged vines may turn bright red and roll downward as a result of the accumulation of surplus sugar in the leaves (Figure 7.1). Inflorescence differentiation as well as pollen germination and pollen tube growth decrease with decreasing soil O_2 concentration, so flooding can be detrimental to bud fruitfulness and fruit set; the latter may be reduced to zero if the soil O_2 concentration decreases below 5% (Kobayashi et al., 1964; Stevens et al., 1999).

Flooding during the dormant winter season ostensibly has little effect on grapevines (Williams et al., 1994). Moreover, many wild grapes that have evolved alongside or near riverbanks frequently experience flooding during or soon after budbreak, when the melting snow in the mountains upstream leads to an annual peak in river runoff (Mastin, 2008). Beyond periodic spring floods, overly abundant water supply during the growing season is a predicament that mainly concerns

FIGURE 7.1

Merlot leaves in late summer, showing symptoms of drought stress (left) or waterlogging (right), with an unstressed leaf in the center.

Photo by M. Keller.

coastal regions and areas on the windward side of mountain ranges, as well as some subtropical and tropical regions that experience frequent summer rainfall. Most Chinese grape growing areas, for example, suffer from the fact that the rainy season coincides with the growing season. Because rainfall requires a cloud cover, such regions also often experience low light intensities that curtail photosynthesis (see Section 5.2). In Bordeaux, France, soils with a propensity to waterlogging are associated with poor wine quality compared with free-draining soils, and interannual variation in water availability under these conditions may be more important for wine quality than is variation in growing-season temperature (Renouf et al., 2010; van Leeuwen et al., 2009). Furthermore, grapevines grown in pots are quite often at risk from hypoxia. The soil water content after drainage of irrigation water from pots is usually much higher ($\Psi_{soil} \approx -0.001$ to -0.003 MPa near the surface) than that in a vineyard at field capacity and depends on the height of the pot, which is why (frequently irrigated) potted plants are more likely to suffer from waterlogging than from drought (Passioura, 2006).

The majority of the world's grapes, however, are grown in more or less Mediterranean climates, which experience cool, moist winters and warm, dry summers with high evaporative demand from the atmosphere. Thus, natural water supply and carbon supply for growth and fruit production are often inversely related (Bloom et al., 1985). In Mediterranean regions and regions with continental climates on the leeward (rainshadow) side of mountain ranges, winter rains often, but by no means always (Davenport et al., 2008b), saturate the soil profile to field capacity. Such refill of soil water stores can be enhanced by cover crops that reduce runoff and improve infiltration, even though the cover crop may later compete with grapevines for water (Battany and Grismer, 2000; Celette et al., 2008; Klik et al., 1998). The soil then progressively dries down as temperatures rise and water evaporates, and the growing season is often characterized by low soil moisture and high vapor pressure deficit (i.e., warm, dry air). The resulting increase in air spaces in the soil and water retention by successively smaller pores leads to an increase in surface tension, which in turn is responsible for the development of increasingly negative pressure. In arid soils, Ψ_{soil} can decline to -3 MPa, compared with -0.01 to -0.03 MPa in a soil at field capacity. Water in pores less than 0.2 μm wide cannot be extracted by the roots (extraction from these pores would require a suction of ~1.5 MPa to overcome the soil Ψ_M in these pores) and thus constitutes the lower limit (termed permanent wilting point) of plant-available water (Watt et al., 2006).

The water balance of grapevines and their cells is determined by the amount of water lost in evaporation to the atmosphere (transpiration; see Section 3.2) and the amount of water absorbed from the soil. A vine can become water stressed as a result of both decreased Ψ_{soil}, which generally occurs progressively over time, and fluctuating transpiration rate (E), which occurs with daily and seasonal changes (due to radiation and humidity) in vapor pressure deficit of the air (Williams et al., 1994). The relationship between plant (leaf) water potential (Ψ_{leaf}) and these variables can be expressed as follows:

$$\Psi_{leaf} = \Psi_{soil} - Er_h$$

where r_h is the total resistance to water flow in the vine (see Section 3.3).

When transpiration exceeds water absorption by a vine's root system, cell turgor (P), relative water content (RWC; percentage of tissue water relative to the water content of the same tissue at full turgor) and cell volume decrease, whereas the concentration of solutes in the cell increases, reducing the cell's osmotic potential (Ψ_π) and Ψ. Consequently, the response of the cells to water deficit involves a response to osmotic stress. Mild plant water stress has been defined as a decrease of Ψ_{leaf} by several bars or of RWC by 8–10% below corresponding values in well-watered plants under mild evaporative demand (Hsiao, 1973). Moderate water stress corresponds to a decrease of Ψ_{leaf} by more than several bars but less than 1.2–1.5 MPa or of RWC by 10–20%, and severe water stress corresponds to a lowering of Ψ_{leaf} by more than 1.5 MPa or of RWC by more than 20%. Initially, the decline in P and RWC reduces shoot growth and stomatal conductance (g_s) for water vapor (Hsiao, 1973). Hence, the severity of water stress has also been classified according to the decrease in g_s (Lovisolo et al., 2010): g_s of well-watered grapevines is usually in the range 200–500 mmol H_2O m^{-2} s^{-1}; under mild water stress, g_s declines to ~150 mmol H_2O m^{-2} s^{-1}, under moderate water stress to 50–150 mmol H_2O m^{-2} s^{-1}, and under severe stress to <50 mmol H_2O m^{-2} s^{-1}. Under nonstress conditions, g_s <50 mmol H_2O m^{-2} s^{-1} is typically observed only at night in grapevine leaves (Rogiers et al., 2009; Rogiers and Clarke, 2013). Shoot elongation and leaf expansion in grapevines decline linearly with decreasing Ψ and stop completely at $\Psi = -1.0$ to -1.2 MPa, even though leaf expansion is relatively independent of P (Schultz and Matthews, 1988b; Shackel et al., 1987; Williams et al., 2010).

The primary function of stomata is to avoid damaging water deficits that would cause xylem cavitation (Brodribb and Holbrook, 2003; Jones, 1998; see also Section 3.3). Nonetheless, cavitation is thought to be quite common in grapevines and might be important in inhibiting shoot growth at moderate water deficits (Lovisolo et al., 2008; Lovisolo and Schubert, 1998; Schultz and Matthews, 1988a). Many of the resulting embolisms probably occur in the leaves and petioles rather than in the shoots or trunks. The proportion of cavitated vessels in a shoot apparently does not exceed about one-third until the xylem water potential (Ψ_x) declines below -2 MPa, whereas in the petioles this proportion may surpass 50% at $\Psi_x \approx -1$ MPa and approaches 100% at $\Psi_x = -2$ MPa (Choat et al., 2010; Zufferey et al., 2011). Thus, *V. vinifera* is rather drought-tolerant and can be grown without irrigation in areas that receive as little as 300 mm annual precipitation, although growth and crop yield are very limited under such conditions (Huglin and Schneider, 1998). However, due to their vulnerability to cavitation, grapevines are usually regarded as isohydric species whose sensitive stomata rapidly decrease g_s and decrease transpiration in response to low Ψ_{soil}, which enables them to maintain relatively constant Ψ_{leaf} throughout the day regardless of Ψ_{soil} (Düring, 1987; Galmés et al., 2007). This effect can override the influence of high light intensity on stomatal opening so that soil water deficit in a vineyard often leads to a midday decrease in g_s (Correia et al., 1990, 1995; Loveys and Düring, 1984; Lovisolo et al., 2010). Nonetheless, pronounced diurnal changes in Ψ_{leaf} have also been reported; with dry soil, the difference between predawn and midday Ψ_{leaf} may sometimes exceed 1 MPa (Williams and Matthews, 1990). Because grapevine species and cultivars differ in their vulnerability to cavitation, they also vary in their stomatal sensitivity to drought (Chaves et al., 2010; Currle et al., 1983; Escalona et al., 1999; Liu et al., 1978; Lovisolo et al., 2010; Schultz, 2003; Soar et al., 2006b; Winkel and Rambal, 1993). So-called isohydric species or cultivars have more sensitive stomata than anisohydric species or cultivars. The latter maintain higher g_s and higher transpiration rates, and they markedly decrease Ψ_{leaf} during the day and in response to soil water deficit or atmospheric moisture stress (high vapor pressure deficit). These differences among cultivars are also reflected in the radial hydraulic resistance (r_h) and activity of aquaporins in the roots, which are regulated to adjust water flow to transpiration rate (Perrone et al., 2012; Tyerman et al., 2009; Vandeleur et al., 2009). Note, however, that the distinction is mostly a matter of degree and not nearly as clear-cut as the terminology implies. Because Ψ_{leaf} never remains absolutely constant, it may be more appropriate to "divide" grapevines into near-isohydric or pseudo-isohydric and near-anisohydric or pseudo-anisohydric species and cultivars. The former include *V. berlandieri*, *V. rupestris*, and some *V. vinifera* cultivars, whereas the latter include *V. californica* and other *V. vinifera* cultivars (Table 7.1). Some cultivars, such as Cabernet Sauvignon, Pinot noir, Tempranillo, and Concord have been variously described as isohydric or

Table 7.1 Some Near-Isohydric and Near-Anisohydric Grapevine Cultivars

Near-Isohydric	Near-Anisohydric
Grenache	Cabernet franc
Tempranillo	Carignan
Trincadeira	Chardonnay
	Merlot
	Muscat
	Nebbiolo
	Riesling
	Sangiovese
	Sémillon
	Syrah
	Thompson Seedless
	Touriga Nacional

From Chaves et al. (2010) and Lovisolo et al. (2010).

anisohydric, depending on environmental conditions and/or developmental status (Lovisolo et al., 2010). Only a few species and cultivars have been studied to date, and it is likely that the boundaries will become increasingly blurred as more information becomes available.

Isohydric species may have evolved in wet habitats; they are more susceptible to xylem cavitation and can be regarded as "pessimists" that modify their behavior to conserve resources. Although this conservative strategy keeps water demand of the transpiring leaves well below levels that would challenge the xylem's supply capacity, it is associated with a decline in photosynthesis and hence a decrease in CO_2 uptake (McDowell, 2011; Palliotti et al., 2014; Sperry, 2004; Tarara et al., 2011). In other words, the stomatal restriction of water loss during drought comes at the cost of a loss in carbon gain. Anisohydric plants, on the other hand, probably originate mostly from more arid regions, are less susceptible to xylem cavitation, and can be viewed as "optimists" using available resources in expectation of more arriving. To them, the soil water "glass" is still half full when it is already half empty to their isohydric counterparts. Due to the narrow safety margin from hydraulic failure, where xylem cavitation becomes irreversible, the anisohydric approach is riskier, and such vines may pay by shedding leaves when a drought becomes too severe (Palliotti et al., 2014). However, it probably enables the plant to more efficiently match water supply with demand so that it can maximize CO_2 uptake by pushing the xylem to its carrying capacity, at least as long as the plant can retain its leaves (Sack and Holbrook, 2006; Sperry, 2004). Indeed, the anisohydric Syrah maintains vigorous shoot growth under water-deficit conditions that would completely inhibit shoot growth in the isohydric Cabernet Sauvignon. Moreover, if left to its own devices, rather than being constrained by a trellis system and foliage wires, the droopy growth habit of Syrah shoots also differs from that of Cabernet Sauvignon or Grenache, whose shoots are much more erect. Anisohydric plants also would be expected to have a larger root:shoot ratio than isohydric plants, which may be a drought-avoidance strategy.

In addition to *V. vinifera* cultivars, rootstocks derived from American *Vitis* species also differ in their susceptibility and response to drought and may alter Ψ_{leaf}, g_s, and photosynthesis of the scion leaves via the production and export in the xylem of ABA and via differences in root growth (Alsina et al., 2011; Bauerle et al., 2008b; Soar et al., 2006a). It furthermore appears that own-rooted *V. vinifera* vines tend to be better able to tolerate water deficit than are *V. vinifera* scions grafted to American rootstocks, especially drought-intolerant *V. rupestris* rootstocks such as St. George (Ezzahouani and Williams, 1995; Keller et al., 2012; Kidman et al., 2014).

In addition to, and because of, its obvious consequences for Ψ, dehydration of plant cells also stimulates the synthesis of ABA from the carotenoid zeaxanthin in the parenchyma cells, especially near the cambium, of the root, shoot, and leaf vascular system (Endo et al., 2008; Nambara and Marion-Poll, 2005). The main function of ABA is to regulate plant water balance and osmotic stress tolerance. The role of ABA in water balance is mainly through the regulation of g_s (see Section 3.2), whereas its role in osmotic stress tolerance is through the production of dehydration-tolerance proteins in most cells. In addition, ABA may induce closure of aquaporins in the bundle-sheath cells of the leaves, which increases r_h and decreases water flow to the mesophyll cells (Shatil-Cohen et al., 2011). This hydraulic reinforcement of the ABA effect on g_s might provide an explanation for the isohydric behavior discussed above (Pantin et al., 2013). It is possible that plants that perceive ABA as a signal to both their stomata and their leaf aquaporins may be more ABA-sensitive and respond to water deficit by closing their stomata sooner and more tightly than plants that perceive ABA only as a stomatal signal.

Another function of ABA is to act as an antagonist of auxin to inhibit cell wall loosening necessary for cell expansion during growth (see Section 3.1). It is possible that ABA prevents auxin from exerting its influence on wall loosening. ABA also directly interferes with the cell cycle machinery, thus inhibiting cell division. In addition to the ABA produced in drying leaves, the amount of root sourced ABA in the xylem sap can increase substantially as a function of reduced soil water availability (Loveys, 1984; Lovisolo et al., 2010). In contrast to ABA, the production of cytokinins in the roots and their transport in the xylem sap decrease in response to soil drying (Davies et al., 2002; Ha et al., 2012; Yang et al., 2002). Both ABA and cytokinins can act as stress signals that mediate conditions in the rootzone via the xylem to the shoots, leaves, and clusters. The production of ABA in the roots is stimulated rapidly in response to water stress, whereas cytokinin production decreases much more gradually. Nevertheless, when water stress occurs suddenly, production and transmission of the root ABA signal may be too slow to enable the stomata to close in time to avoid leaf dehydration. Under these conditions, ABA is produced directly in the shoot and the leaves in response to a hydraulic signal from the roots (Christmann et al., 2007; Soar et al., 2004). Changes in xylem pressure in the roots are transmitted virtually instantaneously to the shoot, where they alter the turgor pressure of the leaf cells. A decrease in turgor may be sufficient to trigger ABA synthesis in the leaves; thus, leaf cell turgor is a key signal of water stress.

Whereas ABA may be involved in inhibiting shoot growth of water-stressed plants, it appears to have the opposite function in roots. Moreover, the pH of the xylem sap and other apoplastic (extracellular) sap generally increases in response to soil drying (Hartung and Slovik, 1991; Li et al., 2011; Wilkinson and Davies, 2002). The pH of xylem sap from well-watered plants is approximately 6.0 (see Section 5.1), whereas the pH of sap from water-stressed plants is approximately 7.0. This increase in xylem sap pH alone could constitute a root-sourced signal to the leaf, inducing stomatal closure and reducing the rate of cell expansion (and hence growth). ABA synthesized in the roots can be leached out into the soil solution, especially in alkaline soils (Wilkinson and Davies, 2002). Consequently, a high soil pH could weaken the root-sourced ABA signal.

Roots must absorb more water than what is lost in transpiration to enable plant growth because growth is caused mainly by cell expansion due to water import (see Section 3.3). Water uptake by cells is driven by accumulation of solutes (e.g., sucrose) inside the cell and thus depends on an osmotically generated $\Delta\Psi$ between cell interior and exterior (see Section 3.1). Because water stress reduces Ψ_{xylem}, a major cause for growth inhibition under water deficit may be the smaller $\Delta\Psi$, which reduces cell water uptake and consequently hampers the generation of the necessary turgor pressure (Nonami and Boyer, 1987; Nonami et al., 1997; Hsiao and Xu, 2000). In addition, cell expansion involves cell wall loosening, which requires acidification by protons (H^+) pumped from the cell interior to the cell wall in exchange for K^+ maintaining the cell's electrical charge balance (Stiles and Van Volkenburgh, 2004). It is possible that the increase in apoplast pH due to water deficit also directly interferes with cell expansion (Bacon et al., 1998). Therefore, growth rates (e.g., shoot elongation, leaf expansion) change rapidly with fluctuating Ψ_{xylem} and pH_{xylem}, and a decline in shoot growth and leaf appearance is the first visible sign of water deficit; shoot growth is even more sensitive to water stress than is photosynthesis (Hsiao and Xu, 2000; Stevens et al., 1995; Sweet et al., 1990; Williams et al., 1994). Moreover, tendrils are also a sensitive indicator of vine water status: On nonstressed, vigorously growing shoots, the uppermost tendrils extend beyond the shoot tip. As water stress sets in and growth begins to slow, new tendrils remain small so that the shoot tip catches up with them. With more severe stress, growth stops and the youngest leaf expands beyond the shoot tip. Young, green tendrils are highly sensitive to water stress and start wilting before the leaves do.

Deliberate application of water deficit is an important management strategy in premium wine grape production. The time between fruit set and veraison is the period when shoot growth and berry size can be most effectively controlled by water deficit. This principle is exploited by regulated deficit irrigation (RDI), whereby water deficit is imposed as soon as possible after fruit set (Dry et al., 2001; Keller, 2005; Kriedemann and Goodwin, 2003). Deficit irrigation means that less water is applied than is necessary to balance vineyard ET, so that soil moisture and plant water status decline. To help growers estimate required irrigation water amounts, weather station networks often provide measures of daily ET, but these are valid only for so-called reference crops such as grass or alfalfa. This baseline or reference ET is referred to as ET_0, and a crop factor (K_c) is applied to estimate crop or vineyard water use as follows:

$$ET_c = K_c ET_0$$

The K_c usually increases from less than 0.1 to about 0.8 in wine grapes, and up to 1.3 in large table grapes, depending on seasonal shoot growth and canopy size (Evans et al., 1993; Williams and Ayars, 2005; Williams, 2014). By applying amounts of water that are lower than ET_c, the soil profile is allowed to dry down during the RDI period until control of shoot growth has been achieved. Obviously, this is possible only in areas with sufficiently low seasonal rainfall or high evaporative demand and on soils with limited water storage capacity. In addition to limiting berry size, RDI also tends to be associated with reduced canopy density and hence higher sun exposure of the clusters (Romero et al., 2010). Once shoot growth stops and especially after veraison, vines are stressed only sufficiently to discourage new shoot growth. In regions with cold winters, growers irrigate to refill the rootzone to near field capacity at the end of the growing season in order to avoid cold injury to roots (see Section 7.4).

Rootzone conditions are generally heterogeneous, and soil moisture is no exception to this rule. When roots experience both wet and dry soil, the reaction of shoot growth depends on whether the difference in soil moisture is perceived by the same or different roots. When the surface soil is dry while the subsoil is still wet, there is no reduction in shoot growth as long as the roots have access to the subsoil water (Phillips and Riha, 1994). However, when separate roots of the same vine experience dry and wet soil columns, shoot growth is suppressed (Dry and Loveys, 1999; Dry et al., 2001; Lovisolo et al., 2002a). This can be exploited in another irrigation strategy termed partial rootzone drying (PRD), in which water is supplied alternately to only one side of a vine while the other side is allowed to dry down (Dry and Loveys, 1998). This strategy attempts to separate the physiological responses to water stress (e.g., ABA production and decreased g_s) from the physical effects of water stress, such as low Ψ_{leaf} (Davies et al., 2002; Dry et al., 2001; Kriedemann and Goodwin, 2003). The ABA produced by the drying roots induces partial stomatal closure and reduced shoot growth while the fully hydrated roots maintain a favorable plant water status (i.e., high Ψ_{leaf}). Water delivery to the shoots by the wet roots may increase during PRD (Kang et al., 2003), and the wet roots also sustain the drying roots by supplying water to them (Stoll et al., 2000). Eventually, however, water flow from the dry roots becomes so slow that ABA delivery from these roots ceases and the stomata begin to reopen (Dodd et al., 2008; Dry et al., 2000a). To prevent such adaptation by the vines, irrigation is alternated every week or two between the two sides of the root system. In addition to increased ABA production, the drying roots release lower amounts of cytokinins into the xylem, which inhibits lateral shoot growth (Stoll et al., 2000). Most of the reduction in canopy size of vines subjected to PRD may be due to inhibition of lateral shoot growth (dos Santos et al., 2003; Dry et al., 2001). Provided Ψ_{leaf} is maintained under PRD, berry size and yield are also maintained, whereas the lower canopy density often results in improved fruit composition. The fundamental difference between RDI and PRD is that RDI imposes a soil water deficit over time, whereas PRD imposes a deficit over space (Keller, 2005). If rainfall during the growing season is low enough, RDI always results in a plant water deficit, whereas PRD usually does not if managed properly (de Souza et al., 2003; Dry et al., 2001; Kriedemann and Goodwin, 2003). Nonetheless, whether or not the effects of PRD differ from those of RDI may depend on the total amount of available soil water (Romero and Martinez-Cutillas, 2012). The distinct PRD effect may be lost if the switching interval is too long, the wet soil volume is too small for a given vine size, two discrete root systems cannot be established (e.g., following conversion of furrow- or sprinkler-irrigated vineyards), where vines have access to a relatively high water table, or in poorly structured clay soils. An additional response to PRD seems to be that grapevines shift root growth to deeper soil layers (Dry et al., 2000b, 2001), whereas under standard drip irrigation roots are often concentrated in the surface soil. Although this adaptive behavior makes vines more drought resistant, it also interferes with the application of the PRD principle on soils with a deep rootzone and ready access of the roots to subsurface water.

Root growth of water-stressed vines declines due to a combination of plant water deficit and increased penetration resistance of the drying soil (Bengough et al., 2011). But the decrease in root growth is less marked than that in shoot growth, because roots can grow at lower Ψ than shoots, partly because higher activity of expansin proteins may increase cell wall extensibility in the root tips (Erlenwein, 1965a; Sharp et al., 2004; Wu and Cosgrove, 2000). In addition, the effect of water stress is more pronounced on the roots' diameter than on their length: Stressed roots become thinner (Mapfumo et al., 1994a). These adaptations increase the root:shoot ratio and maintain water and nutrient supply to the shoots (Hsiao and Xu, 2000). By inducing a blockage of ion channels in the stele (see Section 3.3), ABA reduces the release of K^+ (and possibly other nutrient ions, such as $H_2PO_4^-$) from the root cortex into the xylem conduits (De Boer and Volkov, 2003; Roberts and Snowman, 2000). This traps K^+ taken up from the soil (ABA does not affect K^+ uptake) and that delivered from the shoots in the phloem, which leads to K^+ accumulation in the roots. The osmotic activity of K^+ lowers Ψ_{root}, which may enhance phloem water import (see Sections 3.2 and 5.1). This helps the vine to maintain root growth while avoiding K^+ transport to the leaves, where it would only worsen the water deficit by lowering Ψ_{leaf}. However, at higher concentrations (i.e., more severe stress) ABA inhibits root growth (Zhang et al., 2010b). In addition, the roots, especially the growing root tips, also osmotically adjust, or osmoregulate, by accumulating sugars and amino acids in order to lower Ψ_{root} to favor water uptake and continued growth, and possibly to reduce the risk of xylem cavitation (Düring and Dry, 1995; Schultz and Matthews, 1988a). During more severe water stress, however, the vessels cavitate, which interrupts water flow in the affected xylem conduits altogether.

Soil drying also reduces the microbial activity (which is responsible for transforming mineral nutrients locked up in organic matter into plant-available forms) in the soil and the rate of nutrient transport to the roots. This leads to a decrease in overall nutrient availability and uptake, especially if the water deficit is sufficient to slow root growth (Freeman and Kliewer, 1983; Hsiao, 1973; McDowell, 2011; see also Section 7.3). Thus, decreasing soil moisture makes nutrient uptake

FIGURE 7.2

Relationship between stomatal conductance and transpiration (left) and photosynthesis (right) of mature leaves of two grapevine cultivars during soil drying (Keller, unpublished data). Note the steeper decline of photosynthesis compared to transpiration at low stomatal conductance.

increasingly difficult, and a greater proportion of the nutrient demand of the developing canopy and especially the ripening fruit has to be supplied from stored reserves in the woody parts of the vine. A decline in soil moisture is also associated with a decrease in root respiration and, below approximately 5% volumetric soil moisture, leads to a concomitant loss of membrane integrity so that roots begin to die back under severe drought conditions (Huang et al., 2005b).

The stomatal closure (leading to low g_s), discussed previously, diminishes water loss by transpiration but also limits photosynthesis (Figure 7.2) because CO_2 diffusion into grape leaves is much more dependent on open stomata than is H_2O vapor diffusion out of the leaves (Boyer et al., 1997; Escalona et al., 1999; Flexas et al., 1998, 2002; see also Section 4.2). This is called stomatal regulation or stomatal limitation of photosynthesis and is accompanied by a decrease in ATP and NADPH consumption because less energy is needed for CO_2 fixation. Not surprisingly, the effect on photosynthesis and assimilate export of short episodes of high vapor pressure deficit is similar to the effect of longer periods of soil water deficit (Shirke and Pathre, 2004). However, even a stress that is mild enough not to diminish photosynthesis can reduce shoot growth (i.e., cell expansion), and hence canopy development, because the reduction in growth usually occurs before the stomata begin to close (Hsiao, 1973; McDowell, 2011; Muller et al., 2011). The combination of smaller total leaf area and decreased photosynthesis will result in reduced daily assimilate production of the grapevine canopy (Perez Peña and Tarara, 2004; Poni et al., 2008; Tarara et al., 2011). Because the net assimilation rate of the canopy determines the export rate from the canopy, a reduction in canopy photosynthesis will reduce the amount of photosynthate available for export to the vine's sink organs. Nonetheless, as long as the water deficit is mild enough to restrict shoot (and root) growth more than photosynthesis, there is relatively more sugar available for other sinks (Lawlor and Cornic, 2002; McDowell, 2011; Muller et al., 2011), which may benefit fruit ripening and carbohydrate storage in the vine's perennial organs. More severe drought stress, however, prevents replenishment of storage reserves, especially in the roots (Holzapfel et al., 2010).

Whereas leaf sugar concentrations may remain high for some time, starch becomes depleted, and assimilate export decreases progressively with increasing severity of water stress (Düring, 1984; Quick et al., 1992). Although the sugar concentration in phloem sap increases as water availability declines, more severe water stress may lead to phloem transport failure, because the phloem may no longer be able to maintain turgor (Sala et al., 2010). In addition, when faced with drought stress, expanding leaves and growing root tips commonly accumulate solutes, especially inorganic ions such as K^+ and Ca^{2+}, but also sugars such glucose and fructose and, to a lesser extent, amino acids such as proline and organic acids such as malate (Cramer et al., 2007; Patakas et al., 2002; Sharp et al., 2004). This osmotic adjustment, or osmoregulation, maintains turgor, although this may not be sufficient to permit continued growth (Düring, 1984; Morgan, 1984; Schultz and Matthews, 1993a). Thus leaf expansion is inhibited by water deficit (Sweet et al., 1990). When expansion is complete (e.g., in fully grown leaves), tissues gradually lose the ability to osmoregulate (Patakas et al., 1997). However, leaves developing later in the season can achieve a greater capacity for osmotic adjustment than leaves that are formed earlier.

Under severe stress—that is, when Ψ_{leaf} decreases below approximately -1.5 MPa (depending on cultivar), RWC decreases below approximately 75%, and stomatal conductance (g_s) declines to less than 20% (~ 50 mmol H_2O m^{-2} s^{-1}, according to Flexas et al., 2002) of g_s in well-watered vines—growth slows strongly or stops. Moreover, photosynthetic metabolism is progressively impaired, initially by a reduction of electron transport and ATP synthesis that inhibits the Calvin cycle and, at very low RWC, by inhibition of light harvesting and photosystem (especially photosystem II function (Escalona et al., 1999; Hsiao, 1973; Tezara et al., 1999; see also Section 4.1). Although the amount of rubisco protein in the leaves also declines under severe drought stress (Bota et al., 2004), this might be a consequence rather than a cause of reduced photosynthesis. This depression of photosynthesis is called metabolic regulation or metabolic limitation, and in contrast to stomatal limitation, it is irreversible (Escalona et al., 1999; Lawlor and Cornic, 2002). The decrease in photosynthesis is accompanied by only a slight decrease in respiration (see Section 4.4) and, possibly, an increase in the proportion of photorespiration relative to photosynthesis, which may serve to counter an energy overload by "consuming" excess electrons (Cramer et al., 2007; Düring, 1988; Lawlor and Cornic, 2002; Lawlor and Tezara, 2009; see also Section 4.3). Thus, as water stress intensifies, an increasing proportion of fixed CO_2 is lost from the leaf. The more severe the water stress, the more of this CO_2 is derived from stored carbohydrates (Lawlor and Cornic, 2002; McDowell, 2011). As internal reserves are depleted, amino acids, particularly proline and glutamate, accumulate because their production (supply) exceeds consumption (demand). This in turn seems to lead to nitrate accumulation in leaves, possibly due to feedback inhibition of nitrate reductase (Patakas et al., 2002; see also Section 5.3). The carbon skeletons for the production of amino acids in water-stressed leaves are provided by glycolysis and organic acids from the citric acid cycle, rather than assimilates directly from the chloroplast (as under nonstress conditions), so that there is a decrease in organic acids (Lawlor and Cornic, 2002).

When the leaves lose turgor due to severe water stress, they will wilt; the leaf surface area decreases, and leaves hang down and become parallel to the solar rays (Lawlor and Cornic, 2002; Smart, 1974). Because water stress reduces photosynthesis so that fewer electrons are needed, wilting is an effective mechanism for reducing light absorption (i.e., the number of photons captured by the leaf; see Section 4.1). This can substantially reduce a potential "energy overload" that would be damaging to the photosynthetic system (Flexas et al., 1999b; Lawlor and Cornic, 2002).

Nevertheless, water stress can be especially damaging at high light intensity with large photon fluxes that are associated with a relative energy surplus in the chloroplasts (Lawlor and Tezara, 2009). Moreover, because the cuticle allows small amounts of water vapor to pass through by diffusion, leaves cannot avoid water loss completely, no matter how tightly the stomata are closed (Boyer et al., 1997; see Figure 7.1). Such cuticular transpiration increases exponentially with rising temperature because both the water permeability and the vapor pressure deficit that provides the driving force for water loss increase as the temperature rises (Riederer and Schreiber, 2001). The problem is particularly acute when drying wind reduces r_b and stimulates transpirational water loss. Although young leaves may be more sun-exposed and thus experience higher evaporative demand than older leaves, the old leaves wilt at higher Ψ_{leaf} than the young leaves because the latter are better able to osmoregulate (Patakas et al., 1997). The wilting older leaves produce plenty of ABA, which is exported to the young leaves to reduce water loss.

In addition, ABA can also result in oxidative stress via increased production of active oxygen species that, when produced in excess, damage and eventually kill the plant cells due to uncontrolled oxidation (see Section 7.1). This stimulates the vine's antioxidant defense system to prevent oxidative damage (Cramer et al., 2007), leading to increased foliar concentrations of antioxidants such as carotenoids (especially xanthophylls), glutathione, ascorbate (vitamin C), and α-tocopherol (vitamin E). If the water stress is not relieved by rainfall or irrigation, the increased ABA and reduced cytokinin contents accelerate leaf aging and lead to senescence of older leaves (Romero et al., 2010; Yang et al., 2002). Senescence, which is initiated when leaf cytokinins fall below a threshold level, involves the disassembly of the chloroplasts with their photosynthetic machinery and is accompanied by a decline in chlorophyll and remobilization of carbon, proteins (e.g., degradation of rubisco), and nutrients from these leaves. It culminates in leaf abscission, which can also be induced by excessive xylem cavitation, whereas the meristems inside the buds remain alive. This drought-induced deciduous behavior also promotes early inception of bud dormancy and enables vines to survive severe drought by drastically reducing evaporation and conserving resources. Once initiated, the rate of senescence accelerates with increasing temperature so that a heat wave following drought stress can lead to severe defoliation of vines. The cytosolic glutamine synthetase (GS1) in the phloem may be involved in the production of glutamine used for nitrogen export and of proline used for storage (see Section 5.3) during water stress or senescence. Some of the sugars, amino acids, and phloem-mobile mineral nutrients (see Section 7.3) can be recycled to the fruit, which may partially sustain ripening. However, because nutrients are "swept" into the fruit by the sugar-driven mass flow in the phloem (see Section 5.1), this nutrient recycling may lead to an undesirable increase in fruit K^+ content that may subsequently increase juice and wine pH (see Section 6.2). This effect is probably most pronounced if severe water stress occurs during a period in which the clusters' sink strength is at its maximum (i.e., during and immediately after veraison). Even under drought conditions leading to complete defoliation, the perennial parts of grapevines rarely die. Restoring water supply to defoliated plants usually results in budbreak and regrowth, presumably aided by the generation of root pressure (see Section 3.3). Nevertheless, partial or complete dieback, beginning in the youngest (and smallest) portion of the shoots and roots, can occur due to complete cavitation of the xylem (i.e., hydraulic failure) for prolonged periods.

It is often stated that the reduction in vegetative growth due to water stress is more severe than that of fruit growth (Williams, 1996; Williams et al., 1994). Water deficit, similar to nutrient deficiency, typically reduces yield, particularly if the deficit occurs early in the growing season

(Williams and Matthews, 1990). The sensitivity of reproductive growth, and hence fruit production and yield formation, depends on the developmental stage of the vine. Soon after budbreak, the developing flower clusters begin competing successfully with the growing shoots for limited water and assimilates (Hale and Weaver, 1962). Once inflorescences contribute assimilates via their own gas exchange (Leyhe and Blanke, 1989), they may survive even relatively severe drought sufficient to stunt shoot growth. Closer to and during bloom, however, the flowers' metabolic requirements far exceed their own contribution—after all, the majority of the stomata are located in the cap that is shed at anthesis (Blanke, 1990b; Blanke and Leyhe, 1988, 1989a; Vivin et al., 2003)—but the flower clusters continue to be weak sinks until fruit set. Whereas the female floral organs are thought to be relatively insensitive to water stress (although smaller ovaries are more susceptible to stress than are larger ovaries within the same fruit cluster), water deficit during meiosis in the anthers (which occurs 1–2 weeks before anthesis) may cause pollen sterility (Dorion et al., 1996; Saini, 1997). Even moderate water stress for a brief period during male meiosis can hamper pollen development because water stress may inhibit sugar transport to and starch accumulation in the pollen grains (De Storme and Geelen, 2014; Dorion et al., 1996; Saini, 1997); however, this has yet to be investigated in grapes. Starch is required to fuel respiration for pollen development and germination and for pollen tube growth, so their low "fuel" status prevents the pollen tubes from reaching the ovary in time for fertilization, if they are able to germinate at all (see Section 2.3). Reduced pollination and fertilization result in poor fruit set or, in more severe cases, abscission of inflorescences (Callis, 1995; Smart and Coombe, 1983). In other words, reproductive development is most sensitive to water stress from meiosis to fruit set. The effects of water deficit are likely mediated by a combination of low $\Psi_{inflorescence}$ and changes in xylem sap pH, ABA, and cytokinins in addition to supply of photosynthate via the phloem. For instance, although ABA stimulates vacuolar invertase in the leaves in order to maintain turgor to sustain photosynthetic activity and assimilate export, it suppresses the same enzyme in flowers and young fruits, which are subsequently aborted (Roitsch and González, 2004). An involvement of ABA and cytokinin is consistent with the observation that PRD, which maintains high Ψ_{leaf}, may be associated with poor fruit set (Rogiers et al., 2004b). Moreover, it is possible that when embolisms develop in the inflorescence xylem, they may not be as easy to repair as they are in leaves. Note that it is unknown whether deliberate prebloom water stress could be used in breeding programs to induce male sterility in the female breeding partner.

In addition to reducing berry number, drought stress during bloom could also have implications for a vine's yield potential in the following growing season. Water deficit normally decreases cluster initiation, which can reduce bud fruitfulness and thus result in low numbers of clusters per shoot (Alleweldt and Hofäcker, 1975; Buttrose, 1974b). In contrast, if the water deficit is just mild enough to decrease canopy density, this may actually enhance cluster initiation due to the improved canopy microclimate (Smart and Coombe, 1983; Williams and Matthews, 1990; Williams et al., 1994).

Once fruit set has passed and the vine has made a significant investment in seed production, vines experiencing water stress generally maintain fruit development at the expense of shoot and root growth and replenishment of storage reserves in the roots (Eibach and Alleweldt, 1985; Mullins et al., 1992; Williams et al., 1994). Of course, this can have long-term implications because the vine is heavily dependent on stored reserves for budbreak the following spring. Moreover, a weak root system can predispose vines to drought stress when there is insufficient water available. A small root system supplying water to a large canopy would result in a very steep water potential gradient

($\Delta\Psi = \Psi_{soil} - \Psi_{leaf}$; see Section 3.3), which could render vines vulnerable to xylem cavitation and, consequently, canopy desiccation (Sperry et al., 2002). To balance water loss by transpiration and supply by root uptake, such vines have to either lower Ψ_{leaf} (e.g., by osmotic adjustment) or close their stomata, especially around midday when evaporative demand is at its maximum.

Under dry conditions, moderate nitrogen (N) supply via fertilizer application may conserve soil water in the short term. However, a decrease in the root:shoot ratio due to fertilizer (especially N) addition can further increase a vine's susceptibility to drought, particularly in irrigated vineyards should irrigation be withheld (Erlenwein, 1965a; Sperry et al., 2002). Vigorous growth associated with high N availability leads to an increase in the diameter of xylem vessels and, consequently, a decrease in r_h that comes at the cost of heightened vulnerability to cavitation (Plavcová et al., 2012). The increase in nitrate concentration in the xylem sap that accompanies abundant N supply (see Section 5.3) seems to be associated with a decrease in xylem sap pH, which in turn reduces the sensitivity of the stomata to ABA (Jia and Davies, 2007; Wilkinson and Davies, 2002). This means that the stomata close later and more slowly when a soil with high N status dries compared with a relatively N-poor soil. In nonirrigated plants, however, fertilizer application is thought to improve the capacity for osmotic adjustment and resistance to xylem cavitation, which tends to counter the lower root:shoot ratio (Bucci et al., 2006; Sperry et al., 2002), although excessive N supply can reduce water uptake due to the buildup of solutes in the soil water, similar to the effect of salinity discussed in Section 7.3 (Keller and Koblet, 1994).

Although fruit growth may be less sensitive to drought stress compared with shoot growth, limited water supply during the period of berry cell division and cell expansion restricts berry enlargement, which limits berry size (Roby and Matthews, 2004; Romero and Martinez-Cutillas, 2012; Shellie, 2014; Williams and Matthews, 1990). Whereas cell division in young grape berries is relatively insensitive to water deficit, cell expansion responds readily to changes in water supply (Ojeda et al., 2001). In keeping with the effect on shoot growth, reduced Ψ_{xylem} and pH_{xylem} could also be partly responsible for the smaller berry size of water-stressed vines, particularly if the stress occurs before veraison. Since the limitation in berry size imposed by early-season stress is irreversible, it seems likely that it involves changes in the composition of (skin) cell walls (Ojeda et al., 2001). Water deficit curtails cellulose production and cell wall extensibility in expanding leaves (Schultz and Matthews, 1993a; Sweet et al., 1990), and the same changes probably occur in berries too. Yield reductions due to drought stress can still be severe after fruit set, whereas after veraison berries seem to become increasingly insensitive to water deficit. Thus, the same extent of water deficit occurring during the preveraison phase of berry growth normally reduces berry size much more than if it occurs after veraison (Currle et al., 1983; Hardie and Considine, 1976; Hofäcker et al., 1976; Matthews and Anderson, 1989; McCarthy, 1997; Williams and Matthews, 1990). Applying more water later in the growing season cannot compensate for the decrease in berry size due to early-season deficit (see Figure 6.2), possibly because increased phloem flow into the berry, combined with the stiffening skin, after veraison strongly decrease the sensitivity of berry water status to soil and plant water status (Creasy and Lombard, 1993; Düring et al., 1987; Findlay et al., 1987; Greenspan et al., 1994, 1996; Ollat et al., 2002; Rogiers et al., 2001). Despite the fears of many winemakers, therefore, drip or flood irrigation close to harvest is ineffective at increasing berry size and "diluting" fruit composition (Coombe and Monk, 1979; Keller et al., 2006). The same cannot be said for rainfall or overhead irrigation, however, because uptake of water through the berry skin may induce berry cracking (see Section 6.2). Nevertheless, due to the reversal of xylem flow at veraison, postveraison water deficit, especially prolonged deficit, can lead to berry shrinkage by dehydration when xylem efflux plus berry transpiration exceed phloem influx (Keller et al., 2006). Thus, the extent of preharvest shrinkage is inversely related to the water status of the vine at that time.

The response of berry growth to water deficit also depends on the crop load. Berry size may be curtailed more on plants bearing a heavy crop (i.e., many berries per vine) than on vines with a light crop load because lower sugar import by each berry also limits water uptake by those berries (Fishman and Génard, 1998; Santesteban and Royo, 2006). If a stress occurs early enough to reduce fruit set, the size of the remaining berries may increase and partially compensate for the loss in yield potential, which would offset prospective benefits of smaller berry size for wine quality (see Section 6.2). Water stress, however, may prevent compensatory berry growth on vines that start the growing season with a small crop, whereas it does not seem to have an additional effect on heavily cropped vines, where berry size is already limited (Keller et al., 2008). Although plant water stress symptoms become apparent when more than 50% of plant-available water has been extracted from the soil, grape berries (both before and after veraison) do not begin to shrink until 80% of the available water has been transpired by the vine (Keller et al., 2006).

7.3 NUTRIENTS: DEFICIENCY AND EXCESS

In a growing grapevine, each new cell has a relatively specific requirement for a well-defined set of inorganic mineral nutrients (often abbreviated as "minerals") in addition to water and carbon. Some of these nutrients, especially during budbreak and during leaf fall, are supplied from reserves stored in the permanent structure of the vine or from senescing leaves. However, the majority of nutrients throughout the growing season must be acquired by the roots in the form of ions dissolved in the soil water or "soil solution." Roots can either grow toward nutrient-rich areas or patches in the soil in a process called root interception or wait for the nutrients to arrive with the soil water. In the latter case, the nutrients must move through the soil either by diffusion or by mass flow. Diffusion describes the movement of dissolved nutrients toward the root down a concentration gradient that is generated by nutrient uptake at the root surface (see Section 3.3). Mass flow, on the other hand, is the process of convective nutrient transport toward the root in the "stream" of water (i.e., down a water potential gradient) generated by the transpiration stream, which results in water uptake by the root (Chapman et al., 2012). Diffusion is often the dominant process, but mass flow is more important when plants transpire rapidly and thus remove copious amounts of soil water. In fact, one adaptive advantage of transpiration may be its capacity to enhance mass flow of nutrients to the root surface, which may be important in soils with relatively low nutrient availability—provided water is not limiting (Cramer et al., 2008). Accordingly, grapevines seem to transpire more rapidly when the nutrient concentration in the soil solution is low than when abundant nutrients are present, but the leaves of the high-nutrient vines wilt more readily when they run out of water (Keller and Koblet, 1994; Scienza and Düring, 1980).

Soils vary not only in their capacity to store water but also in the amount and composition of mineral nutrients they contain and in the extent to which these nutrients are available for uptake by the roots. Nutrient storage capacity and accessibility are influenced by soil texture, rooting

CHAPTER 7 ENVIRONMENTAL CONSTRAINTS AND STRESS PHYSIOLOGY

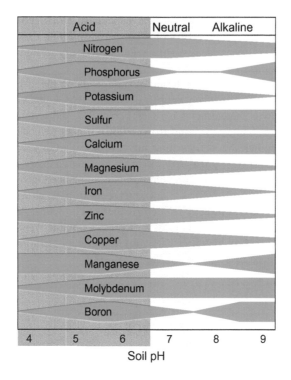

FIGURE 7.3

Nutrient availability (indicated by width of green bands) in the soil is dependent on soil pH.

Illustration by M. Keller, after Lucas and Davis (1961).

depth, and organic matter (humus) content, but availability is modified by soil moisture and pH (Figure 7.3). For instance, just as a loamy soil can store considerably more water than a sandy soil (see Section 7.2), it also adsorbs more nutrient ions than sandy soils where the nutrients are leached away by percolating water. Organic matter, on the other hand, increases water and nutrient storage capacity in both loamy and sandy soils due to its beneficial effect on soil structure via the formation of stable aggregates. Although the nutrient concentration in a dry soil is usually higher than that in a wet soil, these nutrients are less available in the dry soil because the lack of water slows diffusion and mass flow (Marschner, 1995); the diffusion rate decreases roughly with the square of the soil water content. Moreover, mineralization of organic matter and nitrification ($NH_4^+ \rightarrow NO_3^-$) also slow down in drying soil. The concentration of exchangeable cations (e.g., calcium) rises at the expense of anions (e.g., phosphate), and roots may lose contact with soil particles due to shrinkage so that nutrient availability to the roots declines steeply as the soil dries out (Bloom et al., 1985; Jackson et al., 2008). Although this diminishes nutrient uptake and delivery of nutrients to the shoots (Davies et al., 2002; Mpelasoka et al., 2003), the tissue nutrient concentration may nevertheless increase because water stress curtails carbon assimilation and shoot growth more than nutrient uptake (Bloom et al., 1985). Nonetheless, both the concentration and the total amount of nitrogen in the leaves and shoots tend to be lower in grapevines whose growth is slowed by competition with cover crops for access to both water and nutrients (Celette et al., 2009; Guerra and Steenwerth, 2012; Tesic et al., 2007).

Nitrogen availability fluctuates widely throughout the growing season, whereas the availability of the relatively mobile calcium (Ca) and magnesium (Mg), and of the nearly immobile phosphorus (P) and potassium (K), is much more constant (Nord and Lynch, 2009). Potassium and phosphate ($H_2PO_4^-$) can be taken up by mature root sections, whereas absorption of Ca^{2+} and Mg^{2+} seems to be restricted to the young fine roots. Iron (Fe^{2+}) uptake is confined to the growing root tips, whereas sulfate (SO_4^{2-}) uptake is concentrated in the elongation zone immediately behind the meristematic region. In addition to the fibrous fine-root system most effective in nutrient uptake, the older regions of grapevine roots usually form symbiotic (i.e., mutually beneficial) units with mycorrhizal fungi (Gebbing et al., 1977; Menge et al., 1983; Possingham and Groot Obbink, 1971; Schreiner, 2005; Stahl, 1900). These microbes are attracted to the roots by carotenoid-derived strigolactones that are exuded by the roots and induce branching of the fungal hyphae (Akiyama et al., 2005; Gomez-Roldan et al., 2008; Ruyter-Spira et al., 2013). The fungi infect the roots, generally behind the root-hair zone, and provide an additional link between the vine and the soil, living partly within the root cortex, whereby the fungal and plant cytoplasms remain separated by the cell membranes and cell walls of the two partners, and partly within the surrounding soil (Bucher, 2007; Smith and Smith, 2011). The fungi can absorb water and nutrients at considerable distance from the root and thereby greatly extend the soil volume exploited by the roots by forming an extensive and interconnected hyphal network that can even link different plants together. This increases the roots' absorbing capacity well beyond the depletion zones caused by the growing root tips. The fungi transfer most of the absorbed nutrients to the plant roots in exchange for sugar, especially glucose, which is the energy source for mycelium growth, spore production, and active nutrient uptake. The symbiotic microorganisms are mainly known for their ability to enhance $H_2PO_4^-$ and zinc (Zn^{2+}) uptake, but they also transfer sulfur (S) and copper (Cu^{2+}) (Marschner, 1995; Menge et al., 1983; Smith et al., 2001). They may also facilitate acquisition of N, K, Ca, Mg, and iron (Fe) under certain conditions. Furthermore, these fungi grow readily toward organic, and hence nutrient-rich, patches in the soil, where their mycelium proliferates for the benefit of their host plant. Some mycorrhizae can apparently tap directly into plant-unavailable organic material, such as decaying plant or animal remains, where N predominantly occurs as proteins. The fungi promote their decomposition and mineralization, and then they capture and transfer to the roots the resulting inorganic N in addition to P (Hodge, 2006; Hodge et al., 2001; Leigh et al., 2009). However, their carbon requirement makes the mycorrhizae sinks for photosynthetic assimilates (Hall and Williams, 2000; Smith et al., 2001). The supply of sugar to the fungi represents the price paid by the vine for enhanced nutrient uptake. Once established, the fungi ostensibly suppress lateral root growth, making the plant more dependent on nutrients supplied by the mycorrhizae (Osmont et al., 2007). Because of the carbon cost to the plant of establishing and maintaining the symbiosis, young plants initially tend to grow more slowly in mycorrhiza-"contaminated" soil than in fumigated soil, especially if the soil is adequately supplied with nutrients (Clarkson, 1985). In soils with abundant nutrient availability, root colonization by mycorrhizae is usually much slower because in an environment with plentiful resources there is no need for a plant to "waste" carbon on the symbionts.

Nutrient ion uptake is mainly controlled by the demands of the vine and varies according to growth requirements, even as soil nutrients differ in concentrations over several orders of magnitude. Thus, growth is the "pacemaker" for nutrient uptake, and the rates of uptake and root growth reflect the demand created by plant growth (Clarkson, 1985; Clarkson and Hanson, 1980; Gastal and Lemaire, 2002; Tester and Leigh, 2001). Such demand also includes the growing fruit, so that seasonal nutrient uptake by grapevines varies according to their crop load (Pradubsuk and Davenport, 2010, 2011). Correlations between plant growth and nutrient supply may simply indicate that nutrient availability limits grapevine growth. Both nutrient uptake and shoot growth of grapevines show saturation-type responses to increasing soil nitrogen (Bell and Robson, 1999; Keller and Koblet, 1995a; Keller et al., 1995; Linsenmeier et al., 2008; Spayd et al., 1993; Williams and Matthews, 1990). Recycling in the phloem of xylem-delivered nutrients or of their assimilated versions from the leaves to the roots acts as a feedback signaling system to regulate nutrient uptake. For example, nitrate transporters in the roots can be "shut" in response to high concentrations of amino acids (e.g., glutamate and glutamine) or sugars, and sulfate transporters are repressed by cysteine and glutathione (Amtmann and Blatt, 2009; Forde, 2002; Grossman and Takahashi, 2001; Tsay et al., 2011). Such feedback regulation also fine-tunes potassium uptake according to shoot demands (Tester and Leigh, 2001; Véry and Sentenac, 2003). Furthermore, a plant's carbon status impacts nutrient uptake, at least partly because of the need for energy to run ATP pumps. Under conditions of carbon depletion—for instance, when photosynthesis is limited by environmental constraints such as low light—vines reduce nutrient uptake and transport in the xylem (Keller et al., 1995). A glucose phosphate is thought to be the sugar signal that synchronizes nutrient uptake to photosynthesis (Lejay et al., 2008). The decrease in vine growth due to water deficit (see Section 7.2) also decreases the nutrient requirements of grapevines. In addition, the phloem can recycle nutrients from ageing leaves back to the shoot xylem, which helps to buffer short-term changes in xylem nutrient concentrations (Metzner et al., 2010b).

Nevertheless, nutrient uptake can exceed growth requirements when nutrients are abundantly available in the soil solution—for example, after fertilizer application or incorporation and decomposition of cover crops. Although growth and crop yield may not increase further, the extra supply leads to accumulation inside the vine (Delas and Pouget, 1984; Keller et al., 1995; Pouget, 1984). Because grapevines cannot uproot themselves and relocate to search for new resources, this inappropriately named "luxury consumption" can be useful for temporary storage of nutrients, such as nitrate or phosphate, that are subject to wide temporal and spatial fluctuations in the soil (Clarkson, 1985; see also Section 5.3). Storage within plant cells occurs in the vacuoles, whereas the cytoplasm is well buffered against changes in nutrient concentration (Forde, 2002; Tester and Leigh, 2001). As a result, the concentration of nutrient ions may be three to four orders of magnitude lower in the cytoplasm than in either the vacuole or the apoplast. The storage pools serve as an insurance against temporary shortages in supply; they can be accessed to sustain growth when nutrient availability is low or to permit regrowth in spring or following catastrophic events such as drought, fire, or cold damage (Bloom et al., 1985; Pradubsuk and Davenport, 2010, 2011). However, even if their availability increases in unison, all nutrients are not necessarily taken up and accumulated equally; for example, N, P, and K may accumulate throughout the vine, including the fruit, at the expense of Ca and Mg (Delas and Pouget, 1984).

When vacuolar nutrients reach a critical minimum concentration, the cytoplasmic concentration cannot be sustained and metabolism is disturbed. Insufficient supply of a nutrient ion slows shoot growth to a rate consistent with supply (Clarkson, 1985; Gastal and Lemaire, 2002). Leaf expansion, for example, is particularly sensitive to fluctuations in N supply (Forde, 2002). Nutrient deficiency triggers a range of responses in the vine, some of which are general stress responses and some of which are specific for the nutrient in question (Grossman and Takahashi, 2001). General responses include cessation of cell division and cell expansion (i.e., reduced growth), changes in vine morphology (e.g., increased root:shoot ratio), carbohydrate accumulation, decrease in photosynthesis, and modification of metabolism to adapt to the limited nutrient supply. Therefore, the decline in growth (and thus sink activity) when nutrients are deficient is responsible for the accumulation of sugar and feedback inhibition of photosynthesis in the leaves, rather than the decrease in photosynthesis limiting growth (Peuke, 2010). Accumulation of reactive oxygen species also appears to be a general response that is common to at least N, K, P, and S deficiency (Schachtman and Shin, 2007). Another frequent reaction to mineral nutrient deficiency is the accumulation of amino acids such as arginine, glutamine, and asparagine and the polyamine putrescine (Rabe, 1990). As a general rule, it appears that high sugar concentration in the phloem is indicative of nutrient deficiency, whereas high nutrient concentration in the phloem indicates abundant nutrient supply; thus phloem sap composition might be used as a marker of plant nutrient status (Peuke, 2010).

Specific responses to limited supply of a particular nutrient include induction of transport systems to enhance nutrient uptake and remobilization of stored reserves of that nutrient. In addition, because plants can make optimum use of a particular nutrient only if no other nutrient is limiting, deficiency in one nutrient often also impacts the uptake and transport of other nutrients (Amtmann and Blatt, 2009; Keller et al., 2001b). When a portion of a root system is in soil that is depleted of a particular nutrient, plants often enhance growth and uptake mechanisms in another portion of the roots growing in soil where supply of that nutrient is still adequate. Although such local increases in nutrient acquisition can compensate for differences in nutrient supply arising from highly heterogeneous soil conditions (Nibau et al., 2008; Scheible et al., 1997), a decline in shoot growth similar to that in response to PRD (see Section 7.2) often occurs when nutrients such as N or P are available to only a portion of the roots (Baker and Milburn, 1965; Robinson, 1994). Nutrient deficiency, particularly N and P deficiency, can also trigger other plant responses that are very similar to the effects of soil water deficit, namely increased ABA and reduced cytokinin production, increased pH_{xylem} which increases guard cell sensitivity to ABA, increased r_h and r_s, decreased transpiration (which has no effect on Ψ_{leaf}), restricted leaf expansion and senescence of older leaves, and increased root:shoot ratio (Clarkson et al., 2000; Wilkinson and Davies, 2002). These responses lead to a decrease in water uptake and transport, which is probably mediated by closure of aquaporins (Maurel et al., 2008). If different roots of the same vine experience differences in nutrient availability, it makes sense for the plant to reduce water uptake from the depleted soil and compensate by increasing uptake from the nutrient-rich areas.

Nutrient limitation interferes with cell division by preventing the normal progression of the cell cycle so that the time it takes for a cell to divide increases (Gastal and Lemaire, 2002; Grossman and Takahashi, 2001; Lemaire and Millard, 1999). Moreover, meristematic cells are not directly connected to the vine's vascular tissues, and the very small vacuoles of these small, dividing cells permit little nutrient storage. Therefore, these cells are more sensitive to fluctuations in nutrient delivery via the transpiration stream than are mature cells. In developing leaves, both cell division and cell expansion decline rapidly in response to N deficiency. This reaction is, at least in part, caused by a decrease in the production of cytokinin from isoprenoids in the root tips (nitrate and phosphate stimulate cytokinin production) and its transport in the xylem to the shoots, in addition to reduced cytokinin production in the young, expanding leaves (Coruzzi and

Zhou, 2001; Forde, 2002; Kakimoto, 2003; Takei et al., 2002). Lack of cytokinin also seems to inhibit lateral bud outgrowth (Ferguson and Beveridge, 2009) so that low-N vines produce fewer and shorter lateral shoots (Keller and Koblet, 1995a). Root growth, on the other hand, may well increase, notably in response to limiting supplies of N (Keller and Koblet, 1995a) and P but probably also for K and S. In contrast to their action in shoot organs, cytokinins inhibit lateral root growth (Ha et al., 2012; Kakimoto, 2003; Werner et al., 2003), perhaps by stimulating the production of the growth-inhibiting gas ethylene (Dugardeyn and Van Der Straeten, 2008). As the amount of cytokinins declines, root growth increases. This means that when nutrients limit overall plant growth, roots become relatively stronger sinks than shoots for assimilates in order to alleviate the deficiency by improving nutrient uptake from previously untapped soil regions where higher nutrient concentrations may be found (Clarkson, 1985). Additionally, cytokinins seem to hinder the uptake of nutrient ions (e.g., nitrate, phosphate, sulfate, or iron) by repressing the ions' transport proteins in the root epidermis and cortex (Kiba et al., 2011). The local production and movement of other hormones, such as auxin and ABA, also change in response to alterations in nutrient availability and uptake. Transport of these diverse hormones from the roots to the shoots in the xylem or from the shoots to the roots in the phloem mediate and integrate information on nutrient status in plants that is then translated in each organ into a response specific to each nutrient ion (Kiba et al., 2011; Krouk et al., 2011).

Slow growth and the appearance of chlorosis (chlorophyll breakdown leading to leaf yellowing), necrosis (cell death), and senescence of leaves are typical symptoms of mineral nutrient deficiency in grapevines. The shedding of older leaves is often accompanied by remobilization of nutrients in those leaves and redistribution to developing organs such as young leaves and grape berries. This is particularly true of the phloem-mobile ions of N, P, K, S, Mg, Na, and chlorine (Cl), whereas phloem-immobile nutrients, such as Ca and boron (B), cannot readily be redistributed. Consequently, deficiency symptoms of phloem-mobile nutrients are first apparent on older leaves, whereas deficiency symptoms of phloem-immobile nutrients are usually confined to young, growing organs.

The amount of minerals accumulated by different *Vitis* genotypes may vary two- or threefold. Moreover, rootstocks may influence the concentration of nutrients in the xylem sap and leaves of the scion (Delas and Pouget, 1984; Grant and Matthews, 1996a; Keller et al., 2001b; Trieb and Becker, 1969), which has implications for fertilizer recommendations based on tissue nutrient tests. Species, cultivars, and rootstocks that are more efficient than others in nutrient uptake can produce more growth from a given amount of absorbed nutrient ion (Clarkson and Hanson, 1980; Grant and Matthews, 1996a; Scienza et al., 1986). This is especially important when nutrient availability is limited. Nutrient use efficiency is the reciprocal of the nutrient cost of growth (Bloom et al., 1985). Therefore, efficient species or cultivars frequently have lower average tissue concentrations of nutrients than do inefficient species or cultivars. However, the former often produce larger root systems and hence a larger root:shoot ratio (Clarkson and Hanson, 1980). When grafted grapevines are planted, rootstocks with efficient nutrient uptake could be chosen to minimize fertilizer input. On the other hand, inefficient rootstocks might reduce vigor on fertile sites.

Among mineral nutrients required for plant growth, N, P, K, Ca, Mg, and S are usually classed as macronutrients, and the remaining essential nutrients Cl, B, Fe, manganese (Mn), zinc (Zn), copper (Cu), molybdenum (Mo), and nickel (Ni) are classed as micronutrients. This division is arbitrary and merely reflects the fact that different nutrients are required in vastly different amounts to satisfy vine growth. The tissue concentrations of both macro- and micronutrients vary severalfold across different grape cultivars (Christensen, 1984). Depending on cultivar, rootstock, and seasonal weather conditions, the amounts of macronutrients removed from a vineyard per ton of harvested grapes are in the range 1–3 kg N, 0.2–0.4 kg P, 1.5–4 kg K, 0.2–1 kg Ca, and 0.05–0.2 kg Mg (Bettner, 1988; Conradie, 1981a; Currle et al., 1983; Mullins et al., 1992; Schreiner et al., 2006), with additional amounts being incorporated into the growing permanent structure of the vines. The lower end of these ranges applies only where stalks and pomace are recycled to the vineyard. Deficiency of micronutrients, especially B, Fe, Mn, and Zn, is relatively common in vineyards throughout the world (Mullins et al., 1992). Although grapevines naturally acquire nutrients via root uptake from the soil solution, it is possible to alleviate nutrient deficiencies in the short term with foliar sprays. Nutrient ions that are dissolved in water can then (and only then) diffuse through plant cuticles, probably via aqueous polar pores in the cuticular membrane, and can then be taken up by leaf cells and distributed via the apoplast (Schönherr, 2000, 2006). Diffusion increases in proportion to the amount of nutrient ions applied and is independent of temperature but increases with increasing air humidity (Schönherr, 2000; Schreiber et al., 2001). In addition, nutrients may also be taken up through the stomata (Eichert and Goldbach, 2008).

7.3.1 MACRONUTRIENTS

Nitrogen

The role of N and its uptake, assimilation, and transport in the vine are discussed in Section 5.3. Grapevines normally acquire N in the form of inorganic nitrate (NO_3^-), although most of the N in a typical vineyard soil is bound in organic matter, which is unavailable for direct uptake by the roots. Such organic material is decomposed and mineralized, or converted to plant-available NO_3^-, by soil microbes. In addition, rain and lightning may deposit some N that is present in the atmosphere due to fossil fuel and volcanic emissions. However, the distribution of both organic matter and available inorganic N in the soil is extremely heterogeneous in terms of both time and space. Nitrate concentrations in the soil water can vary over several orders of magnitude, from a few micromoles to approximately 100 mM, even over short distances and time spans (Keller et al., 1995, 2001b). Because of the rapid shoot growth in spring, a vine's N demand is greatest between budbreak and bloom, even though most of the N uptake from the soil occurs after bloom, provided there is sufficient soil moisture (Currle et al., 1983; Peacock et al., 1989). During this period, the vine is heavily dependent on the N reserves stored in the permanent structure because the roots absorb very little N from the soil before five or six leaves have unfolded on the shoots (Löhnertz et al., 1989; Pradubsuk and Davenport, 2010; Treeby and Wheatley, 2006). In fact, nutrient application has little effect on bleeding sap composition (Andersen and Brodbeck, 1991), and the bleeding sap of grapevines fertilized with a whopping 672 kg N ha^{-1} contained similar amounts of NO_3^- to that exuded during budbreak of unfertilized vines (Roubelakis-Angelakis and Kliewer, 1979). Even newly planted rooted cuttings initially grow almost exclusively from stored reserves (Groot Obbink et al., 1973). Remobilization from the reserve pool seems to be independent of soil N availability, and poor N reserve status due to inadequate refilling in the previous growing season can restrict early shoot growth and canopy development and probably flower development as well

(Keller and Koblet, 1995a). Uptake increases progressively through bloom, fruit set, and the first phase of berry growth and may increase further after veraison. In some cases, uptake was found to pause once shoot growth has ceased (Löhnertz et al., 1989), but in other cases the uptake rate was at its peak during this preveraison period and was sometimes—especially in warm climates—followed by another rise after harvest (Conradie, 1980, 1986; Peacock et al., 1989). It seems likely that such discrepancies reflect differences in soil moisture during summer because both shoot growth and nutrient uptake slow in drying soil (Keller, 2005). When water is not limiting, maximum N uptake may occur during the warmest portion of the growing season. In any case, uptake until the end of bloom constitutes less than 30% of the seasonal demand, although the leaves may accumulate up to 60% of their seasonal N requirement. Thus, storage reserves reach a minimum around bloom time or even later (Schaller et al., 1989; Weyand and Schultz, 2006b), which makes vines vulnerable to deficiency if insufficient N is available in the soil. However, the strong dependence on reserve N after budbreak also means that fertilizer application is ineffective at the start of the growing season to alleviate an existing deficiency that arises from low reserve status, whereas the ability of vines to absorb N after harvest in sufficiently warm climates permits enhanced refilling of storage reserves through late-season fertilizer supply (Conradie, 1980, 1986; Holzapfel and Treeby, 2007; Peacock et al., 1989; Treeby and Wheatley, 2006).

Overall high soil N status inhibits lateral root growth and leads to a reduction in the root:shoot ratio (Erlenwein, 1965a; Keller et al., 1995; Zerihun and Treeby, 2002). The inhibition of lateral root growth is mediated by ethylene, whose production increases when certain root nitrate transporters sense high nitrate concentrations (Wang et al., 2012). The depression of root growth also occurs when cuttings are rooted that are derived from vines with a very high N status; this effect is exacerbated by high water status of the "mother" vine (Alleweldt, 1964). Nevertheless, localized N supply (e.g., through a drip irrigation system) to a low N soil stimulates lateral root growth in the new high N zone to enhance N uptake there (López-Bucio et al., 2003; Nibau et al., 2008; Scheible et al., 1997). While new lateral roots proliferate in N-rich soil regions, the outgrowth of existing lateral roots is stimulated under N-limiting conditions, which allows the roots to search for N further away from the plant (Gifford et al., 2008). This response is mediated by the assimilated form of N, glutamine: a high glutamine status leads to lateral root proliferation, whereas low glutamine leads to lateral root elongation. In addition, an increase in auxin in the root tip inhibits its growth, whereas auxin seems to stimulate growth of lateral roots (Tsay et al., 2011; Wang et al., 2012). Despite the increase in root growth under low N, water flow through the roots generally decreases due to higher hydraulic resistance (Chapin, 1991). The lower r_h results from a combination of decreased cell expansion, which leads to narrower xylem vessels, and delayed lignification and death of xylem cells (Plavcová et al., 2012). Like other plants, grapevines with an optimum supply of N and water make longer use of the growing season by extending the effective life of their leaves (Alleweldt et al., 1984a; Lawlor, 2002). When vines are grown on N-rich soil, the chlorophyll degradation that accompanies autumn senescence of the leaves is delayed (Keller et al., 2001b), perhaps at least partly because nitrate uptake leads to increased production and export from the roots of the antisenescence hormone cytokinin (Hwang and Sakakibara, 2006; Sakakibara et al., 2006; Takei et al., 2002). Conversely, general N starvation results in reduced plant growth and yield, gradual chlorosis of older leaves (i.e., they turn pale green and later yellow), and early abscission of those leaves (Scienza and Düring, 1980; Schreiner et al., 2013). It seems likely that such premature leaf senescence is triggered by a combination of sugar

FIGURE 7.4

Association between leaf photosynthesis and chlorophyll content in Müller-Thurgau grapevines grown under two long-term N fertilizer regimes (left; from Keller, 2005), and increased lateral shoot growth and delayed leaf senescence on a vine with high N status (left plant on right) compared with a low-N vine (right plant on right; photo by M. Keller).

accumulation and low N status in the leaves (Figure 7.4) (Wingler et al., 2009). Old leaves become an alternative source of N for the roots and other sink organs when N uptake by the roots fails to meet sink demand (Hikosaka, 2005).

Nitrogen starvation increases ABA production in the roots and decreases the amount of cytokinins, whereas N resupply increases cytokinins (Chapin, 1991; Wilkinson and Davies, 2002). Although cytokinins are predominantly produced in the roots, they can also be produced in the shoots and leaves, especially in the phloem (Kiba et al., 2011). Additionally, N deficit leads to an increase in auxin flow from the shoots to the roots (Krouk et al., 2011). Enhanced cytokinin delivery to the shoots may be one possible reason for the delayed lignification that is sometimes found in vines with high N status; cytokinins reduce xylem formation and lignification (Groover and Robischon, 2006). Because cytokinins stimulate cell division, the growth-depressing effect of N deficiency could also be related, at least in part, to the effect of N on cytokinin synthesis: All cytokinin molecules contain five N atoms within the structure formed by the attachment of an isoprene functional group to ATP or ADP. Through the suppression of cell division, N deficiency can inhibit shoot growth at any time during the growing season (Gastal and Lemaire, 2002). Whereas cytokinins stimulate cell division in the shoot apical meristem and promote lateral bud outgrowth (Ferguson and Beveridge, 2009), they inhibit cell division in the root apical meristem. A decrease in cytokinin production thus decreases main and lateral shoot growth and enhances root growth, and an increase in auxin supply stimulates the outgrowth of lateral roots. The sensitivity of leaf initiation and expansion to fluctuations in N supply results in marked changes in the leaf area: plant weight ratio. In addition, cytokinins may function to inhibit nitrate uptake by the roots (Kiba et al., 2011).

Approximately half of the leaves' N is invested in photosynthetic hardware, especially rubisco and other enzymes, and photosynthetic capacity is strongly correlated with N supply (see Section 5.2 and Figure 5.5). By limiting the amount and activity of these proteins, N deficiency restricts photosynthesis (Chen and Cheng, 2003b). This enzymatic inhibition in turn increases r_s and also results in a potential "energy overload." To avoid excessive light absorption that would damage the photosynthetic apparatus, leaves disassemble a portion of their chlorophylls (i.e., they become chlorotic) and activate their thermal energy-dissipation and antioxidant systems (Chen and Cheng, 2003a; Keller, 2005). In other words, the reduced chlorophyll content of N-starved vines (Keller et al., 2001b; Spayd et al., 1993) is a consequence, not a cause, of the decrease in photosynthesis (see Figure 7.4). The N contained in the leaf proteins, as well as any nitrate remaining in the vacuoles, are recycled to young leaves (see Section 5.3) so that older leaves are the first to become chlorotic. Because inadequate N supply decreases the demand for carbon skeletons for N metabolism and generally suppresses growth more than photosynthesis, sugar export declines and starch accumulates in the leaves (Chapin, 1991; Chen and Cheng, 2003b; Gastal and Lemaire, 2002; Lawlor, 2002). At least part of the decrease in photosynthesis of N-deficient leaves may be caused by feedback inhibition as a consequence of carbohydrate accumulation (Hermans et al., 2006; Schreiner et al., 2013). When this is accompanied by oxidative stress arising from too much light absorption, the leaves—even those of white cultivars—also produce photoprotective anthocyanins in the petioles, leaf veins, and sometimes in the leaf blade and cluster peduncles, which therefore turn red (Figure 7.5) (Currle et al., 1983; Gärtel, 1993; Lillo et al., 2008). Incidentally, although malvidin-based pigments tend to dominate the anthocyanin profile in the berries of many dark-skinned cultivars (see Section 6.2), cyanidin and peonidin derivates are much more abundant in the (red) leaves (Kobayashi et al., 2009; Wenzel et al., 1987; Keller, unpublished data). Conversely, the yellowing of deficient and senescent leaves is caused by

FIGURE 7.5

The angle between leaf blade and petiole in a nitrogen-sufficient (left) and a nitrogen-deficient Müller-Thurgau grapevine (right). Also note the absence of lateral shoot growth and the red pigmentation of the petiole on the right.

Photos by M. Keller.

unmasking and partial retention of carotenoids (see Section 4.1) that accompanies chlorophyll breakdown rather than by new production of yellow pigments.

Surplus sugar in the leaves may become available for export to the roots (Lemaire and Millard, 1999). In contrast to the response to water deficit, the leaf sugar is not needed for osmotic adjustment. Similar to their reaction to water stress, however, leaves decrease the angle between leaf blade and petiole (see Figure 7.5), which further curbs light absorption to avoid an energy surplus. In addition, leaves of N-deficient grapevines tend to have a thicker epidermis (Rühl and Imgraben, 1985). Another similarity with water stress is that under prolonged N stress grapevines start shedding older leaves following the remobilization of nutrients in those leaves and redistribution to developing organs such as young leaves and growing fruit (Currle et al., 1983; Gärtel, 1993; Lawlor, 2002). Taken together, the similarity of vine responses to water and N deficit suggests that the combined effects of these two stress factors are predominantly additive rather than interactive. Thus, vines growing under low soil moisture, present during a drought or deficit irrigation, may require a somewhat higher N supply to achieve photosynthetic rates as high as those of fully irrigated vines (Alleweldt et al., 1984a; Rogiers et al., 2004b). Insufficient N also appears to sensitize plants to subsequent water stress, possibly because the increased ABA production induces the stomata to close at higher Ψ_{leaf}, and accelerates nutrient remobilization followed by leaf senescence and abscission during water stress. On the other hand, abundant water and/or N may make vines more susceptible to stress when one of the two resources suddenly is in short supply.

Insufficient N availability during bloom interferes with flowering and reduces fruit set and cluster initiation in the buds, although too much N also decreases cluster initiation (Ewart and Kliewer, 1977; Keller and Koblet, 1995a; Keller et al., 2001a; Spayd et al., 1993). Low plant N status also seems to aggravate the vulnerability of fruit set to water deficit (Keller et al., 1998). Low N status may sometimes (Hilbert et al., 2003; Keller et al., 1998; Rogiers et al., 2004b), although not always (Keller et al., 2001a), be associated with berries growing larger. This is probably a consequence of diminished sink number due to poor fruit set. A restriction in crop load at the beginning of berry development tends to result in compensatory growth of the remaining berries, unless such compensation is prohibited by the effects of water deficit (Keller et al., 2008). Compensatory berry growth might be possible partly because of limited competition from shoot meristems, whose growth is curtailed by N deficiency. This may leave more photosynthates available for export to the grapes, which in turn would also increase water uptake by the berries. In any case, grapes with a low N content often result in musts that are prone to sluggish or stuck fermentations (Bell and Henschke, 2005; Spayd et al., 1995; see also Section 6.2).

Excessive application of N fertilizer can pose problems too, especially in young vines whose small root systems may temporarily be exposed to very high concentrations of soil N following fertilizer application. Very high NO_3^- concentrations may decrease the osmotic potential (Ψ_π) of the soil solution, which can inhibit growth and decrease stomatal conductance due to insufficient water uptake (Keller and Koblet, 1994), similar to the impact of salinity, discussed later. High N also leads to larger xylem vessel and fiber cells with thinner cell walls whose lignification is delayed and reduced (Plavcová et al., 2012). Though this decreases r_h, it increases the vulnerability to cavitation and may weaken the structural support that the fibers lend the wood. In bearing vines, the vigorous growth and often dense canopies that accompany high soil N status can have undesirable consequences for fruit composition (see Section 6.2) and vulnerability to diseases (see Section 8.2). In addition, vigorous vines with high N status ostensibly are also more vulnerable to physiological disorders such as bunch stem necrosis (Christensen and Boggero, 1985; Currle et al., 1983; Keller et al., 2001a). In some cases, however, N deficiency may cause a higher incidence of bunch stem necrosis (Capps and Wolf, 2000).

Soil N status may also have implications for the availability of other nutrients. Higher amounts of available N can enhance the amount of cations such as K^+, Ca^{2+}, and Mg^{2+} in the soil solution and reduce that of anions such as $H_2PO_4^-$ (Keller et al., 1995). Higher cation availability may be coupled with enhanced uptake by grapevines (Keller et al., 2001b), but it may also increase the potential for rainfall-driven leaching of these nutrients (Perret and Roth, 1996). Over time, excessive N application to vineyard soils might therefore increase the likelihood of nutrient deficiencies developing. This may explain the observation that leaf Ca, Mg, Mn, and B are sometimes increased in N-deficient vines (Schreiner et al., 2013). In contrast, abundant N supply sometimes (Spayd et al., 1993), but not always (Hilbert et al., 2003; Keller et al., 1995, 2001b), decreases P uptake, perhaps because an N-induced reduction in root carbohydrate status occasionally limits carbon availability for mycorrhizal fungi. Furthermore, because NO_3^- uptake is coupled to influx of protons (H^+) and release of OH^-, the pH in the rhizosphere rises with increasing N uptake. In extreme cases, such alkalinization can lead to iron-deficiency-induced leaf chlorosis.

Phosphorus

The anion phosphate ($H_2PO_4^-$; often abbreviated P_i for inorganic phosphate) is a major substrate in energy metabolism (ATP contains three phosphate groups; see Sections 4.1 and 4.4) and synthesis of nucleic acids (as phosphate groups in DNA and RNA), cell membranes (as a constituent of phospholipids), carotenoids, and gibberellins. It also plays an important role in photosynthesis (via photophosphorylation), respiration, and the regulation of a number of enzymes and signal receptors. Reversible protein phosphorylation occurs through addition or removal of a phosphate group, which is carried out by kinases or phosphatases, respectively, and activates ($+P_i$) or deactivates ($-P_i$) proteins (i.e., switches proteins on or off); other proteins are labeled by phosphorylation for later disassembly. One of the most ubiquitous groups of these proteins are the H^+-ATPases that are essential for cell expansion (see Section 3.1) and nutrient uptake and transport (see Section 3.3). Furthermore, phosphorylation of aquaporins regulates water transport across membranes: Phosphorylated aquaporins are active—that is, open (Maurel et al., 2008). In addition, inorganic pyrophosphate (PP_i; $P_2O_7^{4-}$), a by-product of starch production, is essential for phloem function (Koch, 2004). Nevertheless, about 90% of the P in a cell is located in the vacuole (Shen et al., 2011). Because of its importance, the leaf P content increases roughly as the four-thirds power of the leaf N content: $P \sim N^{4/3}$ (Niklas, 2006; Niklas and Cobb, 2006). This relationship also holds approximately true across different grape cultivars (calculated from data in Christensen, 1984). Thus, the seasonal pattern of P uptake by grapevines from the soil more or less mirrors the pattern for N uptake (Conradie, 1980, 1981b; Pradubsuk and Davenport, 2010).

Although most soils contain abundant P, it is one of the least available of all essential nutrients. This "phosphorus paradox" arises because P is extremely insoluble, and hence immobile, due to its affinity to cations such as Ca^{2+}, Mg^{2+}, and Al^{3+}, with which it forms insoluble complexes, and its conversion into organic forms by soil microorganisms (Marschner, 1995; Raghothama, 1999; Shen et al., 2011). In addition, the negative charge of $H_2PO_4^-$ means that it is readily leached out. Its concentration in the soil solution is generally below that of many micronutrients; the typical P concentration is less than 10 μM, whereas the concentration in the cell cytosol is 5—20 mM. Many of the heavily leached and often coarse-textured Australian soils are especially poor in P. Under weakly alkaline conditions, P occurs mostly in the form of HPO_4^{2-}, whereas under weakly acidic conditions the predominant form becomes $H_2PO_4^-$. In acid soils (pH ≤ 5.5), which are common in Bordeaux, France, and in the northeastern United States, P occurs as H_3PO_4 and is the major limiting nutrient (Kochian et al., 2004). This is one reason why such soils are often ameliorated during vineyard establishment by the application of lime ($CaCO_3$) to increase the soil pH. Moreover, P availability is normally highest in the topsoil and decreases sharply with soil depth. As a consequence, fertilizers with high phosphate concentration are often applied to improve P availability. Unfortunately, however, a considerable portion of this supplementary P can leach from vineyards into nearby rivers and lakes, where it may cause eutrophication and lead to proliferation of algae (so-called algal blooms), oxygen depletion, and death of fish.

Roots take up P mainly as $H_2PO_4^-$, using high-affinity transporters (Gojon et al., 2009; Grossman and Takahashi, 2001; Shen et al., 2011). Growing roots initially absorb P in the root tips and root hairs, but this uptake rapidly depletes available P in the rhizosphere (Bucher, 2007; Gilroy and Jones, 2000). Decreasing P availability leads to a rapid decrease in the growth of main roots in the depleted soil area and an increase in the growth of lateral roots and the formation and length of root hairs (Hermans et al., 2006; López-Bucio et al., 2003; Nibau et al., 2008; Osmont et al., 2007; Raghothama, 1999). This shift in root growth pattern may be mediated by sucrose and microRNA pieces imported from the leaves via the phloem (Hammond and White, 2011; Lei et al., 2011). Upon arrival in the roots these molecules may induce accumulation of auxin and ethylene and depletion of cytokinin and gibberellin, which ultimately results in a denser but shallower root system and an increase in the root:shoot ratio (Chiou and Lin, 2011; Grant and Matthews, 1996a; Malamy, 2005; Potters et al., 2009). Although this adaptive strategy enables vines to better explore the rather more P-rich surface soil, it also makes them more vulnerable to water stress when the soil dries out. This may be why the response is limited to situations in which the entire root system, rather than just portions of it, experiences P deficiency (Schachtman and Shin, 2007). The roots of some P-deficient species also exude organic acids such as citrate and malate that acidify the rhizosphere and act as anion exchangers that chelate metal cations and solubilize P from rock phosphate (Dakora and Phillips, 2002; Hammond and White, 2008; Marschner, 1995). Such organic acid extrusion, however, comes at a high cost: it can drain more than 10% of a plant's dry matter (Plaxton and Tran, 2011). In addition, P deficiency results in proton extrusion to acidify the soil, and the lower soil pH can greatly enhance the P available to the roots (see Figure 7.3), as well as increase the availability of potassium (K^+) and manganese (Mn^{2+}). Grapevines also rely heavily on mycorrhizal fungi for P uptake, and P-starved plants ostensibly encourage mycorrhizal colonization of their roots by exudation of strigolactone sesquiterpenes that help the fungi to establish root contact (Akiyama et al., 2005; Bais et al., 2006). By modulating auxin flux, the strigolactones also stimulate lateral root growth in P-deficient plants, while they suppress lateral root growth under P-sufficient conditions (Ruyter-Spira et al., 2011, 2013). At the same time, strigolactones are also transported via the xylem to the shoots, where they inhibit lateral shoot growth (Gomez-Roldan et al., 2008; Kohlen et al., 2011; Umehara et al., 2008). A restriction of leaf number and size is one of the earliest and most reliable symptoms of P deficiency. Thus, the increase in carbon partitioning to the roots at the expense of partitioning to the shoots results in a change in plant architecture (i.e., increase in root:shoot ratio) that may be coordinated by strigolactones and favors P uptake while reducing its use for canopy construction (Grant and Matthews, 1996a; Ruyter-Spira et al., 2013).

Low P supply limits cell division, restricting leaf initiation in the shoot apical meristem and expansion of newly developed leaves (Chiera et al., 2002). Reduced cell division may also be responsible for the inhibition of flower cluster initiation or differentiation in P-deficient vines (Gärtel, 1993; Grant and Matthews, 1996b; Skinner et al., 1988). Phosphate deficiency can also increase r_s and r_h so that the reduction in water supply to growing organs restricts cell expansion, which in turn strongly limits leaf expansion (Clarkson et al., 2000). Therefore, although as a phloem-mobile nutrient P can be remobilized and recycled from old leaves and woody structures to support growing organs, P deficiency primarily restricts the sink activity of shoot meristems, and the resulting lack of demand for assimilates in growth eventually leads to sugar and starch accumulation in source leaves (Grant and Matthews, 1996a; Hermans et al., 2006; Kochian et al., 2004). Phosphorus deficiency reduces source activity by decreasing leaf photosynthesis and assimilate partitioning, partly because there is insufficient P_i for ATP synthesis and partly because the leaves do not fully activate their rubisco (Paul and Foyer, 2001; Woodrow and Berry, 1988). The latter could be due to feedback inhibition by excess sugar and seems to be especially pronounced at temperatures below approximately 15°C (Hendrickson et al., 2004b). Nonetheless, sugar export via the phloem to the roots may continue or even increase to stimulate and support lateral root branching (Hammond and White, 2008, 2011; Lei et al., 2011). Deficiency also leads to salvage of P_i from nucleic acids and phospholipids throughout the plant, and the latter are replaced by galacto- and sulfolipids (Plaxton and Tran, 2011; Schachtman and Shin, 2007). Moreover, P-deficient leaves remain small and dark grayish-green rather than turning chlorotic as in other cases of nutrient starvation (Gärtel, 1993; Hammond and White, 2008). Severely P-deficient leaves may turn red due to anthocyanin production from accumulated sugars that are not exported from the leaves (Currle et al., 1983; Grossman and Takahashi, 2001; Lillo et al., 2008). This pigmentation is probably a photoprotective reaction because the decline in photosynthesis is associated with surplus energy (i.e., light absorption and electron transfer proceed even as CO_2 fixation declines) that results in oxidative stress (Feild et al., 2001; Hoch et al., 2003). In addition, insufficient P supply appears to restrict Mg transport in the xylem, which can lead to symptoms of Mg deficiency (Skinner and Matthews, 1990). Whereas P starvation may be associated with increased iron accumulation in plants, excess P may lead to zinc and iron deficiency through complexation (Gärtel, 1993; Lei et al., 2011).

Potassium

Potassium (K^+) is the most abundant cation (a molecule with a positive charge) in plant cells. Its concentration in grape leaves is approximately half that of N (Gärtel, 1993). In contrast with all other macronutrients except calcium, K is not incorporated into organic compounds but remains in its ionic form (Robinson, 1994). As such, it is one of the cells' major osmotic solutes and plays a key role in cell expansion (see Section 3.1) and stomatal movement (see Section 3.2). Thus, rapid plant growth and development require large K^+ fluxes to provide this ion to the growing tissues. This means that growth in general is sensitive to the vine's K^+ status. Cells also use K^+ to neutralize the negative charge of anions, which helps maintain membrane potential and counterbalance the movement of other cations such as protons (H^+). The protons are needed for the activity of enzymes such as ATP synthase, as well as H^+-ATPases and other transport proteins (Amtmann and Blatt, 2009; Clarkson and Hanson, 1980; Terrier et al., 1998). In addition to its role in membrane transport, K^+ also stimulates the loading of sucrose into the phloem for export, perhaps because it helps maintain the electrical neutrality that is necessary to generate pH gradients by H^+-ATPases (Gajdanowicz et al., 2011; Lalonde et al., 2004; Marten et al., 1999; see also Section 5.1). The K^+ concentration in the soil solution varies from approximately 0.1–6 mM, whereas its concentration in the cell cytosol is approximately 100 mM (Ashley et al., 2006). Transport of K^+ in the xylem affects r_h and, thus, sap flow rate (see Section 3.3). Whereas initial growth of grapevines after budbreak is dependent on K reserves mobilized from the permanent structure of the vine, uptake from the soil increases well before bloom and, provided soil moisture is not limiting, proceeds more or less linearly throughout the growing season until a few weeks before leaf fall (Conradie, 1981b; Pradubsuk and Davenport, 2010; Williams and Biscay, 1991). Developing and ripening grape berries are a particularly strong sink for K.

Inadequate K^+ supply strongly reduces xylem sap flow even under well-watered conditions, which limits shoot and fruit growth and greatly increases the risk of drought stress (Currle et al., 1983). Potassium deficiency also suppresses sugar transport in the phloem and can result in sucrose (but not starch) accumulation in the leaves to substitute for the missing K^+ as osmoticum (Cakmak et al., 1994; Hermans et al., 2006; Tsay et al., 2011). This explains not only why root growth cannot increase (and often stops) in K-deficient plants but also why deficiency leads to feedback inhibition of photosynthesis. With sucrose being "trapped" in the leaves and photosynthesis declining, the consequences for fruit production and ripening can be severe. At the same time, its mobility in the phloem enables K^+ to be recycled from older to younger organs and to move preferentially toward growing tissues or organs, including the fruit and the permanent vine structure (Conradie, 1981b; Currle et al., 1983; Mpelasoka et al., 2003). While growth of primary and lateral roots is inhibited, ethylene production that is induced by K deficiency stimulates the outgrowth of root hairs (Tsay et al., 2011).

Visual deficiency symptoms include glossy leaves, especially under high solar irradiance, often followed by bronzelike discoloration from dying epidermis cells and pale-yellow leaf margins (Figure 7.6). In more severe cases, the leaf margins roll upward ("cupping") and eventually become necrotic before the leaves are shed (Currle et al., 1983; Gärtel, 1993). These symptoms develop

FIGURE 7.6

Leaf symptoms of potassium deficiency in Syrah.

Photo by M. Keller.

more quickly under high irradiance because the excited "surplus" electrons lead to oxidative stress (Marschner and Cakmak, 1989). Leaf symptoms are also more severe on heavily cropped vines due to the translocation of K to the growing berries (Currle et al., 1983; Gärtel, 1993) that, between bloom and harvest, accumulate far more K than any other nutrient. Indeed, mature grapes contain almost twice as much K as N and approximately 10 times as much as P and Ca, which are the next most abundant minerals in grape berries. Depending on crop level, the harvested fruit contains 50–75% of a vine's total amount of K (Conradie, 1981b; Rogiers et al., 2006). If the deficiency is not relieved, shoot vigor declines gradually over several years (Gärtel, 1993).

Deficient vines become more susceptible to powdery mildew and to cold injury in winter (Currle et al., 1983; Gärtel, 1993). Moreover, due to the importance of K^+ as an osmoticum for pollen hydration and germination, and for pollen tube growth, K deficiency interferes with pollination and fertilization, which results in poor fruit set. Deficiency symptoms are usually associated with an accumulation of polyamines (especially putrescine) in the shoots and leaves (Geny et al., 1997b). It has been proposed that it is this putrescine buildup that induces the visible leaf symptoms (see Figure 7.6). Nonetheless, putrescine application can improve fruit set (Geny et al., 1997a, 1998), perhaps because a putrescine metabolite (named hypusine-eukaryotic translation initiation factor 5A) is required for the transfer of RNA from the cell nucleus to the cytoplasm during embryo development (Takahashi and Kakehi, 2010). Polyamines are ultimately breakdown products of the amino acid arginine (via ornithine or agmatine; Kusano et al., 2008; Takahashi and Kakehi, 2010), and as K deficiency grows more severe, putrescine may become the main organic soluble N compound. The cause and/or purpose of the accumulation of polyamines during K deficiency is unknown. However, given that polyamines can block cation channels and enhance active transporters (H^+-ATPases) in plant cells (Alcázar et al., 2010; Kusano et al., 2008; Takahashi and Kakehi, 2010), it is possible that they serve to prevent compensatory but detrimental accumulation of Na^+ in the absence of K^+. Despite the decrease in arginine that accompanies polyamine accumulation, grapes from K-deficient vines may not result in sluggish or stuck fermentations because yeast can readily metabolize polyamines.

Putrescine also accumulates in leaves, especially those of vigorously growing young vines, during the stop-and-go growth caused by widely fluctuating spring temperatures (Adams et al., 1990). This can lead to a syndrome referred to as "spring fever," whose symptoms are virtually identical to those of K deficiency, even though the leaves have almost normal K concentrations (but elevated N concentrations). It is possible that the accumulation of putrescine is part of a cold acclimation response that enhances the leaves' ability to survive freezing temperatures (Cuevas et al., 2008). The link between putrescine and low temperature may also be the reason why K deficiency symptoms appear to be more severe under cool conditions. In contrast to K deficiency, however, spring fever symptoms are confined to the oldest leaves, and the shoots usually outgrow the condition within a few weeks. Nevertheless, the berries often remain small, and ripening is delayed.

Potassium deficiency can result from competition for root uptake by other cations present at high concentration in the soil solution (Wang and Wu, 2013). This is possible under saline conditions, where sodium (Na^+) is the predominant dissolved cation but also where NH_4^+ is the predominant nitrogen form. This is typical of acid soils (NH_4^+-based fertilizers in turn contribute to soil acidification), defined as soils with a pH ≤ 5.5 (Kochian et al., 2004). The roots often respond to inadequate K^+ by growing sideways; that is, they deviate from their normal gravitropic growth, which may help them to explore (and exploit) previously untapped soil patches that contain more

K$^+$ (Ashley et al., 2006). Rootstocks differ in their ability to take up K$^+$ and to release K$^+$ into the xylem; for instance, it appears that rootstocks derived from *V. berlandieri* do not take up K$^+$ as readily as those derived from *V. champinii*.

Excess K$^+$ can decrease Ca^{2+} and Mg^{2+} concentrations in the leaves and perennial organs of the vine and induce symptoms of Mg deficiency due to competition for uptake by the roots (Morris et al., 1980, 1983). This is particularly pronounced in young, nonfruiting vines, especially on sandy soils with a low pH (Gärtel, 1993). At very high concentrations (100−200 mM), K$^+$ can even inhibit enzymes, thus becoming toxic.

Sulfur

Sulfur is an essential constituent of the amino acids cysteine and methionine (and hence of proteins), lipids, intermediary metabolites (acetyl-CoA) in energy (ATP) generation and electron transport, and molecules (e.g., the cysteine-containing peptide glutathione) involved in the protection of tissues against oxidative stress (Kopriva, 2006; Takahashi et al., 2011). Sulfur-containing compounds (e.g., cysteine and glutathione) are used by cells to counteract reactive oxygen species brought on by such environmental stressors as drought, cold, heat, high light, and fungal attack (Noctor and Foyer, 1998). Attachment of a sulfate (SO_4^{2-}) group deactivates the hormones brassinosteroid and jasmonate (Takahashi et al., 2011). In addition, S-containing proteins in concert with iron-containing proteins (via the so-called ferredoxin/thioredoxin system) regulate starch production and remobilization in response to light and sugar supply. Following its uptake into the roots by SO_4^{2-}/H^+ cotransport, SO_4^{2-} can be transported in both the xylem and phloem and temporarily stored in the cell vacuoles before it is reduced in the plastids to sulfite (SO_3^{2-}) and further to sulfide (S^{2-}) which is then incorporated, mostly in the cytosol, into cysteine, which can be converted to methionine (Amâncio et al., 2009; Takahashi et al., 2011). Both uptake and reduction of SO_4^{2-} require energy in the form of ATP; reduction additionally requires iron in the form of ferredoxin. Glutathione, produced from cysteine, is the major transport and storage form of reduced S; storage occurs mostly in the bark (Cooper and Williams, 2004; Kopriva, 2006). Nevertheless, plant growth seems to depend on sustained uptake, usually as SO_4^{2-}, from the soil and transport in the xylem to the leaves, where it is reduced and assimilated into cysteine, methionine, and other organic compounds or partially reexported in the phloem as SO_4^{2-} or glutathione (Gojon et al., 2009; Grossman and Takahashi, 2001; Leustek and Saito, 1999).

Feedback inhibition of SO_4^{2-} uptake ensures that uptake is in step with S demand (Takahashi et al., 2011). When S availability in the soil is high, the roots decrease the activity of their high-affinity transporters, whereas they increase such activity when soil S becomes low (Amâncio et al., 2009). Insufficient S supply, however, leads to a depletion of cysteine and glutathione pools, which is associated with an increase in reactive oxygen species (Schachtman and Shin, 2007). This is especially important because the production or release (possibly from glutathione or cysteine degradation) of elemental S or hydrogen sulfide (H$_2$S) may be involved in the vine's defense strategy against certain fungal and bacterial pathogens (Cooper and Williams, 2004; see also Section 8.1). Sulfur deficiency could thus result in increased susceptibility to pathogen attack because only plants growing at high SO_4^{2-} availability are able to release sufficient S to counter invasion. Insufficient S supply also leads to lower photosynthesis as a result of a reduction in the capacity to assimilate S, and to an increase in photorespiration and flavonoid production and breakdown of lipids (Takahashi et al., 2011). Deficiency symptoms include yellowing (chlorosis) of young leaves and stunted shoot growth. Sulfur-starved plants normally accelerate root elongation, form prolific lateral roots close to the root tip to enhance uptake, and decrease shoot growth so that the root:shoot ratio increases, perhaps in response to increased auxin production (López-Bucio et al., 2003; Schachtman and Shin, 2007). Application of S-based fungicides (predominantly elemental S; e.g., for powdery mildew control) may have the added benefits of enhancing vine defense and at the same time countering S deficiency (Cooper and Williams, 2004). It is thought that the protective benefits of S applications in vineyards were discovered in Roman times over 2000 years ago (Rausch, 2007). Despite the inhibition of microbial growth by S, residues on berries decrease rapidly following treatment, so that applications, at least up to veraison, are not usually a concern for fermentation (Thomas et al., 1993). However, foliar S application during hot conditions ($>32°C$) can induce burn symptoms on leaves, shoots, and berries.

Calcium

Calcium is unique among macronutrients in that a high proportion of a plant's Ca^{2+} is located in the cell walls, where it serves as a reinforcing agent. In fact, cell walls probably owe their rigidity to the cross-linking of pectins by Ca^{2+} (Ferguson, 1984; Hamant and Traas, 2010; Jarvis, 1984), which also holds adjacent cells together (i.e., prevents them from sliding). Indeed, the Ca^{2+} concentration in the apoplast is 300–1000 times higher than that in the cytoplasm, and most of this Ca^{2+} is tightly bound to pectic acids (as Ca-pectate) in the cell walls (Boyer, 2009). Although, like K, Ca is not actually incorporated into organic molecules (Amtmann and Blatt, 2009), the ionic bonding of Ca^{2+} contrasts with the mostly soluble apoplastic K^+. Another important site of Ca^{2+} accumulation and storage, in the form of Ca-oxalate, is the cell vacuole. Crystals of Ca-oxalate, especially the needle-shaped raphides, may serve a defense role by deterring herbivores. Because of its ability to connect lipids and proteins at membrane surfaces, Ca^{2+} is important in maintaining membrane integrity; it stabilizes membranes by bridging lipid and protein phosphate and carboxylate groups and prevents membrane damage and leakiness to solutes (Clarkson and Hanson, 1980; Hirschi, 2004). Because membrane leakage (i.e., loss of semipermeability) results in cell death, Ca delays senescence and organ abscission (Poovaiah and Leopold, 1973). Ca^{2+} ions inhibit the Mg^{2+}-dependent reactions of respiratory and intermediary metabolism (Clarkson and Hanson, 1980) and alter the water permeability of membranes by modulating aquaporin activity (Johansson et al., 2000). Plants also require Ca to produce electrons, protons, and oxygen during photosynthesis (see Section 4.1). In addition, Ca is involved in intracellular signaling in response to osmotic stress that arises from many environmental impacts (Plieth, 2005; Zhu, 2002). Calcium serves as a second messenger, whereby an external stimulus leads to an increase in reactive oxygen species that results in temporary oscillations in the cytosolic Ca^{2+} concentration; the period, frequency, and amplitude of these oscillations amount to a characteristic "Ca^{2+} signature" that encodes information about the nature and strength of the stimulus (Hetherington and Brownlee, 2004; McAinsh and Pittman, 2009). For instance, ABA promotes release of Ca^{2+} from the guard cell vacuoles and influx from the apoplast to induce stomatal closure (Allen et al., 2001; Amtmann and Blatt, 2009), and auxin and gibberellin coordinate cell expansion by inducing a rise in cytosolic Ca^{2+} released from intra- and extracellular stores. Calcium binds to a small protein termed calmodulin to modulate the activity of other proteins by inducing their phosphorylation (Scrase-Field and Knight, 2003; Snedden and Fromm, 2001; Veluthambi and Poovaiah, 1984). Moreover, Ca^{2+} influx and a cytosolic Ca^{2+} gradient with increasing concentration toward the tip are necessary for the directional growth of root hairs and pollen tubes (Dutta and Robinson, 2004; Holdaway-Clarke and Hepler, 2003; Véry and Davies, 2000).

At high concentration, Ca^{2+} can disrupt metabolism by precipitating phosphate as $CaHPO_4$; thus, high amounts of Ca are extremely toxic to cells (He et al., 2014; Plieth, 2005). Plants utilize ATP-driven pumps (often regulated by calmodulin) to keep the concentration of cytosolic Ca three or four orders of magnitude lower (at <0.1 μM) than that of apoplastic and vacuolar Ca, allowing only small Ca oscillations potentially generated as signals by the pumps (Hetherington and Brownlee, 2004; Hirschi, 2004; McAinsh and Pittman, 2009). Uptake of Ca occurs mostly near the root tips across nonselective cation channels, increases with Ca^{2+} availability in the soil, and varies along with plant transpiration (Gilliham et al., 2011; Karley and White, 2009; Pradubsuk and Davenport, 2010). As plants have little control over Ca uptake, they sequester Ca as Ca-oxalate crystals inside specialized cells called idioblasts to prevent toxicity (Franceschi and Nakata, 2005; He et al., 2014; Storey et al., 2003b). The idioblasts are located in roots, leaves (especially along veins), petioles, and fruit. This biomineralization takes the soluble Ca out of "circulation," and the crystals may serve as a storage form that can be rapidly mobilized during Ca starvation and during early spring growth. Accumulation, in the form of raphide crystals, in the root cortex occurs especially near the root tip, from where Ca may be supplied to the xylem (Storey et al., 2003b).

The Ca concentration in grape leaves and other organs is similar to or higher than that of N, but in contrast to N, its concentration increases as a leaf ages (Gärtel, 1993; Pradubsuk and Davenport, 2010). Because Ca^{2+} is poorly phloem-mobile, it is preferentially supplied to rapidly transpiring organs such as mature leaves and mostly becomes immobile once deposited in a particular organ; in addition to the leaves, the bark seems to be a major sink for Ca in grapevines (Conradie, 1981b; Gilliham et al., 2011; Karley and White, 2009; Peuke, 2010).

Plants on a low-Ca "diet" grow smaller and fewer leaves than plants with abundant Ca supply (Suárez, 2010), and deficiency symptoms appear mainly in young, expanding leaves. Symptoms of Ca deficiency develop when Ca supply to cells cannot keep pace with Ca movement to the vacuoles. Deficiency causes cell walls to disintegrate and membranes to become leaky, which causes cell death and can lead to the collapse of affected tissues (Hirschi, 2004). This can lead to marginal leaf necrosis and berry shrinkage beginning at the tip of fruit clusters, which resembles symptoms of bunch stem necrosis (Gärtel, 1993). Indeed, the appearance of bunch stem necrosis has been associated with insufficient Ca^{2+} (and often Mg^{2+}) relative to K^+ in rachis tissues, especially in the epidermal and subepidermal layers (Currle et al., 1983; Feucht et al., 1975). Due to its importance for pollen tube growth, Ca deficiency is also detrimental to fertilization and fruit set. Although xylem-supplied Ca tends to accumulate in the stomatal regions of grape berries (Blanke et al., 1999), local deficiency in the fruit may be promoted by excessive canopy transpiration and rapid shoot growth diverting xylem flow, and thus Ca supply, away from the fruit. This could be especially pronounced during periods of rapid berry cell expansion that requires Ca^{2+} for incorporation in new cell walls and membranes and as an intracellular signal. Excessive Ca supply, on the other hand, may retard the rate of fruit ripening, possibly by preventing cell wall disassembly (Ferguson, 1984). However, high plant Ca status can also improve survival after freezing (see Section 7.2).

A shortage in Ca supply, similar to salinity, leads to downregulation of aquaporins, which in turn increases the hydraulic resistance in roots and interferes with water flow to the xylem (Luu and Maurel, 2005). In addition, Ca deficiency makes plants more susceptible to damage by salinity or low soil pH (Hirschi, 2004; Plieth, 2005). Natural Ca deficiency is rare and occurs primarily on soils with very low pH because the highly toxic aluminum (Al^{3+}) becomes very soluble at pH <5 and blocks Ca^{2+} uptake. Excessive soil K^+ (Delas and Pouget, 1984) and/or Mg^{2+} and also salinity can induce Ca deficiency. However, the availability of Ca^{2+} (and Mg^{2+}) at the root surface often greatly exceeds the demand of grapevines, especially in high-pH soils (Storey et al., 2003b). Therefore, vines grown on calcareous soils generally have considerably higher tissue Ca contents (and often higher Mg but lower K contents) than vines grown on other soils. On such soils, the amount of Ca in a vine (especially in its permanent structure) can even exceed its content of N, which is normally the most abundant mineral element in grapevines. Moreover, in soils with a pH >7.5, the soils' calcium carbonate ($CaCO_3$) can lead to precipitation of phosphorus, iron, zinc, copper, and manganese, making these nutrients less plant-available. Different *Vitis* species are adapted to a wide range of Ca^{2+} in different soils and are often grouped as being either calcicole (lime-tolerant or adapted to calcareous or alkaline soils) or calcifuge (lime-sensitive or adapted to neutral to acidic soils). For instance, *V. labrusca* and *V. riparia* are lime sensitive, whereas *V. vinifera* and *V. berlandieri* are lime tolerant.

Magnesium

Although Mg^{2+} is a structural component of chlorophyll, less than 20% of a vine's total Mg content is bound in chlorophyll. As much as 75% of the Mg in leaves may be involved in the manufacture of proteins (Karley and White, 2009). As a divalent cation, Mg^{2+} is an important cofactor required to activate many enzymes (including the key photosynthetic protein rubisco and the N-assimilatory enzyme glutamine synthetase; see also Sections 4.2 and 5.3) and transport proteins (including the ATPase proton pumps responsible for phloem loading; see Section 5.1) (Cakmak and Kirkby, 2008; Clarkson and Hanson, 1980). In addition, Mg seems to be involved in protecting anthocyanins from degradation in cell vacuoles (Sinilal et al., 2011).

In contrast with most other nutrient ions, little is known about how roots take up Mg^{2+} and how they regulate this uptake and subsequent transport. Uptake and transport to the xylem do seem to be mostly active, via transport proteins, while Mg delivery to the leaves varies with transpiration rates (Karley and White, 2009). The pattern of Mg^{2+} uptake by grapevine roots appears to be similar to that of K^+, starting soon after budbreak and continuing until shortly before leaf fall (Conradie, 1981b; Pradubsuk and Davenport, 2010). Grape leaves contain approximately 10-fold less Mg than N (Gärtel, 1993; Pradubsuk and Davenport, 2010).

Magnesium deficiency is a quite common predicament of vines growing in sandy, very acidic (pH <4.5) soils, where high concentrations of aluminum (Al^{3+}), ammonium (NH_4^+), and hydrogen ions (H_3O^+) tend to inhibit Mg uptake (Gärtel, 1993). However, high Ca^{2+} (e.g., in calcareous soils with high pH) and/or K^+ availability can also curb Mg uptake and induce Mg deficiency due to competition among these cations for root uptake (Delas and Pouget, 1984); the same applies to competition with Na^+ in saline soils (Shaul, 2002). Grapevines grafted on rootstocks derived from American *Vitis* species may be more prone than own-rooted *V. vinifera* cultivars to such interference of high soil K with Mg uptake (Mullins et al., 1992). In addition, although application of N fertilizers can enhance the (short-term) availability of Mg in the soil solution and its uptake by the vine, it may also increase the (long-term) risk for leaching and depletion of the surface soil (Keller et al., 1995, 2001b). Because Mg is much more phloem-mobile than Ca, Mg deficiency symptoms first become apparent as chlorotic discoloration of the interveinal areas of old leaves as chlorophylls are being dismantled (Currle et al., 1983; Karley and White, 2009). However, even before any symptoms become visible, the inhibitory effect of Mg starvation on phloem loading leads to accumulation of sucrose and starch in the leaves, and the export of sucrose and amino acids declines (Cakmak and Kirkby, 2008; Cakmak et al., 1994; Hermans et al., 2006). This explains why the roots of Mg-starved plants stop growing, similar to the situation with K and Zn deficiency but in stark contrast to the response to N and P deficiency. The link between Mg status and sucrose export may also be the reason for the observed correlation between petiole Mg content and berry sugar (van Leeuwen et al., 2004). The buildup of sugar in the leaves, in turn, leads to feedback inhibition of photosynthesis and results in the degradation of chlorophyll, probably to mitigate excess light absorption leading to oxidative stress (see Section 7.1). In fact, Mg-deficient plants are extremely light-sensitive; high light intensity accelerates the appearance of the characteristic interveinal chlorosis of Mg-deficient leaves, and the deficiency symptoms are more apparent on sun-exposed than on shaded leaves (Marschner and Cakmak, 1989). The surplus sugar is used to produce anthocyanins as a photoprotectant in the interveinal areas of the leaves of some red but not white cultivars (Figure 7.7) (Currle et al., 1983). In addition, even before the destruction of chlorophyll, the leaves activate their antioxidant system so that ascorbate and glutathione concentrations increase (Cakmak and Kirkby, 2008). It therefore appears that vines grown in areas that typically experience high irradiance may require more Mg to avoid the risk of enduring oxidative stress (Cakmak and Kirkby, 2008). Some *V. vinifera* cultivars (e.g., Barbera, Zinfandel) seem to be more vulnerable to Mg deficiency than are other cultivars, but such differences are probably often masked by rootstocks.

In addition to the detrimental consequences for yield formation and fruit ripening due to the decrease in phloem export from the leaves, insufficient Mg availability (sometimes in conjunction with low Ca and high K supply) has also been implicated in the development of a physiological disorder termed bunch stem necrosis that may develop during ripening (Cocucci et al., 1988; Keller and Koblet, 1995b). The rachis of affected fruit clusters develops reddish-brown to black necrotic

FIGURE 7.7

Leaf symptoms of magnesium deficiency in the white cultivar Müller-Thurgau (left) and the red cultivar Pinot noir (center), and cluster abscising due to severe bunch stem necrosis (right).

Photos by M. Keller.

lesions, beginning with some stomatal guard cells or a few subepidermal cells following collapse of their cell walls (Hall et al., 2011; Hifny and Alleweldt, 1972; Jähnl, 1967, 1971; Theiler, 1970). The lesions probably reflect the oxidation of phenolic compounds and rapidly spread into, along, and/or around the stem and in the process may girdle the phloem, which effectively interrupts import of assimilates into the berries distal to the necrosis, while initially leaving the xylem intact. The girdling effect stops sugar accumulation in the berries, whereas the degradation of malate continues, which can lead to poor fruit quality and berry shrinkage (Bondada and Keller, 2012). The rachis eventually dries up, and in severe cases the distal portion of affected clusters may be abscised. Clusters appear to be most susceptible to this syndrome before and during bloom, at which time the disorder is also termed inflorescence necrosis (see Section 6.1), and after the beginning of sugar accumulation at veraison (Jackson and Coombe, 1988; Keller et al., 2001a). Cabernet Sauvignon, Riesling, and Gewürztraminer are particularly sensitive to bunch stem necrosis, whereas the Pinots and Chardonnay are relatively insensitive (Currle et al., 1983). Affected rachis tissues are vulnerable to infections by saprophytic pathogenic fungi such as *Botrytis cinerea*. It is conceivable that the role of Mg in sugar export from the leaves may be to blame for this disorder, especially because symptom development is often related to adverse environmental conditions that limit photosynthesis and assimilate supply, and the incidence tends to be higher on cool compared with warm vineyard sites (Jackson, 1991; Jackson and Coombe, 1988; Keller and Koblet, 1994, 1995b; Pérez Harvey and Gaete, 1986). In addition, loss of leaf area during the ripening period also seems to favor the appearance of bunch stem necrosis (Redl, 1984). Finally, drying winds and high vapor pressure deficit following even brief episodes of rainfall, as well as high light intensity, seem to favor the sudden appearance of necrotic lesions on the rachis. Whether irrigation can substitute for rainfall in dry climates is unknown. In Europe, applying Mg sprays, typically in the form of dissolved Mg-sulfate or Epsom salt ($MgSO_4 \cdot 7H_2O$), directly to the fruit zone at veraison has successfully increased rachis Mg contents and partially alleviated the incidence of bunch stem necrosis.

Excessive soil Mg^{2+}, defined as greater than 40% of the soil's cation exchange capacity, is rare but occurs in serpentine soils that are poor in silicates but rich in ferromagnesium minerals. Such soils occur mainly in the North American Pacific coastal ranges and California's Sierra Nevada foothills and are often associated with high pH; very low K^+, Ca^{2+}, and P; and high Fe, B, cobalt (Co), and Ni. Although no symptoms of Mg toxicity are known, it is thought that very high amounts of Mg in leaves may impair photosynthesis (Shaul, 2002). Moreover, excessive Mg can lead to poor soil structure and may induce K and, sometimes, P deficiency, especially toward the end of the growing season. Conversely, in high Na soils, high Mg contents may ameliorate some of the adverse effects of Na on soil structure, provided there is also sufficient Ca in the soil.

7.3.2 TRANSITION METALS AND MICRONUTRIENTS

Iron

Iron is the most abundant metal on Earth but is extremely insoluble in oxygen-rich environments, which includes most soils (Schmidt, 2003). It is generally the most abundant micronutrient in grapevines (Gärtel, 1993). Due to its ability to occur in two different ionic states (Fe^{2+} and Fe^{3+}), it is a cofactor or ingredient of proteins involved in electron transfer (e.g., ferredoxin) and many enzymes that catalyze reduction/oxidation (redox) reactions (Clarkson and Hanson, 1980; Curie and Briat, 2003; Curie et al., 2009). As such, it is involved in chlorophyll synthesis (although the chlorophyll molecule does not contain Fe), photosynthesis, and respiration and, via enzyme activation, in C, N, and S assimilation, lipid and hormone (e.g., ABA) synthesis and degradation, and DNA synthesis and repair, as well as detoxification of reactive oxygen species. In addition to leaves, Fe is also abundant in seeds and pollen and may be necessary for pollen production (Bertoldi et al., 2011; Curie et al., 2009). Its profusion in soil particles is the source of many erroneously high apparent Fe concentrations, and consequently inaccurate fertilizer recommendations, resulting from analysis of leaf samples "contaminated" with dust.

In the presence of oxygen (i.e., in the upper soil layers), iron generally occurs in the oxidized form Fe^{3+} (ferric iron), although the reduced form Fe^{2+} (ferrous iron) is the soluble cation and is the form required by plants. However, Fe^{3+} forms soluble complexes (chelates) with many organic and inorganic molecules, including humic and fulvic acids (which are important components of soil organic matter), tannins, and phosphate. Like other plants termed "strategy I" species, *Vitis* species absorb iron molecules as Fe^{3+}-chelates, which the enzyme ferric chelate reductase (FCR) reduce to Fe^{2+} on the root plasma membranes (Bavaresco et al., 1991; Briat and Lobréaux, 1997; Kobayashi and Nishizawa, 2012; Schmidt, 2003; Varanini and Maggioni, 1982). The poorly soluble yet highly reactive Fe^{2+} ions are moved across the membranes by Fe^{2+} transporters and are "picked up" in the root apoplast by citrate to be transported to the leaves in the xylem sap mostly as soluble and metabolically inactive Fe^{3+}−citrate complexes (Curie and Briat, 2003; Kobayashi and Nishizawa, 2012). Upon arrival in the leaves, they are reactivated to Fe^{2+} by FCR before they can be taken up into the leaf mesophyll. The FCR protein is highly sensitive to pH, and enzyme activity declines as the apoplast pH increases (Nikolic et al., 2000). Inside the cells, unused Fe is again chelated, this time by a nonprotein amino acid called nicotianamine that apparently serves to keep Fe in a soluble form and ensures its correct distribution among the various cell organelles and in the phloem (Curie and Briat, 2003; Curie et al., 2009; Schmidt, 2003).

Although Fe^{3+} can be stored in a protein termed ferritin, approximately 80−90% of the Fe in a leaf is located in the chloroplast as part of the photosynthetic machinery (Marschner, 1995). Therefore, Fe deficiency impairs electron transfer, which markedly decreases photosynthesis and sugar production for export to the vine's sink organs (Bertamini and Nedunchezhian, 2005). This in turn strongly limits vine growth and yield (Bavaresco et al., 2003; Gruber and Kosegarten, 2002) and results in oxidative stress from an excess of absorbed light. It is possible that such oxidative stress is the reason for the typical decline in leaf chlorophyll. Thus, deficiency symptoms include leaf chlorosis (from gradual loss of chlorophyll or lack of its production; Figure 7.8) beginning at the margins and progressing to the interveinal areas of young leaves, followed by marginal necrosis and, finally, abscission. Developing lateral shoots are stunted, often show pink internodes, and have small, chlorotic leaves that may fail to unfold, although Fe is phloem-mobile and can be remobilized from older and senescing leaves (Curie and Briat, 2003). Consequently, when a vine's crop load is high, developing grape berries may satisfy some of their Fe demand by importing Fe remobilized from leaves and even from the permanent structure of the vine (Pradubsuk and Davenport, 2011). Under severe deficiency, the tendrils and inflorescences also become chlorotic, and fruit set is poor (Gärtel, 1993), probably because of the Fe requirement for pollen development.

Conversely, excess Fe can also cause oxidative stress, reduce leaf chlorophyll, and result in chlorotic or necrotic spots on the leaves (Curie and Briat, 2003). Because Fe^{2+} reacts rapidly with hydrogen peroxide (H_2O_2) to form reactive oxygen species in the Fenton reaction, plants tightly control the availability of both Fe^{2+} and H_2O_2 (Curie et al., 2009; Halliwell, 2006; Lane, 2002).

FIGURE 7.8

Leaf symptoms of lime-induced chlorosis in Concord.

Photos by M. Keller.

Excess Fe can be stored in vacuoles, where it is chelated by nicotianamine or citrate, or in plastids, where it is bound up by ferritin and from where it can also be retrieved on demand to keep cytosolic Fe concentrations stable. However, Fe toxicity is very rare under natural conditions and is mainly limited to acid soils (Kochian et al., 2004). Toxicity may also occur as a result of incorrect applications of foliar nutrient sprays (e.g., due to errors in dose calculations). Leaf chlorosis due to Fe deficiency, on the other hand, frequently develops when grapes are grown on calcareous soils, which are prevalent, for example, in Burgundy and Champagne in France, much of Hungary, and eastern Washington. Diagnosis of Fe chlorosis (e.g., by tissue analysis) is difficult because the total Fe concentration of chlorotic leaves is often similar to or even higher than that of "healthy," green leaves, partly because the chlorotic leaves are usually smaller (Sattelmacher, 2001). However, the Fe content of a leaf is not closely related to its physiologically active Fe content. Furthermore, Fe-deficient plants also have a tendency to accumulate Zn, Mn, Co, and cadmium (Cd) due to the general increase in the activity of divalent metal transporters under Fe deficiency (Curie and Briat, 2003; Kobayashi and Nishizawa, 2012).

Although Fe may be abundant in calcareous soils, it is often precipitated as insoluble Fe^{3+} oxides and hydroxides, making it unavailable for the roots. A one-unit increase in soil pH can decrease Fe solubility 1000-fold (see Figure 7.3). The ability of grapevines to cope with such Fe deficiency depends on species and cultivar. Species that have evolved in calcareous conditions (e.g., *V. vinifera* and, to a lesser extent, *V. berlandieri* or *V. champinii*) can pump out protons (H^+) and organic acids (malate and citrate) from the roots, which acidifies the soil solution and improves Fe solubilization and uptake (Brancadoro et al., 1995; Jiménez et al., 2007; Mengel and Malissiovas, 1982). Iron-inefficient species (e.g., *V. labrusca*, *V. riparia*, and *Muscadinia rotundifolia*) are unable or less able to release H^+, but they also produce organic acids under Fe deficiency (Ollat et al., 2003). The H^+-release strategy works only for species or cultivars that are not sensitive to high soil pH. Although release of H^+ may enhance Fe uptake and transport to the leaves in sensitive species as well, and although the apoplast pH seems to be unaffected by soil pH, uptake of bicarbonate (HCO_3^-) from calcareous soils (which contain high concentrations of HCO_3^- that forms from the soil's $CaCO_3$) leads to changes in the apoplast that inhibit conversion of the inactive form (Fe^{3+}) to the active form of iron (Fe^{2+}). As a consequence, Fe becomes "locked up" in the apoplast (Mengel and Bübl, 1983; Mengel et al., 1984). Therefore, the Fe cannot enter the mesophyll cells and is unavailable for metabolism, which leads to yellowing between the veins of young leaves and eventually whitening of the whole leaf. Nevertheless, because the affected leaves ostensibly signal the roots to increase Fe uptake, the total leaf Fe concentration may be as high as or higher than that in nonchlorotic plants; Fe may also accumulate in the roots (Currle et al., 1983; Gruber and Kosegarten, 2002). This "iron chlorosis paradox" is frequent in own-rooted Concord grapes (whose soil pH optimum is ~5.5) or scion cultivars grafted to *V. riparia* rootstocks grown on high-pH soils. Hence, the condition is also termed lime-induced chlorosis, although it should probably more appropriately be called bicarbonate-induced chlorosis (Currle et al., 1983; Gärtel, 1993; Mengel, 1994; Ollat et al., 2003). Due to their diverse origin, however, *Vitis* species and thus rootstocks differ in their tolerance of calcareous soils (see Section 1.2). Therefore, the chlorosis problem may be overcome by grafting susceptible cultivars to tolerant rootstocks, such as those derived from *V. berlandieri*, *V. vinifera* (which, however, is susceptible to phylloxera), and, to a lesser extent, *V. rupestris* (Bavaresco et al., 2003; Schumann and Frieß, 1976).

Excessive soil moisture, especially waterlogging, also reduces Fe availability (as well as the availability of N, K, and Mn) in addition to restricting root growth. In addition, the poor aeration in waterlogged soils (e.g., due to abundant rainfall or overirrigation) and in compacted soils also increases the soil HCO_3^- content due to reduced diffusion of CO_2 from the soil water ($CO_2 + H_2O \leftrightarrow H_2CO_3 \leftrightarrow HCO_3^- + H^+$). These effects may combine to induce chlorosis, especially in calcareous soils (Currle et al., 1983; Williams et al., 1994) and soils with a high P content: P_i reacts with soluble Fe to form insoluble Fe-phosphates. Moreover, chlorosis is exacerbated if excessive soil moisture coincides with low soil temperatures (Davenport and Stevens, 2006).

Zinc

Zinc (Zn^{2+}) is the only metal ion that is present in all six classes of enzymes, and much of the Zn in plants is strongly bound in proteins (Broadley et al., 2007). The majority of Zn-binding proteins are involved in the regulation of gene transcription via, among others, effects on DNA and RNA binding and RNA metabolism. Zinc, along with copper, is an integral part of one form of the antioxidant enzyme superoxide dismutase, which may also help protect plant tissues against the production of reactive oxygen species that lead to oxidative stress that accompanies many environmental stresses (Apel and Hirt, 2004). The concentration of Zn in grape leaves is approximately one-third to one-half that of Fe, but the concentrations of these two micronutrients may be similar in grape berries, where the concentration, though not the absolute amount, of the phloem-immobile Zn declines during ripening (Bertoldi et al., 2011; Gärtel, 1993; Rogiers et al., 2006). Most (>90%) of the Zn in a soil is insoluble and hence not plant-available; up to half of the soluble fraction is in the main plant-available form Zn^{2+}, although some Zn may be taken up by the roots as organic complexes (Broadley et al., 2007). Solubility is strongly dependent on pH, and similar to Fe, Zn availability is low in calcareous soils with a pH >7 (see Figure 7.3) and high bicarbonate content. Prolonged flooding, high organic matter content, and high Mg:Ca ratios also seem to decrease Zn availability (Broadley et al., 2007). Moreover, excessive P supply can immobilize Zn due to the formation of insoluble Zn phosphate ($Zn_3(PO_4)_2$) (Currle et al., 1983; Gärtel, 1993). Conversely, Zn deficiency can lead to P accumulation in the plant to the point of toxicity if

P availability is high (Cakmak and Marschner, 1986). Although insufficient Zn availability is the most widespread micronutrient deficiency, especially in sandy soils with high pH, plants respond readily to foliar Zn application (Broadley et al., 2007).

Zinc deficiency results in a decline in protein and starch production, accumulation of sugar in the leaves, and stunted shoots (short internodes) with small, malformed leaves (asymmetrical with a wide petiolar sinus and sharply toothed margins) that develop a mosaic-like chlorotic pattern between the veins ("mottle leaf"); the veins may become clear with green borders (Currle et al., 1983; Gärtel, 1993). High light intensity seems to accelerate symptom development (Marschner and Cakmak, 1989), perhaps as a result of oxidative stress. More severe deficiency delays shoot maturation (periderm formation) and stimulates lateral shoot growth, whereas the leaves' interveinal areas become reddish-brown or bronze and then necrotic, leading to rolling of the leaf blades (Broadley et al., 2007; Currle et al., 1983). In addition, insufficient Zn supply can inhibit pollen formation and, therefore, pollination, leading to poor fruit set and a "hens and chicks" appearance of the clusters, mimicking what occurs during B deficiency. Severe Zn deficiency leads to necrotic root tips, which is lethal. Zinc toxicity, on the other hand, is rare and occurs predominantly in low pH soils treated with sewage sludge or contaminated with other man-made Zn-containing (waste) products, such as corroding galvanized objects (Broadley et al., 2007). It decreases uptake of P, Mg, and Mn, leads to stunted growth and reduced yield, and can result in chlorosis caused by Fe deficiency.

Copper

As a component of copper proteins, such as plastocyanin, cytochrome oxidase, ascorbate oxidase, superoxide dismutase, polyphenol oxidase, and laccase, copper (Cu^+ and Cu^{2+}) participates in photosynthetic and respiratory electron transfer, lignification, ethylene perception, cell wall metabolism, oxidative stress protection (in concert with Zn), and molybdenum cofactor synthesis (Burkhead et al., 2009). Its ability to "switch" between Cu^{2+} and Cu^+ not only makes copper an essential cofactor in many oxidase enzymes that catalyze redox reactions but also contributes to its inherent toxicity (Clarkson and Hanson, 1980; Yruela, 2009). Concentrations of Cu in grape leaves are approximately 15 times lower than those of Fe (Gärtel, 1993), but in grape berries they are in the same range, and the content of both metals continues to increase as the grapes ripen (Bertoldi et al., 2011; Rogiers et al., 2006). Developing berries are strong sinks for Cu, and by harvest up to one-third of a vine's Cu may be in the fruit clusters (Pradubsuk and Davenport, 2011). Roots absorb Cu^{2+} using high-affinity transporters and probably transport it in the xylem in the form of Cu-nicotianamine chelate (Curie et al., 2009; Martins et al., 2014). High plant N status appears to increase the demand for Cu (Yruela, 2009). The phloem-mobile Cu can be remobilized from old leaves for redistribution to sinks in need of extra Cu, although such recycling does not appear to be very efficient so that young leaves are more impacted by Cu deficiency than old leaves (Burkhead et al., 2009; Marschner, 1995).

Deficiency symptoms include inhibition of root growth; stunted shoots; small, misshapen, pale-green or chlorotic leaves, often with curled leaf margins; and decreased fruit set due to poor pollen and embryo viability (Marschner, 1995). However, Cu deficiency is rare and is mainly restricted to soils that are very high in organic matter that strongly binds Cu. Moreover, Fe may substitute for Cu because the tasks of many Cu proteins can also be carried out by equivalent Fe proteins (Yruela, 2009). In contrast, when present in excess, Cu can be toxic and, despite its linkage with antioxidant enzymes, may cause oxidative damage due to the production of reactive oxygen species. Due to the damage it causes to the photosynthetic machinery, Cu can exacerbate the impact of high light intensity in leaves. Excess Cu may also inhibit root elongation while stimulating lateral root growth, and it may cause leaf chlorosis or even necrosis (Yruela, 2009). Copper availability increases at low soil pH (see Figure 7.3), and long-term application of Cu-based fungicides can lead to Cu accumulation in the surface soil (Gärtel, 1993). For instance, downy mildew (see Section 8.2) has been controlled since the late nineteenth century by the copper sulfate−lime blend ($CuSO_4 \cdot 3Cu(OH)_2 \cdot 3CaSO_4$) called Bordeaux mixture and, more recently, by copper oxychloride ($CuCl_2 \cdot 3Cu(OH)_2$) and copper hydroxide ($Cu(OH)_2$), which has often led to more than 10-fold increases in soil Cu. Such Cu accumulation may diminish uptake of P, Fe, and, in sandy soils, Mg and Ca. It can also lead to toxicity (Toselli et al., 2009), especially in vineyards planted on sandy and acid soils (frequent application of S-based fungicides also increases soil acidity). Whereas Cu toxicity is extremely rare in established vineyards because of the older vines' deeper root system, it can result in stunted growth of newly planted vines, whose roots are concentrated in the surface soil (Toselli et al., 2009). In extreme cases, young vines may die after replanting of old vineyard sites or in nurseries. Moreover, the fungicidal activity of Cu may have other undesirable side effects: High Cu content in grapes at harvest can inhibit yeast growth, leading to sluggish or stuck fermentations and, sometimes, poor wine quality (Tromp and De Klerk, 1988).

Manganese

Like other transition metals, such as Fe and Cu, Mn exists in various oxidation states, but only Mn^{2+} can be taken up by roots and transported throughout the plant in both the xylem and the phloem (Pittman, 2005). Leaf Mn concentrations are similar to those of Zn (Gärtel, 1993). In ripening grape berries, the Mn concentration, but not its content, may decline during ripening, similar to that of the phloem-immobile elements Ca and Zn (Bertoldi et al., 2011; Rogiers et al., 2006). Like Zn, Mn has antioxidative effects in plant tissues. It can serve as an antioxidant by being oxidized from Mn^{2+} to Mn^{3+} and is also a structural component of antioxidant enzymes (e.g., superoxide dismutase and catalase) as well as of the water-splitting (i.e., electron-, proton-, and oxygen-producing) enzyme in photosynthesis (Pittman, 2005). In addition, Mn is required for the function of enzymes such as glucosyltransferases, which attach a glucose molecule to phenolics and other compounds (Marschner, 1995). Therefore, Mn deficiency may increase tissue sensitivity to oxidative stress, which can be caused by a variety of environmental stresses (see Section 7.1). Deficiency symptoms are consequently more severe on sun-exposed leaves (energy "overload"). Chlorotic leaves in response to insufficient Mn availability occur first in the basal portion of the shoots, soon after budbreak. Chlorosis appears mosaic-like, and the leaves later acquire a reddish or bronze color, whereas lateral leaves often remain green, and fruit ripening is delayed (Currle et al., 1983; Gärtel, 1993). The availability of Mn is highly pH dependent (see Figure 7.3) and is minimal around pH 7, at which deficiency can be a problem, especially on sandy soils with high amounts of organic matter. However, Mn deficiency is often masked in such soils by the simultaneous development of lime-induced Fe chlorosis (Gärtel, 1993). On the other hand, plants often seem to take up much more Mn than they require, and excess Mn can be extremely toxic (Pittman, 2005). Despite its antioxidative properties at "normal" concentrations, excess Mn induces oxidative stress and results in symptoms such as stunted growth, leaf chlorosis, and necrotic lesions (Kochian et al., 2004). Plants therefore sequester and store Mn^{2+} ions and Mn-chelate (i.e., Mn−organic acid complexes) in the vacuoles in addition to the Mn required in the chloroplasts. However, Mn toxicity is rare and occurs predominantly on acid soils (pH <5.5) with high Mn availability (Kochian et al., 2004). Uptake of Mn may also be increased under conditions of Fe and Zn deficiency, whereas excess Fe inhibits Mn uptake (Pittman, 2005).

Molybdenum

Molybdenum is available to grapevine roots mostly as molybdate oxyanion (MoO_4^{2-}). Although it is an essential element, it is required only in minute amounts by grapevines, mainly as a cofactor that forms the active site of a few proteins called molybdoenzymes (Schwarz and Mendel, 2006). Nitrate reductase is the main molybdoenzyme in plants, and nitrate induces Mo uptake (Clarkson and Hanson, 1980; Kaiser et al., 2005; Reid, 2001). Therefore, a major effect of Mo deficiency is the accumulation of nitrate due to reduced nitrate reductase activity and lack of amino acid synthesis (Currle et al., 1983; see also Section 5.3). Consequently, the visual symptoms are similar to those of N deficiency: reduced growth and yield. Another molybdoenzyme is aldehyde oxidase, which is involved in the production of both auxin and ABA (Schwarz and Mendel, 2006). Shortage in Mo supply may result in reduced fruit set and clusters displaying "hens and chicks," especially in susceptible cultivars such as Merlot (Kaiser et al., 2005; Longbottom et al., 2010; Williams et al., 2004). Other deficiency symptoms include short internodes, zigzag growth, and pale-green leaves with necrotic margins. Mo-deficient Merlot may furthermore show flaccid and cupped leaves that wilt easily due to excessive transpiration (Kaiser et al., 2005). This is because the hampered ABA production renders Mo-starved vines vulnerable to losing stomatal control (see Sections 3.2 and 7.2). Merlot appears to be less able to take up Mo from the soil than other cultivars, but grafting on rootstocks can overcome this problem. Although Mo deficiency is relatively rare, it can occur on acid soils (Mo availability strongly declines below a soil pH of ~5.5) and soils with high Fe content, especially in cool climates. During a cool, wet spring, release of MoO_4^{2-} in the soil may be restricted, as is root growth. In addition, high sulfate availability can inhibit molybdate uptake; the two anions compete for root uptake due to their similar size. However, MoO_4^{2-} is highly mobile in plants, so foliar applications (e.g., in the form of sodium molybdate, $Na_2MoO_4 \times 2H_2O$) can be very effective in distributing Mo throughout the plant, alleviating deficiency symptoms, and improving fruit set (Kaiser et al., 2005; Longbottom et al., 2010; Williams et al., 2004). Toxicity due to excess Mo is even rarer than deficiency; symptoms can include purple leaves due to anthocyanin accumulation.

Boron

Although boron is an essential element for plants, its functions are very poorly understood. As borate ($B(OH)_4^-$), it binds strongly to pectic polysaccharides and thus is involved in the formation or cross-linking of the pectin network in the primary cell walls, which supports the function of Ca^{2+} in cell walls and is essential for normal cell and leaf expansion (O'Neill et al., 2004). Therefore, almost all of the B in plants is located in the apoplast, and B must be continually transported to the plant's growing regions to sustain cell wall structure and growth. Moreover, the B requirement of pollen tubes is especially high because they are rich in pectin, particularly in the tip. Boron is available in the soil solution as boric acid ($B(OH)_3$). Thus, it differs from other micronutrients in that it exists as a neutral molecule at physiological pH and appears to be taken up into plants mostly by simple diffusion, especially at high external concentration (Clarkson and Hanson, 1980; Reid, 2001). Hence, the roots may not be

FIGURE 7.9

Leaf symptoms of boron deficiency in Chardonnay (left) and boron toxicity in Cabernet Sauvignon (right).

Photos by M. Keller.

able to exclude excess B, which increases the risk of B toxicity that arises from the ability of B to bind to ATP and NAD(P)H. In addition to passive diffusion, active transport into the roots and the xylem probably occurs when the B concentration in the soil water is low (Reid, 2001; Takano et al., 2008).

The symptoms of B toxicity resemble those of B deficiency (Figure 7.9), and the tissue concentration range between deficiency and toxicity is very narrow (Currle et al., 1983; Takano et al., 2008). Moreover, B may continue to build up in the leaves, and even in the petioles, as the growing season progresses (Christensen, 1984). Therefore, in contrast to other nutrients, the concentration of B is often highest at the end of the season and in the oldest leaves. Accumulation also continues in grape berries throughout their development and ripening (Bertoldi et al., 2011; Rogiers et al., 2006). Indeed, the developing berries appear to be strong sinks for B, so that by harvest more than one-third of a vine's B may be in the grape clusters (Pradubsuk and Davenport, 2011). Boron toxicity can be induced by undue B fertilization and is a particular peril of arid regions. Insufficient B availability, on the other hand, is frequent on very acid soils (pH <4.5), especially under dry conditions, and rapidly results in cessation of cell division, whereas cell expansion is thought to be unaffected (Clarkson and Hanson, 1980; Currle et al., 1983). The inhibition of meristem activity leads to stunted shoot and root growth, with shoots developing "swollen" internodes and often displaying a zigzag appearance and petioles remaining short and thickened (Currle et al., 1983; Gärtel, 1993). Symptoms are usually confined to young tissues; although B is phloem mobile, it apparently cannot be recycled from old leaves. Bushy, branched shoot growth, resembling symptoms of fanleaf virus infection, and poor bud fruitfulness may be a carryover effect of insufficient B available for proper primordium formation during the previous growing season (Gärtel, 1993). In addition, reduced N uptake in B-deficient vines leads to low leaf N status and sugar and starch accumulation in the leaves (Camacho-Cristóbal and González-Fontes, 1999). Root growth also ceases, and roots may swell and crack. Because B is important for pollen germination and pollen tube growth, B deficiency interferes with fertilization (Currle et al., 1983; May, 2004; O'Neill et al., 2004).

This results in poor fruit set and a condition termed "hens and chicks," with clusters containing small, seedless berries among normal berries (Gärtel, 1993). In addition, seed development is impaired, and the seeds remain smaller than in nondeficient vines (Hardie and Aggenbach, 1996). Boron deprivation probably also induces oxidative damage to cells, and high light intensity consequently exacerbates the symptoms of B deficiency (Marschner and Cakmak, 1989).

Nickel

Compared with the minute amount required by plants (<0.1 μg/g dry weight), nickel (Ni^{2+}) is relatively abundant in almost all soils. Although many proteins contain Ni, very little is known about the element's function in plants. Several enzymes are activated by Ni; one of them, urease, is involved with nitrogen metabolism that is required to process urea, for example, during arginine mobilization (Witte et al., 2005). Another such enzyme is superoxide dismutase, which plays an important role in the defense against oxidative stress. Following uptake of Ni by the roots, it is probably transported in the xylem in the form of Ni-nicotianamine chelate (Curie et al., 2009). Due to its partial mobility in the phloem, Ni can also be remobilized from old leaves and redistributed to high-priority sinks. Nonetheless, the Zn concentration in grape berries declines during ripening, although its total amount remains constant after veraison (Bertoldi et al., 2011; Rogiers et al., 2006). Nickel deficiency results in reduced shoot vigor following budbreak; dwarfed, thick leaves with cupped tips (perhaps caused by oxalate accumulation); loss of apical dominance; and brittle wood due to poor lignification.

Silicon

Silicon (Si) is the second most abundant element (oxygen is the most abundant) in the soil, making up approximately 28% of the Earth's surface. As a component of minerals, it is present in clay, sand, and rocks in the form of silicon dioxide (SiO_2, the building block of quartz and glass), which accounts for 50–70% of the soil mass (Ma and Yamaji, 2006). Roots, but apparently not root hairs, absorb Si passively or via transporters as soluble silicic acid ($Si(OH)_4$), which is then released to and transported in the xylem sap (Liang et al., 2005; Ma and Yamaji, 2006). Thus, Si is taken up and distributed with the transpiration stream as an undissociated and neutral molecule, similar to boric acid ($B(OH)_3$). Because $Si(OH)_4$ polymerizes at concentrations greater than 2 mM, it is ultimately precipitated throughout the plant as amorphous, hydrated silica bodies ($SiO_2\text{-}nH_2O$), which are also called silica gel, plant opal, or phytoliths (Ma and Yamaji, 2006). Due to its immobility in the phloem, xylem-supplied Si accumulates in older tissues, including the stomatal regions of grape berries (Blanke et al., 1999). Its existence as an undissociated molecule and ability to polymerize make Si the only element for which excessive uptake and accumulation does not result in detrimental effects to plants.

Plants use polymeric silicates to impregnate the cell walls of epidermis and vascular tissues, thereby strengthening the tissues, reducing water loss, and retarding invasion by fungal pathogens (Clarkson and Hanson, 1980; Ma and Yamaji, 2006). For example, penetration of the powdery mildew fungus *Erysiphe necator* ostensibly leads to localized accumulation of Si in the cell walls (Blaich and Wind, 1989). In addition, Si also stimulates the antioxidant systems, which may avert stress-induced oxidative damage (see Section 7.1). Accordingly, application of Si sprays to the canopy can reduce powdery mildew infections, perhaps by providing a physical barrier and/or by enhancing the production of phenolic compounds. Thus, although deemed a nonessential nutrient (i.e., one that is not required to complete the plant's life cycle), Si provides many benefits, such as improved resistance to pests and diseases; tolerance of drought, salinity, heavy metals, and high temperatures (Currie and Perry, 2007; Epstein, 1999; Ma and Yamaji, 2006); and even improved yield and fruit quality by enhancing the production of flavonoids. For this reason, Si is integrated in many fertilizers. Most of these stress-alleviating effects result from the strengthening of cell walls by Si and by its ability to enhance the binding of cations (e.g., Na^+ and Mn^{2+}) to the cell walls, which may prevent their buildup to toxic concentrations inside the cells (Saqib et al., 2008). In addition, plants high in Si may be able to reduce uptake of Na^+ from saline soils and to transport less of it to the leaves.

7.3.3 SALINITY

The term salinity describes the occurrence of high concentrations of soluble salts (i.e., in ionic form) in water and soils. The development of salinity is termed salinization and occurs in regions where water evaporation from the soil exceeds precipitation so that salts dissolved in the soil solution tend to become concentrated at the soil surface. Whereas this process is characteristic of arid environments, the opposite process, called acidification, occurs in regions where rainfall consistently exceeds evaporation (i.e., especially in the tropics and subtropics) and thereby leaches cations such as K^+, Ca^{2+}, and Mg^{2+} so that the soil pH decreases and the highly toxic Al^{3+}, along with Mn^{2+} and Fe^{2+}, becomes soluble in the soil solution. Rainwater also contains some salt (≤ 50 mg NaCl L^{-1}), especially in coastal areas (Munns and Tester, 2008). Wind may blow seawater inland and lead to salt deposits on the soil surface. However, irrigated vineyards are at much greater risk from salinization than nonirrigated vineyards because irrigation water is relatively rich in dissolved salts and because irrigation tends to raise water tables. Therefore, irrigation in arid and semiarid regions over prolonged periods can lead to a buildup of salt near the soil surface. Most table and raisin grapes are grown in rather dry and warm climatic regions, such as southwestern Asia, California, Chile, or Australia, and are thus especially threatened by salinity.

The dominant soil salts are cations, such as Na^+, K^+, Ca^{2+}, and Mg^{2+}, and their associated anions, such as chloride (Cl^-), sulfate (SO_4^{2-}), carbonate (CO_3^{2-}), and bicarbonate (HCO_3^-). Small amounts of other ions are also present in the soil solution. The relative amounts of different ions vary between water sources and soil types, but the ions most often associated with the effects of salinity on grapevines are Na^+ and Cl^-. Dissolved ions increase the electrical conductivity of water, and thus salinity of irrigation water or water extracts of soils is expressed in electrical conductivity units (measured in decisiemens per meter or dS m^{-1}). The threshold above which salinity begins to affect *V. vinifera* growth and yield formation is approximately 2 dS m^{-1}, and above 16 dS m^{-1} vines cannot survive (Zhang et al., 2002). Because dissolved ions decrease the osmotic potential (Ψ_π) of water, electrical conductivity is also a measure of Ψ_π: 2 dS m^{-1} corresponds to approximately 20 mM NaCl generating a $\Psi_\pi \approx -0.1$ MPa.

Sodicity is related to salinity and refers to the presence of sodium relative to calcium and magnesium in the soil. As a measure of sodium hazard, sodicity is expressed as the sodium adsorption ratio (SAR) because most cations in the soil are attracted to the negative charges of clays. The sodicity of irrigation water or soil water extracts is estimated as follows (Richards, 1954):

$$\text{SAR} = \frac{[\text{Na}^+]}{\sqrt{\frac{1}{2}([\text{Ca}^{2+}] + [\text{Mg}^{2+}])}}$$

where [...] denotes the concentration of an ion in millimoles per liter (mM).

In addition to their elevated Na^+ content, sodic soils are afflicted with a deterioration of structure due to clay dispersion and a rise in hydraulic resistance. Saline and sodic soils are usually classed together as "salt-affected soils." Such soils contain a sufficiently high concentration of soluble salts or exchangeable Na^+ to interfere with plant growth. But while salinity impacts plants directly through the effect of ions on plant physiology, the influence of sodicity is mostly indirect due to its deleterious effects on the soil's physical properties.

The most common cause of salt stress is a high concentration of Na^+ and Cl^- in the soil solution. Both of these are essential plant nutrients but become toxic at much lower concentrations than other nutrients. Plant damage due to salt-affected soils is the outcome of a combination of hyperosmotic stress and hyperionic stress due to a disruption of homeostasis (Greek $homois$ = similar, $stasis$ = stand still, steady) in water status and ion distribution (Hasegawa et al., 2000; Zhu, 2001). Initially, buildup of salt ions in the soil decreases Ψ_π of the soil solution; the Ψ_π in "normal" soils is generally approximately -0.01 MPa but can drop to less than -0.2 MPa in saline soils. The resulting decrease in Ψ_{soil} impedes water uptake by the roots (Shani et al., 1993; see also Chapter 3), increases root hydraulic resistance due to closure of aquaporins (Luu and Maurel, 2005), and results in water deficit for the vine and a decline in Ψ_{leaf} (Cramer et al., 2007; Downton and Loveys, 1981; Walker et al., 1981). Thus, the initial effects of a rise in soil salinity are identical to the effects of drought stress (see Section 7.2). The ensuing collapse of the water potential gradient necessary for cell expansion curtails shoot growth and leaf expansion. Therefore, the first visible sign of salt stress is an inhibition of shoot and leaf growth (Munns and Tester, 2008; Walker et al., 1981).

The decrease in Ψ_{leaf} and the associated increase in root-derived and locally produced ABA also induce closure of the stomata, which in turn decreases transpiration and photosynthesis and consequently the production of sugar for export to other plant parts (Downton et al., 1990; Shani and Ben-Gal, 2005). Rising Cl^- concentration in the leaves further reduces stomatal conductance (Walker et al., 1981), although Na^+ tends to counter this by replacing K^+ in the guard cells and thereby keeping the stomata partially open. However, the rate of photosynthesis declines as leaf Cl^- increases well before any visible symptoms of salt damage become apparent (Downton, 1977b). Thus the reduction in growth is in part due to the decrease in photosynthesis and in part due to inhibition of cell division and cell expansion (Zhu, 2001). Growth may also slow through deactivation of gibberellins under salt stress (Yamaguchi, 2008). More severe salinity arrests lateral shoot growth and induces necrotic leaf margins ("marginal burn" or "salt burn," see Figure 7.10) in older leaves, followed by progression of the necrotic symptoms toward the petiole, whereas the main veins remain green (Williams and Matthews, 1990; Williams et al., 1994). Such salt injury is the result of the accumulation of Na^+ and/or Cl^- in the transpiring leaves (i.e., ion concentrations gradually increase as the transpiration stream deposits salt ions in the leaves) to the point where the vacuoles can no longer contain these ions. Accumulation in the cytoplasm then leads to enzyme

FIGURE 7.10

Leaf symptoms of salt injury on Merlot.

Photo by M. Keller.

inhibition (salt poisoning), whereas accumulation in the cell walls leads to dehydration of the cells; both outcomes result in cell death (Munns, 2002). The threshold Cl^- content for marginal necrosis seems to be around 2.5% of the leaf dry weight (Walker et al., 1981).

Root growth is usually less sensitive to salt stress than is shoot growth (Munns and Tester, 2008). Yet roots are the first and most important organs to experience salinity, which decreases their ability to explore the soil for water and nutrients, and in some instances, root growth of grapevines was found to be more sensitive to salinity than was shoot growth (Hawker and Walker, 1978). One reason for the decrease in root growth may be that the decline in soil Ψ_π limits the turgor pressure of the expanding cells in the root tip, so that the roots can no longer push their way through the soil (Bengough et al., 2011). Impaired root growth, respiration, and water uptake further restrict vine growth, yield, and fruit quality (Shani and Ben-Gal, 2005; Shani et al., 1993). In some cases, salinity may be associated with changes that are typical of mild water deficit, such as earlier veraison, higher fruit sugar, proline, K^+ and Cl^-, and greater decline in acidity during ripening (Downton and Loveys, 1978; Walker et al., 2000). As the salinity becomes more severe, however, fruit set, berry size, and sugar and anthocyanin accumulation are increasingly curtailed (Hawker and Walker, 1978). Moreover, Cl^- and Na^+ ions also accumulate in the fruit, especially in the skin, and from there are readily extracted into wine during fermentation (Downton, 1977a; Gong et al., 2010; Walker et al., 2010).

The challenge for grapevines growing in saline environments is that their roots must take up nutrient ions while keeping out the toxic Na and Cl. "Toxic," as usual, is a relative term; it is important to remember that plants require some Cl^- for the water-splitting reaction that produces electrons, protons, and oxygen during photosynthesis (see Section 4.1). Roots effectively "pick" the nutrient ions from the toxic ions in the soil solution or pump the toxic ions that have been taken up back out again so that more than 95% of Na^+ and Cl^- is prevented from entering the xylem (Munns, 2002; Munns and Tester, 2008). Nonetheless, although Na^+ is not an essential nutrient, the ion is taken up into cells down the electrochemical gradient, competing with K^+ for uptake

(Hasegawa et al., 2000). The hypodermal and endodermal cells of salt-stressed grapevine roots appear to selectively accumulate K^+ over Na^+ and Cl^- compared with cortical and pericycle cells (Storey et al., 2003a). In contrast, the cortex and pericycle cells sequester large amounts of Na^+ and Cl^- in their vacuoles (Storey et al., 2003a). However, grapevines take up more Cl^- than Na^+ from saline soils that have equivalent concentrations of both ions (Walker et al., 1981), and a small portion of each ion ends up in the xylem and is transported to the shoot with the transpiration stream. Consequently, Cl^- and, to a lesser extent, Na^+ accumulate in the older leaves, building up with increasing concentration in the soil solution, and they continue to do so as the growing season progresses (Downton, 1985; Munns, 2002; Shani and Ben-Gal, 2005; Stevens and Walker, 2002). Over time, the ions may accumulate to toxic concentrations; this ion-specific phase of salinity stress is associated with premature death of older leaves (Munns and Tester, 2008).

Excessive Cl^- uptake interferes with nitrogen nutrition because NO_3^- uptake responds to the concentration of $NO_3^- + Cl^-$ rather than to NO_3^- alone (Clarkson, 1985). This competition for uptake between Cl^- and NO_3^- requires application of abundant nitrogen fertilizer to improve plant N status on saline soils. High Na^+ concentration, on the other hand, is toxic to plants because it interferes with K^+ nutrition, which reduces K^+-stimulated enzyme activities, metabolism, and photosynthesis. At concentrations exceeding approximately 100 mM, Na^+ and Cl^- also directly inhibit many enzymes (Munns, 2002; Munns and Tester, 2008; Zhu, 2001). High salinity eventually overwhelms the leaf vacuoles' capacity to sequester Cl^- and Na^+, leading to toxic concentrations in the cytoplasm and disturbing the cells' ionic balance; Na^+ accumulates at the expense of K^+ and can even result in a loss of K^+ and Ca^{2+} from the cells. Sodium can compete with Ca^{2+} and displace it from the cell wall, thus affecting cell wall properties. Because Ca^{2+} in turn can reduce Na^+ uptake and increase K^+ and Ca^{2+} uptake, addition of Ca^{2+} can somewhat alleviate the toxic effects of salinity (Hasegawa et al., 2000; Plieth, 2005). However, prolonged exposure to high soil Ca^{2+} may itself be stressful for the plant.

Oxidative stress is another characteristic of salinity-induced injuries to plant tissues (Munns and Tester, 2008; Zhu, 2001). It is a secondary stress that results from the effects of ion imbalance and hyperosmotic stress and from the decline in photosynthesis (see Section 7.1). Excess free oxygen radicals (especially $^1O_2$) and hydrogen peroxide (H_2O_2) oxidize membrane lipids and other cellular components, which eventually leads to membrane leakage and tissue deterioration (Møller et al., 2007). However, even before such irreversible damage ensues, oxidative stress may increase the vine's light sensitivity because more photons are being absorbed than can be used by the declining photosynthesis. One way grapevines cope with such an energy overload appears to be an increase in photorespiration (Cramer et al., 2007; Downton, 1977b; Downton et al., 1990; Walker et al., 1981), which dissipates some of the excess energy but comes at the cost of lower photosynthetic efficiency. Another defense strategy is to boost the antioxidant systems (i.e., glutathione and xanthophyll cycles) that capture and inactivate some of the reactive oxygen species (Cramer et al., 2007). Because Mn^{2+} acts as an antioxidant in plant tissues, (foliar) application of Mn-chelates (and possibly Zn-chelates) might alleviate effects of oxidative stress in plants grown on saline sites or irrigated with saline irrigation water (Aktas et al., 2005).

Although they vary somewhat in the extent of Na^+ and Cl^- uptake and accumulation in the leaves (Groot Obbink and Alexander, 1973), most cultivars of *V. vinifera* are moderately sensitive to salt. American *Vitis* species, especially *V. riparia*, *V. berlandieri*, and, to a lesser extent, *V. candicans* and *V. champinii*, are more tolerant (Williams et al., 1994). Some rootstocks derived from these species (e.g., Ramsey, 1103 Paulsen, 110 Richter, 140 Ruggeri, and 101−14 Mgt) are able to exclude much of the salt from root uptake and xylem transport (Antcliff et al., 1983; Sauer, 1968; Tregeagle et al., 2010; Walker et al., 2000, 2010). Therefore, these rootstocks and the scions grafted to them are only marginally affected by high salt concentrations in the soil (Downton, 1985; Stevens and Walker, 2002; Zhang et al., 2002). However, it appears that at least some of them (Ramsey, 1103 Paulsen, and 101−14 Mgt) progressively lose this salt-exclusion ability (Tregeagle et al., 2006). Under long-term exposure to saline conditions, which tends to lead to salt buildup in the soil over time, these rootstocks may become less salt tolerant.

The impact of salinity on vines seems to be more severe on heavier clay loam soils than on lighter loamy sands. Moreover, irrigation and soil management also affect the extent of physical degradation of salt-affected soils. Irrigation is a common cause of agricultural land degradation because salt dissolved in the irrigation water is left in the soil following evaporation. Excessive irrigation, particularly with saline water, as well as frequent cultivation (tillage) and intense trafficking are a good recipe for rapid loss of soil fertility. However, salts can also build up under highly efficient drip irrigation (Stevens and Walker, 2002), when ions move down the soil profile beneath the emitters and then move laterally and rise again to the soil surface with the evaporating water. The resulting high-salt zone around the edges of the wetting zone can restrict root growth in a way similar to the restriction imposed by a pot. The "pot" size is smaller in sandy soils than in loam soils. Waterlogging due to the formation of impermeable soil layers or as a result of excessive irrigation also increases the risk of salt damage because waterlogged grapevine roots lose the ability to exclude Na^+ and Cl^- from uptake. Waterlogging also appears to increase the amount of Na^+ in the soil solution relative to other ions so that Na^+ uptake is often favored over Cl^- uptake (Stevens and Walker, 2002). Even if only a portion of the root system is exposed to saline conditions while other portions continue to have access to freshwater, the latter ostensibly do not compensate for the decline in water uptake by the former (Shani et al., 1993). Prolonged exposure to saline soil water ultimately results in vine death (Shani and Ben-Gal, 2005). On the other hand, where soil salts can be leached out of the rootzone by using a fresh source of irrigation water, the harmful physiological effects can often be quickly reversed. Thus, as long as no irreversible damage has been caused, grapevines restore root functionality, growth, and water uptake to drain the excess ions from the leaves, and growth and gas exchange recover rapidly (Shani et al., 1993; Walker et al., 1981).

7.4 TEMPERATURE: TOO COLD OR TOO WARM

Grapevines growing in cool, continental climates are exposed to a large daily temperature range and often experience widely fluctuating temperatures during spring and autumn. Low temperature may limit growth by decreasing the rate of protein production or cell wall extensibility. Restricting cell wall extensibility also inhibits cell division by preventing cell expansion (see Section 3.1). Furthermore, cold temperatures increase the rigidity of the normally fluid cell membranes (Chinnusamy et al., 2007). Because low temperatures restrict cell division more than photosynthesis (the duration of cell division increases exponentially with decreasing temperature), sugar and starch tend to accumulate in the leaves during a cool episode (Wardlaw, 1990). When the temperature drops too low, it can result in damage to plant tissues. The type and extent of damage depend on whether or not the temperature drops below the freezing point and on the developmental status of the plant. By contrast, higher temperatures tend to accelerate plant growth and development so that phenological stages occur in more rapid succession than under cooler conditions (see Section 2.2). In other words, an increase in temperature accelerates and compresses the temporal program of plant development—up to an optimum—and provided no other factors (e.g., water deficit, which is often coupled with high temperatures) are limiting growth. Temperature also has a strong effect on photosynthesis and respiration (see Section 5.2). Changes in the rate of photosynthetic CO_2 assimilation are reversible over the physiological range of about 10-35°C, but higher and lower temperatures can cause injury to the photosynthetic apparatus. Excessive heat and cold tend to result in symptoms that are similar to those caused by drought stress (see Section 7.2). This is not surprising because heat and cold stress have a major effect on plant water relations, decreasing the RWC of tissues and inducing oxidative stress.

7.4.1 HEAT ACCLIMATION AND DAMAGE

Although the majority of the world's grape crop is produced in temperate climates, production, especially of table and raisin grapes, extends to some hot areas where summertime afternoon temperatures often rise above 35°C or even 40°C (Williams et al., 1994). The optimal daytime temperature for grapevine growth, photosynthesis, yield formation, and fruit ripening is below 30°C, and heat stress is usually defined as temperatures exceeding 5°C above those associated with optimal growing conditions. Very high temperatures, above approximately 40°C, cause a sharp decline in photosynthesis due to the disruption of the functional integrity of the photosynthetic machinery in the chloroplasts (Zsófi et al., 2009). Heat increases the physical distance between the light-harvesting antenna pigments and the reaction center in photosystem II (PSII; see Section 4.1). In addition, the water-splitting system of PSII is very heat sensitive; thus, heat inhibits PSII-driven electron transport (Berry and Björkman, 1980). The activity of PSI, on the other hand, is much more heat stable. The same applies to the photosynthetic enzymes involved in CO_2 assimilation, which are stable to over 50°C, although heat stress leads to protein denaturation and misfolding of newly assembled proteins (Berry and Björkman, 1980). For instance, rubisco inactivation increases exponentially with increasing temperature, and rubisco activity strongly declines above 35°C. Therefore, although stomata start closing above approximately 35°C, this seems to be a response to the reduced photosynthesis rather than the cause of it.

Leaf temperatures exceeding 45°C for brief periods trigger severe and lasting declines in stomatal conductance (g_s) and photosynthesis, and if such high temperatures occur over prolonged periods they will kill grapevine leaves (Gamon and Pearcy, 1989, 1990a). Moreover, high temperatures are often associated with low air humidity so that stomata close in response to the high vapor pressure deficit (see Section 3.2) and not because of heat. The negative impact of high temperature on photosynthesis is also much more pronounced at high light intensity (Gamon and Pearcy, 1989, 1990a,b). Thus, leaves are more susceptible to heat waves under a blue sky or during periods of water deficit. Leaves on the hotter south (northern hemisphere) or west side of a canopy may suffer more than their somewhat cooler northern or eastern counterparts, so that the former are more likely to show symptoms of leaf yellowing and senescence (Palliotti et al., 2009).

Heat-stressed plant cells seem to accumulate calcium ions (Ca^{2+}) in the cytosol, which helps to reduce the permeability (and thus leaking) of membranes and to maintain membrane integrity.

Maintenance of membrane integrity occurs at least in part through changes in the fatty acid composition of the membranes: Warm temperatures lead to higher contents in saturated and monounsaturated fatty acids (Penfield, 2008; Sage and Kubien, 2007). On the other hand, however, these changes in membrane composition may increase the tissue susceptibility to chilling (Iba, 2002). Calcium also enhances heat tolerance by stimulating the plant's antioxidant system in an attempt to protect the photosynthetic machinery from oxidative damage. Heat stress moreover appears to result in breakdown of starch to sugars and accumulation of amino acids (perhaps because they are not being used for protein production or because proteins are being degraded), especially glutamine, but not the "normal" compatible solute proline (Guy et al., 2008; Rizhsky et al., 2004). In addition, emission of volatile isoprenoids, such as isoprene (which is a liquid at room temperature but vaporizes above 34°C) or monoterpenes, helps the leaves to recover from brief episodes of excessive temperatures (>40°C) by stabilizing the thylakoid membranes and by scavenging reactive oxygen species (Loreto and Schnitzler, 2010; Pichersky and Gershenzon, 2002; Sharkey et al., 2008). This antioxidant activity provides thermotolerance against heat spikes and strongly decreases oxidative damage in the leaves. Therefore, despite the injury to the photosynthetic system, leaves can recover (i.e., the inactivation is reversible) within a few days after temperatures return to physiological levels, as long as the heat stress was not so severe as to cause membrane leakage and necrosis; grape leaves appear to be killed when the leaf temperature exceeds approximately 45°C (Abass and Rajashekar, 1991). Emission of isoprenoids also ceases above 45°C, probably because the supply of assimilates for their production becomes insufficient due to the impairment of photosynthesis (Loreto and Schnitzler, 2010).

As is the case with other stresses, the capacity of grapevines to recover from heat stress depends on the intensity and duration of the stress and the growth stage at which it occurs. Brief episodes of extreme heat are worse than longer periods of moderately high temperature. On the other hand, photosynthetic acclimation to high temperatures leads to decreased performance at low temperatures and vice versa. Therefore, although vines can adapt to seasonal changes in temperature regimes (Zsófi et al., 2009), short-term temperature fluctuations over a wide range can result in insufficient photosynthesis for growth and/or fruit ripening. Moreover, heat stress also accelerates leaf aging and senescence so that chlorophyll content often decreases along with photosynthesis (Thomas and Stoddart, 1980).

There is variation among *Vitis* species and cultivars in terms of heat sensitivity. Even under relatively modest light intensities, photosynthesis in heat-stressed (40°C) *V. aestivalis* leaves may decline more than in *V. vinifera*, and this decrease is associated with damage to PSII and subsequent chlorophyll degradation and leaf senescence (Kadir, 2006). During an unusually hot summer in Italy, Sangiovese lost more than half of its basal leaves, especially on the west side of the canopy that was heated by the afternoon sun, whereas Montepulciano did not lose any leaves (Palliotti et al., 2009). Photosynthetic recovery of the leaves that survive a heat episode may be slow, extending over several weeks, and is incomplete (Gamon and Pearcy, 1989; Kadir et al., 2007). When heat-susceptible cultivars such as Concord (*V. labruscana*) are grown in warm climates or hot growing seasons with intense solar radiation, they often develop a condition termed "blackleaf," whereby the adaxial epidermis cells of sun-exposed leaves become discolored, turning bronze-brown, purple, or black (Figure 7.11). The discoloration usually begins in midsummer on the leaves' sun-facing ridges and is exacerbated by water deficit. It is accompanied by chlorophyll degradation and a decline in photosynthesis and may result in necrosis and leaf abscission

FIGURE 7.11

Blackleaf symptoms initially appear on south-facing ridges of Concord leaves (left) and eventually cover most of the exposed leaf surfaces (right).

Photos by M. Keller.

(Lang et al., 1998; Smithyman et al., 2001). These blackleaf symptoms look suspiciously like the UV damage symptoms described in other plants and may be a consequence of the polymerization of phenolic compounds in the epidermal cells that results from oxidative stress (Kakani et al., 2003; Yamasaki et al., 1997). In other words, blackleaf symptoms probably arise from injury caused by UV radiation rather than from heat damage *per se*.

Growth and fruit ripening are suppressed as a consequence of the decrease in photosynthesis and/or a decrease in sink activity. In addition, high temperature shifts carbon partitioning to favor vegetative growth at the expense of fruit growth and ripening (Greer and Weston, 2010b; Sepúlveda et al., 1986). Thus, postveraison heat stress can markedly delay or even inhibit fruit ripening by decreasing the amount of assimilates available for export to the clusters and/or the ability of the berries to import assimilates. Because this may lead to uneven ripening or an overall delay of fruit maturation (Cawthon and Morris, 1982), it has implications for grape production in an increasingly warmer climate. In addition to a rise in average temperatures and an associated increase in cumulative heat units and growing-season length, meteorologists and climate modelers predict that shifts will also occur in the number and extent of extreme weather events (IPCC, 2013). This conclusion follows directly from the statistical normal distribution of temperature data around a mean, which follows a so-called bell curve. Therefore, summer heat waves will probably increase in both frequency and severity. An analysis for southern Australia shows that the probability of a run of 5 consecutive days over 35°C doubles for a 1°C warming and increases by a factor of five for a 3°C warming (Hennessy and Pittock, 1995). As model calculations for the United States demonstrate, such heat episodes could be detrimental to fruit quality in areas that already experience warm or hot climates (White et al., 2006), which encompass many of the current grape production regions in the world.

Water evaporation at the leaf surface cools the leaf tissues so that the leaf temperature is generally close to air temperature (see Section 3.2). Evaporative cooling of the leaves thus provides some heat resistance. The beneficial effect of evaporative cooling is sometimes exploited by growers who use intermittent irrigation, for example, by microsprinklers beneath the canopy or overhead sprinklers above it, to cool leaves in the heat of the midday sun in warm climates. Such "hydrocooling" enables the vines to maintain higher photosynthesis rates and higher rates of berry growth and sugar accumulation (Greer and Weedon, 2014). Contrary to widespread opinion, water droplets on leaves do not cause sunburn, because their ellipsoidal shape focuses the solar rays far below the leaf and because they directly cool leaves through their thermal mass (Egri et al., 2010).

By favoring rapid transpiration rates, high g_s lowers leaf temperature and appears to reduce the deleterious effects of heat stress. This is particularly important during the critical flowering and fruit development stages, in which heat stress would result in lower crop yields. The upper temperature limit for maximum yield formation in grapevines seems to be about 35°C (Keller, 2010). Higher g_s also increases CO_2 diffusion into the leaf and favors higher photosynthetic rates. Higher photosynthetic rates could in turn lead to more biomass and higher crop yields. However, higher g_s appears to favor higher yields by a mechanism not directly related to photosynthesis, perhaps by reducing respiration. Respiration is strongly stimulated by increasing temperature and results in a net loss of assimilated carbon. High yields are generally an important goal for the production of table, raisin, or juice grapes, whereas moderate to low yields are preferred in wine grape production to maximize fruit quality. Nevertheless, in areas where vegetative growth can be controlled by deficit irrigation and crop load can be adjusted manually, management practices (including varying the timing of deficit irrigation) aimed at favoring high g_s could be used to improve quality (e.g., see the discussion of organic acid synthesis in Section 6.2). Because water deficit increases a vine's susceptibility to heat stress, drought during the bloom period is especially detrimental to fruit set if it coincides with a heat episode. But heat stress alone may also impair fruit set, probably mostly by interfering with pollen maturation and viability (De Storme and Geelen, 2014; Zinn et al., 2010). Heat stress reduces auxin production in the anthers, which leads to a blockage of cell division in the pollen. The resulting premature abortion of pollen development induces male sterility and thereby prevents fertilization.

Just like other developmental processes, grape ripening is accelerated by high temperatures (Keller, 2010). This is especially true for sugar accumulation and malate degradation. However, because grape berries are designed to minimize water loss by transpiration, especially after veraison (see Section 6.2), they cannot take advantage of the evaporative cooling mechanism that usually protects leaves from overheating. Too much heat can inhibit or even denature proteins in the berries. When plant organs, such as previously shaded clusters following leaf removal, are suddenly exposed to high temperature (38–40°C), they produce so-called heat-shock proteins that function as chaperones by helping to fold other proteins into a suitable structure, preventing accumulation of proteins that have been denatured by heat, and helping to reactivate them (Iba, 2002). Nevertheless, sunburn symptoms can develop as an expression of oxidative damage that usually results from a combination of high light intensity and high temperature. Heat injury usually results from the disintegration of cell membranes due to lipid phase changes and denaturation of membrane proteins. The disruption of membrane structures is irreversible and leads to leakage of cell solutes and breakdown of cell

FIGURE 7.12

Delayed ripening of the sun-exposed side of a Cabernet Sauvignon cluster in a warm climate (left), symptoms of sunburn on grape berries (center), and shriveling of sunburnt berries (right).

Photos by M. Keller.

compartmentation, which is normally lethal for the affected cells. The poor capacity of grape berries for transpirational cooling makes them relatively susceptible to overheating. Indeed, sun-exposed berries can heat up to 15°C above air temperature, whereas shaded berries are usually at approximately ambient temperature (Spayd et al., 2002). This has implications for the choice of vineyard row direction, trellis design and training system, canopy management, and even vineyard floor management. Because during a sunny day the air temperature in the early afternoon is generally higher than in the morning, exposed grape berries on the "afternoon side" of the canopy can reach considerably higher absolute temperatures than their counterparts on the "morning side." Therefore, whereas cultural practices that maximize sun exposure are usually beneficial in a cool climate (where they can accelerate ripening), they may be detrimental in a warm, sunny climate (see Section 6.2). A bare soil surface, especially of a light-colored soil, reflects a major portion of the incident light back to the canopy, and this can aggravate heat injury on grapes during hot days.

Symptoms of sunburn or sunscald on clusters include initial reddening or browning of the exposed portion of the skin on both red and white grapes (especially before veraison), a change in epicuticular wax structure from crystalline to amorphous (giving red postveraison berries, in particular, a glossy appearance), followed by sinking in of the affected area or dehydration (shriveling) of the whole berry (Bondada and Keller, 2012; see also Figure 7.12). Sunburn symptoms can become particularly acute following leaf removal to enhance sun exposure of the fruit late in the growing season because, in contrast to berries that have been exposed since the beginning of the season, the previously shaded berries have not accumulated sufficient amounts of "sunscreen" such as flavonols and xanthophylls (Düring and Davtyan, 2002; Kolb et al., 2003).

7.4.2 CHILLING STRESS

Damage to nondormant plant tissues caused by low but above freezing temperatures (typically in the range of 0–10°C) is referred to as chilling stress. Note that the same temperature range acting on dormant tissues is not normally viewed as being stressful but is necessary to release dormancy (see Section 2.2). The photosynthetic cell organelles, the chloroplasts, are particularly sensitive to chilling stress (Kratsch and Wise, 2000). Swelling of chloroplasts, distortion of thylakoid membranes, and starch depletion (decrease in number and size of starch granules) inside the chloroplasts are usually the first microscopically visible signs of chilling injury. At the same time, chilling also decreases phloem loading and phloem transport. Accumulation of sugar in the chloroplasts (lowering Ψ_π) due to reduced export and continued starch degradation may be responsible for chloroplast swelling by osmotic water influx. Of course, damage to the photosynthetic "hardware" usually has severe consequences for photosynthesis, although the reduction in photosynthesis could also be caused by feedback inhibition due to sugar accumulation. Sugar may accumulate in the leaves because cell division, and hence growth, ceases at low temperature, which decreases sink demand for assimilates and may result in an oversupply of fixed carbon (Körner, 2003; Koroleva et al., 2002). With prolonged chilling, the chloroplasts may disintegrate completely so that the leaves become chlorotic (Kratsch and Wise, 2000). Leaves also turn chlorotic when they develop under chilling conditions, but this is probably due to an inability to produce thylakoid proteins. Chilling-induced chlorosis is normally reversible; thus leaves may recover and regreen upon exposure to warmer temperatures if chilling injury is not too severe. However, the longer the plants are exposed to low temperatures, the more extensive and irreversible is the damage (Kratsch and Wise, 2000). Under prolonged exposure to low temperature, vines activate an organized cell suicide process (so-called programmed cell death) that involves the systematic, energy-intensive (i.e., ATP-dependent) dismantling of the cells accompanied by recovery of cell components (e.g., proteins and DNA). The rescued breakdown products may be exported to other plant parts to enable the vine to survive the unfavorable conditions.

The more chilling sensitive a grape species or cultivar, the sooner and the more extensive is the development of the ultrastructural changes. The most resistant plants may not suffer injury unless they are simultaneously exposed to some other stress factor, such as high light intensity or water stress. Light greatly exacerbates chilling injury due to energy "overload" because light absorption decreases less than carbon fixation, which leads to generation of reactive oxygen species and, thus, oxidative stress (Kratsch and Wise, 2000). Under such conditions, reactive oxygen species can trigger the degradation of the two photosystems, PSI and PSII, and the photosynthetic enzyme rubisco (Møller et al., 2007). Therefore, symptoms develop sooner and are more severe during cold days than during cold nights. Nevertheless, cold nights seem to make the leaves more susceptible to high light intensity (Bertamini et al., 2006). When cold nights are followed by bright, warm days, grapevine leaf photosynthesis can be inhibited completely within a few days. The chilling-induced depression of photosynthesis appears to be similar to the photosynthetic depression due to water deficit (Bálo et al., 1986; Flexas et al., 1999a; see also Section 7.2), although the closure of stomata below 15°C is thought to be a response to the reduced photosynthesis rather than the reverse (Hendrickson et al., 2004a). Photosynthesis recovers only after several warm nights. Photooxidative stress can trigger anthocyanin accumulation in the exterior leaves of dark-skinned cultivars so that the leaves may turn red, whereas leaves in the canopy interior usually do not accumulate anthocyanins. Because anthocyanins probably act as photoprotectants, autumn coloration often becomes much more intense during cool weather, especially if the light intensity remains high (Feild et al., 2001; Hoch et al., 2003). It also appears that anthocyanins fluctuate according to changes in the weather; they may be degraded or reformed depending on the temperature and light conditions (Keskitalo et al., 2005).

High humidity protects plant tissues against injury, especially if the low temperatures occur during the night (Kratsch and Wise, 2000). Moreover, if temperatures remain relatively low for longer periods of time, the leaves may acclimate to those conditions by reemitting some of the excess energy as fluorescent light (see Section 4.1) and increasing the activity of enzymes involved in sugar production (see Section 4.2). Fluorescence emission occurs especially in mature leaves that suddenly experience a "cold shock," whereas enhanced sugar production may be more important in young leaves developing during cool conditions. The acclimation to low temperature, termed chilling acclimation, is very similar to the photoacclimation to high light intensity (Ensminger et al., 2006; Theocharis et al., 2012). Both processes aim to limit photoinhibition—that is, to balance energy supply and demand. If chilling conditions persist, the leaves that had grown under warm conditions will eventually be shed and will progressively be replaced by new, acclimated leaves. These adjustments lead to higher photosynthetic rates at lower temperatures and a downward shift of the optimum temperature for photosynthesis. This response also results in improved water-use efficiency (i.e., a higher photosynthesis:transpiration ratio), and hence drought tolerance, and accelerates the acquisition of freezing tolerance (Ensminger et al., 2006). Perhaps this is because the maturing shoots and perennial parts of the vine act as alternative sinks for the surplus assimilates that are available due to the cessation of growth. Therefore, repeated exposure to chilling temperatures can also induce adaptive changes in a grapevine that result in subsequent tolerance of freezing temperatures. This process is referred to as cold acclimation or hardening and is discussed later. However, under some continental climate conditions the temperature fluctuates too much between night and day for acclimation to protect the leaves from inhibition of photosynthesis.

Chilling during the period leading up to bloom impairs pollen formation in a way that is very similar to the effect of drought (see Section 7.2); both stresses disturb sugar metabolism in the anthers, whereas ovule fertility appears to be rather "immune" to low temperature (De Storme and Geelen, 2014; Oliver et al., 2005). The cold-induced inhibition of invertase activity in the anthers interferes with sugar supply (unloading) from the phloem even if the leaves supply abundant assimilates. This renders the pollen grains unable to accumulate starch, and they may instead temporarily accumulate sucrose. If this blockage occurs during meiosis (i.e., about 1−2 weeks before anthesis), it causes pollen sterility (De Storme and Geelen, 2014; Koblet, 1966; Oliver et al., 2005). Even brief episodes of chilling during male meiosis, such as two or three consecutive cool nights, can irreversibly inhibit pollen development. This decreases pollination and fertilization, which results in poor fruit set or, in more severe cases, abscission of inflorescences.

Chilling stress during grape ripening may lead to carbon shortage in the vine—for example, due to photoinhibition induced by cold nights followed by warm, sunny days. Because fruit sink activity is more sensitive to temperature than is shoot sink activity, this may induce a syndrome termed bunch stem necrosis. This physiological disorder essentially constitutes abandonment of some of the vine's clusters and tends to be more severe on heavily cropped vines. The vascular system in the peduncle and/or rachis, sometimes only on a shoulder or the tip of a cluster, becomes dysfunctional, the berries on affected clusters stop ripening, and the cluster or its affected portion is eventually shed by the vine (Hall et al., 2011). Thus, reproductive development is most vulnerable to chilling stress during the periods leading up to anthesis (i.e., before the vine has invested heavily in reproduction) and after veraison (i.e., when the seeds are mature). Between fruit set and the onset of ripening, the berries' sink strength is high enough to attract carbon from remobilized storage reserves when necessary (Candolfi-Vasconcelos et al., 1994).

7.4.3 COLD ACCLIMATION AND FREEZE DAMAGE

Ice melts at 0°C, but that does not imply that liquid water freezes at 0°C to form ice crystals. Ice formation (freezing) occurs only if no heat is removed from the water. If, however, heat is lost as the temperature decreases (e.g., the heat loss from plant tissues to the environment), then water will not freeze unless some nucleation event occurs (e.g., vibration of dust or other tiny particles such as bacterial proteins). This phenomenon is commonly referred to as "supercooling," which is followed by a sudden release of heat at the point where water finally freezes. At this point, chemical energy (the so-called latent heat, which is the energy needed to break the hydrogen bonds that hold together individual water molecules in ice crystals) is converted to (i.e., released as) sensible heat, which is heat that we measure with a thermometer. In the absence of ice nucleation, highly purified water can be supercooled to almost −40°C, but supercooling does not occur in plants to such an extreme because water inside plants and on their surfaces is never absolutely pure (Wisniewski et al., 2014). However, the presence of osmotically active solutes (e.g., sugars and nutrient ions) in plant tissues does decrease the freezing temperature below 0°C. Like osmotic pressure, freezing-point depression (more accurately, melting-point depression) is a so-called colligative property of solutions. As a general rule, every mole of dissolved solutes per kilogram of water lowers the melting point of the solution by 1.86°C (Zachariassen and Kristiansen, 2000); thus, the extent of this osmotic freezing-point depression is greater with larger amounts of solutes present in the solution or tissue. Plant tissues freeze when they cannot avoid nucleation and prevent ice growth (Pearce, 2001). The explosive force of water is considerable because the volume of freezing water expands by approximately 10% (Meiering et al., 1980). Incidentally, supercooled rainwater that freezes instantly when it hits a cold (<0°C) surface causes the phenomenon known as "freezing rain" or "ice storm."

During a freezing event, grapevine tissues lose heat and temporarily supercool due to the presence of osmotically active solutes. A tissue is defined as supercooled when it remains unfrozen below its freezing temperature. As the temperature drops, water freezes first on an organ's surface and in the intercellular spaces, cell walls, and tracheary elements (apoplast), where the concentration of solutes is much lower, and hence the freezing point is higher, than inside the cells (symplast) (Müller-Thurgau, 1886; Wisniewski et al., 2014; Xin and Browse, 2000). The formation of ice (i.e., H_2O crystals) leads to exclusion of solutes, which therefore become more concentrated in the apoplast. This concentration effect lowers the apoplast water potential (by -1.16 MPa °C^{-1}), which "pulls" water out of the symplast osmotically (Guy, 1990; Pearce, 2001; Steponkus, 1984; Thomashow, 1999). The resulting increase in the cells' solute concentration in turn lowers their freezing point by another 2°C or 3°C. As a practical aside, the same concentration mechanism is exploited in the production of ice wine, whereby frozen grapes are harvested and pressed immediately. Liquid water can move freely across intact membranes, but ice crystals cannot, so that the symplast is supercooled (Steponkus, 1984). Thus, ice formation in the apoplast, termed extracellular freezing, is usually not lethal. When the ice crystals melt, the water simply diffuses back into the cells and they resume their metabolism. However, in nonacclimated cells of leaves, for example, the cell membranes shrink irreversibly during extracellular freezing, and when the melted water moves back into the cells the resulting turgor increase can make the cells burst (Xin and Browse, 2000). While membrane shrinkage in cold acclimated cells of buds and woody tissues is reversible, the apoplast contains only approximately 10−15% of the total tissue water. Consequently, extended exposure to freezing temperatures dehydrates the symplast and causes a water deficit, which interferes with metabolism and puts cells under physical stress due to shrinkage (Browse and Xin, 2001; Müller-Thurgau, 1886; Thomashow, 2001). The amount of water that is lost by a cell depends on its solute concentration (determining π_{cell}) and the freezing temperature (determining $\Psi_{apoplast}$). For instance, if freezing occurs at $-10°C$, then $\Psi_{apoplast} = -11.6$ MPa, which pulls 90% of the water out of a cell at $\pi_{cell} = 1$ MPa but only 80% from an acclimated cell with $\pi_{cell} = 2$ MPa (Xin and Browse, 2000). Very high solute concentrations can be toxic to the cells, and if the cells shrink beyond their minimum tolerated volume, osmotic stress and membrane rupture ensue (Zachariassen and Kristiansen, 2000). Thus, dehydration of the symplast by extracellular freezing is the most common cause of freezing-induced injury, and tissue survival depends on the extent of the cells' dehydration tolerance (Guy, 1990; Thomashow, 1999; Xin and Browse, 2000). Because it may take several days for the symplast to reach water potential equilibrium with the apoplast, such injury by dehydration is mainly a problem if low temperatures persist for prolonged periods (Gusta and Wisniewski, 2013). Under such conditions of chronic stress the cells may eventually die but the temperature is not low enough to kill them instantaneously; instant death occurs during acute stress that leads to intracellular ice formation.

In contrast with other tissues, xylem parenchyma cells somehow seem to avoid dehydration during extracellular freezing, at least for a while. This mechanism, which is termed deep supercooling, enables these xylem cells to endure lower temperatures than other tissues for extended periods (Quamme, 1991). However, prolonged exposure to freezing temperatures can lead to the coalescence of small ice crystals into large crystals—a process called recrystallization (Pearce, 2001). This formation of large ice masses can deform cells and damage plant tissues by interfering with their structure (or the structure of entire organs), for example, by forcing cell layers apart. Although this does not necessarily kill the vine, the resulting tension in the wood can cause splitting or cracking of woody organs either immediately or during the subsequent growing season, which may provide infection sites for crown gall bacteria (Paroschy et al., 1980; see also Section 8.2). Moreover, the forcing of gases out of solution during ice formation inside the xylem vessels can lead to cavitation (i.e., air embolisms; see Section 3.3). The vulnerability to such embolisms increases with increasing conduit diameter and with water stress that decreases xylem pressure. Therefore, dry soil conditions during winter generally result in more severe xylem cavitation, especially in vines that had grown vigorously during the preceding years and hence have large xylem vessels. Although this presents quite a challenge to grapevines experiencing repeated freeze–thaw cycles at higher latitudes, they are normally very efficient at repairing such cavitation in spring. Such repair cannot occur, however, if unusually dry soil prevents the buildup of root pressure needed to dissolve and push out air bubbles (see Section 2.2). It is also possible that conditions that lead to depletion of root storage reserves (e.g., severe drought stress, loss of functional leaf area) render the roots unable to restore xylem function in spring, which may result in plant death (Galvez et al., 2013).

As the temperature drops, the apoplastic water initially remains unfrozen, and the tissues cool until they reach their ice nucleation temperature. Ice formation in the apoplast is accompanied by a release of heat, called the high-temperature exotherm or freezing exotherm, which results in a rapid but transient rise in tissue temperature (Xin and Browse, 2000). This temperature is the (nonlethal) freezing temperature of the tissue. During freezing, the tissue temperature remains constant and above ambient until all the extracellular water is frozen. Once this has occurred, the tissue temperature returns to ambient and continues to drop. As the temperature continues to decline, a second phase of heat release occurs when the cell protoplasts freeze, and this is termed the low-temperature exotherm (Mills et al., 2006; Ristic and Ashworth, 1993). Intracellular freezing kills cells instantaneously due to the combined effects of membrane damage, symplast dehydration, and protein denaturation. The extent of intracellular ice formation also depends on the rate of tissue cooling; very rapid cooling seems to induce ice formation within the cells even at temperatures that would normally induce only extracellular freezing (Ashworth, 1982; Müller-Thurgau, 1886). Fortunately, under field conditions cooling rates are seldom fast enough to favor intracellular freezing, except in situations in which the sun heats trunks and canes that face it before the temperature plummets during a clear night (Guy, 1990).

Individual vine tissues and organs differ in their ability to tolerate freezing temperatures. Growing organs that have a high water content are very sensitive to frost, and early spring growth after budbreak is especially frost susceptible (Fuller and Telli, 1999). Ambient temperatures of −2°C or −3°C or lower can damage leaves, shoots, and green buds (Fennell, 2004). The lethal temperature also depends on the presence of moisture and ice nucleating agents (especially certain bacteria, such as *Pseudomonas syringae*) on the surface of an organ (Luisetti et al., 1991). For instance, dry leaves can supercool to relatively low temperatures, whereas wet leaves (e.g., due to condensation of water during a radiation frost) freeze at just a few degrees below 0°C due to the presence of nucleating agents on the leaf surface. Ice first forms on the leaf surface and then penetrates through stomata, hydathodes, wounds, or cracks in the cuticle (although the cuticle normally acts as a barrier to ice growth) to initiate freezing in the apoplast (Pearce, 2001; Wisniewski et al., 2014). Severe infections by mites, insects, or pathogens may create lesions through which ice can grow into the leaf.

Cold hardiness, or freezing tolerance, remains low during the growing season but increases in late summer and fall during a process known as cold acclimation and reaches a peak in midwinter. As a consequence, damage to buds, canes, cordons, and trunks in midwinter (Figure 7.13) occurs at much lower temperature than during the growing season and depends on the temperature experienced during the period preceding the cold event. Grapevines can survive freezing temperatures

FIGURE 7.13

Young Merlot cordon (left) with lethal freeze injury to the phloem (brown area), while the xylem parenchyma remains viable (green area), and frozen primary bud with surviving secondary and tertiary buds (right).

Photos by L. Mills.

either by avoiding them (e.g., via late budbreak and early fruit and shoot maturation) or by tolerating them (e.g., by deep supercooling in midwinter). They also have "ice sites" or "ice sinks" to accommodate ice formation. For example, the scales of buds with their down (see Section 1.3) can cope with ice crystals that grow from water moving out of the supercooling meristem tissue in the buds' interior. The formation of ice in the bud scales is termed extraorgan freezing and may be accompanied by progressive dehydration of the interior tissues as temperature decreases (Pearce, 2001). In addition, as they acquire freezing tolerance, that is during cold acclimation, the buds become hydraulically isolated from the subtending cane (i.e., disconnected from its vascular tissues) by tightly packed parenchyma cells with relatively impermeable cell walls that form a diffusion barrier for ice forming in the cane and enable the buds to supercool (Fennell, 2004; Jones et al., 2000). The primordia in such acclimated buds have no functional xylem connection to the subtending bud axis (Ashworth, 1982; see also Figure 2.8). This structural ice barrier protects the dormant buds from the ice spreading through the cane xylem, which can occur at rates of several centimeters per second (Wisniewski et al., 2014).

Although the most conspicuous sign of cold acclimation in grapevines is the abscission of leaves at the end of the growing season, acclimation is a gradual process. Leaf abscission is preceded by periderm formation and thus browning of the shoots, entering into dormancy of the winter buds, and withdrawal of nutrients from the leaves and redistribution to the permanent organs of the vine—a process termed resorption. The "wave" of brown periderm moving up the shoot in late summer is associated with a massive increase in storage carbohydrates (Eifert et al., 1961; Winkler and Williams, 1945) and a decline in water content, and browning correlates closely with the ability to cold acclimate when the temperature decreases (Fennell and Hoover, 1991; Howell and Shaulis, 1980; Salzman et al., 1996; Wolpert and Howell, 1985, 1986). Because they are unable to acclimate, the green ends of the canes usually die back in winter, which is a "self-pruning" effect that can be exploited in lightly (e.g., mechanically) pruned or nonpruned (so-called minimally pruned) vines with high bud and shoot numbers (Clingeleffer, 1984; Clingeleffer and Krake, 1992; Downton and Grant, 1992; Keller and Mills, 2007).

Both periderm formation and cold acclimation are initiated by decreasing day length in late summer, followed by cool nonfreezing (night) temperatures (0–5°C) (Alleweldt, 1957; Schnabel and Wample, 1987; Wake and Fennell, 2000). It is possible that the lengthening twilight periods at dawn and dusk are at least partly responsible for the day-length effect that is detected by the leaves; the low ratio of red to far-red light in twilight may exert its influence via the vine's phytochrome system (Franklin and Whitelam, 2007; see also Section 5.2). Shorter days alone, while inducing dormancy, are insufficient for grapevines to cold acclimate fully (Fennell and Hoover, 1991; Salzman et al., 1996). Dormancy is, however, a prerequisite for the subsequent acquisition of full cold hardiness (van der Schoot and Rinne, 2011): green buds cannot cold acclimate. Even brown buds need exposure to low temperatures in order to cold acclimate (Salzman et al., 1996). Chilling temperatures, or more accurately rapid rates of temperature decline, trigger a massive temporary redistribution of calcium (Ca^{2+}) from the apoplast to the symplast, and it appears to be this Ca^{2+} signal that activates the cold acclimation process by which plants acquire freezing tolerance (Browse and Xin, 2001; Knight and Knight, 2012; Plieth, 2005; Thomashow, 1999, 2001). The Ca^{2+} signal is locally generated in the meristematic tissues and cannot be imported from other plant parts; yet, if the temperature drops fast enough, the signal is strong enough to induce cold acclimation even at high temperatures. Although it is not known whether foliar Ca^{2+} applications can improve cold hardiness, high Ca^{2+} supply in the soil, which is typical of many high-pH soils, may indeed be associated with better freezing tolerance (Percival et al., 1999).

The Ca^{2+} signal is complemented by the action of plant hormones. Low temperatures favor the conversion of bioactive forms of gibberellin to inactive forms. Disappearance of active gibberellins in turn permits accumulation of growth-suppressing DELLA proteins (see Section 2.2), which increases freezing tolerance by an as yet unknown mechanism (Achard et al., 2008). Another hormone also participates in cold acclimation. In addition to increasing temporarily in response to low temperature, ABA rapidly (i.e., within days) reduces shoot growth and promotes periderm formation, bud dormancy, and freezing tolerance even at temperatures that are normally too high to induce cold acclimation (Guy, 1990; Thomashow, 1999; Zhang et al., 2011; Zhu, 2002). This response can be exploited by applying ABA sprays to the canopy during grape ripening (Zhang and Dami, 2012) or controlled water deficit (which triggers ABA production; see Section 3.2) to vines during an unusually warm autumn. In general, acclimation to water deficit is associated with earlier periderm formation (Williams and Matthews, 1990) and predisposes vines to hardening against cold, probably because the stresses caused by extracellular freezing and drought are similar. Conversely, any factor that causes a delay in the cessation of shoot growth tends to delay cold acclimation as well. Examples include late planting of young vines during vineyard establishment or late field grafting. Factors that promote vigor, such as excessive soil moisture or nitrogen availability, as well as overcropping, late summer pruning, defoliation, or damage caused by mechanical harvesters, also may interfere with reserve accumulation and thus slow cold acclimation (Currle et al., 1983; Zufferey et al., 2012). Nonetheless, application of as much as 224 kg N ha^{-1} rarely compromised bud hardiness of Riesling in a study conducted in eastern Washington, an area known for its cold winters (Wample et al., 1993).

It seems that once activated, acclimation accelerates following brief episodes of 0°C, and cold hardiness reaches a maximum after the vine has experienced temperatures around −5°C (Fennell, 2004). Maximum freezing tolerance is thus acquired in sequential phases that are initiated by short days, chilling temperatures and, finally, freezing temperatures (van der Schoot and Rinne, 2011). The rate of cold acclimation is inversely correlated with temperature so that vines harden off faster at cooler sites or in years with cooler autumn temperatures (Ferguson et al., 2011, 2014; Keller et al., 2008; Stergios and Howell, 1977). Therefore, cold hardiness increases gradually during early autumn and then accelerates in late autumn to approach maximum freezing tolerance.

But even though low temperatures lead to cold acclimation, they will not maintain vines in a very cold hardy condition. In fact, grapevines gradually lose hardiness when they are exposed to low but nonfreezing temperatures for long periods. The process of cold acclimation is cumulative; it can be stopped, reversed, and restarted, depending on temperature fluctuations (Kalberer et al., 2006; Ferguson et al., 2014). Warm episodes during the acclimation period induce rapid deacclimation, and the greater the loss of cold hardiness, the longer it takes for the vines to reacclimate. This may spell trouble when an unseasonable freeze event follows a mild autumn (Keller and Mills, 2007) and has implications with respect to the consequences of global climate change. The influence of climate change is expected to be more pronounced at the lower limits of temperature, because temperature increases will probably be higher at night, at higher latitudes, and in the winter, thus continuing the trend that has already occurred since the middle of the twentieth century (IPCC, 2013; Jones, 2005). This asymmetrical warming could mitigate some of the adverse effects of warmer autumn temperatures on cold acclimation. Moreover, as the average temperature creeps

FIGURE 7.14

Dynamics of grapevine cane carbohydrates with changing winter temperature (left): P, start of periderm formation; LF, leaf fall; B, start of bleeding; BB, budbreak (modified with permission from Eifert et al., 1961). Association between cane sugar concentration and the mean temperature over the preceding 5 days (top right; calculated from data in Eifert et al., 1961), and starch grains (St) deposited in the cane xylem (bottom right; reproduced with permission from Plank and Wolkinger, 1976).

upward, so will the extremes (i.e., outliers) on either end of the temperature distribution curve. Therefore, although winter killing freezes and spring and fall frosts will still occur in a generally warmer world, they might become less frequent and less severe.

Osmotic adjustment, or accumulation of so-called cryoprotectants, especially osmotically active sugars, is essential for proper acclimation because of their effects on freezing-point depression and membrane stabilization (Xin and Browse, 2000). These compounds provide cold tolerance by reducing the extent of cell dehydration and inhibiting the nucleation and growth of ice crystals. Therefore, intracellular freezing occurs at much lower temperatures in acclimated tissues than in nonacclimated tissues. Carbohydrates that are produced in the leaves are translocated in the phloem as sucrose and accumulated in the woody organs as starch (Figure 7.14; see also Section 5.1). This accumulation starts soon after bloom, when the rate of shoot growth slows, but peaks only during grape ripening and, sometimes, after harvest. Starch accumulation constitutes a replenishment of nutrient reserves that have been remobilized from the vine's permanent parts during budbreak and the initial establishment of the canopy (Eifert et al., 1961; Holzapfel et al., 2010; Winkler and Williams, 1945; Zapata et al., 2004; Zufferey et al., 2012). Of course, partitioning to the shoots, cordon, trunk, and roots is also necessary for growth and metabolism (e.g., nitrogen assimilation in the roots) of these organs. The middle portion of the shoot appears to be "refilled" first, followed by the basal and apical portion, the cordon and trunk, and, finally, the roots (Winkler and Williams, 1945).

Whereas the developing grape clusters dominate the hierarchy of photosynthate distribution, partitioning to the perennial parts intensifies soon after veraison, when the grape seeds are mature and ready to germinate. Thus, toward the end of the growing season, the permanent organs become the dominant sink of the grapevine (Candolfi-Vasconcelos et al., 1994). This period is concomitant with the final stages of fruit ripening and may sometimes interfere with winemakers' quest to maximize fruit quality. This is especially so when environmental stress factors limit the amount of sugar available for distribution, which can occur due to even a few unusually cool or cloudy autumn days during and after which ripening can stall. The change in relative sink priorities is also the reason why harvest date is usually not a significant factor in reserve carbohydrate accumulation and cold hardiness (Hamman et al., 1996; Keller et al., 2014; Koussa et al., 1998; Wample and Bary, 1992), which permits the recurring and often extreme delay in harvest of grapes destined for ice wine production near the latitudinal and altitudinal margins of grape growing. It is possible that late harvest compromises cold acclimation only in situations of overcropping, especially under cool, cloudy conditions, when storage reserves are used to support fruit ripening.

Starch does not increase cold hardiness of the canes, cordons, and trunks because it is osmotically inert. Acclimation requires conversion of the starch stored in phloem and xylem parenchyma cells back to sugars (mainly sucrose, but also glucose and fructose, and some galactose, raffinose, and stachyose), which decrease the cells' osmotic potential and act as cryoprotectants (Currle et al., 1983; Fennell, 2004; Guy, 1990; Hamman et al., 1996; Jones et al., 1999). This conversion seems to be initiated by short days in mid-September in the northern hemisphere (and thus *before* grapes are usually harvested in cooler climates), then follows the decreasing temperatures in late autumn (below $\sim 5°C$) and continues until midwinter (Eifert et al., 1961; Hamman et al., 1996; Müller-Thurgau, 1882; Wample and Bary, 1992; Winkler and Williams, 1945; see also Figure 7.13). A similar interconversion of starch to sugars occurs simultaneously in the buds (Hamman et al., 1996; Jones et al., 1999; Koussa et al., 1998). Accumulation of soluble sugars also accelerates because cool temperatures reduce respiration rates (see Section 5.2). Nevertheless, the simultaneous decline of photosynthesis due to the cooling temperatures and shortening days requires sugars produced from stored starch to be used to generate energy in addition to providing the substrates for the formation of cryoprotective components (Druart et al., 2007). Moreover, starch is also needed to produce fatty acids because during the transition to dormancy the vacuole in the cambium cells splits into several smaller vacuoles, which requires synthesis of new membrane lipids (Druart et al., 2007). After reaching its maximum near leaf fall, the total amount of carbohydrates in the buds and the perennial organs decreases only slightly (primarily due to maintenance respiration) during the dormant period, but the sugar:starch ratio increases as mean daily temperatures decline (Eifert et al., 1961; Koussa et al., 1998; Wample et al., 1993; Winkler and Williams, 1945).

Once sugar translocation ceases, the pores of the sieve plates separating the phloem's sieve tubes in the vines' stem portion are sealed by callose (see Section 1.3), the cork cells are waterproofed by suberization, and the perennial organs are drained of any water that is not absolutely essential to maintain cell function (e.g., maintenance respiration). This tissue dehydration reduces the water content to 42–45% and is necessary during acclimation because it is the rupture of cell membranes by ice crystals that causes freezing injury in nonacclimated plants. There is a strong association between cold acclimation and declining water content in the buds and woody parts of the vine (Wolpert and Howell, 1985, 1986). Callose is also deposited in the apical meristems (more precisely, in the plasmodesmata that connect individual cells) while they prepare for dormancy inside the buds (van der Schoot and Rinne, 2011; van der Schoot et al., 2014). This process is accompanied by the impregnation of cell walls and may effectively seal the meristems off from the supply of water, nutrients, and hormones from the rest of the plant until warmer temperatures induce callose degradation.

In addition to sugars and tissue water, proteins and amino acids are also involved in cold acclimation. Soluble proteins (especially glycoproteins; i.e., proteins attached to sugar molecules) are secreted to the apoplast, particularly in the bark during autumn, with the nitrogen for their synthesis being provided from remobilization in the senescing leaves (Griffith and Yaish, 2004; Guy, 1990; Pearce, 2001). In addition to reinforcing the cell walls and making them more elastic, the high viscosity of these so-called cryoprotective or "antifreeze" proteins prevents the growth of ice crystals and depresses the freezing point. Incidentally, polar fishes use the same strategy to avoid freezing in subzero saltwater temperatures, and many hibernating insects add glycerol or even ethylene glycol to their supercooling arsenal, enabling them to survive temperatures as low as $-50°C$ (Zachariassen and Kristiansen, 2000). Antifreeze proteins, which are particularly strong inhibitors of ice recrystallization, are very similar to the so-called pathogenesis-related (PR) proteins that grapevines secrete into the apoplast upon fungal attack during the growing season (see Section 8.1). The antifreeze proteins have probably evolved from PR proteins, and they enhance the resistance of plant tissues to both cold-tolerant (psychrophilic) pathogens and freezing temperatures (Seo et al., 2010; Zachariassen and Kristiansen, 2000). Nevertheless, the PR proteins accumulated by nonacclimated tissues (e.g., in shoots following powdery mildew infection) lack antifreeze activity so that late-season fungal infections do not enhance cold tolerance (Griffith and Yaish, 2004). Dehydrins are another group of glycoproteins that accumulate in acclimating buds, perhaps in response to the declining water content (Salzman et al., 1996). Given that dehydrins are involved in desiccation tolerance, they may serve to protect cell membranes when water freezes in the apoplast.

Organic acids (particularly malate) and free amino acids (particularly proline, arginine, alanine, asparagine, and serine) may be accumulated in parallel with sugars during acclimation, but little is known about their contribution to cold hardiness (Currle et al., 1983). It could be that the cryoprotective properties of proline may be due to its function as protein and membrane stabilizer during water stress (Thomashow, 2001; see also Section 7.2). Deep-supercooling xylem parenchyma cells also seem to accumulate flavonols (glycosides of quercetin and kämpferol) that help them suppress ice nucleation (Kasuga et al., 2008). In addition to the effect of osmotic potential, the extent of cold tolerance is also modified by adapting the size of individual cells, thickening of cell walls, and changes in membrane properties that reduce the sensitivity of the membranes to mechanical stress due to contraction and expansion (Steponkus, 1984). Finally, activation of the antioxidant system during the cold acclimation phase enhances the protection of cell membranes from damage due to oxidative stress associated with low temperatures (Druart et al., 2007). The combination of all of these processes can either delay the onset of freezing or diminish its adverse consequences.

During the winter, the aboveground organs of grapevines continually adjust their degree of cold hardiness in response to fluctuating temperatures (Figure 7.15) (Düring, 1997; Ferguson et al., 2011, 2014; Mills et al., 2006). After a minimum of approximately $-10°C$ has passed, rising temperatures induce a reversal of the starch to sugar conversion so that starch again accumulates (Currle et al., 1983; Eifert et al., 1961; Koussa et al., 1998; Winkler and Williams, 1945). This change goes along with a concomitant decrease in cold hardiness as the vine deacclimates and prepares for budbreak in spring. Once the buds' endodormancy has been released (i.e., the chilling

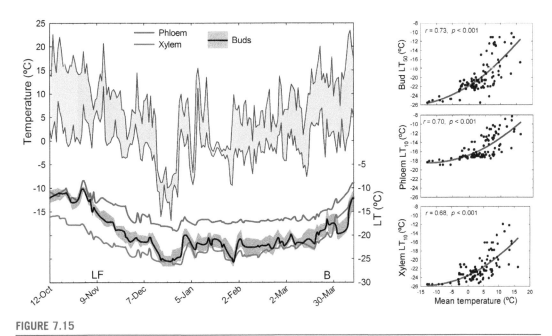

FIGURE 7.15

Dynamics of Merlot cold hardiness with changing winter temperature (left): LF, leaf fall; B, start of bleeding; LT, lethal temperature; lines indicate LT for 10% of the phloem and xylem and 50% of the buds with gray area indicating the 10–90% window. Association among bud, cane phloem, and cane xylem hardiness and the mean daily temperature (right). Mills and Keller, unpublished data.

requirement has been fulfilled), temperature is the main driver of deacclimation; the temperature threshold that induces deacclimation is about 5°C but varies by cultivar and species (Ferguson et al., 2011, 2014; Pagter and Arora, 2013; Stergios and Howell, 1977; Wolf and Cook, 1992). Although the sugar:starch ratio also reflects temperature fluctuations during winter, the conversion of starch back to sugar following colder temperatures gradually slows toward the end of winter (Currle et al., 1983; Wample and Bary, 1992). Moreover, the concentration of antifreeze proteins declines rapidly upon exposure to warm temperatures. Cold hardiness can be partially lost within a day during unseasonable warm spells in winter (Ferguson et al., 2011, 2014; Guy, 1990; Mills et al., 2006; Pagter and Arora, 2013; Wolf and Cook, 1992). The side of a trunk that is exposed to the afternoon sun is particularly vulnerable when it warms up due to reflection of solar radiation from snow covers. Although hardiness is restored when cool temperatures return, the deacclimation behavior predisposes grapevines for cold injury when the temperature drops suddenly after a warm episode. Moreover, the sensitivity to temperature fluctuations increases and deacclimation occurs more readily as spring approaches (Wolf and Cook, 1992), making vines more vulnerable to sudden cold spells during late winter. Once root pressure begins to push water up the trunk before budbreak, the rapid rehydration of the vine tissues is associated with an equally rapid and irreversible loss of cold hardiness that continues as the buds swell and then break (Johnson and Howell, 1981; Meiering et al., 1980; Paroschy et al., 1980). It is during this budswell phase that the vascular

FIGURE 7.16

Small gradients in temperature down a slope can mean the difference between frost damage and no damage. Note the grow tubes around replanted vines at the lower end of the slope and the wind machine on top of the hill.

Photo by M. Keller.

connection is established between the bud primordia and the subtending canes, thus breaking down the buds' ice barrier to the xylem (Ashworth, 1982; Wisniewski et al., 2014).

The location of a vineyard site is an important determinant of the risk of freeze and frost injury. As cold air is heavier than warm air, vineyard sites that are somewhat elevated above a valley floor and permit free air drainage down a slope are at lower risk than vineyards on the valley floor and/or where air drainage is obstructed (Figure 7.16). In cool regions, vineyards are often located near lakes that moderate nighttime temperatures (Bowen et al., 2005). Many growers in particularly vulnerable areas (e.g., New Zealand, Washington, or Ontario) use wind machines in an attempt to increase the temperature around the vines. Wind machines are most effective during clear, calm nights, when heat is lost to the sky by long-wave radiation (Atwell et al., 1999; Battany, 2012; Schultz, 1961). Due to the resulting temperature inversion, the temperature 20 m above the ground can be $1-10°C$ warmer than that at ground level because cold air settles near the ground after sunset. At wind speeds greater than 1.5 m s^{-1}, however, the temperature gradient collapses so that artificial mixing of air is useless.

Although they are an important storage organ for carbohydrates and other nutrients, the roots may be less effective than the trunk, cordons, and canes in terms of the interconversion of starch and sugars during the winter season (Holzapfel et al., 2010; Winkler and Williams, 1945; Zufferey et al., 2012). In contrast to many tree species (Galvez et al., 2013), the bulk of storage carbohydrates in *V. vinifera* roots are thought to remain in the form of starch throughout winter. Moreover, roots do not seal their phloem with callose (Esau, 1948; Pouget, 1963), nor do they experience dormancy in the way the aboveground organs do (see Section 2.2). Consequently, the roots may be unable to cold acclimate and are much more vulnerable to cold temperatures than is the stem portion of grapevines. Roots of *V. vinifera* and *V. labruscana* cultivars may be killed at $-6°C$, whereas *V. amurensis* roots, which have smaller cells, may withstand temperatures as low as $-15°C$, with many rootstocks derived from American *Vitis* species experiencing root death in the $-8°C$ to $-10°C$ range (Ahmedullah and Kawakami, 1986; Guo et al., 1987). Cold injury to roots, even if not lethal, may compromise their ability to generate root pressure for xylem and bud rehydration (see Section 2.2), which may lead to uneven budbreak or canopy collapse after budbreak (Pagter and Arora, 2013). Because water has a greater heat capacity than air, temperatures fluctuate more in dry than in wet soils, and insufficient soil moisture in winter can predispose grapevine roots to cold injury. This is why growers in regions that regularly experience dry summers in combination with cold winters usually plant vines as deep as possible (with the trunk extending up to 30 cm below ground) and irrigate their vineyards at the end of the growing season to rewater the rootzone to field capacity.

Survival of a grapevine following cold injury depends on the type of tissue that is damaged, the extent of damage, and the ability to recover enough to (at least partially) resume proper function of the affected organs. Note that the term crop recovery refers to the crop produced from new growth following injury by spring frost. During the cold acclimation and deacclimation phases, grapevine buds are usually slightly hardier than canes or trunks, whereas in midwinter bud damage occurs at somewhat higher temperature than damage to the woody tissues (Mills et al., 2006). However, during calm nights the temperature near the soil surface is usually lower than at the canopy level (the temperature gradient is in the range $2-5°C\ m^{-1}$), which can sometimes result in trunk injury when canes and buds may just escape the critical temperature. Within the compound winter bud, the primary bud is generally less hardy than the secondary bud, which in turn is less hardy than the tertiary bud (see Figure 7.13). This is probably a reflection of the degree of bud differentiation; increasing differentiation leads to increased susceptibility to ice nucleation. Nonetheless, even primary buds in close proximity are not all killed at exactly the same temperature: The difference between 10% bud damage and 90% damage can be up to $7°C$ during the acclimation period in fall and the deacclimation period in spring, but this range is often smaller ($\sim 3°C$) in midwinter when the buds are fully acclimated (Mills et al., 2006). Swelling buds in spring lose cold hardiness very rapidly as their water content increases within a few days from $<50\%$ before swelling to $\sim 80\%$ by the woolly bud stage (Ferguson et al., 2014; Fuller and Telli, 1999; Hamman et al., 1990; Johnson and Howell, 1981; Lavee and May, 1997; Pouget, 1963). Because late winter pruning, close to the time of budbreak, delays budbreak and hence irreversible deacclimation in spur-pruned vines, this practice is often used in climates with a high risk of spring frosts (Friend et al., 2011). If done earlier, however, winter pruning does not impact either bud cold hardiness or budbreak timing (Wample, 1994).

Within woody organs, the phloem is usually the first tissue to be damaged by a severe cold event (Figures 7.13 and 7.15), even though the dilute water inside the xylem conduits freezes at much higher temperatures (above $-10°C$) than in the live phloem cells. This is at least in part because the phloem cannot accommodate apoplastic ice formation in the same way that ice forms rather harmlessly in the xylem vessels (Meiering et al., 1980). Phloem injury can interrupt the transport of assimilates from the permanent parts of the vine to the developing buds in spring. However, even when the phloem and vascular cambium are completely destroyed, starch can also be remobilized from the xylem parenchyma cells and released as sugars into the xylem conduits along with water (Améglio et al., 2004; Sakr et al., 2003; Winkler and Williams, 1945). This generates root pressure that drives out air bubbles and reactivates the buds (see Sections 2.2

FIGURE 7.17

Cold injury to the phloem and cambium of a young Gewürztraminer trunk above the soil surface, with callus forming above and around the site of injury, and suckers growing from below the soil surface (left). Riesling vine with similar cold injury, with aborted clusters and leaf symptoms of sugar accumulation, chlorophyll degradation, light avoidance, and nutrient deficiency (right); note that the vines in the background experienced no cold injury.

Photos by M. Keller.

and 3.3). Moreover, although dead tissues cannot be repaired, they can be replaced by new tissues (Pratt and Pool, 1981). Phloem can be readily restored if the cambium survives. Death of the cambium is more serious, although the extent of damage depends on the position of the dead cambium. Dead cambium surrounding the trunk effectively "girdles" it (Figure 7.17) and can be lethal to the vine because any surviving buds will break and begin to grow in spring. By blocking the movement of sucrose and auxin, girdling disrupts the communication between buds/leaves and roots.

If the cambium and phloem cannot be restored rapidly in spring and communication reestablished, the shoots and inflorescences that emerge from any surviving buds may receive insufficient nutrient reserves from the trunk and the roots. Because inflorescences are poor competitors for access to the remaining reserves, they are often aborted to permit continued shoot growth, as long as the xylem is able to supply soil water. However, the leaves may soon begin to accumulate carbohydrates and suffer from feedback inhibition of photosynthesis, because the sugar produced cannot be transported to the roots and clusters (although some may be used to produce callus in an attempt to heal the injured trunk). The roots in turn will starve from photosynthate deficiency, which will decrease nutrient uptake, so that the leaves may begin to show secondary symptoms of nutrient deficiency (Figure 7.17). In severe cases, the vine can wilt and collapse.

By producing and releasing auxin, the swelling and breaking buds in spring are responsible for reactivation of the phloem and, subsequently, renewed cambium activity (i.e., cell division) and production of new vascular tissues (see Section 1.3). This implies that at least some buds must be alive for recovery to occur, and this includes latent buds on the cordon or trunk that may break even when all the dormant buds on the canes have been killed (Pratt and Pool, 1981).

In contrast to the phloem, the xylem can accommodate the presence of ice, which can expand inside the vessels. Damage to the xylem is not by itself detrimental to the vine, but it normally occurs only after the phloem and cambium have already been killed so that xylem injury is often a sign of disaster. Remarkably, however, as long as a sufficient number of immature (i.e., relatively undifferentiated) xylem parenchyma cells survive, auxin may induce these cells to dedifferentiate and resume cell division to directly form new phloem cells, complete with sieve elements and companion cells, and new cambium cells (Pang et al., 2008; Zhang et al., 2011). At the same time, the dedifferentiated xylem cells also form callus on the wound surface; the callus, however, does not seem to participate in the regeneration of the vascular tissues. Thus, restoration may occur via the division of surviving xylem cells, provided some bud-derived auxin can diffuse down to the "site of action." However, although more parenchyma cells may differentiate into vascular cells when auxin transport is interrupted (especially near surviving buds acting as auxin sources and above log jams of auxin flow), these vascular tissues may remain discontinuous because they cannot become properly aligned in the absence of polar auxin flow. Because the production of new cells depends on the supply of sucrose and other nutrients stored in the parenchyma cells, it seems probable that a high reserve status promotes recovery from cold injury.

The maximal extent to which vines become cold hardy varies with species and cultivar (Pierquet and Stushnoff, 1980; Williams et al., 1994). Hardier species or cultivars ordinarily can survive with more of their extracellular water frozen than can less hardy ones. The eastern Asian species *V. amurensis* and the northeastern North American species *V. riparia* can survive midwinter temperatures as low as $-40°C$, whereas the southeastern North American species *M. rotundifolia*, at the other end of the spectrum, suffers severe damage at $-12°C$ (buds even at $-5°C$). The majority of the other American species are not as hardy as *V. riparia*, but they are hardier than *V. vinifera*. And yet, there is no conclusive evidence to suggest that rootstocks derived from American species systematically alter the hardiness of their *V. vinifera* grafting partners (Miller et al., 1988; Wolf and Pool, 1988). Most *V. vinifera* cultivars are killed at midwinter temperatures below approximately $-25°C$, but individual cultivars vary by several degrees (Davenport et al., 2008a; Ferguson et al., 2014; Mills et al., 2006). However, the relative order of cultivar differences can be modified by location, cultural practices, and timing of a cold event. For example, Concord (*V. labruscana*) is considered to be significantly hardier than Riesling (*V. vinifera*), and Chardonnay (*V. vinifera*) is considered to be almost as hardy as Riesling, but the early budbreak and, consequently, early deacclimation of Concord and Chardonnay in late winter induce a substantial loss in hardiness. As a general rule, species and cultivars with a short endodormancy period (i.e., low chilling requirement) and/or a low temperature requirement for budbreak (see Section 2.2) deacclimate earlier than species and cultivars with a high chilling requirement and/or higher temperature threshold and are more cold sensitive toward the end of winter. Species or cultivars that originate from higher latitudes or from regions with continental climates tend to be hardier in midwinter than are their lower-latitude or coastal counterparts (Ferguson et al., 2014; Kovács et al., 2003; Wolf and Cook, 1992). When such cultivars are grown side by side, however, the former are more vulnerable to spring frost damage, because their earlier deacclimation is associated with earlier budbreak.

CHAPTER 8

LIVING WITH OTHER ORGANISMS

CHAPTER OUTLINE

8.1 Biotic Stress and Evolutionary Arms Races 343
8.2 Pathogens: Defense and Damage.. 352
    8.2.1 Bunch Rot 353
    8.2.2 Powdery Mildew 357
    8.2.3 Downy Mildew... 359
    8.2.4 Bacteria.. 361
    8.2.5 Viruses 364

ABSTRACT

Most organisms do not impact grapevine growth and fruit production, but some organisms enter into a symbiotic relationship with grapevines or render other services, while other organisms compete with the vines for resources, and yet others eat them or invade them as disease-causing pathogens. Weeds use water and soil nutrients, but some plants can fix atmospheric nitrogen, and all plants stabilize the soil. Some birds, insects, and mites may feed on grape berries, leaves, or roots or vector symbiotic or pathogenic microorganisms. Insects and, especially, microorganisms multiply very rapidly, enabling them to evolve rapidly. Wild grapevines have coevolved with these species and thus produced an arsenal of preformed and induced defense strategies, but vegetatively propagated plants brought to new environments are mostly deprived of this ability. Biotrophic pathogens, such as those causing powdery or downy mildew, can survive only on living host tissues, whereas necrotrophic pathogens, such as bunch-rot fungi, kill their host's tissues and feed on dead material. Some bacteria spread through the plant xylem and slowly kill grapevines. Viruses do not kill their grapevine hosts, but some spread through the phloem and reduce growth, fruit production, and quality.

Keywords: Bacterium; biotroph; disease resistance; fungus; infection; necrotroph; PR protein; stilbene; virus

8.1 BIOTIC STRESS AND EVOLUTIONARY ARMS RACES

Grapevines share their living quarters, both above- and belowground, with a myriad of other organisms, mainly arthropods (spiders, mites, and insects) and microorganisms (fungi, oomycetes, bacteria, and viruses), in addition to some nematodes, birds and mammals, and other plants. Some of these organisms are considered beneficial because they provide useful "services" to the vines for which growers would otherwise have to pay. Earthworms, for instance, enhance soil fertility by converting organic matter into humus, grinding up and mixing mineral particles, and creating biopores through which roots can grow easily and which aid in soil drainage and aeration. A host of microorganisms are involved in the mineralization of soil organic matter to inorganic nutrients that can then be taken up by the roots. Other microorganisms (e.g., mycorrhizal fungi) actively participate in this nutrient uptake in a symbiosis with plant roots (see Section 7.3). Most of the remaining organisms normally leave grapevines alone, do little damage, or are unable to overcome the vines' defense strategies (the latter is called an incompatible interaction). For example, grape berries are typically heavily colonized by various yeast species, other fungi, and bacteria, especially in the vicinity of stomata and small cracks in the cuticle that allow exudation of sap from the mesocarp (Barata et al., 2012b; Belin, 1972; Bokulich et al., 2014). Yet this is very rarely a concern, unless berries are damaged or harvested fruit is left standing too long before processing so that uncontrolled fermentation sets in. Similarly, among the many fungal species in the order Erysiphales that cause powdery mildew in plants, only one is able to infect *Vitis* species (see Section 8.2)—and that one is unable to infect any other plant species. This is one reason, apart from aesthetic considerations, that roses are planted around some vineyards: because the two plant species are susceptible to different species of fungi that thrive under similar weather conditions, the roses may be used to indicate conditions that might favor powdery mildew on grapevines. Some organisms, however, compete with the vines for resources (especially other plants, which are therefore called weeds) or make a living feeding on various grapevine structures, which makes them pests (in the case of arthropods) or pathogens (in the case of microorganisms, when we speak of a compatible interaction). Grapevines therefore perceive such organisms as biotic stress.

Plants sharing a vineyard with grapevines, whether they are called weeds or resident vegetation or are planted deliberately as cover crops or green floor covers, use water and tend to reduce nutrient availability for the vines (Celette et al., 2009; Guerra and Steenwerth, 2012; Keller, 2005; Morlat and Jacquet, 2003; Tesic et al., 2007). Performance of vines with a permanent floor cover in cool and humid climates can be remarkably similar to that often observed under deficit irrigation (see Section 7.2) in warm and dry climates; vines have diminished vigor, smaller berry size, improved fruit composition, and reduced bunch rot (Keller et al., 2001a,b; Valdés-Gómez et al., 2008). Vineyard floor vegetation is desirable for many reasons but may compete with vines for resources if not managed carefully, especially in warm and dry climates or in dry years (Celette et al., 2009; Pool et al., 1990; Tesic et al., 2007). The impact on yield formation arising from such potential competition may be averted by judicious irrigation management that takes into account both the vines and the additional vegetation (Ingels et al., 2005; Steenwerth et al., 2013).

Due to preferential growth of vine roots beneath emitters in drip-irrigated vineyards in dry climates (Stevens and Douglas, 1994), even if interrow cover crops are mowed or incorporated into the soil by tilling, the nutrients released during their microbial decomposition may be mostly available to cover crop regrowth rather than available to the grapevines. However, even in humid climates, competition from cover crops may increase vineyard fertilizer requirements if vine productivity is to be maintained (Keller et al., 2001a,b). Nitrogen, at least, can also be supplied using leguminous cover crops that employ symbiotic *Rhizobium* bacteria to fix atmospheric nitrogen ($N_2 \rightarrow NH_4^+$), which mitigates the need for synthetic fertilizers (Guerra and Steenwerth, 2012; Jackson et al., 2008;

Patrick King and Berry, 2005). Legumes can moreover acidify their rhizosphere, which may improve P availability (Shen et al., 2011). Consequently, cover crop management should aim to synchronize nutrient availability in the soil to the seasonal demand of the vines (Keller, 2005; Perret et al., 1993).

Birds can be important pests of wine grapes, because grapes have evolved to have their seeds dispersed by birds (see Section 2.3). This often interferes with a grower's or winemaker's desire to mature the fruit well beyond the stage at which seeds are ready to germinate. The presence in the green, immature berry of malic acid, which is an effective bird deterrent, and its degradation during ripening are probably adaptive mechanisms evolved by grapes to ensure protection of the developing seeds and their dispersal once they have reached maturity. The astringent tannins and the toxic, needle-like calcium oxalate crystals present in unripe grapes could be another defense against fruit-eating birds and mammals.

A variety of insects and mites (along with the crabs collectively grouped into the arthropods or the phylum Arthropoda, from Greek *arthron* = joint, *pous* = foot) feed on grapevine organs. Among them is the most devastating of grapevine pests—the root-feeding, aphid-like phylloxera, *Daktulosphaira vitifoliae* Fitch (also called *Viteus vitifolii*), which can multiply in all but very sandy soils and spreads rapidly except in very dry climates. Ironically, however, the sandy soils that halt phylloxera are very conducive to the multiplication of another group of root parasites, namely the root-knot nematodes from the genus *Meloidogyne* (Huglin and Schneider, 1998). Phylloxera (Greek *phyllon* = leaf, *xeros* = dry) is indigenous to the eastern and southwestern United States and Central America and was introduced to Europe on infested plant material (probably as eggs on rooted cuttings) in the middle of the nineteenth century. The subsequent epidemic across the continent wiped out most native wild vines and nearly destroyed the European wine industry, before grafting to resistant (more accurately tolerant) American rootstocks was introduced and accepted as the only feasible remedy. On American *Vitis* species, phylloxera has a complex life cycle, alternating between root-feeding and leaf-feeding forms (Forneck and Huber, 2009; Galet, 1996; Granett et al., 2001; Huglin and Schneider, 1998). The leaf-feeding form usually cannot multiply on the leaves of European *Vitis vinifera*, but the root-feeding form is extremely damaging. Once introduced in a new region, the insect, especially its first instar larva or nymph (a.k.a. crawler), can hitchhike on people and vineyard machinery to invade new vineyards (King and Buchanan, 1986). Following infestation, the insects and secondary infections by fungal pathogens can destroy the roots of susceptible *V. vinifera* cultivars. Infested vines gradually decline, often displaying symptoms of potassium deficiency, and eventually die as they run out of water and nutrients. The decline is rapid on heavy clay soils but can be slow on sandy soils, especially in very cool or hot climates (Granett et al., 2001). Because damage to the roots of American *Vitis* and *Muscadinia* species is very limited, these phylloxera-tolerant species or interspecific crosses are used throughout the world as rootstocks that protect vineyards from succumbing to phylloxera (Galet, 1996; Granett et al., 2001). The mechanisms of resistance or tolerance are not well understood, but many grape species may be able to form a cork layer around areas where phylloxera feeds on the roots, which may cut off the food supply to the insects. Despite this knowledge gap, the breeding and use of rootstocks, many of which continue to prevent damage after many decades of use, is an outstanding example of a highly successful biological pest control strategy. Many of these rootstocks also have at least partial resistance to or tolerance of nematodes (see Section 1.2).

Diverse other insect species feed on grapevine leaves or fruit. Many of these insects are able to use the mix of volatile compounds emitted by the leaves or berries to distinguish undesirable materials from edible ones, such as ripe berries (Rodríguez et al., 2013). All arthropods have natural enemies (predators or parasites), which we consider beneficial organisms; many of these are small wasps that parasitize insect eggs, and others are bugs, lady beetles, or spiders that eat other insects or mites (Mullins et al., 1992). However, one insect, the multicolored Asian lady beetle *Harmonia axyridis* Pallas, which was originally introduced to Europe and North America as a biological control agent for aphids and scale insects, may directly influence wine and grape juice aroma. When the beetles hiding in grape clusters are disturbed or crushed during juice processing, they release a yellowish fluid from their legs in a process termed "reflex bleeding." The methoxypyrazines in this fluid lead to off-flavors, termed "ladybug taint," because they are extremely potent at imparting vegetal, green pepper, and herbaceous aromas to grape juice and wine (Pickering et al., 2004, 2005; see also Section 6.2). This lady beetle is therefore now considered an important contaminant pest.

In addition to insects, several mite species (mites are related to spiders rather than insects) also can be potent grapevine pests. For instance, grape rust mites (*Calepitrimerus vitis* Nalepa) and grape bud mites (*Colomerus vitis* Pagenstecher) are responsible for a phenomenon termed "restricted spring growth" (Bernard et al., 2005). Symptoms include retarded budbreak, stunted shoots and distorted leaves in spring until fruit set (with poor fruit set or loss of inflorescences), intense growth of lateral shoots, and bronze discoloration (russetting) of mature leaves later in the growing season. The damage is caused by overwintering populations of rust mites, which feed on the newly emerging shoots and are dispersed mainly by movement of shoots (e.g., due to wind or machinery). Rust mites are particularly destructive when cool spring temperatures slow shoot growth, whereas rapidly growing shoots during warm conditions quickly outgrow the feeding mites. In contrast, bud mites live and feed inside the developing and dormant buds, which leads to stunted shoots (often with zigzag growth) that are missing one or all of their inflorescences or the apical meristem (Bernard et al., 2005). The bud mites migrate to newly forming buds after budbreak; they are spread mainly via infested planting material. In extreme cases, bud necrosis (i.e., death of the primordia inside the bud) from mite feeding can prevent budbreak altogether. In biologically diverse vineyards, both of these (as well as other damaging mite species) are usually kept in check by predatory mites.

Organisms also interact in other ways. Social wasps (*Vespa* and *Polistes* species) that feed on grapes, for instance, inadvertently ingest yeast cells. Because the wasps can harbor these cells in their gut during the winter and pass them on to their larvae in spring, these wasps may maintain and promote relatively site-specific yeast communities in vineyards (Stefanini et al., 2012; Vezinhet et al., 1992). Some pathogens also rely on so-called vectors to carry them from one plant to another. For example, certain viruses are transmitted by nematodes or insects, and phytoplasmas (specialized bacteria living in the phloem) require phloem-sucking insects (e.g., leafhoppers) for their transmission (Bovey et al., 1980; Oliver and Fuchs, 2011). Of course, vegetative propagation is also a very effective means to disperse such pathogens (which is why they are often collectively termed graft-transmissible diseases). The evolution of grapevines was until recently tightly linked to the evolution of other organisms that happen to like fruits and leaves. However, such coevolution has been greatly hindered since the introduction of vegetative propagation that deprived cultivated vines of the opportunity to adapt by sexual reproduction, leaving open only the much less effective avenue of somatic mutations (see Section 2.3).

Because mutations in somatic cells during cell division or mitosis (Greek *mitos* = string, loop) occur several orders of magnitude less frequently than in germ cells during meiosis, there is very little genetic variation within most modern grape cultivars compared with their wild relatives. Consequently, the influence of novel environments or newly introduced pathogens will cause more problems for some cultivars than for others, depending on their specific genotype. Moreover, a population that comprises many genetically identical individuals, such as is typical of the monocultures that are found in many modern vineyards, is vulnerable to any pathogen that discovers the key to exploiting this susceptibility. This puts cultivated grapevines at great risk from organisms—especially microorganisms—that like to eat their fruit, leaves, or roots. Microorganisms such as fungi, bacteria (especially bacteria), and viruses are extremely abundant in the environment, are exceedingly diverse, and have unnervingly short generation times. Not surprisingly, microbes are the most copious fruit consumers (Rodríguez et al., 2013). Large numbers, variation, and rapid reproduction mean rapid evolution. Because random mutations happen every so often, the laws of probability demand that in a sufficiently large or sufficiently fast-reproducing population, even unlikely changes are bound to occur. Nonrandom natural selection does the rest, quickly weeding out unfavorable mutations, ignoring irrelevant changes, and favoring material that benefits reproduction. Consequently, even the most unthinkable mutants may evolve, and any such fungal, bacterial, or viral strain with just a minuscule advantage over the others will soon dominate the population. These traits enable microbes to evolve extremely rapidly, under sufficient selection pressure sometimes within mere days. In addition, bacteria have neither true biological species nor sexual reproduction (Mayr, 2001). They instead readily exchange DNA with one another across apparent "species" boundaries, when two not necessarily related bacteria meet and connect with each other by growing a temporary tube between them across which genes flow from one bacterial cell to its neighbor. This so-called conjugation is analogous to copying systems data files directly from one computer to another. Vegetatively propagated grapevines, by comparison, evolve only very slowly, often remaining essentially unchanged for hundreds or even thousands of years (see Section 1.2). This puts microorganisms at a tremendous competitive advantage over cultivated vines and is a major cause of disease epidemics and of the rapid emergence of strains that are resistant to pesticides.

Even insects can multiply at astonishing rates; a single phylloxera egg hatching in spring may result in five billion parthenogenetically generated descendants by midsummer (Battey and Simmonds, 2005). A pathogen or pest introduced to a new area or made virulent through a mutation to which local cultivars have no defense could have devastating consequences from which wild and diverse plant communities may be somewhat protected, because they have coevolved with the pathogens. Such coevolution essentially constitutes a genetic "arms race," in which mutations occur in and natural selection acts on both the pathogens and the hosts. Moreover, mixing genes in new combinations during sexual reproduction (see Section 2.3) presents a moving (and, ideally, elusive) target for would-be pathogens and pests. Coevolution and sexual reproduction, of course, can do nothing to improve plant resistance, if a pathogen and its potential host evolved in geographic isolation from one another. Both the cultivated and wild forms of the European *V. vinifera*, when confronted with phylloxera and the mildew-causing pathogens introduced from North America in the second half of the nineteenth century, quickly succumbed for precisely the same reason native American peoples succumbed to European diseases after the Spaniards had inadvertently introduced the pathogens to the Americas; they had not coevolved with the pathogens and therefore had not had a chance to "invent" effective genetic resistance strategies. At the beginning of the twenty-first century history seems to be repeating itself in China, where phylloxera is wreaking havoc among native populations of wild and cultivated grapes (Du et al., 2009).

When leaves are damaged by arthropod feeding or other mechanical injuries, they release a blend of volatiles that is composed of ethylene, terpenoids, and other compounds derived from fatty acids in the cell membranes, such as methyl salicylate or methyl jasmonate. The emission of such volatiles is an indirect defense response and serves to attract beneficial arthropods that are predators or parasites of the herbivorous pests but do not feed on grapevines (James and Price, 2004; Pichersky and Gershenzon, 2002; Poelman et al., 2008; Takabayashi and Dicke, 1996; Van den Boom et al., 2004). In addition to this indirect effect, some volatiles also appear to have a direct defensive function. Hexanals and hexenals, for instance, may be toxic to insects, such as certain aphids (which imbibe phloem sap), whereas other volatiles deter some insects (especially Lepidoptera, whose caterpillars chew on the leaves) from laying eggs on the leaves. Moreover, grapevines that have been attacked by pests may be able to use volatile emission as a signal of imminent threat to nearby vines, which can detect this signal and respond to it by activating their own defenses. There is even evidence of plants under herbivore attack releasing root exudates that then induce neighboring plants to release volatiles from their leaves (Bais et al., 2006), but it is unknown whether grapevines can do this. Such plant-to-arthropod and plant-to-plant communication by use of air-transmissible signals may be very important in minimizing herbivore-induced damage because plants cannot hide or run away from pest attack. The number and relative abundance of beneficial arthropods often increases as the botanical diversity increases in a vineyard and its vicinity (e.g., due to diverse cover crops, resident floor vegetation, or adjacent brush or hedgerow vegetation), and so does the number of indifferent or neutral arthropod species (i.e., those that are neither harmful nor beneficial), which results in an overall increase in biodiversity and ecosystem stability and keeps pest populations in check (Altieri et al., 2005; Remund et al., 1989). Added benefits of cover crops compared with bare soil are their ability to reduce the splash dispersal of fungal spores during rainfall and their ability to improve vineyard water-use efficiency (Ntahimpera et al., 1998; Schultz and Stoll, 2010).

Any prospective pathogen that attempts to penetrate the epidermis first has to overcome the waxy cuticle and thick outer cell walls on plant organs before it can cause disease (Yeats and Rose, 2013). Accordingly, there seems to be an inverse association between the thickness of the cuticle and outer cell wall of different *Vitis* cultivars and their susceptibility to powdery mildew (Heintz and Blaich, 1989). Wounding of leaves and grape berries caused by herbivore, bird, or arthropod feeding or by other mechanical damage (e.g., from wind or machinery) can provide possible "enhanced-access" points for pathogens. In addition, shedding of plant organs (e.g., of flower caps during anthesis) leads to exposed fracture surfaces that provide ideal sites for pathogen invasion (Roberts et al., 2002; Viret et al., 2004). This is why plants respond to physical damage by mechanisms that aim at healing wounds and preventing invasion, such as reinforcement of cell walls by deposition of callose, lignin glycoproteins, and phenolics and production of so-called pathogenesis-related (PR) proteins such as chitinases and glucanases. Cell separation is also accompanied by accumulation of PR proteins to prevent infection; for example, these proteins accumulate at the base of anthers and calyptras at the time of shedding (Roberts et al., 2002). If this defense response is unsuccessful and pathogens penetrate into the tissues, the vine can shed the entire infected organ to prevent the spread of the pathogen throughout the plant. However, plants have evolved an array of strategies to resist fungal infection. These strategies can be constitutive or induced. Constitutive resistance strategies are passive and are present regardless of infection (i.e., preinfection). They include physical barriers (e.g., cuticle and cell walls) and chemicals with antimicrobial activity (the so-called phytoanticipins such as phenolics), which are generally accumulated in the cell vacuoles (often as inactive precursors) and provide nonspecific protection against a wide range of would-be invaders. Induced strategies, on the other hand, are actively initiated in response to pathogen invasion (i.e., postinfection) and specifically target pathogens that have overcome the constitutive barriers. They include the production of reactive oxygen species, fortification of cell walls (lignification, suberization, or incorporation of callose, proteins, silicon, and calcium), and production of antimicrobial compounds (e.g., proteins and the so-called phytoalexins). Active defenses are usually restricted to the site of invasion. The infected and neighboring cells accumulate the antimicrobial chemicals to high concentrations in an attempt to restrict the spreading of the invading pathogen. Cell wall reinforcement also requires the release of metabolic precursors to the apoplast before they can be incorporated in the cell walls, and this is usually accompanied by the production of reactive oxygen species that induce cross-linking of the cell walls (Field et al., 2006).

Grapevines, like other plants, have special receptor proteins that recognize invading pathogens by some of their microbial enzymes or complex carbohydrates, proteins, and lipids, especially those in the fungal cell walls, such as chitin (Shibuya and Minami, 2001). In addition, they also seem to be able to interpret as intruder signals the breakdown products of their own cuticle and cell walls (Yeats and Rose, 2013). These compounds are collectively termed elicitors because they elicit a defense response by the plant. In fact, the defense response results from activation of various biochemical pathways by a series of signaling cascades that are triggered by the detection of a pathogen. One of the first signals is the so-called oxidative burst on the cell membrane (Apel and Hirt, 2004; Jones and Dangl, 1996; Mittler, 2002; Smith, 1996; Somssich and Hahlbrock, 1998). Within minutes of an attempted infection by a foreign invader, there is a rapid rise in reactive oxygen species (mainly superoxide, O_2^-) in the apoplast. The superoxide is rapidly converted to hydrogen peroxide (H_2O_2), which diffuses into the cells and is perceived by the vine as a signal of an imminent threat. In contrast to the situation during abiotic stress (see Section 7.1), this oxidative burst is premeditated (O_2^- is produced enzymatically by NADPH-oxidase, peroxidase, and amine oxidase), and the antioxidant systems remain silent to boost the amount of H_2O_2. In response to this signal, the surrounding plant cells mount structural barriers; they reinforce (by callose deposition) and often lignify (H_2O_2 stimulates lignification by peroxidase and also has antimicrobial effects) their cell walls at the sites of attempted entry and produce PR proteins. These proteins owe their antifungal activity to their ability to degrade chitin and glucans, which are important components of the cell walls of fungi (Selitrennikoff, 2001).

Secondary signaling molecules, including salicylic acid, jasmonic acid, and ethylene, then augment the early defense response and may even activate defenses in distant healthy tissues (Heil and Ton, 2008; Thatcher et al., 2005). In other words, the secondary signals exert a kind of remote control; that is, they act systemically. Although salicylic and jasmonic acid are not volatile, their methylated forms—methyl salicylate and methyl jasmonate—can be emitted by the infected leaves within hours of an infection and taken up by distant uninfected leaves, where they may be transformed back into the nonmethylated forms to prime or sensitize those distant leaves for defense. The salicylic and jasmonic acids arriving days to weeks after an infection via the vascular pathway may then amplify the induced defense response in those leaves (Heil and Ton, 2008). The activation of defense responses by such signaling molecules may be exploited in disease control in vineyards; application of methyl jasmonate, it seems, not only triggers accumulation of defense compounds (i.e., PR proteins and stilbenes) in the leaves but also reduces powdery mildew infection (Belhadj et al., 2006; Larronde et al., 2003).

In some instances, H_2O_2 in concert with secondary signals induces the infected cells and those surrounding the infection site to commit suicide in a process termed hypersensitive response. This limits food supply to the pathogen, starving and confining the invaders in a localized area (Apel and Hirt, 2004; Jones and Dangl, 1996). In addition, nearby vascular tissues may be occluded by the deposition of callose, which limits the spread of the invader or its toxins.

Induced structural barriers may fail to contain a pathogen or may work for some but not other pathogens. Several hours after an infection, the vine therefore also activates a second line of defense, this time of a more direct, biochemical nature. This induced biochemical defense includes accumulation of antimicrobial compounds, including the phytoalexins and PR proteins. Another defense strategy, especially in response to xylem-invading fungal and bacterial pathogens, is the accumulation of elemental sulfur in the vessel walls and xylem parenchyma cells (Cooper and Williams, 2004). Unfortunately, most bacteria (e.g., the Pierce's disease−causing *Xylella fastidiosa*) are not sensitive to S. In contrast, the nonvascular pathogen *Erysiphe necator* and other fungi are very sensitive. Accordingly, S dust and S−lime mixtures have been in use as fungicides for thousands of years; indeed, they were the world's first fungicides, and their effectiveness against powdery mildew was discovered soon after the introduction of the fungus in Europe. Moreover, applications of S fertilizers can enhance the resistance to *E. necator* (see Section 7.3). The disadvantage of S application, however, is the risk of spider mite infestations, especially when S is applied as dust: Mites like dust.

Phytoalexins (Greek *phyton* = plant, *alexein* = to ward off) are a diverse group of fungitoxic (i.e., antimicrobial), low-molecular-weight metabolites that plants produce in response to various kinds of stresses, such as attack by pathogens or wounding (Kuc, 1995; Smith, 1996). These "natural fungicides" include more than 300 chemically diverse compounds (e.g., phenolics, terpenoids, polyacetylenes, fatty acid derivatives, and others) and have been found in more than 100 plant species from more than 20 families. The phytoalexins characteristic of grapevines are phenolic compounds named stilbenes and include resveratrol and its glucoside piceid, pterostilbene, and several resveratrol oligomers, the viniferins (Jeandet et al., 2002; see also Section 6.2). They are produced at or near sites of infection and can inhibit spore germination and mycelium growth of a variety of fungi and oomycetes. The production of these phytoalexins is triggered by elicitors comprising high-molecular-weight microbial compounds and components of plant cell walls that are released during infection (Blaich and Bachmann, 1980; Darvill and Albersheim, 1984). In addition, stilbenes also accumulate, along with PR proteins, in response to the air pollutant ozone (O_3) that forms as part of photochemical smog and is highly phytotoxic, inducing leaf symptoms such as numerous yellowish to reddish spots ("stipple") or blotches ("mottling"), bleaching, bronzing, or, in severe cases, premature senescence (Sandermann et al., 1998; Schubert et al., 1997; Williams et al., 1994).

The rate at which phytoalexins accumulate at the site of infection determines whether or not the pathogen attack is successful (Kuc, 1995; Smith, 1996). If accumulation is too slow, the intruder has grown far beyond the inoculation site by the time phytoalexin concentrations are high enough to inhibit its growth. In this natural arms race, many microorganisms have found ways to degrade and thus detoxify phytoalexins. This detoxification is an important trait conferring pathogenicity (also termed virulence), enabling the pathogen to infect the plant tissue (Darvill and Albersheim, 1984; Jeandet et al., 2002; Sbaghi et al., 1996; Smith, 1996; VanEtten et al., 1989). For instance, by employing the all-purpose phenol-oxidizing enzyme laccase (a.k.a. stilbene oxidase), *Botrytis cinerea* is able to degrade resveratrol and pterostilbene (Breuil et al., 1999; Hoos and Blaich, 1990; Pezet et al., 1991), whereas fungi regarded as not normally pathogenic for grape berries ostensibly lack this ability. Moreover, it is possible that accumulation of the fungitoxic viniferins comes too late. They may be formed only after the infected tissues have already become necrotic (Keller et al., 2003b). Despite their inducible nature in green plant organs, stilbenes also seem to be present as constitutive compounds in the vine's woody organs, including canes, trunks, and roots, as well as in peduncles. In concert with other phenolics and even terpenoids, they may contribute to the general pathogen resistance and durability of wood (Kemp and Burden, 1986; Langcake and Pryce, 1976; Pawlus et al., 2013; Pool et al., 1981).

Similar to stilbenes, the PR proteins, which are mostly accumulated in the cell walls, also have the capacity to directly inhibit spore germination and/or hyphal growth of certain pathogens (Giannakis et al., 1998; Jacobs et al., 1999; Monteiro et al., 2003). These antimicrobial proteins include β-1,3-glucanases (members of the so-called PR-2 protein family), chitinases (PR-3 family), osmotin/thaumatin-like proteins (PR-5 family), and perhaps some nonspecific lipid transfer proteins (nsLTPs; PR-14 family). Some of these antifungal proteins are active constitutively (i.e., present at some base concentration), and all accumulate to high concentrations in response to fungal attack of both leaves and berries. Whereas the PR-5 proteins are not produced constitutively in preveraison grape berries and other organs, they and the PR-3 (but not PR-2) proteins begin to accumulate in the berries at veraison (Negri et al., 2008; Pocock et al., 2000; Robinson et al., 1997; Salzman et al., 1998; Tattersall et al., 1997). This induction is probably a consequence of the osmotic stress caused by sugar accumulation and makes the ripening berries increasingly resistant to powdery mildew, downy mildew, and black rot. In addition, whereas moderate amounts of sugars can boost the colonization by some fungal pathogens (seemingly because sugars are important carbon sources for the microbes), higher sugar concentrations may reverse this effect, leading to lower susceptibility. Indeed, *E. necator* appears to be unable to establish new infections on grape berries exceeding approximately 7°Brix, a threshold that is normally reached just before veraison (see Section 6.2). It is possible that this phenomenon, which is termed "high-sugar resistance," may be at least partly due to the osmotic stress the high sugar concentration imposes on the fungal cells. Nevertheless, the rachis supporting the ripening berries remains susceptible as long as it remains green.

Some members of the PR-3, PR-5, and PR-14 families also seem to inhibit spore germination and growth of *Phomopsis viticola* Sacc., which causes cane and leaf spot disease, and *Elsinoe ampelina* (de Bary) Shear, which causes grapevine anthracnose (Monteiro et al., 2003). The *Vitis labrusca*–derived Concord grapes accumulate very high concentrations of these or closely related PR proteins, which also contributes to their strong *B. cinerea* resistance compared with the susceptible *V. vinifera* cultivars (Salzman et al., 1998). Although the latter produce much lower amounts, their PR-3 and PR-5 proteins can make up half of the total soluble protein in mature grape berries and are major contributors to protein instability in white (but usually not red) wines (see Section 6.2).

Because of the immense economic importance of grapevine diseases, stringent phytosanitary measures are generally imposed to control disease outbreaks. Cultural practices aimed at reducing disease inoculums and spread of infections include canopy management (open canopies facilitate rapid drying after rain, minimize shade, and maximize spray penetration and coverage) as well as nutrient and irrigation management (judicious fertilizer and water application avoids excess vigor). In addition, successful disease management often requires the intensive use of costly pesticides to optimize control and to manage pesticide resistance of pathogens. Nonetheless, although such strategies usually work reasonably well for fungal and fungi-like pathogens, they fail completely for most bacteria and viruses that insert themselves within their hosts' cells. Once present in a victim, vegetative propagation ensures their dissemination with new planting material (typically cuttings or buds) taken from the host. Some viruses are further distributed to healthy plants (e.g., in neighboring vineyards) by insect or nematode vectors (Oliver and Fuchs, 2011). In contrast with other disease agents, bacteria and viruses cannot be eradicated or contained with any curative or therapeutic control measures short of destroying infected plants (Rowhani et al., 2005).

8.2 PATHOGENS: DEFENSE AND DAMAGE

Pathogens can be roughly divided into necrotrophs and biotrophs. A necrotroph (Greek *nekrois* = death; *trophé* = nutrition) is defined as a pathogen that kills its host's cells and then obtains its food supply from these dead cells. The typical necrotroph attacking grapevines is *Botrytis cinerea* Pers.:Fr., which can infect all green plant organs but is best known for causing gray mold in grapes and many other plant species (Williamson et al., 2007). In a devious twist, rather than being killed by the post-infection oxidative burst inside the host tissues, such necrotrophs may actually exploit (and thus subvert) the plant's defense response by promoting tissue senescence which, in turn, favors disease progression (Elad and Evensen, 1995; von Tiedemann, 1997). Necrotrophs, as their name suggests, also benefit from rather than being inhibited by the cell death induced during a hypersensitive response which, by contrast, is very effective against the spread of biotrophs, again as their name suggests (Poland et al., 2009). Therefore, although killing the cells around sites of attempted infection is indeed a useful strategy to contain biotrophs, this approach comes at the cost of increased vulnerability to invasion by necrotrophs, whose mode of life evolved on a platform of thriving on dead tissue and which is only encouraged by suicidal plant cells.

A biotroph (Greek *bios* = life) is an organism that can live and multiply only on another living organism (i.e., it cannot be grown in culture), making it an obligate parasite or symbiont, depending on whether its presence is damaging or beneficial to its host. Examples of biotrophs include most viruses, the powdery mildew fungus *Erysiphe necator* [Schwein.] Burr. (formerly *Uncinula necator*), and the downy-mildew–causing oomycete *Plasmopara viticola* (Berk. & Curtis) Berl. & de Toni, both of which develop on green plant parts, but also the mycorrhizal fungi that enter into a symbiotic relationship with the roots of their hosts (see Section 7.3). Biotrophs typically tap into their host's epidermis cells to extract sugar (mainly glucose) and other nutrients (especially amino acids) to support their hyphae growing on the surface and act as sinks that alter carbon distribution in the host to divert nutrients for their own metabolism (Brem et al., 1986; Hall and Williams, 2000; Hayes et al., 2010; Smith et al., 2001). In other words, biotrophs compete with the sinks of their hosts for assimilate supply. They rely on a strategy that attempts to invade host tissues with minimal damage to their host's cells and suppress or bypass its defense system. Infections by biotrophs are usually limited to the epidermis. In fact, after breaking through the cell wall, using a high-precision chemical drilling approach (i.e., they penetrate a cell by digesting a pore through its wall; see Cantu et al., 2008), they may even exploit the host's defense strategies to accommodate the establishment of elaborate intracellular feeding structures. These structures are termed haustoria (singular haustorium) in the case of the mildews or arbuscules in the case of mycorrhiza; the host cell membrane becomes wrapped around the mycelium but is not penetrated by it (Hall and Williams, 2000; Heintz and Blaich, 1990; Pearson and Goheen, 1998; Smith et al., 2001). This maximizes the intruder's surface area inside the host for active nutrient absorption, which is a one-way street for mildews and a two-way street for mycorrhiza.

The remainder of this chapter discusses examples of some of the major pathogens, or groups of pathogens, that afflict vineyards throughout the world, their disease symptoms, and interaction with their hosts.

8.2.1 BUNCH ROT

The term bunch rot refers to the disintegration of ripening grape clusters due to infection by a range of different pathogens. The most important of these is the fungus *B. cinerea*, also known by the name of its sexual form *Botryotinia fuckeliana*. In fact, *B. cinerea* is one of the most ubiquitous plant pathogens in the world, with a wide range of host species. Its dry dispersal spores, called conidia, can survive temperatures as low as $-80°C$ for several months, can be blown by the wind over enormous distances, and can germinate (making them infective) over a temperature range of $1-30°C$, as long as the relative humidity exceeds 90% (Pearson and Goheen, 1998; Pezet et al., 2004b). Most vineyards, even in isolated areas, are under permanent pressure from wind-dispersed conidia, so *B. cinerea* can be regarded as part of a vineyard's environmental microflora. Nevertheless, its requirement for high humidity provides some protection to many vineyards, as long as they are not compromised by overly dense canopies (Valdés-Gómez et al., 2008; see also Section 5.2).

The *B. cinerea* fungus is responsible for one of the worst fungal diseases of grapes, namely gray mold, which can cause serious crop losses and reductions in fruit quality. The fungus is able to use the otherwise stable tartrate as a carbon source (in addition to grape sugars), converting some of the acid's degradation products into small amounts of malate and other organic acids. Enzymes such as the polyphenol oxidase, termed laccase, secreted by the fungus can readily oxidize phenolic compounds in grapes and continue to do so in the fermenting juice and wine made from these grapes (Dittrich, 1989; Dubernet, 1977; Ky et al., 2012; Macheix et al., 1991; Pezet et al., 2004b; Ribéreau-Gayon, 1982). When phenolics are oxidized, they turn into quinonens, which in turn can form brown polymers; this leads to discoloration of red wines and browning of white wines. The intruder also reduces amino acid concentrations and degrades aroma compounds (e.g., terpenoids), which shifts a wine's aroma from fruity to "phenol" or "iodine." Moreover, *B. cinerea* and other fungi that appear as secondary colonizers of infected grapes may produce volatiles with undesirable "mushroom" or "earthy" odors (La Guerche et al., 2006).

Ironically, however, some of the world's most highly prized dessert wines, particularly the French Sauternes, German Trockenbeerenauslesen, and Hungarian Tokaji aszús, are produced from grapes infected with *B. cinerea*. This so-called "noble rot" appears to be due to *B. cinerea* growing mainly in the epidermis of grape berries, which leads to desiccation due to increased water permeability of the skin and to concentration of sugars (especially fructose, due to the "taste" of the fungus for glucose) and, to a lesser extent, acids in the berry (Dittrich, 1989; Pezet et al., 2004b). Additional beneficial changes induced by the fungus include the accumulation of glycerol in the berries, which contributes to the sweetness of the resulting wine; of cysteine-precursors of volatile thiols with an aroma of grapefruit and passion fruit; and of fatty acid−derived lactones with a prune-like aroma (Ribéreau-Gayon et al., 2006; Robinson et al., 2014; Thibon et al., 2009). Early morning fog followed by warm, dry days is thought to favor the development of noble rot during ripening, but rain can quickly result in the detrimental bunch rot (Mullins et al., 1992). Winemakers inadvertently discovered the benefits of noble rot for wine quality when a delay in the harvest permission in the German Rheingau in 1775 and the invasion of France by the German−Russian armies in 1815 both led to late harvests (Dittrich, 1989; Pezet et al., 2004b).

The fungus opportunistically invades all green organs of the vine through wounds or via senescent or dead tissues and also invades young organs directly through the cuticle and outer cell wall, both of which it degrades in the process, causing necrosis (Elad and Evensen, 1995; Martelli et al., 1986; Salinas et al., 1986). Whereas prebloom flowers are relatively well protected by the calyptra, grape flowers immediately after anthesis are particularly vulnerable to infection, especially when senescent anthers and calyptras remain stuck in the inflorescence (Bulit and Dubos, 1982; Pearson and Goheen, 1998). The fungus can penetrate flowers at the receptacle end, possibly through the scar left behind after capfall, and may remain as latent mycelium inside the developing berries without causing disease symptoms until the berries begin to ripen (Cadle-Davidson, 2008; Keller et al., 2003b; Pezet et al., 2003, 2004b; Viret et al., 2004). Many other fungi also seem to be able to establish latent infections in grape berries, some of which (especially *Alternaria*, *Cladosporium*, and *Penicillium* species) may also cause postharvest rot in table grapes (Dugan et al., 2002). Damage to the berries by cracking of the cuticle from pressure within the berry and physical damage from insects, hail, and wind predispose clusters to berry infection (Kretschmer et al., 2007) and, together with wet conditions, leads to disease expression and fruit loss (Figure 8.1). Cracking of berries in susceptible cultivars can ensue during rainfall or during humid nights, particularly in combination with high soil moisture (see Section 6.2). In addition, small pores in the berry surface may also provide entry sites for *B. cinerea* infection (Mlikota Gabler et al., 2003). Once a disease outbreak has occurred, conducive conditions (especially rainfall during the ripening phase) will quickly lead to secondary infections via sporulating conidia. Such secondary infections spread much more readily throughout tight clusters whose berries touch each

FIGURE 8.1

B. cinerea infection in ripening grape cluster (left), *E. necator* blotches on abaxial side of leaf (center), and *P. viticola* "oilspots" on adaxial side of leaf (right).

Photos by M. Keller.

other, retaining surface water and rubbing off epicuticular wax, than through loose clusters whose berries are separated by gaps and dry rapidly after rain (Marois et al., 1986; Rosenquist and Morrison, 1989; Vail and Marois 1991; Vail et al., 1999). The importance of surface water for the spread of the fungus is a major reason why removal of leaves around the fruit clusters is a critical canopy management practice in regions that are prone to rainfall during the ripening period (English et al., 1990; Zoecklein et al., 1992). The increased air flow and, once the sun shines again, the higher berry surface temperature favor speedy drying of the clusters after a rainfall event. High sun exposure, in addition to loose clusters, associated with minimally pruned vines also strongly decreases *B. cinerea* infection (Intrieri et al., 2001).

In contrast with open flowers, young, unripe grape berries are highly resistant to *B. cinerea*, whereas ripening fruit again drop their defenses against fungal attack and become increasingly susceptible to infection (Blaich et al., 1984; Bulit and Dubos, 1982). This relaxation of defenses is partly related to the gradual modification of the grape cuticle and cell walls and the disappearance or modification of several preformed (i.e., constitutive) defensive substances. Once the berries have begun to dismantle their cell walls, they are no match for the necrotroph, which secretes a whole arsenal of wall-degrading proteins to digest the pectins and other polysaccharides in its passing host's cell walls (Cantu et al., 2008). Such cell wall degradation appears to be associated with the release of bound Ca^{2+}, which, due to its toxicity, may help the invader to kill the host cells (Kaile et al., 1991). Although the constitutively produced glycolate and phenolic compounds, especially tannins and their subunits but also hydroxycinnamic acids and flavonols, provide a protective barrier by inhibiting the macerating fungal enzymes as well as laccase (Goetz et al., 1999; Tabacchi, 1994), the increasing degree of polymerization of skin tannins toward fruit maturity (see Section 6.2) may decrease their ability to bind and denature the fungal proteins (Jersch et al., 1989). The continued accumulation in ripening grape berries of oxalate may also contribute to the decline of resistance because oxalate is suspected to facilitate degradation of cell walls and repress plant defenses; in fact, *B. cinerea* secretes oxalate during infection (Williamson et al., 2007; van Kan, 2006). Plants apparently retaliate by producing oxalate oxidase to partly disarm this fungal weapon (Walz et al., 2008). In addition to these constitutive barriers, green berries produce stilbene phytoalexins after infection, which may slow the spread of the fungus, although resveratrol is a rather ineffective botryticide and is easily degraded by fungal laccase (stilbene oxidase) enzymes (Hoos and Blaich, 1990; Pezet et al., 1991). The berries' ability to produce stilbenes also seems to decline over the course of berry development and ripening (Bais et al., 2000; Jeandet et al., 1991).

Accumulation of stilbenes in response to infection by *B. cinerea* might occur at the cost of reduced production of flavonoids such as tannins, anthocyanins, and flavonols. After all, healthy, nonstressed leaves and berries do not produce stilbenes, which is why these compounds are also termed stress metabolites. Conversely, it is possible that the accumulation of anthocyanins in red grape berries at veraison decreases the berries' capacity to produce stilbenes. Like anthocyanins, resveratrol is accumulated predominantly in the berry skin, although it also seems to be present in the seeds. Induction of stilbene synthase can inhibit the activity of chalcone synthase, and it has been argued that the two enzymes compete for substrates (Fischer et al., 1997; Gleitz et al., 1991). However, this does not explain why white grapes also produce lower amounts of stilbenes (Bais et al., 2000) and become more susceptible after veraison, or the fact that vulnerability remains high after the cessation of pigment accumulation in red grapes and is a major cause of damage from storage rot in table grapes (Pearson and Goheen, 1998). If one considers this phenomenon from a plant reproduction standpoint, the picture becomes clearer. The pigmentation of grapes provides visual cues that attract seed dispersers to ripened fruit, which become edible after veraison. Simultaneously, the barriers erected against microbial intruders that would be detrimental during seed development are dismantled when the seeds are viable and ready to be dispersed. Therefore, both the increase in anthocyanins and the decrease in stilbene-synthesizing capacity during ripening may serve the same purpose—to ensure seed dispersal. After all, when a grape berry rots away, its seeds generally remain free of infection and simply drop to the ground, where they may germinate. It seems at least plausible that this might constitute something like an insurance policy in case there are no birds or mammals present to eat the fruit. If this explanation is correct, *B. cinerea* may be regarded as a mutualistic symbiont in grape berries rather than as a pathogen, or at least as a fungus able to switch its lifestyle from pathogenic (in flowers and young leaves) to mutualistic (in developing berries).

In contrast to grape berries, expanding leaves gradually become more resistant to fungal infection (Langcake, 1981; Langcake and McCarthy, 1979). This may partly be caused by the accumulation of the two major phenolic biopolymers—lignin and tannin. Lignin is deposited in mature cell walls, resulting in physical strengthening similar to the role of steel rebars in ferroconcrete. Tannins are also produced at variable concentrations and inhibit the cell wall—degrading enzymes secreted by *B. cinerea* (Bachmann and Blaich, 1979; Goetz et al., 1999). In addition, mature leaves also contain high concentrations of an array of flavonols, hydroxycinnamic acids, and phenolic acids (Bachmann, 1978; Egger et al., 1976; Rapp and Ziegler, 1973), many of which inhibit fungal laccases (Goetz et al., 1999; Tabacchi, 1994), and produce PR proteins in response to infection (Renault et al., 1996). Furthermore, stilbenes accumulate much more rapidly in leaves, especially older leaves, than in fruit. Leaves of the *B. cinerea*—resistant American *Vitis* species (e.g., *V. riparia* and *V. rupestris*) and interspecific hybrids generally have a greater capacity for stilbene synthesis than the susceptible European *V. vinifera* cultivars (Stein and Blaich, 1985). Different cultivars, and perhaps different clones of the same cultivar, also vary in the amount of resveratrol they accumulate following fungal attack. For instance, Cabernet Sauvignon leaves appear to be able to produce twice the amount of stilbenes that accumulate in Pinot noir or Chardonnay. However, it seems unlikely that their greater capacity for stilbene production is the only reason Cabernet Sauvignon, Merlot, Mourvèdre, or Sauvignon blanc have relatively resistant berries, whereas the Pinots and Muscats, Cabernet franc, Grenache, Sémillon, and Thompson Seedless are among the most susceptible cultivars.

Part of the variation in berry susceptibility among cultivars and clones can be attributed to differences in cluster compactness rather than stilbenes; more compact clusters with berries rubbing against each other are more easily infected by *B. cinerea* than loose clusters because rubbing may compromise the integrity of the cuticular wax on the berries. Pores in the berry skin, as well as the thickness of the cuticle and the amount of wax it contains also vary among cultivars. The rather resistant cultivars derived from American *Vitis* species and their interspecific crosses tend to have far fewer pores, a thicker cuticle, and more cuticular wax than do the vulnerable *V. vinifera* cultivars (Mlikota Gabler et al., 2003). Such variation among cultivars in a variety of developmental, morphological, and biochemical traits, each conferring partial resistance, demonstrates the existence of quantitative (or incomplete), rather than qualitative (or complete), disease resistance (Poland et al., 2009) within the genus *Vitis* and within the species *V. vinifera*. This results in a wide range of disease severity among cultivars under similar environmental conditions.

Even the rootstock seems to be able to influence how much resveratrol the tissues of an infected scion produce (e.g., Teleki 5BB > 1103 Paulsen). It is possible that rootstocks exert their effect via differences in nutrient uptake and transport because plant nutrition also plays a role in disease resistance. However, although vines with high nitrogen status tend to have a higher incidence of bunch rot (Delas et al., 1982; Keller et al., 2001a), research has failed to establish a correlation between the nitrogen content of grape berries and their susceptibility to *B. cinerea* (Delas, 1972). On the other hand, the production of both constitutive phenolics, especially flavonols and hydroxycinnamic acids, and inducible stilbenes declines as soil nitrogen availability rises, which may render leaves and berries less able to resist fungal invasion. Of course, the unfavorable canopy microclimate that is often associated with high vine nitrogen status also favors disease development.

In addition to gray mold, grapes can also succumb to sour rot, which is caused by a complex of various microorganisms that include bacteria, yeasts, and other fungi (Barata et al., 2012a). Most of these are secondary pathogens that can only penetrate into grape berries through preformed wounds (e.g., from powdery mildew infection or hail damage). Fungi of the genera *Penicillium*, *Aspergillus*, *Alternaria*, and others participate in the infection, in addition to yeasts (which are single-celled fungi) belonging to *Hanseniaspora*, *Candida*, *Kloeckera*, and others. These yeasts in turn attract fruit flies (genus *Drosophila*) that are vectors of *Acetobacter* (acetic acid bacteria), which causes the typical "vinegar" smell of grapes infected with sour rot.

8.2.2 POWDERY MILDEW

Unlike *B. cinerea*, the fungus *E. necator* that causes powdery mildew (also termed oidium) on grapes is unable to infect plant species that do not belong to the Vitaceae—and it seems to be the only fungus of its family that is able to infect grapes (Dry et al., 2010). Nonetheless, it is arguably the most widespread and most consistently damaging pathogen of grapevines. This is because, in contrast to *B. cinerea* or *P. viticola*, it requires only 40% relative humidity to germinate, a threshold that is easily reached on the lower surface of transpiring leaves, even if the surrounding air is much drier (Keller et al., 2003a; Pearson and Goheen, 1998). The rate of development of the fungus accelerates up to an optimum of approximately 85% humidity and 25°C, but frequent rainfall, especially heavy rain, and temperatures outside the range 10–32°C limit its development, with temperatures greater than 35°C killing it outright (Carroll and Wilcox, 2003; Gadoury and Pearson, 1990; OEPP/EPPO, 2002). Mild rainfall, in contrast, seems to benefit *E. necator* by enhancing spore dispersal. The fungus can colonize all green plant surfaces (Figure 8.1) but thrives in shade and often develops in the interior of dense canopies (Nicholas et al., 1998). In addition to the influence of air humidity, this shade effect seems to be related to the absence of UV light, which is effectively filtered out by the leaves' epidermis (Keller et al., 2003a). Consequently, the severity of disease expression on grape clusters tends to increase as their exposure to sunlight declines, for example as a result of higher canopy density (Austin et al., 2011). The environmental factors that favor disease development are of particular concern for grapevines grown in greenhouses and for young vines growing inside plastic sleeves or "grow tubes" that are used to accelerate growth and facilitate weed control. This is because both greenhouses and grow tubes create a microclimate that is characterized by reduced light, especially UV light, and by elevated relative humidity and, in the case of grow tubes and uncooled greenhouses, elevated daytime temperature. Nevertheless, in climates where grow tubes tend to increase the temperature above 35°C, they may inhibit, rather than enhance, powdery mildew development (Hall and Mahaffee, 2001).

The typical whitish "powdery" symptoms on leaf surfaces are due to mycelium and sporulating bodies of the fungus, whereas infected shoots become necrotic, and petioles or peduncles become brittle and break easily later in the growing season. Spores germinate on the surface of plant organs, penetrate the cuticle and cell walls, and rapidly establish haustoria inside the epidermis cells (Heintz and Blaich, 1990; Pearson and Goheen, 1998). Like all biotrophic pathogens, *E. necator* depends on the living host plant for assimilate supply; that is, it acts as a sink. Therefore, the fungus does not kill its host's cells but suppresses their defense responses in susceptible cultivars. Infected leaves generally have increased concentrations of sugars (especially hexoses, which are the preferred carbohydrates for uptake by the fungus) due to import of sucrose from uninfected plant parts and subsequent breakdown by invertase in the cell walls (Brem et al., 1986; Hall and Williams, 2000; Hayes et al., 2010). This is a rare example of the conversion of source leaves to sinks and a demonstration of the pathological nature of this conversion, which appears to be caused by an "injection" of cytokinin from the pathogen (cytokinins induce invertase activity) and also involves amino acid import (Hayes et al., 2010; Walters and McRoberts, 2006). Some of the sugar may be diverted to produce callose that sometimes seals off the fungal haustoria in older infections (Heintz and Blaich, 1990). Infection—or the sugar accumulation resulting from it—will decrease photosynthesis and starch storage in and assimilate export from infected leaves (Brem et al., 1986; Hall and Williams, 2000; Lakso et al., 1982; Nail and Howell, 2005). Consequently, this powerful extra sink (i.e., the fungus) alters assimilate partitioning in the vine at the expense of other sinks, such as fruit, roots, and storage reserves. Although the fungus-derived cytokinins can sometimes delay leaf senescence (Walters and McRoberts, 2006), the vine may shed severely infected leaves, which can further depress yield formation, fruit ripening, replenishment of storage reserves, and cold acclimation. The close relationship between leaf sugar and fungal growth suggests that events that lead to higher sugar concentration in leaves, such as intense fruit removal (e.g., by early cluster thinning) or girdling, might also increase the leaves' vulnerability to infection.

Infection of flowers very early in the growing season can result in poor fruit set and consequently low yield, whereas infection after fruit set may reduce berry size, perhaps because imported sugar is diverted to and consumed by the fungus instead of being available for berry growth. Berries that are heavily infected early in development usually shrivel up or drop off, whereas later infections damage the epidermis so that the berries may split (crack) upon expansion during ripening and may attract insects that inflict further damage. Cracking, as well as the necrotic epidermal cells that develop at the site of fungal appressoria even on berries that show no visible signs of infection (i.e., they are sparsely covered with diffuse, nonsporulating mycelium), open the door for secondary infections by *B. cinerea* and other bunch rot opportunists. In addition, such surface damage increases colonization by non-*Saccharomyces* yeasts and bacteria (Gadoury et al., 2007). The characteristic "mushroom" (from 1-octen-3-one) and "geranium-like" (from 1,5-octadien-3-one) odor of powdery mildew renders infected grapes unpalatable and can impart off-odors in wine, even though yeasts degrade most of these compounds during fermentation (Darriet et al., 2002; Stummer et al., 2005). However, even less severe infections may blemish the berry surface, which can render table grapes unmarketable.

Whereas American and Chinese *Vitis* and *Muscadinia* species exhibit various degrees of resistance to the fungus, the European *V. vinifera* cultivars (i.e., most wine and table grape cultivars) are readily infected because, unlike their American relatives, they did not coevolve with the pathogen and produce lower amounts of PR proteins (Eibach, 1994; Fung et al., 2008). In addition, *E. necator* times the release of spores from its overwintering bodies, the cleistothecia, to coincide with budbreak of the native *V. riparia* grapes. Because this occurs earlier than budbreak of *V. vinifera*, there are lots of spores around by the time the wine grape shoots emerge. Nevertheless, there is some variation among *V. vinifera* cultivars in the degree of susceptibility (Doster and Schnathorst, 1985; Gaforio et al., 2011; Roy and Ramming, 1990). For instance, Chardonnay, Tempranillo, Grenache, and, to a somewhat lesser degree, Cabernet Sauvignon are among the most susceptible cultivars, whereas Pinot noir, Riesling, Malbec, and Mourvèdre seem to be much better able to resist or tolerate infection. Surprisingly, some *V. vinifera* table grape cultivars (e.g., Kishmish vatkana) from the Near East are apparently as resistant as are American *Vitis* species (Hoffmann et al., 2007).

Although some stilbene phytoalexins are also effective against *E. necator* (at least *in vitro*), infection of *V. vinifera* tissues does not normally trigger their production (Keller et al., 2003a) or triggers it only in the infected cells themselves (Schnee et al., 2008), perhaps because the biotrophic fungus minimizes cell damage so as not to threaten its own survival (Mendgen and Hahn, 2002). In contrast, the resistant American *Vitis* species do accumulate stilbenes in response to infection. In addition, flavonols, which accumulate in the epidermis and cuticular wax in response to UV (especially UVB) radiation and provide a "sunscreen" for plant tissues, may be involved in partial *V. vinifera* resistance against *E. necator*. It is conceivable that enhanced synthesis of such constitutive phenolic compounds in sun-exposed leaves and berries contributes to an unfavorable environment for powdery mildew colonization. Flavonol production, however, is strongly reduced by high soil nitrogen availability, and high plant N status makes vines more susceptible to colonization by *E. necator* (Bavaresco and Eibach, 1987; Keller et al., 2003a). An additional resistance mechanism may be vitrification of penetrating mycelium by the localized accumulation of silicates in the cell walls (Blaich and Wind, 1989).

Grape berries, even those of *V. vinifera* cultivars, become increasingly resistant to *E. necator* over the course of their development, which contrasts with the berries' increasing vulnerability to *B. cinerea*. Although the powdery mildew fungus may continue to colonize the berry surface, it seems to be unable to infect berries as early as 3–5 weeks after bloom (Gadoury et al., 2003, 2007). At least in *V. vinifera* this is likely to be a by-product of the accumulation of PR proteins for other reasons (e.g., sugar accumulation) because this species has not had time to evolve specific resistance mechanisms against this comparatively novel pathogen. The increase in PR protein content following an infection may raise the amount of protein in the juice and wine made from such grapes, which has the undesirable consequence of reducing protein stability (i.e., increasing haze formation) in white (but probably not red) wines (Girbau et al., 2004). In contrast with the berries, the rachis remains susceptible to infection throughout the growing season.

8.2.3 DOWNY MILDEW

Although usually regarded as a fungus because it looks like one and produces spores, the causal agent *P. viticola* is in fact more closely related to marine algae (such as kelps and diatoms). Also, in contrast to "true" fungi, its cell walls contain cellulose instead of chitin, and its cell nuclei are diploid, not haploid like those of fungi. It belongs to the class of Oomycetes (also called water molds) and is much less related to the powdery mildew fungus than the latter is to us. Because its spores can germinate at greater than 95% relative humidity, it thrives under humid, shady conditions, especially with frequent rainfall and temperatures of approximately 20–25°C, which is common in coastal areas and other regions with summer rainfall (Galet, 1996; OEPP/EPPO, 2002; Pearson and Goheen, 1998). Although Mediterranean climates are less favorable for the pathogen native to the southeastern United States, it is widespread throughout the world, except in regions with very low spring and summer rainfall, such as central California, eastern Washington, northern Chile, and Western Australia. Because the little rain that falls in the Mendoza region in Argentina mostly comes in the summer, downy mildew is a major problem even in this otherwise arid region. Along with other North American natives (the fungi *E. necator*, which was introduced earlier, and *G. bidwellii*), *P. viticola* wreaked havoc in the European wine industry when it spread from infested grape material imported in the 1870s from the United States for use as rootstocks (see Section 1.2) resistant to the previously introduced insect phylloxera to which the susceptible *V. vinifera* varieties were succumbing. During the event, the discovery of the so-called Bordeaux mixture (copper sulfate mixed with lime suspended in water) saved the industry. Incidentally, however, the widespread appearance of inferior wines, made by desperate producers who lacked sufficient *V. vinifera* grapes and instead used fruit from hybrids resulting from resistance-breeding programs with *V. vinifera* and American *Vitis* species during the phylloxera-and-mildew crisis, indirectly spurred the advent of the rigid legislation that has since become a hallmark of the European wine industry.

Like the fungi discussed previously, *P. viticola* can infest all green plant parts, but unlike those fungi it ordinarily colonizes young leaves or young berries by entering through the stomata rather than by penetrating the epidermis (Gindro et al., 2003; Langcake and Lovell, 1980; Martelli et al., 1986; Pearson and Goheen, 1998). Its mycelium then develops an intercellular network in the leaf mesophyll and feeds off these cells by means of cell wall–penetrating haustoria to cause downy mildew (also called peronospora). Initial disease symptoms appear on the adaxial sides of leaves as yellow or, in some cultivars, red oily spots (Figure 8.1), which spread and later become angular necrotic patches restricted by the leaf veins. The typical whitish "downy" mildew symptoms arise from sporulation of the pathogen through the stomata on the abaxial leaf surface. Even before any disease symptoms appear, however, the invading pathogen prevents the stomata from closing at night or in response to water deficit so that the unrestrained transpiration may lead to water loss and wilting of infected leaves (Allègre et al., 2007). Although unlike *E. necator*, *P. viticola* does not stimulate sugar accumulation in infected leaves, it also leads to reprogramming of sugar metabolism akin to that in a sink (Hayes et al., 2010; Palmieri et al., 2012), as well as to repression of photosynthesis and shedding of severely damaged leaves. This can adversely impact yield formation, fruit ripening, replenishment of storage reserves, and cold hardiness. Moreover, infected shoot tips, petioles, tendrils, and inflorescences often become necrotic and are abscised, whereas the highly susceptible young grape berries become covered with a grayish "felt." Ripening berries that are infected with downy mildew drastically increase the production of volatile thiol precursors, probably from fatty acid degradation products derived from damaged membranes (Kobayashi et al., 2011).

Having coevolved with the pathogen, most of the North American *Vitis* species are partly resistant (e.g., *V. labrusca*, *V. rupestris*, *V. berlandieri*, and *V. aestivalis*) or fully resistant (e.g., *M. rotundifolia*, *V. riparia*, and *V. cinerea*) to downy mildew. Some Asian species (e.g., *V. amurensis*) also have partial resistance, whereas *V. vinifera* is highly susceptible. It seems that *V. vinifera* has trouble recognizing *P. viticola* as a pathogen and consequently fails to mount effective defense strategies (Palmieri et al., 2012). Perhaps the intruder, whose own survival relies on live host tissues, actively suppresses the defense response (Milli et al., 2012). Nonetheless, even among the European grape cultivars there are various degrees of susceptibility. For instance, Riesling, the Pinot family, and especially Cabernet Sauvignon are among the less susceptible cultivars, whereas Tempranillo and Albarino are among the most susceptible (Boso and Kassemeyer, 2008; Boso et al., 2011). Some of the resistant species defend themselves against infection of their leaves by rapidly secreting callose that plugs their stomata and coats the pathogen spores (Gindro et al., 2003; Langcake and Lovell, 1980). This coating probably kills the germinating spores and stops mycelial growth while also reducing water loss from the leaves. This postinfection defense response can be activated, even in *V. vinifera* cultivars, by soil fungi belonging to the *Trichoderma* species, which may thus be used as biocontrol agents (Palmieri et al., 2012; Perazzolli et al., 2011). In addition, the leaves produce stilbenes (resveratrol, pterostilbene, and viniferins) that kill the cells surrounding infected stomata and hinder sporulation of the pathogen (Dai et al., 1995; Langcake, 1981; Langcake et al., 1979; Malacarne et al., 2011; Pezet et al., 2004a). High plant N status seems to compromise the leaves' ability to produce stilbenes and leads to higher vulnerability to infection (Bavaresco and Eibach, 1987).

Young grape berries, even those of susceptible species and cultivars, also rapidly accumulate stilbenes around sites of infection (Schmidlin et al., 2008), but the berries appear to lose this ability around veraison. Nonetheless, the berries become increasingly resistant to infection during development (Kennelly et al., 2005). The appearance of necrotic cells near invasion sites is termed hypersensitive response and is an example of localized programmed cell death that stops the pathogen spreading to healthy tissues (Busam et al., 1997; Kortekamp, 2006; Langcake and Lovell, 1980). Again, *Vitis* species differ in their ability to accumulate stilbenes in response to *P. viticola* infection (Dai et al., 1995; Pezet et al., 2004a). For instance, stilbenes appear earlier and reach much higher concentrations in leaves of the resistant *V. riparia* than in the susceptible *V. vinifera*, whereas *V. rupestris* seems to be intermediate in both resistance and stilbene production. The resistant *V. cinerea* and *V. champinii*, however, are also poor stilbene producers, which indicates that these phytoalexins are not the only actors on the stage of downy mildew resistance. Indeed, in addition to the constitutive chitinase in their leaves, resistant species also activate glucanase upon challenge by *P. viticola* (Kortekamp, 2006). Greater glucanase activity, and perhaps peroxidase activity, in old rather than in young leaves also may partly account for the decrease in vulnerability to the pathogen as leaves age (Reuveni, 1998).

As in the case of *E. necator*, the American *P. viticola* and the European *V. vinifera* met only recently by evolutionary timescales, so it is probable that the ability to produce stilbenes evolved in *V. vinifera* as a defense against intruders other than *P. viticola*. The formation of necrotic spots on the leaves of at least partly resistant *Vitis* species and of oilspots on susceptible species is, of course, a result of the way the different species respond to the attacking pathogen. Such variation demonstrates the principle that when a pathogen "discovers" and invades a new host species, the symptoms it provokes are often different and usually more severe than those that occur with the original species, which has "honed" its skills at fending off the invader through the long arms race of coevolution.

8.2.4 BACTERIA

Although the majority of all bacteria live in the soil, a rare few have succeeded at becoming plant pathogens, suggesting that plants in general have evolved powerful resistance strategies

(Bais et al., 2006). One of the few perpetrators is *Rhizobium vitis* (a.k.a. *Agrobacterium vitis*, formerly known as *Agrobacterium tumefaciens* biotype 3), the causal agent of crown gall, which is mainly confined to and spreads via xylem vessels—both up and down the trunk and roots (Tarbah and Goodman, 1987; Young et al., 2001). It may live for years within the vascular system of infected vines without any outward expression of disease, which typically develops only at sites of physical injury. Although commonly regarded as a soil-borne pathogen, the bacterium survives only in *Vitis* tissues and is usually introduced into vineyards with contaminated planting material (which can be symptomless) that then serves as a source of inoculum (Burr and Otten, 1999; Burr et al., 1998). Infection often occurs during the propagation process in nurseries. The stage of transplanting freshly callused rootlings into nursery beds, during which many of the soft callus cells may be injured due to friction at the callus-soil interface, seems to be an especially vulnerable period (Goodman et al., 1993). In established vineyards, the bacterium enters the plant predominantly through wounds caused by cold injury but also at sites of other injuries such as those caused by machinery or grafting/topworking. Therefore, the disease thrives in climates that experience frequent freeze injury, and damage caused by galls often surpasses that of the initial injury.

The bacteria, it seems, induce the tumors called crown gall when mechanical wounds rupture the xylem vessels and release the bacteria to the adjacent parenchyma cells, where they insert some of their own DNA into the genome of the infected plant (Burr and Otten, 1999; Martelli et al., 1986; Tarbah and Goodman, 1987). These bacterial DNA pieces are called tumor-inducing plasmids or Ti plasmids. It is this exceptional ability to transfer genetic material to plants that also makes *Agrobacterium* species the organism of choice for plant genetic engineering. In this context it is important to realize that plants have, during their evolution, repeatedly acquired genes from bacteria, as well as from fungi and viruses, and even from other plants (Bock, 2010), and continue to do so. Such asexual integration of foreign genes into the plant genome is termed horizontal gene transfer or lateral gene transfer. The fact that this process occurs naturally indicates that genetic engineering is nothing more than man-made horizontal gene transfer (Bock, 2010).

The bacterial genes induce the now genetically modified host cells to degrade tartrate, to divert auxin away from its normal basipetal movement, and to locally enhance auxin, cytokinin, and ethylene production in order to stimulate cell proliferation that leads to the typical trunk tumors associated with crown gall (Aloni, 2013; Burr and Otten, 1999; Sakakibara, 2006). Although no *Vitis* species have thus far been found to be immune to crown gall, the species vary in their vulnerability to *R. vitis*. For instance, *V. vinifera* cultivars are highly susceptible, whereas *V. labrusca* cultivars are somewhat resistant, the rootstocks of *V. riparia* and *V. rupestris* parentage are relatively resistant, and *V. amurensis* is quite resistant (Burr and Otten, 1999; Burr et al., 1998; Stover et al., 1997; Szegedi et al., 1984). On susceptible species, galls normally develop close to the base of the trunk, from slightly above to slightly below the soil surface (Figure 8.2), but they can also form up to approximately 1 m above the soil surface, and even on canes. Training up suckers to form new trunks is a technique that can be used to replace injured trunks, provided the suckers arise from the asymptomatic base of the trunk and not from a rootstock. On vines grafted to resistant rootstocks, galls usually form at or above the graft union (Burr and Otten, 1999). Although infected roots do not usually form galls, they may develop necrotic lesions.

In addition to inducing gall formation, auxin, cytokinin, and ethylene also trigger the differentiation of new vascular bundles, which are then connected to the vascular system of the vine to provide water and nutrients to the growing tumor at the expense of the shoots (Aloni, 2001;

FIGURE 8.2

Young grapevine with severe symptoms of crown gall (left) and tumor formation above the soil surface (right).

Photos by M. Keller.

Efetova et al., 2007; Veselov et al., 2003). The lack of an epidermis around the tumor leads to water loss through its large surface area, which further promotes water flow to the tumor (Aloni, 2013). In response to this threat, the plant cells may produce osmoprotectants (e.g., proline; see Section 7.1), as well as suberin for incorporation in the exposed cell walls as a sealing agent (Efetova et al., 2007; Veselov et al., 2003).

As developing galls expand, they increasingly restrict movement of water and nutrients with drastic consequences for plant vigor, longevity, and yield (Schroth et al., 1988). Because the girdling effect of the tumors inhibits phloem transport to the roots, the leaves may turn bright red due to the conversion of accumulating sugars into anthocyanins. In severe cases, the portion of the vine above the gall may die, which is a particular problem during vineyard establishment (Figure 8.2). In addition to the striking disease symptoms associated with gall formation, *R. vitis* also causes root lesions (Rodriguez-Palenzuela et al., 1991), thereby compromising plant vigor even in the absence of galls. Upon removal of a vineyard, the bacteria can persist in root debris for several years, which may provide a source of inoculum for newly planted vines. In contrast, the bacteria seem to be unable to move to the tip of growing shoots. This can be exploited to eliminate *R. vitis* from propagation material through the process of shoot-tip culturing (Burr and Otten, 1999; Burr et al., 1988, 1998).

The bacterium *X. fastidiosa*, the causal agent of Pierce's disease, contrasts with *R. vitis* in that it causes damage mainly in areas with mild winters that permit survival of its vectors (Mullins et al., 1992). A harmless endophyte (i.e., a symbiotic microorganism living inside the plant) in most of its many host plant species, the bacterium causes severe damage and even death in *V. vinifera* (Hopkins and Purcell, 2002). It lives exclusively inside xylem vessels and uses them to spread throughout its host plants after it has been deposited there by xylem-sucking insects, such as sharpshooter leafhoppers (Chatelet et al., 2006; Stevenson et al., 2004). The pathogen may facilitate its own migration through the xylem by releasing cell wall–degrading enzymes (e.g., polygalacturonase and cellulase) that digest some of the pit membranes that get in its way (Pérez-Donoso et al., 2010; Roper et al., 2007; Sun et al., 2011). Bacteria populations build up over time, and severe infection can sometimes lead to localized plugging of vessels. Plugging seems to result mostly from dislodged cell wall pectin fragments and from the production of tyloses and gums by the infected vine (Pérez-Donoso et al., 2010; Stevenson et al., 2004; Sun et al., 2013). Xylem occlusion could be a response to the release of ethylene by the injured vine (Sun et al., 2007). The disruption of water flow, especially in the petioles where the bacteria can induce cavitation, may lead to leaf dehydration (McElrone et al., 2008). Water stress may or may not be associated with the symptoms characteristic of Pierce's disease: marginal leaf necrosis that resembles salt injury, abscission of leaves but not the petiole (leaving the so-called "matchsticks" on the shoot), and green islands on otherwise maturing (i.e., browning) shoots (Stevenson et al., 2005; Thorne et al., 2006). The berries of severely infected plants sometimes shrivel up and dry, and vines, too, ultimately die from infection. Species that have coevolved with the bacterium, such as *M. rotundifolia* and *V. arizonica*, are resistant to the bacterium, but the resistance varies according to the home range of local populations of wild vines (Fritschi et al., 2007; Ruel and Walker, 2006).

8.2.5 VIRUSES

In addition to bacteria, grapevines are also vulnerable to infection by a variety of graft-transmissible viral diseases. Grapevine viruses became widespread only in the twentieth century after, in response to the spread of phylloxera, European growers were forced to change their propagation methods from simple rooting of cuttings to grafting onto phylloxera-tolerant rootstocks. Cultivated grapes throughout the world now appear to be infected with more viruses than any other woody species. These viruses are typically introduced into vineyards by infected planting material, although some can also be spread by transport agents called vectors, such as phloem-feeding insects (e.g., leafroll viruses by aphids or mealybugs, often aided by wind, people, or machinery) or root-feeding nematodes (e.g., fanleaf nepovirus by the dagger nematode *Xiphinema index*) (Bovey et al., 1980; OEPP/EPPO, 2002; Oliver and Fuchs, 2011; Olmo, 1986). The latter, of course, also cause direct damage through feeding on the roots in addition to delivering their virus load. Moreover, although some grapevine genotypes, such as certain rootstocks, can tolerate very high nematode numbers, these nematodes can still feed on the roots and transmit viruses (Mullins et al., 1992). In contrast to most *Vitis* species (*V. arizonica* may be an exception), *Muscadinia* species are resistant to *X. index* feeding; yet even *Muscadinia* can be infected with fanleaf virus, albeit very slowly (Oliver and Fuchs, 2011). Once infected, all rootstocks will transmit the virus to their grafting partners.

Viruses can be classed as biotrophic pathogens because they depend on the metabolism of their hosts for multiplication (replication), which leads to systemic infections. All known viruses infecting *Vitis* species have genomes that consist of RNA (rather than DNA, as in their hosts) and recruit host ribosomes to manufacture their proteins. This property enables such viruses to mutate readily because it dodges the host cells' ordinary "proofreading" machinery that is designed to detect and repair errors on DNA. This leads to rapid evolution of novel virus genotypes, although one or a few stable genotypes typically dominate infections in plants (Coetzee et al., 2010; García-Arenal et al., 2001; Karasev, 2000). Despite the virus's RNA genome, however, grapevines have incorporated fragments of leafroll-associated virus genes by reverse-transcribing them into the genome of their own mitochondria (Goremykin et al., 2009). To further their own needs, many viruses suppress the expression of some of their host's genes (a phenomenon called host gene shutoff) while triggering the expression of other genes (Havelda et al., 2008). These viruses thus manipulate the host's metabolism, which ultimately leads to the development of disease symptoms. Once a virus has colonized a grapevine, it remains there for good; a cure is not possible. Vegetative propagation material (e.g., cuttings) obtained from such vines is therefore also compromised. Moreover, infections with a potpourri of different viruses are common, and the association of specific symptoms with particular viruses is often not clear-cut. Intriguingly, many viruses appear to induce the production by their hosts of heat shock proteins (which seem to facilitate virus infection and replication) in addition to stilbenes and PR proteins, among other disease resistance mechanisms (Whitham et al., 2006), which might make virus-infected vines less susceptible to attack by fungal pathogens.

Lower temperatures seem to strongly increase a plant's susceptibility to virus infections and also lead to more pronounced disease symptoms, possibly because cool conditions may favor the multiplication of viruses and slow the degradation of their RNA (Samach and Wigge, 2005; Wang et al., 2008). Many plant viruses induce localized or general yellowing of leaves. The virus particles normally spread symplastically in infected tissues and organs, using special movement proteins that enable them to travel via plasmodesmata from cell to cell and via the phloem to other organs (Lough and Lucas, 2006; Oparka and Santa Cruz, 2000; Roberts and Oparka, 2003).

The leafroll virus (actually a genetically diverse bunch of the so-called ampeloviruses and closteroviruses causing similar symptoms; see Karasev, 2000; Martelli et al., 2002; Oliver and Fuchs, 2011) is arguably the most destructive of all grape viruses because it interferes with phloem transport and curtails root growth and shoot vigor. Though the virus is mostly confined to the phloem and adjacent vascular tissues, phloem blockage occurs not by the virus itself but due to callose deposition, which may be an attempt by the vine to limit the spread of virus particles (Martelli et al., 1986). Unfortunately, this defense response restricts sugar export from the leaves and leads to sugar accumulation inside the leaves, which in turn induces feedback inhibition of photosynthesis (Ayre, 2011; Espinoza et al., 2007) and, often, the characteristic downward rolling and interveinal reddening in red cultivars (Figure 8.3) or yellowing in white cultivars. The red leaf coloration is due to accumulation of anthocyanins (Gutha et al., 2010), which is another consequence of sugar accumulation in the leaves and sometimes starts even before veraison. Given that sugar and abscisic acid act synergistically in inducing anthocyanin production (Pirie and Mullins, 1976), one would expect water deficit to enhance the expression of red leaf symptoms. Sugar accumulation not only triggers anthocyanin production but also leads to an increase in the osmotic pressure of the leaf

FIGURE 8.3

Symptoms of leafroll virus infection in Cabernet Sauvignon: bright red areas appear on the lamina of a mature leaf, while the veins remain green.

Photo by M. Keller.

tissues (Koroleva et al., 2002). Thus, the downward rolling of the leaves might be a result of osmotically induced cell expansion, which would affect the densely packed upper palisade parenchyma more than the more loosely packed lower spongy parenchyma (see also Figure 1.15).

In contrast with *V. vinifera*, the American *Vitis* species and their hybrids (including Concord) typically do not develop visual leaf symptoms, but they still suffer from reduced yield and fruit quality (Oliver and Fuchs, 2011). No grapevine species or cultivar is resistant to or tolerant of leafroll virus infection. The so-called "graft incompatibility" of certain scion–rootstock combinations is often associated with one or both of the grafting partners being infected with the leafroll virus. But the viruses infest own-rooted vines just as readily, and symptoms are identical to, albeit often less pronounced than, those developing on grafted vines.

In addition to causing reduced growth and plant longevity, some viruses, especially the leafroll-associated and fanleaf viruses, result in significant losses in both yield and quality of the fruit (Bovey et al., 1980; Cabaleiro et al., 1999; Goheen and Cook, 1959; Hewitt et al., 1962; Lider et al., 1975; Mannini et al., 2012; Woodham et al., 1983, 1984). This is because diminished sugar transport to the fruit clusters interferes with yield formation, grape berry development, and ripening. Because infected vines become highly sensitive to adverse environmental conditions that limit photosynthesis, clusters often appear very loose due to poor fruit set, berry size and ripening are uneven, and grapes commonly lag behind in maturity, with poor sugar and color and high acidity. Whereas accumulation of sugars, some amino acids (e.g., proline), and anthocyanins is reduced in the fruit of leafroll-infected vines, accumulation of organic acids (tartrate and malate), other amino acids (e.g., arginine), and K^+ is increased (Kliewer and Lider, 1976; Vega et al., 2011). Because of restricted sugar export from the leaves, root growth and replenishment of storage reserves in the perennial parts of the vine also may be reduced (Rühl and Clingeleffer, 1993). This weakens the plant and leads to gradual decline and, finally, death of the plant, especially if multiple infections with different viruses occur. In addition, the leafroll virus also induces the plant to produce volatiles that attract the insect vectors of the disease. This facilitates spreading of the virus from infected to healthy vines and neighboring vineyards.

Despite this list of adverse consequences of virus infection, it is possible that viruses may not be all bad for grapevines. There is some evidence that virus infection may improve drought tolerance and cold hardiness of plants, although this has yet to be tested in grapes (Xu et al., 2008). This outcome seems to be a consequence of the accumulation of sugars in the leaves, which leads to feedback inhibition of photosynthesis resulting in partial closure of the stomata, which in turn decreases transpiration. The lower vigor and, hence, leaf area of infected plants further reduces transpiration. Other metabolites (e.g., proline and polyamines) accumulating in the leaves may confer some protection against oxidative stress that typically results from environmental stress (see Section 7.1). Thus, viruses may further their own survival by "helping" their hosts to survive adverse environmental conditions.

Abbreviations and Symbols

ABA	Abscisic acid
ADP	Adenosine diphosphate
ATP	Adenosine triphosphate
c	Solute concentration
C_a	Atmospheric CO_2 concentration
CHS	Chalcone synthase
C_i	Substomatal CO_2 concentration
DMS	Dimethyl sulfide
DNA	Deoxyribonucleic acid
Δ	Difference or gradient
E	Transpiration rate
ET	Evapotranspiration
F	Volume flow rate
FCR	Ferric chelate reductase
Fd_{ox}	Oxidized ferredoxin
Fd_{red}	Reduced ferredoxin
FR	Far-red light
g_b	Boundary layer conductance
GDD	Growing degree days
GDH	Glutamate dehydrogenase
g_l	Leaf conductance
GOGAT	Glutamine oxoglutarate aminotransferase or glutamate synthase
g_s	Stomatal conductance
GS	Glutamine synthetase
J	Flux or flow rate per unit area
K_c	Crop factor
l_h	Hydraulic conductance
MEP	Methyl-erythritol 4-phosphate
NADH	Nicotinamide adenine dinucleotide
NADPH	Nicotinamide adenine dinucleotide phosphate
NiR	Nitrite reductase
NR	Nitrate reductase
P	Hydrostatic pressure or turgor pressure
P_a	Atmospheric pressure
PAL	Phenylalanine ammonia lyase
PAR	Photosynthetically active radiation
P_{fr}	Active phytochrome
P_i	Inorganic phosphate
P_r	Inactive phytochrome
PR	Pathogenesis-related
PRD	Partial rootzone drying
PSI	Photosystem I
PSII	Photosystem II
P_x	Xylem pressure

Q_{10}	Temperature coefficient
R	Red light
R	Universal gas constant
RDI	Regulated deficit irrigation
RH	Relative humidity
r_b	Boundary layer resistance
r_h	Hydraulic resistance
RNA	Ribonucleic acid
RQ	Respiratory quotient
r_s	Stomatal resistance
Rubisco	Ribulose bisphosphate carboxylase/oxygenase
RWC	Relative water content
σ	Reflection coefficient
SAR	Sodium adsorption ratio
π	Osmotic pressure
T	Temperature
TCA	Tricarboxylic acid, also 2,4,6-trichloroanisole
TDN	Trimethyl-dihydronaphthalene
UDP	Uridine diphosphate
UV	Ultraviolet
VPD	Vapor pressure deficit
WUE	Water use efficiency
Ψ	Water potential
Ψ_G	Gravitational potential
Ψ_M	Matrix potential
Ψ_π	Osmotic potential

GLOSSARY

This glossary provides definitions or explanations of some chemical, anatomical, physiological, and viticultural terms that appear throughout the text. An arrow (→) in front of a word indicates that this word has its own entry in the glossary.

Abscisic acid (ABA) Plant hormone that indicates stress ("stress hormone"), induces stomatal closure and leaf fall (→abscission), inhibits growth, and controls seed dormancy. The carboxylic acid ABA is a breakdown product of →carotenoids and is produced when plant cells lose →turgor. ABA acts antagonistically to →gibberellin.

Abscission (Latin *abscissus* = to cut off) The shedding of leaves, flowers, fruits, or other plant parts, usually following the formation of an abscission zone and preceded by →senescence.

Absorption (Latin *absorbere* = to swallow) Uptake, e.g., of light photons by leaves or of nutrient ions by roots.

Acid A substance that dissociates in water, releasing hydrogen ions called protons (H^+) and thus causing a relative increase in the concentration of these ions. The most important (organic) acids in grapes are tartaric and malic acids. Acid also means having a →pH less than 7 due to the presence of more protons than hydroxyl ions (OH^-) in a solution.

Adaptation An anatomical or physiological trait that has been changed by →evolution, so that it more or less matches a particular set of environmental conditions. Such traits often appear to be "designed" for a specific purpose or environment.

Adsorption (Latin *ad* = to, at; *sorbere* = to suck) Binding of gaseous or dissolved compounds to the surface of a solid object.

Adventitious root (Latin *adventicius* = not properly belonging to) A→root that develops from the →cambium of the stem, as in grapevine cuttings.

Aglycon Organic component of a →glycoside that remains after removal of the sugar.

Allele One particular version of a →gene that has several different forms arising from sexual reproduction or mutation. Alleles can be present on the →chromosome pairs of an individual plant in single dose (→heterozygous) or double dose (→homozygous). Different alleles generally result in differences among →phenotypes in a population.

Amino acid (Greek *Ammon*, the Egyptian Sun god, near whose temple ammonium salts were first prepared from camel dung) Organic acid made up of a chain of carbon atoms to which hydrogen, oxygen, nitrogen, and, sometimes, sulfur atoms are attached; comes in 20 different forms. The units or "building blocks" from which →protein molecules are built. An important food source for yeast during fermentation.

Anther (Greek *anthos* = flower) The part of the stamen that develops and contains pollen and is usually borne on a stalk (filament).

Anthocyanin (Greek *anthos* = flower; *kyanos* = blue) A water-soluble blue or red pigment, belonging to the phenolic class of flavonoids, found in the cell vacuoles of the skin (and sometimes pulp) of dark grapes. Anthocyanins are responsible for the red-purple color of dark grapes, leaves A→root, and wine.

Apical dominance (Latin *apex* = tip; *dominari* = to rule) The phenomenon in annual shoots whereby the growing shoot apex inhibits outgrowth of lateral shoots.

Apoplast (Greek *apo* = away, distant; *plastos* = formed) The totality of spaces within a plant that are outside its cell →membranes, including cell walls, intercellular spaces, and →xylem conduits.

Arginine An amino acid; the main nitrogen storage substance of grapevines.

Assimilate stream The flow of photosynthetic assimilates, or food materials, in the phloem; moves from →source to →sink down a pressure gradient.

Assimilation (Latin *assimilare* = to align, to integrate) The conversion of inorganic chemicals acquired from outside the plant body into organic compounds inside the plant.

ATP Adenosine triphosphate; the almost universal energy currency of biological organisms; functions as a donor of phosphate groups. It consists of the purine base adenine linked to the sugar ribose (adenine + ribose = adenosine) and three phosphate molecules and is produced during →respiration. ATP also is a constituent of →RNA and, following removal of oxygen, of →DNA.

Auxin (Greek *auxanomai* = to grow) Plant →hormone that mainly stimulates the pattern of cell division and differentiation ("growth hormone"), and controls organ formation and →apical dominance. Very generally, auxin is transported to tissues to induce them to grow. The major auxin is indole-3-acetic acid (IAA), a carboxylic acid produced both from the →amino acid tryptophan and from one of its precursors.

Bark Tough external covering of a woody stem or root external to the →vascular cambium; composed mostly of old →phloem tissue.

Basal bud →Bud at the base of a shoot or cane, not classified as a bud at a count node. Such buds do not normally break in the same season as buds at count nodes but may remain latent on old wood for many years; if stimulated by severe pruning, they may produce shoots that may or may not be fruitful, depending on cultivar.

Blade The broad, expanded portion of a leaf; the lamina.

Bloom Flowering as indicated by shedding of the →calyptras covering the reproductive organs of grapevine flowers. Bloom also refers to the waxy coating on grape berries that gives a frosted appearance to dark-colored grape varieties.

Bleeding Sap flow. The exudation of sap from pruning wounds before and during →budbreak; serves to dissolve and push out air bubbles in the →xylem and to rehydrate the buds.

Brassinosteroids Group of plant →hormones that may induce cell expansion ("growth hormone"). Brassinosteroids are produced locally, mostly in the epidermis. The major active brassinosteroid of grapevines is castasterone, produced from a cholesterol-like compound.

Bud Rounded organ (winter bud or eye) at the node of a cane or shoot formed in the axil of the leaf, containing an undeveloped (embryonic) shoot protected by overlapping scales. Each bud appears to be single but comprises at least three "true buds": a more developed primary bud between two less prominent secondary buds.

Budbreak, budburst The stage of bud development when green tissue becomes visible; the emergence of a new shoot from a bud during spring.

Bud fruitfulness A "fruitful bud" has one or more inflorescence primordia, which give rise to clusters. One measure of fruitfulness is the number of clusters per shoot. Fruitfulness is an inherited characteristic that is also influenced by environmental factors at the time of inflorescence primordium initiation.

Bunch Australian term for grape →cluster.

Callus (Latin *callos* = hard skin) Mass of undifferentiated →somatic cells. Parenchyma tissue that grows over a wound or graft and protects it against drying or other injury, or cells that develop from plant pieces in tissue culture.

Calyptra (Greek *kalyptra* = cover, veil) The fused petals (cap) of the grape flower that fall off at bloom. Also, the root cap.

Cambium (Latin *cambiare* = to exchange) Meristematic (dividing) tissue that produces parallel rows of cells.

    Vascular cambium (Latin *vasculum* = small vessel) A thin, cylindrical sheath of undifferentiated meristem between the bark and wood. When active, it produces secondary xylem (to the inside) and phloem (to the outside), resulting in growth of the diameter of stems (shoots, trunks) and roots.

    Cork cambium (=phellogen) Meristem, which is in part responsible for the development of the bark. Produces cork cells, the outer walls of which are impregnated with fatty substances impervious to water and gases.

Cane A mature, woody 1-year-old →shoot from the previous growing season, which generally has more than six count nodes. A shoot becomes a cane after →periderm formation.
Canopy The aboveground portion of a grapevine, comprising the →trunk, →cordon, shoots, leaves, and fruit.
Capacity The total growth of a grapevine, including the production of crop, leaves, shoots, and roots in one growing season. Capacity can be estimated from a vine's total weight of fruit and shoots.
Carbohydrate (Latin *carbo* = ember; *hydro* = water) Organic compound made up of sugar units. Consists of a chain of carbon atoms to which hydrogen and oxygen atoms are attached in a 2:1 ratio; examples are sugars, →starch, and →cellulose.
Carotenoid Yellow, orange, or red photosynthetic →pigment in plant cell →chloroplasts. Carotenoids become visible in leaves and grape berries when →chlorophyll breaks down and are responsible for the yellowish-orange fall colors of leaves.
Cell The structural and functional unit of all living organisms, enclosed in a plasma membrane and containing the →cytoplasm, →nucleus, and other organelles (such as →chloroplasts, →mitochondria, and →vacuoles). Plant (and animal) cells are typically 10–30 µm in diameter, but many bacteria are only 1–2 µm long.
Cellulose A linear, water-insoluble →carbohydrate (a homopolysaccharide), composed of 10,000–15,000 glucose units arranged in β1→4-glucan chemical bonds, forming straight fibrous chains held together by hydrogen bonds; the main cell wall substance of plants and some fungi.
Chilling Low but above-freezing temperature. Required to release →dormancy. Chilling stress refers to damage or injury caused by low temperatures (usually <15°C) to nondormant plant tissues.
Chimera A plant consisting of more than one genetically distinct population of →cells, usually due to a →mutation in one of the two cell layers of the shoot apical →meristem.
Chitin A linear →carbohydrate (a homopolysaccharide), composed of thousands of *N*-acetyl-D-glucosamine units arranged in β1→4-glucan chemical bonds, forming extended rigid fibers similar to those of →cellulose. It is the main cell wall substance of fungi and of all the hard parts of arthropods but is not found anywhere in plants.
Chloroplast →A chlorophyll-containing photosynthetic →cell organelle. Chloroplasts descended from symbiotic microorganisms that invaded (or were "swallowed" by) larger cells and still have their own →genes.
Chlorophyll (Greek *chloros* = green, yellow; *phyllon* = leaf) Green photosynthetic pigment in plant cell →chloroplasts; the receptor of light energy in →photosynthesis.
Chlorosis Loss of →chlorophyll and yellowing of some or all the cells of a tissue or an organ.
Chromosome A segment made up of chromatin, which consists of the histone →proteins acting as spools around which a single large →DNA molecule winds that contains many →genes in the →cell →nucleus of all higher organisms (other than bacteria and archaea).
Clone (Greek *klon* = twig) A group of vines of a uniform, genetically identical type, derived by →vegetative propagation from a single original mother vine.
Cluster Fruiting (reproductive) structure of the grapevine with grape berries.
Coenzyme Organic cofactor (often containing a vitamin) required for certain →enzymes to be active.
Cold hardiness The ability of a grapevine to survive the extreme cold and drying effects of winter weather.
Competition Interaction between organs of the same organism (e.g., shoot tips and inflorescences), members of the same population (e.g., grapevines in a vineyard), or of two or more populations (e.g., grapevines and weeds) to obtain a resource that both require and that is available in limited supply.
Cordon The permanent arm arising from the trunk to form a part of the vine framework and which supports the fruiting 1-year-old wood. Cordons are usually horizontal, but may be inclined or vertical.
Correlative inhibition The phenomenon whereby distal (apical) buds on a cane, i.e., at the apex, break prior to and inhibit the growth of the proximal (basal) buds. Such inhibition is most obvious in early spring when, in cane-pruned vines, the most distal buds of the cane break and grow vigorously, while those toward the base of the cane grow weakly or not at all.

Cork A secondary tissue produced by a cork →cambium in woody stems and roots; made up of nonliving cells with suberized cell walls, which resist the passage of gases and water vapor.

Cropping The formation of crop yield on a grapevine.

   Crop The amount of fruit borne on a vine.

   Crop level The amount of fruit (or number of clusters retained) per shoot or per meter of canopy length. Also used as a synonym for →crop size.

   Crop load The crop size in relation to vine size (measured as pruning weight or leaf surface).

   Crop recovery The crop produced from new growth following frost injury.

   Crop size The yield per vine or per unit of land area. Also used as synonym for →crop level.

Cultivar A →variety of plant found only under cultivation. A group of plants that has been selected for a particular attribute or combination of attributes that is clearly distinct, uniform, and stable in these characteristics, and that retains those characteristics when propagated.

Cuticle (Latin *cutis* = skin) Waxy or fatty layer on the outer wall of epidermal cells, formed of cutin and wax; largely impervious to water vapor and carbon dioxide; prevents uncontrolled water loss.

Cutting A portion of cane usually 30–40 cm long, used for propagation.

Cytokinins (Greek *kytos* = hollow; *kinein* = to move) Group of plant hormones that stimulate cell division and formation of →meristems in shoots but reduce it in roots. Zeatin is the main active cytokinin. Cytokinins are produced from →ATP (and ADP) and isoprenoids in the root tips, shoot meristems, and immature seeds and transported throughout the plant in both the xylem and phloem.

Cytoplasm The internal, highly heterogeneous volume of a →cell, composed of the →cytosol and various insoluble, suspended particles and organelles arranged in a porous, elastic network similar to a sponge filled with liquid.

Cytosol An aqueous →cell solution with a complex composition (e.g., →enzymes, →RNA, →amino acids, metabolites, ions) and a gel-like consistency.

Differentiation Developmental process by which a relatively unspecialized cell undergoes a progressive change to a more specialized cell. The increasing specialization of cells and tissues for particular structures or functions during development.

Diffusion Net movement of molecules in the direction of lower concentration.

Diploid Containing two copies of the same set of →chromosomes in each cell, normally arising from →fertilization.

DNA Deoxyribonucleic acid; the helical chain molecule (polymer) that composes the →genes. DNA is the storage form of genetic information located in the →chromosomes. It consists of long chains of nucleotide building blocks composed of a purine or pyrimidine base, deoxyribose sugar, and phosphate. One pair of bases located on opposite strands of DNA constitutes a letter of the genetic code. Groups of three consecutive bases (triplets) specify (i.e., are the "recipe" for) the different →amino acids in addition to "start" and "stop" signals (i.e., "punctuation marks" that determine beginning and end of a protein). The sequence of triplets determines the various →proteins.

Domestication Selection of wild plants (or animals) for adaptation to cultivation and use by humans. The process usually involves selection of traits regarded as beneficial. Such traits may be present in wild plants or arise via spontaneous or induced →mutations.

Dormancy (Latin *dormire* = to sleep) Temporary suspension of visible growth of plants, buds, or seeds. Growth cannot resume without special environmental cues.

Ecosystem (Greek *oikos* = house) A major interacting system that involves both living organisms and their physical environment.

Enzyme (Greek *en* = into; *zyme* = yeast, sourdough) A →protein (or →RNA) capable of speeding up (catalyzing) specific biochemical reactions by lowering the required activation energy, but which is unaltered itself in the process; a biological catalyst, the "workhorse" of a biochemical reaction.

Ethylene H$_2$C=CH$_2$; The only gaseous plant hormone. Ethylene induces organ abscission, dormancy, and root hair formation. Ethylene acts by increasing the tissue's sensitivity to →gibberellin.

Evapotranspiration The evaporation of water from a land surface, including →transpiration from plants and evaporation from the soil and surface water bodies.

Evolution Change in the genetic makeup of a population of organisms over time due to →mutations and subsequent unequal reproduction (natural selection) of individual organisms. →Species often evolve by giving an existing →gene a new job, rather than by "inventing" new genes.

Fatty acid Organic molecule made up of a polar (hydrophilic, i.e., water-attracting) head with two oxygen atoms and a hydrophobic (i.e., water-repellent) chain ("tail") of carbon atoms to which hydrogen atoms are bound. The hydrocarbon tail can be saturated (i.e., all carbon atoms are joined in single bonds) or unsaturated (i.e., some carbon atoms are joined in double bonds). Important constituent of cell →membranes, fats, oils, and waxes.

Feedback Regulation of the rate of a biochemical pathway by its own products; can be positive (a product speeds up the pathway) or negative (a product slows down the pathway by inhibiting an earlier step).

Fertilization Fusion of the sperm cells (male) from the pollen tube with the egg cells (female) of the ovule, resulting in the →zygote and leading to →fruit set. Also, the application of mineral plant nutrients to the soil or the leaves.

Field capacity The percentage of water a particular soil can hold against the action of gravity, i.e., the water remaining in the soil immediately after all free water has drained away and before the soil begins to dry down, which occurs at a soil water potential of about -10 kPa.

Fluorescence Light emission by an excited (unstable, higher energy) molecule (such as →chlorophyll upon light absorption) upon its reversion to the ground (normal, stable, lower energy) state. The emitted light is always of a longer wavelength (lower energy) than the absorbed light.

Fruit A mature ovary (berry) or cluster of mature ovaries.

Fruit set A stage of berry development 1–3 weeks after →bloom when many of the flowers have fallen and those remaining have set and developed into berries. Also called berry set stage or shatter stage; fruit set is preferred.

Gamete Female or male reproductive (sex) →cell (e.g., egg and sperm), produced by →meiosis.

Gene The fundamental hereditary unit, composed of multiple pairs of →DNA bases and usually located within a segment of a →chromosome. The "recipe" for a →protein; also, a developmental "on/off switch" in specific tissues at specific times; also a "device" to extract information from the environment. Genes can be switched on/off (i.e., "expressed"/"repressed") by acetylation/deacetylation (i.e., addition/removal of a COCH$_3$ group) of the supporting histone protein, or by demethylation/methylation (i.e., removal/addition of a CH$_3$ group) of the DNA itself.

Genome The complete set of →DNA of a particular organism. Grapevines, like all plants, have three distinct genomes: one in the →nucleus (where genome refers to the entire $\sim$500 million base-pair DNA sequence of one set of →chromosomes), one in the →mitochondria, and one in the plastids (e.g., →chloroplasts). The range of genetic variation in a population of individual organisms (i.e., in a →species) forms its gene pool.

Genotype The genetic makeup (set of →genes) of an individual organism.

Gibberellins Also termed gibberellic acids. Group of over 100 structurally similar plant hormones, a few of which are biologically active and mainly induce cell elongation and cell differentiation in meristematic tissues. Gibberellins are cyclic diterpenoids and are produced in meristematic shoot and root tips and leaf vascular tissues.

Girdling The removal of a complete ring of outer and inner bark (outward of the →cambium) from a shoot, cane, or trunk. Also called cincturing or ringing. Girdling temporarily interrupts the transport of assimilates in the →phloem.

Glutamine An →amino acid; the primary form in which assimilated nitrogen is transported in the →xylem and →phloem of grapevines.

Glycoside (Greek *glykys* = sweet) Compound consisting of a sugar bound to another molecule belonging to a variety of chemical classes.

Gradient Difference in concentration or other physical entity per unit distance or across a →membrane.

Grafting The (usually deliberate) fusion of parts from two or more different plants. A union of different individuals in which a portion, called the →scion, of one individual is inserted into a stem or root, called the stock or →rootstock, of the other individual.

Gravitropism (Latin *gravis* = heavy; Greek *tropos* = direction) Orientation of plants or plant organs in relation to gravity.

Guttation (Latin *gutta* = drop) The exudation of liquid water from leaves, caused by root pressure resulting from osmotic water uptake.

Haploid Containing only one copy of a →chromosome set in the cell; →gametes are haploid.

Heterozygous Having two →alleles (i.e., two distinct copies of a →gene, often one dominant and one recessive) at the same chromosome position (locus) but on different →chromosomes within a chromosome pair of a →diploid →cell. The dominant alleles determine the →phenotype, because their instructions are preferred over those of the recessive alleles.

Homozygous Having the same →allele (i.e., two identical copies of a →gene) on both →chromosomes of →diploid cells.

Hormone (Greek *hormaein* = to stimulate) Organic substance produced in minute amounts in one plant part (tissue or organ), from which it is transported to another part of the same plant on which it has a specific effect. Hormones function as highly specific chemical signals between cells.

Humus Decomposing organic matter in the soil.

Hybrid The result of a cross between differing plants or taxonomic units. Most modern grape cultivars are intraspecific hybrids (obtained by crossing vines of the same →species), but some are interspecific hybrids (obtained by crossing vines belonging to different species).

Inflorescence The flowering cluster of a grapevine.

Internode The section of a shoot or cane between two adjacent nodes.

Latent bud A bud that has remained undeveloped for a year or longer; usually the source of a →water shoot.

Lateral A branch or side shoot arising from a prompt bud in the axil of a true leaf on a main shoot. Lateral growth is promoted by topping the main shoot; lateral shoots may vary in length from a few mm to greater than 1 m. Also, a branch of the main axis of a cluster.

Leaf A flat, thin expanded organ growing from the shoot, consisting of a broad →blade, a →petiole, and two inconspicuous stipules (leaf-like structures) at the base of the petiole.

Leaf scar The scar left on the shoot after a leaf falls.

Lenticel Porous region of the →periderm with loosely packed cells; permits gas exchange.

Lipid Small, water-insoluble (i.e., hydrophobic) molecule made up of →fatty acids, sterols, or isoprenoids; important component of cell →membranes.

Macronutrient (Greek *macros* = large; Latin *nutrire* = to nourish) Inorganic chemical element that is required in large amounts for plant growth, such as nitrogen, potassium, phosphorus, calcium, magnesium, and sulfur.

Matrix The aqueous content of a →cell or cell organelle with its dissolved solutes.

Maturity Stage of fruit development when it has reached satisfactory quality for which it is intended.

Meiosis (Greek *meioun* = to make small) Special type of →cell division that occurs in anthers and ovaries and reduces a diploid cell to four haploid cells to become →gametes. Random halving of pairs of →chromosomes and random exchange (mixing) of maternal and paternal →genes before meiotic cell division produces genetic variation and repairs →DNA damage in sexually reproducing organisms.

Membrane (Latin *membrana* = skin) Thin, tough, hydrophobic (bi-)layer consisting of →lipids and →proteins that surrounds →cells and cell organelles, separating their contents from the surroundings. Membranes are barriers to the free passage of ions and other molecules.

Meristem (Greek *meristes* = divider) Tissue in which cell division predominates. The undifferentiated, perpetually young plant tissue from which new cells arise. See also →cambium.

  Apical meristem The meristem at the tip of a root or shoot.

  Intercalary meristem (Latin *intercalare* = to insert) Meristem that produces new primary tissues between more differentiated tissues, as in an internode of a shoot or the interveinal part of a leaf.

Micronutrient (Greek *mikros* = small; Latin *nutrire* = to nourish) Inorganic chemical element that is required only in very small, or trace, amounts for plant growth, such as iron, copper, manganese, zinc, molybdenum, nickel, and boron.

Mitochondrion, *pl.* **mitochondria** →Cell organelle responsible for →respiration; the "power plant" providing energy for metabolic processes. Mitochondria descended from symbiotic microorganisms that invaded (or were "swallowed" by) larger cells and still have their own →genes.

Mitosis (Greek *mitos* = string, loop) The process of →DNA duplication, →chromosome replication (by lengthwise splitting), and →cell division during growth. Each of two identical daughter cells receives one set of chromosomes from the mother cell.

Mutation (Latin *mutatio* = change) A spontaneous or engineered change in the →genotype which may alter the →phenotype. Mutations generally result from copying errors during →mitosis or →meiosis and provide the raw material for →evolution and plant breeding.

NAD, NADP Nicotinamide adenine dinucleotide, nicotinamide adenine dinucleotide phosphate; →coenzymes that serve as carriers of hydrogen atoms and electrons.

Necrosis (Greek *nekrois* = death) Uncontrolled, sudden death of some or all of the cells of a tissue or an organ due to loss of →membrane integrity and flooding of cells with water.

Node (Latin *nodus* = knot) The enlarged portion of a shoot or cane at which leaves, clusters, tendrils, and/or buds are located at regular intervals.

Nucleus The dense central organelle of the cell; contains the →chromosomes and is enclosed by a nuclear →membrane.

Osmosis (Greek *osmos* = impulse, thrust) The diffusion of water across a selectively permeable →membrane. In the absence of other forces, osmotic water movement is in the direction of higher solute concentration (lower water concentration or →water potential).

Overcropping The production of more crop than a grapevine can bring to maturity by normal harvest time.

Own-rooted A vine grown from a cutting that develops its own root system as opposed to a vine grafted or budded onto a →rootstock.

Pathogen A disease-causing agent or organism.

Pedicel The stalk of an individual flower or berry in a grape cluster.

Peduncle The stalk attaching a →cluster to the →shoot; the branched axis of a cluster apart from the →pedicels of individual flowers.

Perennial (Latin *per* = through; *annuus* = year) A plant, such as the grapevine, in which the vegetative structures live through multiple growing seasons.

Periderm The →cork cambium and the waterproof tissues it produces; outer bark.

Permanent wilting point The percentage of water remaining in the soil when it has dried down so much that a plant growing in that soil fails to recover from wilting even if placed in a humid chamber. Occurs at a soil water potential of about -1.5 MPa.

Pest Species, strain or biotype of any plant, animal, or pathogen harmful to plants or plant products.

Petiole The stalk attaching the →leaf blade to the shoot.

pH A symbol denoting the relative concentration of hydronium ions (H_3O^+) or protons (H^+, in mol/L) in a solution; $pH = -\log[H^+] \rightarrow [H^+] = 10^{-pH}$. pH values run from 0 to 14, and the lower the value the more acidic is a solution, i.e., the more protons it contains; pH 7 is neutral (e.g., pure water and →cytoplasm), less than 7 is acidic (e.g., grape juice), and more than 7 is alkaline (e.g., lime soil).

Phenols, phenolics A broad range of ultraviolet light–absorbing organic compounds, all of which have a hydroxyl (OH) group attached to an aromatic ring (a ring of six carbon atoms containing three double bonds); includes cinnamic acids, flavonoids, stilbenes, tannins, and lignins. Phenolics are important determinants of wine quality, mainly through →tannins and →anthocyanins, and contribute to disease resistance of grapevines.

Phenotype The whole of all the characteristics of an individual organism; the physical expression, resulting from interaction with the environment, of the →genotype during its development.

Phloem (Greek *phloos* = bark) The food-conducting and storing tissue of plants, which is composed of living cells. In grapevines old phloem comprises most of the bark.

Photosynthesis (Greek *photos* = light; *syn* = together; *tithenai* = to place) The conversion of light energy to chemical energy, and the process by which a plant converts carbon dioxide and water into carbohydrates. Solar radiation is the energy source for this process.

Phytochrome (Greek *phyton* = plant; *chroma* = color) A pigment →protein located in the →cytoplasm of plants that acts as a light sensor. Photoreceptor for red and far-red light; involved in a number of timing processes, such as leaf formation, flowering, dormancy, and seed germination, as well as shade-avoidance responses.

Phytotoxic Causing injury or death of plants or portions of plants.

Pigment A substance that absorbs light, often selectively (i.e., only particular bands of wavelengths), giving the visual appearance of a specific color; →anthocyanin, →chlorophyll, →phytochrome.

Pomace Solid material remaining after pressing of grape juice or wine.

Pollination (Latin *pollen* = fine dust) The transfer of pollen from the male part of a flower (anther) to the female part of a flower (stigma).

Primordium (Latin *primus* = first; *ordiri* = to begin to weave) A cell or an organ in its earliest stage of differentiation.

Prompt bud Axillary bud that develops into a →lateral shoot in the same season that the bud is produced.

Protein (Greek *proteios* = primary) A complex organic compound, a polypeptide, composed of a string of just a few to several thousand →amino acids. Proteins are folded into a plethora of three-dimensional shapes that define their function. See also →enzyme.

Pruning The process of cutting away of a portion of the annual vegetative growth of a grapevine to maintain the desired number and spacing of →buds or →nodes per vine.

    Summer pruning Trimming or topping of growing shoots during the growing season to regulate shoot length, reduce →cluster shading, and maintain canopy shape.

    Winter pruning Also called "dormant pruning." Carried out during the dormant period to regulate the node number and maintain vine form.

Q_{10} The proportional change in the rate of a (bio)chemical reaction per 10°C increase in temperature; e.g., for $Q_{10} = 2$, respiration will double for each 10°C rise in temperature.

Rachis (Greek *rachis* = backbone) The central axis of a →cluster stem.

Raisin A dried grape.

Receptacle The part of the axis of the flower or fruit stalk that bears the floral organs or grape berry.

Respiration An intracellular process in which molecules, particularly pyruvate in the Krebs cycle, are oxidized (i.e., have electrons removed) to produce energy (→ATP). The complete breakdown of sugar or other organic compounds, such as malate, to carbon dioxide (CO_2) and water.

RNA Ribonucleic acid; the chain molecule that translates the genetic code (→DNA) into →amino acids and hence →proteins. RNA is a working copy and transport form of genetic information (i.e., the link between

DNA and protein). It consists of long chains of nucleotide building blocks composed of a purine or pyrimidine base, ribose sugar, and phosphate and is produced from →ATP and other nucleotide triphosphates.

Root The belowground axis of the grapevine, which serves to anchor the plant and to absorb and conduct water and nutrients; it bears only lateral roots. The true root develops from the hypocotyl of the embryo; the grapevine root is a taproot, an axis with lateral roots initiated near the tip.

Rootstock Specialized planting material to which fruiting varieties of grapes are grafted as →scions to produce a commercially acceptable vine. Grape rootstock varieties are used for their tolerance of or resistance to root parasites, such as phylloxera and nematodes, for their vigor, or for their tolerance of adverse soil conditions.

Sap The fluid contents of the →xylem or the sieve elements of the →phloem. The fluid contents of the →vacuole are called cell sap.

Scion The →shoot or →cane portion that is grafted or budded onto a →rootstock.

Senescence (Latin *senescere* = to grow old) Aging. The ordered degradation of cell constituents that ends in death of the cell.

Sexual reproduction Propagation of plants by means of seeds following fertilization.

Shoot The current growing season's green stem growth. It becomes a →cane when over half of its length becomes woody with a tan or brown color.

Sink An assimilate-importing plant organ. All plant organs unable to meet their own nutritional demands may act as sinks, i.e., expanding leaves or growing fruit or storage organs when they are being replenished.

Sinus The gap in the margin between two individual lobes of a →leaf blade. The petiolar sinus is the gap at the attachment of the →petiole.

Somatic cell (Greek *soma* = body) A body →cell. All cells of a plant are somatic except the cells (spores) produced by →meiosis that develop into →gametes (although even these are derived from somatic cells, unlike in animals where somatic and germ line cells are kept separate). Somatic cells can be forced through tissue culture to form somatic embryos that can develop into plants that are →clones of their mother plant.

Source An assimilate-exporting plant organ. Photosynthesizing leaves are the primary sources, but storage organs, such as the trunk or roots, may also serve as important sources, especially during budbreak.

Species Population or groups of populations of sexually reproducing organisms that interbreed with each other but are reproductively isolated from other populations.

Spur A short fruiting unit of 1-year growth; the basal section of a →cane that is cut back to one to three →buds. These are used in spur →pruning to provide fruiting shoots for the current year and in cane pruning to provide replacement canes for the following year.

Starch A complex insoluble carbohydrate, composed of 1000 or more glucose units arranged in $\alpha 1 \to 4$- and $\alpha 1 \to 6$-glucan chemical bonds; the main food storage substance of grapevines.

Stoma, *pl.* **stomata** (Greek *stoma* = mouth) A minute opening bordered by guard cells in the epidermis of leaves, stems, and grapes through which gases pass. The stoma regulates the exchange of carbon dioxide used in photosynthesis and water used in transpiration between the plant and the atmosphere.

Sucker A →water shoot that emerges from the base of the established trunk of the vine, at or belowground level.

Sucrose A disaccharide (glucose plus fructose); the primary form in which sugar produced by photosynthesis is transported in the →phloem of grapevines.

Symbiosis (Greek *syn* = together, at the same time) The mutually beneficial living together of two or more species in a long-term relationship.

Symplast (Greek *plastos* = formed) The totality of protoplasts; i.e., everything inside a plant that is bound inside a membrane. The continuum of cytoplasm interconnected by plasmodesmata and excluding the →vacuoles.

Tannin A macromolecule, belonging to the →phenolic class of flavonoids, found in the cell vacuoles of the skin and seeds of grape berries. Tannins are responsible for the astringency and bitterness of grapes and wine.

Tendril (Latin *tendere* = to extend) A part of the stem modified into a slender twining organ on a shoot opposite a leaf that can coil around an object for support.

Terpene An organic hydrocarbon or one of its derivatives, assembled from multiple isoprenoid units; responsible for certain odors or flavors of grape berries (and of flowers, fruits, and leaves of other plants).

Tissue A group of similar cells organized into a structural and functional unit.

Transpiration (Latin *trans* = over, beyond; *spirare* = to breathe) The loss of water vapor by plant parts; most transpiration occurs through →stomata. Transpiration is required for cooling of plant tissues.

Trellis A permanent structure that supports the vine framework, consisting of stakes, posts and, usually, at least one wire.

Trunk The main thickened stem of a vine between the →roots and the →head or →cordon of the vine.

Turgor Outward pressure of cell sap due to water absorption by →osmosis across a →membrane, balanced by the rigid cell wall; turgor pressure is necessary for cell expansion during growth.

Vacuole Organelle of plant →cells for storage, degradation, and recycling of a wide variety of compounds and waste products, surrounded by a membrane called tonoplast. Different types of specialized vacuoles can be present within the same cell; may represent up to 90% of the volume of a mature cell. Osmotic water absorption creates outward pressure (→turgor).

Variety A group of closely related plants of common origin and similar characteristics within a species. The so-called grape varieties are botanically →cultivars.

Vascular bundle (Latin *vasculum* = small vessel) A strand of tissue containing →xylem and →phloem; used for conducting water, nutrients, and assimilates.

Vegetative propagation Asexual propagation; the propagation of grapevines (and other organisms) without fertilization but by means of cuttings, layering, meristem or shoot-tip culture, and micropropagation.

Vein A strand of vascular tissues in a flat organ such as a leaf.

Veraison (French *vérer* = to change) Onset of ripening of a grape berry as indicated by a change in color, which is preceded by softening and the beginning of sugar accumulation.

Vigor The rate of shoot growth measured by the change in shoot length over time.

Virus Self-replicating complex of nucleic acids (either →DNA or →RNA) and →proteins that infects living organisms and requires an intact host →cell for replication.

Water shoot Rapidly growing shoot arising from a latent bud on a →trunk or →cordon. A noncount shoot that arises on wood older than 1 year and originates from a →basal bud.

Water potential A measure of water concentration. The free energy of an aqueous solution per unit molar volume of liquid water; typically the sum of the component potentials arising from the effects of →turgor pressure and dissolved molecules (→osmosis).

Weed Generally an herbaceous plant not valued for use or beauty, growing wild, and regarded as intruding on space or hindering the growth of grapevines.

Wood Secondary →xylem. The tissue produced on the inside of the vascular →cambium.

Xylem (Greek *ksúlon* = wood) The dead, woody portion of conducting tissues whose function is the transport of water and minerals and support of the roots, trunk, and shoots.

Zygote Fertilized egg resulting from the fusion of a male and a female →gamete (→fertilization) that forms a zygotic embryo which develops into a seed.

Bibliography

Abass, M., Rajashekar, C.B., 1991. Characterization of heat injury in grapes using ^{1}H nuclear magnetic resonance methods. Plant Physiol. 96, 957–961.

Achard, P., Genschik, P., 2009. Releasing the brakes of plant growth: how GAs shutdown DELLA proteins. J. Exp. Bot. 60, 1085–1092.

Achard, P., Gong, F., Cheminant, S., Alioua, M., Hedden, P., Genschik, P., 2008. The cold-inducible CBF1 factor-dependent signaling pathway modulates the accumulation of the growth-repressing DELLA proteins via its effect on gibberellin metabolism. Plant Cell 20, 2117–2212.

Adam-Blondon, A.F., Lahogue-Esnault, F., Bouquet, A., Boursiquot, J.M., This, P., 2001. Usefulness of two SCAR markers for marker-assisted selection of seedless grapevine cultivars. Vitis 40, 147–155.

Adams, D.O., 2006. Phenolics and ripening in grape berries. Am. J. Enol. Vitic. 57, 249–256.

Adams, D.O., Liyanage, C., 1993. Glutathione increases in grape berries at the onset of ripening. Am. J. Enol. Vitic. 44, 333–338.

Adams, D.O., Franke, K.E., Christensen, L.P., 1990. Elevated putrescine levels in grapevine leaves that display symptoms of potassium deficiency. Am. J. Enol. Vitic. 41, 121–125.

Agaoglu, Y.S., 1971. A study on the differentiation and the development of floral parts in grapes (*Vitis vinifera* L. var.). Vitis 10, 20–26.

Agasse, A., Vignault, C., Kappel, C., Conde, C., Gerós, H., Delrot, S., 2009. Sugar transport and sugar sensing in grape. In: Roubelakis-Angelakis, K.A. (Ed.), Grapevine Molecular Physiology and Biotechnology, second ed. Springer, Dordrecht, The Netherlands, pp. 105–139.

Agati, G., Tattini, M., 2010. Multiple functional roles of flavonoids in photoprotection. New Phytol. 186, 786–793.

Ageorges, A., Issaly, N., Picaud, S., Delrot, S., Romieu, C., 2000. Identification and functional expression in yeast of a grape berry sucrose carrier. Plant Physiol. Biochem. 38, 177–185.

Agusti, J., Herold, S., Schwarz, M., Sanchez, P., Ljung, K., Dun, E.A., et al., 2011. Strigolactone signaling is required for auxin-dependent stimulation of secondary growth in plants. Proc. Natl. Acad. Sci. USA 108, 20242–20247.

Agüero, C.B., Rodríguez, J.G., Martínez, L.E., Dangl, G.S., Meredith, C.P., 2003. Identity and parentage of Torrontés cultivars in Argentina. Am. J. Enol. Vitic. 54, 318–321.

Ahmedullah, M., Kawakami, A., 1986. Evaluation of laboratory tests for determining the lethal temperature of *Vitis labruscana* Bailey Concord roots exposed to subzero temperatures. Vitis 25, 142–150.

Ainsworth, E.A., Rogers, A., 2007. The response of photosynthesis and stomatal conductance to rising [CO_2]: mechanisms and environmental interactions. Plant Cell Environ. 30, 258–270.

Akiyama, K., Matsuzaki, K., Hayashi, H., 2005. Plant sesquiterpenes induce hyphal branching in arbuscular mycorrhizal fungi. Nature 435, 824–827.

Aktas, H., Karni, L., Chang, D.C., Turhan, E., Bar-Tal, A., Aloni, B., 2005. The suppression of salinity-associated oxygen radicals production, in pepper (*Capsicum annuum*) fruit, by manganese, zinc and calcium in relation to its sensitivity to blossom-end rot. Physiol. Plant. 123, 67–74.

Alboresi, A., Gestin, C., Leydecker, M.T., Bedu, M., Meyer, C., Truong, H.N., 2005. Nitrate, a signal relieving seed dormancy in *Arabidopsis*. Plant Cell Environ. 28, 500–512.

Alcázar, R., Altabella, T., Marco, F., Bortolotti, C., Reymond, M., Koncz, C., et al., 2010. Polyamines: molecules with regulatory functions in plant abiotic stress tolerance. Planta 231, 1237–1249.

Allègre, M., Daire, X., Héloir, M.C., Trouvelot, S., Mercier, L., Adrian, M., et al., 2007. Stomatal deregulation in *Plasmopara viticola*-infected grapevine leaves. New Phytol. 173, 832–840.

Allen, G.J., Chu, S.P., Harrington, C.L., Schumacher, K., Hoffman, Y., Tang, Y.Y., et al., 2001. A defined range of guard cell calcium oscillation parameters encodes stomatal movements. Nature 411, 1053–1057.

Alleweldt, G., 1957. Der Einfluß von Photoperiode und Temperatur auf Wachstum und Entwicklung von Holzpflanzen unter besonderer Berücksichtigung der Gattung *Vitis*. Vitis 1, 159–180.
Alleweldt, G., 1958. Eine Frühdiagnose zur Bestimmung der Fruchtbarkeit von Reben. Vitis 4, 230–236.
Alleweldt, G., 1959. Förderung des Infloreszenzwachstums der Reben durch Gibberellinsäure. Vitis 2, 71–78.
Alleweldt, G., 1960. Untersuchungen über den Austrieb der Winterknospen von Reben. Vitis 2, 134–152.
Alleweldt, G., 1961. Hemmung der Blütenbildung von *Vitis rupestris* durch Gibberellin. Naturwissenschaften 45, 628–629.
Alleweldt, G., 1964. Über die Nachwirkung von Umweltfaktoren auf das vegetative Wachstum von Rebenstecklingen im Folgejahr. Z. Acker- Pflanzenbau 119, 178–194.
Alleweldt, G., 1965. Über den Einfluss der Temperatur auf die Blutung von Reben. Vitis 5, 10–16.
Alleweldt, G., 1966. Die Differenzierung der Blütenorgane der Rebe. Wein-Wiss. 21, 393–402.
Alleweldt, G., Düring, H., 1972. Der Einfluss der Photoperiode auf Wachstum und Abscisinsäuregehalt der Rebe. Vitis 11, 280–288.
Alleweldt, G., Hifny, H.A.A., 1972. Zur Stiellähme der Reben. II. Kausalanalytische Untersuchungen. Vitis 11, 10–28.
Alleweldt, G., Hofäcker, W., 1975. Einfluss von Umweltfaktoren auf Austrieb, Blüte, Fruchtbarkeit und Triebwachstum bei der Rebe. Vitis 14, 103–115.
Alleweldt, G., Ilter, E., 1969. Untersuchungen über die Beziehungen zwischen Blütenbildung und Triebwachstum bei Reben. Vitis 8, 286–313.
Alleweldt, G., Istar, A., 1969. Über die apikale Dominanz bei Reben. Vitis 8, 94–104.
Alleweldt, G., Merkt, N., 1992. Der Stickstoffexport der Wurzel und die Zusammensetzung des Xylemexsudats. Teil 1: Der Einfluss einer zunehmenden Stickstoffdüngung. Vitis 31, 121–130.
Alleweldt, G., Possingham, J.V., 1988. Progress in grapevine breeding. Theor. Appl. Genet. 75, 669–673.
Alleweldt, G., Düring, H., Waitz, G., 1975. Untersuchungen zum Mechanismus der Zuckereinlagerung in die wachsenden Weinbeeren. Angew. Bot. 49, 65–73.
Alleweldt, G., Engel, M., Gebbing, H., 1981. Histologische Untersuchungen an Weinbeeren. Vitis 20, 1–7.
Alleweldt, G., Düring, H., El-Sese, A.M.A., 1984a. The influence of nitrogen fertilization and water supply on photosynthesis, transpiration and dry matter production in grapevines (*Vitis vinifera*). Plant Res. Dev. 20, 45–58.
Alleweldt, G., Düring, H., Jung, K.H., 1984b. Zum Einfluß des Klimas auf Beerenentwicklung, Ertrag und Qualität bei Reben: Ergebnisse einer siebenjährigen Faktorenanalyse. Vitis 23, 127–142.
Aloni, R., 2001. Foliar and axial aspects of vascular differentiation: hypotheses and evidence. J. Plant Growth Regul. 20, 22–34.
Aloni, R., 2013. Role of hormones in controlling vascular differentiation and the mechanism of lateral root initiation. Planta 238, 819–830.
Aloni, R., Schwalm, K., Langhans, M., Ullrich, C.I., 2003. Gradual shifts in sites of free-auxin production during leaf-primordium development and their role in vascular differentiation and leaf morphogenesis in *Arabidopsis*. Planta 216, 841–853.
Aloni, R., Langhans, M., Aloni, E., Dreieicher, E., Ullrich, C.I., 2005. Root-synthesized cytokinin in *Arabidopsis* is distributed in the shoot by the transpiration stream. J. Exp. Bot. 56, 1535–1544.
Aloni, R., Aloni, E., Langhans, M., Ullrich, C.I., 2006a. Role of auxin in regulating *Arabidopsis* flower development. Planta 223, 315–328.
Aloni, R., Aloni, E., Langhans, M., Ullrich, C.I., 2006b. Role of cytokinin and auxin in shaping root architecture: regulating vascular differentiation, lateral root initiation, root apical dominance and root gravitropism. Ann. Bot. 97, 883–893.
Alsina, M.M., Smart, D.R., Bauerle, T., de Herralde, F., Biel, C., Stockert, C., et al., 2011. Seasonal changes of whole root system conductance by a drought-tolerant grape root system. J. Exp. Bot. 62, 99–109.

Altieri, M.A., Ponti, L., Nicholls, C.I., 2005. Manipulating vineyard biodiversity for improved insect pest management: case studies from northern California. Int. J. Biodivers. Sci. Manage. 1, 191–203.

Amâncio, S., Tavares, S., Fernandes, J.C., Sousa, C., 2009. Grapevine and sulfur: old partners, new achievements. In: Roubelakis-Angelakis, K.A. (Ed.), Grapevine Molecular Physiology and Biotechnology, second ed. Springer, Dordrecht, The Netherlands, pp. 31–52.

Améglio, T., Ewers, F.W., Cochard, H., Martignac, M., Vandame, M., Bodet, C., et al., 2001. Winter stem xylem pressure in walnut trees: effects of carbohydrates, cooling and freezing. Tree Physiol. 21, 387–394.

Améglio, T., Decourteix, M., Alves, G., Valentin, V., Sakr, S., Julien, J.L., et al., 2004. Temperature effects on xylem sap osmolarity in walnut trees: evidence for a vitalistic model of winter embolism repair. Tree Physiol. 24, 785–793.

Amerine, M.A., Bailey, C.B., 1959. Carbohydrate content of various parts of the grape cluster. Am. J. Enol. Vitic. 10, 196–198.

Amerine, M.A., Root, G.A., 1960. Carbohydrate content of various parts of the grape cluster. II. Am. J. Enol. Vitic. 11, 137–139.

Amerine, M.A., Winkler, A.J., 1944. Composition and quality of musts and wines of California grapes. Hilgardia 15, 493–675.

Amrani Joutei, K., Glories, Y., Mercier, M., 1994. Localisation des tanins dans la pellicule de baie de raisin. Vitis 33, 133–138.

Amthor, J.S., 2000. The McCree – de Wit – Penning de Vries – Thornley respiration paradigms: 30 years later. Ann. Bot. 86, 1–20.

Amtmann, A., Blatt, M.R., 2009. Regulation of macronutrient transport. New Phytol. 181, 35–52.

Anastasiou, E., Lenhard, M., 2007. Growing up to one's standard. Curr. Opin. Plant Biol. 10, 63–69.

Andersen, P.C., Brodbeck, B.V., 1989a. Diurnal and temporal changes in the chemical profile of xylem exudate from *Vitis rotundifolia*. Physiol. Plant. 75, 63–70.

Andersen, P.C., Brodbeck, B.V., 1989b. Temperature and temperature preconditioning on flux and chemical composition of xylem exudate from muscadine grapevines. J. Am. Soc. Hortic. Sci. 114, 440–444.

Andersen, P.C., Brodbeck, B.V., 1991. Influence of fertilization on xylem fluid chemistry of *Vitis rotundifolia* Noble and *Vitis* hybrid Suwannee. Am. J. Enol. Vitic. 42, 245–251.

Anderson, L.J., Comas, L.H., Lakso, A.N., Eissenstat, D.M., 2003. Multiple risk factors in root survivorship: a 4-year study in Concord grape. New Phytol. 158, 489–501.

Antcliff, A.J., May, P., 1961. Dormancy and bud burst in Sultana vines. Vitis 3, 1–14.

Antcliff, A.J., Newman, H.P., Barrett, H.C., 1983. Variation in chloride accumulation in some American species of grapevine. Vitis 22, 357–362.

Antolín, M.C., Baigorri, H., De Luis, I., Aguirrezábal, F., Gény, L., Broquedis, M., et al., 2003. ABA during reproductive development in non-irrigated grapevines (*Vitis vinifera* L. cv. Tempranillo). Aust. J. Grape Wine Res. 9, 169–176.

Apel, K., Hirt, H., 2004. Reactive oxygen species: metabolism, oxidative stress, and signal transduction. Annu. Rev. Plant Biol. 55, 373–399.

Aradhya, M.K., Dangl, G.S., Prins, B.H., Boursiquot, J.M., Walker, A.M., Meredith, C.P., et al., 2003. Genetic structure and differentiation in cultivated grape, *Vitis vinifera* L. Genet. Res. 81, 179–192.

Araujo, F.J., Williams, L.E., 1988. Dry matter and nitrogen partitioning and root growth of young field-grown Thompson Seedless grapevines. Vitis 27, 21–32.

Araújo, W.L., Tohge, T., Ishizaki, K., Leaver, C.J., Fernie, A.R., 2011. Protein degradation—an alternative respiratory substrate for stressed plants. Trends Plant Sci. 16, 489–498.

Arnold, C., Gillet, F., Gobat, J.M., 1998. Situation de la vigne sauvage *Vitis vinifera* ssp. *silvestris* en Europe. Vitis 37, 159–170.

Aroca, R., Porcel, R., Ruiz-Lozano, J.M., 2012. Regulation of root water uptake under abiotic stress conditions. J. Exp. Bot. 63, 43–57.

Arrigo, N., Arnold, C., 2007. Naturalised *Vitis* rootstocks in Europe and consequences to native wild grapevine. PLoS One 2, e521, http://dx.doi.org/10.1371/journal.pone.0000521.

Arroyo-García, R., Ruiz-García, L., Bolling, L., Ocete, R., López, M.A., Arnold, C., et al., 2006. Multiple origins of cultivated grapevine (*Vitis vinifera* L. ssp. *sativa*) based on chloroplast DNA polymorphisms. Mol. Ecol. 15, 3707–3714.

Arvanitoyannis, I.S., Ladas, D., Mavromatis, A., 2006. Potential uses and applications of treated wine waste: a review. Int. J. Food Sci. Technol. 41, 475–487.

Aryan, A.P., Wilson, B., Strauss, C.R., Williams, P.J., 1987. The properties of glycosidases of *Vitis vinifera* and a comparison of their β-glucosidase activity with that of exogenous enzymes. An assessment of possible applications in enology. Am. J. Enol. Vitic. 38, 182–188.

Ashley, M.K., Grant, M., Grabov, A., 2006. Plant responses to potassium deficiencies: a role for potassium transport proteins. J. Exp. Bot. 57, 425–436.

Ashworth, E.N., 1982. Properties of peach flower buds which facilitate supercooling. Plant Physiol. 70, 1475–1479.

Atkin, O.K., Tjoelker, M.G., 2003. Thermal acclimation and the dynamic response of plant respiration to temperature. Trends Plant Sci. 8, 343–351.

Atkin, O.K., Bruhn, D., Hurry, V.M., Tjoelker, M.G., 2005. The hot and the cold: unravelling the variable response of plant respiration to temperature. Funct. Plant Biol. 32, 87–105.

Atwell, B., Kriedemann, P., Turnbull, C., 1999. Plants in Action. Adaptation in Nature, Performance in Cultivation. Macmillan Education, South Yarra, Australia.

Aubert, S., Gout, E., Bligny, R., Marty-Mazars, D., Barrieu, F., Alabouvette, J., et al., 1996. Ultrastructural and biochemical characterization of autophagy in higher plant cells subjected to carbon deprivation: control by the supply of mitochondria with respiratory substrates. J. Cell Biol. 133, 1251–1263.

Aubert, S., Bligny, R., Douce, R., Gout, E., Ratcliffe, R.G., Roberts, J.K.M., 2001. Contribution of glutamate dehydrogenase to mitochondrial glutamate metabolism studied by ^{13}C and ^{31}P nuclear magnetic resonance. J. Exp. Bot. 52, 37–45.

Austin, C.N., Grove, G.G., Meyers, J.M., Wilcox, W.F., 2011. Powdery mildew severity as a function of canopy density: associated impacts on sunlight penetration and spray coverage. Am. J. Enol. Vitic. 62, 23–31.

Ayre, B.G., 2011. Membrane-transport systems for sucrose in relation to whole-plant carbon partitioning. Mol. Plant 4, 377–394.

Bachmann, O., 1978. Verbreitung von Phenolcarbonsäuren und Flavonoiden bei Vitaceen. Vitis 17, 234–257.

Bachmann, O., Blaich, R., 1979. Vorkommen und Eigenschaften kondensierter Tannine in Vitaceen. Vitis 18, 106–116.

Bacilieri, R., Lacombe, T., Le Cunff, L., Di Vecchi-Staraz, M., Laucou, V., Genna, B., et al., 2013. Genetic structure in cultivated grapevines is linked to geography and human selection. BMC Plant Biol. 13, 25, http://dx.doi.org/10.1186/1471-2229-13-25.

Bacon, M.A., Wilkinson, S., Davies, W.J., 1998. pH-Regulated leaf cell expansion in droughted plants is abscisic acid dependent. Plant Physiol. 118, 1507–1515.

Bae, G., Choi, G., 2008. Decoding of light signals by plant phytochromes and their interacting proteins. Annu. Rev. Plant Biol. 59, 281–311.

Bailey-Serres, J., Voesenek, L.A.C.J., 2008. Flooding stress: acclimations and genetic diversity. Annu. Rev. Plant Biol. 59, 313–339.

Baillod, M., Baggiolini, M., 1993. Les stades repères de la vigne. Rev. Suisse Vitic. Arboric. Hortic. 25, 7–9.

Bair, K.E., Davenport, J.R., Stevens, R.G., 2008. Release of available nitrogen after incorporation of a legume cover crop in Concord grape. HortScience 43, 875–880.

Bais, A.J., Murphy, P.J., Dry, I.B., 2000. The molecular regulation of stilbene phytoalexin biosynthesis in *Vitis vinifera* during grape berry development. Aust. J. Plant Physiol. 27, 425–433.

Bais, H.P., Weir, T.L., Perry, L.G., Gilroy, S., Vivanco, J.M., 2006. The role of root exudates in rhizosphere interactions with plants and other organisms. Annu. Rev. Plant Biol. 57, 233–266.

Baker, D.A., Milburn, J.A., 1965. Lateral movement of inorganic solutes in plants. Nature 205, 306–307.

Baker, N.R., 2008. Chlorophyll fluorescence: a probe of photosynthesis *in vivo*. Annu. Rev. Plant Biol. 59, 89–113.

Balint, G., Reynolds, A.G., 2013. Impact of exogenous abscisic acid on vine physiology and grape composition of Cabernet Sauvignon. Am. J. Enol. Vitic. 64, 74–87.

Bálo, B., Mustárdy, L.A., Hideg, E., Faludi-Dániel, F., 1986. Studies on the effect of chilling on the photosynthesis of grapevine. Vitis 25, 1–7.

Bammi, R.K., Randhawa, G.S., 1968. Viticulture in the tropical regions of India. Vitis 7, 124–129.

Banowetz, G.M., Ammar, K., Chen, D.D., 1999. Postanthesis temperatures influence cytokinin accumulation and wheat kernel weight. Plant Cell Environ. 22, 309–316.

Barata, A., Malfeito-Ferreira, M., Loureiro, V., 2012a. Changes in sour rotten grape berry microbiota during ripening and wine fermentation. Int. J. Food Microbiol. 154, 152–161.

Barata, A., Malfeito-Ferreira, M., Loureiro, V., 2012b. The microbial ecology of wine grape berries. Int. J. Food Microbiol. 153, 243–259.

Barbier-Brygoo, H., De Angeli, A., Filleur, S., Frachisse, J.M., Gambale, F., Thomine, S., et al., 2011. Anion channels/transporters in plants: from molecular bases to regulatory networks. Annu. Rev. Plant Biol. 62, 25–51.

Bargel, H., Neinhuis, C., 2005. Tomato (*Lycopersicon esculentum* Mill.) fruit growth and ripening as related to the biomechanical properties of fruit skin and isolated cuticle. J. Exp. Bot. 56, 1049–1060.

Bargel, H., Koch, K., Cerman, Z., Neinhuis, C., 2006. Structure–function relationships of the plant cuticle and cuticular waxes—a smart material? Funct. Plant Biol. 33, 893–910.

Barlass, M., Skene, K.G.M., Woodham, R.C., Krake, L.R., 1982. Regeneration of virus-free grapevines using *in vitro* apical culture. Ann. Appl. Biol. 101, 291–295.

Barnard, H., Dooley, A.N., Areshian, G., Gasparyan, B., Faull, K.F., 2011. Chemical evidence for wine production around 4000 BCE in the Late Chalcolithic Near Eastern highlands. J. Archaeol. Sci. 38, 977–984.

Barnard, H.R., Ryan, M.G., 2003. A test of the hydraulic limitation hypothesis in fast-growing *Eucalyptus saligna*. Plant Cell Environ. 26, 1235–1245.

Barnavon, L., Doco, T., Terrier, N., Ageorges, A., Romieu, C., Pellerin, P., 2000. Analysis of cell wall neutral sugar composition, β-galactosidase activity and a related cDNA clone throughout the development of *Vitis vinifera* grape berries. Plant Physiol. Biochem. 38, 289–300.

Barratt, D.H.P., Kölling, K., Graf, A., Pike, M., Calder, G., Findlay, K., et al., 2011. Callose synthase GSL7 is necessary for normal phloem transport and inflorescence growth in *Arabidopsis*. Plant Physiol. 155, 328–341.

Barritt, B.H., 1970. Ovule development in seeded and seedless grapes. Vitis 9, 7–14.

Bartley, G.E., Scolnik, P.A., 1995. Plant carotenoids: pigments for photoprotection, visual attraction, and human health. Plant Cell 7, 1027–1038.

Bass, P., Vuittenez, A., 1977. Amélioration de la thermothérapie des vignes virosées au moyen de la culture d'apex sur milieux nutritifs ou par greffage sur des vignes de semis obtenues aseptiquement *in vitro*. Ann. Phytopathol. 9, 539–540.

Battany, M.C., 2012. Vineyard frost protection with upward-blowing wind machines. Agric. For. Meteorol. 157, 39–48.

Battany, M.C., Grismer, M.E., 2000. Rainfall runoff and erosion in Napa Valley vineyards: effects of slope, cover and surface roughness. Hydrol. Processes 14, 1289–1304.

Battey, N.H., Simmonds, P.E., 2005. Phylloxera and the grapevine: a sense of common purpose? J. Exp. Bot. 56, 3029–3031.

Bauerle, T.L., Richards, J.H., Smart, D.R., Eissenstat, D.M., 2008a. Importance of internal hydraulic redistribution for prolonging the lifespan of roots in dry soil. Plant Cell Environ. 31, 177–186.

Bauerle, T.L., Smart, D.R., Bauerle, W.L., Stockert, C., Eissenstat, D.M., 2008b. Root foraging in response to heterogeneous soil moisture in two grapevines that differ in potential growth rate. New Phytol. 179, 857–866.

Baumes, R., Wirth, J., Bureau, S., Gunata, Y., Razungles, A., 2002. Biogeneration of C_{13}-norisoprenoid compounds: experiments supportive for an apo-carotenoid pathway in grapevines. Anal. Chim. Acta 458, 3–14.

Baur, J.A., Sinclair, D.A., 2006. Therapeutic potential of resveratrol: the *in vivo* evidence. Nat. Rev. Drug. Discov. 5, 493–506.

Bautista, J., Dangl, G.S., Yang, J., Reisch, B., Stover, E., 2008. Use of genetic markers to assess pedigrees of grape cultivars and breeding program selections. Am. J. Enol. Vitic. 59, 248–254.

Bauwe, H., Hagemann, M., Fernie, A.R., 2010. Photorespiration: players, partners and origin. Trends Plant Sci. 15, 330–336.

Bavaresco, L., Eibach, R., 1987. Investigations on the influence of N fertilizer on resistance to powdery mildew (*Oidium tuckeri*), downy mildew (*Plasmopara viticola*) and on phytoalexin synthesis in different grapevine varieties. Vitis 26, 192–200.

Bavaresco, L., Fregoni, M., Fraschini, P., 1991. Investigations on iron uptake and reduction by excised roots of different grapevine rootstocks and a *V. vinifera* cultivar. Plant Soil 130, 109–113.

Bavaresco, L., Giachino, E., Pezzutto, S., 2003. Grapevine rootstock effects on lime-induced chlorosis, nutrient uptake, and source–sink relationships. J. Plant Nutr. 26, 1451–1465.

Bayonove, C., Cordonnier, R., Dubois, P., 1975a. Etude d'une fraction caractéristique de l'arome du raisin de la variété Cabernet-Sauvignon; mise en évidence de la 2-methoxy-3-isobutylpyrazine. C.R. Hébd. Séances Acad. Sci. Sér. D, Sci. Nat. 281, 75–78.

Bayonove, C., Cordonnier, R., Ratier, R., 1975b. Localisation de l'arome dans la baie de raisin: variétés Muscat d'Aléxandrie et Cabernet-Sauvignon. C.R. Séances Acad. Agric. Fr. 60, 1321–1328.

Becker, H., 1976. Genetische Konstitution, züchterische Bearbeitung und Leistung der Rebsorte Müller-Thurgau. Wein-Wiss. 31, 26–37.

Becker, T., Knoche, M., 2011. Water movement through the surfaces of the grape berry and its stem. Am. J. Enol. Vitic. 62, 340–350.

Becker, T., Grimm, E., Knoche, M., 2012. Substantial water uptake into detached grape berries occurs through the stem surface. Aust. J. Grape Wine Res. 18, 109–114.

Begum, S., Nakaba, S., Oribe, Y., Kubo, T., Funada, R., 2007. Induction of cambial reactivation by localized heating in a deciduous hardwood hybrid poplar (*Populus sieboldii x P. grandidentata*). Ann. Bot. 100, 439–447.

Belhadj, A., Saigne, C., Telef, N., Cluzet, S., Bouscaut, J., Corio-Costet, M.F., et al., 2006. Methyl jasmonate induces defense responses in grapevine and triggers protection against *Erysiphe necator*. J. Agric. Food Chem. 54, 9119–9125.

Belin, J.M., 1972. Recherches sur la répartition des levures à la surface de la grappe de raisin. Vitis 11, 135–145.

Bell, A.A., Ough, C.S., Kliewer, W.M., 1979. Effects on must and wine composition, rates of fermentation, and wine quality of nitrogen fertilization of *Vitis vinifera* var. Thompson Seedless grapevines. Am. J. Enol. Vitic. 30, 124–129.

Bell, S.J., Henschke, P.A., 2005. Implications of nitrogen nutrition for grapes, fermentation and wine. Aust. J. Grape Wine Res. 11, 242–295.

Bell, S.J., Robson, A., 1999. Effect of nitrogen fertilization on growth, canopy density, and yield of *Vitis vinifera* L. cv. Cabernet Sauvignon. Am. J. Enol. Vitic. 50, 351–358.

Bengough, G.A., McKenzie, B.M., Hallett, P.D., Valentine, T.A., 2011. Root elongation, water stress, and mechanical impedance: a review of limiting stresses and beneficial root tip traits. J. Exp. Bot. 62, 59–68.

Benjak, A., Forneck, A., Casacuberta, J.M., 2008. Genome wide analysis of the "cut-and-paste" transposons of grapevine. PLoS One 3, e3107, http://dx.doi.org/10.1371/journal.pone.0003107.

Benjamins, R., Scheres, B., 2008. Auxin: the looping star in plant development. Annu. Rev. Plant Biol. 59, 443–465.

Bennett, J., Jarvis, P., Creasy, G.L., Trought, M.C.T., 2005. Influence of defoliation on overwintering carbohydrate reserves, return bloom, and yield of mature Chardonnay grapevines. Am. J. Enol. Vitic. 56, 386–393.

Berger, F., Grini, P.E., Schnittger, A., 2006. Endosperm: an integrator of seed growth and development. Curr. Opin. Plant Biol. 9, 664–670.

Berger, F., Hamamura, Y., Ingouff, M., Higashiyama, T., 2008. Double fertilization—caught in the act. Trends Plant Sci. 13, 437–443.

Bergqvist, J., Dokoozlian, N., Ebisuda, N., 2001. Sunlight exposure and temperature effects on berry growth and composition of Cabernet Sauvignon and Grenache in the central San Joaquin Valley of California. Am. J. Enol. Vitic. 52, 1–7.

Berleth, T., Scarpella, E., Prusinkiewicz, P., 2007. Towards the systems biology of auxin-transport-mediated patterning. Trends Plant Sci. 12, 151–159.

Berli, F., D'Angelo, J., Cavagnaro, B., Bottini, R., Wuilloud, R., Silva, M.F., 2008. Phenolic composition in grape (*Vitis vinifera* L. cv. Malbec) ripened with different solar UV-B radiation levels by capillary zone electrophoresis. J. Agric. Food Chem. 56, 2892–2898.

Berli, F.J., Fanzone, M., Piccoli, P., Bottini, R., 2011. Solar UV-B and ABA are involved in phenol metabolism of *Vitis vinifera* L. increasing biosynthesis of berry skin polyphenols. J. Agric. Food Chem. 59, 4874–4884.

Bernard, M.B., Horne, P.A., Hoffmann, A.A., 2005. Eriophyoid mite damage in *Vitis vinifera* (grapevine) in Australia: *Calepitrimerus vitis* and *Colomerus vitis* (Acari: Eriophyidae) as the common cause of the widespread 'Restricted Spring Growth' syndrome. Exp. Appl. Acarol. 35, 83–109.

Bernizzoni, F., Civardi, S., van Zeller, M., Gatti, M., Poni, S., 2011. Shoot thinning effects on seasonal whole-canopy photosynthesis and vine performance in *Vitis vinifera* L. cv. Barbera. Aust. J. Grape Wine Res. 17, 351–357.

Bernstein, Z., Lustig, I., 1985. Hydrostatic methods of measurement of firmness and turgor pressure of grape berries (*Vitis vinifera* L.). Sci. Hortic. 25, 129–136.

Berry, J., Björkman, O., 1980. Photosynthetic response and adaptation to temperature in higher plants. Annu. Rev. Plant Physiol. 31, 491–543.

Bertamini, M., Nedunchezhian, N., 2001. Decline of photosynthetic pigments, ribulose-1,5-bisphosphate carboxylase and soluble protein contents, nitrate reductase and photosynthetic activities, and changes in thylakoid membrane protein pattern in canopy shade grapevine (*Vitis vinifera* L. cv. Moscato giallo) leaves. Photosynthetica 39, 529–537.

Bertamini, M., Nedunchezhian, N., 2003. Photosynthetic fuctioning of individual grapevine leaves (*Vitis vinifera* L. cv. Pinot noir) during ontogeny in the field. Vitis 42, 13–17.

Bertamini, M., Nedunchezhian, N., 2005. Grapevine growth and physiological responses to iron deficiency. J. Plant Nutr. 28, 737–749.

Bertamini, M., Muthuchelian, K., Rubinigg, M., Zorer, R., Velasco, R., Nedunchezhian, N., 2006. Low-night temperature increased the photoinhibition of photosynthesis in grapevine (*Vitis vinifera* L. cv. Riesling) leaves. Environ. Exp. Bot. 57, 25–31.

Bertoldi, D., Larcher, R., Nicolini, G., Bertamini, M., Concheri, G., 2009. Distribution of rare earth elements in *Vitis vinifera* L. 'Chardonnay' berries. Vitis 48, 49–51.

Bertoldi, D., Larcher, R., Bertamini, M., Otto, S., Concheri, G., Nicolini, G., 2011. Accumulation and distribution pattern of macro- and microelements and trace elements in *Vitis vinifera* L. cv. Chardonnay berries. J. Agric. Food Chem. 59, 7224–7236.

Bessis, R., 1972a. Etude de l'évolution des caractères morphologiques des cires cuticulaires au cours de la vie du fruit de la vigne. C.R. Hébd. Séances Acad. Sci. Sér. D, Sci. Nat. 274, 1911–1914.

Bessis, R., 1972b. Etude de l'évolution des stomates et des tissus péristomatiques du fruit de la vigne. C.R. Hébd. Séances Acad. Sci. Sér. D, Sci. Nat. 274, 2158–2161.

Bessis, R., 2007. Evolution of the grapevine (*Vitis vinifera* L.) imprinted by natural and human factors. Can. J. Bot. 85, 679–690.

Bettner, W., 1988. Nährstoffgehalte in den Organen des grünen Rebtriebs und deren Nährstoffentzüge. Mitt. Klosterneuburg 38, 130–137.

Beveridge, C.A., 2006. Axillary bud outgrowth: sending a message. Curr. Opin. Plant Biol. 9, 35–40.

Beyer, M., Knoche, M., 2002. Studies on water transport through the sweet cherry fruit surface: V. Conductance for water uptake. J. Am. Soc. Hortic. Sci. 127, 325–332.

Bindi, M., Fibbi, L., Gozzini, B., Orlandini, S., Miglietta, F., 1996. Modelling the impact of future climate scenarios on yield and yield variability of grapevine. Clim. Res. 7, 213–224.

Bindon, K., Dry, P., Loveys, B., 2008. Influence of partial rootzone drying on the composition and accumulation of anthocyanins in grape berries (*Vitis vinifera* cv. Cabernet Sauvignon). Aust. J. Grape Wine Res. 14, 91–103.

Bindon, K.A., Kennedy, J.A., 2011. Ripening-induced changes in grape skin proanthocyanidins modify their interaction with cell walls. J. Agric. Food Chem. 59, 2696–2707.

Bindon, K.A., Bacic, A., Kennedy, J.A., 2012. Tissue-specific and developmental modifications of grape cell walls influence the adsorption of proanthocyanidins. J. Agric. Food Chem. 60, 9249–9260.

Bindon, K.A., Varela, C., Kennedy, J., Holt, H., Herderich, M., 2013. Relationships between harvest time and wine composition in *Vitis vinifera* L. Cabernet Sauvignon 1. Grape and wine chemistry. Food Chem. 138, 1696–1705.

Birky Jr., C.W., 1995. Uniparental inheritance of mitochondrial and chloroplast genes: mechanisms and evolution. Proc. Natl. Acad. Sci. USA 92, 11331–11338.

Blaich, R., Bachmann, O., 1980. Die Resveratrolsynthese bei Vitaceen. Induktion und zytologische Beobachtungen. Vitis 19, 230–240.

Blaich, R., Wind, R., 1989. Inducible silica incrusts in cell walls of *Vitis* leaves. Vitis 28, 73–80.

Blaich, R., Stein, U., Wind, R., 1984. Perforationen in der Cuticula von Weinbeeren als morphologischer Faktor der Botrytisresistenz. Vitis 23, 242–256.

Blaich, R., Konradi, J., Rühl, E., Forneck, A., 2007. Assessing genetic variation among Pinot noir (*Vitis vinifera* L.) clones with AFLP markers. Am. J. Enol. Vitic. 58, 526–529.

Blanke, M.M., 1990a. Carbon economy of the grape inflorescence. 4. Light transmission into grape berries. Wein-Wiss. 45, 21–23.

Blanke, M.M., 1990b. Carbon economy of the grape inflorescence. 5. Energy requirement of the flower bud of grape. Wein-Wiss. 45, 33–35.

Blanke, M.M., Leyhe, A., 1987. Stomatal activity of the grape berry cv. Riesling, Müller-Thurgau and Ehrenfelser. J. Plant Physiol. 127, 451–460.

Blanke, M.M., Leyhe, A., 1988. Stomatal and cuticular transpiration of the cap and berry of grape. J. Plant Physiol. 132, 250–253.

Blanke, M.M., Leyhe, A., 1989a. Carbon economy of the grape inflorescence. 1. Carbon economy in flower buds of grape. Wein-Wiss. 44, 33–36.

Blanke, M.M., Leyhe, A., 1989b. Carbon economy of the grape inflorescence. 3. Chlorophyll in the grape inflorescence. Wein-Wiss. 44, 188–191.

Blanke, M.M., Pring, R.J., Baker, E.A., 1999. Structure and elemental composition of grape berry stomata. J. Plant Physiol. 154, 477–481.

Bleby, T.M., McElrone, A.J., Jackson, R.B., 2010. Water uptake and hydraulic redistribution across large woody root systems to 20 m depth. Plant Cell Environ. 33, 2132–2148.

Bloom, A.J., Chapin III, F.S., Mooney, H.A., 1985. Resource limitation in plants—an economic analogy. Annu. Rev. Ecol. Syst. 16, 363–392.

Bock, R., 2010. The give-and-take of DNA: horizontal gene transfer in plants. Trends Plant Sci. 15, 11–22.

Boerjan, W., Ralph, J., Baucher, M., 2003. Lignin biosynthesis. Annu. Rev. Plant Biol. 54, 519–546.

Bogs, J., Downey, M.O., Harvey, J.S., Ashton, A.R., Tanner, G.J., Robinson, S.P., 2005. Proanthocyanidin synthesis and expression of genes encoding leucoanthocyanidin reductase and anthocyanidin reductase in developing grape berries and grapevine leaves. Plant Physiol. 139, 652–663.

Böhlenius, H., Huang, T., Charbonnel-Campaa, L., Brunner, A.M., Jansson, S., Strauss, S.H., et al., 2006. *CO/FT* regulatory module controls timing of flowering and seasonal growth cessation in trees. Science 312, 1040–1043.

Bohnert, H.J., Nelson, D.E., Jensen, R.G., 1995. Adaptations to environmental stresses. Plant Cell 7, 1099–1111.

Bokulich, N.A., Thorngate, J.H., Richardson, P.M., Mills, D.A., 2014. Microbial biogeography of wine grapes is conditioned by cultivar, vintage, and climate. Proc. Natl. Acad. Sci. USA 111, E139–E148.

Bonanomi, G., Incerti, G., Barile, E., Capodilupo, M., Antignani, V., Mingo, A., et al., 2011. Phytotoxicity, not nitrogen immobilization, explains plant litter inhibitory effects: evidence from solid-state ^{13}C NMR spectroscopy. New Phytol. 191, 1018–1030.

Bond, B.J., Czarnomski, N.M., Cooper, C., Day, M.E., Greenwood, M.S., 2007. Developmental decline in height growth in Douglas-fir. Tree Physiol. 27, 441–453.

Bondada, B., 2011. Micromorpho-anatomical examination of 2,4-D phytotoxicity in grapevine (*Vitis vinifera* L.) leaves. J. Plant Growth Regul. 30, 185–198.

Bondada, B.R., 2012. An array of simple, fast, and safe approaches to visualizing fine cellular structures in free-hand sections of stem, leaf, and fruit using optical microscopy. Curr. Bot. 3, 11–22.

Bondada, B.R., Keller, M., 2012. Not all shrivels are created equal—morpho-anatomical and compositional characteristics differ among different shrivel types that develop during ripening of grape (*Vitis vinifera* L.) berries. Am. J. Plant Sci. 3, 879–898.

Bondada, B.R., Matthews, M.A., Shackel, K.A., 2005. Functional xylem in the post-veraison grape berry. J. Exp. Bot. 56, 2949–2957.

Bongers, A.J., Hinsch, R.T., Bus, V.G., 1991. Physical and chemical characteristics of raisins from several countries. Am. J. Enol. Vitic. 42, 76–78.

Bönisch, F., Frotscher, J., Stanitzek, S., Rühl, E., Wüst, M., Bitz, O., et al., 2014. A UDP-glucose:monoterpenol glucosyltransferase adds to the chemical diversity of the grapevine metabolome. Plant Physiol. 165, 461–581.

Boonman, A., Prinsen, E., Gilmer, F., Schurr, U., Peeters, A.J.M., Voesenek, L.A.C.J., et al., 2007. Cytokinin import rate as a signal for photosynthetic acclimation to canopy light gradients. Plant Physiol. 143, 1841–1852.

Boonman, A., Prinsen, E., Voesenek, L.A.C.J., Pons, T.L., 2009. Redundant roles of photoreceptors and cytokinins in regulating photosynthetic acclimation to canopy density. J. Exp. Bot. 60, 1179–1190.

Boso, S., Kassemeyer, H.H., 2008. Different susceptibility of European grapevine cultivars for downy mildew. Vitis 47, 33–49.

Boso, S., Alonso-Villaverde, V., Gago, P., Santiago, J.L., Martínez, M.C., 2011. Susceptibility of 44 grapevine (*Vitis vinifera* L.) varieties to downy mildew in the field. Aust. J. Grape Wine Res. 17, 394–400.

Boss, P.K., Davies, C., 2009. Molecular biology of anthocyanin accumulation in grape berries. In: Roubelakis-Angelakis, K.A. (Ed.), Grapevine Molecular Physiology and Biotechnology, second ed. Springer, Dordrecht, The Netherlands, pp. 263–292.

Boss, P.K., Thomas, M.R., 2002. Association of dwarfism and floral induction with a grape 'green revolution' mutation. Nature 416, 847–850.
Boss, P.K., Davies, C., Robinson, S.P., 1996. Analysis of the expression of anthocyanin pathway genes in developing *Vitis vinifera* L. cv Shiraz grape berries and the implications for pathway regulation. Plant Physiol. 111, 1059–1066.
Boss, P.K., Buckeridge, E.J., Poole, A., Thomas, M.R., 2003. New insights into grapevine flowering. Funct. Plant Biol. 30, 593–606.
Bota, J., Medrano, H., Flexas, J., 2004. Is photosynthesis limited by decreased Rubisco activity and RuBP content under progressive water stress? New Phytol. 162, 671–681.
Böttcher, C., Boss, P.K., Davies, C., 2013. Increase in cytokinin levels during ripening in developing *Vitis vinifera* cv. Shiraz berries. Am. J. Enol. Vitic. 64, 527–531.
Boudaoud, A., 2010. An introduction to the mechanics of morphogenesis for plant biologists. Trends Plant Sci. 15, 353–360.
Boulton, R., 1980a. The general relationship between potassium, sodium and pH in grape juice and wine. Am. J. Enol. Vitic. 31, 182–186.
Boulton, R., 1980b. The relationships between total acidity, titratable acidity and pH in grape tissues. Vitis 19, 113–120.
Boulton, R., 2001. The copigmentation of anthocyanins and its role in the color of red wine: a critical review. Am. J. Enol. Vitic. 52, 67–87.
Boursiquot, J.M., Dessup, M., Rennes, C., 1995. Distribution des principaux caractères phénologiques, agronomiques et technologiques chez *Vitis vinifera* L. Vitis 34, 31–35.
Boursiquot, J.M., Lacombe, T., Laucou, V., Julliard, S., Perrin, F.X., Lanier, N., et al., 2009. Parentage of Merlot and related winegrape cultivars of southwestern France: discovery of the missing link. Aust. J. Grape Wine Res. 15, 144–155.
Bovey, R., Gärtel, W., Hewitt, W.B., Martelli, G.P., Vuittenez, A., 1980. Virus and Virus-like Diseases of Grapevines. Colour Atlas of Symptoms. Editions Payot, Lausanne, Switzerland.
Bowen, P., Bogdanoff, C., Usher, K., Estergaard, B., Watson, M., 2011. Effects of irrigation and crop load on leaf gas exchange and fruit composition in red winegrapes grown on a loamy sand. Am. J. Enol. Vitic. 62, 9–22.
Bowen, P., Bogdanoff, C., Estergaard, B., 2012. Effects of converting from sprinkler to drip irrigation on water conservation and the performance of Merlot grown on a loamy sand. Am. J. Enol. Vitic. 63, 385–393.
Bowen, P.A., Kliewer, W.M., 1990. Relationships between the yield and vegetative characteristics of individual shoots of 'Cabernet Sauvignon' grapevines. J. Am. Soc. Hortic. Sci. 115, 534–539.
Bowen, P.A., Bogdanoff, C.P., Estergaard, B.F., Marsh, S.G., Usher, K.B., Smith, C.A.S., et al., 2005. Geology and wine 10: use of geographic information system technology to assess viticulture performance in the Okanagan and Similkameen valleys, British Columbia. Geosci. Can. 32, 161–176.
Bowers, J., Boursiquot, J.M., This, P., Chu, K., Johansson, H., Meredith, C., 1999. Historical genetics: the parentage of Chardonnay, Gamay, and other wine grapes of northeastern France. Science 285, 1562–1565.
Bowers, J.E., Meredith, C.P., 1997. The parentage of a classic wine grape, Cabernet Sauvignon. Nat. Genet. 16, 84–87.
Bowles, D., Lim, E.K., Poppenberger, B., Vaistij, F.E., 2006. Glycosyltransferases of lipophilic small molecules. Annu. Rev. Plant Biol. 57, 567–597.
Boyer, J.S., 1969. Measurement of the water status of plants. Annu. Rev. Plant Physiol. 20, 351–364.
Boyer, J.S., 1985. Water transport. Annu. Rev. Plant Physiol. 36, 473–516.
Boyer, J.S., 2009. Cell wall biosynthesis and the molecular mechanism of plant enlargement. Funct. Plant Biol. 36, 383–394.
Boyer, J.S., Silk, W.K., 2004. Hydraulics of plant growth. Funct. Plant Biol. 31, 761–773.

Boyer, J.S., Wong, S.C., Farquhar, G.D., 1997. CO_2 and water vapor exchange across leaf cuticle (epidermis) at various water potentials. Plant Physiol. 114, 185–191.

Boyer, J.S., Silk, W.K., Watt, M., 2010. Path of water for root growth. Funct. Plant Biol. 37, 1105–1116.

Braam, J., 2005. In touch: plant responses to mechanical stimuli. New Phytol. 165, 373–389.

Bradford, K.J., 1994. Water stress and the water relations of seed development: a critical review. Crop Sci. 34, 1–11.

Braidot, E., Zancani, M., Petrussa, E., Peresson, C., Bertolini, A., Patui, S., et al., 2008. Transport and accumulation of flavonoids in grapevine (*Vitis vinifera* L.). Plant Signal. Behav. 3, 626–632.

Braidwood, L., Breuer, C., Sugimoto, K., 2014. My body is a cage: mechanisms and modulation of plant cell growth. New Phytol. 201, 388–402.

Brancadoro, L., Rabotti, G., Scienza, A., Zocchi, G., 1995. Mechanisms of Fe-efficiency in roots of *Vitis* spp. in response to iron deficiency stress. Plant Soil 171, 229–234.

Brantjes, N.B.M., 1978. Pollinator attraction of *Vitis vinifera* subsp. *silvestris*. Vitis 17, 229–233.

Bravdo, B., Hepner, Y., Loinger, C., Cohen, S., Tabacman, H., 1984. Effect of crop level on growth, yield and wine quality of a high yielding Carignane vineyard. Am. J. Enol. Vitic. 35, 247–252.

Bravdo, B., Hepner, Y., Loinger, C., Cohen, S., Tabacman, H., 1985a. Effect of crop level and crop load on growth, yield, must and wine composition, and quality of Cabernet Sauvignon. Am. J. Enol. Vitic. 36, 125–131.

Bravdo, B., Hepner, Y., Loinger, C., Cohen, S., Tabacman, H., 1985b. Effect of irrigation and crop level on growth, yield and wine quality of Cabernet Sauvignon. Am. J. Enol. Vitic. 36, 132–139.

Bray, E.A., 1997. Plant responses to water deficit. Trends Plant Sci. 2, 48–54.

Brem, S., Rast, D.M., Ruffner, H.P., 1986. Partitioning of photosynthate in leaves of *Vitis vinifera* infected with *Uncinula necator* or *Plasmopara viticola*. Physiol. Mol. Plant Pathol. 29, 285–291.

Breuil, A.C., Jeandet, P., Adrian, M., Chopin, F., Pirio, N., Meunier, P., et al., 1999. Characterization of a pterostilbene dehydrodimer produced by laccase of *Botrytis cinerea*. Phytopathology 89, 298–302.

Briat, J.F., Lobréaux, S., 1997. Iron transport and storage in plants. Trends Plant Sci. 2, 187–193.

Brickell, C.D., Alexander, C., David, J.C., Hetterscheid, W.L.A., Leslie, A.C., Malecot, V., et al., 2009. International Code of Nomenclature for Cultivated Plants, eighth ed. Scripta Horticulturae, vol. 10, pp. 184.

Brillouet, J.M., Romieu, C., Schoefs, B., Solymosi, K., Cheynier, V., Fulcrand, H., et al., 2013. The tannosome is an organelle forming condensed tannins in the chlorophyllous organs of Tracheophyta. Ann. Bot. 112, 1003–1014.

Britto, D.T., Kronzucker, H.J., 2006. Futile cycling at the plasma membrane: a hallmark of low-affinity nutrient transport. Trends Plant Sci. 11, 529–534.

Broadley, M.R., White, P.J., Hammond, J.P., Zelko, I., Lux, A., 2007. Zinc in plants. New Phytol. 173, 677–702.

Brodersen, C.R., McElrone, A.J., Choat, B., Matthews, M.A., Shackel, K.A., 2010. The dynamics of embolism repair in xylem: *in vivo* visualizations using high-resolution computed tomography. Plant Physiol. 154, 1088–1095.

Brodersen, C.R., Lee, E.F., Choat, B., Jansen, S., Phillips, R.J., Shackel, K.A., et al., 2011. Automated analysis of three-dimensional xylem networks using high-resolution computed tomography. New Phytol. 191, 1168–1179.

Brodersen, C.R., McElrone, A.J., Choat, B., Lee, E.F., Shackel, K.A., Matthews, M.A., 2013. *In vivo* visualization of drought-induced embolism spread in *Vitis vinifera*. Plant Physiol. 161, 1820–1829.

Brodribb, T.J., Holbrook, N.M., 2003. Stomatal closure during leaf dehydration, correlation with other leaf physiological traits. Plant Physiol. 132, 2166–2173.

Brodribb, T.J., Field, T.S., Jordan, G.J., 2007. Leaf maximum photosynthetic rate and venation are linked by hydraulics. Plant Physiol. 144, 1890–1898.

Brossaud, F., Cheynier, V., Noble, A.C., 2001. Bitterness and astringency of grape and wine polyphenols. Aust. J. Grape Wine Res. 7, 33–39.
Browse, J., Xin, Z., 2001. Temperature sensing and cold acclimation. Curr. Opin. Plant Biol. 4, 241–246.
Brummell, D.A., 2006. Cell wall disassembly in ripening fruit. Funct. Plant Biol. 33, 103–119.
Bucci, S.J., Scholz, F.G., Goldstein, G., Meinzer, F.C., Franco, A.C., Campanello, P.I., et al., 2006. Nutrient availability constrains the hydraulic architecture and water relations of savannah trees. Plant Cell Environ. 29, 2153–2167.
Buchanan, B.B., Gruissem, W., Jones, R.L., 2000. Biochemistry and Molecular Biology of Plants. ASPP, Rockville, MD.
Buchanan-Wollaston, V., 1997. The molecular biology of leaf senescence. J. Exp. Bot. 48, 181–199.
Bucher, M., 2007. Functional biology of plant phosphate uptake at root and mycorrhiza interfaces. New Phytol. 173, 11–26.
Buer, C.S., Muday, G.K., Djordjevic, M.A., 2007. Flavonoids are differentially taken up and transported long distances in *Arabidopsis*. Plant Physiol. 145, 478–490.
Bulit, J., Dubos, B., 1982. Epidémiologie de la pourriture grise. Bull. OEPP 12, 37–48.
Burch-Smith, T.M., Zambryski, P.C., 2012. Plasmodesmata paradigm shift: regulation from without versus within. Annu. Rev. Plant Biol. 63, 239–260.
Burkhead, J.L., Gogolin Reynolds, K.A., Abdel-Ghany, S.E., Cohu, C.M., Pilon, M., 2009. Copper homeostasis. New Phytol. 182, 799–816.
Burr, T.J., Otten, L., 1999. Crown gall of grape: biology and disease management. Annu. Rev. Phytopathol. 37, 53–80.
Burr, T.J., Katz, B.H., Bishop, A.L., Meyers, C.A., Mittak, V.L., 1988. Effect of shoot age and tip culture propagation of grapes on systemic infestations by *Agrobacterium tumefaciens* Biovar 3. Am. J. Enol. Vitic. 39, 67–70.
Burr, T.J., Bazzi, C., Süle, S., Otten, L., 1998. Crown gall of grape: biology of *Agrobacterium vitis* and the development of disease control strategies. Plant Dis. 82, 1288–1297.
Busam, G., Kassemeyer, H.H., Matern, U., 1997. Differential expression of chitinases in *Vitis vinifera* L. responding to systemic acquired resistance activators or fungal challenge. Plant Physiol. 115, 1029–1038.
Buttrose, M.S., 1966a. The effect of reducing leaf area on the growth of roots, stems and berries of Gordo grapevines. Vitis 5, 455–464.
Buttrose, M.S., 1966b. Use of carbohydrate reserves during growth from cuttings of grape vine. Aust. J. Biol. Sci. 19, 247–256.
Buttrose, M.S., 1969a. Fruitfulness in grapevines: effects of light intensity and temperature. Bot. Gaz. 130, 166–173.
Buttrose, M.S., 1969b. The dissolution and reaccumulation of starch granules in grape vine cane. Aust. J. Biol. Sci. 22, 1297–1303.
Buttrose, M.S., 1969c. Vegetative growth of grape-vine varieties under controlled temperature and light intensity. Vitis 8, 280–285.
Buttrose, M.S., 1970a. Fruitfulness in grapevines: development of leaf primordia in buds in relation to bud fruitfulness. Bot. Gaz. 131, 78–83.
Buttrose, M.S., 1970b. Fruitfulness in grape-vines: the response of different cultivars to light, temperature and daylength. Vitis 9, 121–125.
Buttrose, M.S., 1974a. Climatic factors and fruitfulness in grapevines. Hortic. Abstr. 44, 319–325.
Buttrose, M.S., 1974b. Fruitfulness in grape-vines: effect of water stress. Vitis 12, 299–305.
Buttrose, M.S., Hale, C.R., 1973. Effect of temperature on development of the grapevine inflorescence after bud burst. Am. J. Enol. Vitic. 24, 4–16.
Buttrose, M.S., Mullins, M.G., 1968. Proportional reduction in shoot growth of grapevine with root systems maintained at constant relative volumes by repeated pruning. Aust. J. Biol. Sci. 21, 1095–1101.

Buttrose, M.S., Hale, C.R., Kliewer, W.M., 1971. Effect of temperature on the composition of 'Cabernet Sauvignon' berries. Am. J. Enol. Vitic. 22, 71–75.
Cabaleiro, C., Segura, A., García-Berrios, J.J., 1999. Effects of grapevine leafroll-associated virus 3 on the physiology and must of *Vitis vinifera* L. cv. Albariño following contamination in the field. Am. J. Enol. Vitic. 50, 40–44.
Cabezas, J.A., Cervera, M.T., Arroyo-García, R., Ibáñez, J., Rodríguez-Torres, I., Borrego, J., et al., 2003. Garnacha and Garnacha Tintorera: genetic relationships and the origin of teinturier varieties cultivated in Spain. Am. J. Enol. Vitic. 54, 237–245.
Cabezas, J.A., Cervera, M.T., Ruiz-García, L., Carreño, J., Martínez-Zapater, J.M., 2006. A genetic analysis of seed and berry weight in grapevine. Genome 49, 1574–1585.
Cacho, J., Fernández, P., Ferreira, V., Castells, J.E., 1992. Evolution of five anthocyanidin-3-glucosides in the skin of the Tempranillo, Moristel, and Garnacha grape varieties and influence of climatological variables. Am. J. Enol. Vitic. 43, 244–248.
Cadle-Davidson, L., 2008. Monitoring pathogenesis of natural *Botrytis cinerea* infections in developing grape berries. Am. J. Enol. Vitic. 59, 387–395.
Cadle-Davidson, M.M., Owens, C.L., 2008. Genomic amplification of the *Gret1* retroelement in white-fruited accessions of wild *Vitis* and interspecific hybrids. Theor. Appl. Genet. 116, 1079–1094.
Cadot, Y., Miñana-Castelló, M.T., Chevalier, M., 2006. Anatomical, histological, and histochemical changes in grape seeds from *Vitis vinifera* L. cv Cabernet franc during fruit development. J. Agric. Food Chem. 54, 9206–9215.
Cain, D.W., Emershad, R.L., Tarailo, R.E., 1983. In-ovulo embryo culture and seedling development of seeded and seedless grapes (*Vitis vinifera* L.). Vitis 22, 9–14.
Cakmak, I., Kirkby, E.A., 2008. Role of magnesium in carbon partitioning and alleviating photooxidative damage. Physiol. Plant. 133, 692–704.
Cakmak, I., Marschner, H., 1986. Mechanism of phosphorus-induced zinc deficiency in cotton: I. Zinc deficiency-enhanced uptake rate of phosphorus. Physiol. Plant. 68, 483–490.
Cakmak, I., Hengeler, C., Marschner, H., 1994. Changes in phloem export of sucrose in leaves in response to phosphorus, potassium and magnesium deficiency in bean plants. J. Exp. Bot. 45, 1251–1257.
Calderón, A.A., Zapata, J.M., Ros Barceló, A., 1994. Differential expression of a cell wall-localized peroxidase isoenzyme capable of oxidizing 4-hydroxystilbenes during the cell culture of grapevine (*Vitis vinifera* cv. Airen and Monastrell). Plant Cell Tissue. Organ Cult. 37, 121–127.
Calderon-Orellana, A., Matthews, M.A., Drayton, W.M., Shackel, K.A., 2014. Uniformity of ripeness and size in Cabernet Sauvignon berries from vineyards with contrasting crop price. Am. J. Enol. Vitic. 65, 81–88.
Caldwell, M.M., Björn, L.O., Bornman, J.F., Flint, S.D., Kulandaivelu, G., Teramura, A.H., et al., 1998. Effects of increased solar ultraviolet radiation on terrestrial ecosystems. J. Photochem. Photobiol. B 46, 40–52.
Callis, J., 1995. Regulation of protein degradation. Plant Cell 7, 845–857.
Calonje, M., Cubas, P., Martínez-Zapater, J.M., Carmona, M.J., 2004. Floral meristem identity genes are expressed during tendril development in grapevine. Plant Physiol. 135, 1491–1501.
Camacho-Cristóbal, J.J., González-Fontes, A., 1999. Boron deficiency causes a drastic decrease in nitrate content and nitrate reductase activity, and increases the content of carbohydrates in leaves from tobacco plants. Planta 209, 528–536.
Campbell-Clause, J.M., 1998. Stomatal response of grapevines to wind. Aust. J. Exp. Agric. 38, 77–82.
Candolfi-Vasconcelos, M.C., Koblet, W., 1990. Yield, fruit quality, bud fertility and starch reserves of the wood as a function of leaf removal in *Vitis vinifera*—evidence of compensation and stress recovering. Vitis 29, 199–221.

Candolfi-Vasconcelos, M.C., Koblet, W., 1991. Influence of partial defoliation on gas exchange parameters and chlorophyll content of field grown grapevines—mechanisms and limitations of the compensation capacity. Vitis 30, 129–141.

Candolfi-Vasconcelos, M.C., Candolfi, M.P., Koblet, W., 1994. Retranslocation of carbon reserves from the woody storage tissues into the fruit as a response to defoliation stress during the ripening period in *Vitis vinifera* L. Planta 192, 567–573.

Canny, M.J., 1993. The transpiration stream in the leaf apoplast: water and solutes. Philos. Trans. Biol. Sci. 341, 87–100.

Cantos, E., Espín, J.C., Tomás-Barberán, F.A., 2002. Varietal differences among the polyphenol profiles of seven table grape cultivars studied by LC-DAD-MS-MS. J. Agric. Food Chem. 50, 5691–5696.

Cantu, D., Vicente, A.R., Labavitch, J.M., Bennett, A.B., Powell, A.L.T., 2008. Strangers in the matrix: plant cell walls and pathogen susceptibility. Trends Pant Sci. 13, 610–617.

Caporali, E., Spada, A., Marziani, G., Failla, O., Scienza, A., 2003. The arrest of development of abortive reproductive organs in the unisexual flower of *Vitis vinifera* ssp. *silvestris*. Sex. Plant Reprod. 15, 291–300.

Capps, E.R., Wolf, T.K., 2000. Reduction of bunch stem necrosis of Cabernet Sauvignon by increased tissue nitrogen concentration. Am. J. Enol. Vitic. 51, 319–328.

Caputi, L., Carlin, S., Ghiglieno, I., Stefanini, M., Valenti, L., Vrhovsek, U., et al., 2011. Relationship of changes in rotundone content during grape ripening and winemaking to manipulation of the 'peppery' character of wine. J. Agric. Food Chem. 59, 5565–5571.

Carbonneau, A., de Loth, C., 1985. Influence du régime d'éclairement journalier sur la résistance stomatique et la photosynthèse brute chez *Vitis vinifera* L. cv. «Cabernet-Sauvignon». Agronomie 5, 631–638.

Carmona, M.J., Cubas, P., Martínez-Zapater, J.M., 2002. VFL, the grapevine *FLORICAULA/LEAFY* ortholog, is expressed in meristematic regions independently of their fate. Plant Physiol. 130, 68–77.

Carmona, M.J., Cubas, P., Calonje, M., Martínez-Zapater, J.M., 2007. Flowering transition in grapevine (*Vitis vinifera* L.). Can. J. Bot. 85, 701–711.

Carmona, M.J., Chaïb, J., Martínez-Zapater, J.M., Thomas, M.R., 2008. A molecular genetic perspective of reproductive development in grapevine. J. Exp. Bot. 59, 2579–2596.

Carpaneto, A., Geiger, D., Bamberg, E., Sauer, N., Fromm, J., Hedrich, R., 2005. Phloem-localized, proton-coupled sucrose carrier ZmSUT1 mediates sucrose efflux under the control of the sucrose gradient and the proton motive force. J. Biol. Chem. 280, 21437–21443.

Carpita, N.C., Gibeaut, D.M., 1993. Structural models of primary cell walls in flowering plants: consistency of molecular structure with the physical properties of the walls during growth. Plant J. 3, 1–30.

Carreño, J., Almela, L., Martínez, A., Fernández-López, J.A., 1997. Chemotaxonomical classification of red table grapes based on anthocyanin profile and external colour. Lebensm.-Wiss. Technol. 30, 259–265.

Carrier, G., Le Cunff, L., Dereeper, A., Legrand, D., Sabot, F., Bouchez, O., et al., 2012. Transposable elements are a major cause of somatic polymorphism in *Vitis vinifera* L. PLoS One 7, e32973, http://dx.doi.org/10.1371/journal.pone.0032973.

Carroll, D.E., Marcy, J.E., 1982. Chemical and physical changes during maturation of Muscadine grapes (*Vitis rotundifolia*). Am. J. Enol. Vitic. 33, 168–172.

Carroll, D.E., Hoover, M.W., Nesbitt, W.B., 1971. Sugar and organic acid concentrations in cultivars of muscadine grapes. J. Am. Soc. Hortic. Sci. 96, 737–740.

Carroll, J.E., Wilcox, W.F., 2003. Effects of humidity on the development of grapevine powdery mildew. Phytopathology 93, 1137–1144.

Cartechini, A., Palliotti, A., 1995. Effect of shading on vine morphology and productivity and leaf gas exchange characteristics in grapevines in the field. Am. J. Enol. Vitic. 46, 227–234.

Casado, C.G., Heredia, A., 2001. Ultrastructure of the cuticle during growth of the grape berry (*Vitis vinifera*). Physiol. Plant. 111, 220–224.

Casassa, L.F., Larsen, R.C., Beaver, C.W., Mireles, M.S., Keller, M., Riley, W.R., et al., 2013. Impact of extended maceration and regulated deficit irrigation (RDI) in Cabernet Sauvignon wines: characterization of proanthocyanidin distribution, anthocyanin extraction, and chromatic properties. J. Agric. Food Chem. 61, 6446–6457.

Caspari, H.W., Lang, A., Alspach, P., 1998. Effects of girdling and leaf removal on fruit set and vegetative growth in grape. Am. J. Enol. Vitic. 49, 359–366.

Castelan-Estrada, M., Vivin, P., Gaudillère, J.P., 2002. Allometric relationships to estimate seasonal above-ground vegetative and reproductive biomass in *Vitis vinifera* L. Ann. Bot. 89, 401–408.

Castellarin, S.D., Di Gaspero, G., 2007. Transcriptional control of anthocyanin biosynthetic genes in extreme phenotypes for berry pigmentation of naturally occurring grapevines. BMC Plant Biol. 7, 46, http://dx.doi.org/10.1186/1471-2229-7-46.

Castellarin, S.D., Matthews, M.A., Di Gaspero, G., Gambetta, G.A., 2007a. Water deficits accelerate ripening and induce changes in gene expression regulating flavonoid biosynthesis in grape berries. Planta 227, 101–112.

Castellarin, S.D., Pfeiffer, A., Sivilotti, P., Degan, M., Peterlunger, E., Di Gaspero, G., 2007b. Transcriptional regulation of anthocyanin biosynthesis in ripening fruits of grapevine under seasonal water deficit. Plant Cell Environ. 30, 1381–1399.

Castellarin, S.D., Gambetta, G.A., Wada, H., Shackel, K.A., Matthews, M.A., 2011. Fruit ripening in *Vitis vinifera*: spatiotemporal relationships among turgor, sugar accumulation, and anthocyanin biosynthesis. J. Exp. Bot. 62, 4345–4354.

Cawthon, D.L., Morris, J.R., 1982. Relationship of seed number and maturity to berry development, fruit maturation, hormonal changes, and uneven ripening of 'Concord' (*Vitis labrusca* L.) grapes. J. Am. Soc. Hortic. Sci. 107, 1097–1104.

Cazzonelli, C.I., Pogson, B.J., 2010. Source to sink: regulation of carotenoid biosynthesis in plants. Trends Plant Sci. 15, 266–274.

Celette, F., Gaudin, R., Gary, C., 2008. Spatial and temporal changes to the water regime of a Mediterranean vineyard due to the adoption of cover cropping. Eur. J. Agron. 29, 153–162.

Celette, F., Findeling, A., Gary, C., 2009. Competition for nitrogen in an unfertilized intercropping system: the case of an association of grapevine and grass cover in a Mediterranean climate. Eur. J. Agron. 30, 41–51.

Cerpa-Calderón, F.K., Kennedy, J.A., 2008. Berry integrity and extraction of skin and seed proanthocyanidins during red wine fermentation. J. Agric. Food Chem. 56, 9006–9014.

Chacko, E.K., Saidha, T., Swamy, R.D., Reddy, Y.N., Kohli, R.R., 1976. Studies on the cytokinins in fruits. I. Occurrence and levels of cytokinin-like substances in grape berries at different developmental stages. Vitis 15, 221–226.

Chalker-Scott, L., 1999. Environmental significance of anthocyanins in plant stress responses. Photochem. Photobiol. 70, 1–9.

Chandler, J.W., 2011. Founder cell specification. Trends Plant Sci. 16, 607–613.

Chapin III, F.S., 1991. Integrated response of plants to stress. BioScience 41, 29–36.

Chapman, D.M., Matthews, M.A., Guinard, J.X., 2004a. Sensory attributes of Cabernet Sauvignon wines made from vines with different crop yields. Am. J. Enol. Vitic. 55, 325–334.

Chapman, D.M., Thorngate, J.H., Matthews, M.A., Guinard, J.X., Ebeler, S.E., 2004b. Yield effects on 2-methoxy-3-isobutylpyrazine concentration in Cabernet Sauvignon using a solid phase microextraction gas chromatography/mass spectrometry method. J. Agric. Food Chem. 52, 5431–5435.

Chapman, N., Miller, A.J., Lindsey, K., Whalley, W.R., 2012. Roots, water, and nutrient acquisition: let's get physical. Trends Plant Sci. 17, 701–710.

Chatelet, D.S., Matthews, M.A., Rost, T.L., 2006. Xylem structure and connectivity in grapevine (*Vitis vinifera*) shoots provides a passive mechanism for the spread of bacteria in grape plants. Ann. Bot. 98, 483–494.

Chatelet, D.S., Rost, T.L., Matthews, M.A., Shackel, K.A., 2008a. The peripheral xylem of grapevine (*Vitis vinifera*). 2. Anatomy and development. J. Exp. Bot. 59, 1997–2007.

Chatelet, D.S., Rost, T.L., Shackel, K.A., Matthews, M.A., 2008b. The peripheral xylem of grapevine (*Vitis vinifera*). 1. Structural integrity in post-veraison berries. J. Exp. Bot. 59, 1987–1996.

Chater, C.C.C., Oliver, J., Casson, S., Gray, J.E., 2014. Putting the brakes on: abscisic acid as a central environmental regulator of stomatal development. New Phytol. 202, 376–391.

Chatonnet, P., Dubourdieu, D., 1998. Comparative study of the characteristics of American white oak (*Quercus alba*) and European oak (*Quercus petraea* and *Q. robur*) for production of barrels used in barrel aging of wines. Am. J. Enol. Vitic. 49, 79–85.

Chatonnet, P., Dubourdieu, D., Boidron, J.N., Pons, M., 1992. The origin of ethylphenols in wines. J. Sci. Food Agric. 60, 165–178.

Chaves, M.M., Harley, P.C., Tehunen, J.D., Lange, O.L., 1987. Gas exchange studies in two Portuguese grapevine cultivars. Physiol. Plant. 70, 639–647.

Chaves, M.M., Zarrouk, O., Francisco, R., Costa, J.M., Santos, T., Regalado, A.P., et al., 2010. Grapevine under deficit irrigation: hints from physiological and molecular data. Ann. Bot. 105, 661–676.

Chen, L.Q., Qu, X.Q., Hou, B.H., Sosso, D., Osorio, S., Fernie, A.R., et al., 2012. Sucrose efflux mediated by SWEET proteins as a key step for phloem transport. Science 335, 207–211.

Chen, L.S., Cheng, L., 2003a. Both xanthophyll cycle-dependent thermal dissipation and the antioxidant system are up-regulated in grape (*Vitis labrusca* L. cv. Concord) leaves in response to N limitation. J. Exp. Bot. 54, 2165–2175.

Chen, L.S., Cheng, L., 2003b. Carbon assimilation and carbohydrate metabolism of 'Concord' grape (*Vitis labrusca* L.) leaves in response to nitrogen supply. J. Am. Soc. Hortic. Sci. 128, 754–760.

Cheng, H., Qin, L., Lee, S., Fu, X., Richards, D.E., Cao, D., et al., 2004a. Gibberellin regulates *Arabidopsis* floral development via suppression of DELLA protein function. Development 131, 1055–1064.

Cheng, L., Xia, G., Bates, T., 2004b. Growth and fruiting of young 'Concord' grapevines in relation to reserve nitrogen and carbohydrates. J. Am. Soc. Hortic. Sci. 129, 660–666.

Chervin, C., El-Kereamy, A., Roustan, J.P., Latché, A., Lamon, J., Bouzayen, M., 2004. Ethylene seems required for the berry development and ripening in grape, a non-climacteric fruit. Plant Sci. 167, 1301–1305.

Chervin, C., Tira-Umphon, A., Terrier, N., Zouine, M., Severac, D., Roustan, J.P., 2008. Stimulation of the grape berry expansion by ethylene and effects on related gene transcripts, over the ripening phase. Physiol. Plant. 134, 534–546.

Cheung, A.Y., Boavida, L.C., Aggarwal, M., Wu, H.M., Feijó, J.A., 2010. The pollen tube journey in the pistil and imaging the *in vivo* process by two-photon microscopy. J. Exp. Bot. 61, 1907–1915.

Chevalier, É., Loubert-Hudon, A., Zimmerman, E.L., Matton, D.P., 2011. Cell–cell communication and signalling pathways within the ovule: from its inception to fertilization. New Phytol. 192, 13–28.

Cheynier, V., Rigaud, J., 1986. HPLC separation and characterization of flavonols in the skins of *Vitis vinifera* var. Cinsault. Am. J. Enol. Vitic. 37, 248–252.

Cheynier, V., Dueñas-Paton, M., Salas, E., Maury, C., Souquet, J.M., Sarni-Manchado, P., et al., 2006. Structure and properties of wine pigments and tannins. Am. J. Enol. Vitic. 57, 298–305.

Chiera, J., Thomas, J., Rufty, T., 2002. Leaf initiation and development in soybean under phosphorus stress. J. Exp. Bot. 53, 473–481.

Chinnusamy, V., Zhu, J., Zhu, J.K., 2007. Cold stress regulation of gene expression in plants. Trends Plant Sci. 12, 444–451.

Chiou, T.J., Lin, S.I., 2011. Signaling network in sensing phosphate availability in plants. Annu. Rev. Plant Biol. 62, 185–206.

Chitwood, D.H., Ranjan, A., Martinez, C.C., Headland, L.R., Thiem, T., Kumar, R., et al., 2014. A modern ampelography: a genetic basis for leaf shape and venation patterning in grape. Plant Physiol. 164, 259–272.

Choat, B., Lahr, E.C., Melcher, P.J., Zwieniecki, M.A., Holbrook, N.M., 2005. The spatial pattern of air seeding thresholds in mature sugar maple trees. Plant Cell Environ. 28, 1082–1089.
Choat, B., Cobb, A.R., Jansen, S., 2008. Structure and function of bordered pits: new discoveries and impacts on whole-plant hydraulic function. New Phytol. 177, 608–626.
Choat, B., Drayton, W.M., Brodersen, C., Matthews, M.A., Shackel, K.A., Wada, H., et al., 2010. Measurement of vulnerability to water stress-induced cavitation in grapevine: a comparison of four techniques applied to a long-vesseled species. Plant Cell Environ. 33, 1502–1512.
Choné, X., Lavigne-Cruège, V., Tominaga, T., Van Leeuwen, C., Castagnède, C., Saucier, C., et al., 2006. Effect of vine nitrogen status on grape aromatic potential: flavor precursors (*S*-cysteine conjugates), glutathione and phenolic content in *Vitis vinifera* L. cv. Sauvignon blanc grape juice. J. Int. Sci. Vigne Vin 40, 1–6.
Chory, J., 1997. Light modulation of vegetative development. Plant Cell 9, 1225–1234.
Christensen, L.P., Boggero, J.D., 1985. A study of mineral nutrition relationships of waterberry in Thompson Seedless. Am. J. Enol. Vitic. 36, 57–64.
Christensen, L.P., Bianchi, M.L., Peacock, W.L., Hirschfelt, D.J., 1994. Effect of nitrogen fertilizer timing and rate on inorganic nitrogen status, fruit composition, and yield of grapevines. Am. J. Enol. Vitic. 45, 377–387.
Christensen, L.P., Bianchi, M.L., Miller, M.W., Kasimatis, A.N., Lynn, C.D., 1995. The effects of harvest date on Thompson Seedless grapes and raisins. II. Relationships of fruit quality factors. Am. J. Enol. Vitic. 46, 493–498.
Christensen, N.M., Nicolaisen, M., Hansen, M., Schulz, A., 2004. Distribution of phytoplasmas in infected plants as revealed by real time PCR and bioimaging. Mol. Plant Microbe. Interact. 17, 1175–1184.
Christensen, P., 1975. Response of 'Thompson Seedless' grapevines to the timing of preharvest irrigation cut-off. Am. J. Enol. Vitic. 26, 188–194.
Christensen, P., 1984. Nutrient level comparisons of leaf petioles and blades in twenty-six grape cultivars over three years (1979 through 1981). Am. J. Enol. Vitic. 35, 124–133.
Christmann, A., Weiler, E.W., Steudle, E., Grill, E., 2007. A hydraulic signal in root-to-shoot signalling of water shortage. Plant J. 52, 167–174.
Chuine, I., Yiou, P., Viovy, N., Seguin, B., Daux, V., Le Roy Ladurie, E., 2004. Grape ripening as a past climate indicator. Nature 432, 289–290.
Cipriani, G., Spadotto, A., Jurman, I., Di Gaspero, G., Crespan, M., Meneghetti, S., et al., 2010. The SSR-based molecular profile of 1005 grapevine (*Vitis vinifera* L.) accessions uncovers new synonymy and parentages, and reveals a large admixture amongst varieties of different geographic origin. Theor. Appl. Genet. 121, 1569–1585.
Claeys, H., De Bodt, S., Inzé, D., 2014. Gibberellins and DELLAs: central nodes in growth regulatory networks. Trends Plant Sci. 19, 231–239.
Clark, S.E., 1997. Organ formation at the vegetative shoot meristem. Plant Cell 9, 1067–1076.
Clarke, S.J., Hardie, W.J., Rogiers, S.Y., 2010. Changes in susceptibility of grape berries to splitting are related to impaired osmotic water uptake associated with losses in cell vitality. Aust. J. Grape Wine Res. 16, 469–476.
Clarkson, D.T., 1985. Factors affecting mineral nutrient acquisition by plants. Annu. Rev. Plant Physiol. 36, 77–115.
Clarkson, D.T., 1993. Roots and the delivery of solutes to the xylem. Philos. Trans. Biol. Sci. 341, 5–17.
Clarkson, D.T., Hanson, J.B., 1980. The mineral nutrition of higher plants. Annu. Rev. Plant Physiol. 31, 239–298.
Clarkson, D.T., Carvajal, M., Henzler, T., Waterhouse, R.N., Smyth, A.J., Cooke, D.T., et al., 2000. Root hydraulic conductance: diurnal aquaporin expression and the effects of nutrient stress. J. Exp. Bot. 51, 61–70.
Clingeleffer, P.R., 1984. Production and growth of minimal pruned Sultana vines. Vitis 23, 42–54.
Clingeleffer, P.R., 1989. Effect of varying node number per bearer on yield and juice composition of Cabernet Sauvignon grapevines. Aust. J. Exp. Agric. 29, 701–705.

Clingeleffer, P.R., Krake, L.R., 1992. Responses of Cabernet franc grapevines to minimal pruning and virus infection. Am. J. Enol. Vitic. 43, 31–37.

Cocucci, S., Morgutti, S., Cocucci, M., Scienza, A., 1988. A possible relationship between stalk necrosis and membrane transport in grapevine cultivars. Sci. Hortic. 34, 67–74.

Coetzee, B., Freeborough, M.J., Maree, H.J., Celton, J.M., Rees, D.J.G., Burger, J.T., 2010. Deep sequencing analysis of viruses infecting grapevines: virome of a vineyard. Virology 400, 157–163.

Cola, G., Failla, O., Mariani, L., 2009. BerryTone—a simulation model for the daily course of grape berry temperature. Agric. For. Meteorol. 149, 1215–1228.

Columella, L.I.M. (4–ca. 70 AD). De Re Rustica. Available from: <http://www.thelatinlibrary.com/columella.html>.

Comas, L.H., Eissenstat, D.M., Lakso, A.N., 2000. Assessing root death and root system dynamics in a study of grape canopy pruning. New Phytol. 147, 171–178.

Comas, L.H., Anderson, L.J., Dunst, R.M., Lakso, A.N., Eissenstat, D.M., 2005. Canopy and environmental control of root dynamics in a long-term study of Concord grape. New Phytol. 167, 829–840.

Comas, L.H., Bauerle, T.L., Eissenstat, D.M., 2010. Biological and environmental factors controlling root dynamics and function: effects of root ageing and soil moisture. Aust. J. Grape Wine Res. 16, 131–137.

Comstock, J.P., 2002. Hydraulic and chemical signalling in the control of stomatal conductance and transpiration. J. Exp. Bot. 53, 195–200.

Conde, C., Silva, P., Fontes, N., Dias, A.C.P., Tavares, R.M., Sousa, M.J., et al., 2007. Biochemical changes throughout grape berry development and fruit and wine quality. Food 1, 1–22 (Global Science Books).

Coniberti, A., Ferrari, V., Fariña, L., Carrau, F., Dellacassa, E., Boido, E., et al., 2012. Role of canopy management in controlling high pH in Tannat grapes and wines. Am. J. Enol. Vitic. 63, 554–558.

Conn, S., Zhang, W., Franco, C., 2003. Anthocyanic vacuolar inclusions (AVIs) selectively bind acylated anthocyanins in *Vitis vinifera* L. (grapevine) suspension culture. Biotechnol. Lett. 25, 835–839.

Conn, S., Franco, C., Zhang, W., 2010. Characterization of anthocyanic vacuolar inclusions in *Vitis vinifera* L. cell suspension cultures. Planta 231, 1343–1360.

Conradie, W.J., 1980. Seasonal uptake of nutrients by Chenin blanc in sand culture: I. Nitrogen. S. Afr. J. Enol. Vitic. 1, 59–65.

Conradie, W.J., 1981a. Nutrient consumption by Chenin blanc grown in sand culture and seasonal changes in the chemical composition of leaf blades and petioles. S. Afr. J. Enol. Vitic. 2, 15–18.

Conradie, W.J., 1981b. Seasonal uptake of nutrients by Chenin blanc in sand culture: II. Phosphorus, potassium, calcium and magnesium. S. Afr. J. Enol. Vitic. 2, 7–13.

Conradie, W.J., 1986. Utilisation of nitrogen by the grapevine as affected by time of application and soil type. S. Afr. J. Enol. Vitic. 7, 76–83.

Conradie, W.J., Saayman, D., 1989. Effects of long term nitrogen, phosphorus, and potassium fertilization on Chenin blanc vines. II. Leaf analyses and grape composition. Am. J. Enol. Vitic. 40, 91–98.

Considine, J., Brown, K., 1981. Physical aspects of fruit growth: theoretical analysis of distribution of surface growth forces in fruit in relation to cracking and splitting. Plant Physiol. 68, 371–376.

Considine, J.A., 1981. Correlation of resistance to physical stress with fruit structure in the grape *Vitis vinifera* L. Aust. J. Bot. 29, 475–482.

Considine, J.A., Knox, R.B., 1979a. Development and histochemistry of the pistil of the grape, *Vitis vinifera*. Ann. Bot. 43, 11–22.

Considine, J.A., Knox, R.B., 1979b. Development and histochemistry of the cells, cell walls, and cuticle of the dermal system of fruit of the grape, *Vitis vinifera* L. Protoplasma 99, 347–365.

Considine, J.A., Knox, R.B., 1981. Tissue origins, cell lineages and patterns of cell division in the developing dermal system of the fruit of *Vitis vinifera* L. Planta 151, 403–412.

Considine, J.A., Kriedemann, P.E., 1972. Fruit splitting in grapes: determination of the critical turgor pressure. Aust. J. Agric. Res. 23, 17–24.
Cooke, J.E.K., Eriksson, M.E., Junttila, O., 2012. The dynamic nature of bud dormancy in trees: environmental control and molecular mechanisms. Plant Cell Environ. 35, 1707–1728.
Cookson, S.J., Clemente Moreno, M.J., Hevin, C., Nyamba Mendome, L.Z., Delrot, S., Trossat-Magnin, C., et al., 2013. Graft union formation in grapevine induces transcriptional changes related to cell wall modification, wounding, hormone signalling, and secondary metabolism. J. Exp. Bot. 64, 2997–3008.
Cookson, S.J., Clemente Moreno, M.J., Hevin, C., Nyamba Mendome, L.Z., Delrot, S., Trossat-Magnin, C., et al., 2014. Heterografting with nonself rootstocks induces genes involved in stress responses at the graft interface when compared with autografted controls. J. Exp. Bot. 65, 2473–2481.
Coombe, B.G., 1959. Fruit set and development in seeded grape varieties as affected by defoliation, topping, girdling, and other treatments. Am. J. Enol. Vitic. 10, 85–100.
Coombe, B.G., 1960. Relationship of growth and development to changes in sugars, auxins and gibberellins in fruit of seeded and seedless varieties of *Vitis vinifera* L. Plant Physiol. 35, 241–250.
Coombe, B.G., 1962. The effect of removing leaves, flowers and shoot tips on fruit-set in *Vitis vinifera* L. J. Hortic. Sci. 37, 1–15.
Coombe, B.G., 1976. The development of fleshy fruits. Annu. Rev. Plant Physiol. 27, 507–528.
Coombe, B.G., 1987. Distribution of solutes within the developing grape berry in relation to its morphology. Am. J. Enol. Vitic. 38, 120–127.
Coombe, B.G., 1992. Research on development and ripening of the grape berry. Am. J. Enol. Vitic. 43, 101–110.
Coombe, B.G., 1995. Adoption of a system for identifying grapevine growth stages. Aust. J. Grape Wine Res. 1, 104–110.
Coombe, B.G., 2001. Ripening berries—a critical issue. Aust. Vitic. 5, 28–34.
Coombe, B.G., Bishop, G.R., 1980. Development of the grape berry. II. Changes in diameter and deformability during veraison. Aust. J. Agric. Res. 31, 499–509.
Coombe, B.G., Dry, P.R., 2001. Viticulture, vol. 2. Practices, seventh ed. Winetitles, Adelaide, Australia.
Coombe, B.G., Hale, C.R., 1973. The hormone content of ripening grape berries and the effect of growth substance treatments. Plant Physiol. 51, 629–634.
Coombe, B.G., Monk, P.R., 1979. Proline and abscisic acid content of the juice of ripe Riesling grape berries: effect of irrigation during harvest. Am. J. Enol. Vitic. 30, 64–67.
Cooper, R.M., Williams, J.S., 2004. Elemental sulphur as an induced antifungal substance in plant defence. J. Exp. Bot. 55, 1947–1953.
Correia, M.J., Chaves, M.M., Pereira, J.S., 1990. Afternoon depression in photosynthesis in grapevine leaves—evidence for a high light stress effect. J. Exp. Bot. 41, 417–426.
Correia, M.J., Pereira, J.S., Chaves, M.M., Rodrigues, M.L., Pacheco, C.A., 1995. ABA xylem concentrations determine maximum daily leaf conductance of field-grown *Vitis vinifera* L. plants. Plant Cell Environ. 18, 511–521.
Cortell, J.M., Kennedy, J.A., 2006. Effect of shading on accumulation of flavonoid compounds in (*Vitis vinifera* L.) Pinot noir fruit and extraction in a model system. J. Agric. Food Chem. 54, 8510–8520.
Cortell, J.M., Halbleib, M., Gallagher, A.V., Righetti, T.L., Kennedy, J.A., 2005. Influence of vine vigor on grape (*Vitis vinifera* L. cv. Pinot noir) and wine proanthocyanidins. J. Agric. Food Chem. 53, 5798–5808.
Cortell, J.M., Halbleib, M., Gallagher, A.V., Righetti, T.L., Kennedy, J.A., 2007a. Influence of vine vigor on grape (*Vitis vinifera* L. cv. Pinot noir) anthocyanins. 1. Anthocyanin concentration and composition in fruit. J. Agric. Food Chem. 55, 6575–6584.
Cortell, J.M., Halbleib, M., Gallagher, A.V., Righetti, T.L., Kennedy, J.A., 2007b. Influence of vine vigor on grape (*Vitis vinifera* L. cv. Pinot noir) anthocyanins. 2. Anthocyanins and pigmented polymers in wine. J. Agric. Food Chem. 55, 6585–6595.

Coruzzi, G.M., Zhou, L., 2001. Carbon and nitrogen sensing and signaling in plants: emerging 'matrix effects'. Curr. Opin. Plant Biol. 4, 247–253.

Cosgrove, D.J., 1997. Relaxation in a high-stress environment: the molecular bases of extensible cell walls and cell enlargement. Plant Cell 9, 1031–1041.

Cosgrove, D.J., 2000. Loosening of plant cell walls by expansins. Nature 407, 321–326.

Cosgrove, D.J., 2005. Growth of the plant cell wall. Nat. Rev. Mol. Cell. Biol. 6, 850–861.

Costantini, E., Landi, L., Silvestroni, O., Pandolfini, T., Spena, A., Mezzetti, B., 2007. Auxin synthesis-encoding transgene enhances grape fecundity. Plant Physiol. 143, 1689–1694.

Costantini, V., Bellincontro, A., De Santis, D., Botondi, R., Mencarelli, F., 2006. Metabolic changes of Malvasia grapes for wine production during postharvest drying. J. Agric. Food Chem. 54, 3334–3340.

Courtois-Moreau, C.L., Pesquet, E., Sjödin, A., Muñiz, L., Bollhöner, B., Kaneda, M., et al., 2009. A unique program for cell death in xylem fibers of *Populus* stem. Plant J. 58, 260–274.

Coutand, C., 2010. Mechanosensing and thigmomorphogenesis, a physiological and biomechanical point of view. Plant J. 179, 168–182.

Cramer, G.R., Ergül, A., Grimplet, J., Tillett, R.L., Tattersall, E.A.R., Bohlman, M.C., et al., 2007. Water and salinity stress in grapevines: early and late changes in transcript and metabolite profiles. Funct. Integr. Genomics 7, 111–134.

Cramer, M.D., Hoffmann, V., Verboom, G.A., 2008. Nutrient availability moderates transpiration in *Ehrharta calycina*. New Phytol. 179, 1048–1057.

Crane, O., Halaly, T., Pang, X., Lavee, S., Perl, A., Vankova, R., et al., 2012. Cytokinin-induced *VvTFL1A* expression may be involved in the control of grapevine fruitfulness. Planta 235, 181–192.

Cravero, M.C., Guidoni, S., Schneider, A., Di Stefano, R., 1994. Caractérisation variétale de cépages musqués à raisin coloré au moyen de paramètres ampélographiques descriptifs et biochimiques. Vitis 33, 75–80.

Crawford, N.M., 1995. Nitrate: nutrient and signal for plant growth. Plant Cell 7, 859–868.

Crawford, N.M., Glass, A.D.M., 1998. Molecular and physiological aspects of nitrate uptake in plants. Trends Plant Sci. 3, 389–395.

Crawford, S., Shinohara, N., Sieberer, T., Williamson, L., George, G., Hepworth, J., et al., 2010. Strigolactones enhance competition between shoot branches by dampening auxin transport. Development 137, 2905–2913.

Creasy, G.L., Lombard, P.B., 1993. Vine water stress and peduncle girdling effects on pre- and post-veraison grape berry growth and deformability. Am. J. Enol. Vitic. 44, 193–197.

Creasy, G.L., Price, S.F., Lombard, P.B., 1993. Evidence for xylem discontinuity in Pinot noir and Merlot grapes: dye uptake and mineral composition during berry maturation. Am. J. Enol. Vitic. 44, 187–192.

Crespan, M., Milani, N., 2001. The Muscats: a molecular analysis of synonyms, homonyms and genetic relationships within a large family of grapevine cultivars. Vitis 40, 23–30.

Crippen Jr., D.D., Morrison, J.C., 1986. The effects of sun exposure on the compositional development of Cabernet Sauvignon berries. Am. J. Enol. Vitic. 37, 235–242.

Cuéllar, T., Azeem, F., Andrianteranagna, M., Pascaud, F., Verdeil, J.L., Sentenac, H., et al., 2013. Potassium transport in developing fleshy fruits: the grapevine inward K^+ channel VvK1.2 is activated by CIPK-CBL complexes and induced in ripening berry flesh cells. Plant J. 73, 1006–1018.

Cuevas, J.C., López-Cobollo, R., Alcázar, R., Zarza, X., Koncz, C., Altabella, T., et al., 2008. Putrescine is involved in *Arabidopsis* freezing tolerance and cold acclimation by regulating abscisic acid levels in response to low temperature. Plant Physiol. 148, 1094–1105.

Curie, C., Briat, J.F., 2003. Iron transport and signaling in plants. Annu. Rev. Plant Biol. 54, 183–206.

Curie, C., Cassin, G., Couch, D., Divol, F., Higuchi, K., Le Jean, M., et al., 2009. Metal movement within the plant: contribution of nicotianamine and yellow stripe 1-like transporters. Ann. Bot. 103, 1–11.

Currie, H.A., Perry, C.C., 2007. Silica in plants: biological, biochemical and chemical studies. Ann. Bot. 100, 1383–1389.

Currle, O., Bauer, O., Hofäcker, W., Schumann, F., Frisch, W., 1983. Biologie der Rebe. Neustadt an der Weinstrasse. Meininger, Germany.
Dai, G.H., Andary, C., Mondolot-Cosson, L., Boubals, D., 1995. Histochemical studies on the interaction between three species of grapevine, *Vitis vinifera*, *V. rupestris* and *V. rotundifolia* and the downy mildew fungus, *Plasmopara viticola*. Physiol. Mol. Plant Pathol. 46, 177–188.
Dai, Z.W., Ollat, N., Gomès, E., Decroocq, S., Tandonnet, J.P., Bordenave, L., et al., 2011. Ecophysiological, genetic, and molecular causes of variation in grape berry weight and composition: a review. Am. J. Enol. Vitic. 62, 413–425.
Dakora, F.D., Phillips, D.A., 2002. Root exudates as mediators of mineral acquisition in low-nutrient environments. Plant Soil 245, 35–47.
Daly, C., Halbleib, M., Smith, J.I., Gibson, W.P., Doggett, M.K., Taylor, G.H., et al., 2008. Physiographically sensitive mapping of climatological temperature and precipitation across the conterminous Unites States. Int. J. Climatology 28, 2031–2064.
Danilewicz, J.C., 2003. Review of reaction mechanisms of oxygen and proposed intermediate reduction products in wine: central role of iron and copper. Am. J. Enol. Vitic. 54, 73–85.
Darriet, P., Pons, M., Henry, R., Dumont, O., Findeling, V., Cartolaro, P., et al., 2002. Impact odorants contributing to the fungus type aroma from grape berries contaminated by powdery mildew (*Uncinula necator*): incidence of enzymatic activities of the yeast *Saccharomyces cerevisiae*. J. Agric. Food Chem. 50, 3277–3282.
Darvill, A.G., Albersheim, P., 1984. Phytoalexins and their elicitors—a defense against microbial infection in plants. Annu. Rev. Plant Physiol. 35, 243–275.
Darwin, C., 1875. The Movements and Habits of Climbing Plants. Murray, London.
Darwin, C., 2004. In: Levine, G. (Ed.), The Origin of Species. Barnes & Noble Classics, New York, NY, (Original work published in 1859).
Davenport, J.R., Stevens, R.G., 2006. High soil moisture and low soil temperature are associated with chlorosis occurrence in Concord grape. HortScience 41, 418–422.
Davenport, J.R., Keller, M., Mills, L.J., 2008a. How cold can you go? Frost and winter protection for grape. HortScience 43, 1966–1969.
Davenport, J.R., Stevens, R.G., Whitley, K.M., 2008b. Spatial and temporal distribution of soil moisture in drip-irrigated vineyards. HortScience 43, 229–235.
Davidovich-Rikanati, R., Sitrit, Y., Tadmor, Y., Iijiama, Y., Bilenko, N., Bar, E., et al., 2007. Enrichment of tomato flavor by diversion of the early plastidial terpenoid pathway. Nat. Biotechnol. 25, 899–901.
Davies, C., Böttcher, C., 2009. Hormonal control of grape berry ripening. In: Roubelakis-Angelakis, K.A. (Ed.), Grapevine Molecular Physiology and Biotechnology, second ed. Springer, Dordrecht, The Netherlands, pp. 229–261.
Davies, C., Robinson, S.P., 2000. Differential screening indicates a dramatic change in mRNA profiles during grape berry ripening. Cloning and characterization of cDNAs encoding putative cell wall and stress response proteins. Plant Physiol. 122, 803–812.
Davies, C., Boss, P.K., Robinson, S.P., 1997. Treatment of grape berries, a nonclimacteric fruit with a synthetic auxin, retards ripening and alters the expression of developmentally regulated genes. Plant Physiol. 115, 1155–1161.
Davies, C., Wolf, T., Robinson, S.P., 1999. Three putative sucrose transporters are differentially expressed in grapevine tissues. Plant Sci. 147, 93–100.
Davies, C., Shin, R., Liu, W., Thomas, M.R., Schachtman, D.P., 2006. Transporters expressed during grape berry (*Vitis vinifera* L.) development are associated with an increase in berry size and berry potassium accumulation. J. Exp. Bot. 57, 3209–3216.
Davies, W.J., Zhang, J., 1991. Root signals and the regulation of growth and development of plants in drying soil. Annu. Rev. Plant Physiol. Plant Mol. Biol. 42, 55–76.

Davies, W.J., Wilkinson, S., Loveys, B., 2002. Stomatal control by chemical signalling and the exploitation of this mechanism to increase water use efficiency in agriculture. New Phytol. 153, 449–460.

Davis, J.D., Evert, R.F., 1970. Seasonal cycle of phloem development in woody vines. Bot. Gaz. 131, 128–138.

De Andrés, M.T., Benito, A., Peréz-Rivera, G., Ocete, R., Lopez, A., Gaforio, L., et al., 2012. Genetic diversity of wild grapevine populations in Spain and their genetic relationships with cultivated grapevines. Mol. Ecol. 21, 800–816.

De Boer, A.H., Volkov, V., 2003. Logistics of water and salt transport through the plant: structure and functioning of the xylem. Plant Cell Environ. 26, 87–101.

De Freitas, V.A.P., Glories, Y., 1999. Concentration and compositional changes of procyanidins in grape seeds and skin of white *Vitis vinifera* varieties. J. Sci. Food Agric. 79, 1601–1606.

De Freitas, S.T., Handa, A.K., Wu, Q., Park, S., Mitcham, E.J., 2012. Role of pectin methylesterases in cellular calcium distribution and blossom-end rot development in tomato fruit. Plant J. 71, 824–835.

De Mora, S.J., Eschenbruch, R., Knowles, S.J., Spedding, D.J., 1986. The formation of dimethyl sulphide during fermentation using a wine yeast. Food Microbiol. 3, 27–32.

De Schepper, V., De Swaef, T., Bauweraerts, I., Steppe, K., 2013. Phloem transport: a review of mechanisms and controls. J. Exp. Bot. 64, 4839–4850.

De Smet, I., Zhang, H., Inzé, D., Beeckman, T., 2006. A novel role for abscisic acid emerges from underground. Trends Plant Sci. 11, 434–439.

De Souza, C.R., Maroco, J.P., dos Santos, T.P., Rodrigues, M.L., Lopes, C.M., Pereira, J.S., et al., 2003. Partial rootzone drying: regulation of stomatal aperture and carbon assimilation in field-grown grapevines (*Vitis vinifera* cv. Moscatel). Funct. Plant Biol. 30, 653–662.

De Storme, N., Geelen, D., 2014. The impact of environmental stress on male reproductive development in plants: biological processes and molecular mechanisms. Plant Cell Environ. 37, 1–18.

DeBolt, S., Hardie, J., Tyerman, S., Ford, C.M., 2004. Composition and synthesis of raphide crystals and druse crystals in berries of *Vitis vinifera* L. cv. Cabernet Sauvignon: ascorbic acid as precursor for both oxalic and tartaric acids as revealed by radiolabelling studies. Aust. J. Grape Wine Res. 10, 134–142.

DeBolt, S., Cook, D.R., Ford, C.M., 2006. L-Tartaric acid synthesis from vitamin C in higher plants. Proc. Natl. Acad. Sci. USA 103, 5608–5613.

DeBolt, S., Melino, V., Ford, C.M., 2007. Ascorbate as a biosynthetic precursor in plants. Ann. Bot. 99, 3–8.

Dechorgnat, J., Nguyen, C.T., Armengaud, P., Jossier, M., Diatloff, E., Filleur, S., et al., 2011. From the soil to the seeds: the long journey of nitrate in plants. J. Exp. Bot. 62, 1349–1359.

Del Bianco, M., Giustini, L., Sabatini, S., 2013. Spatiotemporal changes in the role of cytokinin during root development. New Phytol. 199, 324–338.

Del Pozo, J.C., Lopez-Matas, M.A., Ramirez-Parra, E., Gutierrez, C., 2005. Hormonal control of the plant cell cycle. Physiol. Plant. 123, 173–183.

Delas, J., 1972. Effets de la fertilisation de la vigne sur le développement de "Botrytis cinerea". Rev. Zool. Agric. Pathol. Vég. 71, 11–17.

Delas, J., Pouget, R., 1984. Action de la concentration de la solution nutritive sur quelques caractéristiques physiologiques et technologiques chez *Vitis vinifera* L. cv. Cabernet-Sauvignon. II. Composition minérale des organes végétatifs, du moût et du vin. Agronomie 4, 350–443.

Delas, J., Molot, C., Soyer, J.P., 1982. Influence d'une fertilisation azotée excessive, du porte-greffe et de la charge sur la sensibilité du cépage Merlot à *Botrytis cinerea*. Bull. OEPP 12, 177–182.

Della Penna, D., Pogson, B.J., 2006. Vitamin synthesis in plants: tocopherols and carotenoids. Annu. Rev. Plant Biol. 57, 711–738.

Deluc, L.G., Grimplet, J., Wheatley, M.D., Tillett, R.L., Quilici, D.R., Osborne, C., et al., 2007. Transcriptomic and metabolite analysis of Cabernet Sauvignon grape berry development. BMC Genomics 8, 429, http://dx.doi.org/10.1186/1471-2164-8-429.

Delwiche, J.F., 2003. Impact of color on perceived wine flavor. Foods Food Ingredients J. Jpn. 208, 349–352.
Demmig-Adams, B., Adams III, W.W., 1996. The role of xanthophyll cycle carotenoids in the protection of photosynthesis. Trends Plant Sci. 1, 21–26.
Dengler, N.G., 2001. Regulation of vascular development. J. Plant Growth Regul. 20, 1–13.
Dettweiler, E., Jung, A., Zyprian, E., Töpfer, R., 2000. Grapevine cultivar Müller-Thurgau and its true to type descent. Vitis 39, 63–65.
Dewitte, W., Chiappetta, A., Azmi, A., Witters, E., Strnad, M., Rembur, J., et al., 1999. Dynamics of cytokinins in apical shoot meristems of a day-neutral tobacco during floral transition and flower formation. Plant Physiol. 119, 111–121.
Deytieux, C., Gény, L., Lapaillerie, D., Claverol, S., Bonneu, M., Donéche, B., 2007. Proteome analysis of grape skins during ripening. J. Exp. Bot. 58, 1851–1862.
Di Filippo, M., Vila, H., 2011. Influence of different rootstocks on the vegetative and reproductive performance of *Vitis vinifera* L. Malbec under irrigated conditions. J. Int. Sci. Vigne Vin 45, 75–84.
Di Vecchi Staraz, M., Bandinelli, R., Boselli, M., This, P., Boursiquot, J.M., Laucou, V., et al., 2007. Genetic structuring and parentage analysis for evolutionary studies in grapevine: kin group and origin of the cultivar Sangiovese revealed. J. Am. Soc. Hortic. Sci. 132, 514–524.
Di Vecchi-Staraz, M., Laucou, V., Bruno, G., Lacombe, T., Gerber, S., Bourse, T., et al., 2009. Low level of pollen-mediated gene flow from cultivated to wild grapevine: consequences for the evolution of the endangered subspecies *Vitis vinifera* L. subsp. *silvestris*. J. Hered. 100, 66–75.
Diakou, P., Carde, J.P., 2001. *In situ* fixation of grape berries. Protoplasma 218, 225–235.
Díaz-Riquelme, J., Lijavetzky, D., Martínez-Zapater, J.M., Carmona, M.J., 2009. Genome-wide analysis of MIKC^C-type MADS box genes in grapevine. Plant Physiol. 149, 354–369.
Dintscheff, D., Stoeff, K., Peschakoff, G., Kolarowa, E., 1964. Untersuchungen über die Stickstoffernährung der Weinrebe unter Anwendung des stabilen Stickstoffisotops ^{15}N. Vitis 4, 347–356.
Dittrich, H.H., 1989. Die Veränderungen der Beereninhaltsstoffe und der Weinqualität durch *Botrytis cinerea*—Übersichtsreferat. Wein-Wiss. 44, 105–131.
Dixon, H.H., Joly, J., 1895. On the ascent of sap. Philos. Trans. R. Soc. B 186, 563–576.
Dixon, R.A., Xie, D.Y., Sharma, S.B., 2005. Proanthocyanidins—a final frontier in flavonoid research? New Phytol. 165, 9–28.
Do, P.T., Prudent, M., Sulpice, R., Causse, M., Fernie, A.R., 2010. The influence of fruit load on the tomato pericarp metabolome in a *Solanum chmielewskii* introgression line population. Plant Physiol. 154, 1128–1142.
Doblin, M.S., Pettolino, F., Bacic, A., 2010. Plant cell walls: the skeleton of the plant world. Funct. Plant Biol. 37, 357–381.
Doco, T., Williams, P., Pauly, M., O'Neill, M.A., Pellerin, P., 2003. Polysaccharides from grape berry cell walls. Part II. Structural characterization of the xyloglucan polysaccharides. Carbohydr. Polym. 53, 253–261.
Dodd, I.C., Egea, G., Davies, W.J., 2008. Abscisic acid signalling when soil moisture is heterogeneous: decreased photoperiod sap flow from drying roots limits abscisic acid export to the shoots. Plant Cell Environ. 31, 1263–1274.
Dokoozlian, N.K., 1999. Chilling temperature and duration interact on the budbreak of 'Perlette' grapevine cuttings. HortScience 34, 1054–1056.
Dokoozlian, N.K., Hirschfelt, D.J., 1995. The influence of cluster thinning at various stages of fruit development on Flame Seedless table grapes. Am. J. Enol. Vitic. 46, 429–436.
Dokoozlian, N.K., Kliewer, W.M., 1995a. The light environment within grapevine canopies. I. Description and seasonal changes during fruit development. Am. J. Enol. Vitic. 46, 209–218.
Dokoozlian, N.K., Kliewer, W.M., 1995b. The light environment within grapevine canopies. II. Influence of leaf area density on fruit zone light environment and some canopy assessment parameters. Am. J. Enol. Vitic. 46, 219–226.

Dokoozlian, N.K., Kliewer, W.M., 1996. Influence of light on grape berry growth and composition varies during fruit development. J. Am. Soc. Hortic. Sci. 121, 869–874.
Dokoozlian, N.K., Williams, L.E., Neja, R.A., 1995. Chilling exposure and hydrogen cyanamide interact in breaking dormancy of grape buds. HortScience 30, 1244–1247.
Domínguez, E., España, L., López-Casado, G., Cuartero, J., Heredia, A., 2009. Biomechanics of isolated tomato (Solanum lycopersicum) fruit cuticles during ripening: the role of flavonoids. Funct. Plant Biol. 36, 613–620.
Domínguez, E., Heredia-Guerrero, J.A., Heredia, A., 2011. The biophysical design of plant cuticles: an overview. New Phytol. 189, 938–949.
Donèche, B., Chardonnet, C., 1992. Evolution et localisation des principaux cations au cours du développement du raisin. Vitis 31, 175–181.
Dorcey, E., Urbez, C., Blázquez, M.A., Carbonell, J., Perez-Amador, M.A., 2009. Fertilization-dependent auxin response in ovules triggers fruit development through the modulation of gibberellin metabolism in Arabidopsis. Plant J. 58, 318–332.
Dorion, S., Lalonde, S., Saini, H.S., 1996. Induction of male sterility in wheat by meiotic-stage water deficit is preceded by a decline in invertase activity and changes in carbohydrate metabolism in anthers. Plant Physiol. 111, 137–145.
dos Santos, T.P., Lopes, C.M., Rodrigues, M.L., de Souza, C.R., Maroco, J.P., Pereira, J.S., et al., 2003. Partial rootzone drying: effects on growth and fruit quality of field-grown grapevines (Vitis vinifera). Funct. Plant Biol. 30, 663–671.
Doster, M.A., Schnathorst, W.C., 1985. Comparative susceptibility of various grapevine cultivars to the powdery mildew fungus Uncinula necator. Am. J. Enol. Vitic. 36, 101–104.
Downey, M.O., Rochfort, S., 2008. Simultaneous separation by reversed-phase high-performance liquid chromatography and mass spectral identification of anthocyanins and flavonols in Shiraz grape skin. J. Chromatogr. A 1201, 43–47.
Downey, M.O., Harvey, J.S., Robinson, S.P., 2003a. Analysis of tannins in seeds and skins of Shiraz grapes throughout berry development. Aust. J. Grape Wine Res. 9, 15–27.
Downey, M.O., Harvey, J.S., Robinson, S.P., 2003b. Synthesis of flavonols and expression of flavonol synthase genes in the developing grape berries of Shiraz and Chardonnay (Vitis vinifera L.). Aust. J. Grape Wine Res. 9, 110–121.
Downey, M.O., Harvey, J.S., Robinson, S.P., 2004. The effect of bunch shading on berry development and flavonoid accumulation in Shiraz grapes. Aust. J. Grape Wine Res. 10, 55–73.
Downey, M.O., Dokoozlian, N.K., Krstic, M.P., 2006. Cultural practice and environmental impacts on the flavonoid composition of grapes and wine: a review of recent research. Am. J. Enol. Vitic. 57, 257–268.
Downton, W.J.S., 1977a. Salinity effects on the ion composition of fruiting Cabernet Sauvignon vines. Am. J. Enol. Vitic. 28, 210–214.
Downton, W.J.S., 1977b. Photosynthesis in salt-stressed grapevines. Aust. J. Plant Physiol. 4, 183–192.
Downton, W.J.S., 1985. Growth and mineral composition of the Sultana grapevine as influenced by salinity and rootstock. Aust. J. Agric. Res. 36, 425–434.
Downton, W.J.S., Grant, W.J.R., 1992. Photosynthetic physiology of spur pruned and minimal pruned grapevines. Aust. J. Plant Physiol. 19, 309–316.
Downton, W.J.S., Loveys, B.R., 1978. Compositional changes during grape berry development in relation to abscisic acid and salinity. Aust. J. Plant Physiol. 5, 415–423.
Downton, W.J.S., Loveys, B.R., 1981. Abscisic acid content and osmotic relations of salt-stressed grapevine leaves. Aust. J. Plant Physiol. 8, 443–452.
Downton, W.J.S., Grant, W.J.R., Loveys, B.R., 1987. Diurnal changes in the photosynthesis of field-grown grape vines. New Phytol. 105, 71–80.

Downton, W.J.S., Loveys, B.R., Grant, W.J.R., 1990. Salinity effects on the stomatal behaviour of grapevine. New Phytol. 116, 499–503.

Dragoni, D., Lakso, A.N., Piccioni, R.M., Tarara, J.M., 2006. Transpiration of grapevines in the humid northeastern United States. Am. J. Enol. Vitic. 57, 160–167.

Drawert, F., Steffan, H., 1966a. Biochemisch-physiologische Untersuchungen an Traubenbeeren. II. Distribution and respiration of the added ^{14}C compounds. Vitis 5, 27–34.

Drawert, F., Steffan, H., 1966b. Biochemisch-physiologische Untersuchungen an Traubenbeeren. III. Stoffwechsel von zugeführten ^{14}C-Verbindungen und die Bedeutung des Säure-Zucker-Mechanismus für die Reifung von Traubenbeeren. Vitis 5, 377–384.

Druart, N., Johansson, A., Baba, K., Schrader, J., Sjödin, A., Bhalerao, R.R., et al., 2007. Environmental and hormonal regulation of the activity-dormancy cycle in the cambial meristem involves stage-specific modulation of transcriptional and metabolic networks. Plant J. 50, 557–573.

Dry, I.B., Feechan, A., Anderson, C., Jermakow, A.M., Bouquet, A., Adam-Blondon, A.F., et al., 2010. Molecular strategies to enhance the genetic resistance of grapevines to powdery mildew. Aust. J. Grape Wine Res. 16, 94–105.

Dry, P.D., 2000. Canopy management for fruitfulness. Aust. J. Grape Wine Res. 6, 109–115.

Dry, P.R., Coombe, B.G., 2004. Viticulture. Volume 1—Resources, second ed. Winetitles, Adelaide, Australia.

Dry, P.R., Loveys, B.R., 1998. Factors influencing grapevine vigour and the potential for control with partial rootzone drying. Aust. J. Grape Wine Res. 4, 140–148.

Dry, P.R., Loveys, B.R., 1999. Grapevine shoot growth and stomatal conductance are reduced when part of the root system is dried. Vitis 38, 151–156.

Dry, P.R., Loveys, B.R., Düring, H., 2000a. Partial drying of the root-zone of grape. I. Transient changes in shoot growth and gas exchange. Vitis 39, 3–8.

Dry, P.R., Loveys, B.R., Düring, H., 2000b. Partial drying of the root-zone of grape. II. Changes in the pattern of root development. Vitis 39, 9–12.

Dry, P.R., Loveys, B.R., McCarthy, M.G., Stoll, M., 2001. Strategic irrigation management in Australian vineyards. J. Int. Sci. Vigne Vin 35, 45–61.

Dry, P.R., Longbottom, M.L., McLoughlin, S., Johnson, T.E., Collins, C., 2010. Classification of reproductive performance of ten winegrape varieties. Aust. J. Grape Wine Res. 16, 47–55.

Du, Y.P., Zhai, H., Sun, Q.H., Wang, Z.S., 2009. Susceptibility of Chinese grapes to grape Phylloxera. Vitis 48, 57–58.

Dubernet, M., Ribereau-Gayon, P., Lerner, H.R., Harel, E., Mayer, A.M., 1977. Purification and properties of laccase from *Botrytis cinerea*. Phytochemistry 16, 191–193.

Dubourdieu, D., Tominaga, T., Masneuf, I., Peyrot des Gachons, C., Murat, M.L., 2006. The role of yeasts in grape flavor development during fermentation: the example of Sauvignon blanc. Am. J. Enol. Vitic. 57, 81–88.

Duchêne, E., Schneider, C., 2005. Grapevine and climatic changes: a glance at the situation in Alsace. Agron. Sustainable Dev. 25, 93–99.

Dudareva, N., Klempien, A., Muhlemann, J.K., Kaplan, I., 2013. Biosynthesis, function and metabolic engineering of plant volatile organic compounds. New Phytol. 198, 16–32.

Dudley, R., 2004. Ethanol, fruit ripening, and the historical origins of human alcoholism in primate frugivory. Integr. Comp. Biol. 44, 315–323.

Dugan, F.M., Lupien, S.L., Grove, G.G., 2002. Incidence, aggressiveness and *in planta* interactions of *Botrytis cinerea* and other filamentous fungi quiescent in grape berries and dormant buds in central Washington state. J. Phytopathol. 150, 375–381.

Dugardeyn, J., Van Der Straeten, D., 2008. Ethylene: fine-tuning plant growth and development by stimulation and inhibition of elongation. Plant Sci. 175, 59–70.

Dun, E.A., Ferguson, B.J., Beveridge, C.A., 2006. Apical dominance and shoot branching. Divergent opinions or divergent mechanisms? Plant Physiol. 142, 812–819.

Dun, E.A., de Saint Germain, A., Rameau, C., Beveridge, C.A., 2012. Antagonistic action of strigolactone and cytokinin in bud outgrowth control. Plant Physiol. 158, 487–498.

Dundon, C.G., Smart, R.E., 1984. Effects of water relations on the potassium status of Shiraz vines. Am. J. Enol. Vitic. 35, 40–45.

Dunlevy, J.D., Kalua, C.M., Keyzers, R.A., Boss, P.K., 2009. The production of flavour and aroma compounds in grape berries. In: Roubelakis-Angelakis, K.A. (Ed.), Grapevine Molecular Physiology and Biotechnology, second ed. Springer, Dordrecht, The Netherlands, pp. 293–340.

Dunlevy, J.D., Soole, K.L., Perkins, M.V., Nicholson, E.L., Maffei, S.M., Boss, P.K., 2013. Determining the methoxypyrazine biosynthesis variables affected by light exposure and crop level in Cabernet Sauvignon. Am. J. Enol. Vitic. 64, 450–458.

Düring, H., 1980. Stomatafrequenz bei Blättern von Vitis-Arten und -Sorten. Vitis 19, 91–98.

Düring, H., 1984. Evidence for osmotic adjustment to drought in grapevines (*Vitis vinifera* L.). Vitis 23, 1–10.

Düring, H., 1987. Stomatal response to alteration of soil and air humidity in grapevines. Vitis 26, 9–18.

Düring, H., 1988. CO_2 assimilation and photorespiration of grapevine leaves: responses to light and drought. Vitis 27, 199–208.

Düring, H., 1997. Potential frost resistance of grape: kinetics of temperature-induced hardening of Riesling and Silvaner buds. Vitis 36, 213–214.

Düring, H., 1999. Photoprotection in leaves of grapevines: responses of the xanthophyll cycle to alterations of light intensity. Vitis 38, 21–24.

Düring, H., 2003. Stomatal and mesophyll conductances control CO_2 transfer to chloroplasts in leaves of grapevine (*Vitis vinifera* L.). Vitis 42, 65–68.

Düring, H., Alleweldt, G., 1973. Der Jahresgang der Abscisinsäure in vegetativen Organen von Reben. Vitis 12, 26–32.

Düring, H., Bachmann, O., 1975. Abscisic acid analysis in *Vitis vinifera* in the period of endogenous bud dormancy by high pressure liquid chromatography. Physiol. Plant. 34, 201–203.

Düring, H., Davtyan, A., 2002. Developmental changes of primary processes of photosynthesis in sun- and shade-adapted berries of two grapevine cultivars. Vitis 41, 63–67.

Düring, H., Dry, P.R., 1995. Osmoregulation in water stressed roots: responses of leaf conductance and photosynthesis. Vitis 34, 15–17.

Düring, H., Loveys, B.R., 1982. Diurnal changes in water relations and abscisic acid in field grown *Vitis vinifera* cvs. I. Leaf water potential components and leaf conductance under humid temperate and semiarid conditions. Vitis 21, 223–232.

Düring, H., Oggionni, F., 1986. Transpiration und Mineralstoffeinlagerung der Weinbeere. Vitis 25, 59–66.

Düring, H., Lang, A., Oggionni, F., 1987. Patterns of water flow in Riesling berries in relation to developmental changes in their xylem morphology. Vitis 26, 123–131.

Dutta, R., Robinson, K.R., 2004. Identification and characterization of stretch-activated ion channels in pollen protoplasts. Plant Physiol. 135, 1398–1406.

Eapen, D., Barroso, M.L., Ponce, G., Campos, M.E., Cassab, G.I., 2005. Hydrotropism: root growth responses to water. Trends Plant Sci. 10, 44–50.

Ebadi, A., Sedgley, M., May, P., Coombe, B.G., 1996. Seed development and abortion in *Vitis vinifera* L., cv. Chardonnay. Int. J. Plant Sci. 157, 703–712.

Edson, C.E., Howell, G.S., Flore, J.A., 1993. Influence of crop load on photosynthesis and dry matter partitioning of Seyval grapevines I. Single leaf and whole vine response pre- and post-harvest. Am. J. Enol. Vitic. 44, 139–147.

Edson, C.E., Howell, G.S., Flore, J.A., 1995a. Influence of crop load on photosynthesis and dry matter partitioning of Seyval grapevines II. Seasonal changes in single leaf and whole vine photosynthesis. Am. J. Enol. Vitic. 46, 469–477.

Edson, C.E., Howell, G.S., Flore, J.A., 1995b. Influence of crop load on photosynthesis and dry matter partitioning of Seyval grapevines III. Seasonal changes in dry matter partitioning, vine morphology, yield, and fruit composition. Am. J. Enol. Vitic. 46, 478–485.

Edye, L.A., Richards, G.N., 1991. Analysis of condensates from wood smoke: components derived from polysaccharides and lignins. Environ. Sci. Technol. 25, 1133–1137.

Efetova, M., Zeier, J., Riederer, M., Lee, C.W., Stingl, N., Mueller, M., et al., 2007. A central role of abscisic acid in drought stress protection of *Agrobacterium*-induced tumors on *Arabidopsis*. Plant Physiol. 145, 853–862.

Egerton-Warburton, L.M., Querejeta, J.I., Allen, M.F., 2007. Common mycorrhizal networks provide a potential pathway for the transfer of hydraulically lifted water between plants. New Phytol. 58, 1473–1483.

Egger, K., Reichling, J., Ammann-Schweizer, R., 1976. Flavonol-Derivate in Formen der Gattung Vitis. Vitis 15, 24–28.

Egri, Á., Horváth, Á., Kriska, G., Horváth, G., 2010. Optics of sunlit water drops on leaves: conditions under which sunburn is possible. New Phytol. 185, 979–987.

Ehret, D.L., Helmer, T., Hall, J.W., 1993. Cuticle cracking in tomato fruit. J. Hortic. Sci. 68, 195–201.

Eibach, R., 1994. Investigations about the genetic resources of grapes with regard to resistance characteristics to powdery mildew (*Oidium tuckeri*). Vitis 33, 143–150.

Eibach, R., Alleweldt, G., 1984. Einfluss der Wasserversorgung auf Wachstum, Gaswechsel und Substanzproduktion traubentragender Reben. II. Der Gaswechsel. Vitis 23, 11–20.

Eibach, R., Alleweldt, G., 1985. Einfluss der Wasserversorgung auf Wachstum, Gaswechsel und Substanzproduktion traubentragender Reben. III. Die Substanzproduktion. Vitis 24, 183–198.

Eichert, T., Goldbach, H.E., 2008. Equivalent pore radii of hydrophilic foliar uptake routes in stomatous and astomatous leaf surfaces—further evidence for a stomatal pathway. Physiol. Plant. 132, 491–502.

Eifert, J., Pánczél, M., Eifert, A., 1961. Änderung des Stärke- und Zuckergehaltes der Rebe während der Ruheperiode. Vitis 2, 257–265.

Elad, Y., Evensen, K., 1995. Physiological aspects of resistance to *Botrytis cinerea*. Phytopathology 85, 637–643.

Else, M.A., Stankiewicz-Davies, A.P., Crisp, C.M., Atkinson, C.J., 2004. The role of polar auxin transport through pedicels of *Prunus avium* L. in relation to fruit development and retention. J. Exp. Bot. 55, 2099–2109.

Else, M.A., Janowiak, F., Atkinson, C.J., Jackson, M.B., 2009. Root signals and stomatal closure in relation to photosynthesis, chlorophyll *a* fluorescence and adventitious rooting of flooded tomato plants. Ann. Bot. 103, 313–323.

Emershad, R.L., Ramming, D.W., 1984. *In-ovulo* embryo culture of *Vitis vinifera* L. c.v. 'Thompson Seedless'. Am. J. Bot. 71, 873–877.

Emershad, R.L., Ramming, D.W., Serpe, M.D., 1989. *In ovulo* embryo development and plant formation from stenospermic genotypes of *Vitis vinifera*. Am. J. Bot. 76, 397–402.

Emes, M.J., Neuhaus, H.E., 1997. Metabolism and transport in non-photosynthetic plastids. J. Exp. Bot. 48, 1995–2005.

Endo, A., Sawada, Y., Takahashi, H., Okamoto, M., Ikegami, K., Koiwai, H., et al., 2008. Drought induction of *Arabidopsis* 9-cis-epoxycarotenoid dioxygenase occurs in vascular parenchyma cells. Plant Physiol. 147, 1984–1993.

English, J.T., Bledsoe, A.M., Marois, J.J., Kliewer, W.M., 1990. Influence of grapevine canopy management on evaporative potential in the fruit zone. Am. J. Enol. Vitic. 41, 137–141.

Enquist, B.J., Niklas, K.J., 2002. Global allocation rules for patterns of biomass partitioning in seed plants. Science 295, 1517–1520.

Ensminger, I., Busch, F., Huner, N.P.A., 2006. Photostasis and cold acclimation: sensing low temperature through photosynthesis. Physiol. Plant. 126, 28–44.

Epstein, E., 1999. Silicon. Annu. Rev. Plant Physiol. Plant Mol. Biol. 50, 641–664.

Erlenwein, H., 1965a. Einfluss der Ernährung und des Pfropfpartners auf das Wurzelwachstum von Vitis-Arten und -Sorten. Vitis 5, 161–186.

Erlenwein, H., 1965b. Einfluß von Klimafaktoren auf das Wachstum von Vitis-Arten und-Sorten. Vitis 5, 94–109.

Esau, K., 1948. Phloem structure in the grapevine, and its seasonal changes. Hilgardia 18, 217–296.

Escalona, J.M., Flexas, J., Medrano, H., 1999. Stomatal and non-stomatal limitations of photosynthesis under water stress in field-grown grapevines. Aust. J. Plant Physiol. 26, 421–433.

Eschenbruch, R., Kleynhans, P.H., 1974. The influence of copper-containing fungicides on the copper content of grape juice and on hydrogen sulphide formation. Vitis 12, 320–324.

Escudero, A., Campo, E., Fariña, L., Cacho, J., Ferreira, V., 2007. Analytical characterization of the aroma of five premium red wines. Insights into the role of odor families and the concept of fruitiness of wines. J. Agric. Food Chem. 55, 4501–4510.

Espinoza, C., Vega, A., Medina, C., Schlauch, K., Cramer, G., Arce-Johnson, P., 2007. Gene expression associated with compatible viral diseases in grapevine cultivars. Funct. Integr. Genomics 7, 95–110.

Esteban, A., Villanueva, J., Lissarrague, J.R., 1999. Effect of irrigation on changes in berry composition of Tempranillo during maturation. Sugars, organic acids, and mineral elements. Am. J. Enol. Vitic. 50, 418–434.

Etchebarne, F., Ojeda, H., Deloire, A., 2009. Influence of water status on mineral composition of berries in 'Grenache Noir' (*Vitis vinifera* L.). Vitis 48, 63–68.

Etchebarne, F., Ojeda, H., Hunter, J.J., 2010. Leaf:fruit ratio and vine water status effects on Grenache noir (*Vitis vinfera* L.) berry composition: water, sugar, organic acids and cations. S. Afr. J. Enol. Vitic. 31, 106–115.

Etienne, A., Génard, M., Lobit, P., Mbeguié-A-Mbéguié, D., Bugaud, C., 2013. What controls fleshy fruit acidity? A review of malate and citrate accumulation in fruit cells. J. Exp. Bot. 64, 1451–1469.

Evans, J.R., 1989. Photosynthesis and nitrogen relationships in leaves of C_3 plants. Oecologia 78, 9–19.

Evans, J.R., Kaldenhoff, R., Genty, B., Terashima, I., 2009. Resistances along the CO_2 diffusion pathway inside leaves. J. Exp. Bot. 60, 2235–2248.

Evans, R.G., Spayd, S.E., Wample, R.L., Kroeger, M.W., Mahan, M.O., 1993. Water use of *Vitis vinifera* grapes in Washington. Agric. Water Manage. 23, 109–124.

Evert, R.F., 2006. Esau's Plant Anatomy. Meristems, Cells, and Tissues of the Plant Body—Their Structure, Function, and Development, third ed. John Wiley & Sons, Hoboken, NJ.

Ewart, A., Kliewer, W.M., 1977. Effects of controlled day and night temperatures and nitrogen on fruit-set, ovule fertility, and fruit composition of several wine grape cultivars. Am. J. Enol. Vitic. 28, 88–95.

Ewers, F.W., Fisher, J.B., 1991. Why vines have narrow stems: histological trends in *Bauhinia* (Fabaceae). Oecologia 88, 233–237.

Ezzahouani, A., Williams, L.E., 1995. The influence of rootstock on leaf water potential, yield, and berry composition of Ruby Seedless grapevines. Am. J. Enol. Vitic. 46, 559–563.

Failla, O., Mariani, L., Brancadoro, L., Minelli, R., Scienza, A., Murada, G., et al., 2004. Spatial distribution of solar radiation and its effects on vine phenology and grape ripening in an alpine environment. Am. J. Enol. Vitic. 55, 128–138.

Fait, A., Hanhineva, K., Beleggia, R., Dai, N., Rogachev, I., Nikiforova, V.J., et al., 2008. Reconfiguration of the achene and receptacle metabolic networks during strawberry fruit development. Plant Physiol. 148, 730–750.

Falcetti, M., Scienza, A., 1989. Influence de la densité de plantation et du porte-greffe sur la production et la qualité du moût de Pinot blanc et de Chardonnay cultivés en Italie dans le Trentin. J. Int. Sci. Vigne Vin 23, 151–164.

Falginella, L., Di Gaspero, G., Castellarin, S.D., 2012. Expression of flavonoid genes in the red grape berry of 'Alicante Bouschet' varies with the histological distribution of anthocyanins and their chemical composition. Planta 236, 1037–1051.

Famiani, F., Walker, R.P., Técsi, L., Chen, Z.H., Proietti, P., Leegood, R.C., 2000. An immunohistochemical study of the compartmentation of metabolism during the development of grape (*Vitis vinifera* L.) berries. J. Exp. Bot. 51, 675–683.

Fan, S.C., Lin, C.S., Hsu, P.K., Lin, S.H., Tsay, Y.F., 2009. The *Arabidopsis* nitrate transporter NRT1.7, expressed in phloem, is responsible for source-to-sink remobilization of nitrate. Plant Cell 21, 2750–2761.

Fanciullino, A.L., Bidel, L.P.R., Urban, L., 2014. Carotenoid responses to environmental stimuli: integrating redox and carbon controls into a fruit model. Plant Cell Environ. 37, 273–289.

Fang, Y., Qian, M.C., 2006. Quantification of selected aroma-active compounds in Pinot noir wines from different grape maturities. J. Agric. Food Chem. 54, 8567–8573.

Fanizza, G., Lamaj, F., Resta, P., Ricciardi, L., Savino, V., 2005. Grapevine cvs Primitivo, Zinfandel and Crljenak kastelanski: molecular analysis by AFLP. Vitis 44, 147–148.

Fankhauser, C., Staiger, D., 2002. Photoreceptors in *Arabidopsis thaliana*: light perception, signal transduction and entrainment of the endogenous clock. Planta 216, 1–16.

Farquhar, G.D., Wetselaar, R., Firth, P.M., 1979. Ammonia volatilization from senescing leaves of maize. Science 203, 1257–1258.

Favero, A.C., Angelucci de Amorim, D., Vieira da Mota, R., Soares, A.M., de Souza, C.R., de Albuquerque Regina, M., 2011. Double-pruning of 'Syrah' grapevines: a management strategy to harvest wine grapes during the winter in the Brazilian Southeast. Vitis 50, 151–158.

Feild, T.S., Lee, D.W., Holbrook, N.M., 2001. Why leaves turn red in autumn. The role of anthocyanins in senescing leaves of red-osier dogwood. Plant Physiol. 127, 566–574.

Feldman, L.J., 1984. Regulation of root development. Annu. Rev. Plant Physiol. 35, 223–242.

Fennell, A., 2004. Freezing tolerance and injury in grapevines. In: Arora, R. (Ed.), Adaptations and Responses of Woody Plants to Environmental Stresses. Food Products Press, Binghamton, NY, pp. 201–235.

Fennell, A., Hoover, E., 1991. Photoperiod influences growth, bud dormancy, and cold acclimation in *Vitis labruscana* and *V. riparia*. J. Am. Soc. Hortic. Sci. 116, 270–273.

Ferguson, B.J., Beveridge, C.A., 2009. Roles for auxin, cytokinin, and strigolactone in regulating shoot branching. Plant Physiol. 149, 1929–1944.

Ferguson, I.B., 1984. Calcium in plant senescence and fruit ripening. Plant Cell Environ. 7, 477–489.

Ferguson, J.C., Tarara, J.M., Mills, L.J., Grove, G.G., Keller, M., 2011. Dynamic thermal time model of cold hardiness for dormant grapevine buds. Ann. Bot. 107, 389–396.

Ferguson, J.C., Moyer, M.M., Mills, L.J., Hoogenboom, G., Keller, M., 2014. Modeling dormant bud cold hardiness and budbreak in twenty-three *Vitis* genotypes reveals variation by region of origin. Am. J. Enol. Vitic. 65, 59–71.

Fernandez, L., Pradal, M., Lopez, G., Berud, F., Romieu, C., Torregrosa, L., 2006. Berry size variability in *Vitis vinifera* L. Vitis 45, 53–55.

Fernandez, L., Torregrosa, L., Segura, V., Bouquet, A., Martinez-Zapater, J.M., 2010. Transposon-induced gene activation as a mechanism generating cluster shape somatic variation in grapevine. Plant J. 61, 545–557.

Fernie, A.R., Carrari, F., Sweetlove, L.J., 2004. Respiratory metabolism: glycolysis, the TCA cycle and mitochondrial electron transport. Curr. Opin. Plant Biol. 7, 254–261.

Ferrandino, A., Carra, A., Rolle, L., Schneider, A., Schubert, A., 2012. Profiling of hydroxycinnamoyl tartrates and acylated anthocyanins in the skin of 34 *Vitis vinifera* genotypes. J. Agric. Food Chem. 60, 4931–4945.

Feucht, W., Forche, E., Porstendörfer, J., 1975. Ermittlung der Kalium- und Calciumverteilung in Traubenachsen von *Vitis vinifera* mit Hilfe der Röntgenstrahlenmikroanalyse am Rasterelektronenmikroskop. Vitis 14, 190–197.

Field, B., Jordán, F., Osbourn, A., 2006. First encounters—deployment of defence-related natural products by plants. New Phytol. 172, 193–207.

Field, S.K., Smith, J.P., Holzapfel, B.P., Hardie, W.J., Emery, R.J.N., 2009. Grapevine response to soil temperature: xylem cytokinins and carbohydrate reserve mobilization from budbreak to anthesis. Am. J. Enol. Vitic. 60, 164–172.

Filippetti, I., Intrieri, C., Cemtinari, M., Bucchetti, B., Pastore, C., 2005. Molecular characterization of officially registered Sangiovese clones and of other Sangiovese-like biotypes in Tuscany, Corsica and Emilia-Romagna. Vitis 44, 167–172.

Fillion, L., Ageorges, A., Picaud, S., Coutos-Thévenot, P., Lemoine, R., Romieu, C., et al., 1999. Cloning and expression of a hexose transporter gene expressed during the ripening of grape berry. Plant Physiol. 120, 1083–1093.

Findlay, N., Oliver, K.J., Nii, N., Coombe, B.G., 1987. Solute accumulation by grape pericarp cells. IV. Perfusion of pericarp apoplast via the pedicel and evidence for xylem malfunction in ripening berries. J. Exp. Bot. 38, 668–679.

Fine, P.M., Cass, G.R., Simoneit, B.R.T., 2001. Chemical characterization of fine particle emissions from fireplace combustion of woods grown in the northeastern United States. Environ. Sci. Technol. 35, 2665–2675.

Firn, R.D., Jones, C.G., 2009. A Darwinian view of metabolism: molecular properties determine fitness. J. Exp. Bot. 60, 719–726.

Fischer, R., Budde, I., Hain, R., 1997. Stilbene synthase gene expression causes changes in flower colour and male sterility in tobacco. Plant J. 11, 489–498.

Fischer, T.C., Mirbeth, B., Rentsch, J., Sutter, C., Ring, L., Flachowsky, H., et al., 2014. Premature and ectopic anthocyanin formation by silencing of anthocyanidin reductase in strawberry (*Fragaria* × *ananassa*). New Phytol. 201, 440–451.

Fischer, W.N., André, B., Rentsch, D., Krolkiewicz, S., Tegeder, M., Breitkreuz, K., et al., 1998. Amino acid transport in plants. Trends Plant Sci. 3, 188–195.

Fishman, S., Génard, M., 1998. A biophysical model of fruit growth: simulation of seasonal and diurnal dynamics of mass. Plant Cell Environ. 21, 739–752.

Flexas, J., Escalona, J.M., Medrano, H., 1998. Down-regulation of photosynthesis by drought under field conditions in grapevine leaves. Aust. J. Plant Physiol. 25, 893–900.

Flexas, J., Badger, M., Chow, W.S., Medrano, H., Osmond, C.B., 1999a. Analysis of the relative increase in photosynthetic O_2 uptake when photosynthesis in grapevine leaves is inhibited following low night temperatures and/or water stress. Plant Physiol. 121, 675–684.

Flexas, J., Escalona, J.M., Medrano, H., 1999b. Water stress induces different levels of photosynthesis and electron transport rate regulation in grapevines. Plant Cell Environ. 22, 39–48.

Flexas, J., Bota, J., Escalona, J.M., Sampol, B., Medrano, H., 2002. Effects of drought on photosynthesis in grapevines under field conditions: an evaluation of stomatal and mesophyll limitations. Funct. Plant Biol. 29, 461–471.

Flore, J.A., Lakso, A.N., 1989. Environmental and physiological regulation of photosynthesis in fruit crops. Hortic. Rev. (Am. Soc. Hortic. Sci) 11, 111–157.

Fontes, N., Côrte-Real, M., Gerós, H., 2011a. New observations on the integrity, structure, and physiology of flesh cells from fully ripened grape berry. Am. J. Enol. Vitic. 62, 279–284.

Fontes, N., Gerós, H., Delrot, S., 2011b. Grape berry vacuole: a complex and heterogeneous membrane system specialized in the accumulation of solutes. Am. J. Enol. Vitic. 62, 270–278.

Fonti, P., Solomonoff, N., García-González, I., 2007. Earlywood vessels of *Castanea sativa* record temperature before their formation. New Phytol. 173, 562–570.

Forde, B.G., 2002. Local and long-range signaling pathways regulating plant responses to nitrate. Annu. Rev. Plant Biol. 53, 203–224.

Forde, C.G., Cox, A., Williams, E.R., Boss, P.K., 2011. Associations between the sensory attributes and volatile composition of Cabernet Sauvignon wines and the volatile composition of the grapes used for their production. J. Agric. Food Chem. 59, 2573–2583.

Forkmann, G., 1991. Flavonoids as flower pigments: the formation of the natural spectrum and its extension by genetic engineering. Plant Breed. 106, 1–26.

Forneck, A., Huber, L., 2009. (A)sexual reproduction—a review of life cycles of grape phylloxera, *Daktulosphaira vitifoliae*. Entomol. Exper. Appl. 131, 1–10.

Fougère-Rifot, M., Park, H.S., Bouard, J., 1995. Données nouvelles sur l'hypoderme et la pulpe des baies normales et des baies millerandées d'une variété de *Vitis vinifera* L., le Merlot noir. Vitis 34, 1–7.

Fournand, D., Vicens, A., Sidhoum, L., Souquet, J.M., Moutounet, M., Cheynier, V., 2006. Accumulation and extractability of grape skin tannins and anthocyanins at different advanced physiological stages. J. Agric. Food Chem. 54, 7331–7338.

Fournier-Level, A., Le Cunff, L., Gomez, C., Doligez, A., Ageorges, A., Roux, C., et al., 2009. Quantitative genetic bases of anthocyanin variation in grape (*Vitis vinifera* L. ssp. *sativa*) berry: a quantitative trait locus to quantitative trait nucleotide integrated study. Genetics 183, 1127–1139.

Fournier-Level, A., Lacombe, T., Le Cunff, L., Boursiquot, J.M., This, P., 2010. Evolution of the *VvMybA* gene family, the major determinant of berry colour in cultivated grapevine (*Vitis vinifera* L.). Heredity 104, 351–362.

Foyer, C.H., Noctor, G., 2005. Oxidant and antioxidant signalling in plants: a re-evaluation of the concept of oxidative stress in a physiological context. Plant Cell Environ. 28, 1056–1071.

Foyer, C.H., Parry, M., Noctor, G., 2003. Markers and signals associated with nitrogen assimilation in higher plants. J. Exp. Bot. 54, 585–593.

Foyer, C.H., Bloom, A.J., Queval, G., Noctor, G., 2009. Photorespiratory metabolism: genes, mutants, energetics, and redox signaling. Annu. Rev. Plant Biol. 60, 455–484.

Fracheboud, Y., Luquez, V., Björkén, L., Sjödin, A., Tuominen, H., Jansson, S., 2009. The control of autumn senescence in European aspen. Plant Physiol. 149, 1982–1991.

Franceschi, V.R., Nakata, P.A., 2005. Calcium oxalate in plants: formation and function. Annu. Rev. Plant Biol. 56, 41–71.

Franck, N., Morales, J.P., Arancibia-Avendaño, D., García de Cortázar, V., Perez-Quezada, J.F., Zurita-Silva, A., et al., 2011. Seasonal fluctuations in *Vitis vinifera* root respiration in the field. New Phytol. 192, 939–951.

Frank, D., Gould, I., Millikan, M., 2004. Browning reactions during storage of low-moisture Australian sultanas: evidence for arginine-mediated Maillard reactions. Aust. J. Grape Wine Res. 10, 151–163.

Franke, R., Schreiber, L., 2007. Suberin—a biopolyester forming apoplastic plant interfaces. Curr. Opin. Plant Biol. 10, 252–259.

Franklin, K.A., 2008. Shade avoidance. New Phytol. 179, 930–944.

Franklin, K.A., Whitelam, G.C., 2007. Light-quality regulation of freezing tolerance in *Arabidopsis thaliana*. Nat. Genet. 39, 1410–1413.

Franks, P.J., Adams, M.A., Amthor, J.S., Barbour, M.M., Berry, J.A., Ellsworth, D.S., et al., 2013. Sensitivity of plants to changing atmospheric CO_2 concentration: from the geological past to the next century. New Phytol. 197, 1077–1094.

Franks, P.J., Farquhar, G.D., 2007. The mechanical diversity of stomata and its significance in gas-exchange control. Plant Physiol. 143, 78–87.

Franks, T., Botta, R., Thomas, M.R., 2002. Chimerism in grapevines: implications for cultivar identity, ancestry and genetic improvement. Theor. Appl. Genet. 104, 192–199.

Fraser, P.D., Truesdale, M.R., Bird, C.R., Schuch, W., Bramley, P.M., 1994. Carotenoid biosynthesis during tomato fruit development. Plant Physiol. 105, 405–413.

Freeman, B.M., 1983. Effects of irrigation and pruning of Shiraz grapevines on subsequent red wine pigments. Am. J. Enol. Vitic. 34, 23–26.

Freeman, B.M., Kliewer, W.M., 1983. Effect of irrigation, crop level and potassium fertilization on Carignane vines. II. Grape and wine quality. Am. J. Enol. Vitic. 34, 197–207.

Freeman, B.M., Lee, T.H., Turkington, C.R., 1979. Interaction of irrigation and pruning level on growth and yield of Shiraz vines. Am. J. Enol. Vitic. 30, 218–223.

Freeman, B.M., Kliewer, W.M., Stern, P., 1982. Influence of windbreaks and climatic region on diurnal fluctuation of leaf water potential, stomatal conductance, and leaf temperature of grapevines. Am. J. Enol. Vitic. 33, 233–236.

Friend, A.P., Trought, M.C.T., Stushnoff, C., Wells, G.H., 2011. Effect of delaying budburst on shoot development and yield of *Vitis vinifera* L. Chardonnay 'Mendoza' after a spring freeze event. Aust. J. Grape Wine Res. 17, 378–382.

Frigerio, M., Alabadí, D., Pérez-Gómez, J., García-Cárcel, L., Phillips, A.L., Hedden, P., et al., 2006. Transcriptional regulation of gibberellin metabolism genes by auxin signaling in *Arabidopsis*. Plant Physiol. 142, 553–563.

Friml, J., 2003. Auxin transport—shaping the plant. Curr. Opin. Plant Biol. 6, 7–12.

Fritschi, F.B., Lin, H., Walker, M.A., 2007. *Xylella fastidiosa* population dynamics in grapevine genotypes differing in susceptibility to Pierce's disease. Am. J. Enol. Vitic. 58, 326–332.

Fritz, C., Palacios-Rojas, N., Feil, R., Stitt, M., 2006. Regulation of secondary metabolism by the carbon-nitrogen status in tobacco: nitrate inhibits large sectors of phenylpropanoid metabolism. Plant J. 46, 533–548.

Fromm, J., Lautner, S., 2007. Electrical signals and their physiological significance in plants. Plant Cell Environ. 30, 249–257.

Fu, Q., Cheng, L., Guo, Y., Turgeon, R., 2011. Phloem loading strategies and water relations in trees and herbaceous plants. Plant Physiol. 157, 1518–1527.

Fu, X., Harberd, N.P., 2003. Auxin promotes *Arabidopsis* root growth by modulating gibberellin response. Nature 421, 740–743.

Fujioka, S., Yokota, T., 2003. Biosynthesis and metabolism of brassinosteroids. Annu. Rev. Plant Biol. 54, 137–164.

Fukui, M., Yokotsuka, K., 2003. Content and origin of protein in white and red wines: changes during fermentation and maturation. Am. J. Enol. Vitic. 54, 178–188.

Fulcrand, H., Dueñas, M., Salas, E., Cheynier, V., 2006. Phenolic reactions during winemaking and aging. Am. J. Enol. Vitic. 57, 289–297.

Fuller, M.P., Telli, G., 1999. An investigation of the frost hardiness of grapevine (*Vitis vinifera*) during bud break. Ann. Appl. Biol. 135, 589–595.

Fung, R.W.M., Gonzalo, M., Fekete, C., Kovacs, L.G., He, Y., Marsh, E., et al., 2008. Powdery mildew induces defense-oriented reprogramming of the transcriptome in a susceptible but not in a resistant grapevine. Plant Physiol. 146, 236–249.

Furch, A.C.U., Hafke, J.B., Schulz, A., van Bel, A.J.E., 2007. Ca^{2+}-mediated remote control of reversible sieve tube occlusion in *Vicia faba*. J. Exp. Bot. 58, 2827–2838.

Furiya, T., Suzuki, S., Sueta, T., Takayanagi, T., 2009. Molecular characterization of a bud sport of Pinot gris bearing white berries. Am. J. Enol. Vitic. 60, 66–73.

Gadoury, D.M., Pearson, R.C., 1990. Germination of ascospores and infection of *Vitis* by *Uncinula necator*. Phytopathology 80, 1198–1203.

Gadoury, D.M., Seem, R.C., Ficke, A., Wilcox, W.F., 2003. Ontogenic resistance to powdery mildew in grape berries. Phytopathology 93, 547–555.

Gadoury, D.M., Seem, R.C., Wilcox, W.F., Henick-Kling, T., Conterno, L., Day, A., et al., 2007. Effects of diffuse colonization of grape berries by *Uncinula necator* on bunch rots, berry microflora, and juice and wine quality. Phytopathology 97, 1356–1365.

Gaforio, L., García-Muñoz, S., Cabello, F., Muñoz-Organero, G., 2011. Evaluation of susceptibility to powdery mildew (*Erysiphe necator*) in *Vitis vinifera* varieties. Vitis 50, 123–126.

Gagné, S., Saucier, C., Gény, L., 2006. Composition and cellular localization of tannins in Cabernet Sauvignon skins during growth. J. Agric. Food Chem. 54, 9465–9471.

Gago, P., Santiago, J.L., Boso, S., Alonso-Villaverde, V., Grando, M.S., Martínez, M.C., 2009. Biodiversity and characterization of twenty-two *Vitis vinifera* L. cultivars in the northwestern Iberian Peninsula. Am. J. Enol. Vitic. 60, 293–301.

Gajdanowicz, P., Michard, E., Sandmann, M., Rocha, M., Guedes Corrêa, L.G., Ramírez-Aguilar, S.J., et al., 2011. Potassium (K^+) gradients serve as a mobile energy source in plant vascular tissues. Proc. Natl. Acad. Sci. USA 108, 864–869.

Galet, P., 1985. Précis d'Ampélographie Pratique. fifth ed. Déhan, Montpellier, France.

Galet, P., 1996. Grape Diseases (J. Smith, Trans.). Oenoplurimédia, Chaintré, France.

Galet, P., 1998. Grape Varieties and Rootstock Varieties (J. Smith, Trans.). Oenoplurimédia, Chaintré, France.

Galet, P., 2000. General Viticulture (J. Smith, Trans.). Oenoplurimédia, Chaintré, France.

Gallavotti, A., 2013. The role of auxin in shaping shoot architecture. J. Exp. Bot. 64, 2593–2608.

Galmés, J., Pou, A., Alsina, M.M., Tomàs, M., Medrano, H., Flexas, J., 2007. Aquaporin expression in response to different water stress intensities and recovery in Richter-110 (*Vitis* sp.): relationship with ecophysiological status. Planta 226, 671–681.

Galvez, D.A., Landhäusser, S.M., Tyree, M.T., 2013. Low root reserve accumulation during drought may lead to winter mortality in poplar seedlings. New Phytol. 198, 139–148.

Gamalei, Y., 1989. Structure and function of leaf minor veins in trees and herbs. A taxonomic review. Trees 3, 96–110.

Gambetta, G.A., Matthews, M.A., Shaghasi, T.H., McElrone, A.J., Castellarin, S.D., 2010. Sugar and abscisic acid signaling orthologs are activated at the onset of ripening in grape. Planta 232, 219–234.

Gambetta, G.A., Fei, J., Rost, T.L., Knipfer, T., Matthews, M.A., Shackel, K.A., et al., 2013. Water uptake along the length of grapevine fine roots: developmental anatomy, tissue-specific aquaporin expression, and pathways of water transport. Plant Physiol. 163, 1254–1265.

Gambetta, J.M., Bastian, S.E.P., Cozzolino, D., Jeffery, D.W., 2014. Factors influencing the aroma composition of Chardonnay wines. J. Agric. Food Chem. 62, 6512–6534.

Gamon, J.A., Pearcy, R.W., 1989. Leaf movement, stress avoidance and photosynthesis in *Vitis californica*. Oecologia 79, 475–481.

Gamon, J.A., Pearcy, R.W., 1990a. Photoinhibition in *Vitis californica*: interactive effects of sunlight, temperature and water status. Plant Cell Environ. 13, 267–275.

Gamon, J.A., Pearcy, R.W., 1990b. Photoinhibition in *Vitis californica*. The role of temperature during high-light treatment. Plant Physiol. 92, 487–494.

Gang, D.R., 2005. Evolution of flavors and scents. Annu. Rev. Plant Biol. 56, 301–325.

García-Arenal, F., Fraile, A., Malpica, J.M., 2001. Variability and genetic structure of plant virus populations. Annu. Rev. Phytopathol. 39, 157–186.

Garris, A., Clark, L., Owens, C., McKay, S., Luby, J., Mathiason, K., et al., 2009. Mapping of photoperiod-induced growth cessation in the wild grape *Vitis riparia*. J. Am. Soc. Hortic. Sci. 134, 261–272.

Gärtel, W., 1993. Grapes. In: Bennett, W.F. (Ed.), Nutrient Deficiencies and Toxicities in Crop Plants. APS Press, St. Paul, MN, pp. 177–183.

Gastal, F., Lemaire, G., 2002. N uptake and distribution in crops: an agronomical and ecophysiological perspective. J. Exp. Bot. 53, 789–799.

Gatti, M., Bernizzoni, F., Civardi, S., Poni, S., 2012. Effects of cluster thinning and preflowering leaf removal on growth and grape composition in cv. Sangiovese. Am. J. Enol. Vitic. 63, 325–332.

Gauthier, P.P.G., Bligny, R., Gout, E., Mahé, A., Nogués, S., Hodges, M., et al., 2010. *In folio* isotopic tracing demonstrates that nitrogen assimilation into glutamate is mostly independent from current CO_2 assimilation in illuminated leaves of *Brassica napus*. New Phytol. 185, 988–999.

Gawel, R., 1998. Red wine astringency: a review. Aust. J. Grape Wine Res. 4, 74–95.

Gebbing, H., Schwab, A., Alleweldt, G., 1977. Mykorrhiza der Rebe. Vitis 16, 279–285.

Geiger, D., 2011. Plant sucrose transporters from a biophysical point of view. Mol. Plant 4, 395–406.

Geiger, D.R., Servaites, J.C., 1991. Carbon allocation and responses to stress. In: Mooney, H.A., Winner, W.E., Pell, E.J. (Eds.), Response of Plants to Multiple Stresses. Academic Press, San Diego, CA, pp. 103–127.

Geiger, D.R., Servaites, J.C., Fuchs, M.A., 2000. Role of starch in carbon translocation and partitioning at the plant level. Aust. J. Plant Physiol. 27, 571–582.

Geisler, G., Radler, R., 1963. Entwicklungs- und Reifevorgänge an Trauben von *Vitis*. Ber. Dtsch. Bot. Ges. 76, 112–119.

Geisler, G., Staab, J., 1958. Versuchsanstellung im Weinbau. Vitis 1, 257–281.

Geldner, N., 2013. The endodermis. Annu. Rev. Plant Biol. 64, 531–558.

Gény, L., Broquedis, M., Martin-Tanguy, J., Bouard, J., 1997a. Free, conjugated, and wall-bound polyamines in various organs of fruiting cuttings of *Vitis vinifera* L. cv. Cabernet Sauvignon. Am. J. Enol. Vitic. 48, 80–84.

Gény, L., Broquedis, M., Martin-Tanguy, J., Soyer, J.P., Bouard, J., 1997b. Effects of potassium nutrition on polyamine content of various organs of fruiting cuttings of *Vitis vinifera* L. cv. Cabernet Sauvignon. Am. J. Enol. Vitic. 48, 85–92.

Gény, L., Ollat, N., Soyer, J.P., 1998. Les boutures fructifères de vigne: validation d'un modèle d'étude de la physiologie de la vigne. II. Étude de développement de la grappe. J. Int. Sci. Vigne Vin 32, 83–90.

Gény, L., Saucier, C., Bracco, S., Daviaud, F., Glories, Y., 2003. Composition and cellular localization of tannins in grape seeds during maturation. J. Agric. Food Chem. 51, 8051–8054.

German, J.B., Walzem, R.L., 2000. The health benefits of wine. Annu. Rev. Nutr. 20, 561–593.

Gerrath, J.M., 1992. Developmental morphology and anatomy of grape flowers. Hortic. Rev. (Am. Soc. Hortic. Sci). 13, 315–337.

Gerrath, J.M., Posluszny, U., 1988a. Morphological and anatomical development in the Vitaceae. I. Vegetative development in *Vitis riparia*. Can. J. Bot. 66, 209–224.

Gerrath, J.M., Posluszny, U., 1988b. Morphological and anatomical development in the Vitaceae. II. Floral development in *Vitis riparia*. Can. J. Bot. 66, 1334–1351.

Gerrath, J.M., Posluszny, U., 2007. Shoot architecture in the Vitaceae. Can. J. Bot. 85, 691–700.

Gerrath, J.M., Posluszny, U., Dengler, N.G., 2001. Primary vascular patterns in the Vitaceae. Int. J. Plant Sci. 162, 729–745.

Gessler, A., Schultze, M., Schrempp, S., Rennenberg, H., 1998. Interaction of phloem-translocated amino compounds with nitrate net uptake by the roots of beech (*Fagus sylvatica*) seedlings. J. Exp. Bot. 49, 1529–1537.

Ghaffari, S., Hasnaoui, N., Zinelabidine, L.H., Ferchichi, A., Martínez-Zapater, J.M., Ibáñez, J., 2013. Genetic identification and origin of grapevine cultivars (*Vitis vinifera* L.) in Tunisia. Am. J. Enol. Vitic. 64, 538–544.

Gholami, M., Hayasaka, Y., Coombe, B.G., Jackson, J.F., Robinson, S.P., Williams, P.J., 1995. Biosynthesis of flavour compounds in Muscat Gordo Blanco grape berries. Aust. J. Grape Wine Res. 1, 19–24.

Giacomelli, L., Rota-Stabelli, O., Masuero, D., Acheampong, A.K., Moretto, M., Caputi, L., et al., 2013. Gibberellin metabolism in *Vitis vinifera* L. during bloom and fruit-set: functional characterization and evolution of grapevine gibberellin oxidases. J. Exp. Bot. 64, 4403–4419.

Giannakis, C., Bucheli, C.S., Skene, K.G.M., Robinson, S.P., Scott, N.S., 1998. Chitinase and ß-1,3-glucanase in grapevine leaves: a possible defence against powdery mildew infection. Aust. J. Grape Wine Res. 4, 14–22.

Gifford, M.L., Dean, A., Gutierrez, R.A., Coruzzi, G.M., Birnbaum, K.D., 2008. Cell-specific nitrogen responses mediate developmental plasticity. Proc. Natl. Acad. Sci. USA 105, 803–808.
Gillaspy, G., Ben-David, H., Gruissem, W., 1993. Fruits: a developmental perspective. Plant Cell 5, 1439–1451.
Gilliham, M., Dayod, M., Hocking, B.J., Xu, B., Conn, S.J., Kaiser, B.N., et al., 2011. Calcium delivery and storage in plant leaves: exploring the link with water flow. J. Exp. Bot. 62, 2233–2250.
Gilroy, S., Jones, D.L., 2000. Through form to function: root hair development and nutrient uptake. Trends Plant Sci. 5, 56–60.
Gindro, K., Pezet, R., Viret, O., 2003. Histological study of the response of two *Vitis vinifera* cultivars (resistant and susceptible) to *Plasmopara viticola* infections. Plant Physiol. Biochem. 41, 846–853.
Girbau, T., Stummer, B.E., Pocock, K.F., Baldock, G.A., Scott, E.S., Waters, E.J., 2004. The effect of *Uncinula necator* (powdery mildew) and *Botrytis cinerea* infection of grapes on the levels of haze-forming pathogenesis-related proteins in grape juice and wine. Aust. J. Grape Wine Res. 10, 125–133.
Giribaldi, M., Perugini, I., Sauvage, F.X., Schubert, A., 2007. Analysis of protein changes during grape berry ripening by 2-DE and MALDI-TOF. Proteomics 7, 3154–3170.
Glad, C., Regnard, J.L., Querou, Y., Brun, O., Morot-Gaudry, J.F., 1992. Flux and chemical composition of xylem exudates from Chardonnay grapevines: temporal evolution and effect of recut. Am. J. Enol. Vitic. 43, 275–282.
Gladstone, E.A., Dokoozlian, N.K., 2003. Influence of leaf area density and trellis/training system on the light microclimate within grapevine canopies. Vitis 42, 123–131.
Gladstones, J., 1992. Viticulture and Environment. Winetitles, Adelaide, Australia.
Gleitz, J., Schnitzler, J.P., Steimle, D., Seitz, H.U., 1991. Metabolic changes in carrot cells in response to simultaneous treatment with ultraviolet light and a fungal elicitor. Planta 184, 362–367.
Gloser, V., Zwieniecki, M.A., Orians, C.M., Holbrook, N.M., 2007. Dynamic changes in root hydraulic properties in response to nitrate availability. J. Exp. Bot. 58, 2409–2415.
Godinho, S., Paulo, O.S., Morais-Cecílio, L., Rocheta, M., 2012. A new *gypsy*-like retroelement family in *Vitis vinifera*. Vitis 51, 65–72.
Goes da Silva, F., Iandolino, A., Al-Kayal, F., Bohlmann, M.C., Cushman, M.A., Lim, H., et al., 2005. Characterizing the grape transcriptome. Analysis of expressed sequence tags from multiple *Vitis* species and development of a compendium of gene expression during berry development. Plant Physiol. 139, 574–597.
Goetz, G., Fkyerat, A., Métais, N., Kunz, M., Tabacchi, R., Pezet, R., et al., 1999. Resistance factors to grey mould in grape berries: identification of some phenolics inhibitors of *Botrytis cinerea* stilbene oxidase. Phytochemistry 52, 759–767.
Goetz, M., Godt, D.E., Guivarc'h, A., Kahmann, U., Chriqui, D., Roitsch, T., 2001. Induction of male sterility in plants by metabolic engineering of the carbohydrate supply. Proc. Natl. Acad. Sci. USA 98, 6522–6527.
Goff, S.A., Klee, H.J., 2006. Plant volatile compounds: sensory cues for health and nutritional value? Science 311, 815–819.
Goheen, A.C., Cook, J.A., 1959. Leafroll (red-leaf or rougeau) and its effects on vine growth, fruit quality, and yields. Am. J. Enol. Vitic. 10, 173–181.
Gojon, A., Nacry, P., Davidian, J.C., 2009. Root uptake regulation: a central process for NPS homeostasis in plants. Curr. Opin. Plant Biol. 12, 328–338.
Goldberg, R., Morvan, C., Jauneau, A., Jarvis, M.C., 1996. Methyl-esterification, de-esterification and gelation of pectins in the primary cell wall. Prog. Biotechnol. 14, 151–172.
Gollop, R., Even, S., Colova-Tsolova, V., Perl, A., 2002. Expression of the grape dihydroflavonol reductase gene and analysis of its promoter region. J. Exp. Bot. 53, 1397–1409.
Gomez, C., Conejero, G., Torregrosa, L., Cheynier, V., Terrier, N., Ageorges, A., 2011. *In vivo* grapevine anthocyanin transport involves vesicle-mediated trafficking and the contribution of anthoMATE transporters and GST. Plant J. 67, 960–970.

Gomez-Roldan, V., Fermas, S., Brewer, P.B., Puech-Pagès, V., Dun, E.A., Pillot, J.P., et al., 2008. Strigolactone inhibition of shoot branching. Nature 455, 189−194.

Gong, H., Blackmore, D.H., Walker, R.R., 2010. Organic and inorganic anions in Shiraz and Chardonnay grape berries and wine as affected by rootstock under saline conditions. Aust. J. Grape Wine Res. 16, 227−236.

Gonzalez, N., Vanhaeren, H., Inzé, D., 2012. Leaf size control: complex coordination of cell division and expansion. Trends Plant Sci. 17, 332−340.

González-Centeno, M.R., Jourdes, M., Femenia, A., Simal, S., Rosselló, C., Teissedre, P.L., 2013. Characterization of polyphenols and antioxidant potential of white grape pomace byproducts (*Vitis vinifera* L.). J. Agric. Food Chem. 61, 11579−11587.

Goodman, R.N., Grimm, R., Frank, M., 1993. The influence of grape rootstocks on the crown gall infection process and on tumor development. Am. J. Enol. Vitic. 44, 22−26.

Goremykin, V.V., Salamini, F., Velasco, R., Viola, R., 2009. Mitochondrial DNA of *Vitis vinifera* and the issue of rampant horizontal gene transfer. Mol. Biol. Evol. 26, 99−110.

Gorska, A., Ye, Q., Holbrook, N.M., Zwieniecki, M.A., 2008. Nitrate control of root hydraulic properties in plants: translating local information to whole plant response. Plant Physiol. 148, 1159−1167.

Grace, J., 2006. The temperature of buds may be higher than you thought. New Phytol. 170, 1−3.

Graf, A., Smith, A.M., 2011. Starch and the clock: the dark side of plant productivity. Trends Plant Sci. 16, 169−175.

Granett, J., Walker, M.A., Kocsis, L., Omer, A.D., 2001. Biology and management of grape phylloxera. Annu. Rev. Entomol. 46, 387−412.

Grant, O.M., Tronina, L., Jones, H.G., Chaves, M.M., 2007. Exploring thermal imaging variables for the detection of stress responses in grapevine under different irrigation regimes. J. Exp. Bot. 58, 815−825.

Grant, R.S., Matthews, M.A., 1996a. The influence of phosphorus availability and rootstock on root system characteristics, phosphorus uptake, phosphorus partitioning, and growth efficiency. Am. J. Enol. Vitic. 47, 403−409.

Grant, R.S., Matthews, M.A., 1996b. The influence of phosphorus availability, scion, and rootstock on grapevine shoot growth, leaf area, and petiole phosphorus concentration. Am. J. Enol. Vitic. 47, 217−224.

Grassi, F., Labra, M., Imazio, S., Spada, A., Sgorbati, S., Scienza, A., et al., 2003. Evidence of a secondary grapevine domestication centre detected by SSR analysis. Theor. Appl. Genet. 107, 1315−1320.

Grassi, F., Labra, M., Imazio, S., Ocete Rubio, R., Failla, O., Scienza, A., et al., 2006. Phylogeographical structure and conservation genetics of wild grapevine. Conserv. Genet. 7, 837−845.

Gray, J.D., Kolesik, P., Høj, P.B., Coombe, B.G., 1999. Confocal measurement of the three-dimensional size and shape of plant parenchyma cells in a developing fruit tissue. Plant J. 19, 229−236.

Green, B.R., 2011. Chloroplast genomes of photosynthetic eukaryotes. Plant J. 66, 34−44.

Green, M.A., Fry, S.C., 2005. Vitamin C degradation in plant cells via enzymatic hydrolysis of 4-O-oxalyl-L-threonate. Nature 433, 83−87.

Greenspan, M.D., Shackel, K.A., Matthews, M.A., 1994. Developmental changes in the diurnal water budget of the grape berry exposed to water deficits. Plant Cell Environ. 17, 811−820.

Greenspan, M.D., Schultz, H.R., Matthews, M.A., 1996. Field evaluation of water transport in grape berries during water deficits. Physiol. Plant. 97, 55−62.

Greer, D.H., Weedon, M.M., 2012. Modelling photosynthetic responses to temperature of grapevine (*Vitis vinifera* cv. Semillon) leaves on vines grown in a hot climate. Plant Cell Environ. 35, 1050−1064.

Greer, D.H., Weedon, M.M., 2014. Does the hydrocooling of *Vitis vinifera* cv. Semillon vines protect the vegetative and reproductive growth processes and vine performance against high summer temperatures? Funct. Plant Biol. 41, 620−633.

Greer, D.H., Weston, C., 2010a. Effects of fruiting on vegetative growth and development dynamics of grapevines (*Vitis vinifera* cv. Semillon) can be traced back to events at or before budbreak. Funct. Plant Biol. 37, 756–766.

Greer, D.H., Weston, C., 2010b. Heat stress affects flowering, berry growth, sugar accumulation and photosynthesis of *Vitis vinifera* cv. Semillon grapevines grown in a controlled environment. Funct. Plant Biol. 37, 206–214.

Gregan, S.M., Wargent, J.J., Liu, L., Shinkle, J., Hoffmann, R., Winefield, C., et al., 2012. Effects of solar ultraviolet radiation and canopy manipulation on the biochemical composition of Sauvignon blanc grapes. Aust. J. Grape Wine Res. 18, 227–238.

Grenan, S., 1984. Polymorphisme foliaire consécutif à la culture *in vitro* de *Vitis vinifera* L. Vitis 23, 159–174.

Griffith, M., Yaish, M.W.F., 2004. Antifreeze proteins in overwintering plants: a tale of two activities. Trends Plant Sci. 9, 399–405.

Grncarevic, M., Radler, F., 1971. A review of the surface lipids of grapes and their importance in the drying process. Am. J. Enol. Vitic. 22, 80–86.

Groot Obbink, J., Alexander, D. McE., 1973. Response of six grapevine cultivars to a range of chloride concentrations. Am. J. Enol. Vitic. 24, 65–68.

Groot Obbink, J., Alexander, D.McE., Possingham, J.V., 1973. Use of nitrogen and potassium reserves during growth of grape vine cuttings. Vitis 12, 207–213.

Groover, A., Robischon, M., 2006. Developmental mechanisms regulating secondary growth in woody plants. Curr. Opin. Plant Biol. 9, 55–58.

Grossman, A., Takahashi, H., 2001. Macronutrient utilization by photosynthetic eukaryotes and the fabric of interactions. Annu. Rev. Plant Physiol. Plant Mol. Biol. 52, 163–210.

Gruber, B., Kosegarten, H., 2002. Depressed growth of non-chlorotic vine grown in calcareous soil is an iron deficiency symptom prior to leaf chlorosis. J. Plant Nutr. Soil Sci. 165, 111–117.

Grzegorczyk, W., Walker, M.A., 1998. Evaluating resistance to grape phylloxera in *Vitis* species with an *in vitro* dual culture assay. Am. J. Enol. Vitic. 49, 17–22.

Gu, L., Hanson, P.J., Post, W.M., Kaiser, D.P., Yang, B., Nemani, R., et al., 2008. The 2007 eastern US spring freeze: increased cold damage in a warming world? BioScience 58, 253–262.

Gu, S., Jacobs, S., Du, G., 2011. Efficacy, rate and timing of applications of abscisic acid to enhance fruit anthocyanin contents in 'Cabernet Sauvignon' grapes. J. Hortic. Sci. Biotechnol. 86, 505–510.

Gu, S., Jacobs, S.D., McCarthy, B.S., Gohl, H.L., 2012. Forcing vine regrowth and shifting fruit ripening in a warm region to enhance fruit quality in 'Cabernet Sauvignon' grapevine (*Vitis vinifera* L.). J. Hortic. Sci. Biotechnol. 87, 287–292.

Guerra, B., Steenwerth, K., 2012. Influence of floor management technique on grapevine growth, disease pressure, and juice and wine composition: a review. Am. J. Enol. Vitic. 63, 149–164.

Guilford, J.M., Pezzuto, J.M., 2011. Wine and health: a review. Am. J. Enol. Vitic. 62, 471–486.

Guillaumie, S., Ilg, A., Réty, S., Brette, M., Trossat-Magnin, C., Decroocq, S., et al., 2013. Genetic analysis of the biosynthesis of 2-methoxy-3-isobutylpyrazine, a major grape-derived aroma compound impacting wine quality. Plant Physiol. 162, 604–615.

Gunata, Y.Z., Bayonove, C.L., Baumes, R.L., Cordonnier, R.E., 1985. The aroma of grapes. Localisation and evolution of free and bound fractions of some grape aroma components c.v. Muscat during first development and maturation. J. Sci. Food Agric. 36, 857–862.

Guo, C., Guo, R., Xu, X., Gao, M., Li, X., Song, J., et al., 2014. Evolution and expression analysis of the grape (*Vitis vinifera* L.) *WRKY* gene family. J. Exp. Bot. 65, 1513–1528.

Guo, X.W., Fu, W.H., Wang, G.J., 1987. Studies on cold hardiness of grapevine roots. Vitis 26, 161–171.

Gusta, L.V., Wisniewski, M., 2013. Understanding plant cold hardiness: an opinion. Physiol. Plant. 147, 4–14.

Guth, H., 1997a. Identification of character impact odorants of different white wine varieties. J. Agric. Food Chem. 45, 3022–3026.

Guth, H., 1997b. Quantitation and sensory studies of character impact odorants of different white wine cultivars. J. Agric. Food Chem. 45, 3027–3032.

Gutha, L.R., Casassa, L.F., Harbertson, J.F., Naidu, R.A., 2010. Modulation of flavonoid biosynthetic pathway genes and anthocyanins due to virus infection in grapevine (*Vitis vinifera* L.) leaves. BMC Plant Biol. 10, 187, http://dx.doi.org/10.1186/1471-2229-10-187.

Gutierrez, L., Van Wuytswinkel, O., Castelain, M., Bellini, C., 2007. Combined networks regulating seed maturation. Trends Plant Sci. 12, 294–300.

Gutiérrez-Granda, M.J., Morrison, J.C., 1992. Solute distribution and malic enzyme activity in developing grape berries. Am. J. Enol. Vitic. 43, 323–328.

Guy, C., Kaplan, F., Kopka, J., Selbig, J., Hincha, D.K., 2008. Metabolomics of temperature stress. Physiol. Plant. 132, 220–235.

Guy, C.L., 1990. Cold acclimation and freezing stress tolerance: role of protein metabolism. Annu. Rev. Plant Physiol. Plant Mol. Biol. 41, 187–223.

Ha, S., Vankova, R., Yamaguchi-Shinozaki, K., Shinozaki, K., Tran, L.S.P., 2012. Cytokinins: metabolism and function in plant adaptation to environmental stresses. Trends Plant Sci. 17, 172–179.

Hacke, U.G., Sperry, J.S., Wheeler, J.K., Castro, L., 2006. Scaling of angiosperm xylem structure with safety and efficiency. Tree Physiol. 26, 689–701.

Hafke, J.B., van Amerongen, J.K., Kelling, F., Furch, A.C.U., Gaupels, F., van Bel, A.J.E., 2005. Thermodynamic battle for photosynthate acquisition between sieve tubes and adjoining parenchyma in transport phloem. Plant Physiol. 138, 1527–1537.

Hale, C.R., 1962. Synthesis of organic acids in the fruit of the grape. Nature 195, 917–918.

Hale, C.R., 1977. Relation between potassium and the malate and tartrate contents of grape berries. Vitis 16, 9–19.

Hale, C.R., Buttrose, M.S., 1974. Effect of temperature on ontogeny of berries of *Vitis vinifera* L. cv. Cabernet Sauvignon. J. Am. Soc. Hortic. Sci. 99, 390–394.

Hale, C.R., Weaver, R.J., 1962. The effect of developmental stage on direction of translocation of photosynthate in *Vitis vinifera*. Hilgardia 33, 89–131.

Hales, S., 1727. Vegetable Staticks. W. and J. Innys and T. Woodward, London.

Hall, A., Jones, G.V., 2010. Spatial analysis of climate in winegrape-growing regions in Australia. Aust. J. Grape Wine Res. 16, 389–404.

Hall, A., Lamb, D.W., Holzapfel, B., Louis, J., 2002. Optical remote sensing applications in viticulture—a review. Aust. J. Grape Wine Res. 8, 36–47.

Hall, G.E., Bondada, B.R., Keller, M., 2011. Loss of rachis cell viability is associated with ripening disorders in grapes. J. Exp. Bot. 62, 1145–1153.

Hall, J.L., Williams, L.E., 2000. Assimilate transport and partitioning in fungal biotrophic interactions. Aust. J. Plant Physiol. 27, 549–560.

Hall, T.W., Mahaffee, W.F., 2001. Impact of vine shelter use on development of grape powdery mildew. Am. J. Enol. Vitic. 52, 204–209.

Halliwell, B., 2006. Reactive species and antioxidants. Redox biology is a fundamental theme of aerobic life. Plant Physiol. 141, 312–322.

Halliwell, B., 2007. Dietary polyphenols: good, bad, or indifferent for your health? Cardiovasc. Res. 73, 341–347.

Hamant, O., Traas, J., 2010. The mechanics behind plant development. New Phytol. 185, 369–385.

Hamman Jr., R.A., Dami, I.E., Walsh, T.M., Stushnoff, C., 1996. Seasonal carbohydrate changes and cold hardiness of Chardonnay and Riesling grapevines. Am. J. Enol. Vitic. 47, 31–36.

Hamman Jr., R.A., Renquist, A.R., Hughes, H.G., 1990. Pruning effects on cold hardiness and water content during deacclimation of Merlot bud and cane tissues. Am. J. Enol. Vitic. 41, 251–260.

Hammond, J.P., White, P.J., 2008. Sucrose transport in the phloem: integrating root responses to phosphorus starvation. J. Exp. Bot. 59, 93–109.

Hammond, J.P., White, P.J., 2011. Sugar signaling in root responses to low phosphorus availability. Plant Physiol. 156, 1033–1040.

Hanlin, R.L., Downey, M.O., 2009. Condensed tannin accumulation and composition in skin of Shiraz and Cabernet Sauvignon grapes during berry development. Am. J. Enol. Vitic. 60, 13–23.

Hanlin, R.L., Hrmova, M., Harbertson, J.F., Downey, M.O., 2010. Condensed tannin and grape cell wall interactions and their impact on tannin extractability into wine. Aust. J. Grape Wine Res. 16, 173–188.

Hanlin, R.L., Kelm, M.A., Wilkinson, K.L., Downey, M.O., 2011. Detailed characterization of proanthocyanidins in skin, seeds, and wine of Shiraz and Cabernet Sauvignon wine grapes (*Vitis vinifera*). J. Agric. Food Chem. 59, 13265–13276.

Hanson, J., Smeekens, S., 2009. Sugar perception and signaling—an update. Curr. Opin. Plant Biol. 12, 562–567.

Harbertson, J.F., Keller, M., 2012. Rootstock effects on deficit-irrigated winegrapes in a dry climate: grape and wine composition. Am. J. Enol. Vitic. 63, 40–48.

Harbertson, J.F., Kennedy, J.A., Adams, D.O., 2002. Tannin in skins and seeds of Cabernet Sauvignon, Syrah, and Pinot noir berries during ripening. Am. J. Enol. Vitic. 53, 54–59.

Harbertson, J.F., Hodgins, R.E., Thurston, L.N., Schaffer, L.J., Reid, M.S., Landon, J.L., et al., 2008. Variability of tannin concentration in red wines. Am. J. Enol. Vitic. 59, 210–214.

Hardie, W.J., 2000. Grapevine biology and adaptation to viticulture. Aust. J. Grape Wine Res. 6, 74–81.

Hardie, W.J., Aggenbach, S.J., 1996. Effects of site, season and viticultural practices on grape seed development. Aust. J. Grape Wine Res. 2, 21–24.

Hardie, W.J., Considine, J.A., 1976. Response of grapes to water-deficit stress in particular stages of development. Am. J. Enol. Vitic. 27, 55–61.

Hardie, W.J., Aggenbach, S.J., Jaudzems, V.G., 1996a. The plastids of the grape pericarp and their significance in isoprenoid synthesis. Aust. J. Grape Wine Res. 2, 144–154.

Hardie, W.J., O'Brien, T.P., Jaudzems, V.G., 1996b. Morphology, anatomy and development of the pericarp after anthesis in grape, *Vitis vinifera* L. Aust. J. Grape Wine Res. 2, 97–142.

Hardy, P.H., 1969. Selective diffusion of basic and acidic products of CO_2 fixation into the transpiration stream in grapevine. J. Exp. Bot. 20, 856–862.

Harris, J.M., Kriedemann, P.E., Possingham, J.V., 1968. Anatomical aspects of grape berry development. Vitis 7, 106–119.

Harris, J.M., Kriedemann, P.E., Possingham, J.V., 1971. Grape berry respiration: effects of metabolic inhibitors. Vitis 9, 291–298.

Hartung, W., Slovik, S., 1991. Physicochemical properties of plant growth regulators and plant tissues determine their distribution and redistribution: stomatal regulation by abscisic acid in leaves. New Phytol. 119, 361–382.

Hartweck, L.M., 2008. Gibberellin signaling. Planta 229, 1–13.

Hasegawa, P.M., Bressan, R.A., Zhu, J.K., Bohnert, H.J., 2000. Plant cellular and molecular responses to high salinity. Annu. Rev. Plant Physiol. Plant Mol. Biol. 51, 463–499.

Haselgrove, L., Botting, D., van Heeswijck, R., Høj, P.B., Dry, P.R., Ford, C., et al., 2000. Canopy microclimate and berry composition: the effect of bunch exposure on the phenolic composition of *Vitis vinifera* L. cv. Shiraz grape berries. Aust. J. Grape Wine Res. 6, 141–149.

Hashizume, K., Samuta, T., 1997. Green odorants of grape cluster stem and their ability to cause a wine stemmy flavor. J. Agric. Food Chem. 45, 1333–1337.

Hashizume, K., Samuta, T., 1999. Grape maturity and light exposure affect berry methoxypyrazine concentration. Am. J. Enol. Vitic. 50, 194–198.

Havelda, Z., Várallyay, E., Válóczi, A., Burgyán, J., 2008. Plant virus infection-induced persistent host gene downregulation in systemically infected leaves. Plant J. 55, 278–288.

Hawker, J.S., Walker, R.R., 1978. The effect of sodium chloride on the growth and fruiting of Cabernet Sauvignon vines. Am. J. Enol. Vitic. 29, 172–176.

Hawker, J.S., Ruffner, H.P., Walker, R.R., 1976. The sucrose content of some Australian grapes. Am. J. Enol. Vitic. 27, 125–129.

Hawker, J.S., Jenner, C.F., Niemitz, C.M., 1991. Sugar metabolism and compartmentation. Aust. J. Plant Physiol. 18, 227–237.

Hayasaka, Y., Baldock, G.A., Parker, M., Pardon, K.H., Black, C.A., Herderich, M.J., et al., 2010. Glycosylation of smoke-derived volatile phenols in grapes as a consequence of grapevine exposure to bushfire smoke. J. Agric. Food Chem. 58, 10989–10998.

Hayes, M.A., Davies, C., Dry, I.B., 2007. Isolation, functional characterization, and expression analysis of grapevine (*Vitis vinifera* L.) hexose transporters: differential roles in sink and source tissues. J. Exp. Bot. 58, 1985–1997.

Hayes, M.A., Feechan, A., Dry, I.B., 2010. Involvement of abscisic acid in the coordinated response of a stress-inducible hexose transporter (VvHT5) and a cell wall invertase in grapevine in response to biotrophic fungal infection. Plant Physiol. 153, 211–221.

Hayward, A., Stirnberg, P., Beveridge, C., Leyser, O., 2009. Interactions between auxin and strigolactone in shoot branching control. Plant Physiol. 151, 400–412.

He, F., He, J.J., Pan, Q.H., Duan, C.Q., 2010a. Mass-spectrometry evidence confirming the presence of pelargonidin-3-*O*-glucoside in the berry skins of Cabernet Sauvignon and Pinot noir (*Vitis vinifera* L.). Aust. J. Grape Wine Res. 16, 464–468.

He, F., Mu, L., Yan, G.L., Liang, N.N., Pan, Q.H., Wang, J., et al., 2010b. Biosynthesis of anthocyanins and their regulation in colored grapes. Molecules 15, 9057–9091.

He, H., Veneklaas, E.J., Kuo, J., Lambers, H., 2014. Physiological and ecological significance of biomineralization in plants. Trends Plant Sci. 19, 166–174.

Heazlewood, J.E., Wilson, S., 2004. Anthesis, pollination and fruitset in Pinot noir. Vitis 43, 65–68.

Hedrich, R., 2012. Ion channels in plants. Physiol. Rev. 92, 1777–1811.

Heil, M., Ton, J., 2008. Long-distance signalling in plant defence. Trends Plant Sci. 13, 264–272.

Heintz, C., Blaich, R., 1989. Structural characters of epidermal cell walls and resistance to powdery mildew of different grapevine cultivars. Vitis 28, 153–160.

Heintz, C., Blaich, R., 1990. Ultrastructural and histochemical studies on interactions between *Vitis vinifera* L. and *Uncinula necator* (Schw.) Burr. New Phytol. 115, 107–117.

Hellmann, H., Barker, L., Funck, D., Frommer, W.B., 2000. The regulation of assimilate allocation and transport. Aust. J. Plant Physiol. 27, 583–594.

Hendrickson, L., Ball, M.C., Osmond, C.B., Furbank, R.T., Chow, W.S., 2003. Assessment of photoprotection mechanisms of grapevines at low temperature. Funct. Plant Biol. 30, 631–642.

Hendrickson, L., Ball, M.C., Wood, J.T., Chow, W.S., Furbank, R.T., 2004a. Low temperature effects on photosynthesis and growth of grapevine. Plant Cell Environ. 27, 795–809.

Hendrickson, L., Chow, W.S., Furbank, R.T., 2004b. Low temperature effects on grapevine photosynthesis: the role of inorganic phosphate. Funct. Plant Biol. 31, 789–801.

Hennessy, K.J., Pittock, A.B., 1995. The potential impact of greenhouse warming on threshold temperature events in Victoria. Int. J. Climatol. 15, 591–692.

Hepner, Y., Bravdo, B., 1985. Effect of crop level and drip irrigation scheduling on the potassium status of Cabernet Sauvignon and Carignane vines and its influence on must and wine composition and quality. Am. J. Enol. Vitic. 36, 140–147.

Hepner, Y., Bravdo, B., Loinger, C., Cohen, S., Tabacman, H., 1985. Effect of drip irrigation schedules on growth, yield, must composition and wine quality of Cabernet Sauvignon. Am. J. Enol. Vitic. 36, 77−85.

Hermans, C., Hammond, J.P., White, P.J., Verbruggen, N., 2006. How do plants respond to nutrient shortage by biomass allocation? Trends Plant Sci. 11, 610−617.

Hernández, I., Alegre, L., Van Breusegem, F., Munné-Bosch, S., 2009. How relevant are flavonoids as antioxidants in plants? Trends Plant Sci. 14, 125−132.

Hernández-Orte, P., Guitart, A., Cacho, J., 1999. Changes in the concentration of amino acids during the ripening of *Vitis vinifera* Tempranillo variety from the Denomination d'Origine Somontano (Spain). Am. J. Enol. Vitic. 50, 144−154.

Hetherington, A.M., Brownlee, C., 2004. The generation of Ca^{2+} signals in plants. Annu. Rev. Plant Biol. 55, 401−427.

Hewitt, W.B., Goheen, A.C., Raski, D.J., Gooding, G.V., 1962. Studies on virus diseases of grapevines in California. Vitis 3, 57−83.

Heymann, H., Noble, A.C., 1987. Descriptive analysis of commercial Cabernet Sauvignon wines from California. Am. J. Enol. Vitic. 38, 41−44.

Hichri, I., Barrieu, F., Bogs, J., Kappel, C., Delrot, S., Lauvergeat, V., 2011. Recent advances in the transcriptional regulation of the flavonoid biosynthetic pathway. J. Exp. Bot. 62, 2465−2483.

Hifny, H.A.A., Alleweldt, G., 1972. Untersuchungen zur Stiellähme der Reben. I. Die Symptomatologie der Krankheit. Vitis 10, 298−313.

Hikosaka, K., 2005. Leaf canopy as a dynamic system: ecophysiology and optimality in leaf turnover. Ann. Bot. 95, 521−533.

Hikosaka, K., Ishikawa, K., Borjigidai, A., Muller, O., Onoda, Y., 2006. Temperature acclimation of photosynthesis: mechanisms involved in the changes in temperature dependence of photosynthetic rate. J. Exp. Bot. 57, 291−302.

Hilbert, G., Soyer, J.P., Molot, C., Giraudon, J., Milin, S., Gaudillère, J.P., 2003. Effects of nitrogen supply on must quality and anthocyanins accumulation in berries of cv. Merlot. Vitis 42, 69−76.

Hirschberg, J., 2001. Carotenoid biosynthesis in flowering plants. Curr. Opin. Plant Biol. 4, 210−218.

Hirschi, K.D., 2004. The calcium conundrum. Both versatile nutrient and specific signal. Plant Physiol. 136, 2438−2442.

Hoch, W.A., Singsaas, E.L., McCown, B.H., 2003. Resorption protection. Anthocyanins facilitate nutrient recovery in autumn by shielding leaves from potentially damaging light levels. Plant Physiol. 133, 1296−1305.

Hocquigny, S., Pelsy, F., Dumas, V., Kindt, S., Heloir, M.C., Merdinoglu, D., 2004. Diversification within grapevine cultivars goes through chimeric states. Genome 47, 579−589.

Hodge, A., 2006. Plastic plants and patchy soils. J. Exp. Bot. 57, 401−411.

Hodge, A., Campbell, C.D., Fitter, A.H., 2001. An arbuscular mycorrhizal fungus accelerates decomposition and acquires nitrogen directly from organic material. Nature 413, 297−299.

Hofäcker, W., 1978. Untersuchungen zur Photosynthese der Rebe. Einfluß der Entblätterung, der Dekapitierung, der Ringelung und der Entfernung der Traube. Vitis 17, 10−22.

Hofäcker, W., Alleweldt, G., Khader, S., 1976. Einfluss der Umweltfaktoren auf Beerenwachstum und Mostqualität bei der Rebe. Vitis 15, 96−112.

Hoffmann, S., Di Gaspero, G., Kovács, L., Howard, S., Kiss, E., Galbács, Z., et al., 2007. Resistance to *Erysiphe necator* in the grapevine 'Kishmish vatkana' is controlled by a single locus through restriction of hyphal growth. Theor. Appl. Genet. 116, 427−438.

Hogewoning, S.W., Douwstra, P., Trouwborst, G., van Ieperen, W., Harbinson, J., 2010. An artificial solar spectrum substantially alters plant development compared with usual climate room irradiance spectra. J. Exp. Bot. 61, 1267−1276.

Holbrook, N.M., Ahrens, E.T., Burns, M.J., Zwieniecki, M.A., 2001. In vivo observation of cavitation and embolism repair using magnetic resonance imaging. Plant Physiol. 126, 27–31.
Holdaway-Clarke, T.L., Hepler, P.K., 2003. Control of pollen tube growth: role of ion gradients and fluxes. New Phytol. 159, 539–563.
Holländer, H., Amrhein, N., 1980. The site of the inhibition of the shikimate pathway by glyphosate. I. Inhibition by glyphosate of phenylpropanoid synthesis in buckwheat (*Fagopyrum esculentum* Moench). Plant Physiol. 66, 823–829.
Holmes, M.G., Smith, H., 1977a. The function of phytochrome in the natural environment. I. Characterization of daylight for studies in photomorphogenesis and photoperiodism. Photochem. Photobiol. 25, 533–538.
Holmes, M.G., Smith, H., 1977b. The function of phytochrome in the natural environment. II. The influence of vegetation canopies on the spectral energy distribution of natural daylight. Photochem. Photobiol. 25, 539–545.
Holst, T., Mayer, H., Schindler, D., 2004. Microclimate within beech stands—part II: thermal conditions. Eur. J. For. Res. 123, 13–28.
Holst, T., Rost, J., Mayer, H., 2005. Net radiation balance for two forested slopes on opposite sides of a valley. Int. J. Biometeorol. 49, 275–284.
Holt, H.E., Birchmore, W., Herderich, M.J., Iland, P.G., 2010. Berry phenolics in Cabernet Sauvignon (*Vitis vinifera* L.) during late-stage ripening. Am. J. Enol. Vitic. 61, 285–299.
Holzapfel, B.P., Treeby, M.T., 2007. Effects of timing and rate of N supply on leaf nitrogen status, grape yield and juice composition from Shiraz grapevines grafted to one of three different rootstocks. Aust. J. Grape Wine Res. 13, 14–22.
Holzapfel, B.P., Smith, J.P., Mandel, R.M., Keller, M., 2006. Manipulating the postharvest period and its impact on vine productivity of Semillon grapevines. Am. J. Enol. Vitic. 57, 148–157.
Holzapfel, B.P., Smith, J.P., Field, S.K., Hardie, W.J., 2010. Dynamics of carbohydrate reserves in cultivated grapevines. Hortic. Rev. (Am. Soc. Hortic. Sci.) 37, 143–211.
Hoos, G., Blaich, R., 1990. Influence of resveratrol on germination of conidia and mycelial growth of *Botrytis cinerea* and *Phomopsis viticola*. J. Phytopathol. 129, 102–110.
Hopkins, D.L., Purcell, A.H., 2002. *Xylella fastidiosa*: cause of Pierce's disease of grapevine and other emergent diseases. Plant Dis. 86, 1056–1066.
Hörtensteiner, S., 2006. Chlorophyll degradation during senescence. Annu. Rev. Plant Biol. 57, 55–77.
Hörtensteiner, S., 2009. Stay-green regulates chlorophyll and chlorophyll-binding protein degradation during senescence. Trends Plant Sci. 14, 155–162.
Hörtensteiner, S., Feller, U., 2002. Nitrogen metabolism and remobilization during senescence. J. Exp. Bot. 53, 927–937.
Horvath, D.P., Anderson, J.V., Chao, W.S., Foley, M.E., 2003. Knowing when to grow: signals regulating bud dormancy. Trends Plant Sci. 8, 534–540.
Hotta, C.T., Gardner, M.J., Hubbard, K.E., Baek, S.J., Dalchau, N., Suhita, D., et al., 2007. Modulation of environmental responses of plants by circadian clocks. Plant Cell Environ. 30, 333–349.
Houdaille, F., Guillon, J.M., 1895. Contribution à l'étude des pleurs de la vigne. Rev. Vitic. 67, 305–308; 69, 362–364; 71, 408–415; 73, 457–461.
Howell, G.S., Shaulis, N., 1980. Factors influencing within-vine variation in the cold resistance of cane and primary bud tissues. Am. J. Enol. Vitic. 31, 158–161.
Hrazdina, G., 1975. Anthocyanin composition of Concord grapes. Lebensm.-Wiss. Technol. 8, 111–113.
Hrazdina, G., Parsons, G.F., Mattick, L.R., 1984. Physiological and biochemical events during development and maturation of grape berries. Am. J. Enol. Vitic. 35, 220–227.
Hsiao, T.C., 1973. Plant responses to water stress. Annu. Rev. Plant Physiol. 24, 519–570.
Hsiao, T.C., Xu, L.K., 2000. Sensitivity of growth of roots versus leaves to water stress: biophysical analysis and relation to water transport. J. Exp. Bot. 51, 1595–1616.

Huang, S., Pollack, H.N., Shen, P.Y., 2000. Temperature trends over the past five centuries reconstructed from borehole temperatures. Nature 403, 756–758.

Huang, X., Lakso, A.N., Eissenstat, D.M., 2005b. Interactive effects of soil temperature and moisture on Concord grape root respiration. J. Exp. Bot. 56, 2651–2660.

Huang, X.M., Huang, H.B., 2001. Early post-veraison growth in grapes: evidence for a two-step mode of berry enlargement. Aust. J. Grape Wine Res. 7, 132–136.

Huang, X.M., Huang, H.B., Wang, H.C., 2005a. Cell walls of loosening skin in post-veraison grape berries lose structural polysaccharides and calcium while accumulate structural proteins. Sci. Hortic. 104, 249–263.

Huang, Z., Ough, C.S., 1991. Amino acid profiles of commercial grape juices and wines. Am. J. Enol. Vitic. 42, 261–267.

Hubbard, R.M., Ryan, M.G., Stiller, V., Sperry, J.S., 2001. Stomatal conductance and photosynthesis vary linearly with plant hydraulic conductance in ponderosa pine. Plant Cell Environ. 24, 113–121.

Hufnagel, J.C., Hofmann, T., 2008a. Orosensory-directed identification of astringent mouthfeel and bitter-tasting compounds in red wine. J. Agric. Food Chem. 56, 1376–1386.

Hufnagel, J.C., Hofmann, T., 2008b. Quantitative reconstruction of the nonvolatile sensometabolome of a red wine. J. Agric. Food Chem. 56, 9190–9199.

Hughes, N.M., Morley, C.B., Smith, W.K., 2007. Coordination of anthocyanin decline and photosynthetic maturation in juvenile leaves of three deciduous tree species. New Phytol. 175, 675–685.

Huglin, P., Balthazard, J., 1975. Variabilité et fluctuation de la composition des inflorescences et des grappes chez quelques variétés de *Vitis vinifera*. Vitis 14, 6–13.

Huglin, P., Schneider, C., 1998. Biologie et Ecologie de la Vigne, second ed. Lavoisier, Paris.

Humphreys, J.M., Chapple, C., 2002. Rewriting the lignin roadmap. Curr. Opin. Plant Biol. 5, 224–229.

Hunter, J.J., Visser, J.H., 1988. The effect of partial defoliation, leaf position and developmental stage of the vine on the photosynthetic activity of *Vitis vinifera* L. cv Cabernet Sauvignon. S. Afr. J. Enol. Vitic. 9, 9–15.

Hunter, J.J., Skrivan, R., Ruffner, H.P., 1994. Diurnal and seasonal physiological changes in leaves of *Vitis vinifera* L.: CO_2 assimilation rates, sugar levels and sucrolytic enzyme activity. Vitis 33, 189–195.

Huppe, H.C., Turpin, D.H., 1994. Integration of carbon and nitrogen metabolism in plant and algal cells. Annu. Rev. Plant Physiol. Plant Mol. Biol. 45, 577–607.

Hwang, I., Sakakibara, H., 2006. Cytokinin biosynthesis and perception. Physiol. Plant. 126, 528–538.

Iacono, F., Sommer, K.J., 1996. Photoinhibition of photosynthesis and photorespiration in *Vitis vinifera* under field conditions: effects of light climate and leaf position. Aust. J. Grape Wine Res. 2, 1–11.

Iacono, F., Bertamini, M., Scienza, A., Coombe, B.G., 1995. Differential effects of canopy manipulation and shading of *Vitis vinifera* L. cv. Cabernet Sauvignon. Leaf gas exchange, photosynthetic electron transport rate and sugar accumulation in berries. Vitis 34, 201–206.

Iba, K., 2002. Acclimative response to temperature stress in higher plants: approaches of gene engineering for temperature tolerance. Annu. Rev. Plant Biol. 53, 225–245.

Ibáñez, J., Vargas, A.M., Palancar, M., Borrego, J., de Andrés, M.T., 2009. Genetic relationships among table-grape varieties. Am. J. Enol. Vitic. 60, 35–42.

Ibáñez, J., Muñoz-Organero, G., Zinelabidine, L.H., de Andrés, M.T., Cabello, F., Martínez-Zapater, J.M., 2012. Genetic origin of the grapevine cultivar Tempranillo. Am. J. Enol. Vitic. 63, 549–553.

Iland, P., Dry, P., Proffitt, T., Tyerman, S., 2011. The Grapevine: From the Science to the Practice of Growing Vines for Wine. Patrick Iland Wine Promotions, Adelaide, Australia.

Iland, P.G., Coombe, B.G., 1988. Malate, tartrate, potassium, and sodium in flesh and skin of Shiraz grapes during ripening: concentration and compartmentation. Am. J. Enol. Vitic. 39, 71–76.

Inaba, A., Ishida, M., Sobajima, Y., 1976. Changes in endogenous hormone concentrations during berry development in relaton to the ripening of Delaware grapes. J. Jpn. Soc. Hortic. Sci. 45, 245–252.

Ingels, C.A., Scow, K.M., Whisson, D.A., Drenovsky, R.E., 2005. Effects of cover crops on grapevines, yield, juice composition, soil microbial ecology, and gopher activity. Am. J. Enol. Vitic. 56, 19–29.

Intrieri, C., Poni, S., Silvestroni, O., Filippetti, I., 1992. Leaf age, leaf position and photosynthesis in potted grapevines. Adv. Hortic. Sci. 6, 23–27.

Intrieri, C., Zerbi, G., Marchiol, L., Poni, S., Caiado, T., 1995. Physiological response of grapevine leaves to lightflecks. Sci. Hortic. 61, 47–59.

Intrieri, C., Poni, S., Rebucci, B., Magnanini, E., 1997. Effects of canopy manipulations on whole-vine photosynthesis: results from pot and field experiments. Vitis 36, 167–173.

Intrieri, C., Poni, S., Lia, G., Gomez del Campo, M., 2001. Vine performance and leaf physiology of conventionally and minimally pruned Sangiovese grapevines. Vitis 40, 123–130.

Intrigliolo, D.S., Lakso, A.N., Centinari, M., 2009. Effects of the whole vine versus single shoot-crop level on fruit growth in *Vitis labruscana* 'Concord'. Vitis 48, 1–5.

Intergovernmental Panel on Climate Change, 2013. Climate Change 2013: The Physical Science Basis. Contribution of Working Group I to the Fifth Assessment Report of the Intergovernmental Panel on Climate Change. In: Stocker, T.F., Qin, D., Plattner, G.K., Tignor, M., Allen, S.K., Boschung, J. (Eds.), Cambridge University Press, Cambridge, United Kingdom.

Iriti, M., Rossoni, M., Faoro, F., 2006. Melatonin content in grape: myth or panacea? J. Sci. Food Agric. 86, 1432–1438.

Ishimaru, M., Kobayashi, S., 2002. Expression of a xyloglucan endo-transglycosylase gene is closely related to grape berry softening. Plant Sci. 162, 621–628.

Israelsson, M., Sundberg, B., Moritz, T., 2005. Tissue-specific localization of gibberellins and expression of gibberellin-biosynthetic and signaling genes in wood-forming tissues in aspen. Plant J. 44, 494–504.

Ivanchenko, M.G., Muday, G.K., Dubrovsky, J.G., 2008. Ethylene–auxin interactions regulate lateral root initiation and emergence in *Arabidopsis thaliana*. Plant J. 55, 335–347.

Iwahori, S., Weaver, R.J., Pool, R.M., 1968. Gibberellin-like activity in berries of seeded and seedless Tokay grapes. Plant Physiol. 43, 333–337.

Jaakola, L., Hohtola, A., 2010. Effect of latitude on flavonoid biosynthesis in plants. Plant Cell Environ. 33, 1239–1247.

Jackson, D.I., 1991. Environmental and hormonal effects on development of early bunch stem necrosis. Am. J. Enol. Vitic. 42, 290–294.

Jackson, D.I., Cherry, N.J., 1988. Prediction of a district's grape-ripening capacity using a latitude-temperature index (LTI). Am. J. Enol. Vitic. 39, 19–28.

Jackson, D.I., Coombe, B.G., 1988. Early bunchstem necrosis in grapes—a cause of poor fruit set. Vitis 27, 57–61.

Jackson, D.I., Lombard, P.B., 1993. Environmental and management practices affecting grape composition and wine quality—a review. Am. J. Enol. Vitic. 44, 409–430.

Jackson, D.I., Steans, G.F., Hemmings, P.C., 1984. Vine response to increased node numbers. Am. J. Enol. Vitic. 35, 161–163.

Jackson, L.E., Burger, M., Cavagnaro, T.R., 2008. Roots, nitrogen transformations, and ecosystem services. Annu. Rev. Plant Biol. 59, 341–363.

Jackson, R.S., 2008. Wine Science: Principles and Applications, third ed. Academic Press, San Diego, CA.

Jacobs, A.K., Dry, I.B., Robinson, S.P., 1999. Induction of different pathogenesis-related cDNAs in grapevine infected with powdery mildew and treated with ethephon. Plant Pathol. 48, 325–336.

Jähnl, G., 1967. Stiellähme, mikroskopisch betrachtet. Wein-Wiss. 22, 446.

Jähnl, G., 1971. Anatomische Veränderungen durch Stiellähme. Wein-Wiss. 26, 242–252.

Jaillon, O., Aury, J.M., Noel, B., Policriti, A., Clepet, C., Casagrande, A., et al., 2007. The grapevine genome sequence suggests ancestral hexaploidization in major angiosperm phyla. Nature 449, 463–468.

James, D.G., Price, T.S., 2004. Field-testing of methyl salicylate for recruitment and retention of beneficial insects in grapes and hops. J. Chem. Ecol. 30, 1613–1628.

Jansen, M.A.K., Gaba, V., Greenberg, B.M., 1998. Higher plants and UV-B radiation: balancing damage, repair and acclimation. Trends Plant Sci. 3, 131–135.

Jansen, R.K., Kaittanis, C., Saski, C., Lee, S.B., Tomkins, J., Alverson, A.J., et al., 2006. Phylogenetic analyses of *Vitis* (Vitaceae) based on complete chloroplast genome sequences: effects of taxon sampling and phylogenetic methods on resolving relationships among rosids. BMC Evol. Biol. 6, 32, http://dx.doi.org/10.1186/1471-2148-6-32.

Jansen, S.C., van Dusseldorp, M., Bottema, K.C., Dubois, A.E., 2003. Intolerance to dietary biogenic amines: a review. Ann. Allergy Asthma Immunol. 91, 233–240.

Jánváry, L., Hoffmann, T., Pfeiffer, J., Hausmann, L., Töpfer, R., Fischer, T.C., et al., 2009. A double mutation in the anthocyanin 5-*O*-glucosyltransferase gene disrupts enzymatic activity in *Vitis vinifera* L. J. Agric. Food Chem. 57, 3512–3518.

Jarvis, M.C., 1984. Structure and properties of pectin gels in plant cell walls. Plant Cell Environ. 7, 153–164.

Javelle, M., Vernoud, V., Rogowsky, P.M., Ingram, G.C., 2011. Epidermis: the formation and functions of a fundamental plant tissue. New Phytol. 189, 17–39.

Jeandet, P., Bessis, R., Gautheron, B., 1991. The production of resveratrol (3,5,4′-trihydroxystilbene) by grape berries in different developmental stages. Am. J. Enol. Vitic. 42, 41–46.

Jeandet, P., Douillet-Breuil, A.C., Bessis, R., Debord, S., Sbaghi, M., Adrian, M., 2002. Phytoalexins from Vitaceae: biosynthesis, phytoalexin gene expression in transgenic plants, antifungal activity, and metabolism. J. Agric. Food Chem. 50, 2731–2741.

Jenik, P.D., Irish, V.F., 2000. Regulation of cell proliferation patterns by homeotic genes during *Arabidopsis* floral development. Development 127, 1267–1276.

Jeong, S.T., Goto-Yamamoto, N., Kobayashi, S., Esaka, M., 2004. Effects of plant hormones and shading on the accumulation of anthocyanins and the expression of anthocyanin biosynthetic genes in grape berry skins. Plant Sci. 167, 247–252.

Jersch, S., Scherer, C., Huth, G., Schlösser, E., 1989. Proanthocyanidins as basis for quiescence of *Botrytis cinerea* in immature strawberry fruits. Z. Pflanzenkr. Pflanzenschutz 96, 365–378.

Jia, H.F., Chai, Y.M., Li, C.L., Lu, D., Luo, J.J., Qin, L., et al., 2011. Abscisic acid plays an important role in the regulation of strawberry fruit ripening. Plant Physiol. 157, 188–199.

Jia, W., Davies, W.J., 2007. Modification of leaf apoplastic pH in relation to stomatal sensitivity to root-sourced abscisic acid signals. Plant Physiol. 143, 68–77.

Jiménez, S., Gogorcena, Y., Hévin, C., Rombolà, A.D., Ollat, N., 2007. Nitrogen nutrition influences some biochemical responses to iron deficiency in tolerant and sensitive genotypes of *Vitis*. Plant Soil 290, 343–355.

Jin, K., Shen, J., Ashton, R.W., Dodd, I.C., Parry, M.A.J., Whalley, W.R., 2013. How do roots elongate in a structured soil? J. Exp. Bot. 64, 4761–4777.

Johansson, I., Karlsson, M., Johanson, U., Larsson, C., Kjellbom, P., 2000. The role of aquaporins in cellular and whole plant water balance. Biochim. Biophys. Acta 1465, 324–342.

Johnson, D.E., Howell, G.S., 1981. Factors influencing critical temperatures for spring freeze damage to developing primary shoots on Concord grapevines. Am. J. Enol. Vitic. 32, 144–149.

Johnson, D.M., McCulloh, K.A., Woodruff, D.R., Meinzer, F.C., 2012. Evidence for xylem embolism as a primary factor in dehydration-induced declines in leaf hydraulic conductance. Plant Cell Environ. 35, 760–769.

Johnson, J.O., Weaver, R.J., Paige, D.F., 1982. Differences in the mobilization of assimilates of *Vitis vinifera* L. grapevines as influenced by an increased source strength. Am. J. Enol. Vitic. 33, 207–213.

Jones, A.M., Dangl, J.L., 1996. Logjam at the Styx: programmed cell death in plants. Trends Plant Sci. 1, 114–119.

Jones, G.V., 2005. Climate change in the western United States grape growing regions. Acta Hortic. 689, 41–59.

Jones, G.V., Davis, R.E., 2000. Climate influences on grapevine phenology, grape composition, and wine production and quality for Bordeaux, France. Am. J. Enol. Vitic. 51, 249–261.

Jones, G.V., White, M.A., Cooper, O.R., Storchmann, K., 2005. Climate change and global wine quality. Clim. Change 73, 319–343.

Jones, G.V., Duff, A.A., Hall, A., Myers, J.W., 2010. Spatial analysis of climate in winegrape growing regions in the western United States. Am. J. Enol. Vitic. 61, 313–326.

Jones, H.G., 1998. Stomatal control of photosynthesis and transpiration. J. Exp. Bot. 49, 387–398.

Jones, K.S., Paroschy, J., McKersie, B.D., Bowley, S.R., 1999. Carbohydrate composition and freezing tolerance of canes and buds in *Vitis vinifera*. J. Plant Physiol. 155, 101–106.

Jones, K.S., McKersie, B.D., Paroschy, J., 2000. Prevention of ice propagation by permeability barriers in bud axes of *Vitis vinifera*. Can. J. Bot. 78, 3–9.

Jones, R.S., Ough, C.S., 1985. Variations in the percent ethanol (v/v) per Brix conversions of wines from different climatic regions. Am. J. Enol. Vitic. 36, 268–270.

Joslyn, M.A., Dittmar, H.F.K., 1967. The proanthocyanidins of Pinot blanc grapes. Am. J. Enol. Vitic. 18, 1–10.

Kadir, S., 2006. Thermostability of photosynthesis of *Vitis aestivalis* and *V. vinifera*. J. Am. Soc. Hortic. Sci. 131, 476–483.

Kadir, S., Von Weihe, M., Al-Khatib, K., 2007. Photochemical efficiency and recovery of photosystem II in grapes after exposure to sudden and gradual heat stress. J. Am. Soc. Hortic. Sci. 132, 764–769.

Kaile, A., Pitt, D., Kuhn, P.J., 1991. Release of calcium and other ions from various plant host tissues infected by different necrotrophic pathogens with special reference to *Botrytis cinerea* Pers. Physiol. Mol. Plant Pathol. 38, 275–291.

Kaiser, B.N., Gridley, K.L., Ngaire Brady, J., Phillips, T., Tyerman, S.D., 2005. The role of molybdenum in agricultural plant production. Ann. Bot. 96, 745–754.

Kakani, V.G., Reddy, K.R., Zhao, D., Sailaja, K., 2003. Field crop responses to ultraviolet-B radiation: a review. Agric. For. Meteorol. 120, 191–218.

Kakimoto, T., 2003. Perception and signal transduction of cytokinins. Annu. Rev. Plant Biol. 54, 605–627.

Kalberer, S.R., Wisniewski, M., Arora, R., 2006. Deacclimation and reacclimation of cold-hardy plants: current understanding and emerging concepts. Plant Sci. 171, 3–16.

Kalua, C.M., Boss, P.K., 2009. Evolution of volatile compounds during the development of Cabernet Sauvignon grapes (*Vitis vinifera* L.). J. Agric. Food Chem. 57, 3818–3830.

Kammerer, D., Claus, A., Carle, R., Schieber, A., 2004. Polyphenol screening of pomace from red and white grape varieties (*Vitis vinifera* L.) by HPLC-DAD-MS/MS. J. Agric. Food Chem. 52, 4360–4367.

Kanellis, A.K., Roubelakis-Angelakis, K.A., 1993. Grape. In: Seymour, G.B., Taylor, J.E., Tucker, G.A. (Eds.), Biochemistry of Fruit Ripening. Chapman & Hall, London, pp. 189–234.

Kang, S., Hu, T., Jerie, P., Zhang, J., 2003. The effects of partial rootzone drying on root, trunk sap flow and water balance in an irrigated pear (*Pyrus communis* L.) orchard. J. Hydrol. 280, 192–206.

Karabourniotis, G., Bornman, J.F., Liakoura, V., 1999. Different leaf surface characteristics of three grape cultivars affect leaf optical properties as measured with fibre optics: possible implication in stress tolerance. Aust. J. Plant Physiol. 26, 47–53.

Karasev, A.V., 2000. Genetic diversity and evolution of closteroviruses. Annu. Rev. Phytopathol. 38, 293–324.

Karley, A.J., White, P.J., 2009. Moving cationic minerals to edible tissues: potassium, magnesium, calcium. Curr. Opin. Plant Biol. 12, 291–298.

Kasimatis, A.N., Vilas Jr., E.P., Swanson, F.H., Baranek, P.P., 1975. A study of the variability of 'Thompson Seedless' berries for soluble solids and weight. Am. J. Enol. Vitic. 26, 37–42.

Kassemeyer, H.H., Staudt, G., 1981. Über die Entwicklung des Embryosacks und die Befruchtung der Reben. Vitis 20, 202–210.

Kassemeyer, H.H., Staudt, G., 1982. Cytologische Untersuchungen über die Ursachen des Verrieselns bei *Vitis*. Vitis 21, 121 135.

Kassemeyer, H.H., Staudt, G., 1983. Über das Wachstum von Endosperm, Embryo und Samenanlagen von *Vitis vinifera*. Vitis 22, 109–119.

Kasuga, J., Hashidoko, Y., Nishioka, A., Yoshiba, M., Arakawa, K., Fujikawa, S., 2008. Deep supercooling xylem parenchyma cells of katsura tree (*Cercidiphyllum japonicum*) contain flavonol glycosides exhibiting anti-ice nucleation activity. Plant Cell Environ. 31, 1335–1348.

Kataoka, I., Sugiura, A., Utsunomiya, N., Tomana, T., 1982. Effect of abscisic acid and defoliation on anthocyanin accumulation in Kyoho grapes (*Vitis vinifera* L. × *V. labruscana* Bailey). Vitis 21, 325–332.

Katsuhara, M., Hanba, Y.T., Shiratake, K., Maeshima, M., 2008. Expanding roles of plant aquaporins in plasma membranes and cell organelles. Funct. Plant Biol. 35, 1–14.

Kawai, Y., Benz, J., Kliewer, W.M., 1996. Effect of flooding on shoot and root growth of rooted cuttings of four grape rootstocks. J. Jpn. Soc. Hortic. Sci. 65, 455–461.

Kawamura, E., Horiguchi, G., Tsukaya, H., 2010. Mechanisms of leaf tooth formation in *Arabidopsis*. Plant J. 62, 429–441.

Kehr, J., Buhtz, A., 2008. Long distance transport and movement of RNA through the phloem. J. Exp. Bot. 59, 85–92.

Keller, M., 2005. Deficit irrigation and vine mineral nutrition. Am. J. Enol. Vitic. 56, 267–283.

Keller, M., 2010. Managing grapevines to optimise fruit development in a challenging environment: a climate change primer for viticulturists. Aust. J. Grape Wine Res. 16, 56–69.

Keller, M., Hrazdina, G., 1998. Interaction of nitrogen availability during bloom and light intensity during veraison: II. Effects on anthocyanin and phenolic development during grape ripening. Am. J. Enol. Vitic. 49, 341–349.

Keller, M., Koblet, W., 1994. Is carbon starvation rather than excessive nitrogen supply the cause of inflorescence necrosis in *Vitis vinifera* L.? Vitis 33, 81–86.

Keller, M., Koblet, W., 1995a. Dry matter and leaf area partitioning, bud fertility and second-season growth of *Vitis vinifera* L.: responses to nitrogen supply and limiting irradiance. Vitis 34, 77–83.

Keller, M., Koblet, W., 1995b. Stress-induced development of inflorescence necrosis and bunch stem necrosis in *Vitis vinifera* L. in response to environmental and nutritional effects. Vitis 34, 145–150.

Keller, M., Mills, L.J., 2007. Effect of pruning on recovery and productivity of cold-injured Merlot grapevines. Am. J. Enol. Vitic. 58, 351–357.

Keller, M., Shrestha, P.M., 2014. Solute accumulation differs in the vacuoles and apoplast of ripening grape berries. Planta 239, 633–642.

Keller, M., Tarara, J.M., 2010. Warm spring temperatures induce persistent season-long changes in shoot development in grapevines. Ann. Bot. 106, 131–141.

Keller, M., Hess, B., Schwager, H., Schärer, H., Koblet, W., 1995. Carbon and nitrogen partitioning in *Vitis vinifera* L.: responses to nitrogen supply and limiting irradiance. Vitis 34, 19–26.

Keller, M., Arnink, K.J., Hrazdina, G., 1998. Interaction of nitrogen availability during bloom and light intensity during veraison. I. Effects on grapevine growth, fruit development, and ripening. Am. J. Enol. Vitic. 49, 333–340.

Keller, M., Pool, R.M., Henick-Kling, T., 1999. Excessive nitrogen supply and shoot trimming can impair colour development in Pinot noir grapes and wine. Aust. J. Grape Wine Res. 5, 45–55.

Keller, M., Kummer, M., Vasconcelos, M.C., 2001a. Reproductive growth of grapevines in response to nitrogen supply and rootstock. Aust. J. Grape Wine Res. 7, 12–18.

Keller, M., Kummer, M., Vasconcelos, M.C., 2001b. Soil nitrogen utilisation for growth and gas exchange by grapevines in response to nitrogen supply and rootstock. Aust. J. Grape Wine Res. 7, 2–11.

Keller, M., Rogiers, S.Y., Schultz, H.R., 2003a. Nitrogen and ultraviolet radiation modify grapevines' susceptibility to powdery mildew. Vitis 42, 87–94.

Keller, M., Viret, O., Cole, F.M., 2003b. *Botrytis cinerea* infection in grape flowers: defense reaction, latency, and disease expression. Phytopathology 93, 316–322.

Keller, M., Mills, L.J., Wample, R.L., Spayd, S.E., 2005. Cluster thinning effects on three deficit-irrigated *Vitis vinifera* cultivars. Am. J. Enol. Vitic. 56, 91–103.

Keller, M., Smith, J.P., Bondada, B.R., 2006. Ripening grape berries remain hydraulically connected to the shoot. J. Exp. Bot. 57, 2577–2587.

Keller, M., Mills, L.J., Smithyman, R.P., 2008. Interactive effects of deficit irrigation and crop load on Cabernet Sauvignon in an arid climate. Am. J. Enol. Vitic. 59, 221–234.

Keller, M., Tarara, J.M., Mills, L.J., 2010. Spring temperatures alter reproductive development in grapevines. Aust. J. Grape Wine Res. 16, 445–454.

Keller, M., Mills, L.J., Harbertson, J.F., 2012. Rootstock effects on deficit-irrigated winegrapes in a dry climate: vigor, yield formation, and fruit ripening. Am. J. Enol. Vitic. 63, 29–39.

Keller, M., Mills, L.J., Olmstead, M.A., 2014. Fruit ripening has little influence on grapevine cold acclimation. Am. J. Enol. Vitic. 65, 417–423.

Kemp, M.S., Burden, R.S., 1986. Phytoalexins and stress metabolites in the sapwood of trees. Phytochemistry 25, 1261–1269.

Kende, H., Zeevaart, J.A.D., 1997. The five "classical" plant hormones. Plant Cell 9, 1197–1210.

Kennedy, J.A., Troup, G.J., Pilbrow, J.R., Hutton, D.R., Hewitt, D., Hunter, C.R., et al., 2000. Development of seed polyphenols in berries from *Vitis vinifera* L. cv. Shiraz. Aust. J. Grape Wine Res. 6, 244–254.

Kennedy, J.A., Hayasaka, Y., Vidal, S., Waters, E.J., Jones, G.P., 2001. Composition of grape skin proanthocyanidins at different stages of berry development. J. Agric. Food Chem. 49, 5348–5355.

Kennedy, J.A., Matthews, M.A., Waterhouse, A.L., 2002. Effect of maturity and vine water status on grape skin and wine flavonoids. Am. J. Enol. Vitic. 53, 268–274.

Kennelly, M.M., Gadoury, D.M., Wilcox, W.F., Magarey, P.A., Seem, R.C., 2005. Seasonal development of ontogenic resistance to downy mildew in grape berries and rachises. Phytopathology 95, 1445–1452.

Kennison, K.R., Gibberd, M.R., Pollnitz, A.P., Wilkinson, K.L., 2008. Smoke-derived taint in wine: the release of smoke-derived volatile phenols during fermentation of Merlot juice following grapevine exposure to smoke. J. Agric. Food Chem. 56, 7379–7383.

Kennison, K.R., Wilkinson, K.L., Pollnitz, A.P., Williams, H.G., Gibberd, M.R., 2009. Effect of timing and duration of grapevine exposure to smoke on the composition and sensory properties of wine. Aust. J. Grape Wine Res. 15, 228–237.

Kennison, K.R., Wilkinson, K.L., Pollnitz, A.P., Williams, H.G., Gibberd, M.R., 2011. Effect of smoke application to field-grown Merlot grapevines at key phenological growth stages on wine sensory and chemical properties. Aust. J. Grape Wine Res. 17, S5–S12.

Keppler, F., Hamilton, J.T.G., McRoberts, W.C., Vigano, I., Brass, M., Röckmann, T., 2008. Methoxyl groups of plant pectin as a precursor of atmospheric methane: evidence from deuterium labelling studies. New Phytol. 178, 808–814.

Kerstetter, R.A., Hake, S., 1997. Shoot meristem formation in vegetative development. Plant Cell 9, 1001–1010.

Kerstiens, G., 1996. Signalling across the divide: a wider perspective of cuticular structure–function relationships. Trends Plant Sci. 1, 125–129.

Keskitalo, J., Bergquist, G., Gardeström, P., Jansson, S., 2005. A cellular timetable of autumn senescence. Plant Physiol. 139, 1635–1648.

Kevan, P.G., Longair, R.W., Gadawski, R.M., 1985. Dioecy and pollen dimorphism in *Vitis riparia* (Vitaceae). Can. J. Bot. 63, 2263–2267.

Kevan, P.G., Blades, D.C.A., Posluszny, U., Ambrose, J.D., 1988. Pollen dimorphism and dioecy in *Vitis aestivalis*. Vitis 27, 143–146.

Khanal, B.P., Grimm, E., Finger, S., Blume, A., Knoche, M., 2013. Intracuticular wax fixes and restricts strain in leaf and fruit cuticles. New Phytol. 200, 134–143.

Kiba, T., Kudo, T., Kojima, M., Sakakibara, H., 2011. Hormonal control of nitrogen acquisition: roles of auxin, abscisic acid, and cytokinin. J. Exp. Bot. 62, 1399–1409.

Kidman, C.M., Olarte Mantilla, S., Dry, P.R., McCarthy, M.G., Collins, C., 2014. Effect of water stress on the reproductive performance of Shiraz (*Vitis vinifera* L.) grafted to rootstocks. Am. J. Enol. Vitic. 65, 96–108.

Kimura, P.H., Okamoto, G., Hirano, K., 1997. Flower types, pollen morphology and germination, fertilization, and berry set in *Vitis coignetiae* Pulliat. Am. J. Enol. Vitic. 48, 323–327.

King, P.D., Buchanan, G.A., 1986. The dispersal of phylloxera crawlers and spread of phylloxera infestations in New Zealand and Australian vineyards. Am. J. Enol. Vitic. 37, 26–33.

King, P.D., Smart, R.E., McClellan, D.J., 2014. Within-vineyard variability in vine vegetative growth, yield, and fruit and wine composition of Cabernet Sauvignon in Hawke's Bay, New Zealand. Aust. J. Grape Wine Res. 20, 234–246.

Klee, H.J., 2010. Improving the flavor of fresh fruits: genomics, biochemistry, and biotechnology. New Phytol. 187, 44–56.

Klenert, M., Rapp, A., Alleweldt, G., 1978. Einfluß der Traubentemperatur auf Beerenwachstum und Beerenreife der Rebsorte Silvaner. Vitis 17, 350–360.

Kliewer, W.M., 1964. Influence of environment on metabolism of organic acids and carbohydrates in *Vitis vinifera*. I. Temperature. Plant Physiol. 39, 869–880.

Kliewer, W.M., 1965a. Changes in concentration of glucose, fructose, and total soluble solids in flowers and berries of *Vitis vinifera*. Am. J. Enol. Vitic. 16, 101–110.

Kliewer, W.M., 1965b. Changes in the concentration of malates, tartrates, and total free acids in flowers and berries of *Vitis vinifera*. Am. J. Enol. Vitic. 16, 92–100.

Kliewer, W.M., 1966. Sugars and organic acids of *Vitis vinifera*. Plant Physiol. 41, 923–931.

Kliewer, W.M., 1967a. Annual cyclic changes in the concentration of free amino acids in grapevines. Am. J. Enol. Vitic. 18, 126–137.

Kliewer, W.M., 1967b. Concentration of tartrates, malates, glucose and fructose in the fruits of the genus *Vitis*. Am. J. Enol. Vitic. 18, 87–96.

Kliewer, W.M., 1967c. The glucose–fructose ratio of *Vitis vinifera* grapes. Am. J. Enol. Vitic. 18, 33–41.

Kliewer, W.M., 1968. Changes in the concentration of free amino acids in grape berries during maturation. Am. J. Enol. Vitic. 19, 166–174.

Kliewer, W.M., 1970. Free amino acids and other nitrogenous fractions in wine grapes. J. Food Sci. 35, 17–21.

Kliewer, W.M., 1973. Berry composition of *Vitis vinifera* cultivars as influenced by photo- and nycto-temperatures during maturation. J. Am. Soc. Hortic. Sci. 98, 153–159.

Kliewer, W.M., 1975. Effect of root temperature on budbreak, shoot growth, and fruit-set of 'Cabernet Sauvignon' grapevines. Am. J. Enol. Vitic. 26, 82–89.

Kliewer, W.M., 1977a. Effect of high temperatures during the bloom-set period on fruit-set, ovule fertility, and berry growth of several grape cultivars. Am. J. Enol. Vitic. 28, 215–222.

Kliewer, W.M., 1977b. Influence of temperature, solar radiation and nitrogen on coloration and composition of Emperor grapes. Am. J. Enol. Vitic. 28, 96–103.

Kliewer, W.M., Cook, J.A., 1971. Arginine and total free amino acids as indicators of the nitrogen status of grapevines. J. Am. Soc. Hortic. Sci. 96, 581–587.

Kliewer, W.M., Dokoozlian, N.K., 2005. Leaf area/crop weight ratios of grapevines: influence on fruit composition and wine quality. Am. J. Enol. Vitic. 56, 170–181.

Kliewer, W.M., Lider, L.A., 1968. Influence of cluster exposure to the sun on the composition of Thompson Seedless fruit. Am. J. Enol. Vitic. 19, 175–184.

Kliewer, W.M., Lider, L.A., 1976. Influence of leafroll virus on composition of Burger fruits. Am. J. Enol. Vitic. 27, 118–124.

Kliewer, W.M., Ough, C.S., 1970. The effect of leaf area and crop level on the concentration of amino acids and total nitrogen in 'Thompson Seedless' grapes. Vitis 9, 196–206.

Kliewer, W.M., Schultz, H.B., 1964. Influence of environment on metabolism of organic acids and carbohydrates in *Vitis vinifera*. II. Light. Am. J. Enol. Vitic. 15, 119–129.

Kliewer, W.M., Soleimani, A., 1972. Effect of chilling on budbreak in 'Thompson Seedless' and 'Carignane' grapevines. Am. J. Enol. Vitic. 23, 31–34.

Kliewer, W.M., Torres, R.E., 1972. Effect of controlled day and night temperatures on grape coloration. Am. J. Enol. Vitic. 23, 71–77.

Kliewer, W.M., Weaver, R.J., 1971. Effect of crop level and leaf area on growth, composition, and coloration of 'Tokay' grapes. Am. J. Enol. Vitic. 22, 172–177.

Kliewer, W.M., Howarth, L., Omori, M., 1967. Concentrations of tartaric acid and malic acids and their salts in *Vitis vinifera* grapes. Am. J. Enol. Vitic. 18, 42–54.

Kliewer, W.M., Lider, L.A., Ferrari, N., 1972. Effects of controlled temperature and light intensity on growth and carbohydrate levels of 'Thompson Seedless' grapevines. J. Am. Soc. Hortic. Sci. 97, 185–188.

Kliewer, W.M., Freeman, B.M., Hosssom, C., 1983. Effect of irrigation, crop level and potassium fertilization on Carignane vines. I. Degree of water stress and effect on growth and yield. Am. J. Enol. Vitic. 34, 186–196.

Kliewer, W.M., Bowen, P., Benz, M., 1989. Influence of shoot orientation on growth and yield development in Cabernet Sauvignon. Am. J. Enol. Vitic. 40, 259–264.

Klik, A., Rosner, J., Loiskandl, W., 1998. Effects of temporary and permanent soil cover on grape yield and soil chemical and physical properties. J. Soil Water Conserv. 53, 249–253.

Knight, M.R., Knight, H., 2012. Low-temperature perception leading to gene expression and cold tolerance in higher plants. New Phytol. 195, 737–751.

Knoblauch, M., Oparka, K., 2012. The structure of the phloem—still more questions than answers. Plant J. 70, 147–156.

Knoblauch, M., Peters, W.S., 2010. Münch, morphology, microfluidics—our structural problem with the phloem. Plant Cell Environ. 33, 1439–1452.

Kobayashi, A., Iwasaki, K., Teranuma, T., 1964. Pollen germination and berry set, growth and quality of Delaware grapes as affected by soil oxygen concentration. J. Jpn. Soc. Hortic. Sci. 33, 265–272.

Kobayashi, H., Suzuki, S., Tanzawa, F., Takayanagi, T., 2009. Low expression of flavonoid 3′,5′-hydroxylase ($F3'{,}5'H$) associated with cyanidin-based anthocyanins in grape leaf. Am. J. Enol. Vitic. 60, 362–367.

Kobayashi, H., Takase, H., Suzuki, Y., Tanzawa, F., Takata, R., Fujita, K., et al., 2011. Environmental stress enhances biosynthesis of flavor precursors, *S*-3-(hexan-1-ol)-glutathione and *S*-3-(hexan-1-ol)-cysteine, in grapevine through glutathione *S*-transferase activation. J. Exp. Bot. 62, 1325–1336.

Kobayashi, H., Matsuyama, S., Takase, H., Sasaki, K., Suzuki, S., Takata, R., et al., 2012. Impact of harvest timing on the concentration of 3-mercaptohexan-1-ol precursors in *Vitis vinifera* berries. Am. J. Enol. Vitic. 63, 544–548.

Kobayashi, S., Goto-Yamamoto, N., Hirochika, H., 2004. Retrotransposon-induced mutations in grape skin color. Science 304, 982.

Kobayashi, T., Nishizawa, N.K., 2012. Iron uptake, translocation, and regulation in higher plants. Annu. Rev. Plant Biol. 63, 131–152.

Koblet, W., 1966. Fruchtansatz bei Reben in Abhängigkeit von Triebbehandlungen und Klimafaktoren. Wein-Wiss. 21, 297–321 and 346–379.

Koblet, W., 1969. Wanderung von Assimilaten in Rebtrieben und Einfluss der Blattfläche auf Ertrag und Qualität der Trauben. Wein-Wiss. 24, 277–319.

Koblet, W., 1975. Wanderung von Assimilaten aus verschiedenen Rebenblättern während der Reifephase der Trauben. Wein-Wiss. 30, 241–249.
Koblet, W., Perret, P., 1971. Kohlehydratwanderung in Geiztrieben von Reben. Wein-Wiss. 26, 202–211.
Koblet, W., Perret, P., 1972. Wanderung von Assimilaten innerhalb der Rebe. Wein-Wiss. 27, 146–154.
Koch, A., Doyle, C.L., Matthews, M.A., Williams, L.E., Ebeler, S.E., 2010. 2-Methoxy-3-isobutylpyrazine in grape berries and its dependence on genotype. Phytochemistry 71, 2190–2198.
Koch, A., Ebeler, S.E., Williams, L.E., Matthews, M.A., 2012. Fruit ripening in *Vitis vinifera*: light intensity before and not during ripening determines the concentration of 2-methoxy-3-isobutylpyrazine in Cabernet Sauvignon berries. Physiol. Plant. 145, 275–285.
Koch, G.W., Sillett, S.C., Jennings, G.M., Davis, S.D., 2004. The limits to tree height. Nature 428, 851–854.
Koch, K., 2004. Sucrose metabolism: regulatory mechanisms and pivotal roles in sugar sensing and plant development. Curr. Opin. Plant Biol. 7, 235–246.
Koch, R., Alleweldt, G., 1978. Der Gaswechsel reifender Weinbeeren. Vitis 17, 30–44.
Kochian, L.V., Hoekenga, O.A., Piñeros, M.A., 2004. How do crop plants tolerate acid soils? Mechanisms of aluminum tolerance and phosphorous efficiency. Annu. Rev. Plant Biol. 55, 459–493.
Köckenberger, W., Pope, J.M., Xia, Y., Jeffrey, K.R., Komor, E., Callaghan, P.T., 1997. A non-invasive measurement of phloem and xylem water flow in castor bean seedlings by nuclear magnetic resonance microimaging. Planta 201, 53–63.
Kohlen, W., Charnikhova, T., Liu, Q., Bours, R., Domagalska, M.A., Beguerie, S., et al., 2011. Strigolactones are transported through the xylem and play a key role in shoot architectural response to phosphate deficiency in nonarbuscular mycorrhizal host *Arabidopsis*. Plant Physiol. 155, 974–987.
Kolb, C.A., Pfündel, E.E., 2005. Origins of non-linear and dissimilar relationships between epidermal UV absorbance and UV absorbance of extracted phenolics in leaves of grapevine and barley. Plant Cell Environ. 25, 580–590.
Kolb, C.A., Käser, M.A., Kopecký, J., Zotz, G., Riederer, M., Pfündel, E.E., 2001. Effects of natural intensities of visible and ultraviolet radiation on epidermal ultraviolet screening and photosynthesis in grape leaves. Plant Physiol. 127, 863–875.
Kolb, C.A., Kopecký, J., Riederer, M., Pfündel, E.E., 2003. UV screening by phenolics in berries of grapevine (*Vitis vinifera*). Funct. Plant Biol. 30, 1177–1186.
Kopriva, S., 2006. Regulation of sulfate assimilation in *Arabidopsis* and beyond. Ann. Bot. 97, 479–495.
Körner, C., 2003. Carbon limitation in trees. J. Ecol. 91, 4–17.
Koroleva, O.A., Tomos, A.D., Farrar, J., Pollock, C.J., 2002. Changes in osmotic and turgor pressure in response to sugar accumulation in barley source leaves. Planta 215, 210–219.
Kortekamp, A., 2006. Expression analysis of defence-related genes in grapevine leaves after inoculation wtih a host and a non-host pathogen. Plant Physiol. Biochem. 44, 58–67.
Koussa, T., Broquedis, M., Bouard, J., 1994. Importance de l'acide abscissique dans le développement des bourgeons latents de vigne (*Vitis vinifera* L. var. Merlot) et plus particulièrement dans la phase de levée de dormance. Vitis 33, 63–67.
Koussa, T., Cherrad, M., Bertrand, A., Broquedis, M., 1998. Comparaison de la teneur en amidon, en glucides solubles et en acide abscissique des bourgeons latents et des entre-noeuds au cours du cycle végétatif de la vigne. Vitis 37, 5–10.
Kovács, L.G., Byers, P.L., Kaps, M.L., Saenz, J., 2003. Dormancy, cold hardiness, and spring frost hazard in *Vitis amurensis* hybrids under continental climatic conditions. Am. J. Enol. Vitic. 54, 8–14.
Koyama, K., Goto-Yamamoto, N., 2008. Bunch shading during different developmental stages affects the phenolic biosynthesis in berry skins of 'Cabernet Sauvignon' grapes. J. Am. Soc. Hortic. Sci. 133, 743–753.
Koyama, K., Sadamatsu, K., Goto-Yamamoto, N., 2009. Abscisic acid stimulated ripening and gene expression in berry skins of the Cabernet Sauvignon grape. Funct. Integr. Genomics 10, 367–381.

Kramer, E.M., Bennett, M.J., 2006. Auxin transport: a field in flux. Trends Plant Sci. 11, 382–386.
Kramer, E.M., Myers, D.R., 2013. Osmosis is not driven by water dilution. Trends Plant Sci. 18, 195–197.
Kramer, E.M., Frazer, N.L., Baskin, T.I., 2007. Measurement of diffusion within the cell wall in living roots of *Arabidopsis thaliana*. J. Exp. Bot. 58, 3005–3015.
Kramer, P.J., Boyer, J.S., 1995. Water Relations of Plants and Soils. Academic Press, San Diego, CA.
Krasnow, M., Matthews, M., Shackel, K., 2008. Evidence for substantial maintenance of membrane integrity and cell viability in normally developing grape (*Vitis vinifera* L.) berries throughout development. J. Exp. Bot. 59, 849–859.
Kratsch, H.A., Wise, R.R., 2000. The ultrastructure of chilling stress. Plant Cell Environ. 23, 337–350.
Kraus, T.E.C, Yu, Z., Preston, C.M., Dahlgren, R.A., Zasoski, R.J., 2003. Linking chemical reactivity and protein precipitation to structural characteristics of foliar tannins. J. Chem. Ecol. 29, 703–730.
Kretschmer, M., Kassemeyer, H.H., Hahn, M., 2007. Age-dependent grey mould susceptibility and tissue-specific defence gene activation of grapevine berry skins after infection by *Botrytis cinerea*. J. Phytopathol. 155, 258–263.
Kriedemann, P.E., 1968a. Observations on gas exchange in the developing Sultana berry. Aust. J. Biol. Sci. 21, 907–916.
Kriedemann, P.E., 1968b. Photosynthesis in vine leaves as a function of light intensity, temperature, and leaf age. Vitis 7, 213–220.
Kriedemann, P.E., Buttrose, M.S., 1971. Chlorophyll content and photosynthetic activity within woody shoots of *Vitis vinifera* (L.). Photosynthetica 5, 22–27.
Kriedemann, P.E., Goodwin, I., 2003. Regulated Deficit Irrigation and Partial Rootzone Drying. An Overview of Principles and Applications. Land and Water Australia, Canberra.
Kriedemann, P.E., Törökfalvy, E., Smart, R.E., 1973. Natural occurrence and photosynthetic utilisation of sunflecks by grapevine leaves. Photosynthetica 7, 18–27.
Kritzinger, E.C., Bauer, F.F., du Toit, W.J., 2013. Role of glutathione in winemaking: a review. J. Agric. Food Chem. 61, 269–277.
Krouk, G., Ruffel, S., Gutiérrez, R.A., Gojon, A., Crawford, N.M., Coruzzi, G.M., et al., 2011. A framework integrating plant growth with hormones and nutrients. Trends Plant Sci. 16, 178–182.
Kruse, J., Hopmans, P., Adams, M.A., 2008. Temperature responses are a window to the physiology of dark respiration: differences between CO_2 release and O_2 reduction shed light on energy conservation. Plant Cell Environ. 31, 901–914.
Kruse, J., Rennenberg, H., Adams, M.A., 2011. Steps towards a mechanistic understanding of respiratory temperature responses. New Phytol. 189, 659–677.
Kuć, J., 1995. Phytoalexins, stress metabolism, and disease resistance in plants. Annu. Rev. Phytopathol. 33, 275–297.
Kuhlemeier, C., 2007. Phyllotaxis. Trends Plant Sci. 12, 143–150.
Kühn, C., Barker, L., Bürkle, L., Frommer, W.B., 1999. Update on sucrose transport in higher plants. J. Exp. Bot. 50, 935–953.
Kurkdjian, A., Guern, J., 1989. Intracellular pH: measurement and importance in cell activity. Annu. Rev. Plant Physiol. Plant Mol. Biol. 40, 271–303.
Kusano, T., Berberich, T., Tateda, C., Takahashi, Y., 2008. Polyamines: essential factors for growth and survival. Planta 228, 367–381.
Kutschera, U., 2008a. The growing outer epidermal wall: design and physiological role of a composite structure. Ann. Bot. 101, 615–621.
Kutschera, U., 2008b. The pacemaker of plant growth. Trends Plant Sci. 13, 105–107.
Kutschera, U., Briggs, W.R., 2012. Root phototropism: from dogma to the mechanism of blue light perception. Planta 235, 443–452.

Ky, I., Lorrain, B., Jourdes, M., Pasquier, G., Fermaud, M., Gény, L., et al., 2012. Assessment of grey mould (*Botrytis cinerea*) impact on phenolic and sensory quality of Bordeaux grapes, musts and wines for two consecutive vintages. Aust. J. Grape Wine Res. 18, 215–226.

La Guerche, S., Dauphin, B., Pons, M., Blancard, D., Darriet, P., 2006. Characterization of some mushroom and earthy off-odors microbially induced by the development of rot on grapes. J. Agric. Food Chem. 54, 9193–9200.

Lacey, M.J., Allen, M.S., Harris, R.L.N., Brown, W.V., 1991. Methoxypyrazines in Sauvignon blanc grapes and wines. Am. J. Enol. Vitic. 42, 103–108.

Lacombe, T., Boursiquot, J.M., Laucou, V., Dechesne, F., Varès, D., This, P., 2007. Relationships and genetic diversity within the accessions related to Malvasia held in the Domaine de Vassal grape germplasm repository. Am. J. Enol. Vitic. 58, 124–131.

Lacroux, F., Tregoat, O., Van Leeuwen, C., Pons, A., Tominaga, T., Lavigne-Cruège, V., et al., 2008. Effect of foliar nitrogen and sulphur application on aromatic expression of *Vitis vinifera* L. cv. Sauvignon blanc. J. Int. Sci. Vigne Vin 42, 125–132.

Lafon-Lafourcade, S., Guimberteau, G., 1962. Evolution des amino-acides au cours de la maturation des raisins. Vitis 3, 130–135.

Laguna, E., 2003. Sobre las formas naturalizadas de *Vitis* en la Comunidad Valenciana. I. Las especies. Flora Montiberica 23, 46–82.

Laidig, F., Piepho, H.P., Hofäcker, W., 2009. Statistical analysis of 'White Riesling' (*Vitis vinifera* ssp. *sativa* L.) clonal performance at 16 locations in the Rheinland-Pfalz region of Germany between 1971 and 2007. Vitis 48, 77–85.

Lakso, A.N., Kliewer, W.M., 1975. The influence of temperature on malic acid metabolism in rape berries. I. Enzyme responses. Plant Physiol. 56, 370–372.

Lakso, A.N., Kliewer, W.M., 1978. The influence of temperature on malic acid metabolism in grape berries. II. Temperature responses of net dark CO_2 fixation and malic acid pools. Am. J. Enol. Vitic. 29, 145–149.

Lakso, A.N., Pratt, C., Pearson, R.C., Pool, R.M., Seem, R.C., Welser, M.J., 1982. Photosynthesis, transpiration, and water use efficiency of mature grape leaves infected with *Uncinula necator* (powdery mildew). Phytopathology 72, 232–236.

Lalonde, S., Tegeder, M., Throne-Holst, M., Frommer, W.B., Patrick, J.W., 2003. Phloem loading and unloading of sugars and amino acids. Plant Cell Environ. 26, 37–56.

Lalonde, S., Wipf, D., Frommer, W.B., 2004. Transport meachnisms for organic forms of carbon and nitrogen between source and sink. Annu. Rev. Plant Biol. 55, 341–372.

Lamb, D.W., Weedon, M.M., Bramley, R.G.V., 2004. Using remote sensing to predict grape phenolics and colour at harvest in a Cabernet Sauvignon vineyard. Timing observations against vine phenology and optimising image resolution. Aust. J. Grape Wine Res. 10, 46–54.

Lane, N., 2002. Oxygen. The Molecule That Made the World. Oxford University Press, Oxford.

Lang, A., Düring, H., 1990. Grape berry splitting and some mechanical properties of the skin. Vitis 29, 61–70.

Lang, A., Thorpe, M.R., 1989. Xylem, phloem and transpiration flows in a grape: application of a technique for measuring the volume of attached fruits to high resolution using Archimedes' principle. J. Exp. Bot. 40, 1069–1078.

Lang, G.A., Early, J.D., Martin, G.C., Darnell, R.L., 1987. Endo-, para-, and eco-dormancy: physiological terminology and classification for dormancy research. HortScience 22, 371–377.

Lang, N.S., Wample, R.L., Smithyman, R., Mills, L., 1998. Photosynthesis and chlorophyll fluorescence in blackleaf-affected Concord leaves. Am. J. Enol. Vitic. 49, 367–374.

Langcake, P., 1981. Disease resistance of *Vitis* spp. and the production of the stress metabolites resveratrol, ε-viniferin, α-viniferin and pterostilbene. Physiol. Plant Pathol. 18, 213–226.

Langcake, P., Lovell, P.A., 1980. Light and electron microscopical studies of the infection of *Vitis* spp. by *Plasmopara viticola*, the downy mildew pathogen. Vitis 19, 321–337.

Langcake, P., McCarthy, W.V., 1979. The relationship of resveratrol production to infection of grapevine leaves by *Botrytis cinerea*. Vitis 18, 244–253.

Langcake, P., Pryce, R.J., 1976. The production of resveratrol by *Vitis vinifera* and other members of the Vitaceae as a response to infection or injury. Physiol. Plant Pathol. 9, 77–86.

Langcake, P., Cornford, C.A., Pryce, R.J., 1979. Identification of pterostilbene as a phytoalexin from *Vitis vinifera* leaves. Phytochemistry 18, 1025–1027.

Larronde, F., Krisa, S., Decendit, A., Chèze, C., Deffieux, G., Mérillon, J.M., 1998. Regulation of polyphenol production in *Vitis vinifera* cell suspension cultures by sugars. Plant Cell Rep. 17, 946–950.

Larronde, F., Gaudillère, J.P., Krisa, S., Decendit, A., Deffieux, G., Mérillon, J.M., 2003. Airborne methyl jasmonate induces stilbene accumulation in leaves and berries of grapevine plants. Am. J. Enol. Vitic. 54, 63–66.

Laucou, V., Lacombe, T., Dechesne, F., Siret, R., Bruno, J.P., Dessup, M., et al., 2011. High throughput analysis of grape genetic diversity as a tool for germplasm collection management. Theor. Appl. Genet. 122, 1233–1245.

Lavee, S., May, P., 1997. Dormancy of grapevine buds—facts and speculation. Aust. J. Grape Wine Res. 3, 31–46.

Lawlor, D.W., 2002. Carbon and nitrogen assimilation in relation to yield: mechanisms are the key to understanding production systems. J. Exp. Bot. 53, 773–787.

Lawlor, D.W., Cornic, G., 2002. Photosynthetic carbon assimilation and associated metabolism in relation to water deficits in higher plants. Plant Cell Environ. 25, 275–294.

Lawlor, D.W., Tezara, W., 2009. Causes of decreased photosynthetic rate and metabolic capacity in water-deficient leaf cells: a critical evaluation of mechanisms and integration of processes. Ann. Bot. 103, 561–579.

Lawson, T., 2008. Guard cell photosynthesis and stomatal function. New Phytol. 181, 13–34.

Lazar, G., Goodman, H.M., 2006. *MAX1*, a regulator of the flavonoid pathway, controls vegetative axillary bud outgrowth in *Arabidopsis*. Proc. Natl. Acad. Sci. USA 103, 472–476.

Leakey, A.D.B., Xu, F., Gillespie, K.M., McGrath, J.M., Ainsworth, E.A., Ort, D.R., 2009. Genomic basis for stimulated respiration by plants growing under elevated carbon dioxide. Proc. Natl. Acad. Sci. USA 106, 3597–3602.

Lebon, E., Pellegrino, A., Tardieu, F., Lecoeur, J., 2004. Shoot development in grapevine (*Vitis vinifera*) is affected by the modular branching pattern of the stem and intra- and inter-shoot trophic competition. Ann. Bot. 93, 263–274.

Lebon, G., Wojnarowiez, G., Holzapfel, B., Fontaine, F., Vaillant-Gaveau, N., Clément, C., 2008. Sugars and flowering in the grapevine (*Vitis vinifera* L.). J. Exp. Bot. 59, 2565–2578.

Lecas, M., Brillouet, J.M., 1994. Cell wall composition of grape berry skins. Phytochemistry 35, 1241–1243.

Lecourieux, F., Kappel, C., Lecourieux, D., Serrano, A., Torres, E., Arce-Johnson, P., et al., 2014. An update on sugar transport and signalling in grapevine. J. Exp. Bot. 65, 821–832.

Ledbetter, C.A., Ramming, D.W., 1989. Seedlessness in grapes. Hortic. Rev. (Am. Soc. Hortic. Sci) 11, 159–184.

Ledbetter, C.A., Shonnard, C.B., 1991. Berry and seed characteristics associated with stenospermocarpy in *vinifera* grapes. J. Hortic. Sci. 66, 247–252.

Lee, C.Y., Smith, N.L., Nelson, R.R., 1979. Relationship between pectin methylesterase activity and the formation of methanol in Concord grape juice and wine. Food Chem. 4, 143–148.

Lee, S.H., Seo, M.J., Riu, M., Cotta, J.P., Block, D.E., Dokoozlian, N.K., et al., 2007. Vine microclimate and norisoprenoid concentration in Cabernet Sauvignon grapes and wines. Am. J. Enol. Vitic. 58, 291–301.

Lehnart, R., Michel, H., Löhnertz, O., Linsenmeier, A., 2008. Root dynamics and pattern of 'Riesling' on 5C rootstock using minirhizotrons. Vitis 47, 197−200.

Lei, M., Liu, Y., Zhang, B., Zhao, Y., Wang, X., Zhou, Y., et al., 2011. Genetic and genomic evidence that sucrose is a global regulator of plant responses to phosphate starvation in *Arabidopsis*. Plant Physiol. 156, 1116−1130.

Leide, J., Hildebrandt, U., Reussing, K., Riederer, M., Vogg, G., 2007. The developmental pattern of tomato fruit wax accumulation and its impact on cuticular transpiration barrier properties: effects of a deficiency in a β-ketoacyl-coenzyme A synthase (LeCER6). Plant Physiol. 144, 1667−1679.

Leigh, J., Hodge, A., Fitter, A.H., 2009. Arbuscular mycorrhizal fungi can transfer substantial amounts of nitrogen to their host plant from organic material. New Phytol. 181, 199−207.

Lejay, L., Wirth, J., Pervent, M., Cross, J.M.F., Tillard, P., Gojon, A., 2008. Oxidative pentose phosphate pathway-dependent sugar sensing as a mechanism for regulation of root ion transporters by photosynthesis. Plant Physiol. 146, 2036−2053.

Lemaire, G., Millard, P., 1999. An ecophysiological approach to modelling resource fluxes in competing plants. J. Exp. Bot. 50, 15−28.

Lendzian, K.J., 2006. Survival strategies of plants during secondary growth: barrier properties of phellems and lenticels towards water, oxygen, and carbon dioxide. J. Exp. Bot. 57, 2535−2546.

Lenk, S., Buschmann, C., Pfündel, E.E., 2007. *In vivo* assessing flavonols in white grape berries (*Vitis vinifera* L. cv. Pinot Blanc) of different degrees of ripeness using chlorophyll fluorescence imaging. Funct. Plant Biol. 34, 1092−1104.

Lens, F., Sperry, J.S., Christman, M.A., Choat, B., Rabaey, D., Jansen, S., 2011. Testing hypotheses that link wood anatomy to cavitation resistance and hydraulic conductivity in the genus *Acer*. New Phytol. 190, 709−723.

Leustek, T., Saito, K., 1999. Sulfate transport and assimilation in plants. Plant Physiol. 120, 637−644.

Levadoux, L., 1956. Les populations sauvages et cultivées de *Vitis vinifera* L. Ann. Amélior. Plantes 6, 59−118.

Levey, D.J., 2004. The evolutionary ecology of ethanol production and alcoholism. Integr. Comp. Biol. 44, 284−289.

Lewinsohn, E., Sitrit, Y., Bar, E., Azulay, Y., Meir, A., Zamir, D., et al., 2005. Carotenoid pigmentation affects the volatile composition of tomato and watermelon fruits, as revealed by comparative genetic analyses. J. Agric. Food Chem. 53, 3142−3148.

Leyhe, A., Blanke, M.M., 1989. Kohlenstoff-Haushalt der Infloreszenz der Rebe. 2. CO_2-Gaswechsel der Blütenstände und Weinbeere. Wein-Wiss. 44, 147−150.

Leyser, O., 2010. The power of auxin in plants. Plant Physiol. 154, 501−505.

Li, B., Feng, Z., Xie, M., Sun, M., Zhao, Y., Liang, L., et al., 2011. Modulation of the root-sourced ABA signal along its way to the shoot in *Vitis riparia* × *Vitis labrusca* under water deficit. J. Exp. Bot. 62, 1731−1741.

Li, D., Wan, Y., Wang, Y., He, P., 2008. Relatedness of resistance to anthracnose and to white rot in Chinese wild grapes. Vitis 47, 213−215.

Li, Y., Jones, L., McQueen-Mason, S., 2003. Expansins and cell growth. Curr. Opin. Plant Biol. 6, 603−610.

Li, Z., Wakao, S., Fischer, B.B., Niyogi, K.K., 2009. Sensing and responding to excess light. Annu. Rev. Plant Biol. 60, 239−260.

Liakopoulos, G., Nikolopoulos, D., Klouvatou, A., Vekkos, K.A., Manetas, Y., Karabourniotis, G., 2006. The photoprotective role of epidermal anthocyanins and surface pubescence in young leaves of grapevine (*Vitis vinifera*). Ann. Bot. 98, 257−265.

Liakoura, V., Fotelli, M.N., Rennenberg, H., Karabourniotis, G., 2009. Should structure−function relations be considered separately for homobaric vs. heterobaric leaves? Am. J. Bot. 96, 612−619.

Liang, N.N., Pan, Q.H., He, F., Wang, J., Reeves, M.J., Duan, C.Q., 2013. Phenolic profiles of *Vitis davidii* and *Vitis quinquangularis* species native to China. J. Agric. Food Chem. 61, 6016–6027.

Liang, Y., Si, J., Römheld, V., 2005. Silicon uptake and transport is an active process in *Cucumis sativus*. New Phytol. 167, 797–804.

Liang, Z., Owens, C.L., Zhong, G.Y., Cheng, L., 2011. Polyphenolic profiles detected in the ripe berries of *Vitis vinifera* germplasm. Food Chem. 129, 940–950.

Liang, Z., Yang, Y., Cheng, L., Zhong, G.Y., 2012a. Characterization of polyphenolic metabolites in the seeds of *Vitis* germplasm. J. Agric. Food Chem. 60, 1291–1299.

Liang, Z., Yang, Y., Cheng, L., Zhong, G.Y., 2012b. Polyphenolic composition and content in the ripe berries of wild *Vitis* germplasm. Food Chem. 132, 730–738.

Lider, L.A., Goheen, A.C., Ferrari, N.L., 1975. A comparison between healthy and leafroll-affected grapevine planting stocks. Am. J. Enol. Vitic. 26, 144–147.

Lijavetzky, D., Ruiz-García, L., Cabezas, J.A., De Andrés, M.T., Bravo, G., Ibáñez, A., et al., 2006. Molecular genetics of berry colour variation in table grape. Mol. Genet. Genomics 276, 427–435.

Lijavetzky, D., Carbonell-Bejerano, P., Grimplet, J., Bravo, G., Flores, P., Fenoll, J., et al., 2012. Berry flesh and skin ripening features in *Vitis vinifera* as assessed by transcriptional profiling. PLoS One 7, e39547, http://dx.doi.org/10.1371/journal.pone/0039547.

Lillo, C., Lea, U.S., Ruoff, P., 2008. Nutrient depletion as a key factor for manipulating gene expression and product formation in different branches of the flavonoid pathway. Plant Cell Environ. 31, 587–601.

Lilov, D., Andonova, T., 1976. Cytokinins, growth, flower and fruit formation in *Vitis vinifera*. Vitis 15, 160–170.

Lim, P.O., Kim, H.J., Nam, H.G., 2007. Leaf senescence. Annu. Rev. Plant Biol. 58, 115–136.

Lin, C.H., Lin, J.H., Chang, L.R., Lin, H.S., 1985. The regulation of the Golden Muscat grape production season in Taiwan. Am. J. Enol. Vitic. 36, 114–117.

Linsenmeier, A., Rauhut, D., Kürbel, H., Löhnertz, O., Schubert, S., 2007. Untypical ageing off-flavour and masking effects due to long-term nitrogen fertilization. Vitis 46, 33–38.

Linsenmeier, A.W., Loos, U., Löhnertz, O., 2008. Must composition and nitrogen uptake in a long-term trial as affected by timing of nitrogen fertilization in a cool-climate Riesling vineyard. Am. J. Enol. Vitic. 59, 255–264.

Lisch, D., 2009. Epigenetic regulation of transposable elements in plants. Annu. Rev. Plant Biol. 60, 43–66.

Liu, H.F., Wu, B.H., Fan, P.G., Li, S.H., Li, L.S., 2006. Sugar and acid concentrations in 98 grape cultivars analyzed by principal component analysis. J. Sci. Food Agric. 86, 1526–1536.

Liu, W.T., Pool, R., Wenkert, W., Kriedemann, P.E., 1978. Changes in photosynthesis, stomatal resistance and abscisic acid of *Vitis labruscana* through drought and irrigation cycles. Am. J. Enol. Vitic. 29, 239–246.

Llorens, N., Arola, L., Blade, C., Mas, A., 2002. Nitrogen metabolism in a grapevine *in vitro* system. J. Int. Sci. Vigne Vin 36, 157–159.

Loescher, W.H., McCamant, T., Keller, J.D., 1990. Carbohydrate reserves, translocation, and storage in woody plant roots. HortScience 25, 274–281.

Loewus, F.A., 1999. Biosynthesis and metabolism of ascorbic acid in plants and of analogs of ascorbic acid in fungi. Phytochemistry 52, 193–210.

Logan, D.C., 2006. The mitochondrial compartment. J. Exp. Bot. 57, 1225–1243.

Löhnertz, O., Schaller, K., Mengel, K., 1989. Nährstoffdynamik in Reben. III. Mitteilung: Stickstoffkonzentration und Verlauf der Aufnahme in der Vegetation. Wein-Wiss. 44, 192–204.

Long, S.P., Ainsworth, E.A., Rogers, A., Ort, D.R., 2004. Rising atmospheric carbon dioxide: plants FACE the future. Annu. Rev. Plant Biol. 55, 591–628.

Longbottom, M.L., Dry, P.R., Sedgley, M., 2008. Observations on the morphology and development of star flowers of *Vitis vinifera* L. cvs Chardonnay and Shiraz. Aust. J. Grape Wine Res. 14, 203–210.

Longbottom, M.L., Dry, P.R., Sedgley, M., 2010. Effects of sodium molybdate foliar sprays on molybdenum concentration in the vegetative and reproductive structures and on yield components of *Vitis vinifera* cv. Merlot. Aust. J. Grape Wine Res. 16, 477–490.

Longo, E., Cansado, J., Agrelo, D., Villa, T.G., 1991. Effect of climatic conditions on yeast diversity in grape musts from northwest Spain. Am. J. Enol. Vitic. 42, 141–144.

López-Bucio, J., Cruz-Ramírez, A., Herrera-Estrella, L., 2003. The role of nutrient availability in regulating root architecture. Curr. Opin. Plant Biol. 6, 280–287.

Lopez-Casado, G., Salamanca, A., Heredia, A., 2010. Viscoelastic nature of isolated tomato (*Solanum lycopersicum*) fruit cuticles: a mathematical model. Physiol. Plant. 140, 79–88.

Loreto, F., Schnitzler, J.P., 2010. Abiotic stresses and induced BVOCs. Trends Plant Sci. 15, 154–166.

Lott, R.V., Barrett, H.C., 1967. The dextrose, levulose, sucrose, and acid content of the juice from 39 grape clones. Vitis 6, 257–268.

Lough, T.J., Lucas, W.J., 2006. Integrative plant biology: role of phloem long-distance macromolecular trafficking. Annu. Rev. Plant Biol. 57, 203–232.

Loulakakis, K.A., Morot-Gaudry, J.F., Velanis, C.N., Skopelitis, D.S., Moschou, P.N., Hirel, B., et al., 2009. Advancements in nitrogen metabolism in grapevine. In: Roubelakis-Angelakis, K.A. (Ed.), Grapevine Molecular Physiology and Biotechnology, second ed. Springer, Dordrecht, The Netherlands, pp. 161–205.

Loulakakis, K.A., Roubelakis-Angelakis, K.A., 2001. Nitrogen assimilation in grapevine. In: Roubelakis-Angelakis, K.A. (Ed.), Molecular Biology and Biotechnology of the Grapevine. Kluwer, Dordrecht, The Netherlands, pp. 59–85.

Loulakakis, K.A., Primikirios, N.I., Nikolantonakis, M.A., Roubelakis-Angelakis, K.A., 2002. Immunocharacterization of *Vitis vinifera* L. ferredoxin-dependent glutamate synthase, and its spatial and temporal changes during leaf development. Planta 215, 630–638.

Loveys, B.R., 1984. Diurnal changes in water relations and abscisic acid in field-grown *Vitis vinifera* cultivars. III. The influence of xylem-derived abscisic acid on leaf gas exchange. New Phytol. 98, 563–573.

Loveys, B.R., Düring, H., 1984. Diurnal changes in water relations and abscisic acid in field-grown *Vitis vinifera* cultivars. II. Abscisic acid changes under semi-arid conditions. New Phytol. 97, 37–47.

Loveys, B.R., Scheurwater, I., Pons, T.L., Fitter, A.H., Atkin, O.K., 2002. Growth temperature influences the underlying components of relative growth rate: an investigation using inherently fast- and slow-growing plant species. Plant Cell Environ. 25, 975–987.

Lovisolo, C., Schubert, A., 1998. Effects of water stress on vessel size and xylem hydraulic conductivity in *Vitis vinifera* L. J. Exp. Bot. 49, 693–700.

Lovisolo, C., Schubert, A., 2000. Downward shoot positioning affects water transport in field-grown grapevines. Vitis 39, 49–53.

Lovisolo, C., Schubert, A., 2006. Mercury hinders recovery of shoot hydraulic conductivity during grapevine rehydration: evidence from a whole-plant approach. New Phytol. 172, 469–478.

Lovisolo, C., Hartung, W., Schubert, A., 2002a. Whole-plant hydraulic conductance and root-to-shoot flow of abscisic acid are independently affected by water stress in grapevines. Funct. Plant Biol. 29, 1349–1356.

Lovisolo, C., Schubert, A., Sorce, C., 2002b. Are xylem radial development and hydraulic conductivity in downwardly-growing grapevine shoots influenced by perturbed auxin metabolism? New Phytol. 156, 65–74.

Lovisolo, C., Perrone, I., Hartung, W., Schubert, A., 2008. An abscisic acid-related reduced transpiration promotes gradual embolism repair when grapevines are rehydrated after drought. New Phytol. 180, 642–651.

Lovisolo, C., Perrone, I., Carra, A., Ferrandino, A., Flexas, J., Medrano, H., et al., 2010. Drought-induced changes in development and function of grapevine (*Vitis* spp.) organs and in their hydraulic and non-hydraulic interactions at the whole-plant level: a physiological and molecular update. Funct. Plant Biol. 37, 98–116.

Lu, Y., Foo, L.Y., 1999. The polyphenol constituents of grape pomace. Food Chem. 65, 1–8.

Luan, F., Wüst, M., 2002. Differential incorporation of 1-deoxy-D-xylulose into (3S)-linalool and geraniol in grape berry exocap and mesocarp. Phytochemistry 60, 451–459.

Luan, F., Mosandl, A., Münch, A., Wüst, M., 2005. Metabolism of geraniol in grape berry mesocarp of *Vitis vinifera* L. cv. Scheurebe: demonstration of stereoselective reduction, E/Z isomerization, oxidation and glycosylation. Phytochemistry 66, 295–303.

Luan, F., Mosandl, A., Gubesch, M., Wüst, M., 2006. Enantioselective analysis of monoterpenes in different grape varieties during berry ripening using stir bar sorptive extraction- and solid phase extraction-enantioselective-multidimensional gas chromatography-mass spectrometry. J. Chromatogr. A 1112, 369–374.

Lucas, R.E., Davis, J.F., 1961. Relationships between pH values of organic soils and availabilities of 12 plant nutrients. Soil Sci. 92, 177–182.

Luisetti, J., Gaignard, J.L., Devaux, M., 1991. *Pseudomonas syringae* pv. *syringae* as one of the factors affecting the ice nucleation of grapevine buds in controlled conditions. J. Phytopathol. 133, 334–344.

Lund, S.T., Peng, F.Y., Nayar, T., Reid, K.E., Schlosser, J., 2008. Gene expression analyses in individual grape (*Vitis vinifera* L.) berries during ripening initiation reveal that pigmentation intensity is a valid indicator of developmental staging within the cluster. Plant Mol. Biol. 68, 301–315.

Lundgren, M.R., Osborne, C.P., Christin, P.A., 2014. Deconstructing Kranz anatomy to understand C_4 evolution. J. Exp. Bot. 65, 3357–3369.

Lushai, G., Loxdale, H.D., 2002. The biological improbability of a clone. Genet. Res. 79, 1–9.

Lustig, I., Bernstein, Z., 1985. Determination of the mechanical properties of the grape berry skin by hydraulic measurements. Sci. Hortic. 25, 279–285.

Luu, D.T., Maurel, C., 2005. Aquaporins in a challenging environment: molecular gears for adjusting plant water status. Plant Cell Environ. 28, 85–96.

Ma, J.F., Yamaji, N., 2006. Silicon uptake and accumulation in higher plants. Trends Plant Sci. 11, 392–397.

Macheix, J.J., Fleuriet, A., Billot, J., 1990. Fruit Phenolics. CRC Press, Boca Raton, FL.

Macheix, J.J., Sapis, J.C., Fleuriet, A., 1991. Phenolic compounds and polyphenoloxidase in relation to browning in grapes and wines. Crit. Rev. Food Sci. Nutr. 30, 441–486.

Madronich, S., McKenzie, R.L., Björn, L.O., Caldwell, M.M., 1998. Changes in biologically active ultraviolet radiation reaching the Earth's surface. J. Photochem. Photobiol. B 46, 5–19.

Maga, J.A., 1989. The contribution of wood to the flavor of alcoholic beverages. Food Rev. Int. 5, 39–99.

Maghradze, D., Rustioni, L., Turok, J., Scienza, A., Failla, O., 2012. Caucasus and Northern Black Sea Region Ampelography. Vitis Special Issue. Maierdruck, Lingenfeld, Germany.

Malacarne, G., Vrhovsek, U., Zulini, L., Cestaro, A., Stefanini, M., Mattivi, F., et al., 2011. Resistance to *Plasmopara viticola* in a grapevine segregating population is associated with stilbenoid accumulation and with specific host transcriptional responses. BMC Plant Biol. 11, 114, http://dx.doi.org/10.1186/1471-2229-11-114.

Malamy, J.E., 2005. Intrinsic and environmental response pathways that regulate root system architecture. Plant Cell Environ. 28, 67–77.

Malamy, J.E., Benfey, P.N., 1997. Down and out in *Arabidopsis*: the formation of lateral roots. Trends Plant Sci. 2, 390–396.

Maletić, E., Pejić, I., Karoglan Kontić, J., Piljac, J., Dangl, G.S., Vokurka, A., et al., 2004. Zinfandel, Dobričić, and Plavac mali: the genetic relationship among three cultivars of the Dalmatian coast of Croatia. Am. J. Enol. Vitic. 55, 174–180.

Mancuso, S., 1999. Hydraulic and electrical transmission of wound-induced signals in *Vitis vinifera*. Aust. J. Plant Physiol. 26, 55–61.

Mancuso, S., Boselli, M., 2002. Characterisation of the oxygen fluxes in the division, elongation and mature zones of *Vitis* roots: influence of oxygen availability. Planta 214, 767–774.

Mancuso, S., Marras, A.M., 2006. Adaptive response of *Vitis* root to anoxia. Plant Cell Physiol. 47, 401–409.

Mannini, F., Mollo, A., Credi, R., 2012. Field performance and wine quality modification in a clone of *Vitis vinifera* cv. Dolcetto after GLRaV-3 elimination. Am. J. Enol. Vitic. 63, 144–147.

Mapfumo, E., Aspinall, D., Hancock, T.W., 1994a. Growth and development of roots of grapevine (*Vitis vinifera* L.) in relation to water uptake from soil. Ann. Bot. 74, 75−85.
Mapfumo, E., Aspinall, D., Hancock, T.W., 1994b. Vessel-diameter distribution in roots of grapevines (*Vitis vinifera* L. cv. Shiraz). Plant Soil 160, 49−55.
Marais, J., Hunter, J.J., Haasbroek, P.D., 1999. Effect of canopy microclimate, season and region on Sauvignon blanc grape composition and wine quality. S. Afr. J. Enol. Vitic. 20, 19−30.
Marais, J., Calitz, F., Haasbroek, P.D., 2001. Relationship between microclimatic data, aroma component concentrations and wine quality parameters in the prediction of Sauvignon blanc wine quality. S. Afr. J. Enol. Vitic. 22, 22−26.
Marangoni, B., Vitagliano, C., Peterlunger, E., 1986. The effect of defoliation on the composition of xylem sap from Cabernet franc grapevines. Am. J. Enol. Vitic. 37, 259−262.
Marga, F., Pesacreta, T.C., Hasenstein, K.H., 2001. Biochemical analysis of elastic and rigid cuticles of *Cirsium horridulum* Michx. Planta 213, 841−848.
Margna, U., Vainjsärv, T., Laanest, L., 1989. Different L-phenylalanine pools available for the biosynthesis of phenolics in buckwheat seedling tissues. Phytochemistry 28, 469−475.
Markham, K.R., Gould, K.S., Winefield, C.S., Mitchell, K.A., Bloor, S.J., Boase, M.R., 2000. Anthocyanic vacuolar inclusions—their nature and significance in flower colouration. Phytochemistry 55, 327−336.
Marois, J.J., Nelson, J.K., Morrison, J.C., Lile, L.S., Bledsoe, A.M., 1986. The influence of berry contact within grape clusters on the development of *Botrytis cinerea* and epicuticular wax. Am. J. Enol. Vitic. 37, 293−296.
Marschner, H., 1995. Mineral Nutrition of Higher Plants. Academic Press, London.
Marschner, H., Cakmak, I., 1989. High light intensity enhances chlorosis and necrosis in leaves of zinc, potassium, and magnesium deficient bean (*Phaseolus vulgaris* L.) plants. J. Plant Physiol. 134, 308−315.
Marshall, D.A., Spiers, J.M., Stringer, S.J., Curry, K.J., 2007. Laboratory method to estimate rain-induced splitting in cultivated blueberries. HortScience 42, 1551−1553.
Martelli, G.P., Graniti, A., Ercolani, G.L., 1986. Nature and physiological effects of grapevine diseases. Experientia 42, 933−942.
Martelli, G.P., Agranovsky, A.A., Bar-Joseph, M., Boscia, D., Candresse, T., Coutts, R.H.A., et al., 2002. The family *Closteroviridae* revised. Arch. Virol. 147, 2039−2044.
Marten, I., Hoth, S., Deeken, R., Ache, P., Ketchum, K.A., Hoshi, T., et al., 1999. AKT3, a phloem-localized K^+ channel, is blocked by protons. Proc. Natl. Acad. Sci. USA 96, 7581−7586.
Martens, H.J., Roberts, A.G., Oparka, K.J., Schulz, A., 2006. Quantification of plasmodesmatal endoplasmic reticulum coupling between sieve elements and companion cells using fluorescence redistribution after photobleaching. Plant Physiol. 142, 471−480.
Martin, C., Smith, A.M., 1995. Starch biosynthesis. Plant Cell 7, 971−985.
Martin, D.M., Toub, O., Chiang, A., Lo, B.C., Ohse, S., Lund, S.T., et al., 2009. The bouquet of grapevine (*Vitis vinifera* L. cv. Cabernet Sauvignon) flowers arises from the biosynthesis of sesquiterpene volatiles in pollen grains. Proc. Natl. Acad. Sci. USA 106, 7245−7250.
Martinelli, L., Gribaudo, I., 2001. Somatic embryogenesis in grapevine. In: Roubelakis-Angelakis, K.A. (Ed.), Molecular Biology and Biotechnology of the Grapevine. Kluwer, Dordrecht, The Netherlands, pp. 327−351.
Martínez, L., Cavagnaro, P., Boursiquot, J.M., Agüero, C., 2008. Molecular characterization of Bonarda-type grapevine (*Vitis vinifera* L.) cultivars from Argentina, Italy, and France. Am. J. Enol. Vitic. 59, 287−291.
Martínez-Esteso, M.J., Sélles-Marchart, S., Lijavetzky, D., Pedreño, M.A., Bru-Martínez, R., 2011. A DIGE-based quantitative proteomic analysis of grape berry flesh development and ripening reveals key events in sugar and organic acid metabolism. J. Exp. Bot. 62, 2521−2569.
Martinoia, E., Meyer, S., De Angeli, A., Nagy, R., 2012. Vacuolar transporters in their physiological context. Annu. Rev. Plant Biol. 63, 183−213.

Martins, V., Bassil, E., Hanana, M., Blumwald, E., Gerós, H., 2014. Copper homeostasis in grapevine: functional characterization of the *Vitis vinifera* copper transporter. Planta 240, 91–101.

Marty, F., 1999. Plant vacuoles. Plant Cell 11, 587–599.

Masclaux, C., Valadier, M.H., Brugière, N., Morot-Gaudry, J.F., Hirel, B., 2000. Characterization of the sink/source transition in tobacco (*Nicotiana tabacum* L.) shoots in relation to nitrogen management and leaf senescence. Planta 211, 510–518.

Masclaux-Daubresse, C., Daniel-Vedele, F., Dechorgnat, J., Chardon, F., Gaufichon, L., Suzuki, A., 2010. Nitrogen uptake, assimilation and remobilization in plants: challenges for sustainable and productive agriculture. Ann. Bot. 105, 1141–1157.

Mason, J.R., Clark, L., Shah, P.S., 1991. Ortho-aminoacetophenone repellency to birds: similarities to methyl anthranilate. J. Wildl. Manage. 55, 334–340.

Mason, M.G., Ross, J.J., Babst, B.A., Wienclaw, B.N., Beveridge, C.A., 2014. Sugar demand, not auxin, is the initial regulator of apical dominance. Proc. Natl. Acad. Sci. USA 111, 6092–6097.

Mastin, M.C. (2008). Effects of potential future warming on runoff in the Yakima River Basin, Washington. U. S. Geol. Surv. Sci. Invest. Rep. 2008–5124.

Mathiason, K., He, D., Grimplet, J., Vekateswari, J., Galbraith, D.W., Or, E., et al., 2009. Transcript profiling in *Vitis riparia* during chilling requirement fulfillment reveals coordination of gene expression patterns with optimized bud break. Funct. Integr. Genomics 9, 81–96.

Mathieu, S., Terrier, N., Procureur, J., Bigey, F., Günata, Z., 2005. A carotenoid cleavage dioxygenase from *Vitis vinifera* L.: functional characterization and expression during grape berry development in relation to C_{13}-norisoprenoid accumulation. J. Exp. Bot. 56, 2721–2731.

Matsui, S., Ryugo, K., Kliewer, W.M., 1986. Growth inhibition of Thompson Seedless and Napa Gamay berries by heat stress and its partial reversibility by applications of growth regulators. Am. J. Enol. Vitic. 37, 67–71.

Matsumoto-Kitano, M., Kusumoto, T., Tarkowski, P., Kinoshita-Tsujimura, K., Václavíková, K., Miyawaki, K., et al., 2008. Cytokinins are central regulators of cambial activity. Proc. Natl. Acad. Sci. USA 105, 20027–20031.

Matthews, M.A., Anderson, M.M., 1988. Fruit ripening in *Vitis vinifera* L.: responses to seasonal water deficits. Am. J. Enol. Vitic. 39, 313–320.

Matthews, M.A., Anderson, M.M., 1989. Reproductive development in grape (*Vitis vinifera* L.): responses to seasonal water deficits. Am. J. Enol. Vitic. 40, 52–60.

Matthews, M.A., Shackel, K.A., 2005. Growth and water transport in fleshy fruit. In: Holbrook, N.M., Zwieniecki, M.A. (Eds.), Vascular Transport in Plants. Elsevier, Boston, MA, pp. 181–195.

Matthews, M.A., Cheng, G., Weinbaum, S.A., 1987. Changes in water potential and dermal extensibility during grape berry development. J. Am. Soc. Hortic. Sci. 112, 314–319.

Mattivi, F., Guzzon, R., Vrhovsek, U., Stefanini, M., Velasco, R., 2006. Metabolite profiling of grape: flavonols and anthocyanins. J. Agric. Food Chem. 54, 7692–7702.

Mattivi, F., Vrhovsek, U., Masuero, D., Trainotti, D., 2009. Differences in the amount and structure of extractable skin and seed tannins amongst red grape varieties. Aust. J. Grape Wine Res. 15, 27–35.

Maurel, C., Verdoucq, L., Luu, D.T., Santoni, V., 2008. Plant aquaporins: membrane channels with multiple integrated functions. Annu. Rev. Plant Biol. 59, 595–624.

May, P., 1960. Effect of direction of growth on Sultana canes. Nature 185, 394–395.

May, P., 1994. Using Grapevine Rootstocks. The Australian Perspective. Winetitles, Adelaide, Australia.

May, P., 2000. From bud to berry, with special reference to inflorescence and bunch morphology in *Vitis vinifera* L. Aust. J. Grape Wine Res. 6, 82–98.

May, P., 2004. Flowering and Fruitset in Grapevines. Lythrum Press, Adelaide, Australia.

May, P., Antcliff, A.J., 1963. The effect of shading on fruitfulness and yield in the Sultana. J. Hortic. Sci. 38, 85–94.

May, P., Shaulis, N.J., Antcliff, A.J., 1969. The effect of controlled defoliation in the Sultana vine. Am. J. Enol. Vitic. 20, 237–250.

May, P., Clingeleffer, P.R., Brien, C.J., 1976. Sultana (*Vitis vinifera* L.) canes and their exposure to light. Vitis 14, 278 288.

Mayer, H., Holst, T., Schindler, D., 2002. Mikroklima in Buchenbeständen—Teil I: Photosynthetisch aktive Strahlung. Forstwiss. Centralbl. 121, 301–321.

Mayr, E., 2001. What Evolution Is. Basic Books, New York, NY.

Mazza, G., Miniati, E., 1993. Anthocyanins in Fruits, Vegetables, and Grains. CRC Press, Boca Raton, FL.

Mc Intyre, G.N., Kliewer, W.M., Lider, L.A., 1987. Some limitations of the degree day system as used in viticulture in California. Am. J. Enol. Vitic. 38, 128–132.

McAinsh, M.R., Pittman, J.K., 2009. Shaping the calcium signature. New Phytol. 181, 275–294.

McCann, M.C., Wells, B., Roberts, K., 1990. Direct visualization of cross-links in the primary plant cell wall. J. Cell Sci. 96, 323–334.

McCarthy, M.G., 1997. The effect of transient water deficit on berry development of cv. Shiraz (*Vitis vinifera* L.). Aust. J. Grape Wine Res. 3, 102–108.

McCarthy, M.G., Cirami, R.M., 1990. The effect of rootstocks on the performance of Chardonnay from a nematode-infested Barossa Valley vineyard. Am. J. Enol. Vitic. 41, 126–130.

McClung, C.R., 2001. Circadian rhythms in plants. Annu. Rev. Plant Physiol. Plant Mol. Biol. 52, 139–162.

McClung, C.R., 2008. Comes a time. Curr. Opin. Plant Biol. 11, 514–520.

McDowell, N.G., 2011. Mechanisms linking drought, hydraulics, carbon metabolism, and vegetation mortality. Plant Physiol. 155, 1051–1059.

McElrone, A.J., Jackson, S., Habdas, P., 2008. Hydraulic disruption and passive migration by a bacterial pathogen in oak tree xylem. J. Exp. Bot. 59, 2649–2657.

McGarvey, D.J., Croteau, R., 1995. Terpenoid metabolism. Plant Cell 7, 1015–1026.

McGovern, P.E., 2003. Ancient Wine. The Search for the Origins of Viniculture. Princeton University Press, Princeton, NJ.

McGovern, P.E., Glusker, D.L., Exner, L.J., Voigt, M.M., 1996. Neolithic resinated wine. Nature 381, 480–481.

McGovern, P.E., Zhang, J., Tang, J., Zhang, Z., Hall, G.R., Moreau, R.A., et al., 2004. Fermented beverages of pre- and proto-historic China. Proc. Natl. Acad. Sci. USA 101, 17593–17598.

McKenry, M.V., 1984. Grape root phenology relative to control of parasitic nematodes. Am. J. Enol. Vitic. 35, 206–211.

McKenzie, R., Connor, B., Bodeker, G., 1999. Increased summertime UV radiation in New Zealand in response to ozone loss. Science 285, 1709–1711.

McKey, D., Elias, M., Pujol, B., Duputié, A., 2010. The evolutionary ecology of clonally propagated domesticated plants. New Phytol. 186, 318–332.

McLeod, A.R., Fry, S.C., Loake, G.J., Messenger, D.J., Reay, D.S., Smith, K.A., et al., 2008. Ultraviolet radiation drives methane emissions from terrestrial plant pectins. New Phytol. 180, 124–132.

McQueen-Mason, S., 2005. Cell walls: the boundaries of plant development. New Phytol. 166, 717–722.

McSteen, P., Leyser, O., 2005. Shoot branching. Annu. Rev. Plant Biol. 56, 353–374.

Meiering, A.G., Paroschy, J.H., Peterson, R.L., Hostetter, G., Neff, A., 1980. Mechanical freezing injury in grapevine trunks. Am. J. Enol. Vitic. 31, 81–89.

Melino, V.J., Soole, K.L., Ford, C.M., 2009. Ascorbate metabolism and the developmental demand for tartaric and oxalic acids in ripening grape berries. BMC Plant Biol. 9, 145, http://dx.doi.org/10.1186/1471-2229-9-145.

Mendgen, K., Hahn, M., 2002. Plant infection and the establishment of fungal biotrophy. Trends Plant Sci. 7, 1–5.

Meneghetti, S., Gardiman, M., Calò, A., 2006. Flower biology of grapevine: a review. Adv. Hortic. Sci. 20, 317–325.

Menge, J.A., Raski, D.J., Lider, L.A., Johnson, E.L.V., Jones, N.O., Kissler, J.J., et al., 1983. Interactions between mycorrhizal fungi, soil fumigation, and growth of grapes in California. Am. J. Enol. Vitic. 34, 117–121.

Mengel, K., 1994. Iron availability in plant tissues: iron chlorosis on calcareous soils. Plant Soil 165, 275–283.

Mengel, K., Bübl, W., 1983. Verteilung von Eisen in Blättern von Weinreben mit HCO_3^--induzierter Fe-Chlorose. Z. Pflanzenern. Bodenk. 146, 560–571.

Mengel, K., Malissiovas, N., 1982. Light dependent proton excretion by roots of entire vine plants (*Vitis vinifera* L.). Z. Pflanzenern. Bodenk. 145, 261–267.

Mengel, K., Breininge, M.T., Bübl, W., 1984. Bicarbonate, the most important factor inducing iron chlorosis in vine grapes on calcareous soils. Plant Soil 81, 333–344.

Mercurio, M.D., Dambergs, R.G., Cozzolino, D., Herderich, M.J., Smith, P.A., 2010. Relationship between red wine grades and phenolics. 1. Tannin and total phenolics concentrations. J. Agric. Food Chem. 58, 12313–12319.

Meredith, C.P., Bowers, J.E., Riaz, S., Handley, V., Bandman, E.B., Dangl, G.S., 1999. The identity and parentage of the variety known in California as Petite Sirah. Am. J. Enol. Vitic. 50, 236–242.

Merzlyak, M.N., Solovchenko, A.E., Chivkunova, O.B., 2002. Patterns of pigment changes in apple fruits during adaptation to high sunlight and sunscald development. Plant Physiol. Biochem. 40, 679–684.

Messenger, D.J., McLeod, A.R., Fry, S.C., 2009. The role of ultraviolet radiation, photosensitizers, reactive oxygen species and ester groups in mechanisms of methane formation from pectin. Plant Cell Environ. 32, 1–9.

Metzner, R., Schneider, H.U., Breuer, U., Thorpe, M.R., Schurr, U., Schroeder, W.H., 2010a. Tracing cationic nutrients from xylem into stem tissue of French bean by stable isotope tracers and cryo-secondary ion mass spectrometry. Plant Physiol. 152, 1030–1043.

Metzner, R., Thorpe, M.R., Breuer, U., Blümler, P., Schurr, U., Schneider, H.U., et al., 2010b. Contrasting dynamics of water and mineral nutrients in stems shown by stable isotope tracers and cryo-SIMS. Plant Cell Environ. 33, 1393–1407.

Meyer, R.C., Steinfath, M., Lisec, J., Becher, M., Witucka-Wall, H., Törjék, O., et al., 2007. The metabolic signature related to high plant growth rate in *Arabidopsis thaliana*. Proc. Natl Acad. Sci. USA 104, 4759–4764.

Meyer, S., De Angeli, A., Fernie, A.R., Martinoia, E., 2010. Intra- and extra-cellular excretion of carboxylates. Trends Plant Sci. 15, 40–47.

Meynhardt, J.T., 1964. Some studies on berry splitting of Queen of the Vineyard grape. S. Afr. J. Agric. Sci. 7, 179–186.

Meynhardt, J.T., Malan, A.H., 1963. Translocation of sugars in double-stem grape vines. S. Afr. J. Agric. Sci. 6, 337–338.

Miele, A., Bouard, J., Bertrand, A., 1993. Fatty acids from lipid fractions of leaves and different tissues of Cabernet Sauvignon grapes. Am. J. Enol. Vitic. 44, 180–186.

Miguel, C., Mesias, J.L., Maynar, J.I., 1985. Évolution des acides aminés pendant la maturation des raisins des variétés *Cayetana* et *Macabeo* (*Vitis vinifera*). Sci. Aliments 5, 599–605.

Millar, A.A., 1972. Thermal regime of grapevines. Am. J. Enol. Vitic. 23, 173–176.

Millar, A.H., Whelan, J., Soole, K.L., Day, D.A., 2011. Organization and regulation of mitochondrial respiration in plants. Annu. Rev. Plant Biol. 62, 79–104.

Miller, A.J., Fan, X., Shen, Q., Smith, S.J., 2008. Amino acids and nitrate as signals for the regulation of nitrogen acquisition. J. Exp. Bot. 59, 111–119.

Miller, D.P., Howell, G.S., 1998. Influence of vine capacity and crop load on canopy development, morphology, and dry matter partitioning in Concord grapevines. Am. J. Enol. Vitic. 49, 183–190.

Miller, D.P., Howell, G.S., Striegler, R.K., 1988. Cane and bud hardiness of own-rooted White Riesling and scions of White Riesling and Chardonnay grafted to selected rootstocks. Am. J. Enol. Vitic. 39, 60–66.

Milli, A., Cecconi, D., Bortesi, L., Persi, A., Rinalducci, S., Zamboni, A., et al., 2012. Proteomic analysis of the compatible interaction between *Vitis vinifera* and *Plasmopara viticola*. J. Proteomics 75, 1284–1302.

Mills, L.J., Ferguson, J.C., Keller, M., 2006. Cold hardiness evaluation of grapevine buds and cane tissues. Am. J. Enol. Vitic. 57, 194–200.

Minchin, P.E.H., Lacointe, A., 2005. New understanding on phloem physiology and possible consequences for modelling long-distance carbon transport. New Phytol. 166, 771–779.

Mirabet, V., Das, P., Boudaoud, A., Hamant, O., 2011. The role of mechanical forces in plant morphogenesis. Annu. Rev. Plant Biol. 62, 365–385.

Mitani, N., Azuma, A., Fukai, E., Hirochika, H., Kobayashi, S., 2009. A retrotransposon-inserted *VvmybA1a* allele has been spread among cultivars of *Vitis vinifera* but not North American of East Asian *Vitis* species. Vitis 48, 55–56.

Mittler, R., 2002. Oxidative stress, antioxidants and stress tolerance. Trends Plant Sci. 7, 405–410.

Miyashita, Y., Good, A.G., 2008. NAD(H)-dependent glutamate dehydrogenase is essential for the survival of *Arabidopsis thaliana* during dark-induced carbon starvation. J. Exp. Bot. 59, 667–680.

Mlikota Gabler, F., Smilanick, J.L., Mansour, M., Ramming, D.W., Mackey, B.E., 2003. Correlations of morphological, anatomical, and chemical features of grape berries with resistance to *Botrytis cinerea*. Phytopathology 93, 1263–1273.

Mohr, H.D., 1996. Periodicity of root tip growth of vines in the Moselle valley. Wein-Wiss. 51, 83–90.

Mok, D.W.S., Mok, M.C., 2001. Cytokinin metabolism and action. Annu. Rev. Plant Physiol. Plant Mol. Biol. 52, 89–118.

Molitor, D., Behr, M., Hoffmann, L., Evers, D., 2012. Impact of grape cluster division on cluster morphology and bunch rot epidemic. Am. J. Enol. Vitic. 63, 508–514.

Møller, I.M., Jensen, P.E., Hansson, A., 2007. Oxidative modifications to cellular components in plants. Annu. Rev. Plant Biol. 58, 459–481.

Monagas, M., Garrido, I., Bartolomé, B., Gómez-Cordovés, C., 2006. Chemical characterization of commercial dietary ingredients from *Vitis vinifera* L. Anal. Chim. Acta 563, 401–410.

Moncada, X., Pelsy, F., Merdinoglu, D., Hinrichsen, P., 2006. Genetic diversity and geographical dispersal in grapevine clones revealed by microsatellite markers. Genome 49, 1459–1472.

Mongélard, G., Seemann, M., Boisson, A.M., Rohmer, M., Bligny, R., Rivasseau, C., 2011. Measurement of carbon flux through the MEP pathway for isoprenoid synthesis by ^{31}P-NMR spectroscopy after specific inhibition of 2-*C*-methyl-D-erythritol 2,4-cyclodiphosphate reductase. Effect of light and temperature. Plant Cell Environ. 34, 1241–1247.

Montagu, K.D., Conroy, J.P., Atwell, B.J., 2001. The position of localized soil compaction determines root and subsequent shoot growth responses. J. Exp. Bot. 52, 2127–2133.

Monteiro, S., Barakat, M., Piçarra-Pereira, M.A., Teixeira, A.R., Ferreira, R.B., 2003. Osmotin and thaumatin from grape: a putative general defense mechanism against pathogenic fungi. Phytopathology 93, 1505–1512.

Monteiro, S., Piçarra-Pereira, M.A., Loureiro, V.B., Teixeira, A.R., Ferreira, R.B., 2007. The diversity of pathogenesis-related proteins decreases during grape maturation. Phytochemistry 68, 416–425.

Morano, L.D., Walker, M.A., 1995. Soils and plant communities associated with three *Vitis* species. Am. Midl. Nat. 134, 254–263.

Moreau, L., Vinet, E., 1923. Sur la composition des pleurs de vignes. C.R. Acad. Agric. France 9, 554–557.

Morelli, G., Ruberti, I., 2002. Light and shade in the photocontrol of *Arabidopsis* growth. Trends Plant Sci. 7, 399–404.

Morgan, D.C., Stanley, C.J., Warrington, I.J., 1985. The effects of simulated daylight and shadelight on vegetative and reproductive growth in kiwifruit and grapevine. J. Hortic. Sci. 60, 473–484.

Morgan, J.M., 1984. Osmoregulation and water stress in higher plants. Annu. Rev. Plant Physiol. 35, 299–319.

Mori, K., Saito, H., Goto-Yamamoto, N., Kitayama, M., Kobayashi, S., Sugaya, S., et al., 2005. Effects of abscisic acid treatment and night temperatures on anthocyanin composition in Pinot noir grapes. Vitis 44, 161–165.

Mori, K., Goto-Yamamoto, N., Kitayama, M., Hashizume, K., 2007. Loss of anthocyanins in red-wine grape under high temperature. J. Exp. Bot. 58, 1935–1945.

Morinaga, K., Imai, S., Yakushiji, H., Koshita, Y., 2003. Effects of fruit load on partitioning of ^{15}N and ^{13}C, respiration, and growth of grapevine roots at different fruit stages. Sci. Hortic. 97, 239–253.

Morita, M.T., Tasaka, M., 2004. Gravity sensing and signaling. Curr. Opin. Plant Biol. 7, 712–718.

Morlat, R., Jacquet, A., 1993. The soil effects on the grapevine root system in several vineyards of the Loire Valley, France. Vitis 32, 35–42.

Morlat, R., Jacquet, A., 2003. Grapevine root system and soil characteristics in a vineyard maintained long-term with or without interrow sward. Am. J. Enol. Vitic. 54, 1–7.

Morlat, R., Symoneaux, R., 2008. Long-term additions of organic amendments in a Loire Valley vineyard on a calcareous sandy soil. III. Effects on fruit composition and chemical and sensory characteristics of Cabernet franc wine. Am. J. Enol. Vitic. 59, 375–386.

Morris, D.A., 2000. Transmembrane auxin carrier systems—dynamic regulators of polar auxin transport. Plant Growth Regul. 32, 161–172.

Morris, J.R., Cawthon, D.L., Fleming, J.W., 1980. Effects of high rates of potassium fertilization on raw product quality and changes in pH and acidity during storage of Concord grape juice. Am. J. Enol. Vitic. 31, 323–328.

Morris, J.R., Sims, C.A., Cawthon, D.L., 1983. Effects of excessive potassium levels on pH, acidity and color of fresh and stored grape juice. Am. J. Enol. Vitic. 34, 35–39.

Morris, J.R., Sims, C.A., Striegler, R.K., Cackler, S.D., Donley, R.A., 1987. Effects of cultivar, maturity, cluster thinning, and excessive potassium fertilization on yield and quality of Arkansas wine grapes. Am. J. Enol. Vitic. 38, 260–264.

Morrison, J.C., 1991. Bud development in *Vitis vinifera* L. Bot. Gaz. 152, 304–315.

Morrison, J.C., Iodi, M., 1990. The influence of waterberry on the development and composition of Thompson Seedless grapes. Am. J. Enol. Vitic. 41, 301–305.

Morrison, J.C., Noble, A.C., 1990. The effects of leaf and cluster shading on the composition of Cabernet Sauvignon grapes and fruit and wine sensory properties. Am. J. Enol. Vitic. 41, 193–200.

Morrot, G., Brochet, F., Dubourdieu, D., 2001. The color of odors. Brain Lang. 79, 309–320.

Moskowitz, A.H., Hrazdina, G., 1981. Vacuolar contents of fruit subepidermal cells from *Vitis* species. Plant Physiol. 68, 686–692.

Motomura, Y., 1990. Distribution of ^{14}C-assimilates from individual leaves on clusters in grape shoots. Am. J. Enol. Vitic. 41, 306–312.

Motomura, Y., 1993. ^{14}C-Assimilate partitioning in grapevine shoots: effects of shoot pinching, girdling of shoot, and leaf-halving on assimilates partitioning from leaves into clusters. Am. J. Enol. Vitic. 44, 1–7.

Mpelasoka, B.S., Schachtmann, D.P., Treeby, M.T., Thomas, M.R., 2003. A review of potassium nutrition in grapevines with special emphasis on berry accumulation. Aust. J. Grape Wine Res. 9, 154–168.

Muday, G.K., Rahman, A., Binder, B.M., 2012. Auxin and ethylene: collaborators or competitors? Trends Plant Sci. 17, 181–195.

Muganu, M., Bellincontro, A., Barnaba, F.E., Paolocci, M., Bignami, C., Gambellini, G., et al., 2011. Influence of bunch position in the canopy on berry epicuticular wax during ripening and on weight loss during postharvest dehydration. Am. J. Enol. Vitic. 62, 91–98.

Muller, B., Pantin, F., Génard, M., Turc, O., Freixes, S., Piques, M., et al., 2011. Water deficits uncouple growth from photosynthesis, increase C content, and modify the relationships between C and growth in sink organs. J. Exp. Bot. 62, 1715–1729.

Müller, D., Leyser, O., 2011. Auxin, cytokinin and the control of shoot branching. Ann. Bot. 107, 1203–1212.

Müller, K., 1985. Abhängigkeit des pflanzenverfügbaren Stickstoffs (N_{min}) von den Humusgehalten und der mineralischen N-Düngung auf einem fränkischen Weinbaustandort. Z. Pflanzenern. Bodenk. 148, 169–178.

Müller, T., Ulrich, M., Ongania, K.H., Kräutler, B., 2007. In reifen Früchten gefundene farblose tetrapyrrolische Chlorophyll-Katabolitien sind wirksame Antioxidantien. Angew. Chem. 119, 8854–8857.

Müller-Stoll, W.R., 1950. Mutative Färbungsänderungen bei Weintrauben. Züchter 20, 288–291.

Müller-Thurgau, H., 1882. Über Zuckeranhäufung in Pflanzenteilen infolge niederer Temperatur. Landwirtsch. Jahrb. 11, 751–828.

Müller-Thurgau, H., 1883a. Über das Abfallen der Rebenblüten und die Entstehung kernloser Traubenbeeren. Der Weinbau 9, 87–89 and 95–96.

Müller-Thurgau, H., 1883b. Über die Fruchtbarkeit der aus den älteren Theilen der Weinstöcke hervorgehenden Triebe, sowie der sog. Nebentriebe. Botanisches Centralblatt 4, 85–86.

Müller-Thurgau, H., 1886. Über das Gefrieren und Erfrieren der Pflanzen. Landwirtsch. Jahrb. 15, 453–610.

Mullins, M.G., 1980. Regulation of flowering in the grapevine (*Vitis vinifera* L.). In: Skoog, F.K. (Ed.), Plant Growth Substances 1979: Proceedings of the 10th International Conference on Plant Growth Substances, Madison, Wisconsin, July 22–26, 1979. Springer, Berlin, pp. 323–330.

Mullins, M.G., Srinivasan, C., 1976. Somatic embryos and plantlets from an ancient clone of the grapevine (cv. Cabernet-Sauvignon) by apomixis *in vitro*. J. Exp. Bot. 27, 1022–1030.

Mullins, M.G., Bouquet, A., Williams, L.E., 1992. Biology of the Grapevine. Cambridge University Press, Cambridge, UK.

Münch, E., 1930. Die Stoffbewegungen in der Pflanze. Gustav Fischer, Jena, Germany.

Munné-Bosch, S., 2008. Do perennials really senesce? Trends Plant Sci. 13, 216–220.

Munns, R., 2002. Comparative physiology of salt and water stress. Plant Cell Environ. 25, 239–250.

Munns, R., Tester, M., 2008. Mechanisms of salinity tolerance. Annu. Rev. Plant Biol. 59, 651–681.

Muñoz, C., Gomez-Talquenca, S., Chialva, C., Ibáñez, J., Martinez-Zapater, J.M., Peña-Neira, Á., et al., 2014. Relationships among gene expression and anthocyanin composition of Malbec grapevine clones. J. Agric. Food Chem. 62, 6716–6725.

Müntz, K., 2007. Protein dynamics and proteolysis in plant vacuoles. J. Exp. Bot. 58, 2391–2407.

Murphey, J.M., Spayd, S.E., Powers, J.R., 1989. Effect of grape maturation on soluble protein characteristics of Gewürztraminer and White Riesling juice and wine. Am. J. Enol. Vitic. 40, 199–207.

Mwange, K.N.K., Hou, H.W., Wang, Y.Q., He, X.Q., Cui, K.M., 2005. Opposite patterns in the annual distribution and time-course of endogenous abscisic acid and indole-3-acetic acid in relation to the periodicity of cambial activity in *Eucommia ulmoides* Oliv. J. Exp. Bot. 56, 1017–1028.

Myles, S., Chia, J.M., Hurwitz, B., Simon, C., Zhong, G.Y., Buckler, E., et al., 2010. Rapid genomic characterization of the genus *Vitis*. PLoS One 5, e8218, http://dx.doi.org/10.1371/journal.pone.0008219.

Myles, S., Boyko, A.R., Owens, C.L., Brown, P.J., Grassi, F., Aradhya, M.K., et al., 2011. Genetic structure and domestication history of the grape. Proc. Natl. Acad. Sci. USA 108, 3530–3535.

Nail, W.R., Howell, G.S., 2005. Effects of timing of powdery mildew infection on carbon assimilation and subsequent seasonal growth of potted Chardonnay grapevines. Am. J. Enol. Vitic. 56, 220–227.

Nakagawa, S., Nanjo, Y., 1965. A morphological study of Delaware grape berries. J. Jpn. Soc. Hortic. Sci. 34, 85–95.

Nambara, E., Marion-Poll, A., 2005. Abscisic acid biosynthesis and catabolism. Annu. Rev. Plant Biol. 56, 165–185.

Naor, A., Gal, Y., Bravdo, B., 1997. Crop load affects assimilation rate, stomatal conductance, stem water potential and water relations of field-grown Sauvignon blanc grapevines. J. Exp. Bot. 48, 1675–1680.

Nardini, A., Tyree, M.T., Salleo, S., 2001. Xylem cavitation in the leaf of *Prunus laurocerasus* L. and its impact on leaf hydraulics. Plant Physiol. 125, 1700–1709.

Nardini, A., Salleo, S., Jansen, S., 2011. More than just a vulnerable pipeline: xylem physiology in the light of ion-mediated regulation of plant water transport. J. Exp. Bot. 62, 4701–4718.

Negi, S., Ivanchenko, M.G., Muday, G.K., 2008. Ethylene regulates lateral root formation and auxin transport in *Arabidopsis thaliana*. Plant J. 55, 175–187.

Negri, A.S., Prinsi, B., Rossoni, M., Failla, O., Scienza, A., Cocucci, M., et al., 2008. Proteome changes in the skin of the grape cultivar Barbera among different stages of ripening. BMC Genomics 9, 388, http://dx.doi.org/10.1186/1471-2164-9-378.

Negrul, A.M., 1936. Variabilität und Vererbung des Geschlechts bei der Rebe. Gartenbauwissenschaft 10, 215–231.

Nelson, N., Yocum, C.F., 2006. Structure and function of photosystems I and II. Annu. Rev. Plant Biol. 57, 521–565.

Newsham, K.K., Low, M.N.R., McLeod, A.R., Greenslade, P.D., Emmett, B.A., 1997. Ultraviolet-B radiation influences the abundance and distribution of phylloplane fungi on pedunculate oak (*Quercus robur*). New Phytol. 136, 287–297.

Nibau, C., Gibbs, D.J., Coates, J.C., 2008. Branching out in new directions: the control of root architecture by lateral root formation. New Phytol. 179, 595–614.

Nicholas, P., Magarey, P., Wachtel, M., 1998. Diseases and Pests. Grape Production Series Number 1, third ed. Winetitles, Adelaide, Australia.

Nieminen, K., Immanen, J., Laxell, M., Kauppinen, L., Tarkowski, P., Dolezal, K., et al., 2008. Cytokinin signaling regulates cambial development in poplar. Proc. Natl. Acad. Sci. USA 105, 20032–20037.

Niimi, Y., Torikata, H., 1979. Changes in photosynthesis and respiration during berry development in relation to the ripening of Delaware grapes. J. Jpn. Soc. Hortic. Sci. 47, 448–453.

Niklas, K.J., 2006. Plant allometry, leaf nitrogen and phosphorus stoichiometry, and interspecific trends in annual growth rates. Ann. Bot. 97, 155–163.

Niklas, K.J., Cobb, E.D., 2006. Biomass partitioning and leaf N,P—stoichiometry: comparisons between tree and herbaceous current-year shoots. Plant Cell Environ. 29, 2030–2042.

Nikolic, M., Römheld, V., Merkt, N., 2000. Effect of bicarbonate on uptake and translocation of ^{59}Fe in two grapevine rootstocks differing in their resistance to Fe deficiency chlorosis. Vitis 39, 145–149.

Nitsch, J.P., Pratt, C., Nitsch, C., Shaulis, N.J., 1960. Natural growth substances in Concord and Concord Seedless grapes in relation to berry development. Am. J. Bot. 47, 566–576.

Niyogi, K.K., 2000. Safety valves for photosynthesis. Curr. Opin. Plant Biol. 3, 455–460.

Nobel, P.S., 2009. Physicochemical and Environmental Plant Physiology, fourth ed. Academic Press, San Diego, CA.

Noble, A.C., 1979. Evaluation of Chardonnay wines obtained from sites with different soil compositions. Am. J. Enol. Vitic. 30, 214–217.

Noble, A.C., Bursick, G.F., 1984. The contribution of glycerol to perceived viscosity and sweetness in white wine. Am. J. Enol. Vitic. 35, 110–112.

Noble, A.C., Ebeler, S.E., 2002. Use of multivariate statistics in understanding wine flavor. Food Rev. Int. 18, 1–21.

Noble, A.C., Strauss, C.R., Williams, P.J., Wilson, B., 1988. Contribution of terpene glycosides to bitterness in Muscat wines. Am. J. Enol. Vitic. 39, 129–131.

Noctor, G., Foyer, C.H., 1998. Ascorbate and glutathione: keeping active oxygen under control. Annu. Rev. Plant Physiol. Plant Mol. Biol. 49, 249–279.

Nonami, H., Boyer, J.S., 1987. Origin of growth-induced water potential. Solute concentration is low in apoplast of enlarging tissues. Plant Physiol. 83, 596–601.

Nonami, H., Wu, Y., Boyer, J.S., 1997. Decreased growth-induced water potential. A primary cause of growth inhibition at low water potentials. Plant Physiol. 114, 501–509.

Nord, E.A., Lynch, J.P., 2009. Plant phenology: a critical controller of soil resource acquisition. J. Exp. Bot. 60, 1927–1937.

Nordström, A., Tarkowski, P., Tarkowska, D., Norbaek, R., Åstot, C., Dolezal, K., et al., 2004. Auxin regulation of cytokinin biosynthesis in *Arabidopsis thaliana*: a factor of potential importance for auxin-cytokinin-regulated development. Proc. Natl. Acad. Sci. USA 101, 8039–8044.

Ntahimpera, N., Ellis, M.A., Wilson, L.L., Madden, L.V., 1998. Effects of a cover crop on splash dispersal of *Colletotrichum acutatum* conidia. Phytopathology 88, 536–543.

Nunan, K.J., Sims, I.M., Bacic, A., Robinson, S.P., Fincher, G.B., 1997. Isolation and characterization of cell walls from the mesocarp of mature grape berries. Planta 203, 93–100.

Nunan, K.J., Sims, I.M., Bacic, A., Robinson, S.P., Fincher, G.B., 1998. Changes in cell wall composition during ripening of grape berries. Plant Physiol. 118, 783–792.

Nunan, K.J., Davies, C., Robinson, S.P., Fincher, G.B., 2001. Expression patterns of cell wall-modifying enzymes during grape berry development. Planta 214, 257–264.

Nuzzo, V., Matthews, M.A., 2006. Response of fruit growth and ripening to crop level in dry-farmed Cabernet Sauvignon on four rootstocks. Am. J. Enol. Vitic. 57, 314–324.

Oaks, A., Hirel, B., 1985. Nitrogen metabolism in roots. Annu. Rev. Plant Physiol. 36, 345–365.

O'Brien, E.E., Gersani, M., Brown, J.S., 2005. Root proliferation and seed yield in response to spatial heterogeneity of below-ground competition. New Phytol. 168, 401–412.

OEPP/EPPO, 2002. Good plant protection practice. Grapevine. EPPO Bull. 32, 371–392.

Ogren, W.L., 1984. Photorespiration: pathways, regulation, and modification. Annu. Rev. Plant Physiol. 35, 415–442.

Ojeda, H., Deloire, A., Carbonneau, A., Ageorges, A., Romieu, C., 1999. Berry development of grapevines: relations between the growth of berries and their DNA content indicate cell multiplication and enlargement. Vitis 38, 145–150.

Ojeda, H., Deloire, A., Carbonneau, A., 2001. Influence of water deficits on grape berry growth. Vitis 40, 141–145.

Ojeda, H., Andary, C., Kraeva, E., Carbonneau, A., Deloire, A., 2002. Influence of pre- and postveraison water deficit on synthesis and concentration of skin phenolic compounds during berry growth of *Vitis vinifera* cv. Shiraz. Am. J. Enol. Vitic. 53, 261–267.

Okuda, T., Yokotsuka, K., 1999. Levels of glutathione and activities of related enzymes during ripening of Koshu and Cabernet Sauvignon grapes and during winemaking. Am. J. Enol. Vitic. 50, 264–270.

Olien, W.C., 1990. The muscadine grape: botany, viticulture, history, and current industry. HortScience 25, 732–739.

Oliveira, C., Ferreira, A.C., Costa, P., Guerra, J., Guedes de Pinho, P., 2004. Effect of some viticultural parameters on the grape carotenoid profile. J. Agric. Food Chem. 52, 4178–4184.

Oliver, J.E., Fuchs, M., 2011. Tolerance and resistance to viruses and their vectors in *Vitis* sp.: a virologist's perspective of the literature. Am. J. Enol. Vitic. 62, 438–451.

Oliver, S.N., Van Dongen, J.T., Alfred, A.C., Mamun, E.A., Zhao, X., Saini, H.S., et al., 2005. Cold-induced repression of the rice anther-specific cell wall invertase gene *OSINV4* is correlated with sucrose accumulation and pollen sterility. Plant Cell Environ. 28, 1534–1551.

Ollat, N., Gaudillere, J.P., 1998. The effect of limiting leaf area during stage I of berry growth on development and composition of berries of *Vitis vinifera* L. cv. Cabernet Sauvignon. Am. J. Enol. Vitic. 49, 251–258.

Ollat, N., Diakou-Verdin, P., Carde, J.P., Barrieu, F., Gaudillère, J.P., Moing, A., 2002. Grape berry development: a review. J. Int. Sci. Vigne Vin 36, 109–131.

Ollat, N., Laborde, B., Neveux, M., Diakou-Verdin, P., Renaud, C., Moing, A., 2003. Organic acid metabolism in roots of various grapevine (*Vitis*) rootstocks submitted to iron deficiency and bicarbonate nutrition. J. Plant Nutr. 26, 2165–2176.

Olmo, H.P., 1936. Pollination and the setting of fruit in the black Corinth grape. Proc. Am. Soc. Hortic. Sci. 34, 402–404.

Olmo, H.P., 1946. Correlations between seed and berry development in some seeded varieties of *Vitis vinifera*. Proc. Am. Soc. Hortic. Sci. 48, 291–297.

Olmo, H.P., 1952. Wine grape varieties of the future. Am. J. Enol. Vitic. 3, 45–51.

Olmo, H.P., 1986. The potential role of (*vinifera* × *rotundifolia*) hybrids in grape variety improvement. Experientia 42, 921–926.

Olsen, K.M., Slimestad, R., Lea, U.S., Brede, C., Løvdal, T., Ruoff, P., et al., 2009. Temperature and nitrogen effects on regulators and products of the flavonoid pathway: experimental and kinetic model studies. Plant Cell Environ. 32, 286–299.

Olszewski, N., Sun, T., Gubler, F., 2002. Gibberellin signaling: biosynthesis, catabolism, and response pathways. Plant Cell 14, S61–S80.

O'Neill, M.A., Ishii, T., Albersheim, P., Darvill, A.G., 2004. Rhamnogalacturonan II: structure and function of a borate cross-linked cell wall pectic polysaccharide. Annu. Rev. Plant Biol. 55, 109–139.

O'Neill, S.D.O., 1997. Pollination regulation of flower development. Annu. Rev. Plant Physiol. Plant Mol. Biol. 48, 547–574.

Ong, B.Y., Nagel, C.W., 1978. High pressure liquid chromatographic analysis of hydroxycinnamic acid-tartaric acid esters and their glucose esters in *Vitis vinifera*. J. Chromatogr. 157, 345–355.

Ongaro, V., Leyser, O., 2008. Hormonal control of shoot branching. J. Exp. Bot. 59, 67–74.

Oparka, K.J., Prior, D.A.M., 1992. Direct evidence for pressure-generated closure of plasmodesmata. Plant J. 2, 741–750.

Oparka, K.J., Santa Cruz, S., 2000. The great escape: phloem transport and unloading of macromolecules. Annu. Rev. Plant Physiol. Plant Mol. Biol. 51, 323–347.

Orrù, M., Mattana, E., Pritchard, H.W., Bacchetta, G., 2012. Thermal thresholds as predictors of seed dormancy release and germination timing: altitude-related risks from climate warming for the wild grapevine *Vitis vinifera* subsp. *sylvestris*. Ann. Bot. 110, 1651–1660.

Ortega-Regules, A., Romero-Cascales, I., López-Roca, J.M., Ros-García, J.M., Gómez-Plaza, E., 2006a. Anthocyanin fingerprint of grapes: environmental and genetic variations. J. Sci. Food Agric. 86, 1460–1467.

Ortega-Regules, A., Romero-Cascales, I., Ros-García, J.M., López-Roca, J.M., Gómez-Plaza, E., 2006b. A first approach towards the relationship between grape skin cell-wall composition and anthocyanin extractability. Anal. Chim. Acta 563, 26–32.

Ortega-Regules, A., Ros-García, J.M., Bautista-Ortín, A.B., López-Roca, J.M., Gómez-Plaza, E., 2008. Changes in skin cell wall composition during the maturation of four premium wine grape varieties. J. Sci. Food Agric. 88, 420–428.

Ortoidze, T., Düring, H., 2001. Light utilisation and thermal dissipation in light- and shade-adapted leaves of *Vitis* genotypes. Vitis 40, 131–136.

Osmont, K.S., Sibout, R., Hardtke, C.S., 2007. Hidden branches: developments in root system architecture. Annu. Rev. Plant Biol. 58, 93–113.

Ou, C., Du, X., Shellie, K., Ross, C., Qian, M.C., 2010. Volatile compounds and sensory attributes of wine from cv. Merlot (*Vitis vinifera* L.) grown under differential levels of water deficit with or without kaolin-based, foliar reflectant particle film. J. Agric. Food Chem. 58, 12890–12898.

Ough, C.S., Amerine, M.A., 1963. Regional, varietal, and type influences on the degree Brix and alcohol relationship of grape musts and wines. Hilgardia 34, 585–600.

Ough, C.S., Nagaoka, R., 1984. Effect of cluster thinning and vineyard yields on grape and wine composition and wine quality of Cabernet Sauvignon. Am. J. Enol. Vitic. 35, 30–34.

Owen, S.J., Lafond, M.D., Bowen, P., Bogdanoff, C., Usher, K., Abrams, S.R., 2009. Profiles of abscisic acid and its catabolites in developing Merlot grape (*Vitis vinifera*) berries. Am. J. Enol. Vitic. 60, 277–284.

Ozga, J.A., Reinecke, D.M., 2003. Hormonal interactions in fruit development. J. Plant Growth Regul. 22, 73–81.

Pagay, V., Cheng, L., 2010. Variability in berry maturation of Concord and Cabernet franc in a cool climate. Am. J. Enol. Vitic. 61, 61–67.

Pagter, M., Arora, R., 2013. Winter survival and deacclimation of perennials under warming climate: physiological perspectives. Physiol. Plant. 147, 75–87.

Paiva, E.A.S., Buono, R.A., Lombardi, J.A., 2009. Food bodies in *Cissus verticillata* (Vitaceae): ontogenesis, structure and functional aspects. Ann. Bot. 103, 517–524.

Palanivelu, R., Brass, L., Edlund, A.F., Preuss, D., 2003. Pollen tube growth and guidance is regulated by *POP2*, an *Arabidopsis* gene that controls GABA levels. Cell 114, 47–59.

Palejwala, V.A., Parikh, H.R., Modi, V.V., 1985. The role of abscisic acid in the ripening of grapes. Physiol. Plant. 65, 498–502.

Pallas, B., Louarn, G., Christophe, A., Lebon, E., Lecoeur, J., 2008. Influence of intra-shoot trophic competition on shoot development in two grapevine cultivars (*Vitis vinifera*). Physiol. Plant. 134, 49–63.

Palliotti, A., Cartechini, A., 2001. Developmental changes in gas exchange activity in flowers, berries, and tendrils of field-grown Cabernet Sauvignon. Am. J. Enol. Vitic. 52, 317–323.

Palliotti, A., Cartechini, A., Ferranti, F., 2000. Morpho-anatomical and physiological characteristics of primary and lateral shoot leaves of Cabernet franc and Trebbiano toscano grapevines under two irradiance regimes. Am. J. Enol. Vitic. 51, 122–130.

Palliotti, A., Silvestroni, O., Petoumenou, D., 2009. Photosynthetic and photoinhibition behavior of two field-grown grapevine cultivars under multiple summer stresses. Am. J. Enol. Vitic. 60, 189–198.

Palliotti, A., Silvestroni, O., Petoumenou, D., 2010. Seasonal patterns of growth rate and morphophysiological features in green organs of Cabernet Sauvignon grapevines. Am. J. Enol. Vitic. 61, 74–82.

Palliotti, A., Gatti, M., Poni, S., 2011. Early leaf removal to improve vineyard efficiency: gas exchange, source-to-sink balance, and reserve storage responses. Am. J. Enol. Vitic. 62, 219–228.

Palliotti, A., Tombesi, S., Frioni, T., Famiani, F., Silvestroni, O., Zamboni, M., et al., 2014. Morpho-structural and physiological response of container-grown Sangiovese and Montepulciano cvv. (*Vitis vinifera*) to re-watering after pre-veraison limiting water deficit. Funct. Plant Biol. 41, 634–647.

Palmieri, M.C., Perazzolli, M., Matafora, V., Moretto, M., Bacchi, A., Pertot, I., 2012. Proteomic analysis of grapevine resistance induced by *Trichoderma harzianum* T39 reveals specific defence pathways activated against downy mildew. J. Exp. Bot. 63, 6237–6251.

Pan, Q.H., Li, M.J., Peng, C.C., Zhang, N., Zou, X., Zou, K.Q., et al., 2005. Abscisic acid activates acid invertases in developing grape berries. Physiol. Plant. 125, 157–170.

Pandolfini, T., Molesini, B., Spena, A., 2007. Molecular dissection of the role of auxin in fruit initiation. Trends Plant Sci. 12, 327–329.

Pang, X., Halaly, T., Crane, O., Keilin, T., Keren-Keiserman, A., Ogrodovitch, A., et al., 2007. Involvement of calcium signalling in dormancy release of grape buds. J. Exp. Bot. 58, 3249–3262.

Pang, Y., Zhang, J., Cao, J., Yin, S.Y., He, X.Q., Cui, K.M., 2008. Phloem transdifferentiation from immature xylem cells during bark regeneration after girdling in *Eucommia ulmoides* Oliv. J. Exp. Bot. 59, 1341–1351.

Pantin, F., Simonneau, T., Muller, B., 2012. Coming of leaf age: control of growth by hydraulics and metabolics during leaf ontogeny. New Phytol. 196, 349–366.

Pantin, F., Monnet, F., Jannaud, D., Costa, J.M., Renaud, J., Muller, B., et al., 2013. The dual effect of abscisic acid on stomata. New Phytol. 197, 65–72.
Park, S.K., Morrison, J.C., Adams, D.O., Noble, A.C., 1991. Distribution of free and glycosidically bound monoterpenes in the skin and mesocarp of Muscat of Alexandria grapes during development. J. Agric. Food Chem. 39, 514–518.
Parker, A.K., de Cortázar-Atauri, I.G., van Leeuwen, C., Chuine, I., 2011. General phenological model to characterise the timing of flowering and veraison of *Vitis vinifera* L. Aust. J. Grape Wine Res. 17, 206–216.
Paroschy, J.H., Meiering, A.G., Peterson, R.L., Hostetter, G., Neff, A., 1980. Mechanical winter injury in grapevine trunks. Am. J. Enol. Vitic. 31, 227–232.
Parpinello, G.P., Heymann, H., Vasquez, S., Cathline, K.A., Fidelibus, M.W., 2012. Grape maturity, yield, quality, sensory properties, and consumer acceptance of Fiesta and Selma Pete dry-on-vine raisins. Am. J. Enol. Vitic. 63, 212–219.
Passioura, J.B., 2002. Soil conditions and plant growth. Plant Cell Environ. 25, 311–318.
Passioura, J.B., 2006. The perils of pot experiments. Funct. Plant Biol. 33, 1075–1079.
Pastor del Rio, J.L., Kennedy, J.A., 2006. Development of proanthocyanidins in *Vitis vinifera* L. cv. Pinot noir grapes and extraction into wine. Am. J. Enol. Vitic. 57, 125–132.
Patakas, A., Noitsakis, B., Stavrakas, D., 1997. Adaptation of leaves of *Vitis vinifera* L. to seasonal drought as affected by leaf age. Vitis 36, 11–14.
Patakas, A., Nikolaou, N., Zioziou, E., Radoglou, K., Noitsakis, B., 2002. The role of organic solute and ion accumulation in osmotic adjustment in drought-stressed grapevines. Plant Sci. 163, 361–367.
Patakas, A.A., Zotos, A., Beis, A.S., 2010. Production, localisation and possible roles of nitric oxide in drought-stressed grapevines. Aust. J. Grape Wine Res. 16, 203–209.
Pate, J.S., 1980. Transport and partitioning of nitrogenous solutes. Annu. Rev. Plant Physiol. 31, 313–340.
Patrick, J.W., 1997. Phloem unloading: sieve element unloading and post-sieve element transport. Annu. Rev. Plant Physiol. Plant Mol. Biol. 48, 191–222.
Patrick, J.W., Zhang, W., Tyerman, S.D., Offler, C.E., Walker, N.A., 2001. Role of membrane transport in phloem translocation of assimilates and water. Aust. J. Plant Physiol. 28, 695–707.
Patrick King, A., Berry, A.M., 2005. Vineyard $\delta^{15}N$, nitrogen and water status in perennial clover and bunch grass cover crop systems of California's central valley. Agric. Ecosyst. Environ. 109, 262–272.
Pattison, R.J., Catalá, C., 2012. Evaluating auxin distribution in tomato (*Solanum lycopersicum*) through an analysis of the *PIN* and *AUX/LAX* gene families. Plant J. 70, 585–598.
Pattison, R.J., Csukasi, F., Catalá, C., 2014. Mechanisms regulating auxin action during fruit development. Physiol. Plant. 151, 62–72.
Paul, M.J., Foyer, C.H., 2001. Sink regulation of photosynthesis. J. Exp. Bot. 52, 1383–1400.
Pawlus, A.D., Sahli, R., Bisson, J., Rivière, C., Delaunay, J.C., Richard, T., et al., 2013. Stilbenoid profiles of canes from *Vitis* and *Muscadinia* species. J. Agric. Food Chem. 61, 501–511.
Peacock, W.L., Christensen, L.P., Broadbent, F.E., 1989. Uptake, storage, and utilization of soil-applied nitrogen by Thompson Seedless as affected by time of application. Am. J. Enol. Vitic. 40, 16–20.
Pearce, R.S., 2001. Plant freezing and damage. Ann. Bot. 87, 417–424.
Pearcy, R.W., 1990. Sunflecks and photosynthesis in plant canopies. Annu. Rev. Plant Physiol. Plant Mol. Biol. 41, 421–453.
Pearsall, K.R., Williams, L.E., Castorani, S., Bleby, T.M., McElrone, A.J., 2014. Evaluating the potential of a novel dual heat-pulse sensor to measure volumetric water use in grapevines under a range of flow conditions. Funct. Plant Biol. 41, 874–883.
Pearson, H.M., 1932. Parthenocarpy and seed abortion in *Vitis vinifera*. Proc. Am. Soc. Hortic. Sci. 29, 169–175.
Pearson, R.C., Goheen, A.C., 1998. Compendium of Grape Diseases. APS Press, St. Paul, MN.

Pecinka, A., Abdelsamad, A., Vu, G.T.H., 2013. Hidden genetic nature of epigenetic natural variation in plants. Trends Plant Sci. 18, 625–632.
Peer, W.A., Murphy, A.S., 2007. Flavonoids and auxin transport: modulators or regulators? Trends Plant Sci. 12, 556–563.
Pelsy, F., 2010. Molecular and cellular mechanisms of diversity within grapevine varieties. Heredity 104, 331–340.
Pelsy, F., Hocquigny, S., Moncada, X., et al., 2010. An extensive study of the genetic diversity within seven French wine grape variety collections. Theor. Appl. Genet. 120, 1219–1231.
Peña-Gallego, A., Hernández-Orte, P., Cacho, J., Ferreira, V., 2012. S-cysteinylated and S-glutathionylated thiol precursors in grapes. A review. Food Chem. 131, 1–13.
Penfield, S., 2008. Temperature perception and signal transduction in plants. New Phytol. 179, 615–628.
Penning de Vries, F.W.T., 1975. The cost of maintenance processes in plant cells. Ann. Bot. 39, 77–92.
Penning de Vries, F.W.T., Brunsting, A.H.M., Van Laar, H.H., 1974. Products, requirements and efficiency of biosynthesis: a quantitative approach. J. Theor. Biol. 45, 339–377.
Peppi, M.C., Walker, M.A., Fidelibus, M.W., 2008. Application of abscisic acid rapidly upregulated UFGT gene expression and improved color of grape berries. Vitis 47, 11–14.
Perazzolli, M., Roatti, B., Bozza, E., Pertot, I., 2011. *Trichoderma harzianum* T39 induces resistance against downy mildew by priming for defense without costs for grapevine. Biol. Control 58, 74–82.
Percival, G.C., Boyle, C., Baird, L., 1999. The influence of calcium supplementation on the freezing tolerance of woody plants. J. Arboric. 25, 285–291.
Pereira, G.E., Gaudillère, J.P., Pieri, P., Hilbert, G., Maucourt, M., Deborde, C., et al., 2006a. Microclimate influence on mineral and metabolic profiles of grape berries. J. Agric. Food Chem. 54, 6765–6775.
Pereira, G.E., Gaudillere, J.P., van Leeuwen, C., Hilbert, G., Maucourt, M., Deborde, C., et al., 2006b. ^{1}H NMR metabolite fingerprints of grape berry: comparison of vintage and soil effects in Bordeaux grapevine growing areas. Anal. Chim. Acta 563, 346–352.
Pérez, F.J., Viani, C., Retamales, J., 2000. Bioactive gibberellins in seeded and seedless grapes: identification and changes in content during berry development. Am. J. Enol. Vitic. 51, 315–318.
Pérez, F.J., Villegas, D., Mejia, N., 2002. Ascorbic acid and flavonoid-peroxidase reaction as a detoxifying system of H_2O_2 in grapevine leaves. Phytochemistry 60, 573–580.
Pérez, F.J., Rubio, S., Ormeño-Núñez, J., 2007. Is erratic bud-break in grapevines grown in warm winter areas related to disturbances in mitochondrial respiratory capacity and oxidative metabolism? Funct. Plant Biol. 34, 624–632.
Pérez, F.J., Vergara, R., Or, E., 2009. On the mechanism of dormancy release in grapevine buds: a comparative study between hydrogen cyanamide and sodium azide. Plant Growth Regul. 59, 145–152.
Perez, J., Kliewer, W.M., 1990. Effect of shading on bud necrosis and bud fruitfulness of Thompson Seedless grapevines. Am. J. Enol. Vitic. 41, 168–175.
Perez, J.R., Kliewer, W.M., 1982. Influence of light regime and nitrate fertilization on nitrate reductase activity and concentrations of nitrate and arginine in tissues of three cultivars of grapevines. Am. J. Enol. Vitic. 33, 86–93.
Pérez Harvey, J., Gaete, L., 1986. Efecto del microclima luminoso sobre la calidad de la uva Sultanina en sistema de parronal español. II. Desgrane, palo negro y pudrición gris. Cienc. Invest. Agrar. 13, 113–120.
Perez Peña, J., Tarara, J., 2004. A portable whole canopy gas exchange system for several mature field-grown grapevines. Vitis 43, 7–14.
Pérez-Donoso, A.G., Sun, Q., Roper, M.C., Greve, C., Kirkpatrick, B., Labavitch, J.M., 2010. Cell wall-degrading enzymes enlarge the pore size of intervessel pit membranes in healthy and *Xylella fastidiosa*-infected grapevines. Plant Physiol. 152, 1748–1759.
Perret, P., Roth, I., 1996. Nährstoffverluste durch Versickerung in ostschweizerischen Rebbergen. Schweiz. Z. Obst- Weinbau 132, 396–399.

Perret, P., Weissenbach, P., Schwager, H., Heller, W., Koblet, W., 1993. "Adaptive nitrogen-management"—a tool for the optimisation of N-fertilisation in vineyards. Wein-Wiss. 48, 124–126.

Perrone, I., Gambino, G., Chitarra, W., Vitali, M., Pagliarani, C., Riccomagno, N., et al., 2012. The grapevine root-specific aquaporin *VvPIP2;4N* controls root hydraulic conductance and leaf gas exchange under well-watered conditions but not under water stress. Plant Physiol. 160, 965–977.

Petit, J.R., Jouzel, J., Raynaud, D., Barkov, N.I., Barnola, J.M., Basile, I., et al., 1999. Climate and atmospheric history of the past 420,000 years from the Vostok ice core, Antarctica. Nature 399, 429–436.

Petrášek, J., Friml, J., 2009. Auxin transport routes in plant development. Development 136, 2675–2688.

Petrie, P.R., Clingeleffer, P.R., 2005. Effects of temperature and light (before and after budburst) on inflorescence morphology and flower number of Chardonnay grapevines (*Vitis vinifera* L.). Aust. J. Grape Wine Res. 11, 59–65.

Petrie, P.R., Trought, M.C.T., Howell, G.S., 2000a. Fruit composition and ripening of Pinot noir (*Vitis vinifera* L.) in relation to leaf area. Aust. J. Grape Wine Res. 6, 46–51.

Petrie, P.R., Trought, M.C.T., Howell, G.S., 2000b. Growth and dry matter partitioning of Pinot noir (*Vitis vinifera* L.) in relation to leaf area and crop load. Aust. J. Grape Wine Res. 6, 40–45.

Petrie, P.R., Trought, M.C.T., Howell, G.S., Buchan, G.D., Palmer, J.W., 2009. Whole-canopy gas exchange and light interception of vertically trained *Vitis vinifera* L. under direct and diffuse light. Am. J. Enol. Vitic. 60, 173–182.

Peuke, A.D., 2010. Correlations in concentrations, xylem and phloem flows, and partitioning of elements and ions in intact plants. A summary and statistical re-evaluation of modelling experiments in *Ricinus communis*. J. Exp. Bot. 61, 635–655.

Peuke, A.D., Rokitta, M., Zimmermann, U., Schreiber, L., Haase, A., 2001. Simultaneous measurement of water flow velocity and solute transport in xylem and phloem of adult plants of *Ricinus communis* over a daily time course by nuclear magnetic resonance spectrometry. Plant Cell Environ. 24, 491–503.

Peyrot des Gachons, C., Tominaga, T., Dubourdieu, D., 2002a. Localisation in the berry of *S*-cysteine conjugates, aroma precursors of *Vitis vinifera* L. cv. Sauvignon blanc grapes. The effect of skin contact on the aromatic potential of the musts. Am. J. Enol. Vitic. 53, 144–146.

Peyrot des Gachons, C., Tominaga, T., Dubourdieu, D., 2002b. Sulfur aroma precursor present in *S*-glutathione conjugate form: identification of *S*-3-(hexan-1-ol)-glutathione in must from *Vitis vinifera* L. cv. Sauvignon blanc. J. Agric. Food Chem. 50, 4076–4079.

Peyrot des Gachons, C., Van Leeuwen, C., Tominaga, T., Soyer, J.P., Gaudillère, J.P., Dubourdieu, D., 2005. Influence of water and nitrogen deficit on fruit ripening and aroma potential of *Vitis vinifera* L cv Sauvignon blanc in field conditions. J. Sci. Food Agric. 85, 73–85.

Pezet, R., Pont, V., Hoang-Van, K., 1991. Evidence for oxidative detoxication of pterostilbene and resveratrol by a laccase-like stilbene oxidase produced by *Botrytis cinerea*. Physiol. Mol. Plant Pathol. 39, 441–450.

Pezet, R., Viret, O., Perret, C., Tabacchi, R., 2003. Latency of *Botrytis cinerea* Pers.: Fr. and biochemical studies during growth and ripening of two grape berry cultivars, respectively susceptible and resistant to grey mould. J. Phytopathol. 151, 208–214.

Pezet, R., Gindro, K., Viret, O., Richter, H., 2004a. Effects of resveratrol, viniferins and pterostilbene on *Plasmopara viticola* zoospore mobility and disease development. Vitis 43, 145–148.

Pezet, R., Viret, O., Gindro, K., 2004b. Plant–microbe interaction: the *Botrytis* grey mould of grapes—biology, biochemistry, epidemiology and control management. In: Hemantaranjan, A. (Ed.), Molecular Plant Physiology and Plant Microbiology. Advances in Plant Physiology, vol. 7. Scientific Publishers, Jodhpur, India, pp. 71–116.

Pezzuto, J.M., 2008. Grapes and human health: a perspective. J. Agric. Food Chem. 56, 6777–6784.

Pfanz, H., 2008. Bark photosynthesis. Trees 22, 137–138.

Phillips, J.G., Riha, S.J., 1994. Root growth, water uptake and canopy development in *Eucalyptus viminalis* seedlings. Aust. J. Plant Physiol. 21, 69–78.

Piazza, P., Jasinski, S., Tsiantis, M., 2005. Evolution of leaf developmental mechanisms. New Phytol. 167, 693–710.

Pichersky, E., Gershenzon, J., 2002. The formation and function of plant volatiles: perfumes for pollinator attraction and defense. Curr. Opin. Plant Biol. 5, 237–243.

Pickering, G., Lin, J., Riesen, R., Reynolds, A., Brindle, I., Soleas, G., 2004. Influence of *Harmonia axyridis* on the sensory properties of white and red wine. Am. J. Enol. Vitic. 55, 153–159.

Pickering, G.J., Lin, Y., Reynolds, A., Soleas, G., Riesen, R., Brindle, I., 2005. The influence of *Harmonia axyridis* on wine composition and aging. J. Food Sci. 70, 128–135.

Pien, S., Wyrzykowska, J., McQueen-Mason, S., Smart, C., Fleming, A., 2001. Local expression of expansin induces the entire process of leaf development and modifies leaf shape. Proc. Natl. Acad. Sci. USA 98, 11812–11817.

Pierik, R., Cuppens, M.L.C., Voesenek, L.A.C.J., Visser, E.J.W., 2004. Interactions between ethylene and gibberellins in phytochrome-mediated shade avoidance responses in tobacco. Plant Physiol. 136, 2928–2936.

Pierquet, P., Stushnoff, C., 1980. Relationship of low temperature exotherms to cold injury in *Vitis riparia* Michx. Am. J. Enol. Vitic. 31, 1–6.

Pilati, S., Perazzolli, M., Malossini, A., Cestaro, A., Dematté, L., Fontana, P., et al., 2007. Genome-wide transcriptional analysis of grapevine berry ripening reveals a set of genes similarly modulated during three seasons and the occurrence of an oxidative burst at veraison. BMC Genomics 8, 428, http://dx.doi.org/10.1186/1471-2164-8-428.

Pineau, B., Barbe, J.C., Van Leeuwen, C., Dubourdieu, D., 2007. Which impact for β-damascenone on red wines aroma? J. Agric. Food Chem. 55, 4103–4108.

Pinelo, M., Arnous, A., Meyer, A.S., 2006. Upgrading of grape skins: significance of plant cell-wall structural components and extraction techniques for phenol release. Trends Food Sci. Technol. 17, 579–590.

Pirie, A., Mullins, M.G., 1976. Changes in anthocyanin and phenolics content of grapevine leaf and fruit tissue treated with sucrose, nitrate, and abscisic acid. Plant Physiol. 58, 468–472.

Pirie, A., Mullins, M.G., 1977. Interrelationships of sugars, anthocyanins, total phenols and dry weight in the skin of grape berries during ripening. Am. J. Enol. Vitic. 28, 204–209.

Pirie, A.J.G., Mullins, M.G., 1980. Concentration of phenolics in the skin of grape berries during fruit development and ripening. Am. J. Enol. Vitic. 31, 34–36.

Pittman, J.K., 2005. Managing the manganese: molecular mechanisms of manganese transport and homeostasis. New Phytol. 167, 733–742.

Piva, C.R., Lopez Garcia, J.L., Morgan, W., 2006. The ideal table grapes for the Spanish market. Rev. Bras. Frutic. 28, 258–261.

Plank, S., Wolkinger, F., 1976. Holz von *Vitis vinifera* im Raster-Elektronenmikroskop. Vitis 15, 153–159.

Plavcová, L., Hacke, U.G., Almeida-Rodriguez, A.M., Li, E., Douglas, C.J., 2012. Gene expression patterns underlying changes in xylem structure and function in response to increased nitrogen availability in hybrid poplar. Plant Cell Environ. 36, 186–199.

Plaxton, W.C., Tran, H.T., 2011. Metabolic adaptations of phosphate-starved plants. Plant Physiol. 156, 1006–1015.

Plieth, C., 2005. Calcium: just another regulator in the machinery of life? Ann. Bot. 96, 1–8.

Plomion, C., Leprovost, G., Stokes, A., 2001. Wood formation in trees. Plant Physiol. 127, 1513–1523.

Pocock, K.F., Hayasaka, Y., McCarthy, M.G., Waters, E.J., 2000. Thaumatin-like proteins and chitinases, the haze-forming proteins of wine, accumulate during ripening of grape (*Vitis vinifera*) berries and drought stress does not affect the final levels per berry at maturity. J. Agric. Food Chem. 48, 1637–1643.

Pocock, K.F., Hayasaka, Y., Peng, Z., Williams, P.J., Waters, E.J., 1998. The effect of mechanical harvesting and long-distance transport on the concentration of haze-forming proteins in grape juice. Aust. J. Grape Wine Res. 4, 23–29.

Poelman, E.H., van Loon, J.J.A., Dicke, M., 2008. Consequences of variation in plant defense for biodiversity at higher trophic levels. Trends Plant Sci. 13, 534–541.

Poland, J.A., Balint-Kurti, P.J., Wisser, R.J., Pratt, R.C., Nelson, R.J., 2009. Shades of gray: the world of quantitative disease resistance. Trends Plant Sci. 14, 21–29.

Pomar, F., Novo, M., Masa, A., 2005. Varietal differences among the anthocyanin profiles of 50 red table grape cultivars studied by high performance liquid chromatography. J. Chromatogr. A 1094, 34–41.

Pongrácz, D.P., 1983. Rootstocks for Grape-Vines. David Philip, Cape Town, South Africa.

Poni, S., Intrieri, C., Silvestroni, O., 1994a. Interactions of leaf age, fruiting, and exogenous cytokinins in Sangiovese grapevines under non-irrigated conditions. I. Gas exchange. Am. J. Enol. Vitic. 45, 71–78.

Poni, S., Intrieri, C., Silvestroni, O., 1994b. Interactions of leaf age, fruiting, and exogenous cytokinins in Sangiovese grapevines under non-irrigated conditions. II. Chlorophyll and nitrogen content. Am. J. Enol. Vitic. 45, 278–284.

Poni, S., Magnanini, E., Bernizzoni, F., 2003. Degree of correlation between total light interception and whole-canopy net CO_2 exchange rate in two grapevine growth systems. Aust. J. Grape Wine Res. 9, 2–11.

Poni, S., Casalini, L., Bernizzoni, F., Civardi, S., Intrieri, C., 2006. Effects of early defoliation on shoot photosynthesis, yield components, and grape composition. Am. J. Enol. Vitic. 57, 397–407.

Poni, S., Bernizzoni, F., Civardi, S., Gatti, M., Porro, D., Camin, F., 2008. Performance and water-use efficiency (single-leaf vs. whole-canopy) of well-watered and half-stressed split-root Lambrusco grapevines grown in Po Valley (Italy). Agric. Ecosyst. Environ. 129, 97–106.

Poni, S., Bernizzoni, F., Civardi, S., Libelli, N., 2009. Effects of pre-bloom leaf removal on growth of berry tissues and must composition in two red *Vitis vinifera* L. cultivars. Aust. J. Grape Wine Res. 15, 185–193.

Pons, T.L., Jordi, W., Kuiper, D., 2001. Acclimation of plants to light gradients in leaf canopies: evidence for a possible role for cytokinins transported in the transpiration stream. J. Exp. Bot. 52, 1563–1574.

Pool, R., Dunst, R., Lakso, A., 1990. Comparison of sod, mulch, cultivation, and herbicide floor management practices for grape production in non-irrigated vineyards. J. Am. Soc. Hortic. Sci. 115, 872–877.

Pool, R.M., Pratt, C., Hubbard, H.D., 1978. Structure of base buds in relation to yield of grapes. Am. J. Enol. Vitic. 29, 36–41.

Pool, R.M., Creasy, L.L., Frackelton, A.S., 1981. Resveratrol and the viniferins, their application to screening for disease resistance in grape breeding programs. Vitis 20, 136–145.

Poorter, H., Bühler, J., van Dusschoten, D., Climent, J., Postma, J.A., 2012. Pot size matters: a meta-analysis of the effects of rooting volume on plant growth. Funct. Plant Biol. 39, 839–850.

Poorter, H., Nagel, O., 2000. The role of biomass allocation in the growth response of plants to different levels of light, nutrients and water: a quantitative review. Aust. J. Plant Physiol. 27, 595–607.

Poorter, H., Pepin, S., Rijkers, T., de Jong, Y., Evans, J.R., Körner, C., 2006. Construction costs, chemical composition and payback time of high- and low-irradiance leaves. J. Exp. Bot. 57, 355–371.

Poovaiah, B.W., Leopold, A.C., 1973. Inhibition of abscission by calcium. Plant Physiol. 51, 848–851.

Possingham, J.V., 1994. New concepts in pruning grapevines. Hortic. Rev. 16, 235–254.

Possingham, J.V., Groot Obbink, J., 1971. Endotrophic mycorrhiza and the nutrition of grape vines. Vitis 10, 120–130.

Possingham, J.V., Chambers, T.C., Radler, F., Grncarevic, M., 1967. Cuticular transpiration and wax structure and composition of leaves and fruit of *Vitis vinifera*. Aust. J. Biol. Sci. 20, 1149–1153.

Possner, D.R.E., Kliewer, W.M., 1985. The localisation of acids, sugars, potassium and calcium in developing grape berries. Vitis 24, 229–240.

Potters, G., Pasternak, T.P., Guisez, Y., Palme, K.J., Jansen, M.A.K., 2007. Stress-induced morphogenic responses: growing out of trouble? Trends Plant Sci. 12, 98–105.

Potters, G., Pasternak, T.P., Guisez, Y., Jansen, M.A.K., 2009. Different stresses, similar morphogenic responses: integrating a plethora of pathways. Plant Cell Environ. 32, 158−169.

Pouget, R., 1963. Recherches physiologiques sur le repos végétatif de la vigne (*Vitis vinifera* L.): La dormance des bourgeons et le mécanisme de sa disparition. Ann. Amélior. Plantes 13, 1−247.

Pouget, R., 1972. Considérations générales sur le rythme végétatif et la dormance des bourgeons de la vigne. Vitis 11, 198−217.

Pouget, R., 1981. Action de la température sur la différenciation des inflorescences et des fleurs durant les phases de pré-débourrement et de post-débourrement des bourgeons latents de la vigne. Connaiss. Vigne Vin 15, 65−79.

Pouget, R., 1984. Action de la concentration de la solution nutritive sur quelques caractéristiques physiologiques et technologiques chez *Vitis vinifera* L. cv. Cabernet-Sauvignon. I. Vigueur, rendement, qualité du moût et du vin. Agronomie 4, 437−442.

Pouget, R., 1988. Le débourrement des bourgeons de la vigne: méthode de prévision et principes d'établissement d'une échelle de précocité de débourrement. Connaiss. Vigne Vin 22, 105−123.

Pourcel, L., Routaboul, J.M., Cheynier, V., Lepiniec, L., Debeaujon, I., 2007. Flavonoid oxidation in plants: from biochemical properties to physiological functions. Trends Plant Sci. 12, 29−36.

Pourtchev, P., 2003. To the problem of depth of root system penetration of grapevine. Soil Sci. Agrochem. Ecol. 2, 47−52.

Powles, S.B., Yu, Q., 2010. Evolution in action: plants resistant to herbicides. Annu. Rev. Plant Biol. 61, 317−347.

Pradubsuk, S., Davenport, J.R., 2010. Seasonal uptake and partitioning of macronutrients in mature 'Concord' grape. J. Am. Soc. Hortic. Sci. 135, 474−483.

Pradubsuk, S., Davenport, J.R., 2011. Seasonal distribution of micronutrients in mature 'Concord' grape: boron, iron, manganese, copper, and zinc. J. Am. Soc. Hortic. Sci. 136, 69−77.

Pratt, C., 1971. Reproductive anatomy in cultivated grapes—a review. Am. J. Enol. Vitic. 22, 92−109.

Pratt, C., 1973. Reproductive system of "Concord" and two sports (*Vitis labruscana* Bailey). J. Am. Soc. Hortic. Sci. 98, 489−496.

Pratt, C., 1974. Vegetative anatomy of cultivated grapes—a review. Am. J. Enol. Vitic. 25, 131−150.

Pratt, C., 1978. Shoot nodes of *Vitis labruscana* Bailey cv. Concord. Vitis 17, 329−334.

Pratt, C., 1979. Shoot and bud development during the prebloom period of *Vitis*. Vitis 18, 1−5.

Pratt, C., Coombe, B.G., 1978. Shoot growth and anthesis in *Vitis*. Vitis 17, 125−133.

Pratt, C., Pool, R.M., 1981. Anatomy of recovery of canes of *Vitis vinifera* L. from simulated freezing injury. Am. J. Enol. Vitic. 32, 223−227.

Price, C.A., Gilooly, J.F., Allen, A.P., Weitz, J.S., Niklas, K.J., 2010. The metabolic theory of ecology: prospects and challenges for plant biology. New Phytol. 188, 696−710.

Price, S.F., Breen, P.J., Valladao, M., Watson, B.T., 1995. Cluster sun exposure and quercetin in Pinot noir grapes and wine. Am. J. Enol. Vitic. 46, 187−194.

Priestley, J.H., Wormall, A., 1925. On the solutes exuded by root pressure from vines. New Phytol. 24, 24−38.

Prieto, J.A., Louarn, G., Perez Peña, J., Ojeda, H., Simonneau, T., Lebon, E., 2012. A leaf gas exchange model that accounts for intra-canopy variability by considering leaf nitrogen content and local acclimation to radiation in grapevine (*Vitis vinifera* L.). Plant Cell Environ. 35, 1313−1328.

Prieur, C., Rigaud, J., Cheynier, V., Moutounet, M., 1994. Oligomeric and polymeric procyanidins from grape seeds. Phytochemistry 36, 781−784.

Prusinkiewicz, P., Crawford, S., Smith, R.S., Ljung, K., Bennett, T., Ongaro, V., et al., 2009. Control of bud activation by an auxin transport switch. Proc. Natl. Acad. Sci. USA 106, 17431−17436.

Puech, J.L., Feuillat, F., Mosedale, J.R., 1999. The tannins of oak heartwood: structure, properties, and their influence on wine flavor. Am. J. Enol. Vitic. 50, 469−478.

Quamme, H.A., 1991. Application of thermal analysis to breeding fruit crops for increased cold hardiness. HortScience 26, 513–517.

Querejeta, J.I., Egerton-Warburton, L.M., Allen, M.F., 2003. Direct nocturnal water transfer from oaks to their mycorrhizal symbionts during severe soil drying. Oecologia 134, 55–64.

Quick, W.P., Chaves, M.M., Wendler, R., David, M., Rodrigues, M.L., Passaharinho, J.A., et al., 1992. The effect of water stress on photosynthetic carbon metabolism in four species grown under field conditions. Plant Cell Environ. 15, 25–35.

Quijada-Morín, N., Regueiro, J., Simal-Gándara, J., Tomás, E., Rivas-Gonzalo, J.C., Escribano-Bailón, M.T., 2012. Relationship between the sensory-determined astringency and the flavanolic composition of red wines. J. Agric. Food Chem. 60, 12355–12361.

Quinlan, J.D., Weaver, R.J., 1969. Influence of benzyladenine, leaf darkening, and ringing on movement of ^{14}C-labeled assimilates into expanding leaves of *Vitis vinifera* L. Plant Physiol. 44, 1247–1252.

Quinlan, J.D., Weaver, R.J., 1970. Modification of pattern of the photosynthate movement within and between shoots of *Vitis vinifera* L. Plant Physiol. 46, 527–530.

Rabe, E., 1990. Stress physiology: the functional significance of the accumulation of nitrogen-containing compounds. J. Hortic. Sci. 65, 231–243.

Radler, F., 1965. Reduction of loss of moisture by the cuticle wax components of grapes. Nature 207, 1002–1003.

Rafei, M.S., 1941. Anatomical studies in *Vitis* and allied genera. I. Development of the fruit. II. Floral anatomy. PhD thesis, Oregon State College.

Raghavan, V., 2003. Some reflections on double fertilization, from its discovery to the present. New Phytol. 159, 565–583.

Raghothama, K.G., 1999. Phosphate acquisition. Annu. Rev. Plant Physiol. Plant Mol. Biol. 50, 665–693.

Ramming, D.W, 2009. Water loss from fresh berries of raisin cultivars under controlled drying conditions. Am. J. Enol. Vitic. 60, 208–214.

Rankine, B.C., Cellier, K.M., Boehm, E.W., 1962. Studies on grape variability and field sampling. Am. J. Enol. Vitic. 13, 58–72.

Rapp, A., Klenert, M., 1974. Einfluß der Samen auf die Beerenreife bei *Vitis vinifera* L. Vitis 13, 222–232.

Rapp, A., Versini, G., 1996. Influence of nitrogen compounds in grapes on aroma compounds of wines. Wein-Wiss. 51, 193–203.

Rapp, A., Ziegler, A., 1973. Bestimmung der Phenolcarbonsäuren (Hydroxybenzoesäuren und Hydroxyzimtsäuren) in Rebblättern, Weintrauben und Wein mittels Mikro-Polyamid-Dünnschichtchromatographie. Vitis 12, 226–236.

Rascher, U., Nedbal, L., 2006. Dynamics of photosynthesis in fluctuating light. Curr. Opin. Plant Biol. 9, 671–678.

Raschke, K., 1975. Stomatal action. Annu. Rev. Plant Physiol. 26, 309–340.

Rasmussen, A., Mason, M.G., De Cuyper, C., Brewer, P.B., Herold, S., Agusti, J., et al., 2012. Strigolactones suppress adventitious rooting in *Arabidopsis* and pea. Plant Physiol. 158, 1976–1987.

Rasmussen, A., Depuydt, S., Goormachtig, S., Geelen, D., 2013. Strigolactones fine-tune the root system. Planta 238, 615–626.

Rausch, T., 2007. When plant life gets tough sulfur gets going. Plant Biol. 9, 551–555.

Rayle, D.L., Cleland, R.E., 1992. The Acid Growth Theory of auxin-induced cell elongation is alive and well. Plant Physiol. 99, 1271–1274.

Razungles, A., Bayonove, C.L., Cordonnier, R.E., Sapis, J.C., 1988. Grape carotenoids: changes during the maturation period and localization in mature berries. Am. J. Enol. Vitic. 39, 44–48.

Razungles, A.J., Babic, I., Sapis, J.C., Bayonove, C.L., 1996. Particular behavior of epoxy xanthophylls during veraison and maturation of grape. J. Agric. Food Chem. 44, 3821–3825.

Razungles, A.J., Baumes, R.L., Dufour, C., Sznaper, C.N., Bayonove, C.L., 1998. Effect of sun exposure on carotenoids and C-13-norisoprenoid glycosides in Syrah berries (*Vitis vinifera* L.). Sci. Aliments 18, 361–373.

Rebucci, B., Poni, S., Intrieri, C., Magnanini, E., Lakso, A.N., 1997. Effects of manipulated grape berry transpiration and post-veraison sugar accumulation. Aust. J. Grape Wine Res. 3, 57–65.

Redl, H., 1984. Die Entblätterung der Traubenzone als vorbeugende Massnahme gegen die Stiellähme. Wein-Wiss. 39, 75–82.

Regner, F., Stadlbauer, A., Eisenheld, C., Kaserer, H., 2000. Genetic relationships among Pinots and related cultivars. Am. J. Enol. Vitic. 51, 7–14.

Reich, P.B., Falster, D.S., Ellsworth, D.S., Wright, I.J., Westoby, M., Oleksyn, J., et al., 2009. Controls on declining carbon balance with leaf age among 10 woody species in Australian woodland: do leaves have zero daily net carbon balances when they die? New Phytol. 183, 153–166.

Reid, R.J., 2001. Mechanisms of micronutrient uptake in plants. Aust. J. Plant Physiol. 28, 659–666.

Reimers, H., Steinberg, B., Kiefer, W., 1994. Ergebnisse von Wurzeluntersuchungen an Reben bei offenem und begrüntem Boden. Wein-Wiss. 49, 136–145.

Reinhardt, D., Mandel, T., Kuhlemeier, C., 2000. Auxin regulates the initiation and radial position of plant lateral organs. Plant Cell 12, 507–518.

Reisch, B.I., Goodman, R.N., Martens, M.H., Weeden, N.F., 1993. The relationship between Norton and Cynthiana, red wine cultivars derived from *Vitis aestivalis*. Am. J. Enol. Vitic. 44, 441–444.

Reiter, W.D., 2002. Biosynthesis and properties of the plant cell wall. Curr. Opin. Plant Biol. 5, 536–542.

Remund, U., Niggli, U., Boller, E.F., 1989. Faunistische und botanische Erhebungen in einem Rebberg der Ostschweiz. Einfluss der Unterwuchsbewirtschaftung auf das Oekosystem Rebberg. Landwirtsch. Schweiz 2, 393–408.

Renault, A.S., Deloire, A., Bierne, J., 1996. Pathogenesis-related proteins in grapevines induced by salicylic acid and *Botrytis cinerea*. Vitis 35, 49–52.

Rennie, E.A., Turgeon, R., 2009. A comprehensive picture of phloem loading strategies. Proc. Natl. Acad. Sci. USA 106, 14162–14167.

Renouf, V., Tregoat, O., Roby, J.P., Van Leeuwen, C., 2010. Soils, rootstocks and grapevine varieties in prestigious Bordeaux vineyards and their impact on yield and quality. J. Int. Sci. Vigne Vin 44, 1–8.

Reuther, G., Reichhardt, A., 1963. Temperatureinflüsse auf Blutung und Stoffwechsel bei *Vitis vinifera*. Planta 59, 391–410.

Reuveni, M., 1998. Relationships between leaf age, peroxidase and β-1,3-glucanase activity, and resistance to downy mildew in grapevines. J. Phytopathol. 146, 525–530.

Revilla, E., García-Beneytiz, E., Cabello, F., 2009. Anthocyanin fingerprint of clones of Tempranillo grapes and wines made with them. Aust. J. Grape Wine Res. 15, 70–78.

Reynolds, A.G., Vanden Heuvel, J.E., 2009. Influence of grapevine training systems on vine growth and fruit composition: a review. Am. J. Enol. Vitic. 60, 251–268.

Reynolds, A.G., Wardle, D.A., 1989a. Impact of various canopy manipulation techniques on growth, yield, fruit composition, and wine quality of Gewürztraminer. Am. J. Enol. Vitic. 40, 121–129.

Reynolds, A.G., Wardle, D.A., 1989b. Influence of fruit microclimate on monoterpene levels of Gewürztraminer. Am. J. Enol. Vitic. 40, 149–154.

Reynolds, A.G., Pool, R.M., Mattick, L.R., 1986. Influence of cluster exposure on fruit composition and wine quality of Seyval blanc grapes. Vitis 25, 85–95.

Reynolds, A.G., Edwards, C.G., Wardle, D.A., Webster, D.R., Dever, M., 1994a. Shoot density affects 'Riesling' grapevines. I. Vine performance. J. Am. Soc. Hortic. Sci. 119, 874–880.

Reynolds, A.G., Edwards, C.G., Wardle, D.A., Webster, D.R., Dever, M., 1994b. Shoot density affects 'Riesling' grapevines. II. Wine composition and sensory response. J. Am. Soc. Hortic. Sci. 119, 881–892.

Reynolds, A.G., Wardle, D.A., Dever, M.J., 1996. Vine performance, fruit composition, and wine sensory attributes of Gewürztraminer in response to vineyard location and canopy manipulation. Am. J. Enol. Vitic. 47, 77–92.

Reynolds, A.G., Molek, T., De Savigny, C., 2005. Timing of shoot thinning in *Vitis vinifera*: impacts on yield and fruit composition variables. Am. J. Enol. Vitic. 56, 343–356.

Reynolds, A.G., Senchuk, I.V., van der Reest, C., de Savigny, C., 2007. Use of GPS and GIS for elucidation of the basis for terroir: spatial variation in an Ontario Riesling vineyard. Am. J. Enol. Vitic. 58, 145–162.

Reynolds, A.G., Taylor, G., de Savigny, C., 2013. Defining Niagara terroir by chemical and sensory analysis of Chardonnay wines from various soil textures and vine sizes. Am. J. Enol. Vitic. 64, 180–194.

Riaz, S., Garrison, K.E., Dangl, G.S., Boursiquot, J.M., Meredith, C.P., 2002. Genetic divergence and chimerism within ancient asexually propagated winegrape cultivars. J. Am. Soc. Hortic. Sci. 127, 508–514.

Ribéreau-Gayon, J., Ribéreau-Gayon, P., 1958. The anthocyans and leucoanthocyans of grapes and wines. Am. J. Enol. Vitic. 9, 1–9.

Ribéreau-Gayon, P., 1964. Les flavonosides de la baie dans le genre *Vitis*. C.R. Hébd. Séances Acad. Sci. (Paris) 258, 1335–1337.

Ribéreau-Gayon, P., 1973. Interprétation chimique de la couleur des vins rouges. Vitis 12, 119–142.

Ribéreau-Gayon, P., 1982. Incidences oenologiques de la pourriture du raisin. Bull. OEPP 12, 201–214.

Ribéreau-Gayon, P., Boidron, J.N., Terrier, A., 1975. Aroma of Muscat grape varieties. J. Agric. Food Chem. 23, 1042–1047.

Ribéreau-Gayon, P., Dubourdieu, D., Donéche, B., Lonvaud, A., 2006. Handbook of Enology. Volume 1: The Microbiology of Wine and Vinifications, second ed. Wiley & Sons, Chichester, UK, Trans. J.M. Branco.

Rice-Evans, C.A., Miller, N.J., Paganga, G., 1997. Antioxidant properties of phenolic compounds. Trends Plant Sci. 2, 152–159.

Richards, D., 1983. The grape root system. Hortic. Rev. 5, 127–168.

Richards, L.A., 1954. Diagnosis and Improvement of Saline and Sodic Soils. Agriculture Handbook 60. United States Department of Agriculture, Washington, DC.

Richardson, A.C., Marsh, K.B., Boldingh, H.L., Pickering, A.H., Bulley, S.M., Frearson, N.J., et al., 2004. High growing temperatures reduce fruit carbohydrate and vitamin C in kiwifruit. Plant Cell Environ. 27, 423–435.

Riederer, M., Schreiber, L., 2001. Protecting against water loss: analysis of the barrier properties of plant cuticles. J. Exp. Bot. 52, 2023–2032.

Ristic, R., Iland, P.G., 2005. Relationships between seed and berry development of *Vitis vinifera* L. cv Shiraz: developmental changes in seed morphology and phenolic composition. Aust. J. Grape Wine Res. 11, 43–58.

Ristic, R., Bindon, K., Francis, L.I., Herderich, M.J., Iland, P.G., 2010. Flavonoids and C_{13}-norisoprenoids in *Vitis vinifera* L. cv. Shiraz: relationships between grape and wine composition, wine colour and wine sensory properties. Aust. J. Grape Wine Res. 16, 369–388.

Ristic, Z., Ashworth, E.N., 1993. Ultrastructural evidence that intracellular ice formation and possibly cavitation are the sources of freezing injury in supercooling wood tissue of *Cornus florida* L. Plant Physiol. 103, 753–761.

Rivera Núñez, D., Walker, M.J., 1989. A review of palaeobotanical findings of early *Vitis* in the Mediterranean and of the origins of cultivated grape-vines, with special reference to new pointers to prehistoric exploitation in the western Mediterranean. Rev. Palaeobot. Palynol. 61, 205–237.

Rizhsky, L., Liang, H., Shuman, J., Shulaev, V., Davletova, S., Mittler, R., 2004. When defense pathways collide. The response of *Arabidopsis* to a combination of drought and heat stress. Plant Physiol. 134, 1683–1696.

Roberts, A.G., Oparka, K.J., 2003. Plasmodesmata and the control of symplastic transport. Plant Cell Environ. 26, 103–124.

Roberts, J.A., Elliott, K.A., Gonzalez-Carranza, Z.H., 2002. Abscission, dehiscence, and other cell separation processes. Annu. Rev. Plant Biol. 53, 131–158.

Roberts, J.K.M., Callis, J., Jardetzky, O., Walbot, V., Freeling, M., 1984. Cytoplasmic acidosis as a determinant of flooding intolerance in plants. Proc. Natl. Acad. Sci. USA 81, 6029–6033.

Roberts, S.K., Snowman, B.N., 2000. The effects of ABA on channel-mediated K$^+$ transport across higher plant roots. J. Exp. Bot. 51, 1585–1594.

Robertson, G.L., Eschenbruch, R., Cresswell, K.J., 1980. Seasonal changes in the pectic substances of grapes and their implication in juice extraction. Am. J. Enol. Vitic. 31, 162–164.

Robichaud, J.L., Noble, A.C., 1990. Astringency and bitterness of selected phenolics in wine. J. Sci. Food Agric. 53, 343–353.

Robinson, A.L., Boss, P.K., Solomon, P.S., Trengove, R.D., Heymann, H., Ebeler, S.E., 2014. Origins of grape and wine aroma. Part 1. Chemical components and viticultural impacts. Am. J. Enol. Vitic. 65, 1–24.

Robinson, D., 1994. The response of plants to non-uniform supplies of nutrients. New Phytol. 127, 635–674.

Robinson, J., Harding, J., Vouillamoz, J., 2012. Wine Grapes. HarperCollins, New York, NY.

Robinson, S.A., Slade, A.P., Fox, G.G., Phillips, R., Ratcliffe, R.G., Stewart, G.R., 1991. The role of glutamate dehydrogenase in plant nitrogen metabolism. Plant Physiol. 95, 509–516.

Robinson, S.P., Davies, C., 2000. Molecular biology of grape berry ripening. Aust. J. Grape Wine Res. 6, 175–188.

Robinson, S.P., Jacobs, A.K., Dry, I.B., 1997. A class IV chitinase is highly expressed in grape berries during ripening. Plant Physiol. 114, 771–778.

Roby, G., Matthews, M.A., 2004. Relative proportions of seed, skin and flesh, in ripe berries from Cabernet Sauvignon grapevines grown in a vineyard either well irrigated or under water deficit. Aust. J. Grape Wine Res. 10, 74–82.

Rockwell, N.C., Su, Y.S., Lagarias, J.C., 2006. Phytochrome structure and signaling mechanisms. Annu. Rev. Plant Biol. 57, 837–858.

Rodríguez, A., Alquézar, B., Peña, L., 2013. Fruit aromas in mature fleshy fruits as signals of readiness for predation and seed dispersal. New Phytol. 197, 36–48.

Rodríguez Montealegre, R., Romero Peces, R., Chacón Vozmediano, J.L., Martínez Gascueña, J., García Romero, E., 2006. Phenolic compounds in skins and seeds of ten grape *Vitis vinifera* varieties grown in a warm climate. J. Food Compos. Anal. 19, 687–693.

Rodriguez-Lovelle, B., Gaudillére, J.P., 2002. Carbon and nitrogen partitioning in either fruiting or non-fruiting grapevines: effects of nitrogen limitation before and after veraison. Aust. J. Grape Wine Res. 8, 86–94.

Rodríguez-Palenzuela, P., Burr, T.J., Collmer, A., 1991. Polygalacturonase is a virulence factor in *Agrobacterium tumefaciens* biovar 3. J. Bacteriol. 173, 6547–6552.

Roelfsema, M.R.G., Hedrich, R., 2005. In the light of stomatal opening: new insights into 'the Watergate'. New Phytol. 167, 665–691.

Rogiers, S.Y., Clarke, S.J., 2013. Nocturnal and daytime stomatal conductance respond to root-zone temperature in 'Shiraz' grapevines. Ann. Bot. 111, 433–444.

Rogiers, S.Y., Smith, J.A., White, R., Keller, M., Holzapfel, B.P., Virgona, J.M., 2001. Vascular function in berries of *Vitis vinifera* (L) cv. Shiraz. Aust. J. Grape Wine Res. 7, 47–51.

Rogiers, S.Y., Hatfield, J.M., Jaudzems, V.G., White, R.G., Keller, M., 2004a. Grape berry cv. Shiraz epicuticular wax and transpiration during ripening and preharvest weight loss. Am. J. Enol. Vitic. 55, 121–127.

Rogiers, S.Y., Hatfield, J.M., Keller, M., 2004b. Irrigation, nitrogen, and rootstock effects on volume loss of berries from potted Shiraz vines. Vitis 43, 1–6.

Rogiers, S.Y., Greer, D.H., Hatfield, J.M., Orchard, B.A., Keller, M., 2006. Mineral sinks within ripening grape berries (*Vitis vinifera* L.). Vitis 45, 115–123.

Rogiers, S.Y., Greer, D.H., Hutton, R.J., Landsberg, J.J., 2009. Does night-time transpiration contribute to anisohydric behaviour in a *Vitis vinifera* cultivar? J. Exp. Bot. 60, 3751–3763.

Rogiers, S.Y., Smith, J.P., Holzapfel, B.P., Hardie, W.J., 2011. Soil temperature moderates grapevine carbohydrate reserves after bud break and conditions fruit set responses to photoassimilatory stress. Funct. Plant Biol. 38, 899–909.

Rogiers, S.Y., Clarke, S.J., Schmidtke, L.M., 2014. Elevated root-zone temperature hastens vegetative and reproductive development in Shiraz grapevines. Aust. J. Grape Wine Res. 20, 123–133.
Rohde, A., Bhalerao, R.P., 2007. Plant dormancy in the perennial context. Trends Plant Sci. 12, 217–223.
Roitsch, T., 1999. Source-sink regulation by sugar and stress. Curr. Opin. Plant Biol. 2, 198–206.
Roitsch, T., González, M.C., 2004. Function and regulation of plant invertases: sweet sensations. Trends Plant Sci. 9, 606–613.
Rojas-Lara, B.A., Morrison, J.C., 1989. Differential effects of shading fruit or foliage on the development and composition of grape berries. Vitis 28, 199–208.
Rolland, F., Baena-Gonzalez, E., Sheen, J., 2006. Sugar sensing and signaling in plants: conserved and novel mechanisms. Annu. Rev. Plant Biol. 57, 675–709.
Rolle, L., Siret, R., Río Segade, S., Maury, C., Gerbi, V., Jourjon, F., 2012. Instrumental texture analysis parameters as markers of table-grape and winegrape quality: a review. Am. J. Enol. Vitic. 63, 11–28.
Romero, P., Martinez-Cutillas, A., 2012. The effects of partial root-zone irrigation and regulated deficit irrigation on the vegetative and reproductive development of field-grown Monastrell grapevines. Irrig. Sci. 30, 377–396.
Romero, P., Fernández-Fernández, J.I., Martinez-Cutillas, A., 2010. Physiological thresholds for efficient regulated deficit-irrigation management in winegrapes grown under semiarid conditions. Am. J. Enol. Vitic. 61, 300–312.
Romero, P., Gil-Muñoz, R., del Amor, F.M., Valdés, E., Fernández, J.I., Martinez-Cutillas, A., 2013. Regulated deficit irrigation based upon optimum water status improves phenolic composition in Monastrell grapes and wines. Agric. Water Manage. 121, 85–101.
Romero-Pérez, A.I., Ibern-Gómez, M., Lamuela-Raventós, R.M., de la Torre-Boronat, M.C., 1999. Piceid, the major resveratrol derivative in grape juices. J. Agric. Food Chem. 47, 1533–1536.
Roper, M.C., Greve, L.C., Warren, J.G., Labavitch, J.M., Kirkpatrick, B.C., 2007. *Xylella fastidiosa* requires polygalacturonase for colonization and pathogenicity in *Vitis vinifera* grapevines. Mol. Plant Microbe. Interact. 20, 411–419.
Roper, T.R., Williams, L.E., 1989. Net CO_2 assimilation and carbohydrate partitioning of grapevine leaves in response to trunk girdling and gibberellic acid application. Plant Physiol. 89, 1136–1140.
Ros Barceló, A., Pomar, F., López-Serrano, M., Pedreño, M.A., 2003. Peroxidase: a multifunctional enzyme in grapevines. Funct. Plant Biol. 30, 577–591.
Rose, J.K.C., Bennett, A.B., 1999. Cooperative disassembly of the cellulose-xyloglucan network of plant cell walls: parallels between cell expansion and fruit ripening. Trends Plant Sci. 4, 176–183.
Rose, J.K.C., Lee, H.H., Bennett, A.B., 1997. Expression of a divergent expansin gene is fruit-specific and ripening-regulated. Proc. Natl. Acad. Sci. USA 94, 5955–5960.
Rose, J.K.C., Hadfield, K.A., Labavitch, J.M., Bennett, A.B., 1998. Temporal sequence of cell wall disassembly in rapidly ripening melon fruit. Plant Physiol. 117, 345–361.
Rose, J.K.C., Saladié, M., Catalá, C., 2004. The plot thickens: new perspectives of primary cell wall modification. Curr. Opin. Plant Biol. 7, 296–301.
Rosenheim, O., 1920. Observations on anthocyanins. I. The anthocyanins of the young leaves of the grape vine. Biochem. J. 14, 178–188.
Rosenquist, J.K., Morrison, J.C., 1988. The development of the cuticle and epicuticular wax of the grape berry. Vitis 27, 63–70.
Rosenquist, J.K., Morrison, J.C., 1989. Some factors affecting cuticle and wax accumulation on grape berries. Am. J. Enol. Vitic. 40, 241–244.
Rosner, N., Cook, J.A., 1983. Effects of differential pruning on Cabernet Sauvignon grapevines. Am. J. Enol. Vitic. 34, 243–248.
Ross, J.J., Weston, D.E., Davidson, S.E., Reid, J.B., 2011. Plant hormone interactions: how complex are they? Physiol. Plant. 141, 299–309.

Roubelakis, K.A., Kliewer, W.M., 1976. Influence of light intensity and growth regulators on fruit-set and ovule fertilization in grape cultivars under low temperature conditions. Am. J. Enol. Vitic. 27, 163–167.

Roubelakis-Angelakis, K.A., Kliewer, W.M., 1979. The composition of bleeding sap from Thompson Seedless grapevines as affected by nitrogen fertilization. Am. J. Enol. Vitic. 30, 14–18.

Roubelakis-Angelakis, K.A., Kliewer, W.M., 1983. Ammonia assimilation in *Vitis vinifera* L.: II. Leaf and root glutamine synthetase. Vitis 22, 299–305.

Roubelakis-Angelakis, K.A., Kliewer, W.M., 1992. Nitrogen metabolism in grapevine. Hortic. Rev. 14, 407–452.

Roufet, M., Bayonove, C.L., Cordonnier, R.E., 1987. Etude de la composition lipidique du raisin, *Vitis vinifera* L.: evolution au cours de la maturation et localisation dans la baie. Vitis 26, 85–97.

Roujou de Boubée, D., Van Leeuwen, C., Dubourdieu, D., 2000. Organoleptic impact of 2-methoxy-3-isobutylpyrazine on red Bordeaux and Loire wines. Effect of environmental conditions on concentrations in grapes during ripening. J. Agric. Food Chem. 48, 4830–4834.

Roujou de Boubée, D., Cumsille, A.M., Pons, M., Dubourdieu, D., 2002. Location of 2-methoxy-3-isobutylpyrazine in Cabernet Sauvignon grape bunches and its extractability during vinification. Am. J. Enol. Vitic. 53, 1–5.

Rousseaux, M.C., Hall, A.J., Sánchez, R.A., 2000. Basal leaf senescence in a sunflower (*Helianthus annuus*) canopy: responses to increased R/FR ratio. Physiol. Plant. 110, 477–482.

Rowhani, A., Uyemoto, J.K., Golino, D.A., Martelli, G.P., 2005. Pathogen testing and certification of *Vitis* and *Prunus* species. Annu. Rev. Phytopathol. 43, 261–278.

Roy, R.R., Ramming, D.W., 1990. Varietal resistance of grape to the powdery mildew fungus, *Uncinula necator*. Fruit Var. J. 44, 149–155.

Roytchev, V.R., 1998. Microsporogenesis and male gametophyte development in seedless grapevine (*Vitis vinifera* L.) cultivars. Wein-Wiss. 53, 162–167.

Roytchev, V.R., 2000. Megasporogenesis and female gametophyte development in seedless grapevine (*Vitis vinifera* L.) cultivars. Wein-Wiss. 55, 21–28.

Rozema, J., van de Staaij, J., Björn, L.O., Caldwell, M., 1997. UV-B as an environmental factor in plant life: stress and regulation. Tree 12, 22–28.

Ruan, Y.L., Patrick, J.W., Bouzayen, M., Osorio, S., Fernie, A.R., 2012. Molecular regulation of seed and fruit set. Trends Plant Sci. 17, 656–665.

Rubio, M., Alvarez-Ortí, M., Alvarruiz, A., Fernández, E., Pardo, J.E., 2009. Characterization of oil obtained from grape seeds collected during berry development. J. Agric. Food Chem. 57, 2812–2815.

Ruel, J.J., Walker, M.A., 2006. Resistance to Pierce's disease in *Muscadinia rotundifolia* and other native grape species. Am. J. Enol. Vitic. 57, 158–165.

Ruffner, H.P., 1982a. Metabolism of tartaric and malic acids in *Vitis*: a review—Part A. Vitis 21, 247–259.

Ruffner, H.P., 1982b. Metabolism of tartaric and malic acids in *Vitis*: a review—Part B. Vitis 21, 346–358.

Ruffner, H.P., Hawker, J.S., 1977. Control of glycolysis in ripening berries of *Vitis vinifera*. Phytochemistry 16, 1171–1175.

Ruffner, H.P., Koblet, W., Rast, D., 1975. Gluconeogenese in reifenden Beeren von *Vitis vinifera*. Vitis 13, 319–328.

Rufty Jr., T.W., MacKown, C.T., Volk, R., 1989. Effects of altered carbohydrate availability on whole-plant assimilation of $^{15}NO_3^-$. Plant Physiol. 89, 457–463.

Rüger, S., Netzer, Y., Westhoff, M., Zimmermann, D., Reuss, R., Ovadiya, S., et al., 2010. Remote monitoring of leaf turgor pressure of grapevines subjected to different irrigation treatments using the leaf patch clamp pressure probe. Aust. J. Grape Wine Res. 16, 405–412.

Rühl, E., Imgraben, H.J., 1985. Einfluß der Stickstoffversorgung auf die Stomatazahl und den Blattaufbau von Weinreben (*Vitis vinifera* L.). Wein-Wiss. 40, 160–171.

Rühl, E.H., Clingeleffer, P.R., 1993. Effect of minimal pruning and virus inoculation on the carbohydrate and nitrogen accumulation in Cabernet franc vines. Am. J. Enol. Vitic. 44, 81–85.

Rühl, E.H., Alleweldt, G., Hofäcker, W., 1981. Untersuchungen zum Lichtkompensationspunkt der Rebe. Vitis 20, 122–129.

Ruhl, E.H., Clingeleffer, P.R., Nicholas, P.R., Cirami, R.M., McCarthy, M.G., Whiting, J.R., 1988. Effect of rootstocks on berry weight and pH, mineral content and organic acid concentrations of grape juice of some wine varieties. Aust. J. Exp. Agric. 28, 119–125.

Ruhl, E.H., Fuda, A.P., Treeby, M.T., 1992. Effect of potassium, magnesium and nitrogen supply on grape juice composition of Riesling, Chardonnay and Cabernet Sauvignon vines. Aust. J. Exp. Agric. 32, 645–649.

Ruyter-Spira, C., Kohlen, W., Charnikhova, T., van Zeijl, A., van Bezouwen, L., de Ruijter, N., et al., 2011. Physiological effects of the synthetic strigolactone analog GR24 on root system architecture in *Arabidopsis*: another belowground role for strigolactones? Plant Physiol. 155, 721–734.

Ruyter-Spira, C., Al-Babili, S., van der Krol, S., Bouwmeester, H., 2013. The biology of strigolactones. Trends Plant Sci. 18, 72–83.

Ryan, M.G., Yoder, B.J., 1997. Hydraulic limits to tree height and growth. Bioscience 47, 235–242.

Ryan, M.G., Phillips, N., Bond, B.J., 2006. The hydraulic limitation hypothesis revisited. Plant Cell Environ. 29, 367–381.

Ryona, I., Pan, B.S., Intrigliolo, D.S., Lakso, A.N., Sacks, G.L., 2008. Effects of cluster light exposure on 3-isobutyl-2-methoxypyrazine accumulation and degradation patterns in red wine grapes (*Vitis vinifera* L. cv. Cabernet franc). J. Agric. Food Chem. 56, 10838–10846.

Ryona, I., Pan, B.S., Sacks, G.L., 2009. Rapid measurement of 3-alkyl-2-methoxypyrazine content of winegrapes to predict levels in resultant wines. J. Agric. Food Chem. 57, 8250–8257.

Ryona, I., Leclerc, S., Sacks, G.L., 2010. Correlation of 3-isobutyl-2-methoxypyrazine to 3-isobutyl-2-hydroxypyrazine during maturation of bell pepper (*Capsicum annuum*) and wine grapes (*Vitis vinifera*). J. Agric. Food Chem. 58, 9723–9730.

Sack, L., Holbrook, N.M., 2006. Leaf hydraulics. Annu. Rev. Plant Biol. 57, 361–381.

Sack, L., Cowan, P.D., Jaikumar, N., Holbrook, N.M., 2003. The 'hydrology' of leaves: co-ordination of structure and function in temperate woody species. Plant Cell Environ. 26, 1343–1356.

Sadras, V.O., Moran, M.A., 2012. Elevated temperature decouples anthocyanins and sugars in berries of Shiraz and Cabernet franc. Aust. J. Grape Wine Res. 18, 115–122.

Sáenz-Navajas, M.P., Tao, Y.S., Dizy, M., Ferreira, V., Fernández-Zurbano, P., 2010. Relationship between nonvolatile composition and sensory properties of premium Spanish red wines and their correlation to quality perception. J. Agric. Food Chem. 58, 12407–12416.

Sage, R.F., Kubien, D.S., 2007. The temperature response of C_3 and C_4 photosynthesis. Plant Cell Environ. 30, 1086–1106.

Sage, R.F., Sage, T.L., Kocacinar, F., 2012. Photorespiration and the evolution of C_4 photosynthesis. Annu. Rev. Plant Biol. 63, 19–47.

Saini, H.S., 1997. Effects of water stress on male gametophyte development in plants. Sex. Plant Reprod. 10, 67–73.

Saito, K., Ohmoto, J., Kuriha, N., 1997. Incorporation of ^{18}O into oxalic, L-threonic and L-tartaric acids during cleavage of L-ascorbic and 5-keto-D-gluconic acids in plants. Phytochemistry 44, 805–809.

Sakakibara, H., 2006. Cytokinins: activity, biosynthesis, and translocation. Annu. Rev. Plant Biol. 57, 431–449.

Sakakibara, H., Takei, K., Hirose, N., 2006. Interactions between nitrogen and cytokinin in the regulation of metabolism and development. Trends Plant Sci. 11, 440–448.

Sakr, S., Alves, G., Morillon, R., Maurel, K., Decourteix, M., Guilliot, A., et al., 2003. Plasma membrane aquaporins are involved in winter embolism recovery in walnut tree. Plant Physiol. 133, 630–641.

Sala, A., Hoch, G., 2009. Height-related growth declines in ponderosa pine are not due to carbon limitation. Plant Cell Environ. 32, 22–30.

Sala, A., Piper, F., Hoch, G., 2010. Physiological mechanisms of drought-induced tree mortality are far from being resolved. New Phytol. 186, 274–281.

Sala, C., Busto, O., Guasch, J., Zamora, F., 2005. Contents of 3-alkyl-2-methoxypyrazines in musts and wines from *Vitis vinifera* variety Cabernet Sauvignon: influence of irrigation and plantation density. J. Sci. Food Agric. 85, 1131–1136.

Salinas, J., Warnaar, F., Verhoeff, K., 1986. Production of cutin hydrolyzing enzymes by *Botrytis cinerea* in vitro. J. Phytopathol. 116, 299–307.

Salisbury, F.B., Ross, C.W., 1992. Plant Physiology, fourth ed. Wadsworth Publishing, Belmont, CA.

Salmasco, M., Dalla Valle, R., Lucchin, M., 2008. Gene pool variation and phylogenetic relationships of an indigenous northeast Italian grapevine collection revealed by nuclear and chloroplast SSRs. Genome 51, 838–855.

Salomé, P.A., McClung, C.R., 2005. What makes the *Arabidopsis* clock tick on time? A review on entrainment. Plant Cell Environ. 28, 21–38.

Salón, J.L., Chirivella, C., Castel, J.R., 2005. Response of cv. Bobal to timing of deficit irrigation in Requena, Spain: water relations, yield, and wine quality. Am. J. Enol. Vitic. 56, 1–8.

Salzman, R.A., Bressan, R.A., Hasegawa, P.M., Ashworth, E.N., Bordelon, B.P., 1996. Programmed accumulation of LEA-like proteins during desiccation and cold acclimation of overwintering grape buds. Plant Cell Environ. 19, 713–720.

Salzman, R.A., Tikhonova, I., Bordelon, B.P., Hasegawa, P.M., Bressan, R.A., 1998. Coordinate accumulation of antifungal proteins and hexoses constitutes a developmentally controlled defense response during fruit ripening in grape. Plant Physiol. 117, 465–472.

Samach, A., Wigge, P.A., 2005. Ambient temperature perception in plants. Curr. Opin. Plant Biol. 8, 483–486.

Sampson, B., Noffsinger, S., Gupton, C., Magee, J., 2001. Pollination biology of the muscadine grape. HortScience 36, 120–124.

Samuels, L., Kunst, L., Jetter, R., 2008. Sealing plant surfaces: cuticular wax formation by epidermal cells. Annu. Rev. Plant Biol. 59, 683–707.

Sánchez, L.A., Dokoozlian, N.K., 2005. Bud microclimate and fruitfulness in *Vitis vinifera* L. Am. J. Enol. Vitic. 56, 319–329.

Sanchez, P., Nehlin, L., Greb, T., 2012. From thin to thick: major transitions during stem development. Trends Plant Sci. 17, 113–121.

Sandermann, H., Ernst, D., Heller, W., Langebartels, C., 1998. Ozone: an abiotic elicitor of plant defence reactions. Trends Plant Sci. 3, 47–50.

Santana, J.C., Heuertz, M., Arranz, C., Rubio, J.A., Martínez-Zapater, J.M., Hidalgo, E., 2010. Genetic structure, origins, and relationships of grapevine cultivars from the Castilian Plateau of Spain. Am. J. Enol. Vitic. 61, 214–224.

Santesteban, L.G., Royo, J.B., 2006. Water status, leaf area and fruit load influence on berry weight and sugar accumulation of cv. 'Tempranillo' under semiarid conditions. Sci. Hortic. 109, 60–65.

Santner, A., Calderon-Villalobos, L.I.A., Estelle, M., 2009. Plant hormones are versatile chemical regulators of plant growth. Nat. Chem. Biol. 5, 301–307.

Santos, T.P., Lopes, C.M., Rodrigues, M.L., de Souza, C.R., Ricardo-Da-Silva, J.M., Maroco, J.P., et al., 2005. Effects of partial root-zone drying irrigation on cluster microclimate and fruit composition of field-grown Castelão grapevines. Vitis 44, 117–125.

Sapis, J.C., Macheix, J.J., Cordonnier, R.E., 1983. The browning capacity of grapes. II. Browning potential and polyphenol oxidase activities in different mature grape varieties. Am. J. Enol. Vitic. 34, 157–162.

Saqib, M., Zörb, C., Schubert, S., 2008. Silicon-mediated improvement in the salt resistance of wheat (*Triticum aestivum*) results from increased sodium exclusion and resistance to oxidative stress. Funct. Plant Biol. 35, 633–639.

Sarry, J.E., Sommerer, N., Sauvage, F.X., Bergoin, A., Rossignol, M., Albagnac, G., et al., 2004. Grape berry biochemistry revisited upon proteomic analysis of the mesocarp. Proteomics 4, 201–215.

Sartorius, O., 1926. Zur Entwicklung und Physiologie der Rebblüte. Angew. Bot. 8, 29–62.

Sartorius, O., 1968. Die Blütenknospen der Rebe. Wein-Wiss. 23, 309–338.

Sartorius, O., 1973. Holzreserven und Traubenreife. Wein-Wiss. 28, 100–107.

Sassi, M., Vernoux, T., 2013. Auxin and self-regulation at the shoot apical meristem. J. Exp. Bot. 64, 2579–2592.

Sattelmacher, B., 2001. The apoplast and its significance for plant mineral nutrition. New Phytol. 149, 167–192.

Sauer, M.R., 1968. Effects of vine rootstocks on chloride concentration in Sultana scions. Vitis 7, 223–226.

Saulnier, L., Brillouet, J.M., 1989. An arabinogalactan-protein from the pulp of grape berries. Carbohydr. Res. 188, 137–144.

Savaldi-Goldstein, S., Peto, C., Chory, J., 2007. The epidermis both drives and restricts plant shoot growth. Nature 446, 199–202.

Saveyn, A., Steppe, K., Ubierna, N., Dawson, T.E., 2010. Woody tissue photosynthesis and its contribution to trunk growth and bud development in young plants. Plant Cell Environ. 33, 1949–1958.

Sawler, J., Reisch, B., Aradhya, M.K., Prins, B., Zhong, G.Y., Schwaninger, H., et al., 2013. Genomics assisted ancestry deconvolution in grape. PLoS One 8, e80791, http://dx.doi.org/10.1371/journal.pone.0080791.

Sbaghi, M., Jeandet, P., Bessis, R., Leroux, P., 1996. Degradation of stilbene-type phytoalexins in relation to the pathogenicity of *Botrytis cinerea* to grapevines. Plant Pathol. 45, 139–144.

Schachtman, D.P., Shin, R., 2007. Nutrient sensing and signaling: NPKS. Annu. Rev. Plant Biol. 58, 47–69.

Schaefer, H., 1981. Der jahreszeitliche Verlauf des Stoffwechsels der löslichen und unlöslichen Proteine in den verholzten Reborganen. Wein-Wiss. 36, 3–20.

Schaller, K., Löhnertz, O., 1991. N-Stoffwechsel von Reben. 3. Mitteilung: Arginin- und Prolindynamik in vegetativen und generativen Teilen der Müller-Thurgau-Rebe im Verlaufe einer Vegetationsperiode. Wein-Wiss. 46, 151–160.

Schaller, K., Löhnertz, O., Oswald, D., Sprengart, B., 1985. Nitratanreicherung in Reben. 3. Mitteilung: Nitratdynamik in Rappen und Beeren während einer Vegetationsperiode in verschiedenen Rebsorten. Wein-Wiss. 40, 147–159.

Schaller, K., Löhnertz, O., Geiben, R., Breit, N., 1989. N-Stoffwechsel von Reben. 1. Mitteilung: N- und Arginindynamik im Holzkörper der Sorte Müller-Thurgau im Verlaufe einer Vegetationsperiode. Wein-Wiss. 44, 91–101.

Schaller, K., Löhnertz, O., Geiben, R., 1990. N-Stoffwechsel von Reben. 2. Mitteilung: N- und Aminosäuredynamik in vegetativen und generativen Teilen der Müller-Thurgau-Rebe im Verlaufe einer Vegetationsperiode. Wein-Wiss. 45, 160–166.

Scheenen, T.W.J., Vergeldt, F.J., Heemskerk, A.M., Van As, H., 2007. Intact plant magnetic resonance imaging to study dynamics in long-distance sap flow and flow-conducting surface area. Plant Physiol. 144, 1157–1165.

Scheffeldt, P., Hrazdina, G., 1978. Co-pigmentation of anthocyanins under physiological conditions. J. Food Sci. 43, 517–520.

Scheible, W.R., Lauerer, M., Schulze, E.D., Caboche, M., Stitt, M., 1997. Accumulation of nitrate in the shoot acts as a signal to regulate shoot-root allocation in tobacco. Plant J. 11, 671–691.

Scheible, W.R., Morcuende, R., Czechowski, T., Fritz, C., Osuna, D., Palacios-Rojas, N., et al., 2004. Genome-wide reprogramming of primary and secondary metabolism, protein synthesis, cellular growth processes, and the regulatory infrastructure of *Arabidopsis* in response to nitrogen. Plant Physiol. 136, 2483–2499.

Scheiner, J.J., Sacks, G.L., Pan, B., Ennahli, S., Tarlton, L., Wise, A., et al., 2010. Impact of severity and timing of basal leaf removal on 3-isobutyl-2-methoxypyrazine concentrations in red winegrapes. Am. J. Enol. Vitic. 61, 358–364.

Scheiner, J.J., Vanden Heuvel, J.E., Pan, B., Sacks, G.L., 2012. Modeling impacts of viticultural and environmental factors on 3-isobutyl-2-methoxypyrazine in Cabernet franc grapes. Am. J. Enol. Vitic. 63, 94–105.

Scherz, W., 1939. Sind selbstfertile hermaphrodite Weinreben obligat autogam? Züchter 11, 244–249.

Schlosser, J., Olsson, N., Weis, M., Reid, K., Peng, F., Lund, S., et al., 2008. Cellular expansion and gene expression in the developing grape (*Vitis vinifera* L.). Protoplasma 232, 255–265.

Schmidlin, L., Poutaraud, A., Claudel, P., Mestre, P., Prado, E., Santos-Rosa, M., et al., 2008. A stress-inducible resveratrol *O*-methyltransferase involved in the biosynthesis of pterostilbene in grapevine. Plant Physiol. 148, 1630–1639.

Schmidt, W., 2003. Iron solutions: acquisition strategies and signaling pathways in plants. Trends Plant Sci. 8, 188–193.

Schmitt, B., Stadler, R., Sauer, N., 2008. Immunolocalization of solanaceous SUT1 proteins in companion cells and xylem parenchyma: new perspectives for phloem loading and transport. Plant Physiol. 148, 187–199.

Schnabel, B.J., Wample, R.L., 1987. Dormancy and cold hardiness of *Vitis vinifera* L. cv. White Riesling as influenced by photoperiod and temperature. Am. J. Enol. Vitic. 38, 265–272.

Schnee, S., Viret, O., Gindro, K., 2008. Role of stilbenes in the resistance of grapevine to powdery mildew. Physiol. Mol. Plant Pathol. 72, 128–133.

Schneider, W., Staudt, G., 1978. Zur Abhängigkeit des Verrieselns von Umwelt und Genom bei *Vitis vinifera*. Vitis 17, 45–53.

Scholander, P.F., Love, W.E., Kanwisher, J.W., 1955. The rise of sap in tall grapevines. Plant Physiol. 30, 93–104.

Scholefield, P.B., Ward, R.C., 1975. Scanning electron microscopy of the developmental stages of the Sultana inflorescence. Vitis 14, 14–19.

Scholefield, P.B., May, P., Neales, T.F., 1977a. Harvest-pruning and trellising of 'Sultana' vines. 1. Effects on yield and vegetative growth. Sci. Hortic. 7, 115–122.

Scholefield, P.B., May, P., Neales, T.F., 1977b. Harvest-pruning and trellising of Sultana vines. 2. Effects on early spring development. Sci. Hortic. 7, 123–132.

Scholefield, P.B., Neales, T.F., May, P., 1978. Carbon balance of the Sultana vine (*Vitis vinifera* L.) and the effects of autumn defoliation by harvest-pruning. Aust. J. Plant Physiol. 5, 561–570.

Schönherr, J., 2000. Calcium chloride penetrates plant cuticles via aqueous pores. Planta 212, 112–118.

Schönherr, J., 2006. Characterization of aqueous pores in plant cuticles and permeation of ionic solutes. J. Exp. Bot. 57, 2471–2491.

Schönherr, J., Eckl, K., Gruler, H., 1979. Water permeability of plant cuticles: the effect of temperature on diffusion of water. Planta 147, 21–26.

Schopfer, P., 2006. Biomechanics of plant growth. Am. J. Bot. 93, 1415–1425.

Schopp, L.M., Lee, J., Osborne, J.P., Chescheir, S.C., Edwards, C.G., 2013. Metabolism of nonesterified and esterified hydroxycinnamic acids in red wines by *Brettanomyces bruxellensis*. J. Agric. Food Chem. 61, 11610–11617.

Schreiber, L., 2005. Polar paths of diffusion across plant cuticles: new evidence for an old hypothesis. Ann. Bot. 95, 1069–1073.

Schreiber, L., 2010. Transport barriers made of cutin, suberin and associated waxes. Trends Plant Sci. 15, 546–553.

Schreiber, L., Skrabs, M., Hartmann, K.D., Diamantopoulos, P., Simanova, E., Santrucek, J., 2001. Effect of humidity on culticular water permeability of isolated culticular membranes and leaf disks. Planta 214, 274–282.

Schreiner, R.P., 2005. Spatial and temporal variation of roots, arbuscular mycorrhizal fungi, and plant and soil nutrients in a mature Pinot noir (*Vitis vinifera* L.) vineyard in Oregon, USA. Plant Soil 276, 219−234.

Schreiner, R.P., Scagel, C.F., Baham, J., 2006. Nutrient uptake and distribution in a mature 'Pinot noir' vineyard. HortScience 41, 336−345.

Schreiner, R.P., Tarara, J.M., Smithyman, R.P., 2007. Deficit irrigation promotes arbuscular colonization of fine roots by mycorrhizal fungi in grapevines (*Vitis vinifera* L.) in an arid climate. Mycorrhiza 17, 551−562.

Schreiner, R.P., Lee, J., Skinkis, P.A., 2013. N, P, and K supply to Pinot noir grapevines: impact on vine nutrient status, growth, physiology, and yield. Am. J. Enol. Vitic. 64, 26−38.

Schreiner, R.P., Scagel, C.F., Lee, J., 2014. N, P, and K supply to Pinot noir grapevines: impact on berry phenolics and free amino acids. Am. J. Enol. Vitic. 65, 43−49.

Schröder, J., 1997. A family of plant-specific polyketide synthases: facts and predictions. Trends Plant Sci. 2, 373−378.

Schroth, M.N., McCain, A.H., Foott, J.H., Huisman, O.C., 1988. Reduction in yield and vigor of grapevine caused by crown gall disease. Plant Dis. 72, 241−246.

Schubert, A., Restagno, M., Novello, V., Peterlunger, E., 1995. Effects of shoot orientation on growth, net photosynthesis, and hydraulic conductivity of *Vitis vinifera* L. cv. Cortese. Am. J. Enol. Vitic. 46, 324−328.

Schubert, A., Lovisolo, C., Peterlunger, E., 1999. Shoot orientation affects vessel size, shoot hydraulic conductivity and shoot growth rate in *Vitis vinifera* L. Plant Cell Environ. 22, 197−204.

Schubert, R., Fischer, R., Hain, R., Schreier, P.H., Bahnweg, G., Ernst, D., et al., 1997. An ozone-responsive region of the grapevine resveratrol synthase promoter differs from the basal pathogen-responsive sequence. Plant Mol. Biol. 34, 417−426.

Schuetz, M., Smith, R., Ellis, B., 2013. Xylem tissue specification, patterning, and differentiation mechanisms. J. Exp. Bot. 64, 11−31.

Schultz, H.B., 1961. Microclimates on spring frost nights in Napa Valley vineyards. Am. J. Enol. Vitic. 12, 81−87.

Schultz, H.R., 1991. Seasonal and nocturnal changes in leaf age dependent dark respiration of grapevine sun and shade leaves. Modelling the temperature effect. Wein-Wiss. 46, 129−141.

Schultz, H.R., 2000. Climate change and viticulture: a European perspective on climatology, carbon dioxide and UV-B effects. Aust. J. Grape Wine Res. 6, 2−12.

Schultz, H.R., 2003. Differences in hydraulic architecture account for near-isohydric and anisohydric behaviour of two field-grown *Vitis vinifera* L. cultivars during drought. Plant Cell Environ. 26, 1393−1405.

Schultz, H.R., Matthews, M.A., 1988a. Resistance to water transport in shoots of *Vitis vinifera* L. Relation to growth at low water potential. Plant Physiol. 88, 718−724.

Schultz, H.R., Matthews, M.A., 1988b. Vegetative growth distribution during water deficits in *Vitis vinifera* L. Aust. J. Plant Physiol. 15, 641−656.

Schultz, H.R., Matthews, M.A., 1993a. Growth, osmotic adjustment, and cell-wall mechanics of expanding grape leaves during water deficits. Crop Sci. 33, 287−294.

Schultz, H.R., Matthews, M.A., 1993b. Xylem development and hydraulic conductance in sun and shade shoots of grapevine (*Vitis vinifera* L.): evidence that low light uncouples water transport capacity from leaf area. Planta 190, 393−406.

Schultz, H.R., Stoll, M., 2010. Some critical issues in environmental physiology of grapevines: future challenges and current limitations. Aust. J. Grape Wine Res. 16, 4−24.

Schultz, H.R., Kiefer, W., Gruppe, W., 1996. Photosynthetic duration, carboxylation efficiency and stomatal limitation of sun and shade leaves of different ages in field-grown grapevine (*Vitis vinifera* L.). Vitis 35, 169−176.

Schumann, F., 1974. Beziehungen zwischen Edelreis und Unterlagen—langjährige Ergebnisse aus Adaptationsversuchen. Wein-Wiss. 29, 216−229.

Schumann, F., Frieß, H., 1976. Beziehung zwischen Edelreis und Unterlagen. 2. Teil: Unterlagen für kalkreiche Böden. Wein-Wiss. 31, 94−120.

Schwab, W., Davidovich-Rikanati, R., Lewinsohn, E., 2008. Biosynthesis of plant-derived flavor compounds. Plant J. 54, 712 732.

Schwarz, G., Mendel, R.R., 2006. Molybdenum cofactor biosynthesis and molybdenum enzymes. Annu. Rev. Plant Biol. 57, 623−647.

Schütz, M., Kunkee, R.E., 1977. Formation of hydrogen sulfide from elemental sulfur during fermentation by wine yeast. Am. J. Enol. Vitic. 28, 137−144.

Scienza, A., Boselli, M., 1981. Fréquence et caractéristiques biométriques des stomates de certains porte-greffes de vigne. Vitis 20, 281−292.

Scienza, A., Düring, H., 1980. Stickstoffernährung und Wasserhaushalt bei Reben. Vitis 19, 301−307.

Scienza, A., Miravalle, R., Visai, C., Fregoni, M., 1978. Relationships between seed number, gibberellin and abscisic acid levels and ripening in Cabernet Sauvignon grape berries. Vitis 17, 361−368.

Scienza, A., Failla, O., Romano, F., 1986. Untersuchungen zur sortenspezifischen Mineralstoffaufnahme bei Reben. Vitis 25, 160−168.

Scrase-Field, S.A.M.G., Knight, M.R., 2003. Calcium: just a chemical switch? Curr. Opin. Plant Biol. 6, 500−506.

Secchi, F., Zwieniecki, M.A., 2010. Patterns of PIP gene expression in *Populus trichocarpa* during recovery from xylem embolism suggest a major role for the PIP1 aquaporin subfamily as moderators of refilling process. Plant Cell Environ. 33, 1285−1297.

Secchi, F., Zwieniecki, M.A., 2012. Analysis of xylem sap from functional (nonembolized) and nonfunctional (embolized) vessels of *Populus nigra*: chemistry of refilling. Plant Physiol. 160, 955−964.

Seddon, T.J., Downey, M.O., 2008. Comparison of analytical methods for the determination of condensed tannins in grape skin. Aust. J. Grape Wine Res. 14, 54−61.

Seemann, J.R., Sharkey, T.D., Wang, J., Osmond, C.B., 1987. Environmental effects on photosynthesis, nitrogen-use efficiency, and metabolite pools in leaves of sun and shade plants. Plant Physiol. 84, 796−802.

Sefc, K.M., Steinkellner, H., Glössl, J., Kampfer, S., Regner, F., 1998. Reconstruction of a grapevine pedigree by microsatellite analysis. Theor. Appl. Genet. 97, 227−231.

Sefc, K.M., Steinkellner, H., Lefort, F., Botta, R., da Câmara Machado, A., Borrego, J., et al., 2003. Evaluation of the genetic contribution of local wild vines to European grapevine cultivars. Am. J. Enol. Vitic. 54, 15−21.

Sefton, M.A., Francis, I.L., Williams, P.J., 1990. Volatile norisoprenoid compounds as constituents of oak woods used in wine and spirit maturation. J. Agric. Food Chem. 38, 2045−2049.

Sefton, M.A., Francis, I.L., Williams, P.J., 1993. The volatile composition of Chardonnay juices: a study by flavor precursor analysis. Am. J. Enol. Vitic. 44, 359−370.

Sefton, M.A., Skouroumounis, G.K., Elsey, G.M., Taylor, D.K., 2011. Occurrence, sensory impact, formation, and fate of damascenone in grapes, wines, and other foods and beverages. J. Agric. Food Chem. 59, 9717−9746.

Segurel, M.A., Razungles, A.J., Riou, C., Salles, M., Baumes, R.L., 2004. Contribution of dimethyl sulfide to the aroma of Syrah and Grenache noir wines and estimation of its potential in grapes of these varieties. J. Agric. Food Chem. 52, 7084−7093.

Sekse, L., 1995. Fruit cracking in sweet cherries (*Prunus avium* L.). Some physiological aspects—a mini review. Sci. Hortic. 63, 135−141.

Selitrennikoff, C.P., 2001. Antifungal proteins. Appl. Environ. Microbiol. 67, 2883−2894.

Seo, P.J., Kim, M.J., Park, J.Y., Kim, S.Y., Jeon, J., Lee, Y.H., et al., 2010. Cold activation of a plasma membrane-tethered NAC transcription factor induces a pathogen resistance response in *Arabidopsis*. Plant J. 61, 661−671.

Sepúlveda, G., Kliewer, W.M., 1986. Effect of high temperature on grapevines (*Vitis vinifera* L.). II. Distribution of soluble sugars. Am. J. Enol. Vitic. 37, 20–25.

Sepúlveda, G., Kliewer, W.M., Ryugo, K., 1986. Effect of high temperature on grapevines (*Vitis vinifera* L.). I. Translocation of ^{14}C-photosynthates. Am. J. Enol. Vitic. 37, 13–19.

Serrani, J.C., Fos, M., Atarés, A., García-Martínez, J.L., 2007. Effect of gibberellin and auxin on parthenocarpic fruit growth induction in the cv. Micro-Tom of tomato. J. Plant Growth Regul. 26, 211–221.

Serrani, J.C., Ruiz-Rivero, O., Fos, M., García-Martínez, J.L., 2008. Auxin-induced fruit-set in tomato is mediated in part by gibberellins. Plant J. 56, 922–934.

Serratosa, M.P., Lopez-Toledano, A., Merida, J., Medina, M., 2008. Changes in color and phenolic compounds during the raisining of grape cv. Pedro Ximenez. J. Agric. Food Chem. 56, 2810–2816.

Seymour, G.B., Østergaard, L., Chapman, N.H., Knapp, S., Martin, C., 2013. Fruit development and ripening. Annu. Rev. Plant Biol. 64, 219–241.

Shackel, K.A., Matthews, M.A., Morrison, J.C., 1987. Dynamic relation between expansion and cellular turgor in growing grape (*Vitis vinifera* L.) leaves. Plant Physiol. 84, 1166–1171.

Shani, U., Ben-Gal, A., 2005. Long-term response of grapevines to salinity: osmotic effects and ion toxicity. Am. J. Enol. Vitic. 56, 148–154.

Shani, U., Waisel, Y., Eshel, A., Xue, S., Ziv, G., 1993. Responses to salinity of grapevine plants with split root systems. New Phytol. 124, 695–701.

Sharkey, T.D., Wiberley, A.E., Donohue, A.R., 2008. Isoprene emission from plants: why and how. Ann. Bot. 101, 5–18.

Sharp, R.E., LeNoble, M.E., 2002. ABA, ethylene and the control of shoot and root growth under water stress. J. Exp. Bot. 53, 33–37.

Sharp, R.E., Poroyko, V., Hejlek, L.G., Spollen, W.G., Springer, G.K., Bohnert, H.J., et al., 2004. Root growth maintenance during water deficits: physiology to functional genomics. J. Exp. Bot. 55, 2343–2351.

Shatil-Cohen, A., Attia, Z., Moshelion, M., 2011. Bundle-sheath cell regulation of xylem-mesophyll water transport via aquaporins under drought stress: a target of xylem-borne ABA? Plant J. 67, 72–80.

Shaul, O., 2002. Magnesium transport and function in plants: the tip of the iceberg. Biometals 15, 309–323.

Shaulis, N., Amberg, H., Crowe, D., 1966. Response of Concord grapes to light exposure and Geneva double courtain training. Proc. Am. Soc. Hortic. Sci. 89, 268–280.

Shellie, K.C., 2011. Interactive effects of deficit irrigation and berry exposure aspect on Merlot and Cabernet Sauvignon in an arid climate. Am. J. Enol. Vitic. 62, 462–470.

Shellie, K.C., 2014. Water productivity, yield, and berry composition in sustained versus regulated deficit irrigation of Merlot grapevines. Am. J. Enol. Vitic. 65, 197–205.

Shellie, K.C., King, B.A., 2013. Kaolin particle film and water deficit influence red winegrape color under high solar radiation in an arid climate. Am. J. Enol. Vitic. 64, 214–222.

Shen, J., Yuan, L., Zhang, J., Li, H., Bai, Z., Chen, X., et al., 2011. Phosphorus dynamics: from soil to plant. Plant Physiol. 156, 997–1005.

Shepherd, T., Griffiths, D.W., 2006. The effects of stress on plant cuticular waxes. New Phytol. 171, 469–499.

Sheppard, A.E., Ayliffe, M.A., Blatch, L., Day, A., Delaney, S.K., Khairul-Fahmy, N., et al., 2008. Transfer of plastid DNA to the nucleus is elevated during male gametogenesis in tobacco. Plant Physiol. 148, 328–336.

Shibuya, N., Minami, E., 2001. Oligosaccharide signaling for defense responses in plants. Physiol. Mol. Plant Pathol. 59, 223–233.

Shimazaki, K.I., Doi, M., Assmann, S.M., Kinoshita, T., 2007. Light regulation of stomatal movement. Annu. Rev. Plant Biol. 58, 219–247.

Shiratake, K., Martinoia, E., 2007. Transporters in fruit vacuoles. Plant Biotechnol. 24, 127–133.

Shirke, P.A., Pathre, U.V., 2004. Influence of leaf-to-air vapour pressure deficit (VPD) on the biochemistry and physiology of photosynthesis in *Prosopis juliflora*. J. Exp. Bot. 55, 2111–2120.

Shure, K.B., Acree, T.E., 1994. Changes in the odor-active compounds in *Vitis labruscana* cv. Concord during growth and development. J. Agric. Food Chem. 42, 350–353.

Sidlowski, J.J., Phillips, W.S., Kuykendall, J.R., 1971. Phloem regeneration across girdles of grape vines. J. Am. Soc. Hortic. Sci. 96, 97–102.

Silacci, M.W., Morrison, J.C., 1990. Changes in pectin content of Cabernet Sauvignon grape berries during maturation. Am. J. Enol. Vitic. 41, 111–115.

Simpson, R.F., Miller, G.C., 1983. Aroma composition of aged Riesling wine. Vitis 22, 51–63.

Simpson, R.F., Miller, G.C., 1984. Aroma composition of Chardonnay wine. Vitis 23, 143–158.

Singleton, V.L., 1992. Tannins and the qualities of wines. In: Hemingway, R.W., Laks, P.E. (Eds.), Plant Polyphenols. Plenum Press, New York, NY, pp. 859–880.

Singleton, V.L., Trousdale, E., 1983. White wine phenolics: varietal and processing differences as shown by HPLC. Am. J. Enol. Vitic. 34, 27–34.

Singleton, V.L., Ough, C.S., Nelson, K.E., 1966. Density separations of wine grape berries and ripeness distribution. Am. J. Enol. Vitic. 17, 95–105.

Singleton, V.L., Trousdale, E., Zaya, J., 1985. One reason sun-dried raisins brown so much. Am. J. Enol. Vitic. 36, 111–113.

Singleton, V.L., Zaya, J., Trousdale, E., 1986. Compositional changes in ripening grapes: caftaric and coutaric acids. Vitis 25, 107–117.

Sinilal, B., Ovadia, R., Nissim-Levi, A., Perl, A., Carmeli-Weissberg, M., Oren-Shamir, M., 2011. Increased accumulation and decreased catabolism of anthocyanins in red grape cell suspension culture following magnesium treatment. Planta 234, 61–71.

Skene, K.G.M., Antcliff, A.J., 1972. A comparative study of cytokinin levels in bleeding sap of *Vitis vinifera* (L.) and the two grapevine rootstocks, Salt Creek and 1613. J. Exp. Bot. 23, 283–293.

Skene, K.G.M., Kerridge, G.H., 1967. Effect of root temperature on cytokinin activity in root exudate of *Vitis vinifera* L. Plant Physiol. 42, 1131–1139.

Skinkis, P.A., Bordelon, B.P., Wood, K.V., 2008. Comparison of monoterpene constituents in Traminette, Gewürztraminer, and Riesling winegrapes. Am. J. Enol. Vitic. 59, 440–445.

Skinner, P.W., Matthews, M.A., 1990. A novel interaction of magnesium translocation with the supply of phosphorus to roots of grapevine (*Vitis vinifera* L.). Plant Cell Environ. 13, 821–826.

Skinner, P.W., Cook, J.A., Matthews, M.A., 1988. Responses of grapevine cvs. Chenin blanc and Chardonnay to phosphorus fertilizer applications under phosphorus-limited conditions. Vitis 27, 95–109.

Smart, D.R., Kocsis, L., Walker, M.A., Stockert, C., 2003. Dormant buds and adventitious root formation by *Vitis* and other woody plants. J. Plant Growth Regul. 21, 296–314.

Smart, D.R., Carlisle, E., Goebel, M., Núñez, B.A., 2005. Transverse hydraulic redistribution by a grapevine. Plant Cell Environ. 28, 157–166.

Smart, D.R., Schwass, E., Lakso, A., Morano, L., 2006. Grapevine rooting patterns: a comprehensive analysis and a review. Am. J. Enol. Vitic. 57, 89–104.

Smart, R.E., 1974. Aspects of water relations of the grapevine (*Vitis vinifera*). Am. J. Enol. Vitic. 25, 84–91.

Smart, R.E., 1985. Principles of grapevine canopy microclimate manipulation with implications for yield and quality. A review. Am. J. Enol. Vitic. 36, 230–239.

Smart, R.E., 1988. Shoot spacing and canopy light microclimate. Am. J. Enol. Vitic. 39, 325–333.

Smart, R.E., Coombe, B.G., 1983. Water relations in grapevine. In: Kozlowski, T.T. (Ed.), Water Deficits and Plant Growth. Vol. VII. Additional Woody Crop Plants. Academic Press, New York, NY, pp. 137–196.

Smart, R.E., Sinclair, T.R., 1976. Solar heating of grape berries and other spherical fruits. Agric. Meteorol. 17, 241–259.

Smart, R.E., Shaulis, N.J., Lemon, E.R., 1982. The effect of Concord vineyard microclimate on yield. I. The effects of pruning, training, and shoot positioning on radiation microclimate. Am. J. Enol. Vitic. 33, 99–108.

Smart, R.E., Robinson, J.B., Due, G.R., Brien, C.J., 1985a. Canopy microclimate modification for the cultivar Shiraz. I. Definition of canopy microclimate. Vitis 24, 17–31.

Smart, R.E., Robinson, J.B., Due, G.R., Brien, C.J., 1985b. Canopy microclimate modification for the cultivar Shiraz. II. Effects on must and wine composition. Vitis 24, 119–128.

Smart, R.E., Smith, S.M., Winchester, R.V., 1988. Light quality and quantity effects on fruit ripening for Cabernet Sauvignon. Am. J. Enol. Vitic. 39, 250–258.

Smart, R.E., Dick, J.K., Gravett, I.M., Fisher, B.M., 1990. Canopy management to improve grape yield and wine quality—principles and practices. S. Afr. J. Enol. Vitic. 11, 3–17.

Smith, A.M., 2012. Starch in the *Arabidopsis* plant. Starch-Stärke 64, 421–434.

Smith, A.M., Zeeman, S.C., Smith, S.M., 2005. Starch degradation. Annu. Rev. Plant Biol. 56, 73–98.

Smith, C.J., 1996. Accumulation of phytoalexins: defence mechanism and stimulus response system. New Phytol. 132, 1–45.

Smith, F.A., Raven, J.A., 1979. Intracellular pH and its regulation. Annu. Rev. Plant Physiol. 30, 289–311.

Smith, H., 2000. Phytochromes and light signal perception by plants—an emerging synthesis. Nature 407, 585–591.

Smith, H., Whitelam, G.C., 1997. The shade avoidance syndrome: multiple responses mediated by multiple phytochromes. Plant Cell Environ. 20, 840–844.

Smith, S.E., Smith, F.A., 2011. Roles of arbuscular mycorrhizas in plant nutrition and growth: new paradigms from cellular to ecosystem scales. Annu. Rev. Plant Biol. 62, 227–250.

Smith, S.E., Dickson, S., Smith, F.A., 2001. Nutrient transfer in arbuscular mycorrhizas: how are fungal and plant processes integrated? Aust. J. Plant Physiol. 28, 683–694.

Smithyman, R.P., Howell, G.S., Miller, D.P., 1998. The use of competition for carbohydrates among vegetative and reproductive sinks to reduce fruit set and *Botrytis* bunch rot in Seyval blanc grapevines. Am. J. Enol. Vitic. 49, 163–170.

Smithyman, R.P., Wample, R.L., Lang, N.S., 2001. Water deficit and crop level influences on photosynthetic strain and blackleaf symptom development in Concord grapevines. Am. J. Enol. Vitic. 52, 364–375.

Snedden, W.A., Fromm, H., 2001. Calmodulin as a versatile calcium signal transducer in plants. New Phytol. 151, 35–66.

Snyder, J.C., 1933. Flower bud formation in the Concord grape. Bot. Gaz. 94, 771–779.

Soar, C.J., Loveys, B.R., 2007. The effect of changing patterns in soil-moisture availability on grapevine root distribution, and viticultural implications for converting full-cover irrigation into a point-source irrigation system. Aust. J. Grape Wine Res. 13, 2–13.

Soar, C.J., Speirs, J., Maffei, S.M., Loveys, B.R., 2004. Gradients in stomatal conductance, xylem sap ABA and bulk leaf ABA along canes of *Vitis vinifera* cv. Shiraz: molecular and physiological studies investigating their source. Funct. Plant Biol. 31, 659–669.

Soar, C.J., Dry, P.R., Loveys, B.R., 2006a. Scion photosynthesis and leaf gas exchange in *Vitis vinifera* L. cv. Shiraz: mediation of rootstock effects via xylem sap ABA. Aust. J. Grape Wine Res. 12, 82–96.

Soar, C.J., Speirs, J., Maffei, S.M., Penrose, A.B., McCarthy, M.G., Loveys, B.R., 2006b. Grape vine varieties Shiraz and Grenache differ in their stomatal response to VPD: apparent links with ABA physiology and gene expression in leaf tissue. Aust. J. Grape Wine Res. 12, 2–12.

Soares, S., Mateus, N., de Freitas, V., 2012. Carbohydrates inhibit salivary proteins precipitation by condensed tannins. J. Agric. Food Chem. 60, 3966–3972.

Soltis, P.S., Soltis, D.E., 2009. The role of hybridization in plant speciation. Annu. Rev. Plant Biol. 60, 561–588.

Somers, T.C., 1968. Pigment profiles of grapes and of wines. Vitis 7, 303–320.
Somers, T.C., 1971. The polymeric nature of wine pigments. Phytochemistry 10, 2175–2186.
Sommer, K.J., Clingeleffer, P.R., Shulman, Y., 1995. Comparative study of vine morphology, growth, and canopy development in cane-pruned and minimal-pruned Sultana. Aust. J. Exp. Agric. 35, 265–273.
Sommer, K.J., Islam, M.T., Clingeleffer, P.R., 2000. Light and temperature effects on shoot fruitfulness in *Vitis vinifera* L. cv. Sultana: influence of trellis type and grafting. Aust. J. Grape Wine Res. 6, 99–108.
Somssich, I.E., Hahlbrock, K., 1998. Pathogen defence in plants—a paradigm of biological complexity. Trends Plant Sci. 3, 86–90.
Sondergaard, T.E., Schulz, A., Palmgren, M.G., 2004. Energization of transport processes in plants. Roles of the plasma membrane H^+-ATPase. Plant Physiol. 136, 2475–2482.
Souquet, J.M., Cheynier, V., Brossaud, F., Moutounet, M., 1996. Polymeric proanthocyanidins from grape skins. Phytochemistry 43, 509–512.
Souquet, J.M., Labarbe, B., Le Guerneve, C., Cheynier, V., Moutounet, M., 2000. Phenolic composition of grape stems. J. Agric. Food Chem. 48, 1076–1080.
Spalding, E.P., Folta, K.M., 2005. Illuminating topics in plant photobiology. Plant Cell Environ. 28, 39–53.
Spayd, S.E., Nagel, C.W., Hayrynen, L.D., Ahmedullah, M., 1987. Effect of freezing fruit on the composition of musts and wines. Am. J. Enol. Vitic. 38, 243–245.
Spayd, S.E., Wample, R.L., Stevens, R.G., Evans, R.G., Kawakami, A.K., 1993. Nitrogen fertilization of White Riesling in Washington: effects on petiole nutrient concentration, yield, yield components, and vegetative growth. Am. J. Enol. Vitic. 44, 378–386.
Spayd, S.E., Wample, R.L., Evans, R.G., Stevens, R.G., Seymour, B.J., Nagel, C.W., 1994. Nitrogen fertilization of White Riesling grapes in Washington. Must and wine composition. Am. J. Enol. Vitic. 45, 34–42.
Spayd, S.E., Nagel, C.W., Edwards, C.G., 1995. Yeast growth in Riesling juice as affected by vineyard nitrogen fertilization. Am. J. Enol. Vitic. 46, 49–55.
Spayd, S.E., Tarara, J.M., Mee, D.L., Ferguson, J.C., 2002. Separation of sunlight and temperature effects on the composition of *Vitis vinifera* cv. Merlot berries. Am. J. Enol. Vitic. 53, 171–182.
Speirs, J., Binney, A., Collins, M., Edwards, E., Loveys, B., 2013. Expression of ABA synthesis and metabolism genes under different irrigation strategies and atmospheric VPDs is associated with stomatal conductance in grapevine (*Vitis vinifera* L. cv Cabernet Sauvignon). J. Exp. Bot. 64, 1907–1916.
Sperry, J.S., 2004. Coordinating stomatal and xylem functioning—an evolutionary perspective. New Phytol. 162, 568–570.
Sperry, J.S., Holbrook, N.M., Zimmermann, M.H., Tyree, M.T., 1987. Spring filling of xylem vessels in wild grapevine. Plant Physiol. 83, 414–417.
Sperry, J.S., Adler, F.R., Campbell, G.S., Comstock, J.P., 1998. Limitation of plant water use by rhizosphere and xylem conductance: results from a model. Plant Cell Environ. 21, 347–359.
Sperry, J.S., Hacke, U.G., Oren, R., Comstock, J.P., 2002. Water deficits and hydraulic limits to leaf water supply. Plant Cell Environ. 25, 251–263.
Sperry, J.S., Stiller, V., Hacke, U.G., 2003. Xylem hydraulics and the soil-plant-atmosphere continuum: opportunities and unresolved issues. Agron. J. 95, 1362–1370.
Spreitzer, R.J., Salvucci, M.E., 2002. RUBISCO: structure, regulatory interactions, and possibilities for a better enzyme. Annu. Rev. Plant Biol. 53, 449–475.
Srinivasan, C., Mullins, M.G., 1978. Control of flowering in the grapevine (*Vitis vinifera* L.). Plant Physiol. 61, 127–130.
Srinivasan, C., Mullins, M.G., 1980. Effects of temperature and growth regulators on formation of anlagen, tendrils and inflorescences in *Vitis vinifera* L. Ann. Bot. 45, 439–446.
Srinivasan, C., Mullins, M.G., 1981. Physiology of flowering in the grapevine—a review. Am. J. Enol. Vitic. 32, 47–63.

Stafford, H.A., 1988. Proanthocyanidins and the lignin connection. Phytochemistry 27, 1−6.
Stahl, E., 1900. Der Sinn der Mycorrhizenbildung. Jahrb. wiss. Bot. 34, 539−668.
Starkenmann, C., Le Calvé, B., Niclass, Y., Cayeux, I., Beccucci, S., Troccaz, M., 2008. Olfactory perception of cysteine−S-conjugates from fruits and vegetables. J. Agric. Food Chem. 56, 9575−9580.
Staudt, G., 1982. Pollenkeimung und Pollenschlauchwachstum in vivo bei Vitis und die Abhängigkeit von der Temperatur. Vitis 21, 205−216.
Staudt, G., 1999. Opening of flowers and time of anthesis in grapevines, Vitis vinifera L. Vitis 38, 15−20.
Staudt, G., Kassemeyer, H.H., 1984. Entstehen kleine Beeren bei Vitis vinifera durch Parthenocarpie? Vitis 23, 205−213.
Staudt, G., Kassrawi, M., 1973. Untersuchungen über das Rieseln di- und tetraploider Reben. Vitis 12, 1−15.
Staudt, G., Schneider, W., Leidel, J., 1986. Phases of berry growth in Vitis vinifera. Ann. Bot. 58, 789−800.
Steenwerth, K.L., McElrone, A.J., Calderon-Orellana, A., Hanifin, R.C., Storm, C., Collatz, W., et al., 2013. Cover crops and tillage in a mature Merlot vineyard show few effects on grapevines. Am. J. Enol. Vitic. 64, 515−521.
Stefanini, I., Dapporto, L., Legras, J.L., Calabretta, A., Di Paola, M., De Filippo, C., et al., 2012. Role of social wasps in Saccharomyces cerevisiae ecology and evolution. Proc. Natl. Acad. Sci. USA 109, 13398−13403.
Steffan, H., Rapp, A., 1979. Ein Beitrag zum Nachweis unterschiedlicher Malatpools in Beeren der Rebe. Vitis 18, 100−105.
Stegemann, S., Bock, R., 2009. Exchange of genetic material between cells in plant tissue grafts. Science 324, 649−651.
Stein, U., Blaich, R., 1985. Untersuchungen über Stilbenproduktion und Botrytisanfälligkeit bei Vitis-Arten. Vitis 24, 75−87.
Steponkus, P.L., 1984. Role of the plasma membrane in freezing injury and cold acclimation. Annu. Rev. Plant Physiol. 35, 543−584.
Steppe, K., Lemeur, R., 2004. An experimental system for analysis of the dynamic sap-flow characteristics in young trees: results of a beech tree. Funct. Plant Biol. 31, 83−92.
Stergios, B.G., Howell, G.S., 1977. Effect of site on cold acclimation and deacclimation of Concord grapevines. Am. J. Enol. Vitic. 28, 43−48.
Steudle, E., 2000. Water uptake by roots: effects of water deficit. J. Exp. Bot. 51, 1531−1542.
Steudle, E., 2001. The cohesion-tension mechanism and the acquisition of water by plant roots. Annu. Rev. Plant Physiol. Plant Mol. Biol. 52, 847−875.
Steudle, E., Peterson, C.A., 1998. How does water get through roots? J. Exp. Bot. 49, 775−788.
Stevens, R.M., Douglas, T., 1994. Distribution of grapevine roots and salt under drip and full-ground cover microjet irrigation systems. Irrig. Sci. 15, 147−152.
Stevens, R.M., Prior, L.D., 1994. The effect of transient waterlogging on the growth, leaf gas exchange, and mineral composition of potted Sultana grapevines. Am. J. Enol. Vitic. 45, 285−290.
Stevens, R.M., Walker, R.R., 2002. Response of grapevines to irrigation-induced saline−sodic soil conditions. Aust. J. Exp. Agric. 42, 323−331.
Stevens, R.M., Harvey, G., Aspinall, D., 1995. Grapevine growth of shoots and fruit linearly correlate with water stress indices based on root-weighted soil matric potential. Aust. J. Grape Wine Res. 1, 58−66.
Stevens, R.M., Harvey, G., Johns, R.E., 1999. Waterlogging reduces shoot growth and bud fruitfulness in pot-grown grapevines with a split-root system. Aust. J. Grape Wine Res. 5, 99−103.
Stevens, R.M., Pech, J.M., Gibberd, M.R., Jones, J.A., Taylor, J., Nicholas, P.R., 2008. Effect of reduced irrigation on growth, yield, ripening rates and water relations of Chardonnay vines grafted to five rootstocks. Aust. J. Grape Wine Res. 14, 177−190.

Stevens, R.M., Pech, J.M., Gibberd, M.R., Walker, R.R., Nicholas, P.R., 2010. Reduced irrigation and rootstock effects on vegetative growth, yield and its components, and leaf physiological responses of Shiraz. Aust. J. Grape Wine Res. 16, 413–425.

Stevenson, J.F., Matthews, M.A., Greve, L.C., Labavitch, J.M., Rost, T.L., 2004. Grapevine susceptibility to Pierce's disease II: progression of anatomical symptoms. Am. J. Enol. Vitic. 55, 238–245.

Stevenson, J.F., Matthews, M.A., Rost, T.L., 2005. The developmental anatomy of Pierce's disease symptoms in grapevines: green islands and matchsticks. Plant Dis. 89, 543–548.

Stiles, K.A., Van Volkenburgh, E., 2004. Role of K^+ in leaf growth: K^+ uptake is required for light-stimulated H^+ efflux but not solute accumulation. Plant Cell Environ. 27, 315–325.

Stines, A.P., Grubb, J., Gockowiak, H., Henschke, P.A., Høj, P.B., van Heeswijck, R., 2000. Proline and arginine accumulation in developing berries of *Vitis vinifera* L. in Australian vineyards: influence of vine cultivar, berry maturity and tissue type. Aust. J. Grape Wine Res. 6, 150–158.

Stitt, M., 1999. Nitrate regulation of metabolism and growth. Curr. Opin. Plant Biol. 2, 178–186.

Stitt, M., Müller, C., Matt, P., Gibon, Y., Carillo, P., Morcuende, R., et al., 2002. Steps towards an integrated view of nitrogen metabolism. J. Exp. Bot. 53, 959–970.

Stitt, M., Gibon, Y., Lunn, J.E., Piques, M., 2007. Multilevel genomics analysis of carbon signalling during low carbon availability: coordinating the supply and utilisation of carbon in a fluctuating environment. Funct. Plant Biol. 34, 526–549.

Stoev, K.D., Dobreva, S.I., Wosteninez, G., 1966. Über die Synthese von Aminosäuren im Wurzelsystem der Rebe. Vitis 5, 265–287.

Stoll, M., Loveys, B., Dry, P., 2000. Hormonal changes induced by partial rootzone drying of irrigated grapevine. J. Exp. Bot. 51, 1627–1634.

Storey, R., 1987. Potassium localization in the grape berry pericarp by energy-dispersive X-ray microanalysis. Am. J. Enol. Vitic. 38, 301–309.

Storey, R., Schachtman, D.P., Thomas, M.R., 2003a. Root structure and cellular chloride, sodium and potassium distribution in salinized grapevines. Plant Cell Environ. 26, 789–800.

Storey, R., Wyn Jones, R.G., Schachtman, D.P., Treeby, M.T., 2003b. Calcium-accumulating cells in the meristematic region of grapevine root apices. Funct. Plant Biol. 30, 719–727.

Stover, E.W., Swartz, H.J., Burr, T.J., 1997. Crown gall formation in a diverse collection of *Vitis* genotypes inoculated with *Agrobacterium vitis*. Am. J. Enol. Vitic. 48, 26–32.

Strader, L.C., Chen, G.L., Bartel, B., 2010. Ethylene directs auxin to control root cell expansion. Plant J. 64, 874–884.

Strauss, C.R., Wilson, B., Anderson, R., Williams, P.J., 1987. Development of precursors of C_{13} norisoprenoid flavorants in Riesling grapes. Am. J. Enol. Vitic. 38, 23–27.

Strefeler, M.S., Weeden, N.F., Reisch, B.I., 1992. Inheritance of chloroplast DNA in two full-sib *Vitis* populations. Vitis 31, 183–187.

Striegler, R.K., Howell, G.S., Flore, J.A., 1993. Influence of rootstock on the response of Seyval grapevines to flooding stress. Am. J. Enol. Vitic. 44, 313–319.

Stummer, B.E., Francis, I.L., Zanker, T., Lattey, K.A., Scott, E.S., 2005. Effects of powdery mildew on the sensory properties and composition of Chardonnay juice and wine when grape sugar ripeness is standardized. Aust. J. Grape Wine Res. 11, 66–76.

Suárez, N., 2010. Leaf lifetime photosynthetic rate and leaf demography in whole plants of *Ipomoea pescaprae* growing with a low supply of calcium, a 'non-mobile' nutrient. J. Exp. Bot. 61, 843–855.

Sun, B.S., Pinto, T., Leandro, M.C., Ricardo-da-Silva, J.M., Spranger, M.I., 1999. Transfer of catechins and proanthocyanidins from solid parts of the grape cluster into wine. Am. J. Enol. Vitic. 50, 179–184.

Sun, Q., Rost, T.L., Matthews, M.A., 2006. Pruning-induced tylose development in stems of current-year shoots of *Vitis vinifera* (Vitaceae). Am. J. Bot. 93, 1567–1576.

Sun, Q., Rost, T.L., Reid, M.S., Matthews, M.A., 2007. Ethylene and not embolism is required for wound-induced tylose development in stems of grapevines. Plant Physiol. 145, 1629–1636.

Sun, Q., Rost, T.L., Matthews, M.A., 2008. Wound-induced vascular occlusions in *Vitis vinifera* (Vitaceae): tyloses in summer and gels in winter. Am. J. Bot. 95, 1498–1505.

Sun, Q., Greve, L.C., Labavitch, J.M., 2011. Polysaccharide compositions of intervessel pit membranes contribute to Pierce's disease resistance of grapevines. Plant Physiol. 155, 1976–1987.

Sun, Q., Gates, M.J., Lavin, E.H., Acree, T.E., Sacks, G.L., 2011. Comparison of odor-active compounds in grapes and wines from *Vitis vinifera* and non-foxy American grape species. J. Agric. Food Chem. 59, 10657–10664.

Sun, Q., Sun, Y., Walker, M.A., Labavitch, J.M., 2013. Vascular occlusions in grapevines with Pierce's disease make disease symptom development worse. Plant Physiol. 161, 1529–1541.

Swanepoel, J.J., Archer, E., 1988. The ontogeny and development of *Vitis vinifera* L. cv. Chenin blanc inflorescence in relation to phenological stages. Vitis 27, 133–141.

Swanson, C.A., El-Shishiny, E.D.H., 1958. Translocation of sugars in the Concord grape. Plant Physiol. 33, 33–37.

Sweet, W.J., Morrison, J.C., Labavitch, J.M., Matthews, M.A., 1990. Altered synthesis and composition of cell wall of grape (*Vitis vinifera* L.) leaves during expansion and growth-inhibiting water deficits. Plant Cell Physiol. 31, 407–414.

Sweetlove, L.J., Beard, K.F.M., Nunes-Nesi, A., Fernie, A.R., Ratcliffe, R.G., 2010. Not just a circle: flux modes in the plant TCA cycle. Trends Plant Sci. 15, 462–470.

Sweetman, C., Deluc, L.G., Cramer, G.R., Ford, C.M., Soole, K.L., 2009. Regulation of malate metabolism in grape berry and other developing fruits. Phytochemistry 70, 1329–1344.

Swift, J.G., Buttrose, M.S., Possingham, J.V., 1973. Stomata and starch in grape berries. Vitis 12, 38–45.

Symons, G.M., Davies, C., Shavrukov, Y., Dry, I.B., Reid, J.B., Thomas, M.R., 2006. Grapes on steroids. Brassinosteroids are involved in grape berry ripening. Plant Physiol. 140, 150–158.

Szegedi, E., Korbuly, J., Koleda, I., 1984. Crown gall resistance in East-Asian *Vitis* species and in their *V. vinifera* hybrids. Vitis 23, 21–26.

Szyjewicz, E., Rosner, N., Kliewer, W.M., 1984. Ethephon ((2-chloroethyl) phosphonic acid, Ethrel, CEPA) in viticulture—a review. Am. J. Enol. Vitic. 35, 117–123.

Tabacchi, R., 1994. Secondary phytotoxic metabolites from pathogenic fungi: structure, synthesis and activity. Pure Appl. Chem. 66, 2299–2302.

Taiz, L., 1984. Plant cell expansion: regulation of cell wall mechanical properties. Annu. Rev. Plant Physiol. 35, 585–657.

Taiz, L., Zeiger, E., 2006. Plant Physiology, fourth ed. Sinauer, Sunderland, MA.

Takabayashi, J., Dicke, M., 1996. Plant-carnivore mutualism through herbivore-induced carnivore attractants. Trends Plant Sci. 1, 109–113.

Takahashi, H., Kopriva, S., Giordano, M., Saito, K., Hell, R., 2011. Sulfur assimilation in photosynthetic organisms: molecular functions and regulations of transporters and assimilatory enzymes. Annu. Rev. Plant Biol. 62, 157–184.

Takahashi, K., Hayashi, K., Kinoshita, T., 2012. Auxin activates the plasma membrane H^+-ATPase by phosphorylation during hypocotyl elongation in *Arabidopsis*. Plant Physiol. 159, 632–641.

Takahashi, N., Yamazaki, Y., Kobayashi, A., Higashitani, A., Takahaski, H., 2003. Hydrotropism interacts with gravitropism by degrading amyloplasts in seedling roots of *Arabidopsis* and radish. Plant Physiol. 132, 805–810.

Takahashi, S., Badger, M.R., 2011. Photoprotection in plants: a new light on photosystem II damage. Trends Plant Sci. 16, 53–60.

Takahashi, S., Murata, N., 2008. How do environmental stresses accelerate photoinhibition? Trends Plant Sci. 13, 178–182.

Takahashi, T., Kakehi, J.I., 2010. Polyamines: ubiquitous polycations with unique roles in growth and stress responses. Ann. Bot. 105, 1–6.

Takano, J., Miwa, K., Fujiwara, T., 2008. Boron transport mechanisms: collaboration of channels and transporters. Trends Plant Sci. 13, 451–457.

Takei, K., Takahashi, T., Sugiyama, T., Yamaya, T., Sakakibara, H., 2002. Multiple routes communicating nitrogen availability from roots to shoots: a signal transduction pathway mediated by cytokinin. J. Exp. Bot. 53, 971–977.

Takimoto, K., Saito, K., Kasai, Z., 1976. Diurnal change of tartrate dissimilation during the ripening of grapes. Phytochemistry 15, 927–930.

Talbott, L.D., Zeiger, E., 1996. Central roles for potassium and sucrose in guard-cell osmoregulation. Plant Physiol. 111, 1051–1057.

Tallman, G., 2004. Are diurnal patterns of stomatal movement the result of alternating metabolism of endogenous guard cell ABA and accumulation of ABA delivered to the apoplast around guard cells by transpiration? J. Exp. Bot. 55, 1963–1976.

Tanaka, M., Takei, K., Kojima, M., Sakakibara, H., Mori, H., 2006. Auxin controls local cytokinin biosynthesis in the nodal stem in apical dominance. Plant J. 45, 1028–1036.

Tanaka, Y., Sasaki, N., Ohmiya, A., 2008. Biosynthesis of plant pigments: anthocyanins, betalains and carotenoids. Plant J. 54, 733–749.

Tanner, W., Beevers, H., 2001. Transpiration, a prerequisite for long-distance transport of minerals in plants? Proc. Natl. Acad. Sci. USA 98, 9443–9447.

Tarara, J.M., Ferguson, J.C., 2006. Two algorithms for variable power control of heat-balance sap flow gauges under high flow rates. Agron. J. 98, 830–838.

Tarara, J.M., Hellman, E.W., 1991. 'Norton' and 'Cynthiana'—premium native wine grapes. Fruit Var. J. 45, 66–69.

Tarara, J.M., Ferguson, J.C., Hoheisel, G.A., Perez Peña, J.E., 2005. Asymmetrical canopy architecture due to prevailing wind direction and row orientation creates an imbalance in irradiance at the fruiting zone of grapevines. Agric. For. Meteorol. 135, 144–155.

Tarara, J.M., Lee, J., Spayd, S.E., Scagel, C.F., 2008. Berry temperature and solar radiation alter acylation, proportion, and concentration of anthocyanin in Merlot grapes. Am. J. Enol. Vitic. 59, 235–247.

Tarara, J.M., Perez Peña, J.E., Keller, M., Schreiner, R.P., Smithyman, R.P., 2011. Net carbon exchange in grapevine canopies responds rapidly to timing and extent of regulated deficit irrigation. Funct. Plant Biol. 38, 386–400.

Tarbah, F., Goodman, R.N., 1987. Systemic spread of *Agrobacterium tumefaciens* biovar 3 in the vascular system of grapes. Phytopathology 77, 915–920.

Tardaguila, J., Martinez de Toda, F., Poni, S., Diago, M.P., 2010. Impact of early leaf removal on yield and fruit and wine composition of *Vitis vinifera* L. Graciano and Carignan. Am. J. Enol. Vitic. 61, 372–381.

Tardieu, F., Simonneau, T., 1998. Variability among species of stomatal control under fluctuating soil water status and evaporative demand: modelling isohydric and anisohydric behaviours. J. Exp. Bot. 49, 419–432.

Tattersall, D.B., van Heeswijck, R., Høj, P.B., 1997. Identification and characterization of a fruit-specific, thaumatin-like protein that accumulates at very high levels in conjunction with the onset of sugar accumulation and berry softening in grapes. Plant Physiol. 114, 759–769.

Tattersall, D.B., Pocock, K.F., Hayasaka, Y., Adams, K., van Heeswijck, R., Waters, E.J., et al., 2001. Pathogenesis-related proteins—their accumulation in grapes during berry growth and their involvement in white wine heat instability. In: Roubelakis-Angelakis, K.A. (Ed.), Molecular Biology and Biotechnology of the Grapevine. Kluwer, Dordrecht, The Netherlands, pp. 183–196.

Taylor, J.E., Whitelaw, C.A., 2001. Signals in abscission. New Phytol. 151, 323–339.

Temple, S.J., Vance, C.P., Gantt, J.S., 1998. Glutamate synthase and nitrogen assimilation. Trends Plant Sci. 3, 51–56.

Terashima, I., Hanba, Y.T., Tholen, D., Niinemets, Ü., 2011. Leaf functional anatomy in relation to photosynthesis. Plant Physiol. 155, 108–116.

Terral, J.F., Tabard, E., Bouby, L., Ivorra, S., Pastor, T., Figueiral, I., et al., 2010. Evolution and history of grapevine (*Vitis vinifera*) under domestication: new morphometric perspectives to understand seed domestication syndrome and reveal origins of ancient European cultivars. Ann. Bot. 105, 443–455.

Terrier, N., Deguilloux, C., Sauvage, F.X., Martinoia, E., Romieu, C., 1998. Proton pumps and anion transport in *Vitis vinifera*: the inorganic pyrophosphatase plays a predominant role in the energization of the tonoplast. Plant Physiol. Biochem. 36, 367–377.

Terrier, N., Glissant, D., Grimplet, J., Barrieu, F., Abbal, P., Couture, C., et al., 2005. Isogene specific oligo arrays reveal multifaceted changes in gene expression during grape berry (*Vitis vinifera* L.) development. Planta 222, 832–847.

Terrier, N., Ollé, D., Verriés, C., Cheynier, V., 2009. Biochemical and molecular aspects of flavan-3-ol synthesis during berry development. In: Roubelakis-Angelakis, K.A. (Ed.), Grapevine Molecular Physiology and Biotechnology, second ed. Springer, Dordrecht, The Netherlands, pp. 365–388.

Tesic, D., Keller, M., Hutton, R., 2007. Influence of vineyard floor management practices on grapevine growth, yield, and fruit composition. Am. J. Enol. Vitic. 58, 1–11.

Tesniere, C., Davies, C., Sreekantan, L., Bogs, J., Thomas, M., Torregrosa, L., 2006. Analysis of the transcript levels of *VvAdh1*, *VvAdh2* and *VvGrip4*, three genes highly expressed during *Vitis vinifera* L. berry development. Vitis 45, 75–79.

Tester, M., Leigh, R.A., 2001. Partitioning of nutrient transport processes in roots. J. Exp. Bot. 52, 445–457.

Tetlow, I.J., Morell, M.K., Emes, M.J., 2004. Recent developments in understanding the regulation of starch metabolism in higher plants. J. Exp. Bot. 55, 2131–2145.

Tezara, W., Mitchell, V.J., Driscoll, S.D., Lawlor, D.W., 1999. Water stress inhibits plant photosynthesis by decreasing coupling factor and ATP. Nature 401, 914–917.

Thatcher, L.F., Anderson, J.P., Singh, K.B., 2005. Plant defence responses: what have we learnt from *Arabidopsis*? Funct. Plant Biol. 32, 1–19.

Theiler, R., 1970. Anatomische Untersuchungen an Traubenstielen im Zusammenhang mit der Stiellähme. Wein-Wiss. 25, 381–417.

Theiler, R., Müller, H., 1986. Beziehungen zwischen Klimafaktoren und dem Stiellähmebefall bei Riesling × Sylvaner. Vitis 25, 8–20.

Theiler, R., Müller, H., 1987. Beziehung zwischen mittlerer Tagesmaximumtemperatur während der Blüteperiode und Stiellähmebefall für verschiedene Rebsorten und Standorte. Mitt. Klosterneuburg 37, 102–108.

Theocharis, A., Clément, C., Ait Barka, E., 2012. Physiological and molecular changes in plants grown at low temperatures. Planta 235, 1091–1105.

Thibon, C., Dubourdieu, D., Darriet, P., Tominaga, T., 2009. Impact of noble rot on the aroma precursor of 3-sulfanylhexanol content in *Vitis vinifera* L. cv Sauvignon blanc and Semillon grape juice. Food Chem. 114, 1359–1364.

This, P., Lacombe, T., Thomas, M.R., 2006. Historical origins and genetic diversity of wine grapes. Trends Genet. 22, 511–519.

This, P., Lacombe, T., Cadle-Davidson, M., Owens, C.L., 2007. Wine grape (*Vitis vinifera* L.) color associates with allelic variation in the domestication gene *VvmybA1*. Theor. Appl. Genet. 114, 723–730.

Thomas, C.S., Gubler, W.D., Silacci, M.W., Miller, R., 1993. Changes in elemental sulfur residues on Pinot noir and Cabernet Sauvignon grape berries during the growing season. Am. J. Enol. Vitic. 44, 205–210.

Thomas, H., 2013. Senescence, ageing and death of the whole plant. New Phytol. 197, 696–711.

Thomas, H., Stoddart, J.L., 1980. Leaf senescence. Annu. Rev. Plant Physiol. 31, 83–111.

Thomas, M.R., Scott, N.S., 1993. Microsatellite repeats in grapevine reveal DNA polymorphisms when analysed as sequence-tagged sites (STSs). Theor. Appl. Genet. 86, 985–990.

Thomas, T.R., Matthews, M.A., Shackel, K.A., 2006. Direct in situ measurement of cell turgor in grape (*Vitis vinifera* L.) berries during development and in response to plant water deficits. Plant Cell Environ. 29, 993–1001.

Thomas, T.R., Shackel, K.A., Matthews, M.A., 2008. Mesocarp cell turgor in *Vitis vinifera* L. berries throughout development and its relation to firmness, growth, and the onset of ripening. Planta 228, 1067–1076.

Thomashow, M.F., 1999. Plant cold acclimation: freezing tolerance genes and regulatory mechanisms. Annu. Rev. Plant Physiol. Plant Mol. Biol. 50, 571–599.

Thomashow, M.F., 2001. So what's new in the field of plant cold acclimation? Lots!. Plant Physiol. 125, 89–93.

Thompson, D.S., Davies, W.J., Ho, L.C., 1998. Regulation of tomato fruit growth by epidermal cell wall enzymes. Plant Cell Environ. 21, 589–599.

Thompson, M.M., Olmo, H.P., 1963. Cytohistological studies of cytochimeric and tetraploid grapes. Am. J. Bot. 50, 901–906.

Thompson, M.V., 2006. Phloem: the long and the short of it. Trends Plant Sci. 11, 26–32.

Thompson, M.V., Holbrook, N.M., 2003. Scaling phloem transport: water potential equilibrium and osmoregulatory flow. Plant Cell Environ. 26, 1561–1577.

Thorne, E.T., Stevenson, J.F., Rost, T.L., Labavitch, J.M., Matthews, M.A., 2006. Pierce's disease symptoms: comparison with symptoms of water deficit and the impact of water deficits. Am. J. Enol. Vitic. 57, 1–11.

Thorngate, J.H., Singleton, V.L., 1994. Localization of procyanidins in grape seeds. Am. J. Enol. Vitic. 45, 259–262.

Tibbetts, T.J., Ewers, F.W., 2000. Root pressure and specific conductivity in temperate lianas: exotic *Celastrus orbiculatus* (Celastraceae) vs. native *Vitis riparia* (Vitaceae). Am. J. Bot. 87, 1272–1278.

Tilbrook, J., Tyerman, S.D., 2008. Cell death in grape berries: varietal differences linked to xylem pressure and berry weight loss. Funct. Plant Biol. 35, 173–184.

Tilbrook, J., Tyerman, S.D., 2009. Hydraulic connection of grape berries to the vine: varietal differences in water conductance into and out of berries, and potential for backflow. Funct. Plant Biol. 36, 541–550.

Timmis, J.N., Ayliffe, M.A., Huang, C.Y., Martin, W., 2004. Endosymbiotic gene transfer: organelle genomes forge eukaryotic chromosomes. Nat. Rev. Genet. 5, 123–135.

Tognetti, R., Raschi, A., Longobucco, A., Lanini, M., Bindi, M., 2005. Hydraulic properties and water relations of *Vitis vinifera* L. exposed to elevated CO_2 concentrations in a free air CO_2 enrichment (FACE). Phyton 45, 243–256.

Tominaga, T., Peyrot des Gachons, C., Dubourdieu, D., 1998. A new type of flavor precursors in *Vitis vinifera* L. cv. Sauvignon blanc: *S*-cysteine conjugates. J. Agric. Food Chem. 46, 5215–5219.

Tominaga, T., Baltenweck-Guyot, R., Peyrot des Gachons, C., Dubourdieu, D., 2000. Contribution of volatile thiols to the aromas of white wines made from several *Vitis vinifera* grape varieties. Am. J. Enol. Vitic. 51, 178–181.

Torabinejad, J., Caldwell, M.M., Flint, S.D., Durham, S., 1998. Susceptibility of pollen to UV-B radiation: an assay of 34 taxa. Am. J. Bot. 85, 360–369.

Toselli, M., Baldi, E., Marcolini, G., Malaguti, D., Quartieri, M., Sorrenti, G., et al., 2009. Response of potted grapevines to increasing soil copper concentration. Aust. J. Grape Wine Res. 15, 85–92.

Tournaire-Roux, C., Sutka, M., Javot, H., Gout, E., Gerbeau, P., Luu, D.T., et al., 2003. Cytosolic pH regulates root water transport during anoxic stress through gating of aquaporins. Nature 425, 393–397.

Treeby, M.T., Wheatley, D.M., 2006. Effect of nitrogen fertiliser on nitrogen partitioning and pool sizes in irrigated Sultana grapevines. Aust. J. Exp. Agric. 46, 1207–1215.

Treeby, M.T., Holzapfel, B.P., Walker, R.R., Nicholas, P.R., 1998. Profiles of free amino acids in grapes of grafted Chardonnay grapevines. Aust. J. Grape Wine Res. 4, 121–126.

Tregeagle, J.M., Tisdall, J.M., Blackmore, D.H., Walker, R.R., 2006. A diminished capacity for chloride exclusion by grapevine rootstocks following long-term saline irrigation in an inland versus a coastal region of Australia. Aust. J. Grape Wine Res. 12, 178–191.

Tregeagle, J.M., Tisdall, J.M., Tester, M., Walker, R.R., 2010. Cl^- uptake, transport and accumulation in grapevine rootstocks of differing capacity for Cl^- exlusion. Funct. Plant Biol. 37, 665–673.

Triantaphylidès, C., Krischke, M., Hoeberichts, F.A., Ksas, B., Gresser, G., Havaux, M., et al., 2008. Singlet oxygen is the major reactive oxygen species involved in photooxidative damage to plants. Plant Physiol. 148, 960–968.

Trieb, G., Becker, H., 1969. Untersuchungen über den Einfluß verschiedener Unterlagen auf die mineralische Ernährung des Edelreises. Wein-Wiss. 24, 258–266.

Trifilò, P., Gascò, A., Raimondo, F., Nardini, A., Salleo, S., 2003. Kinetics of recovery of leaf hydraulic conductance and vein functionality from cavitation-induced embolism in sunflower. J. Exp. Bot. 54, 2323–2330.

Tromp, A., De Klerk, C.A., 1988. Effect of copperoxychloride on the fermentation of must and on wine quality. S. Afr. J. Enol. Vitic. 9, 31–36.

Tropf, S., Lanz, T., Rensing, S., Schröder, J., Schröder, G., 1994. Evidence that stilbene synthases have developed from chalcone synthases several times in the course of evolution. J. Mol. Evol. 38, 610–618.

Tsay, Y.F., Ho, C.H., Chen, H.Y., Lin, S.H., 2011. Integration of nitrogen and potassium signaling. Annu. Rev. Plant Biol. 62, 207–226.

Tsukaya, H., 2006. Mechanism of leaf-shape determination. Annu. Rev. Plant Biol. 57, 477–496.

Tucker, S.C., Hoefert, L.L., 1968. Ontogeny of the tendril in *Vitis vinifera*. Am. J. Bot. 55, 1110–1119.

Turgeon, R., 1984. Efflux of sucrose from minor veins of tobacco leaves. Planta 161, 120–128.

Turgeon, R., 2000. Plasmodesmata and solute exchange in the phloem. Aust. J. Plant Physiol. 27, 521–529.

Turgeon, R., 2010. The role of phloem loading reconsidered. Plant Physiol. 152, 1817–1823.

Turgeon, R., Wolf, S., 2009. Phloem transport: cellular pathways and molecular trafficking. Annu. Rev. Plant Biol. 60, 207–221.

Turnbull, C.G.N., Lopez-Cobollo, R.M., 2013. Heavy traffic in the fast lane: long-distance signalling by macromolecules. New Phytol. 198, 33–51.

Turner, S., Gallois, P., Brown, D., 2007. Tracheary element differentiation. Annu. Rev. Plant Biol. 58, 407–433.

Tyerman, S.D., Bohnert, H.J., Maurel, C., Steudle, E., Smith, J.A.C., 1999. Plant aquaporins: their molecular biology, biophysics and significance for plant water relations. J. Exp. Bot. 50, 1055–1071.

Tyerman, S.D., Tilbrook, J., Pardo, C., Kotula, L., Sullivan, W., Steudle, E., 2004. Direct measurement of hydraulic properties in developing berries of *Vitis vinifera* L. cv Shiraz and Chardonnay. Aust. J. Grape Wine Res. 10, 170–181.

Tyerman, S.D., Vandeleur, R.K., Shelden, M.C., Tilbrook, J., Mayo, G., Gilliham, M., et al., 2009. Water transport and aquaporins in grapevine. In: Roubelakis-Angelakis, K.A. (Ed.), Grapevine Molecular Physiology and Biotechnology, second ed. Springer, Dordrecht, The Netherlands, pp. 73–104.

Tyree, M.T., Ewers, F.W., 1991. The hydraulic architecture of trees and other woody plants. New Phytol. 119, 345–360.

Tyree, M.T., Zimmermann, M.H., 2002. Xylem Structure and the Ascent of Sap, second ed. Springer, Berlin.

Tyree, M.T., Christy, L., Ferrier, J.M., 1974. A simpler iterative steady state solution of Münch pressure-flow systems applied to long and short translocation paths. Plant Physiol. 54, 589–600.

Uhlig, B.A., Clingeleffer, P.R., 1998. Ripening characteristics of the fruit from *Vitis vinifera* L. drying cultivars Sultana and Merbein seedless under furrow irrigation. Am. J. Enol. Vitic. 49, 375–382.

Umehara, M., Hanada, A., Yoshida, S., Akiyama, K., Arite, T., Takeda-Kamiya, N., et al., 2008. Inhibition of shoot branching by new terpenoid plant hormones. Nature 455, 195–200.

Ureta, F., Boidron, J.N., Bouard, J., 1981. Influence of dessèchement de la rafle on grape quality. Am. J. Enol. Vitic. 32, 90–92.

Vail, M.E., Marois, J.J., 1991. Grape cluster architecture and the susceptibility of berries to *Botrytis cinerea*. Phytopathology 81, 188–191.

Vail, M.E., Gubler, W.D., Rademacher, M.R., 1999. Effect of cluster tightness on *Botrytis* bunch rot in six Chardonnay clones. Plant Dis. 82, 107–109.

Valdés-Gómez, H., Fermaud, M., Roudet, J., Calonnec, A., Gary, C., 2008. Grey mould incidence is reduced on grapevines with lower vegetative and reproductive growth. Crop Prot. 27, 1174–1186.

van Bel, A.J.E., 2003. The phloem, a miracle of ingenuity. Plant Cell Environ. 26, 125–149.

van Bel, A.J.E., Ehlers, K., Knoblauch, M., 2002. Sieve elements caught in the act. Trends Plant Sci. 7, 126–132.

van Dongen, J.T., Schurr, U., Pfister, M., Geigenberger, P., 2003. Phloem metabolism and function have to cope with low internal oxygen. Plant Physiol. 131, 1529–1543.

van Doorn, W.G., Woltering, E.J., 2004. Senescence and programmed cell death: substance or semantics? J. Exp. Bot. 55, 2147–2153.

van Doorn, W.G., Hiemstra, T., Fanourakis, D., 2011. Hydrogel regulation of xylem water flow: an alternative hypothesis. Plant Physiol. 157, 1642–1649.

van Heeswijck, R., Stines, A.P., Grubb, J., Skrumsager Møller, I., Høj, P.B., 2001. Molecular biology and biochemistry of proline accumulation in developing grape berries. In: Roubelakis-Angelakis, K.A. (Ed.), Molecular Biology and Biotechnology of the Grapevine. Kluwer, Dordrecht, The Netherlands, pp. 87–108.

van Ieperen, W., van Meeteren, U., van Gelder, H., 2000. Fluid ionic composition influences hydraulic conductance of xylem conduits. J. Exp. Bot. 51, 769–776.

van Kan, J.A.L., 2006. Licensed to kill: the lifestyle of a necrotrophic plant pathogen. Trends Plant Sci. 11, 247–253.

van Leeuwen, C., Friant, P., Choné, X., Tregoat, O., Koundouras, S., Dubourdieu, D., 2004. Influence of climate, soil, and cultivar on terroir. Am. J. Enol. Vitic. 55, 207–217.

van Leeuwen, C., Tregoat, O., Choné, X., Bois, B., Pernet, D., Gaudillère, J.P., 2009. Vine water status is a key factor in grape ripening and vintage quality for red Bordeaux wine. How can it be assessed for vineyard management purposes? J. Int. Sci. Vigne Vin 43, 121–134.

van Leeuwen, C., Roby, J.P., Alonso-Villaverde, V., Gindro, K., 2013. Impact of clonal variability in *Vitis vinifera* Cabernet franc on grape composition, wine quality, leaf blade stilbene content, and downy mildew resistance. J. Agric. Food Chem. 61, 19–24.

van Volkenburgh, E., 1999. Leaf expansion—an integrating plant behaviour. Plant Cell Environ. 22, 1463–1473.

van den Boom, C.E.M., Van Beek, T.A., Posthumus, M.A., De Groot, A., Dicke, M., 2004. Qualitative and quantitative variation among volatile profiles induced by *Tetranychus urticae* feeding on plants from various families. J. Chem. Ecol. 30, 69–89.

van der Schoot, C., Rinne, P.L.H., 2011. Dormancy cycling at the shoot apical meristem: transitioning between self-organization and self-arrest. Plant Sci. 180, 120–131.

van der Schoot, C., Paul, L.K., Rinne, P.L.H., 2014. The embryonic shoot: a lifeline through winter. J. Exp. Bot. 65, 1699–1712.

VanEtten, H.D., Matthews, D.E., Matthews, P.S., 1989. Phytoalexin detoxification: importance for pathogenicity and practical implications. Annu. Rev. Phytopathol. 27, 143–164.

Vandeleur, R.K., Mayo, G., Shelden, M.C., Gilliham, M., Kaiser, B.N., Tyerman, S.D., 2009. The role of plasma membrane intrinsic protein aquaporins in water transport through roots: diurnal and drought stress responses reveal different strategies between isohydric and anisohydric cultivars of grapevine. Plant Physiol. 149, 445–460.

Vandenbussche, F., Suslov, D., De Grauwe, L., Leroux, O., Vissenberg, K., Van der Straeten, D., 2011. The role of brassinosteroids in shoot gravitropism. Plant Physiol. 156, 1331–1336.

van't Hoff, J.H., 1887. Die Rolle des osmotischen Druckes in der Analogie zwischen Lösungen und Gasen. Z. Phys. Chem. 1, 481–508.

Varanini, Z., Maggioni, A., 1982. Iron reduction and uptake by grapevine roots. J. Plant Nutr. 5, 521–529.

Vasconcelos, M.C., Castagnoli, S., 2000. Leaf canopy structure and vine performance. Am. J. Enol. Vitic. 51, 390–396.

Vasconcelos, M.C., Greven, M., Winefield, C.S., Trought, M.C.T., Raw, V., 2009. The flowering process of *Vitis vinifera*: a review. Am. J. Enol. Vitic. 60, 411–434.

Vasudevan, L., Wolf, T.K., Welbaum, G.G., Wisniewski, M.E., 1998. Anatomical developments and effects of artificial shade on bud necrosis of Riesling grapevines. Am. J. Enol. Vitic. 49, 429–439.

Vega, A., Gutiérrez, R.A., Peña-Neira, A., Cramer, G.R., Arce-Johnson, P., 2011. Compatible GLRaV-3 viral infections affect berry ripening decreasing sugar accumulation and anthocyanin biosynthesis in *Vitis vinifera*. Plant Mol. Biol. 77, 261–274.

Velasco, R., Zharkikh, A., Troggio, M., Cartwright, D.A., Cestaro, A., Pruss, D., et al., 2007. A high quality draft consensus sequence of the genome of a heterozygous grapevine variety. PLoS One 2, e1326, http://dx.doi.org/10.1371/journal.pone.0001326.

Veluthambi, K., Poovaiah, B.W., 1984. Calcium-promoted protein phosphorylation in plants. Science 223, 167–169.

Véry, A.A., Davies, J.M., 2000. Hyperpolarization-activated calcium channels at the tip of *Arabidopsis* root hairs. Proc. Natl. Acad. Sci. USA 97, 9801–9806.

Véry, A.A., Sentenac, H., 2003. Molecular mechanisms and regulation of K^+ transport in higher plants. Annu. Rev. Plant Biol. 54, 575–603.

Veselov, D., Langhans, M., Hartung, W., Aloni, R., Feussner, I., Götz, C., et al., 2003. Development of *Agrobacterium tumefaciens* C58-induced plant tumors and impact on host shoots are controlled by a cascade of jasmonic acid, auxin, cytokinin, ethylene and abscisic acid. Planta 216, 512–522.

Vezinhet, F., Hallet, J.N., Valade, M., Poulard, A., 1992. Ecological survey of wine yeast strains by molecular methods of identification. Am. J. Enol. Vitic. 43, 83–86.

Vezzulli, S., Leonardelli, L., Malossini, U., Stefanini, M., Velasco, R., Moser, C., 2012. Pinot blanc and Pinot gris arose as independent somatic mutations of Pinot noir. J. Exp. Bot. 63, 6359–6369.

Viala, P., Vermorel, V., 1901–1909. Ampélographie. Tomes I-VII. Masson, Paris.

Vicens, A., Fournand, D., Williams, P., Sidhoum, L., Moutounet, M., Doco, T., 2009. Changes in polysaccharide and protein composition of cell walls in grape berry skin (cv. Shiraz) during ripening and over-ripening. J. Agric. Food Chem. 57, 2955–2960.

Vidal, S., Williams, P., O'Neill, M.A., Pellerin, P., 2001. Polysaccharides from grape berry cell walls. Part I: tissue distribution and structural characterization of the pectic polysaccharides. Carbohydr. Polym. 45, 315–323.

Vidal, S., Francis, L., Guyot, S., Marnet, N., Kwiatkowski, M., Gawel, R., et al., 2003. The mouth-feel properties of grape and apple proanthocyanins in a wine-like medium. J. Sci. Food Agric. 83, 564–573.

Vidal, S., Meudec, E., Cheynier, V., Skouroumounis, G., Hayasaka, Y., 2004. Mass spectrometric evidence for the existence of oligomeric anthocyanins in grape skins. J. Agric. Food Chem. 52, 7144–7151.

Vignault, C., Vachaud, M., Cakir, B., Glissant, D., Dédaldéchamp, F., Büttner, M., et al., 2005. *VvHT1* encodes a monosaccharide transporter expressed in the conducting complex of the grape berry phloem. J. Exp. Bot. 56, 1409–1418.

Viret, O., Keller, M., Jaudzems, V.G., Cole, F.M., 2004. *Botrytis cinerea* infection of grape flowers: light and electron microscopical studies of infection sites. Phytopathology 94, 850–857.

Vitrac, X., Larronde, F., Krisa, S., Decendit, A., Deffieux, G., Mérillon, J.M., 2000. Sugar sensing and Ca^{2+}–calmodulin requirement in *Vitis vinifera* cells producing anthocyanins. Phytochemistry 53, 659–665.

Vivin, P., Castelan-Estrada, M., Gaudillére, J.P., 2003. Seasonal changes in chemical composition and construction costs of grapevine tissues. Vitis 42, 5–12.
Vogel, S., 2009. Leaves in the lowest and highest winds: temperature, force and shape. New Phytol. 183, 13–26.
Volder, A., Smart, D.R., Bloom, A.J., Eissenstat, D.M., 2005. Rapid decline in nitrate uptake and respiration with age in fine lateral roots of grape: implications for root efficiency and competitive effectiveness. New Phytol. 165, 493–502.
von Bassermann-Jordan, F., 1923. Geschichte des Weinbaus. Bände I-III, second ed. Frankfurter Verlags-Anstalt, Frankfurt, Germany.
von Goethe, J.W., 1790. Versuch, die Metamorphose der Pflanzen zu erklären. Ettinger, Gotha, Germany.
von Tiedemann, A., 1997. Evidence for a primary role of active oxygen species in induction of host cell death during infection of bean leaves with *Botrytis cinerea*. Physiol. Mol. Plant Pathol. 50, 151–166.
Vouillamoz, J.F., Grando, M.S., 2006. Genealogy of wine grape cultivars: 'Pinot' is related to 'Syrah'. Heredity 97, 102–110.
Vouillamoz, J.F., Schneider, A., Grando, M.S., 2007. Microsatellite analysis of Alpine grape cultivars (*Vitis vinifera* L.): alleged descendants of Pliny the Elder's Raetica are genetically related. Genet. Resour. Crop Evol. 54, 1095–1104.
Vršič, A., Ivančič, A., Šušek, A., Zagradišnik, B., Valdhuber, J., Šiško, M., 2011. The World's oldest living grapevine specimen and its genetic relationships. Vitis 50, 167–171.
Wada, H., Shackel, K.A., Matthews, M.A., 2008. Fruit ripening in *Vitis vinifera*: apoplastic solute accumulation accounts for pre-veraison turgor loss in berries. Planta 227, 1351–1361.
Wada, H., Matthews, M.A., Shackel, K.A., 2009. Seasonal pattern of apoplastic solute accumulation and loss of cell turgor during ripening of *Vitis vinifera* fruit under field conditions. J. Exp. Bot. 60, 1773–1781.
Wake, C.M.F., Fennell, A., 2000. Morphological, physiological and dormancy responses of three *Vitis* genotypes to short photoperiod. Physiol. Plant. 109, 203–210.
Walbot, V., Evans, M.M.S., 2003. Unique features of the plant life cycle and their consequences. Nat. Rev. Genet. 4, 369–379.
Walker, A.R., Lee, E., Bogs, J., McDavid, D.A.J., Thomas, M.R., Robinson, S.P., 2007. White grapes arose through the mutation of two similar and adjacent regulatory genes. Plant J. 49, 772–785.
Walker, R.P., Chen, Z.H., Técsi, L.I., Famiani, F., Lea, P.J., Leegood, R.C., 1999. Phosphoenolpyruvate carboxykinase plays a role in interactions of carbon and nitrogen metabolism during grape seed development. Planta 210, 9–18.
Walker, R.R., Blackmore, D.H., 2012. Potassium concentration and pH inter-relationships in grape juice and wine of Chardonnay and Shiraz from a range of rootstocks in different environments. Aust. J. Grape Wine Res. 18, 183–193.
Walker, R.R., Törökfalvy, E., Scott, N.S., Kriedemann, P.E., 1981. An analysis of photosynthetic response to salt treatment in *Vitis vinifera*. Aust. J. Plant Physiol. 8, 359–374.
Walker, R.R., Clingeleffer, P.R., Kerridge, G.H., Ruhl, E.H., Nicholas, P.R., Blackmore, D.H., 1998. Effects of the rootstock Ramsey (*Vitis champini*) on ion and organic acid composition of grapes and wine, and on wine spectral characteristics. Aust. J. Grape Wine Res. 4, 100–110.
Walker, R.R., Read, P.E., Blackmore, D.H., 2000. Rootstock and salinity effects on rates of berry maturation, ion accumulation and colour development in Shiraz grapes. Aust. J. Grape Wine Res. 6, 227–239.
Walker, R.R., Blackmore, D.H., Clingeleffer, P.R., 2010. Impact of rootstock on yield and ion concentrations in petioles, juice and wine of Shiraz and Chardonnay in different viticultural environments with different irrigation water salinity. Aust. J. Grape Wine Res. 16, 243–257.
Walters, D.R., McRoberts, N., 2006. Plants and biotrophs: a pivotal role for cytokinins? Trends Plant Sci. 11, 581–586.

Walz, A., Zingen-Sell, I., Loeffler, M., Sauer, M., 2008. Expression of an oxalate oxidase gene in tomato and severity of disease caused by *Botrytis cinerea* and *Sclerotinia sclerotiorum*. Plant Pathol. 57, 453–458.
Wample, R.L., 1994. A comparison of short- and long-term effects of mid-winter pruning on cold hardiness of Cabernet Sauvignon and Chardonnay buds. Am. J. Enol. Vitic. 45, 388–392.
Wample, R.L., Bary, A., 1992. Harvest date as a factor in carbohydrate storage and cold hardiness of Cabernet Sauvignon grapevines. J. Am. Soc. Hortic. Sci. 117, 32–36.
Wample, R.L., Spayd, S.E., Evans, R.G., Stevens, R.G., 1993. Nitrogen fertilization of White Riesling grapes in Washington: nitrogen seasonal effects on bud cold hardiness and carbohydrate reserves. Am. J. Enol. Vitic. 44, 159–167.
Wan, Y., Schwaninger, H., He, P., Wang, Y., 2007. Comparison of resistance to powdery mildew and downy mildew in Chinese wild grapes. Vitis 46, 132–136.
Wan, Y., Schwaninger, H., Li, D., Simon, C.J., Wang, Y., He, P., 2008a. The eco-geographic distribution of wild grape germplasm in China. Vitis 47, 77–80.
Wan, Y., Schwaninger, H., Li, D., Simon, C.J., Wang, Y., Zhang, C., 2008b. A review of taxonomic research on Chinese wild grapes. Vitis 47, 81–88.
Wan, Y., Wang, Y., Li, D., He, P., 2008c. Evaluation of agronomic traits in Chinese wild grapes and screening superior accessions for use in a breeding program. Vitis 47, 153–158.
Wang, J., De Luca, V., 2005. The biosynthesis and regulation of biosynthesis of Concord grape fruit esters, including 'foxy' methylanthranilate. Plant J. 44, 606–619.
Wang, Q., Cuellar, W.J., Rajamäki, M.L., Hirata, Y., Valkonen, J.P.T., 2008. Combined thermotherapy and cryotherapy for efficient virus eradication: relation of virus distribution, subcellular changes, cell survival and viral RNA degradation in shoot tips. Mol. Plant Pathol. 9, 237–250.
Wang, S., Okamoto, G., Hirano, K., Lu, J., Zhang, C., 2001. Effects of restricted rooting volume on vine growth and berry development of Kyoho grapevines. Am. J. Enol. Vitic. 52, 248–253.
Wang, W.Q., Song, S.Q., Li, S.H., Gan, Y.Y., Wu, J.H., Cheng, H.Y., 2011. Seed dormancy and germination in *Vitis amurensis* and its variation. Seed Sci. Res. 21, 255–265.
Wang, Y., Li, J., 2008. Molecular basis of plant architecture. Annu. Rev. Plant Biol. 59, 253–279.
Wang, Y., Wu, W.H., 2013. Potassium transport and signaling in higher plants. Annu. Rev. Plant Biol. 64, 451–476.
Wang, Y.Y., Hsu, P.K., Tsay, Y.F., 2012. Uptake, allocation and signaling of nitrate. Trends Plant Sci. 17, 458–467.
Wardlaw, I.F., 1990. The control of carbon partitioning in plants. New Phytol. 116, 341–381.
Wareing, P.F., Nasr, T.A.A., 1961. Gravimorphism in trees. I. Effects of gravity on growth and apical dominance in fruit trees. Ann. Bot. 25, 321–340.
Waterhouse, A.L., Laurie, V.F., 2006. Oxidation of wine phenolics: a critical evaluation and hypotheses. Am. J. Enol. Vitic. 57, 306–313.
Watt, M., Silk, W.K., Passioura, J.B., 2006. Rates of root and organism growth, soil conditions, and temporal and spatial development of the rhizosphere. Ann. Bot. 97, 839–855.
Weaver, L.M., Herrmann, K.M., 1997. Dynamics of the shikimate pathway in plants. Trends Plant Sci. 2, 346–351.
Weaver, R.J., 1960. Toxicity of gibberellin to seedless and seeded varieties of *Vitis vinifera*. Nature 187, 1135–1136.
Weaver, R.J., Ibrahim, I.M., 1968. Effect of thinning and seededness on maturation of *Vitis vinifera* L. grapes. Proc. Am. Soc. Hortic. Sci. 92, 311–318.
Weaver, R.J., Pool, R.M., 1968. Effect of various levels of cropping on *Vitis vinifera* grapevines. Am. J. Enol. Vitic. 19, 185–193.
Weaver, R.J., Pool, R.M., 1971. Chemical thinning of grape clusters (*Vitis vinifera* L.). Vitis 10, 201–209.

Weaver, R.J., Amerine, M.A., Winkler, A.J., 1957. Preliminary report on effect of level of crop on development of color in certain red wine grapes. Am. J. Enol. Vitic. 8, 157–166.
Weaver, R.J., Shindy, W., Kliewer, W.M., 1969. Growth regulator induced movement of photosynthetic products into fruits of 'Black Corinth' grapes. Plant Physiol. 44, 183–188.
Webb, L.B., Whetton, P.H., Barlow, E.W.R., 2007. Modelled impact of future climate change on the phenology of winegrapes in Australia. Aust. J. Grape Wine Res. 13, 165–175.
Weber, B., Hoesch, L., Rast, D.M., 1995. Protocatechualdehyde and other phenols as cell wall components of grapevine leaves. Phytochemistry 40, 433–437.
Wegner, L.H., Zimmermann, U., 2009. Hydraulic conductance and K^+ transport into the xylem depend on radial volume flow, rather than on xylem pressure, in roots of intact, transpiring maize seedlings. New Phytol. 181, 361–373.
Wegscheider, E., Benjak, A., Forneck, A., 2009. Clonal variation in Pinot noir revealed by S-SAP involving universal retrotransposon-based sequences. Am. J. Enol. Vitic. 60, 104–109.
Weir, T.L., Park, S.W., Vivanco, J.M., 2004. Biochemical and physiological mechanisms mediated by allelochemicals. Curr. Opin. Plant Biol. 7, 472–479.
Weiss, A., Allen Jr., L.H., 1976. Air-flow patterns in vineyard rows. Agric. Meteorol. 16, 329–432.
Weiss, D., Ori, N., 2007. Mechanisms of cross talk between gibberellin and other hormones. Plant Physiol. 144, 1240–1246.
Went, F.W., 1944. Plant growth under controlled conditions. II. Thermoperiodicity in growth and fruiting of the tomato. Am. J. Bot. 31, 135–150.
Wenzel, K., Dittrich, H.H., Heimfarth, M., 1987. Die Zusammensetzung der Anthocyane in den Beeren verschiedener Rebsorten. Vitis 26, 65–78.
Wermelinger, B., Koblet, W., 1990. Seasonal growth and nitrogen distribution in grapevine leaves, shoots and grapes. Vitis 29, 15–26.
Werner, T., Motyka, V., Laucou, V., Smets, R., Van Onckelen, H., Schmülling, T., 2003. Cytokinin-deficient transgenic *Arabidopsis* plants show multiple developmental alterations indicating opposite functions of cytokinins in the regulation of shoot and root meristem activity. Plant Cell 15, 2532–2550.
Weyand, K.M., Schultz, H.R., 2006a. Light interception, gas exchange and carbon balance of different canopy zones of minimally and cane-pruned field grown Riesling grapevines. Vitis 45, 105–114.
Weyand, K.M., Schultz, H.R., 2006b. Long-term dynamics of nitrogen and carbohydrate reserves in woody parts of minimally and severely pruned Riesling vines in a cool climate. Am. J. Enol. Vitic. 57, 172–182.
Weyand, K.M., Schultz, H.R., 2006c. Regulating yield and wine quality of minimal pruning systems through the application of gibberellic acid. J. Int. Sci. Vigne Vin 40, 151–163.
White, M.A., Diffenbaugh, N.S., Jones, G.V., Pal, J.S., Giorgi, F., 2006. Extreme heat reduces and shifts United States premium wine production in the 21st century. Proc. Natl. Acad. Sci. USA 103, 11217–11222.
Whitelam, G.C., Devlin, P.F., 1998. Light signalling in *Arabidopsis*. Plant Physiol. Biochem. 36, 125–133.
Whitham, S.A., Yang, C., Goodin, M.M., 2006. Global impact: elucidating plant responses to viral infection. Mol. Plant Microbe. Interact. 19, 1207–1215.
Wiedemann, P., Neinhuis, C., 1998. Biomechanics of isolated plant cuticles. Bot. Acta 111, 28–34.
Wiley, E., Helliker, B., 2012. A re-evaluation of carbon storage in trees lends greater support for carbon limitation to growth. New Phytol. 195, 285–289.
Wilkinson, S., Davies, W.J., 2002. ABA-based chemical signalling: the co-ordination of responses to stress in plants. Plant Cell Environ. 25, 195–210.
Wille, A.C., Lucas, W.J., 1984. Ultrastructural and histochemical studies on guard cells. Planta 160, 129–142.
Williams, C.M.J., Maier, N.A., Bartlett, L., 2004. Effect of molybdenum foliar sprays on yield, berry size, seed formation, and petiolar nutrient composition of "Merlot" grapevines. J. Plant Nutr. 27, 1891–1916.

Williams, L.E., 1996. Grape. In: Zamski, E., Schaffer, A. (Eds.), Photoassimilate Distribution in Plants and Crops: Source-Sink Relationships. Dekker, New York, NY, pp. 851–881.

Williams, L.E., 1987. Growth of 'Thompson Seedless' grapevines: II. Nitrogen distribution. J. Am. Soc. Hortic. Sci. 112, 330–333.

Williams, L.E., 2010. Interaction of rootstock and applied water amounts at various fractions of estimated evapotranspiration (Et_c) on productivity of Cabernet Sauvignon. Aust. J. Grape Wine Res. 16, 434–444.

Williams, L.E., 2014. Determination of evapotranspiration and crop coefficients for a Chardonnay vineyard located in a cool climate. Am. J. Enol. Vitic. 65, 159–169.

Williams, L.E., Ayars, J.E., 2005. Grapevine water use and the crop coefficient are linear functions of the shaded area measured beneath the canopy. Agric. For. Meteorol. 132, 201–211.

Williams, L.E., Baeza, P., 2007. Relationships among ambient temperature and vapor pressure deficit and leaf and stem water potentials of fully irrigated, field-grown grapevines. Am. J. Enol. Vitic. 58, 173–181.

Williams, L.E., Biscay, P.J., 1991. Partitioning of dry weight, nitrogen, and potassium in Cabernet Sauvignon grapevines from anthesis until harvest. Am. J. Enol. Vitic. 42, 113–117.

Williams, L.E., Matthews, M.A., 1990. Grapevine. In: Stewart, B.A., Nielson, N.R. (Eds.), Irrigation of Agricultural Crops, Agronomy Monograph No. 30. ASA-CSSA-SSSA, Madison, WI, pp. 1019–1055.

Williams, L.E., Smith, R.J., 1985. Net CO_2 assimilation rate and nitrogen content of grape leaves subsequent to fruit harvest. J. Am. Soc. Hortic. Sci. 110, 846–850.

Williams, L.E., Dokoozlian, N.K., Wample, R., 1994. Grape. In: Schaffer, B., Andersen, P.C. (Eds.), Handbook of Environmental Physiology of Fruit Crops, vol. I. Temperate Crops. CRC Press, Boca Raton, FL, pp. 85–133.

Williams, L.E., Retzlaff, W.A., Yang, W., Biscay, P.J., Ebisuda, N., 2000. Effect of girdling on leaf gas exchange, water status, and non-structural carbohydrates of field-grown *Vitis vinifera* L. (cv. Flame Seedless). Am. J. Enol. Vitic. 51, 49–54.

Williams, L.E., Phene, C.J., Grimes, D.W., Trout, T.J., 2003. Water use of mature Thompson Seedless grapevines in California. Irrig. Sci. 22, 11–18.

Williams, L.E., Grimes, D.W., Phene, C.J., 2010. The effects of applied water at various fractions of measured evapotranspiration on water relations and vegetative growth of Thompson Seedless grapevines. Irrig. Sci. 28, 221–232.

Williamson, B., Tudzynsk, B., Tudzynski, P., van Kan, J.A.L., 2007. *Botrytis cinerea*: the cause of grey mould disease. Mol. Plant Pathol. 8, 561–580.

Wilson, B., Strauss, C.R., Williams, P.J., 1984. Changes in free and glycosidically bound monoterpenes in developing Muscat grapes. J. Agric. Food Chem. 32, 919–924.

Wilson, B., Strauss, C.R., Williams, P.J., 1986. The distribution of free and glycosidically-bound monoterpenes among skin, juice, and pulp fractions of some white grape varieties. Am. J. Enol. Vitic. 37, 107–111.

Wilson, E.O., Brown Jr., W.L., 1953. The subspecies concept and its taxonomic application. Syst. Zool. 2, 97–111.

Windt, C.W., Vergeldt, F.J., de Jager, P.A., Van As, H., 2006. MRI of long-distance water transport: a comparison of the phloem and xylem flow characteristics and dynamics in poplar, castor bean, tomato and tobacco. Plant Cell Environ. 29, 1715–1729.

Wingler, A., Masclaux-Daubresse, C., Fischer, A.M., 2009. Sugars, senescence, and ageing in plants and heterotrophic organisms. J. Exp. Bot. 60, 1063–1066.

Winkel, T., Rambal, S., 1993. Influence of water stress on grapevines growing in the field: from leaf to whole-plant response. Aust. J. Plant Physiol. 20, 143–157.

Winkel-Shirley, B., 2002. Biosynthesis of flavonoids and effects of stress. Curr. Opin. Plant Biol. 5, 218–223.

Winkler, A.J., 1929. The effect of dormant pruning on the carbohydrate metabolism of *Vitis vinifera*. Hilgardia 4, 153–173.

Winkler, A.J., 1954. Effects of overcropping. Am. J. Enol. Vitic. 5, 4–12.
Winkler, A.J., 1958. The relation of leaf area and climate to vine performance and grape quality. Am. J. Enol. Vitic. 9, 10–23.
Winkler, A.J., 1959. The effect of vine spacing at Oakville on yields, fruit composition, and wine quality. Am. J. Enol. Vitic. 10, 39–43.
Winkler, A.J., 1969. Effect of vine spacing in an unirrigated vineyard on vine physiology, production and wine quality. Am. J. Enol. Vitic. 20, 7–15.
Winkler, A.J., Williams, W.O., 1935. Effect of seed development on the growth of grapes. Proc. Am. Soc. Hortic. Sci. 33, 430–434.
Winkler, A.J., Williams, W.O., 1945. Starch and sugars of *Vitis vinifera*. Plant Physiol. 20, 412–432.
Winkler, A.J., Cook, J.A., Kliewer, W.M., Lider, L.A., 1974. General Viticulture. University of California Press, Berkeley, CA.
Winship, L.J., Obermeyer, G., Geitmann, A., Hepler, P.K., 2010. Under pressure, cell walls set the pace. Trends Plant Sci. 15, 363–369.
Winterhalter, P., Schreier, P., 1994. C_{13}-Norisoprenoid glycosides in plant tissues: an overview on their occurrence, composition and role as flavour precursors. Flavour Fragr. J. 9, 281–287.
Wisniewski, M., Gusta, L., Neuner, G., 2014. Adaptive mechanisms of freeze avoidance in plants: a brief update. Environ. Exp. Bot. 99, 133–140.
Witte, C.P., Rosso, M.G., Romeis, T., 2005. Identification of three urease accessory proteins that are required for urease activation in *Arabidopsis*. Plant Physiol. 139, 1155–1162.
Wobus, U., Weber, H., 1999. Seed maturation: genetic programmes and control signals. Curr. Opin. Plant Biol. 2, 33–38.
Wolf, S., Hématy, K., Höfte, H., 2012. Growth control and cell wall signaling in plants. Annu. Rev. Plant Biol. 63, 381–407.
Wolf, T.K., Cook, M.K., 1992. Seasonal deacclimation patterns of three grape cultivars at constant, warm temperature. Am. J. Enol. Vitic. 43, 171–179.
Wolf, T.K., Pool, R.M., 1988. Nitrogen fertilization and rootstock effects on wood maturation and dormant bud cold hardiness of cv. Chardonnay grapevines. Am. J. Enol. Vitic. 39, 308–312.
Wolfe, D.W., Schwartz, M.D., Lakso, A.N., Otsuki, Y., Pool, R.M., Shaulis, N.J., 2005. Climate change and shifts in spring phenology of three horticultural woody perennials in northeastern USA. Int. J. Biometeorol. 49, 303–309.
Wollmann, N., Hofmann, T., 2013. Compositional and sensory characterization of red wine polymers. J. Agric. Food Chem. 61, 2045–2061.
Wolpert, J.A., Howell, G.S., 1985. Cold acclimation of Concord grapevines. II. Natural acclimation pattern and tissue moisture decline in canes and primary buds of bearing vines. Am. J. Enol. Vitic. 36, 189–194.
Wolpert, J.A., Howell, G.S., 1986. Cold acclimation of Concord grapevines. III. Relationship between cold hardiness, tissue water content, and shoot maturation. Vitis 25, 151–159.
Wood, C., Siebert, T.E., Parker, M., Capone, D.L., Elsey, G.M., Pollnitz, A.P., et al., 2008. From wine to pepper: rotundone, an obscure sesquiterpene, is a potent spicy aroma compound. J. Agric. Food Chem. 56, 3738–3744.
Wood, W.F., Sollers, B.G., Dragoo, G.A., Dragoo, J.W., 2002. Volatile components in defensive spray of the hooded skunk, *Mephitis macroura*. J. Chem. Ecol. 28, 1865–1870.
Woodham, R.C., Alexander, D.M., 1966. The effect of root temperature on development of small fruiting Sultana vines. Vitis 5, 345–350.
Woodham, R.C., Krake, L.R., Cellier, K.M., 1983. The effect of grapevine leafroll plus yellow speckle disease on annual growth, yield and quality of grapes from Cabernet Franc under two pruning systems. Vitis 22, 324–330.

Woodham, R.C., Antcliff, A.J., Krake, L.R., Taylor, R.H., 1984. Yield differences between Sultana clones related to virus status and genetic factors. Vitis 23, 73–83.

Woodrow, I.E., Berry, J.A., 1988. Enzymatic regulation of photosynthetic CO_2 fixation in C_3 plants. Annu. Rev. Plant Physiol. Plant Mol. Biol. 39, 533–594.

Woodward, A.W., Bartel, B., 2005. Auxin: regulation, action, and interaction. Ann. Bot. 95, 707–735.

Woodward, F.I., 1987. Stomatal numbers are sensitive to CO_2 increases from pre-industrial levels. Nature 327, 617–618.

Wu, Y., Cosgrove, D.J., 2000. Adaptation of roots to low water potentials by changes in cell wall extensibility and cell wall proteins. J. Exp. Bot. 51, 1543–1553.

Wyka, T.P., Oleksyn, J., Karolewski, P., Schnitzer, S.A., 2013. Phenotypic correlates of the lianescent growth form: a review. Ann. Bot. 112, 1667–1681.

Xia, G., Cheng, L., 2004. Foliar urea application in the fall affects both nitrogen and carbon storage in young 'Concord' grapevines grown under a wide range of nitrogen supply. J. Am. Soc. Hortic. Sci. 129, 653–659.

Xie, D.Y., Sharma, S.B., Paiva, N.L., Ferreira, D., Dixon, R.A., 2003. Role of anthocyanidin reductase, encoded by *BANYULS* in plant flavonoid biosynthesis. Science 299, 396–399.

Xie, Z., Kapteyn, J., Gang, D.R., 2008. A systems biology investigation of the MEP/terpenoid and shikimate/phenylpropanoid pathways points to multiple levels of metabolic control in sweet basil glandular trichomes. Plant J. 54, 349–361.

Xin, Z., Browse, J., 2000. Cold comfort farm: the acclimation of plants to freezing temperatures. Plant Cell Environ. 23, 893–902.

Xiong, J., Bauer, C.E., 2002. Complex evolution of photosynthesis. Annu. Rev. Plant Biol. 53, 503–521.

Xu, G., Fan, X., Miller, A.J., 2012. Plant nitrogen assimilation and use efficiency. Annu. Rev. Plant Biol. 63, 153–182.

Xu, P., Chen, F., Mannas, J.P., Feldman, T., Sumner, L.W., Roossinck, M.J., 2008. Virus infection improves drought tolerance. New Phytol. 180, 911–921.

Yamaguchi, S., 2008. Gibberellin metabolism and its regulation. Annu. Rev. Plant Biol. 59, 225–251.

Yamane, T., Jeong, S.T., Goto-Yamamoto, N., Koshita, Y., Kobayashi, S., 2006. Effects of temperature on anthocyanin biosynthesis in grape berry skins. Am. J. Enol. Vitic. 57, 54–59.

Yamasaki, H., Sakihama, Y., Ikehara, N., 1997. Flavonoid-peroxidase reaction as a detoxification mechanism of plant cells against H_2O_2. Plant Physiol. 115, 1405–1412.

Yang, J., Zhang, J., Wang, Z., Zhu, Q., Liu, L., 2002. Abscisic acid and cytokinins in the root exudates and leaves and their relationship to senescence and remobilization of carbon reserves in rice subjected to water stress during grain filling. Planta 215, 645–652.

Yang, T., Poovaiah, B.W., 2002. Hydrogen peroxide homeostasis: activation of plant catalase by calcium/calmodulin. Proc. Natl. Acad. Sci. USA 99, 4097–4102.

Yang, Y., Duan, C., Du, H., Tian, J., Pan, Q., 2010. Trace element and rare earth element profiles in berry tissues of three grape cultivars. Am. J. Enol. Vitic. 61, 401–407.

Yang, Y.S., Hori, Y., 1979. Studies on retranslocation of accumulated assimilates in 'Delaware' grapevines. I. Retranslocation of ^{14}C-assimilates in the following spring after ^{14}C feeding in summer and autumn. Tohoku J. Agric. Res. 30, 43–56.

Yang, Y.S., Hori, Y., 1980. Studies on retranslocation of accumulated assimilates in 'Delaware' grapevines. III. Early growth of new shoots as dependent on accumulated and current year assimilates. Tohoku J. Agric. Res. 31, 120–129.

Ye, Q., Wiera, B., Steudle, E., 2004. A cohesion/tension mechanism explains the gating of water channels (aquaporins) in *Chara* internodes by high concentration. J. Exp. Bot. 55, 449–461.

Ye, Z.H., 2002. Vascular tissue differentiation and pattern formation in plants. Annu. Rev. Plant Biol. 53, 183–202.

Yeats, T.H., Rose, J.K.C., 2013. The formation and function of plant cuticles. Plant Physiol. 163, 5−20.

Yin, R., Messner, B., Faus-Kessler, T., Hoffman, T., Schwab, W., Hajirezaei, M.R., et al., 2012. Feedback inhibition of the general phenylpropanoid and flavonol biosynthetic pathways upon a compromised flavonol-3-*O*-glycosylation. J. Exp. Bot. 63, 2465−2478.

Yokotsuka, K., Fukui, M., 2002. Changes in nitrogen compounds in berries of six grape cultivars during ripening over two years. Am. J. Enol. Vitic. 53, 69−77.

Young, J.M., Kuykendall, L.D., Martínez-Romero, E., Kerr, A., Sawada, H., 2001. A revision of *Rhizobium* Frank 1889, with an amended description of the genus, and the inclusion of all species of *Agrobacterium* Conn 1942 and *Allorhizobium undicola* de Lajudie *et al.* 1998 as new combinations: *Rhizobium radiobacter*, *R. rhizogenes*, *R. rubi*, *R. undicola* and *R. vitis*. Int. J. Syst. Evol. Microbiol. 51, 89−103.

Yruela, I., 2009. Copper in plants: acquisition, transport and interactions. Funct. Plant Biol. 36, 409−430.

Yu, O., Jez, J.M., 2008. Nature's assembly line: biosynthesis of simple phenylpropanoids and polyketides. Plant J. 54, 750−762.

Zabalza, A., van Dongen, J.T., Froehlich, A., Oliver, S.N., Faix, B., Jagadis Gupta, K., et al., 2009. Regulation of respiration and fermentation to control the plant internal oxygen concentration. Plant Physiol. 149, 1087−1098.

Zachariassen, K.E., Kristiansen, E., 2000. Ice nucleation and antinucleation in nature. Cryobiology 41, 257−279.

Zamboni, A., Minoia, L., Ferrarini, A., Tornielli, G.B., Zago, E., Delledonne, M., et al., 2008. Molecular analysis of post-harvest withering in grape by AFLP transcriptional profiling. J. Exp. Bot. 59, 4145−4159.

Zapata, C., Deléens, E., Chaillou, S., Magné, C., 2004. Partitioning and mobilization of starch and N reserves in grapevine (*Vitis vinifera* L.). J. Plant Physiol. 161, 1031−1040.

Zeeman, S.C., Kossmann, J., Smith, A.M., 2010. Starch: its metabolism, evolution, and biotechnological modification in plants. Annu. Rev. Plant Biol. 61, 209−234.

Zell, M.B., Fahnenstich, H., Maier, A., Saigo, M., Voznesenskaya, E.V., Edwards, G.E., et al., 2010. Analysis of *Arabidopsis* with highly reduced levels of malate and fumarate sheds light on the role of these organic acids as storage carbon molecules. Plant Physiol. 152, 1251−1262.

Zelleke, A., Düring, H., 1994. Evidence for tertiary buds within latent buds of Müller-Thurgau grapevines. Vitis 33, 96.

Zelleke, A., Kliewer, W.M., 1979. Influence of root temperature and rootstock on budbreak, shoot growth, and fruit composition of Cabernet Sauvignon grapevines grown under controlled conditions. Am. J. Enol. Vitic. 30, 312−317.

Zelleke, A., Kliewer, W.M., 1981. Factors affecting the qualitative and quantitative levels of cytokinins in xylem sap of grapevines. Vitis 20, 93−104.

Zerihun, A., Treeby, M.T., 2002. Biomass distribution and nitrate assimilation in response to N supply for *Vitis vinifera* L. cv. Cabernet Sauvignon on five *Vitis* rootstock genotypes. Aust. J. Grape Wine Res. 8, 157−162.

Zhang, C., Tateishi, N., Tanabe, K., 2010a. Pollen density on the stigma affects endogenous gibberellin metabolism, seed and fruit set, and fruit quality in *Pyrus pyrifolia*. J. Exp. Bot. 61, 4291−4302.

Zhang, H., Han, W., De Smet, I., Talboys, P., Loya, R., Hassan, A., et al., 2010b. ABA promotes quiescence of the quiescent centre and suppresses stem cell differentiation in the *Arabidopsis* primary root meristem. Plant J. 64, 764−774.

Zhang, J., Ma, H., Feng, J., Zeng, L., Wang, Z., Chen, S., 2008. Grape berry plasma membrane proteome analysis and its differential expression during ripening. J. Exp. Bot. 59, 2979−2990.

Zhang, J., Gao, G., Chen, J.J., Taylor, G., Cui, K.M., He, X.Q., 2011. Molecular features of secondary vascular tissue regeneration after bark girdling in *Populus*. New Phytol. 192, 869−884.

Zhang, X., Walker, R.R., Stevens, R.M., Prior, L.D., 2002. Yield—salinity relationships of different grapevine (*Vitis vinifera* L.) scion-rootstock combinations. Aust. J. Grape Wine Res. 8, 150—156.

Zhang, X.Y., Wang, X.L., Wang, X.F., Xia, G.H., Pan, Q.H., Fan, R.C., et al., 2006. A shift of phloem unloading from symplasmic to apoplasmic pathway is involved in developmental onset of ripening in grape berry. Plant Physiol. 142, 220—232.

Zhang, Y., Dami, I.E., 2012. Foliar application of abscisic acid increases freezing tolerance of field-grown *Vitis vinifera* Cabernet franc grapevines. Am. J. Enol. Vitic. 63, 377—384.

Zhang, Y., Mechlin, T., Dami, I., 2011. Foliar application of abscisic acid induces dormancy responses in greenhouse-grown grapevines. HortScience 46, 1271—1277.

Zhao, J., Pang, Y., Dixon, R.A., 2010. The mysteries of proanthocyanidin transport and polymerization. Plant Physiol. 153, 437—443.

Zhu, J.K., 2001. Plant salt tolerance. Trends Plant Sci. 6, 66—71.

Zhu, J.K., 2002. Salt and drought stress signal transduction in plants. Annu. Rev. Plant Biol. 53, 247—273.

Zimman, A., Waterhouse, A.L., 2004. Incorporation of malvidin-3-glucoside into high molecular weight polyphenols during fermentation and wine aging. Am. J. Enol. Vitic. 55, 139—146.

Zinn, K.E., Tunc-Ozdemir, M., Harper, J.F., 2010. Temperature stress and plant sexual reproduction: uncovering the weakest links. J. Exp. Bot. 61, 1959—1968.

Zoecklein, B.W., Wolf, T.K., Duncan, N.W., Judge, J.M., Cook, M.K., 1992. Effects of fruit zone leaf removal on yield, fruit composition, and fruit rot incidence of Chardonnay and White Riesling (*Vitis vinifera* L.) grapes. Am. J. Enol. Vitic. 43, 139—148.

Zohary, D., Hopf, M., 2001. Domestication of Plants in the Old World—The Origin and Spread of Cultivated Plants in West Asia, Europe, and the Nile Valley, third ed. Oxford University Press, Oxford.

Zohary, D., Spiegel-Roy, P., 1975. Beginnings of fruit growing in the old world. Science 187, 319—327.

Zonia, L., Munnik, T., 2007. Life under pressure: hydrostatic pressure in cell growth and function. Trends Plant Sci. 12, 90—97.

Zsófi, Z., Váradi, G., Bálo, B., Marschall, M., Nagy, Z., Dulai, S., 2009. Heat acclimation of grapevine leaf photosynthesis: mezo- and macroclimatic aspects. Funct. Plant Biol. 36, 310—322.

Zufferey, V., Cochard, H., Ameglio, T., Spring, J.L., Viret, O., 2011. Diurnal cycles of embolism formation and repair in petioles of grapevine (*Vitis vinifera* cv. Chasselas). J. Exp. Bot. 62, 3885—3894.

Zufferey, V., Murisier, F., Vivin, P., Belcher, S., Lorenzini, F., Spring, J.L., et al., 2012. Carbohydrate reserves in grapevine (*Vitis vinifera* L. 'Chasselas'): the influence of the leaf to fruit ratio. Vitis 51, 103—110.

Zwieniecki, M.A., Melcher, P.J., Holbrook, N.M., 2001. Hydrogel control of xylem hydraulic resistance in plants. Science 291, 1059—1062.

Internet Resources

American Society for Enology and Viticulture (including *American Journal of Enology and Viticulture*): http://www.asev.org

American Society of Plant Biologists (including various journals): http://my.aspb.org

Australian Society of Viticulture and Oenology (including *Australian Journal of Grape and Wine Research*): http://www.asvo.com.au

Biology of the Grapevine (University of Hohenheim virtual lectures by R. Blaich, in German): http://www.uni-hohenheim.de/lehre370/weinbau

European *Vitis* Database: http://www.eu-vitis.de

French Network of Grapevine Repositories: http://bioweb.ensam.inra.fr/collections_vigne

Grape Genome Browser: http://www.genoscope.cns.fr/externe/GenomeBrowser/Vitis

International Grape Genome Program: http://www.vitaceae.org

International Society for Horticultural Science (including *Acta Horticulturae*): http://www.ishs.org

Kimball's Biology Pages: http://users.rcn.com/jkimball.ma.ultranet/BiologyPages

Molecular Cell Biology Online: http://www.ncbi.nlm.nih.gov/books/NBK21475

National Grape Registry of the USA: http://ngr.ucdavis.edu

Plant Physiology Online: http://5e.plantphys.net

PlantStress: http://www.plantstress.com

The Virtual Vine Doctor (in German): http://www.rebendoktor.de

University of California Integrated Viticulture Online: http://iv.ucdavis.edu

University of California Davis Virtual Crops Grape (*Vitis vinifera*): http://www-plb.ucdavis.edu/Labs/Rost/virtual%20grape%20dreamweaver/index.htm

USDA Natural Resources Conservation Service Plants Profile: http://plants.usda.gov/core/profile?symbol=VITIS

Vitis International Variety Catalogue: http://www.vivc.de

Vitis Journal of Grapevine Research: http://www.jki.bund.de/en/startseite/veroeffentlichungen/vitis.html

Vitis-VEA, Viticulture and Enology Abstracts: http://www.vitis-vea.de

Vitisvinum, The Multilingual Wine Dictionary: http://www.vitisvinum.info

Washington State University Viticulture and Enology: http://wine.wsu.edu

Index

Note: Page numbers followed by "*f*" and "*t*" refer to figures and tables, respectively.

A

Abortion of flowers and ovaries, 93
Abscisic acid (ABA), 26, 61, 141, 154
    ATPases and, 154
    grape ripening and, 205–206
    in inducing anthocyanin production, 365–366
    organ's sensitivity to, 269
    reaction of stomata to, 107–108
Accessory buds, 43
Accessory pigments, 127
Acetyl coenzyme A (acetyl-CoA), 139–140
Acid growth theory, 104–105
Acids. *See also* Organic acids
    grape composition and fruit quality, 216–219, 216*f*
Active (or reactive) oxygen species, 129–130, 270
Adventitious roots, 26–27
Agrobacterium species, 362
    A. tumefaciens, 361–362
    A. vitis, 3
Allocation, photosynthate, 156–163
    assimilate surplus in, 161, 162*f*
    assimilation rate determined by source in, 160–161
    categories, 156
    patterns in, 158–159
    phosphates availability in regulation of, 156
    source-sink relation with, 157–159, 157*f*
    storage in, 156
    transport in, 156
    utilization in, 156
Allorhizic root system, 26–27
Alphonse Lavallée, 16
American group
    Vitis aestivalis Michaux, 7–8
    Vitis berlandieri Planchon, 9
    Vitis candicans Engelmann, 9
    Vitis cinerea Engelmann, 9
    Vitis labrusca L., 7
    Vitis riparia Michaux, 8, 8*f*
    Vitis rupestris Scheele, 8
Amino acids, 2, 13–14, 32–33, 118–119, 141, 147, 151, 167–168, 182–184, 184*f*, 209, 217, 244–245, 256, 281, 283, 289–290, 314, 336, 353, 358, 366–367
    ammonium assimilation into, 182–183, 185–187
    arginine, 302
    aromatic, 221, 223–226
    assembly of, 187
    in berries, 260
    cysteine, 270–271
    free, 219–221, 336
    glutamine and glutamate, 189–190
    in leaf, 262
    mesophyll cell produced, 152
    methionine, 274
    N:C ratio of, 189–190
    nitrate and sulfate assimilation, 143, 188, 190–192, 258–259
    nitrogen-containing methoxypyrazines, 241–242
    nonprotein, 309
    nutrient deficiency and, 290–291
    phenylalanine, 135–136
    during photorespiratory carbon oxidation cycle, 137
    production, 152, 187–188, 259–260, 283
    profile of a wine, 221
    protein breakdown during ripening and, 220–221
    proteolysis production of, 190–191
    root-delivered, 152
    root pressure and, 61–64
    rootstocks and, 265
    S-containing, 242, 259–260
    in shaded berries, 246–247
    during springtime remobilization of storage, 36–37, 121
    in sucrose concentration, 147–149
    sulfur in, 303
    TCA cycle and, 140*f*, 141
    temperature influencing concentration of, 253
    tryptophan, 36–37
    xylem/phloem sap with, 148*t*
Ammonium assimilation, 185–187
    alternative pathway for, 185*f*, 186
    glutamate dehydrogenase (GDH) catalyzation, 186–187
    glutamine produced by GS, 186
    GS/GOGAT cycle, 185–186, 185*f*
    structure and reactions of compounds in, 185*f*
Ampelography, 10–11
Amylopectin, 134
Amylose, 134
Anastomosis, 44
Androecium, 52–54
Anemic cultivars, 17
Angiospermae (Magnoliophyta), 4
Angiosperms, 4, 86–87
    fertilization in, 89–90

Antenna pigments, 127–128
Antheraxanthin, 129–130
Anthesis, 61, 83–84, 86–91
    abortion of flowers and ovaries following, 93
    berry cytokinin concentration decline following, 94–95
    effect of arrested ovule development, 92
    environmental conditions, effect on, 201
    flower caps during, 348–349
    grape flowers, vulnerability of, 354–355
    grape flowers immediately before, 87f
    during meiosis, 284–285, 328
    proportion of flowers that develop into berries following, 93
    skin cell division and, 94–95
    stomatal density and, 203–204
    volume of mesocarp cells and, 94–95
    weather conditions, effect on, 200
Anthocyanidin reductase, 228–229
Anthocyanins, 13, 38, 141, 171–173, 206, 215–216, 224t, 228, 230–231, 243–244, 258, 271, 306, 366–367
    accumulation of, 181, 231–232, 244–247, 327, 365–366
        in exocarp, 97–98
    cloudy or overcast conditions, impact of, 247
    color of, 233
    composition of, 233
    conversion of accumulating sugars into, 363–364
    cyanidin-based, 259
    disappearance of, 68–69
    free, 230–232
    malvidin-based, 252–253, 256–257
    molecular forms, 235
    peonidin-based, 259
    petunidin-based, 256–257
    photoprotective, 234, 296–297, 306–307
    production of, 71–73, 256–257, 355–356
        by crushing of grapes, 235
        glucose, role in, 233
    protection of, 232–233, 235–236
    reaction with acetaldehyde or pyruvate, 235–236
    red, 71–73
    reddish or bluish berry skin, role in, 233–234
    specific metabolic pathway, 233–234
    sun exposure, impact on, 247–248
    transport proteins and, 207–208
    in wine, 235
Apianic cultivars, 17
Apical meristem, 28–29, 34–37
Apoplast, 104, 118–121, 151–155, 184, 207–208, 290, 329–333, 336, 349
    boron concentration in, 314–315
    Ca^{2+} concentration in, 304–305
    Fe concentration in, 310–311
    K^+ concentration in, 304–305
    nonxylem, 121
    pH, 279–280, 309–311
Apoplastic (extracellular) sap, 279
Apoplastic ice formation, 339–340
Apoplastic pathway, 118–119, 154
Apoplastic unloading, 154–155, 207–208, 210
Apoplastic water, 330–331
Aquaporins, 101–102, 108, 111, 114–117, 119–123, 130–131, 147–149, 210, 273–274, 276–278, 291, 298, 318
Arginine, 189–190
Aromatic amino acid phenylalanine, 135–136
Assimilation, 130–136
Assimilation rate of a grapevine canopy, 160–161
ATP to ADP, 133–134
Axillary buds, 36
Axillary meristems, 36

B

Bacteria, 361–364, 363f
Bark, 40–41
Basal buds, 44
Basal inflorescence of a shoot, 86
Basic function of lamina, 45–46
Berry pericarp cells, 155, 155f, 203–204
Berry-pigment mutants, 15–16
Biologically effective day degrees, 74–75
Biopores, 80–81
Biotrophs, 352–353
Bleeding sap, 61–66, 64f, 293–294
Bloom, 56–57, 62f, 63f
    application of gibberellin sprays at, 250
    beginning of, 86–87, 87f
    bloom-time temperature effects, 201–202
    inflorescences with, 86–87, 87f
    yield potential in, 200
        cell divisions before and after bloom, 202–203
        conditions, 198, 200–201
        nitrogen (N) availability, 199
Bonarda, 16
Bordeaux cultivar, 15–16
Boron, 183, 223, 288f, 292, 314–316
    B toxicity, 315–316
    deficiency, 315–316, 315f
Botanical classification, 2–10
    class Dicotyledoneae, 4
    division Angiospermae, 4
    domain Eukarya, 3
    family Vitaceae, 4–5
    genus *Muscadinia*, 5
    genus *Vitis*, 5–7

kingdom Plantae, 3–4
order Rhamnales, 4
order Vitales, 4
Botrytis cinerea, 3–4, 93–94, 353, 358
Box cut, 33–34
Bracts, 37, 41–44, 52, 279
Brassinosteroids, 29–30
Brettanomyces, 227–228
Browning of shoots, 40–41
Bud fruitfulness, 197
Buds, 41–44
    accessory, 43
    apical meristem, 43
    basal, 44
    bracts with, 41–43
    cell division and auxin production in, 65
    distal, 67
    dormant, 43
    duration of chilling and lower temperatures and budbreak, 65–66
    grapevine, 41–43
    higher-order, 43
    latent, 43–44
    lateral, 41–43
    necrosis, 65–66
    secondary, 43
    during shoot's subsequent free-growth phase, 43
    tertiary, 43
    water contentduring budswell phase, 64–65
    winter, 43
Bunch rot, 353–357
Burgundy cultivar, 15–16

C

Cabernet franc, 15–16
Cabernet Sauvignon, 195, 360–361
    berry growth, 95f
    rate of sap flow, 115–116
    tannin concentrations, 230
Calvin, Melvin, 132–133
Calvin cycle, 132–133
    ATP to ADP in, 132f, 133
    carbon into starch production with, 134–135
    carboxylation step, 132f
    first stage: carboxylation, 132–133
    NADPH to NADP in, 133
    3-phosphoglycerate formation, 133
    rubisco in, 132–133
    second stage: reduction, 133
    sucrose in, 132f
    three stages of, 132f

Calyptra, 52–54
Canalization, 69–70
Cane pruning, 33–34
Canopy, defined, 163
Canopy–Environment Interactions, 163–182
    assimilation rate of a grapevine canopy, 160–161
    climatic conditions for resource availability in
        macroclimate, 164
        mesoclimate, 164
        microclimate, 164
    effect of altitude, 164
    grapevines productivity, influence of, 163
    humidity in, 179
        leaf growth with, 179
        transpiration with, 179
        vine water status with, 179
    ideal, 180–182
        canopy height/row width ratio in, 180
        canopy surface area in, 180
        fruit exposure in, 181
        fruit zone in, 181
        lateral shoot growth limited in, 181
        north/south oriented rows in, 180–181
        productivity upper limit with, 180–181
        pruning weight in, 181
        renewal zone in, 181
        shoots 15 nodes long in, 181
        15 shoots per meter of canopy in, 180
        vertically trained foliage in, 180
        yield to pruning weight ratio in, 181
    light in, 165–173
        compensation point, 166–167
        in the fruiting zone, 167
        grapevine adjustment to, 167–168
        importance for plant development of, 165
        leaf evolution for, 165–166
        leaf hairs in scatter of UV, 172–173
        leaf layers alter quantity/quality of, 169–170, 170f
        leaf's net CO_2 assimilation with zero, 166–167
        limited, 167
        photosynthetically active radiation (PAR), 165, 167
        range for vine vigor of, 167
        red/far red leaf absorption of, 169–170
        red/far red variations for day's, 170–171
        UV range with, 172–173
    in regions at higher latitude, 164
    south-facing slopes, 164–165
    temperature in, 174–177
        clouds alter, 176
        CO_2 assimilation influenced by, 174, 176
        global climate change with increase in, 176
        leaf growth with, 177

Canopy—Environment Interactions (*Continued*)
    leaf temperature changes, 174, 175*f*
    photosynthesis stimulated by, 175—176
  wind in, 177—179
    damage from, 177—178
    down row *vs* across row turbulence, 179
    leaf's boundary layer resistance decreased with, 178—179
Canopy photosynthesis, 163
Capfall, 86—87
Carbohydrates, construction costs of, 141—142, 142*t*
Carbon metabolism, 182—192
Carbon uptake, 130—136
  Calvin cycle with, 132—133, 132*f*
    ATP to ADP in, 132*f*, 133
    carbon into starch production with, 134—135
    first stage: carboxylation, 132—133
    NADPH to NADP in, 133
    3-phosphoglycerate molecule formation, 133
    rubisco in, 132—133
    second stage: reduction, 133
    sucrose in, 132*f*
  chloroplast with, 136
    starch production in, 133—134
    sugar production in, 133, 134*f*
  glucose molecules in, 135—136
  H_2O loss with, 131—132, 131*f*
  sucrose production with, 133—134
  uptake of CO_2 by leaves with, 130—131
Cardinal, 16
Catechin, 228—229
Cavitation, 111
"Cell-expansion" protein expansin, 44
Cellulose, 135—136, 141
Cell wall loosening in pulp, 207—208
Chalcone synthase (CHS), 226—227
Chicks, 93
Chilling requirement, 65—66, 77—78
Chlorophyll *a* and *b*, 127, 129—130
Chloroplast, 136
Cinnamic acid, 226—227
Clones, 19—20
  Gewürztraminer, 14
Cofactors, 235—236
Collection phloem, 151
Columella, 27—28
Concord, 16
  rate of sap flow, 115—116
Condensed tannins, 228—229
Convection, 109
Copigmentation, 235—236
Copper (Cu), 312—313
  deficiency, 312—313
  toxicity, 312—313
Cordon-trained grapevines, 33—34
Corolla, 84—85
Coulure, 93
4-coumaryl-CoA, 226—227
C_3 plants, 132—133
C_4 plants, 138
Crop load, 261—264
Crown buds, 44
Cultivar classification, 17—19
  anemic, 17
  apianic, 17
  based on use of grapes, 19
    brandy (distillation) grapes, 19
    juice grapes, 19
    raisin grapes, 19
    table grapes, 19
    wine grapes, 19
  based on viticultural characteristics, 18—19
  on the basis of winemaking characteristics, 19
    grape composition, 19
    production costs, 19
    varietal aroma, 19
  European wine grapes, 17
    Vinum francicum, 17
    Vinum hunicum, 17
  grape cultivars based on relative heat requirement, 18*t*
  grape varieties *vs* grape cultivars, 11
  by local soil, climatic conditions and cultural practices, 18—19, 18*t*
  mutation variation, 14
  natural hybridization, 14—16
  Negrul's grouping, 17—19
  nonanemic, 17
  proles, 17—18
    Proles occidentalis, 18
    Proles orientalis, 18
    Proles pontica, 17
  root growth and distribution is influenced by, 81—82
  V. sylvestris, 18—19
Cultivated grapevines, 11. *See also* Cultivar classification
  code of nomenclature, 11
  commercial, 12
  by hybridization, 12
  with "muscat" aroma, 15
  mutations, 12—13
    berry-pigment mutants, 15—16
    genetic influence, 13
    somatic, 12
  propagation by cuttings, 16—17

for winemaking, 11
  historical evidence, 11
Cytokinins, 26, 38–40, 79, 81, 121, 141, 284, 294–295, 362–363
  action in root tips, 28–31
  in buds, 69–70
  callose deposition, role in, 40–41
  fruit development and, 202–203
  fungus-derived, 358
  isopentenyladenine-type, 38–40
  meristem activity and, 28–29
  production of, 86, 96–98, 279, 291–292, 299
  production of pollen stimulated by, 87–88
  ratio of auxin to, 269
  root-derived, 38–40, 65, 69–70, 159, 167–168, 269, 279
  root growth and, 26–27
  in stimulating cell division and cell expansion, 36, 47, 50, 65, 94–95, 295
  support of cambial activity, 38–40
  synthesis in the shoot, 69–70
  thermosensitivity of, 202–203
  zeatin-type, 38–40
Cytosol, 60, 104–105, 107–108, 154–155, 155f, 207–208, 213–214, 232–233, 239, 273–274, 298–299, 303
  cytosolic Ca^{2+} concentration, 60, 304–305, 322–323
  cytosolic Fe concentrations, 309–310
  cytosolic glutamine synthetase (GS1), 186, 284
  of root epidermis and cortex cells, 184
  sucrose production in, 133, 134f

D

Daktulosphaira vitifoliae Fitch, 3–4
Day length, 59–61
  at higher latitude, 61
  typical responses to, 60
2,4-D (2,4-dichlorophenoxyacetic acid), 46–47
DELLA proteins, 70–71
Developmental physiology
  yield formation, 190–191, 194–205
    components of grapevine yield, 194
    interactions between genotype and environment, 195
    internal and external factors, 194–195
    optimal balance between vine's vegetative and reproductive growth, 195
    production of fruit, 196
    pruning and other cultural practices, 196–197
    reproductive performance, 195
    shoot vigor for, 195–196
    vine capacity, 195
    vine self-regulation for, 196–197
    vine's vegetative and reproductive growth, 195
    vineyard yield, 194
    yield potential in, 197–205
      berry and seed photosynthesis, effect of, 203–204
      bloom occurrence with, 200
      bloom-time temperature effects, 201–202
      bud development in dense canopies, 198
      bud fruitfulness, 198
      cell divisions before and after bloom and, 202–203
      conditions during bloom, 198, 200–201
      effect of light intensity on fruitfulness, 198
      effect of seed number, 202
      final fruit yield, 200
      functional leaf area for inflorescence development, 198–199
      of grapevine bud, 197
      "heat wave" during the bloom–fruit set period, 203
      of inflorescence primordia, 198
      limitation in assimilate supply, effect of, 204
      maximum number of potentially fruitful shoots, 197–198
      need to supply carbon and other nutrients, 199–200
      nitrogen (N) availability during bloom, 199
      period of inflorescence differentiation, 199–200
      phosphorus and potassium deficiency, effects of, 199
      reserve status, 199–200
      temperature effects, 198–201
      time of day and, 204
      timing and rate of water or nutrient application, 204–205
      in viticulture, 197
      water deficit effects, 199
Dicotyledoneae (Magnoliopsida), 4
Dormant buds, 34, 36f, 41–44, 340–341
  adventitious root formation, 79
  chilling requirement to resume growth, 65–66, 77–78
  rehydration of, 64–65
  structural ice barrier protection, 331–332
Dormant-bud side (ventral side), 43
Downy mildew, 312–313, 351–353, 359–361
Drought stress, 116–117
Druses, 50

E

Ecodormancy, 78
Elastic modulus, 167
Electromagnetic spectrum, 126f, 165, 169–170, 172
Electron transport chain, 128–129, 138–140, 176–177
  photosynthetic, 185
  standard, 140–141
  transfer of electrons from NADH to oxygen, 140–141
Elicitors, 350
Embolisms in xylem network, 111–113
Embryonic root, 26–27
Endocarp, 55–57, 247
Endocytosis, 154–155

Endodormancy, 76–78
Endodormant buds, 77f
Endosperm nucleus, 91–92
Energy capture, 125–130
Epicatechin, 228–229
Epicuticular wax, 56–57
Erysiphe necator, 3–4, 154, 357–358
Ethanol production, 137
Eugenol, 227–228
Eukaryota, 3
Eurasian group, 5–7, 6t, 9–10
   Vitis amurensis Ruprecht, 10
   Vitis coignetiae Pulliat, 10
   Vitis davidii, 10
   Vitis romanetii Romanet du Caillaud, 10
   Vitis sylvestris, 9–10
   Vitis vinifera L., 9
Euvitis, 4–5
Evapotranspiration, 106
Exocarp, 55–57, 91–92, 95–98, 213–214, 227

F

Feedback inhibition, 134–135
Ferredoxin, 128–129
Fertilization, 14–15, 54, 88, 92. *See also* Pollination
   in angiosperms, 89–90
   B toxicity and, 315–316
   Ca deficiency, impact of, 305
   division of endosperm nucleus, 91–92
   double, 89–90
   formation of outer and inner integuments, 54–55
   fuel status and, 284–285
   K deficiency, impact of, 302
   ovary to develop into a berry, 93
   rate of pollen tube growth, 89
   in seeded grapes, 94–95
   temperature effects and, 325
Ferulic acid, 227
Flame Tokay, 16
Flavanols, 228–231, 233, 236, 252
Flavonoids, 226–227
Flavonols, 206, 224t, 228, 234–235, 257–258, 355–357, 359
   absorption of UV light, 172–173
   accumulation of, 246–247, 336
   aglycones, 235–236
   in dark-skinned cultivars, 234–235
   hydrolyzation of, 235–236
   inhibition of auxin transport, 172–173
   as "sunscreen," 87–88, 172–173, 234–235, 326
   useful markers of light exposure, 247
   UV-protective, 248
Floaters, 92

Floral organogenesis, 84–85
Flowers, 53f, 157f
   androecium of, 52–54
   anthesis and, 86–87
   from basal inflorescence of a shoot, 86
   calyptra of, 54
   flower initiation, 84–86
   growth stages, 62f, 63f
   gynoecium of, 52–54
   infection of, 358
   morphology/anatomy of, 52–57
   ovule of, 53f, 54–55
   pistil of, 52–54, 53f
   proportion developing into berries, 93
   stamens of, 52–54, 53f
   star, 86–87
   vascular bundles, 55–56
Foliage leaves, 44
Free energy per unit volume, 103
Fruitfulness, 196–197
   of basal buds, 44
   effect of light intensity on, 198

G

Galloylation, 229
Gamma-amino butyrate (GABA), 89
Gating, 101–102
Geneva double-curtain, 75–76
Geographical distribution
   American group
     Vitis aestivalis Michaux, 7–8
     Vitis berlandieri Planchon, 9
     Vitis candicans Engelmann, 9
     Vitis cinerea Engelmann, 9
     Vitis labrusca L., 7
     Vitis riparia Michaux, 8, 8f
     Vitis rupestris Scheele, 8
   Eurasian group, 5–7, 6t, 9–10
     Vitis amurensis Ruprecht, 10
     Vitis coignetiae Pulliat, 10
     Vitis davidii, 10
     Vitis romanetii Romanet du Caillaud, 10
     Vitis sylvestris, 9–10
     Vitis vinifera L., 9
Gewürztraminer clones, 14
Gibberellins, 70–71, 202
Girdling, 202
Global climate change
   consequences, 60, 333–334
     adaptation to a conversion, 99
     increase in photosynthesis, 143
     rise in temperature, 75, 176, 249

photorespiration with, 136–138
reasons
rising atmospheric CO_2 concentration, 137–138
Glucose phosphate molecule, 133–134
Glutamate dehydrogenase (GDH), 185f, 186–187, 189
ability to oxidize glutamate of, 190–191
Glycolate pathway, 137
Glycolysis, 138–141, 140f, 176–177, 213–214, 218–219, 223–226, 283
ATP production via, 273–274
Glycosyltransferases, 228
Goethe, Johann Wolfgang von, 36
Gouais blanc, 15–17
Graft incompatibility, 366
Grafting, 5–7, 21–24, 243–244, 264–265, 310–311, 333, 341, 361–362, 364–365
Grape, 11–12
berries, slow transpiration rate of, 212
brandy, 19
composition, 19
hybridization of, 14–15
juice, 19
mutations in, 12–14
production costs, 19
raisin, 19
table, 19
tannin and anthocyanin extractability during ripening, 231
varietal aroma, 19
wine, 19
Grapevines. *See also* Morphology/anatomy
buds of, 41–43
clones, 19–20
cooling tactics of, 109–110
convection, 109
radiation, 109
transpiration, 109–110
cultivar classification, 17–19, 18t
diffuse-porous, 38–40
family, 4–5
flow rate in field-grown vines, 115–116
flowers, formation and development of, 52–54
fruiting structures of, 52
genus *Vitis*, 5–7
meristems, 163
monitoring of seasons, 60
net assimilation rate of a grapevine canopy, 160–161
order Rhamnales, 4
phytoalexins characteristic of, 350
production of phenolics (A) and terpenoids (B) in, 225f
resistances to water flow through a, 122–123
root xylem of, 113
rootstocks, 21–24, 22t

shoot change and source–sink relations, 157f
tendrils and fruiting clusters of, 50
varieties *vs* cultivars, 11
viticultural traits of American and Eurasian, 6t, 7–10
water balance of, 276
Gravimorphism, 75–76
Gravity-sensing columella cells, 81
Ground state, 127
Growing degree days (GDD), 74–75
Growth cycle
day length, 59–61
reproductive cycle, 83–99
seasons, 59–61
vegetative cycle, 61–83
Guaiacol, 227–228
Guttation, 46
Gynoecium, 52–54

H
Heat dissipation, 109–110
convection, 109
radiation, 109
transpiration, 109–110
Heliothermal index, 74–75
Hemicellulose, 135–136
Hens, 93
Heritability, 10–11
leaf shape in grapevines, 10–11
Higher-order buds, 43
Hydathodes, 46
Hydraulic capacitance, 121–122
Hydraulic conductance, 114
Hydraulic effect of stomatal opening or closing, 107–108
Hydraulic lift, 117–118
Hydraulic redistribution, 117–118
Hydrolysable tannins, 228–229
Hydrotropism, 82–83
Hydroxybenzoic acids, 224t
Hydroxycinnamic acids, 224t, 227–228

I
Idioblasts, 56–57, 218, 305
Imbibition, 26–27
Inflorescence necrosis, 190–191, 201
Inflorescences, 4, 36–38, 43–44, 52, 68f, 93, 190–191, 194, 284–285, 309, 328, 340, 346, 360
abortion of, 200–201, 204
basal, 86
beginning of bloom with, 86–87, 87f
buds with, 84
flower differentiation with, 84–85, 85f

Inflorescences (*Continued*)
    on lateral shoots, 85–86
    N deficiency and N excess, effect of, 199
    number of flowers per, 86, 199–200
    practice of cutting through, 250
    temperature effect on, 198
Interfascicular cambium, 38–40
Intrafascicular cambium, 38–40
Ion channels, 101–102, 119–121, 281
Iron, 308–311
Isoprenoids, 141

J

Jumping genes, 12–14, 90, 241–242

K

α-ketoglutarate, 186
Königin der Weingärten, 16

L

Lamina, 41–43, 366f
    basic function of, 45–46
    growth of, 47
    petiole and, 44
    vein network of, 45–46
Latent buds, 43–44
Lateral buds, 41–43
Lateral meristems, 36
Lateral roots, 26–27
Lateral shoots, 41–43
Latitude-temperature index, 74–75
Leaves, 44–50, 49f
    auxin production in, 46–47
    basic function of lamina, 45–46
    basis of ampelography, 45–46
    bracts, 44
    bundle sheaths, 45–46
    cell division and expansion in formation of, 46–47
    Chardonnay, 45f
    cotyledons, 44
    differentiation of, 44
    epidermis, 48
    foliage, 44
    grapevine, 45–46, 48–49
    hydathodes, 46
    margins of, 46
    Merlot, 45f
    mesophyll, 49–50
        cells of, 45–46
    N concentration in, 189
        rate of photosynthesis and, 191
    nitrogen assimilation in, 183–184, 188–189
    number and length of trichomes in, 47–48
    palisade parenchyma, 49–50
    petiole of, 44
    photosynthetic efficiency of, 50
    primordia, 44
    scales, 44
    shape and basic architecture of its vascular system, 46–47
    stomatal density of, 48–49
    V. vinifera, cross section of, 45f
    vascular bundles of, 44
    vein network, 45–46
    of *Vitis*, 45–46
Leeaceae, 4
Lenticels, 40
Leucoanthocyanidin reductase, 228–229
Light, 165–173
    anthocyanin accumulation in, 246–247
    compensation point, 166–167
    cultivar variation and fruit's susceptibility to shade, 247–248
    DOXP/MEP pathway, role of, 248–249
    in the fruiting zone, 167
    grape berries composition/quality influenced by, 245–249
    grape norisoprenoids and, 248–249
    grapevine adjustment to, 167–168
    importance for plant development of, 165
    leaf evolution for, 165–166
    leaf hairs in scatter of UV, 172–173
    leaf layers alter quantity/quality of, 169–170, 170f
    leaf's net CO_2 assimilation with zero, 166–167
    limited, 167
    methoxypyrazine accumulation and, 248–249
    photosynthetically active radiation (PAR), 165, 167
    range for vine vigor of, 167
    red/far red leaf absorption of, 169–170
    red/far red variations for day's, 170–171
    shaded fruit *vs* sun-exposed fruit, 246–247
    tannin accumulation and, 247–248
    tartrate and, 246–247
    UV range with, 172–173
    within-canopy microclimate, 245–246
    xanthophyll carotenoid production and, 248
    on yeast microflora, effect of, 249
Light absorption, 125–130
    conversion of light energy into chemical energy, 126–127
    electromagnetic spectrum with, 126, 126f
    green chlorophyll, role of, 127
    maximizing, 45–46
    visible light with, 126, 247–248
Light compensation point, 166–167
Lignin, 135–136
Lipids, 48, 73, 97–98, 139, 141, 270, 304–305, 335, 349

construction cost of, 141–142, 142t
grape berries composition/quality influenced by, 236–242
mutations or cell death by oxidizing, 129
"Lizard's tail" strategy, 112–113

M

Macroclimate, 164
Magnesium (Mg), 222, 288f, 289, 306–308
Magnoliophyta, 4
Magnoliopsida, 4
Malate, 61–64, 96–98, 102–103, 107–108, 139, 140f, 147–149, 186–188, 207–208, 213–214, 216–219, 239, 241–242, 244–245, 251–253, 256, 258–261, 283, 299, 310–311, 325–326, 336, 353, 366–367
Malvidin-based pigment, 233–234
Malvidin-3-glucoside, 233–234
Manganese (Mn), 127–128, 130, 223, 288f, 292–293, 299, 313–314
Marginal leaf necrosis, 364
Matrix potential, 103
Merlot, 195
    tannin concentrations, 230
Mesocarp, 91–98, 154–155, 204, 206–207, 213–214, 218–219, 222–223, 227–229, 232–233, 241–242, 255
    cell vacuoles, 207–209, 217–218
    cell walls, 208–211
    expansion, 208
    exudation of sap from, 343–344
    fatty acids, 236–237
    of a mature berry, 215–216
    nitrogen, 219–220
    pH, 221–222
    turgor pressure, 210–211
Mesoclimate, 164
Mesophyll cells, 45–46, 49–50, 68–69, 73, 106–108, 122–123, 130–132, 151–152, 155, 165, 169, 184, 272, 278, 309–311, 360
Microclimate, 164
Micropyle, 54–55
Mild plant water stress, 276
Millerandage, 93
Mineral nutrients, grape berries composition/quality with, 219–223
Mitochondria, 3
Mitochondrial electron transfer chain, 140–141
Molybdenum (Mo), 288f, 292–293, 314
    cofactor synthesis, 312
Monocotyledons, 26–27
Morphology/anatomy, 24–57
    buds, 41–44
        apical meristem, 43
        axillary, 36–37
        basal, 42f, 44
        basal-bud fruitfulness, 44
        crown, 44
        dormant, 41–43
        dormant Cabernet Sauvignon, 42f
        higher-order, 43
        initiation of bracts, 41–43
        latent, 41–44
        lateral, 41–43
        prompt, 41–43
        secondary, 43
    callose deposition, 40–41
    cambial activity, 38–40
    cell differentiation
        cortex, 38
        endodermis, 38
        epidermis, 37–38
        pearls, pearl glands and pearl bodies, 37–38
        phloem, 38
        procambium, 38–40
        specialized protein storage vacuoles, 38
        starch sheath, 38
        vascular bundles, 38
        vascular tissues, 38
        xylem, 38
    clusters, 50–52
        architecture, 52
        auxin and, 52
        bracts in, 52
        hypoclade, 52
        pedicel, 52
        peduncle, 52
        rachis, 52
        vascular bundles of, 52
    flowers, 52–57
        epicotyl, 54–55
        female, 52–54
        gynoecium of, 52–54
        male, 52–54
        Muscat Hamburg, 53f
        ovules of, 54
        pistil, 52–54
        sepals and petals, 52–54
        stamen, 52–54
    grape berries, 52–57
        endocarp of, 57
        epicuticular wax, 56–57
        epidermis of, 56–57
        exocarp, 56–57
        locules, 54
        mesocarp of, 57
        pericarp, 55–56

Morphology/anatomy (*Continued*)
    structure of a ripe grape berry, 55f
    sun-exposed berries, 56–57
    tracheary elements, 55–56
  lateral organs, development of, 36–37
    accumulation of auxin, 36–37
    cytokinins, role of, 36–37
  leaf-opposed tendrils and clusters, production of, 37
  leaves, 44–50, 49f
    auxin production in, 46–47
    basic function of lamina, 45–46
    basis of ampelography, 45–46
    bracts, 44
    bundle sheaths, 45–46
    cell division and expansion in formation of, 46–47
    Chardonnay, 45f
    cotyledons, 44
    differentiation of, 44
    epidermis, 48
    foliage, 44
    grapevine, 45–46, 48–49
    hydathodes, 46
    margins of, 46
    Merlot, 45f
    mesophyll, 49–50
    mesophyll cells of, 45–46
    number and length of trichomes in, 47–48
    palisade parenchyma, 49–50
    petiole of, 44
    photosynthetic efficiency of, 50
    primordia, 44
    scales, 44
    shape and basic architecture of its vascular system, 46–47
    stomatal density of, 48–49
    V. vinifera, cross section of, 45f
    vascular bundles of, 44
    vein network, 45–46
    of *Vitis*, 45–46
  nodes, 41–44
  phloem reactivation, 40–41
  phyllotactic positioning of lateral organs, 36–37
  rings of xylem, 38–40
  root system, 26–33
    absorption zone, 30
    cell differentiation in formation of, 28–29, 28f
    cell division in, 28–29
    elongation zone of, 27–28
    endodermis, 29
    of established grapevines, 27
    horizontal roots, 27
    hypocotyl development of, 26–27
    initiation and elongation of root hairs, 30
    lateral roots, 26–27, 29–30
    lignification and, 32–33
    meristem cells of, 28–29
    Merlot, 28f
    parenchyma cells, 32–33
    perforation plates, 33
    pericycle, 29–30
    phloem, 30–31
    programmed cell death, 32–33
    root cap, 27–28
    root tip or apex, 27–28
    surface area of, 27
    tracheary elements, 32–33
    tracheids, 33
    uptake of nutrients and water, 27–28
    V. vinifera, 32f
    vascular cambium, 30–31
    vessel elements, 33
    xylem cells, 30–31
  shoots, 33–41
    apical meristem, 34–35
    browning of, 40–41
    central zone, 34–35
    Concord shoot tip, 35f
    epicotyl (embryonic shoot), 34
    "fixed" growth and "free" growth of, 34
    internodes, 37–38
    juvenile phase of growth, 34
    lateral, 36, 41–43
    lateral meristems, 34–35
    pearls of, 37–38
    peripheral (or morphogenetic) zone, 34–35
    phyllotaxy of, 34
    pruning of, 33–34
    rib zone, 34–35
    tendrils, 34
    tertiary, 41–43
    three-node pattern of Syrah, 36f
    V. vinifera, 39f
  tendrils, 50–52, 51f
    intercalary growth of, 50–52
    thigmotropic movement, 50–52
    tips, 50–52
  trunks, 33–41
    sucker, 35
  tyloses, development of, 40
Mucigel or mucilage, 27–28
Müller–Thurgau, 201f, 220–221, 295f, 296f, 307f
Muscadinia species, 4–5
  Muscadinia munsoniana Small, 5
  Muscadinia popenoei Fennell, 5
  Muscadinia rotundifolia Small, 5, 52–54, 233–234, 310–311
    resistance to fungus, 358–359
Muscat of Alexandria, 15

N

NADPH to NADP in Calvin cycle, 133
Necrotrophs, 352
Nectaries, 54
Nickel (Ni), 292–293, 316
Nitrite reductase (NiR), 184–185
Nitrogen, 293–298
Nitrogen assimilation, 141–142, 182–192
  ammonium assimilation, 185–187
  from cells to plants, 187–192
    application of N fertilizer, 189–192
    energy availability, 187–188
    environmental factors, 188, 190–192
    leaves in assimilation for, 188–189
    nitrogen as nucleic acid component for, 187
    nitrogen distribution in soil for, 187
    photosynthesis effect on, 190–191
    response of roots, 187
    roots/shoots nitrate assimilation for, 187–188
    sucrose transportation and, 188
    sufficient photosynthate energy for, 189–190
  definition, 182
  in grapevines, 182–185, 184f
    energy-dependent, 188
    with high carbohydrate status, 188
  nitrate uptake/reduction with, 183–185, 184f
    activity and forms of NR, 184
    boron (B) deficiency and, 183
    grapevines store/distribute, 184–185
    nitrate reductase forms for, 183
    protein ferredoxin in, 185
    in roots, 183–185
    vacuolar storage pool for, 184–185
    in *Vitis* species, 183
  nitrogen-assimilating GS/GOGAT cycle, 226
  process, 182–183
Nitrogen deficiency, 188, 192, 293–294
Nitrogenous compounds, grape berries composition/quality with, 219–223
Nodes, 34
  ideal canopy with shoots of 15, 181
  morphology/anatomy of, 41–44
  shoots distinguished from, 189–190
  three-node pattern of a Syrah shoot, 175f
Nomentanic cultivars, 17
Nucleotides, 141
Nutrient assimilation, photosynthesis with, 146, 190
Nutrient remobilization, 190
Nutrient uptake and transport, 110–124, 289–290
  active transport of, 120
  ATP hydrolysis, 119–120
  carriers of macronutrients, 120–121
  Casparian bands with, 118–119, 118f
  cell-to-cell movement of solutes, 118–119
  via endodermis cell membranes, 118–119
  hydraulic lift or hydraulic redistribution, 117–118
  inside xylem conduits, 121
  ion channels for, 119–120
  organic acids, 121
  passive, 120–121
  in phloem, 123–124
  root pressure and, 119
  via roots, 117–118, 118f
  semipermeability of cell membranes, 119
  soil–root–shoot pathway, 117–118
  symplastic and apoplastic pathways, 118–119, 118f
  transpiration stream, 121–122
  transport proteins, role of, 119–120
  water flow dependent, 117–118
Nutrients, 121–122
  grape berries composition/quality with, 257–261
  macronutrients, 293–308
    calcium, 304–306
    magnesium, 306–308
    nitrogen, 293–298
    phosphorus, 298–300
    potassium, 300–303
    sulfur, 303–304
  shoot/root growth influenced by, 83
  transition metal/micronutrients, 308–317
    boron, 314–316
    copper, 312–313
    iron, 308–311
    manganese, 313–314
    molybdenum, 314
    nickel, 316
    silicon, 316–317
    zinc, 311–312

O

Ohm's law, 106, 113–114
Organic acids, 57, 61–64, 74, 96, 102–103, 120–121, 139, 141, 142t, 147–149, 152, 182–183, 186–188, 213, 216–218, 231, 235, 251–252, 258, 283, 299, 310–311, 313–314, 325, 336, 366–367
Osmosis, 101–108
  aquaporins with, 101–102
  cell expansion and, 104–105
  definition, 102
  diffusion with, 103f
  osmotic potential, 102
  osmotic pressure, defined, 102
  osmotic solutes (osmolytes), 102–104
  turgor pressure, 102–103
Osmotic pressure, 102–103

Ovules, 53f, 54–55, 92
Oxidative phosphorylation, 139–140
Oxidative stress, 270, 272, 284, 296–297, 301–302, 306–307, 309, 313–314, 320, 327
Oxidized tannins, 230
Oxygenation, 136–137

P

Palisade parenchyma, 49–50
Panicles, 52
Paradormant buds, 77f
Parthenocarpic fruit development, 92–93
Partial rootzone drying (PRD), 280–281
Partitioning, 156–163
   communication with, 159
   competition with, 159
   connection with, 158
   definition, 157–159
   development with, 159
   environmental variables with, 159–160, 163
   interference with, 158–159
   patterns of, 157–159
   proximity with, 158
Peonidin-based pigment, 233–234
Pericarp, 54–57, 55f, 91–92, 94–98, 154, 208–209, 223
Peroxidases, 227–228
Petiole, 44
Petunidin-3-glucoside, 233–234
Phenolics, 48, 140f, 142t, 205–206, 223–236, 348–349
   amino acids with, 223
   anthocyanins, 231–234
   bonding with cell wall polysaccharides and polymers, 208
   class of phenolic compounds, 223
      in grapes, 224t
   in cutin matrix, 210
   enzymatic oxidation of, 223
   flavonoids, 226–230
   flavonols, 234–235
   glycosylated, 228
   incorporation into leaf cells, 172–173
   nonflavonoid, 227
   perception of astringency from, 229–230
   production of, 223–226
      biosynthetic pathways, 225f
   reactions in winemaking practices, 235
   in seed coat, 218
   smoke-derived, 227–228
   synthesis pathways
      malonate, 223–226
      phenylpropanoid, 226
      shikimate, 223–226
   tannins, 228–231
   ultraviolet (UV)-protecting, 60, 172–173
Phenology, 59–60
Phenylalanine ammonia lyase (PAL), 226
Phloem
   bulk flow rate of solutes, 149–150
   cell differentiation, 38
   collection, 151
   minor-vein, 151–152
   nutrient uptake and transport in, 123–124
   pathways for sugar to vacuole, 155f
   pressure flow mechanism and, 156
   reactivation, 40–41
   release, 151, 153–155
   root system, 30–31
   sequential sectors of, 151
   short pieces of RNA, 147–149
   sucrose concentration in, 153–154
   sucrose loading into, 152–154
   in sugar metabolism, 147–149
   symplastic to apoplastic, 154–155
   transport, 152–153
      rate in, 150–151
   water movement into and out of, 150–151
Phloem sap, 121, 147–149, 148t, 209
   along transport path, 149–150
   composition of, 148t, 149, 290–291
   flow velocities of, 149, 155–156
   pressure flow theory of, 149
   sugar concentration in, 283
Phosphate deficiency, 134–135
Phosphate translocator, 134–135
3-phosphoglycerate, 132–133
Phosphorus, 298–300
   deficiency, effects of, 199
Photochemical quenching, 65
Photochemistry, 64–65
Photoinhibition, 130, 137
Photons, 60
   energy content (E) of, 61
Photoperiodism, 60
Photophosphorylation, 129
Photorespiration, 136–138
Photorespiratory carbon oxidation cycle, 137
Photosynthate, 143
   allocation, 156–163
      assimilate surplus in, 161, 162f
      assimilation rate determined by source in, 160–161
      categories, 156
      patterns in, 158–159
      phosphates availability in regulation of, 156
      source-sink relation with, 157–159, 157f
      storage in, 156

transport in, 156
utilization in, 156
photosynthetic machinery, 146–147
production of, 146
sufficient energy for, nitrogen assimilation with, 189–190
translocation/distribution of, 146–163
    amino acids in, 148t, 152
    flexibility with, 147
    nutrient assimilation with, 146–147
    phloem sap in, 147–149, 148t
    sink with, 146–147, 153–154
    solute unloading in, 153–154
    sucrose in, 147–149, 154–155
    xylem sap in, 148t, 149
in woody organs of the vine, 147
Photosynthesis, 60, 138
    capacity per unit leaf area, 167–168
    carbon uptake/assimilation with
        Calvin cycle with, 132–135, 132f
    from cells to plants, 141–143
    chlorophyll content and, 146–147
    chloroplast with, 61, 136
        starch production in, 133–134
        sugar production in, 133, 134f
    construction costs of compounds with, 141–142, 142t
    conversion of carbon dioxide (CO_2) and water (H_2O) to sugar (carbohydrate), 61
    in C_4 plants, 138
    electromagnetic spectrum, 126f, 165, 169–170, 172
    energy content (E) of photons and, 126
    energy overload with, 130
    energy transfer with, 129
    and export rate of canopy, 160–161
    light absorption/energy capture with, 125–130
    light limited, 167
    light-saturated, 166–167
    nitrogen assimilation with, 190–191
    organic compounds produced during, 146
    photochemical capture of solar energy, 126–127
    photosynthate $(CH_2O)_n$, 133
    processes
        chlorophyll excitation, 127
        conversion of light energy into chemical energy, 129–130
        electron transport chain, 128–129
        energy-releasing mechanisms, 127–128
        multiprotein complex, formation of, 127–128
        photochemical quenching, 128–129
        photoinhibition, 130
        photophosphorylation, 129
        production of ATP from ADP, 129
        water-splitting reaction, 127–128
    product of, 137
    quantum yield or quantum efficiency, 129
    respiration, 141–143
    shade on the leaves and, 246
    sucrose export with, 133–134
        rate, 135
    temperature change and, 175–176
    visible light with, 126
    whole-canopy, 167
Photosynthetic carbon reduction cycle, 132–133
Photosystem II (PSII), 127–130
Phyllotaxy or phyllotaxis, 34
Phytoalexins, 348–350
    detoxification of, 350–351
    production of, 350
    rate of accumulation at the site of infection, 350–351
    stilbene, 355, 359, 361
Phytoanticipins, 348–349
Pierce's disease, 350, 364
Pinot family, 14f, 15–16
    Pinot blanc, 13, 234
    Pinot gris, 13, 233–234
    Pinot noir, 13, 233–234
    tannin concentrations, 230
Pistil, 52–54, 84–85, 89, 91–92
Plantae kingdom, 3–4
Plasmodesmata, 31, 118–119
Plasmopara viticola, 352–353, 359–361
Plastochron, 44
Plugging of vessels, 364
Pollen, 87–88
    production of, 87–88
    tube growth, 88–89
Pollination, 54–55, 87–88, 91–93, 200, 204, 284–285, 302, 312. *See also* Fertilization
    chilling stress and, 328
    cross, 12, 14–15, 88
    insect, 88
    K deficiency, impact of, 302
    self, 12
Polymerization, 230
Polyphenoloxidases, 227–228
Porphyrins, 141
Powdery mildew, 181, 260–261, 302–304, 343–344, 348–350, 357–359
    fungus *E. necator*, 178–179, 316–317, 358
        resistance against, 359
    infection, 316–317, 336, 357
Preprimordium, 35
Primitivo, 15–16
Proanthocyanidins, 228–229
Procambium, 38–40

Prokaryotes, 3
Proles occidentalis, 18–19
Proles orientalis, 18–19
Proles pontica, 17–19
Proline, 192
Prompt buds, 44
Prompt-bud side (dorsal side), 43
Proteolysis, 99
Pruning
   cane, 33–34
   of shoots, 33–34
   spur, 33–34
   strategies in different seasons, 61
   vine capacity depressed with, 196
   weight in canopy–environment interactions, 181
   yield formation and, 196–197
Pseudomonas syringae, 331
Pyruvate, 139–140

Q

Quanta, 125–126
Quantum leap, 127
Quercus suber L., 227

R

Radiation, 109
Radicle, 26–27
Rain shadow effect, 105–106
Raphides, 50
Reaction center, 127–128
Red/far red light, 167–171, 332–333
Reductive pentose phosphate pathway, 132–133
Regeneration, 133
Release phloem, 151, 153–155
Reproductive cycle, 83–99. *See also* Vegetative cycle
   abscission of stamens, 87–88
   anthesis in, 86–87, 87*f*
   beginning of bloom, 86–87, 87*f*
   cell division in berry, rate and duration of, 94–98
   cell expansion during cluster development, 93–94
   chromosome mixing, 90–91
   cluster initiation/differentiation in, 85*f*
   cytokinins and auxins, production of, 86, 94–95
   endosperm development, 89–90
   endosperm nucleus division, 91–92
   extent of branching before and after dormancy period, 86
   fertilization, 89–90, 94–95
   floral organogenesis, 84–85
   flower formation, 83–84
   flower initiation, 84–85
   formation of lateral meristems, 83–84
   fruit development and growth, 91–93
   fertilized berries, 93
   lag phase, 96–97
   parthenocarpic grapes, 92–93
   rapid increase in size of seeds and pericarp, 96
   ripening period, 97–98
   stenospermocarpic grapes, 92–93
   inflorescence and flower differentiation, 84–85, 85*f*
   inflorescences, formation of, 83–84
   inflorescences per bud with, 84
   meiosis process, 90–91, 93
   millerandage, 93
   pollen production, 87–88
   pollen tubes, growth of, 89
   pollination, 87–88
   primordia
      branching, 86
      development of, 83–84
   seed development, 91–92
   seed germination, 98–99
   seed maturation, 98–99
   seedless grape berries, 98
   shoot organogenesis, 90
   somatic embryogenesis, 90–91
   stratification of seeds, 98–99
Resorption, 190
Respiration, 138–143
   ADP to ATP, 140–141
   carbon intermediates from, 141
   citric acid cycle, 139–140, 140*f*
   efficiency, 139
   electron transport chain, 140–141
   glycolysis, 139, 140*f*
   mitochondria and, 139–140
   nitrate and sulfate acquisition, costs of, 143
   oxygen intermediates from, 140–141
   rate of meristems, 143
   of a vineyard soil, total annual, 142–143
Resveratrol, 355–356
Rhamnaceae, 4
Rhamnales, 4
Rhizobia, 182–183
Rhizobium vitis, 361–362
Rhizosphere, 120–121
Ripe grape berry, structure of, 55*f*
   exocarp, 56–57
   mesocarp, 57
Roots, 26–33, 115–116, 143
   ABA derived, ability of, 107–109
   ability to find water, 82–83
   absorption zone, 30
   adventitious, 26–27, 79
   allorhizic root system, 26–27
   biomass of grapevine, 27

cavitation in the surface, 113
cell differentiation in formation of, 28–29, 28f
cell division in, 28–29
cytokinin action
   growth, role in, 26–27
   in root tips, 28–29
delivered amino acids, 152
derived cytokinins, 38–40, 65, 69–70, 167–168, 269, 279
elongation zone of, 27–28
endodermis, 29
of established grapevines, 27
flow of water from the soil toward, 113–114
gravity response, 82–83
growth, 79
   soil temperature, influence of, 61–64
   of tips, 79–80
horizontal, 27
hypocotyl development of, 26–27
information exchange between shoots and, 26
initiation and elongation of root hairs, 30
lateral, 26–27, 29–30
lignification and, 32–33
of mature vines, growth pattern of, 82–83
meristem cells of, 28–29
Merlot, 28f
nitrate uptake/reduction in, 183–185
nutrient uptake and transport in, 117–118, 118f
osmotic pressure of xylem and osmotic water uptake by, 61–64
parenchyma cells, 32–33
perforation plates, 33
pericycle, 29–30
phloem, 30–31
programmed cell death, 32–33
response in nitrogen assimilation, 187–188
root cap, 27–28
root tip or apex, 27–28
starch accumulation in, 134–135
surface area of, 27
tracheary elements, 32–33
tracheids, 33
uptake of nutrients and water, 27–28
V. vinifera, 32f
vascular cambium, 30–31
vessel elements, 33
Vitaceae, 4–5
water uptake and transport, 111–112, 116
whole-plant respiration and growth via, 142–143
woody, 27
xylem cells, 30–31
xylem of grapevines, 113
Rootstocks
   agronomic characteristics of, 22t
   amino acids and, 265
   commonly used, 21
   grafting of, 21
   influence on composition of the fruit, 24
   in management of vine vigor, 21–24
   root growth and distribution is influenced by, 81–82
Rubisco, 127–128, 136–137, 191, 284, 296–297, 300, 306, 327
   Calvin cycle with, 132–133
   carboxylation function of, 137–138
   in C_4 plants, 138
   during drought stress, 283
   effect on photosynthesis, 272
   N content and, 191
   oxygenation rate, 176–177
   temperature changes and, 176–177

S

Saccharomyces cerevisiae, 3–4
Salinity, 317–321
Sauvignon blanc berries during, 205f
Savagnin, 15–16
Scavenging cells, 46
Scion cultivars, 21, 81–82, 310–311
Scott–Henry systems, 75–76
Seasons, 59–61
   endogenous "clock," 60
   pruning strategies and, 61
   reproductive cycle, 83–99
   time of active growth, 61
   vegetative cycle, 61–83
Second crop, 41–43
Secondary meristems, 36
Seedless table grapes, 92–93. *See also* Thompson Seedless
Senescence, 136
Shatter, 93
Shedding, 93
Shoot growth, 67, 123–124
   accumulation of auxin in, 75–76
   in correlative inhibition, 67
   cycle, 71–73
   following budbreak, 67–68
   internodes development, 70–71
   leaf unfolding, 70–71
   N reserves, role of, 189–190
   rapid, 76
   secondary, 76
   shoot apex, sensitivity of, 71
   shoot "maturation," 71
   slowing down of, 76–77
   soil compaction, impact of, 83
   starting of, 67
   temperature effect on, 74–76

Shoot growth (*Continued*)
  time of rapid, 76
  upward and downward, 67–68
  vigor, 74–75
Shoot organogenesis, 90
Shoots, 33–41, 134–135
  browning of, 40–41
  central zone, 34–35
  Concord shoot tip, 35*f*
  epicotyl (embryonic shoot), 34
  "fixed" growth and "free" growth of, 34
  internodes, 37–38
  juvenile phase of growth, 34
  lateral, 36, 41–43
  leaf-opposed tendril production with, 37
  meristems, 143
    apical, 34–35
    lateral, 34–35
  pearls of, 37–38
  peripheral (or morphogenetic) zone, 34–35
  phyllotaxy of, 34
  pruning of, 33–34
  rib zone, 34–35
  tendrils, 34
  tertiary, 41–43
  three-node pattern of Syrah, 36*f*
  V. vinifera, 39*f*
Shot berries, 93, 201
Silicon, 316–317
Sink, 146–147, 153–154
  sink-to-source transition, 146–147, 153–154
  solutes unloaded in, 153–154
  source–sink relation, 157–159, 157*f*
    shoot change, 157*f*
  strength, 157–159
  sucrose imported into tissues of, 155
Sinkers, 92
Sodicity, 317–318
Sodium adsorption ratio (SAR), 317–318
Soil–plant–air continuum, 110–111
Somatic embryogenesis, 90–91
Specific conductivity, 114
Spur pruning, 33–34
Stamens, 52–54, 84–85
Starch accumulation, 134–135
Starch "assembly line," 134
Stenospermocarpic fruit, 92–93
Stenospermocarpy, 92–93
Stilbenes, 224*t*
  accumulation of, 355–356
Stomatal action, 106–110
  guard cells of closed stomata, 107–108
  protection of xylem conduits, 107
  reaction of stomata to ABA, 108–109
  role in regulating diffusional water loss, 107
  stomatal aperture, 107
  stomatal conductance, 107
  stomatal opening and closure, 107–108
Stress physiology
  abiotic stress, 268–272
  biotic stress, 343–352
  carbon depletion, 271
  flooding, 273–274
  nutrients: deficiency or excess in, 287–321
    macronutrients, 293–308
    transition metal/micronutrients, 308–317
  organisms: defense and damage, 272–287
    bacteria, 298–300
    biotrophs, 270
    bunch rot, 287–321
    downy mildew, 293–298
    necrotrophs, 270
    phytoalexins, 268, 272, 278–280
    powdery mildew, 293–308
    PR proteins, 268–269
    viruses, 300–303
  oxidative stress and damages, 270–271
    antioxidant defense systems, 270–271
  responses to abiotic stress, 268–272
    antioxidant defense systems, 270–271
    auxin and cytokinin concentrations, change in ratio of, 269
    compatible solutes, role of, 269–270
    optimum resource allocation, 268–269
    organ sacrifice, 271
    reduction of leaves' photosynthetic rate, 271
    rubisco, role of, 272
    senescence and abscission process, 272
    shoot and root apical meristems, adaptation of, 268–269
  salinity, 317–321
  temperature: too cold or too warm in, 321–341
    chilling stress, 326–328
    cold acclimation/freeze damage with, 329–341
    heat acclimation/damage, 322–326
  waterlogging, 273–274
  Vitis species, susceptibility of, 274
  water stress, 276
Strigolactones, 38–40, 69–70, 79
Sucrose, 133–134
  in amino acids, 147–149
  in Calvin cycle, 132*f*
  carbon uptake and production of, 133–134
  in cytosol, 133, 134*f*
  loading into phloem, 152–154

in phloem, 153−154
phosphate groups (P_i), release of, 133−134
photosynthesis and export of, 133−135
synthesis, 133−134
translocation/distribution of photosynthate, 147−149, 154−155
transportation, 188
transporter (ST) proteins, 154−155
Supercooled tissue, 329−330
Symplast, 107−108, 118−119, 118f, 122−123, 151−155, 329−330, 332−333
Sympodial bundles, 38
Syrah, 15−16
tannin concentrations, 230

T

Tannins, 224t, 229
classification of, 228−229
colorless, 229
concentration in red wines, 231
condensed, 228−229
extractability during ripening, 231
functional properties, 231
genes responsible for biosynthesis of, 230
hydrolysable, 228−229
polymerization of, 229−231
production of flavanols, 230−231
in seeds, 230
subunits in the skin, 230−231
wine's astringency and, 229−230
Tartrate, 227
Temperature, 174−177
clouds alter, 176
CO_2 assimilation influenced by, 174, 176
global climate change with increase in, 176
influencing concentration of amino acids, 253
leaf growth with, 177
leaf temperature changes, 174, 175f
photosynthesis stimulated by, 175−176
Temperature inversions, 164
Tendrils, 50−52, 51f
intercalary growth of, 50−52
thigmotropic movement, 50−52
tips, 50−52
Thigmomorphogenetic response, 177−178
Thigmotropic movement, 50−52
Thompson Seedless, 15, 18−19, 44, 65−66, 92−94, 356
application of gibberellin sprays at bloom, 250
fruitfulness of buds, 83−84
inflorescence initiation and development, 198
maintenance of water flow, 115−116
rate of sap flow, 115−116
vigor of, 74
Thylakoid lumen, 129
Torus, 52−54
Tracheobionta, 3−4
Traminer, 16−17
Transaminases, 189−190
Transpiration, 106−110
boundary-layer resistance and, 106
of canopies, 106
definition, 106
driving water flow up xylem, 109
equation, 107
factors influencing, 106
in grapevine leaf, 107
negative pressure of xylem and, 109−110
rate per unit leaf area, 106
stomatal pore resistance and, 106
temperature as determinant of, 106−107
turnover of water in transpiring leaf, 109−110
in a typical grapevine leaf, 107
water uptake via, 110−111
Transpiration−cohesion−tension theory of xylem sap flow, 110−111
Transpiration stream, 121−122
Trichoblasts, 30
Triose phosphate transporter (TPT), 134−135, 134f
Trunks, 33−41
sucker, 35
Turgor pressure, 102−103
Tyloses, 40

U

Uncinula necator (Schwein.) Burr., 3−4
Uridine triphosphate (UTP), 133−134
UV light, 126f, 172−173
leaf hairs in scatter of, 172−173

V

Vapor pressure deficit (VPD), 106−107
Vascular cambium, 38−40
Vegetative cycle, 61−83. *See also* Reproductive cycle
anthocyanins, action of, 68−69
auxin production, 69−70
bleeding, initiation of, 61−64
soil moisture and, 64−65
budbreak phenomenon, 67−68
bud dormancy, 76−77
canalization, 69−70
cell division and auxin production in buds, 65
chilling requirement, 65−66
cytokinin production, 70

Vegetative cycle (*Continued*)
  DELLA proteins, development of, 70–71
  early canopy development, 80
  ecodormancy, 78
  endodormancy, 77–78
  evolution of tendrils, 67
  gibberellins, action of, 70–71
  grapevine growth stages, 62f, 63f
  leaf senescence, 71–73
  root growth and distribution, 81–82, 82f
    auxin, role of, 79
    root density, 80–81
    root tips, 79–80
  root pressure, 61–65
  shoot growth, 67
    internodes development, 70–71
    leaf unfolding, 70–71
    rapid, 76
    secondary, 76
    shoot apex, sensitivity of, 71
    shoot "maturation," 71
    soil compaction, impact of, 83
    temperature effect on, 74–76
    upward and downward, 67–68
    vigor, 74–75
  strigolactones production, 69–70
  suppression of axillary bud outgrowth, 69
  uptake of water and nutrients, 80–81
Vine capacity, 195–196
Vineyard floor vegetation, 344
  root density and, 81
Vinum francicum, 17
Vinum hunicum, 17
Viognier, 16–17
Violaxanthin, 129–130
Viruses, 300–303
Visible light, 126, 165, 172, 246–248, 253
Vitaceae family, 4–5
Vitaceae roots, 4–5
Vitales, 4
Vitis species, 5–7, 13, 46, 360–361
  leaves of, 45–46
  pith and cortex, 41
  reproductive growth in, 83–84
  root pressure of, 111–112
  susceptibility to waterlogging, 274
  tannin subunits in, 229
  V. aestivalis, 7–8
  V. amurensis, 10, 362
  V. berlandieri, 9, 52–54
  V. candicans, 9
  V. champinii, 361

  V. cinerea, 9, 227–228
  V. coignetiae, 10
  V. davidii, 10
  V. labrusca, 37, 52–56, 362
    seedling, 16
  V. labruscana, 84, 86–87
    shoots of, 86–87
  V. riparia, 8, 8f, 84, 274, 361–362
  V. romanetii, 10
  V. rupestris, 8, 84, 274, 361–362
  V. sylvestris, 9–11, 26–27
  V. vinifera, 9, 11, 13–14, 16–17, 52–56, 60, 86–87, 195, 361–362
    primordia in, 84
    resistance to fungus, 358–359
    salinity effects on, 317
    shoots of, 86–87
Volatiles, grape berries composition/quality with, 236–242
Vugava bijela, 16–17

W

Water potential, 103, 105
  equation, 116–117
Water relations
  aquaporins, 101–102
  cell expansion, 101–106
  matrix potential, 103
  nutrient uptake/transport, 110–124
  osmosis, 101–106
  stomatal action, 106–110
  transpiration, 106–110
  water potential, 101–106
  water uptake/transport, 110–124
Water uptake and transport, 110–124
  aquaporin activity of root cell membranes, 115–116
  canopy size and, 115–116
  cavitation and, 111–112, 112f, 115–116
    by osmotic water "lifting," 113
    root system, 113
    vulnerability to, 113
  direction of movement, 110
    transpiration-driven water flow, 110–111, 115–116, 121
    xylem sap, 110–111
  drought stress and imbalance between, 116–117
  embolisms in xylem network and, 111–113
  friction between water and conduit walls, effect of, 114
  fungi, 113
  gas blockages and, 111
  hydraulic architecture for, 114–115
  long-distance transport, 121–122
  pit permeability, 115
  resistances to water flow through a grapevine, 122–123

root pressure for, 111–112, 116
by suction at the root surface, 113–114
surface area of root system and, 117
surface tension, role of, 111
symplastic and apoplastic pathways, 118–119, 118f
transpiration rate and soil-to-leaf flow rate, 116
vessel diameter, effect of, 113
Vitis species, 111–112
within xylem, 111–112, 114, 122–124
Water vapor concentration, 106–107
Water vapor pressure, 106–107
Wind, 177–179
down row *vs* across row turbulence, 179
leaf's boundary layer resistance, 178–179
physical damages from strong, 177–178, 178f
shoot growth and, 177–178
thigmomorphogenetic responses to, 177–178
Winkler regions, 74–75
Winter buds, 43
Woody roots, 27

X

Xanthophyll cycle, 129–130
Xylella fastidiosa, 3, 364
Xylem
ABA derived in, 109
cell differentiation, 38, 123–124
cells of roots, 30–31, 135–136
embolisms in, 111
of grapevines, 113
negative pressure of, and transpiration, 109–110
nutrient uptake and transport in, 121
osmotic pressure of, and osmotic water uptake by, 61–64
protection of, 107
rings of, 38–40
transpiration driving water flow up, 109
water uptake and transport within, 111–112, 114, 122–124
Xylem sap, 209
sustaining of surface tension, 111
translocation/distribution of photosynthate, 148t, 149
transpiration–cohesion–tension theory of flow of, 110–111
Xylem vessel, 38, 46, 108, 361–362
bacterial infection, 364
cavitation of, 111, 112f, 273, 330
ice formation within, 339–340
lizard's tail strategy protecting, 112–113
nitrogen (N) status and, 286, 294–295, 297–298
scavenging cells of, 46
shoot growth and, 75–76
water and solute uptake, 122–123
water deficit and narrower, 113

Y

Yield component compensation principle, 162, 197, 205
Yield formation, 190–191, 194–205
components of grapevine yield, 194
interactions between genotype and environment, 195
internal and external factors, 194–195
optimal balance between vine's vegetative and reproductive growth, 195
production of fruit, 196
pruning and other cultural practices, 196–197
reproductive performance, 195
shoot vigor for, 195–196
vine capacity, 195
vine self-regulation for, 196–197
vine's vegetative and reproductive growth, 195
vineyard yield, 194
Yield potential, 197–205
berry and seed photosynthesis, effect of, 203–204
bloom occurrence with, 200
bloom-time temperature effects, 201–202
bud development in dense canopies, 198
bud fruitfulness, 198
cell divisions before and after bloom and, 202–203
conditions during bloom, 198, 200–201
effect of light intensity on fruitfulness, 198
effect of seed number, 202
final fruit yield, 200
functional leaf area for inflorescence development, 198–199
of grapevine bud, 197
"heat wave" during the bloom–fruit set period, 203
of inflorescence primordia, 198
limitation in assimilate supply, effect of, 204
maximum number of potentially fruitful shoots, 197–198
need to supply carbon and other nutrients, 199–200
nitrogen (N) availability during bloom, 199
period of inflorescence differentiation, 199–200
phosphorus and potassium deficiency, effects of, 199
reserve status, 199–200
temperature effects, 198–201
time of day and, 204
timing and rate of water or nutrient application, 204–205
in viticulture, 197
water deficit effects, 199

Z

Zeatin-type cytokinins, 38–40
Zeaxanthin, 129–130
Zinc, 311–312
Ziziphus jujuba Mill., 4
Zygotic embryo, 90

Edwards Brothers Malloy
Ann Arbor MI. USA
February 12, 2015